# ADVANCES IN NEUROPHARMACOLOGY

## Drugs and Therapeutics

# ADVANCES IN NEUROPHARMACOLOGY

## Drugs and Therapeutics

*Edited by*

**Md. Sahab Uddin**
**Mamunur Rashid**

AAP APPLE ACADEMIC PRESS

Apple Academic Press Inc.
4164 Lakeshore Road
Burlington ON L7L 1A4
Canada

Apple Academic Press Inc.
1265 Goldenrod Circle NE
Palm Bay, Florida 32905
USA

© 2020 by Apple Academic Press, Inc.

First issued in paperback 2021

*Exclusive worldwide distribution by CRC Press, a member of Taylor & Francis Group*

No claim to original U.S. Government works

ISBN 13: 978-1-77463-471-4 (pbk)
ISBN 13: 978-1-77188-797-7 (hbk)

**Library and Archives Canada Cataloguing in Publication**

Title: Advances in neuropharmacology : drugs and therapeutics / edited by Md. Sahab Uddin, Mamunur Rashid.

Names: Sahab Uddin, Md., 1992- editor. | Rashid, Mamunur (Professor of pharmacology), editor.

Description: Includes bibliographical references and index.

Identifiers: Canadiana (print) 20190154322 | Canadiana (ebook) 20190154330 | ISBN 9781771887977 (hardcover) | ISBN 9780429242717 (PDF)

Subjects: LCSH: Neuropharmacology.

Classification: LCC RM315 .A38 2020 | DDC 615.7/8—dc23

**Library of Congress Cataloging-in-Publication Data**

Names: Sahab Uddin, Md., 1992- editor. | Rashid, Mamunur (Professor of pharmacology), editor.

Title: Advances in neuropharmacology : drugs and therapeutics / edited by Md. Sahab Uddin, Mamunur Rashid.

Description: Oakville, ON : Palm Bay, Florida : Apple Academic Press, [2020] | Includes bibliographical references and index. | Summary: "This new volume, Advances in Neuropharmacology: Drugs and Therapeutics, provides a comprehensive overview of the drugs that act on the central and peripheral nervous systems. It thoroughly describes the diseases that are associated with the nervous system and drugs used their treatment while also looking at the current status of these drugs and their future potential and challenges. This book is divided into three sections that describe the nervous system associated diseases and their treatment as well as current status and future opportunities and challenges. Section 1 focuses on the drugs that affect the functions of the autonomic nervous system to produce therapeutic effects. These drugs may act presynaptically by manipulating the genesis, storage, and secretion, and by blocking the action of neurotransmitters. Some drugs may trigger or impede postsynaptic receptors. Section 2 focuses on drugs that affect the central nervous system, including antianxiety drugs, sedative and hypnotic drugs, antidepressant drugs, antipsychotic drugs, antiepileptic drugs, and many more. It covers the pharmacological management of various diseases, including Alzheimer's, Parkinson's, Huntington's disease, and others. Section 3 offers explanations of neurochemical interactions with the aim to develop drugs that have beneficial effects on neurochemical imbalances. This section demonstrates models to assess the transport of drugs across the blood-brain barrier and nanomedicine to treat brain disorders. This rich compilation provides thorough and extensive research updates on the important advances in neuropharmacological drugs and drug therapy from experienced and eminent academicians, researchers, and scientists from throughout the world. It will be rich resource for professionals, academicians, students, researchers, scientists, and industry professionals around the world in the biomedical, health, and life science fields. Key features: Presents recent advances in neuropharmacology, covering the drugs that act on the central and peripheral nervous systems Provides extensive explanations of various emerging research in this field Explores the intricacy of neurological disorders Looks at existing and forthcoming therapeutic drug strategies"-- Provided by publisher.

Identifiers: LCCN 2019033080 (print) | LCCN 2019033081 (ebook) | ISBN 9781771887977 (hardcover) | ISBN 9780429242717 (ebook)

Subjects: MESH: Central Nervous System Agents | Autonomic Nervous System--drug effects | Brain Diseases--drug therapy | Neuropharmacology

Classification: LCC RM315 (print) | LCC RM315 (ebook) | NLM QV 76.5 | DDC 615.7/8--dc23

LC record available at https://lccn.loc.gov/2019033080

LC ebook record available at https://lccn.loc.gov/2019033081

Apple Academic Press also publishes its books in a variety of electronic formats. Some content that appears in print may not be available in electronic format. For information about Apple Academic Press products, visit our website at **www.appleacademicpress.com** and the CRC Press website at **www.crcpress.com**

# Dedication

---

This book is dedicated to
our beloved parents.

# About the Editors

**Md. Sahab Uddin, RPh**

*Department of Pharmacy, Southeast University, Dhaka, Bangladesh*

Md. Sahab Uddin, RPh, is a Registered Pharmacist and a Research Scholar in the Department of Pharmacy at Southeast University, Dhaka, Bangladesh. He has published copious articles in peer-reviewed international scientific journals. He has also authored and edited several books, including *Handbook of Research on Critical Examinations of Neurodegenerative Disorders*; *Oxidative Stress and Antioxidant Defense: Biomedical Value in Health and Diseases*; *Comprehensive MCQs in Pharmacology*; *Comprehensive MCQs in Physical Pharmacy*; *Tools of Pharmacy: Getting Familiar with the Regular Terms, Words and Abbreviations*; and *Quality Control of Pharmaceuticals: Compendial Standards and Specifications*. Md. Uddin also serves as a guest editor, editorial and reviewer board member of numerous scholarly journals. He is a member of many national and international scientific societies. He has developed Matching Capacity, Dissimilarity Identification, Numeral Finding, Typo Revealing, and Sense Making tests for the estimation of memory, attention, and cognition. Moreover, he is the Founder and Executive Director of the Pharmakon Neuroscience Research Network, an open innovation hub bringing together neuroscientists to advance brain health. He received his BPharm in 2014 securing a first position from the Department of Pharmacy, Southeast University, Bangladesh. Md. Uddin's research interest is how neuronal operation can be restored to abate Alzheimer's dementia.

**Mamunur Rashid, PhD**

*Department of Pharmacy, University of Rajshahi, Rajshahi, Bangladesh*

Mamunur Rashid, PhD, is a Professor in the Department of Pharmacy at the University of Rajshahi, Rajshahi, Bangladesh, where he also served as Chairman of the department. He previously worked as a visiting scientist at the Department of Pharmacology, Niigata University of Pharmacy and Applied Life Sciences, Japan. He was also the Chairman of the Department of Pharmacy, Southeast University, Bangladesh. He is a member of the Bangladesh Pharmaceutical Society. He also works for the development of the pharmaceutical sector in Bangladesh. He has made a vast contribution to the promotion of pharmacy education and the pharmacy profession to create better opportunities for Bangladeshi pharmacists. He is one of the pioneers who are working with the Commonwealth Pharmacists Association and Bangladesh Pharmaceutical Society to develop hospital and community pharmacy in Bangladesh. He has published 75 articles and 23 abstracts in conference proceedings.

He is an editorial and reviewer board member for several international peer-reviewed journals. He has supervised the research of many PhD, MPhil, and MPharm students. His research fields of interest are pharmacology, neuropharmacology, molecular biology, and cardiology. He obtained his PhD degree from the Department of Pharmacology, Niigata University of Pharmacy and Applied Life Sciences, Japan. After completion of his PhD, he was awarded a Japan Society for Promotion of Science postdoctoral research fellowship at the Tohoku University Graduate School of Medicine, Japan.

# Contents

# Contributors

**Abhinav Anand**
Lovely Professional University, Punjab, India

**Elena González Burgos**
University Complutense of Madrid, Madrid, Spain

**Vivek K. Chaturvedi**
University of Allahabad, Uttar Pradesh, India

**Himani Chaurasia**
University of Allahabad, Uttar Pradesh, India

**Muralikrishnan Dhanasekaran**
Auburn University, Auburn, USA

**Leslie B. Essel**
Kwame Nkrumah University of Science and Technology, Kumasi, Ghana
University of Missouri, Missouri, USA

**Luis García-García**
University Complutense of Madrid, Madrid, Spain

**Munish Garg**
Maharshi Dayanand University, Haryana, India

**M. Pilar Gómez-Serranillos**
University Complutense of Madrid, Madrid, Spain

**Manoj Govindarajulu**
Auburn University, Auburn, USA

**Vishal S. Gulecha**
SNJB's SSDJ College of Pharmacy, Maharashtra, India

**Ajay Gupta**
All India Institute of Medical Sciences, Rajasthan, India

**Shallina Gupta**
Guru Nanak Dev University Amritsar, Punjab, India

**Tanya Gupta**
Jaypee Institute of Information Technology, Uttar Pradesh, India

**Seetha Harilal**
Kerala University of Health Sciences, Kerala, India

**Suvarna Ingale**
SCES's Indira College of Pharmacy, Maharashtra, India

**Ellery Jones**
Auburn University, Auburn, USA

**Harleen Kaur**
Amity Institute of Biotechnology, Uttar Pradesh, India

**Ramneek Kaur**
Jaypee Institute of Information Technology, Uttar Pradesh, India

**Navneet Khurana**
Lovely Professional University, Punjab, India

**Chahat Kubba**
Jaypee Institute of Information Technology, Uttar Pradesh, India

**Rajan Kumar**
Lovely Professional University, Punjab, India

**Rakesh Kumar**
Lovely Professional University, Punjab, India

**Sachin Kumar**
Jaypee Institute of Information Technology, Uttar Pradesh, India

**Manoj S. Mahajan**
SNJB's SSDJ College of Pharmacy, Maharashtra, India

**Shalini Mani**
Jaypee Institute of Information Technology, Uttar Pradesh, India

**Pawan Kumar Maurya**
Central University of Haryana, Mahendergarh, India

**Richa Mishra**
University of Allahabad, Uttar Pradesh, India

**Arup Kumar Misra**
All India Institute of Medical Sciences, Rajasthan, India

**Timothy Moore**
Auburn University, Auburn, USA

**Samuel Obeng**
Virginia Commonwealth University, Virginia, USA

**David D. Obiri**
Kwame Nkrumah University of Science and Technology, Kumasi, Ghana

**Francisca Gómez Oliver**
University Complutense of Madrid, Madrid, Spain

**Newman Osafo**
Kwame Nkrumah University of Science and Technology, Kumasi, Ghana

**Rupali Patil**
GES's SDMSG College of Pharmaceutical Education and Research, Maharashtra, India

**K. Pramod**
Government Medical College, Kerala, India

**Rachana**
Jaypee Institute of Information Technology, Uttar Pradesh, India

**Rajesh K**
Kerala University of Health Sciences, Kerala, India

**Rashi Rajput**
Jaypee Institute of Information Technology, Uttar Pradesh, India

**Sindhu Ramesh**
Auburn University, Auburn, USA

**Varsha Rani**
Amity Education Group, New York, USA

**Chinnu Sabu**
Government Medical College, Kerala, India

**Kanishka Sharma**
Amity Education Group, New York, USA

**Neha Sharma**
Lovely Professional University, Punjab, India

**Pramod Kumar Sharma**
All India Institute of Medical Sciences, Rajasthan, India

**Sonia Sharma**
Guru Nanak Dev University Amritsar, Punjab, India

**Sushant Sharma**
University of KwaZulu Natal, Durban, South Africa

**Tanya Sharma**
Jaypee Institute of Information Technology, Uttar Pradesh, India

**Abdulla Sherikar**
SNJB's SSDJ College of Pharmacy, Maharashtra, India

**Aarushi Singh**
Jaypee Institute of Information Technology, Uttar Pradesh, India

**Manisha Singh**
Jaypee Institute of Information Technology, Uttar Pradesh, India

**Surjit Singh**
All India Institute of Medical Sciences, Rajasthan, India

**Vishal K. Singh**
University of Allahabad, Uttar Pradesh, India

**Vishal Srivastava**
University of Allahabad, Uttar Pradesh, India

**Vishnu Suppiramaniam**
Auburn University, Auburn, USA

**Aman Upaganlawar**
SNJB's SSDJ College of Pharmacy, Maharashtra, India

**Chandrashekhar Upasani**
SNJB's SSDJ College of Pharmacy, Maharashtra, India

**Vaibhav Walia**
Maharshi Dayanand University, Haryana, India

**Saumya Yadav**
Jaypee Institute of Information Technology, Uttar Pradesh, India

**Oduro K. Yeboah**
Kwame Nkrumah University of Science and Technology, Kumasi, Ghana

# Abbreviations

| | | | |
|---|---|---|---|
| 2-PAM | 2-pyridine aldoxime methyl chloride/pralidoxime | ALDH | aldehyde dehydrogenase |
| | | ALS | amyotrophic lateral sclerosis |
| 3-OMD | 3-O-methyldopa | AMP | adenosine monophosphate/ |
| 4-DAMP | 1,1-dimethyl-4-diphenylace-toxypiperidinium iodide | | adenosine 3',5'-monophosphate |
| | | AMPA | α-amino-3-hydroxy-5-methyl-4-isoxazolepropionic acid |
| 5-HT | 5-hydroxytryptamine/serotonin | | |
| 8-OHdG | 8-hydroxy-2-deoxyguanosine | AMPH | amphetamines |
| 8-OH-DPAT | 8-hydroxy-2-(dipropylamino) tetralin | ANS | autonomic nervous system |
| | | Anti-ChE | anticholinesterase |
| AAAD | aromatic L-amino acid decarboxylase | APIs | active pharmaceutical ingredients |
| AAGBI | Anesthetists of Great Britain and Ireland | APP | amyloid precursor protein |
| | | ASD | autism spectrum disorder |
| ABCC2 | ATP-binding cassette sub-family C member 2 | ASOs | antisense oligonucleotides |
| | | ATA | atmospheres absolute |
| ABCC3 | ATP-binding cassette sub-family C member 3 | AUC | area under the curve |
| | | AUD | alcohol use disorder |
| AC | adenylyl cyclase | AV | atrioventricular |
| ACE | angiotensin converting enzyme | Aβ | amyloid beta |
| ACET | kainate receptor antagonist | BAC | blood alcohol concentration |
| ACh | acetylcholine | BAT | brown adipose tissue |
| AChE | acetylcholinesterase | BBB | blood–brain barrier |
| ACLS | advance cardiovascular life support | BChE | butyrylcholinesterase |
| | | Bcl-2 | B-cell lymphocyte protein-2 |
| ACNU | nimustine | BD | bipolar disorder |
| ACP | anticholinergic drug used in Parkinson's disease | BDNF | brain-derived neurotrophic factor |
| ACTH | adrenocorticotropic hormone | BDs | brain disorders |
| AD | Alzheimer's disease | BMEC | brain microvascular endothe-lial cells |
| ADH | alcohol dehydrogenase | | |
| ADR | adverse drug reaction | BM-MSC | bone marrow-derived mesen-chymal stem cells |
| ADT | antidepressant treatment | | |
| AEDs | antiepileptic drugs | BPH | benign prostatic hyperplasia |
| AGP | α1-acid glycoprotein | BuChE | butyrylcholinesterase |
| AJ | adhesion junction | BZs/BZDs | benzodiazepines |

| | | | |
|---|---|---|---|
| C6G | codeine-6-glucuronide | CSF | cerebrospinal fluid |
| CA | catecholamines | CT | computed tomography |
| CACT | carnitine-acylcarnitine transferase | CTZ | chemotactic trigger zone |
| | | CVS | cardiovascular system |
| CAG | cysteine adenosine-guanine | CYP | cytochrome/cytochrome P450 |
| cAMP | cyclic adenosine monophos-phate/adenosine-3′,5′-cyclic monophosphate | DA | dopamine |
| | | DAG | diacylglycerol/1,2-diacylglyc-erol |
| CAMs | cell adhesion molecules | DAM | diacetylmonoxime |
| CAT/ChAT | choline acetyltransferase | dbcAMP | adenosine monophosphate |
| CB | cannabinoid receptors | DC | dendritic cell |
| CB1 | cannabinoid type 1 receptors | DCTN1 | p150 subunit of dynactin |
| CB1Rs | CB1 receptors | DFP | di-isopropyl flurophosphonate |
| CBZ | carbamazepine | DHA | docosahexaenoic acid |
| CCK | cholecystokinin | DHB | dihydrobunolol |
| CD | carbidopa | DMPP | dimethylphenylpiperazinium |
| CDAI | Crohn's Disease Activity Index | DNMTs | DNA methyltransferases |
| CDEIS | Crohn's Disease Endoscopy Index Severities Score | DOMA | 3,4-dihydroxymandelic acid |
| | | DOP/DOR | delta opioid peptide receptor |
| CE | cognitive enhancers | DOPA | di-hydroxy phenylalanine |
| cGMP | cyclic guanosine mono-phosphate/cyclic guanosine 3′,5′-monophosphate | DOPAC | 3,4-dihydroxyphenylacetic acid |
| | | DOPE | dioleoyl phosphatidylethanolamine |
| ChAT | choline acetyltransferase | | |
| CHD | coronary heart disease | DSPC | distearoyl phosphatidylcholine |
| CHF | congestive heart failure | DSPE | 1,2-distearoyl-sn-glycero-3-phosphoethanolamine |
| CHO | Chinese hamster ovary | | |
| CLD | Creutzfeldt-Jakob disease | DT | diagnostic testing |
| $C_{max}$ | maximum plasma concentration | DUI | driving under the influence |
| | | EBC | eye blink conditioning |
| CNF | central neurotrophic factor | eCBs | endocannabinoids |
| CO | carbon monoxide/cardiac output | ECF | extracellular fluid |
| | | ECV | electroconvulsive |
| COMT | catechol-O-methyl transferase | EDRF | endothelial derived relaxing factor |
| COPD | chronic obstructive pulmonary disease | | |
| | | EDTA | ethylenediaminetetraacetic acid |
| COX | cyclooxygenase | | |
| CP | cerebral palsy | EEG | electroencephalogram |
| CREB | cAMP response element binding protein | EKC | [$^3$H]-ethylketocyclazocine |
| | | EMG | electromyography |
| CRF | corticotrophin releasing factor | eNOS | endothelial nitric oxide synthase |
| CSD | cortical spreading depression | | |

| | | | |
|---|---|---|---|
| ENT | extraneuronal amine transporter | HATs | histone acetylases |
| EPA | eicosapentaenoic acid | HATs | histone acetyltransferases |
| EPN | epinephrine | HBOT | hyperbaric oxygen therapy |
| EPP | excitatory postsynaptic potential | HD | Huntington's disease |
| | | HDACs | histone deacetylases |
| EPS | extrapyramidal symptoms | HDL | high density lipoprotein |
| EPSP | excitatory postsynaptic potential | HDLC | high density lipoprotein cholesterol |
| ERK | extracellular-signal-regulated kinase | HEK | human embryonic kidney |
| | | HET | human isolated erectile tissue |
| ESC | embryonic stem cell | HIF-1$\alpha$ | hypoxia-inducible factor 1-$\alpha$ |
| ETC | electron transport chain | HMG-CoA | hydroxymethylglutaryl coen-zyme A |
| FAS | fetal alcohol syndrome | | |
| FFI | fatal familial insomnia | HO | heme oxygenase |
| FRs | free radicals | HPNS | high pressure nervous syndrome |
| FVC | forced vital capacity | | |
| GA | glatiramer acetate | Hpsc | human pluripotent stem cells |
| GABA | $\gamma$-aminobutyric acid | HR | heart rate |
| GAD | generalized anxiety disorder/ glutamic acid decarboxylase | HTN | hypertension |
| | | HVA | homovanillic acid |
| GAPDH | glyceraldehyde 3-phosphate dehydrogenase | IBDQ | inflammatory bowel disease questionnaire |
| | | IDA-SLN | idarubicin-loaded solid lipid nanoparticles |
| GAT-1 | sodium- and chloride-depended GABA transporter 1 | | |
| | | IM | intramuscular |
| GBM | glioblastoma multiforme | iNOS | inducible nitric oxide synthase |
| GDH | glutamate dehydrogenase | IOP | intraocular pressure |
| GDP | guanosine diphosphate | IP | inositol monophosphate |
| GH$_3$ | growth-hormone-secreting pituitary | IP$_3$ | inositol triphosphate/inositol 1,4,5-triphosphate |
| GHB | $\gamma$-hydroxybutyrate | IPSP | inhibitory postsynaptic potential |
| GIRK | G protein-coupled inwardly rectifying K$^+$ channels | | |
| | | iRISA | impaired response inhibition and salience attribution |
| GIT | gastrointestinal tract | | |
| GMFM | gross motor function measure | ISA | intrinsic sympathomimetic activity |
| GPCRs | G protein-coupled receptors | | |
| GPx | glutathione peroxidase | IUPHAR | International Union of Basic and Clinical Pharmacology |
| GSK-3b | glycogen synthase kinase-3beta | | |
| | | IVRA | intravenous regional anesthesia |
| GSS | Gerstmann-Straussler-Scheinker syndrome | JAM | junction adhesion molecules |
| | | KOR | kappa opioid peptide receptor |
| GTP | guanosine triphosphate | L-arg | L-arginine |

| | | | |
|---|---|---|---|
| LAs | local anesthetics | MMP9 | matrix metalloproteinase 9 |
| LC | locus coeruleus | MMSE | mini-mental state examination |
| LCM | lacosamide | MMTV | mouse mammary tumor virus |
| LD | levodopa | MOP/MOR | μ-opioid peptide receptor |
| LDL | low density lipoprotein | MPH | methylphenidate |
| LDL-C | low-density lipoprotein cholesterol | MPP$^+$ | N-methyl-4-phenylpyridinium |
| LDR | long duration response | MPTP | 1-methyl-4-phenyl-1,2,3,6-tetrahydropyridine |
| LEV | levetiracetam | MRI | magnetic resonance imaging |
| LOO$^•$ | lipid peroxyl radical | MRT | mean residence time |
| LSD | lysergic acid diethylamide | MSA | membrane stabilizing activity |
| M | muscarinic | MSN | medium spiny neuron |
| M3G | morphine-3-glucuronide | N | nicotinic |
| M6G | morphine-6-glucuronide | NA | non-adrenergic/noradrenaline |
| MAC | minimum alveolar concentration | NAc | nucleus accumbens |
| mAChR | muscarinic acetylcholine receptors | nAChR | nicotinic acetylcholine receptors |
| MAG | myelin associated glycoprotein | NAD | nicotinamide adenine dinucleotide |
| MAGUK | membrane associated guanylatekinase | NASSAs | noradrenergic specific serotonergic agents |
| MAO | monoamine oxidase | NBOT/NBO | normobaric oxygen therapy |
| MAO-B | monoamine oxidase B | NBQX | AMPA receptor antagonist |
| MAOIs | monoamine oxidase inhibitors | NC | non-cholinergic |
| MAPK | mitogen-activated protein kinase | ncRNAs | non-coding RNAs |
| MARCKS | myristoylated alanine-rich PKC-kinase substrate | NDRIs | noradrenaline and dopamine reuptake inhibitors |
| MCAO | middle cerebral artery occlusion | NDs | neurodegenerative disorders |
| MCI | mild cognitive impairment | NE | norepinephrine |
| MDCK | madindarby canine kidney | NET | norepinephrine transporter |
| MDRI | multidrug resistance gene | NF | nanofibers |
| MDZ | midazolam | NFTs | neurofibrillary tangles |
| MEF2 | myocyte enhancing factor-2 | NF-κB | nuclear factor-κB |
| MEOs | microsomal ethanol-oxidizing system | NGF | nerve growth factor |
| MI | myocardial infarction | NHERF | Na$^+$/H$^+$ exchanger regulatory factor |
| MIT | Massachusetts Institute of Technology | NIV | non-invasive ventilation |
| MLAC | minimum local analgesia concentration | NL | nanoliposomes |
| | | NLCs | nanostructured lipid carriers |
| | | NMBDs | neuromuscular blocking drugs |
| | | NMDA | N-methyl-D-aspartate |
| | | NMDAR | N-methyl-D-aspartate receptor |

| | | | |
|---|---|---|---|
| NMJ | neuromuscular junction | PEG | percutaneous endoscopic gastrostomy/polyethylene glycol |
| NMs | nanomaterials | | |
| NMS | neuroleptic malignant syndrome | | |
| | | PET | positron emission tomography |
| nNOS | neuronal nitric oxide synthase | PGB | Pregabalin |
| NOP | nociception/orphanin FQ peptide | PGE | prostaglandin E |
| | | P-gp | P-glycoprotein |
| NOS | nitric oxide synthase | PGs | prostaglandins |
| NPs | nanoparticles | PHT | phenytoin/fosphenytoin |
| NRM | nucleus raphe magnus | PIP | phosphatidylinositol monophosphate |
| NRPG | nucleus reticularis paragigantocellularis | | |
| | | $PIP_2$ | phosphatidylinositol 4,5-bisphosphate |
| NSC | neuronal stem cell | | |
| NT | neuronal tissue | PKA | protein kinase A |
| NTs | neurotrophins | PKC | protein kinase C |
| OAB | overactive bladder syndrome | PKG | cGMP dependent protein kinase |
| OCD | obsessive-compulsive disorder | | |
| OEC | olfactory ensheathing cells | PLC | phospholipase C |
| Omgp | oligodendrocyte-myelin glycoprotein | PLGA | poly(DL-lactic-co-glycolic acid) |
| ONH | optic nerve head | PM | pore module |
| $ONOO^-$ | peroxynitrite | PNS | peripheral nervous system |
| OXC | oxcarbazepine | PO | per oral |
| p53 | tumor supressor protein 53 | $pO_2$ | partial pressure of oxygen |
| PABA | para-amino-benzoic acid | PPARγ | peroxisome proliferator-activated receptor gamma |
| $paCO_2$ | carbon dioxide tension | | |
| PAG | periaqueductal gray | PR | peripheral resistance |
| PAM | pralidoxime | PrD | prion disease |
| PAMPA | parallel artificial membrane permeability assay | PRL | plasma prolactin |
| | | PRO | propofol |
| PANSS | Positive and Negative Syndrome Scale | PrP | prion protein |
| | | PrPSc | prion protease-resistant isoform |
| PARIs | serotonin partial agonist reuptake inhibitors | | |
| | | PS | permeability surface area product |
| PBCA | poly(butyl cyanoacrylate) | | |
| PCL | polycaprolactone | PS | presenilin |
| PD | panic disorder | PT | predictive testing |
| PD | Parkinson's disease | PTB | phenobarbital |
| PDE | phosphodiesterase | PTMA | phenyl trimethyl ammonium |
| PDI | protein-disulphide isomerase | PTSD | post-traumatic stress disorder |
| PDL | poly-D-lysine | QAR | qualitative autoradiography |
| | | QD | quantum dot |

| | | | |
|---|---|---|---|
| REM | rapid-eye movement | TEER | transendothelial electrical |
| RNS | reactive nitrogen species | | resistance |
| ROCs | receptor-operated channels | TEM | transmission electron |
| ROS | reactive oxygen species | | microscopy |
| rTMS | repetitive transcranial | TENS | transcutaneous electrical nerve |
| | magnetic stimulation | | stimulation |
| RXR | retinoid X receptor | TEPA | tetraethylenepentamine |
| SA | sinoatrial | TG | triglyceride |
| SAD | social anxiety disorder | THC | delta-9-tetrahydrocannabinol |
| SAH | aneurysmal subarachnoid | THP | thiopental |
| | hemorrhage | TJ | tight junction |
| SAR | structure activity relationship | TM | thrombomodulin |
| SARIs | serotonin antagonist and reup- | TMA | tetramethylammonium |
| | take inhibitors | TNF-α | tumor necrosis factor-α |
| SC | subcutaneous | TNS | transient neurological |
| SDR | short duration response | | symptoms |
| SF-36 | 36-Item Short Form Survey | TPM | topiramate |
| sGC | soluble guanylate cyclase | TRH | thyroid releasing hormone |
| SMS | serotonin modulator and | TrkB | tropomyosin receptor kinase B |
| | stimulator | TTX | tetrodotoxin |
| S-NO | S-nitrosylation | TZDs | thiazolidinediones |
| SNP | single nucleotide | UDP | glucuronosyltransferase-2B7 |
| | polymorphism | UGT | UDP-glucuronosyltransferase |
| SNpc | substantia nigra pars compacta | UGT | uridine diphosphoglucuronosyl |
| SNr | substantia nigra pars reticulata | | transferase |
| SOCs | store operated channels | UHDRS | Unified *Huntington's* |
| SOD1 | superoxide dismutase-1 | | *Disease* Rating Scale |
| SPECT | single-photon emission | UI | urinary incontinence |
| | computed tomography | VAPB/ALS8 | vesicle associated membrane |
| SSJ | Stevens-Johnson syndrome | | protein |
| SSRIs | selective serotonin reuptake | $V_d$ | volume of distribution |
| | inhibitors | VEGF | vascular endothelial growth |
| STN | subthalamic nucleus | | factor |
| SV2A | synaptic vesicle glycoprotein | VEP | visual evoked potentials |
| | 2A | VGSCs | voltage-gated sodium channels |
| $t_{1/2}$ | half-life | VMAT | vesicular monoamine |
| TBZ | tetrabenazine | | transporter |
| TC | total cholesterol | VNS | vagal nerve stimulation |
| TCAs | tricyclic antidepressants | VNT | vesicular neurotransmitter |
| TCTP | translationally controlled | | transporter |
| | tumor protein | VOCs | voltage-operated channels |
| TEA | tetramethylammonium | VPA | valproic acid |

| | | | |
|---|---|---|---|
| VSDs | voltage sensor domains | WAT | white adipose tissue |
| VSSCs | voltage-sensitive sodium channels | Wnt | wingless-type MMTV integration site family |
| VTA | ventral tegmental area | ZO | zonula occludens |

# Preface

The nervous system is a complex network of specialized cells known as neurons; understanding of the nervous system and how neurons communicate with one another have become one of the most auspicious areas of research to comprehend the neurological disorders. Neuropharmacology is a very rife region of science that involves countless traits of the nervous system from single neuron manipulation to entire parts of the brain, spinal cord, and peripheral nerves. The prime research of neuropharmacology is to analyze the functions of the brain, spinal cord, sensory systems, and peripheral nerves with the intention to reveal mechanisms of neurological disorders and new approaches for the treatment.

*Advances in Neuropharmacology: Drugs and Therapeutics* emphasizes the drugs that act on the central and peripheral nervous systems. This book, divided into three parts that consist of 24 chapters, describes the nervous system associated diseases and their treatment as well as current status and future opportunities and challenges. Part I represents the drugs that affect the functions of the autonomic nervous system to produce the therapeutic effects. This section comprises four chapters. Most of the drugs affecting the central nervous system to produce the therapeutic effects primarily act on the steps of neurotransmission. These drugs may act presynaptically by manipulating the genesis, storage, secretion, or blocking the action of neurotransmitters. Some drugs may trigger or impede postsynaptic receptors. Part II focuses on the drugs that affect the central nervous system and comprises 17 chapters. Part III offers neurochemical interactions, aiming at developing drugs that have beneficial effects on neurochemical imbalances, models to assess the transport of drugs across the blood–brain barrier, and nanomedicine to treat brain disorders. This section comprises three chapters.

This book represents the copious set of specific research updates. All over the world numerous erudite, experienced, and eminent academicians, researchers, and scientists participated to write the text of this book to give a precise and diaphanous understanding of drugs affecting the nervous system at a more advanced level with excellent presentation.

This book is suitable for professionals, academicians, students, researchers, scientists, and industrialists around the world. Biomedical, health, and life science departments can use this book as a crucial textbook. Researchers and scientists from research institutes can use the efficient research information presented here. Pharmacists, physicians, and other healthcare professionals can use this book as a reference book. Furthermore, for interested readers, this book is a storehouse of knowledge to help comprehend the complexity of the nervous system acting drugs.

It is expected that readers shall find this book very informative and enormously useful. Since science is constantly changing; readers are strongly recommended to check for recent updates. The editors are ebulliently ready to accept any comment, suggestion, advice, or critique.

**Md. Sahab Uddin, RPh**
*Department of Pharmacy, Southeast University, Dhaka, Bangladesh*

**Mamunur Rashid, PhD**
*Department of Pharmacy, University of Rajshahi, Rajshahi, Bangladesh*

# Acknowledgments

Editing is a complex process. The editors would like to thank the people involved in this project for their quality time, expertise, and countless efforts. The editors are extremely indebted to the following:

- The authors for the countless time and expertise that they have put into their texts.
- Those authors who submitted their work to contribute to this project but unfortunately were rejected.
- The reviewers for their constructive reviews in improving the quality and presentation of the content.
- The teachers, colleagues, friends, and supporters for their endless inspiration and assistance.
- Especially Apple Academic Press for giving us a great opportunity to edit this book and their aid in the preparation of this edition.
- Last but not least, the editors would like to express the heartiest gratitude to almighty Allah and their parents.

**Md. Sahab Uddin, RPh**
*Department of Pharmacy, Southeast University, Dhaka, Bangladesh*
**Mamunur Rashid, PhD**
*Department of Pharmacy, University of Rajshahi, Rajshahi, Bangladesh*

# PART I
## Drugs Affecting the
## Autonomic Nervous System

# CHAPTER 1

# Cholinergic Agonists

RUPALI PATIL[1*] and AMAN UPAGANLAWAR[2]

[1]GES's SDMSG College of Pharmaceutical Education and Research, Nashik, Maharashtra, India

[2]SNJB's SSDJ College of Pharmacy, Nashik, Maharashtra, India

*Corresponding author. E-mail: ruupalipatil@rediffmail.com

## ABSTRACT

Cholinergic agonists include a wide range of drugs with varied chemical structures and properties. In clinical practice, they are used for the treatment of various diseases, such as glaucoma, myasthenia gravis, belladonna poisoning, paralytic ileus, urinary retention, and reversal of neuromuscular blockade. They also have clinical implication in the treatment of Alzheimer's disease. Use of irreversible anticholinesterase is limited as pesticides or insecticides as they may cause poisoning in humans causing death of a person. This chapter deals with the drugs mimicking cholinergic actions in the periphery at muscarinic receptors. Muscarinic agonists, as a group, or cholinomimetics imitate or mimic the actions of acetylcholine, and hence, such drugs are also called parasympathomimetics. Mainly, they produce effects resembling those of parasympathetic stimulation.

## 1.1 INTRODUCTION

Drugs acting on the autonomic nervous system (ANS) are classified into two types according to the involvement of type of neuron as cholinergic drugs acting on receptors activated by acetylcholine (ACh) and adrenergic drugs acting on receptors stimulated by noradrenaline or adrenaline (Rang et al., 2011). In the middle of the 19th century, the peripheral nervous system, and particularly the ANS, received a great deal of attention. In 1869, it had been shown that muscarine, an exogenous substance, mimics the effects of stimulating the vagus nerve. Actions of muscarine and vagus nerve stimulation could be inhibited by atropine. It was not until 1921, in Germany, that Loewi showed that excitation of the vagosympathetic trunk connected to an isolated and cannulated frog's heart could cause the release into the cannula of a substance (Vagusstoff) that if the cannula fluid was transported from the first heart to a second, it would inhibit the second heart (Rang et al., 2011).

Studies of Loewi first time provided a direct indication in the involvement of release of chemical substances in nerve impulses (Brunton, 2011). He used two isolated frog hearts. He stimulated vagus nerve of first frog heart. The perfusate of the first frog heart was passed to the second frog heart. Second heart of frog was used as an assessment object. The second heart of frog responded in the same way as the first heart after a short lag. This proved that a constituent released from the first heart was responsible to slow second heart's rate. He labeled this constituent as Vagusstoff, also known as parasympathin, a vagus substance (Brunton, 2011). Evidence to identify this substance as ACh was provided by Loewi and Navratil. Loewi also proved that when actions of inhibitory fibers are suppressed than the sympathetic fibers in the frog's vagus nerve, Acceleranstoff, an accelerator substance similar to adrenaline, was liberated into the perfusion fluid. According to studies of Feldberg and Krayer in 1933, the cardiac vagus substance is also ACh in mammals (Brunton, 2011). Loewi's findings can be briefed as to the following:

- Perfusate of first frog heart with stimulated vagus caused the appearance of a substance in the perfusate capable of producing an inhibitory effect resembling vagus stimulation in a second heart.
- Sympathetic nervous system stimulation caused the appearance of a substance capable of accelerating a second heart. By fluorescence measurements, Loewi concluded later that this substance was epinephrine.
- Inhibitory action of the vagus on the heart was prevented by atropine but it did not prevent the release of the transmitter, Vagusstoff.
- When Vagusstoff was incubated with ground-up heart muscle, it became inactivated (Rang et al., 2011) due to the enzymatic destruction of ACh by cholinesterase (ChE).

Physostigmine (eserine) prevented the destruction of Vagusstoff by the heart muscle. Vagus stimulation of heart was also stimulated by physostigmine. This evidenced that the stimulation is associated with the inhibition of enzyme ChE involved in the destruction of the transmitter substance ACh (Rang et al., 2011). Synthesis of ACh from choline occurs in the axon terminal of a cholinergic neuron. Transporter helps in uptake of choline into nerve terminal. In nerve terminal, acetylation of free choline occurs in the presence of choline acetyltransferase (CAT). Transfer of acetyl group from acetyl coenzyme A (CoA) occurs in the presence of CAT (Fig. 1.1).

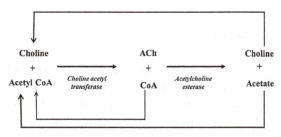

**FIGURE 1.1** Synthesis of acetylcholine in cholinergic neuron.

ChE is located in the nerve terminal. ACh is continually hydrolyzed and resynthesized. ACh is stored in the synaptic vesicles. After fusion of the synaptic membrane with axonal membrane, ACh gets released from synaptic vesicles by the process of exocytosis involving $Ca^{2+}$ influx into the nerve terminal. After release, ACh binds with receptors on pre- or postsynaptic membrane (Rang et al., 2011). This chapter deals with the drugs, which act as cholinomimetics in the periphery at muscarinic receptors (mAChRs). Muscarinic agonists, as a group, or cholinomimetics imitate or mimic the actions of ACh. The main effects they produce resemble those of parasympathetic stimulation; hence, such drugs are also called parasympathomimetics.

## 1.2 CHOLINERGIC RECEPTORS

Muscarinic and nicotinic are mainly two subtypes of cholinergic receptors (Fig. 1.2). The effects of parasympathetic nerve discharge are mimicked by the alkaloid muscarine. Parasympathomimetic effects are effects of muscarine at autonomic neuroeffector junctions and are facilitated by mAChRs (Katzung et al., 2009). Autonomic ganglia and skeletal muscle neuromuscular junctions (NMJs) are stimulated by low concentrations of the alkaloid nicotine due to stimulation of nicotinic receptors (nAChRs). But nicotine does not stimulate autonomic effector cells (Katzung et al., 2009).

### 1.2.1 MUSCARINIC RECEPTORS AND SIGNAL TRANSDUCTION

ACh is the neurotransmitter of the parasympathetic ANS. Effects of mAChRs are potentiated by ACh (Broadley, 1996). mAChRs are located pre- and postsynaptically. They control transmitter release and are responsible for ganglionic excitation. Effects of ACh at postganglionic synapses located in heart, smooth muscles, and glands are mediated by mAChRs. They are also present in many parts of the central nervous system (CNS) (Rang et al., 2011).

Study of the responses of cells and organ systems in the CNS and periphery helps in the initial differentiation of mAChRs. mAChRs were categorized into two types: $M_1$ (also called ganglionic) and $M_2$ (also called effector cell) based on actions of muscarinic agonists, namely, bethanechol and McN-A-343 on the tone of the lower esophageal sphincter (Goyal and Rattan, 1978). mAChRs were identified as $M_1$–$M_5$ based on the cloning of the cDNAs encoding five distinct genes for mAChRs (Bonner et al., 1987).

mAChRs belong to G-protein-coupled receptor (Brunton, 2011). Cellular effectors responsible for pharmacological actions produced by activation of mAChRs include formation of second messengers like inositol triphosphate ($IP_3$) and 1,2-diacylglycerol (DAG) by activation of phospholipase C (PLC), stimulation of potassium ($K^+$) channels,

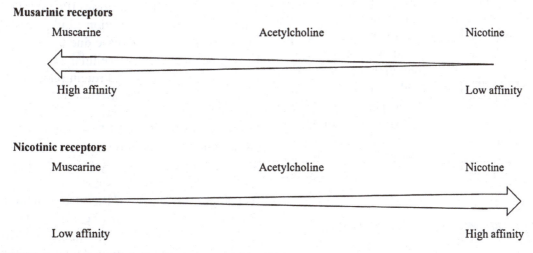

**FIGURE 1.2** Categories of cholinergic receptors and its affinity.
*Source*: Adapted from Harvey et al. (2011).

or inhibition of calcium ($Ca^{2+}$) channels and adenylyl cyclase (Rang et al., 2011).

### 1.2.2   $M_1$ RECEPTORS

$M_1$ receptors (location: CNS, parietal cells, and neurons in the periphery) facilitate stimulatory effect. Slow muscarinic stimulation in sympathetic ganglia and central neurons by ACh is due to membrane depolarization caused by a decrease in $K^+$ conductance. A deficit of such type of effects mediated by ACh in the brain may be responsible for dementia (Wess, 2004). Stimulation of $M_1$ receptors may increase gastric acid secretion following vagal stimulation. Most agonists are nonselective and antagonists show more selectivity, but most of the classic antagonists (e.g., atropine, scopolamine) are nonselective (Rang et al., 2011).

### 1.2.3   $M_2$ RECEPTORS

Inhibitory effects of $M_2$ receptors (location: heart, presynaptic terminals of peripheral and central neurons) are due to an increase in $K^+$ conductance and inhibition of $Ca^{2+}$ channels. Negative inotropic and chronotropic effects are observed due to stimulation of $M_2$ receptors causing cholinergic inhibition of heart. Activation of $M_2$ receptors also shows presynaptic inhibition in the CNS and periphery. Visceral smooth muscles reveal involvement of $M_2$ and $M_3$ in the stimulation of smooth muscles in various organs by muscarinic agonists (Rang et al., 2011).

### 1.2.4   $M_3$, $M_4$, AND $M_5$ RECEPTORS

Excitatory effects such as stimulation of glandular secretions (bronchial, sweat, salivary, etc.) and visceral smooth muscle contraction are produced by $M_3$ receptors. Release of nitric oxide by adjacent endothelial cells is the key factor in relation to smooth muscle (mainly vascular) enriched with the $M_3$ receptors (Wess, 2004).

$M_4$ and $M_5$ are molecular mAChR mainly restricted to the CNS (Rang et al., 2011). Their functional role is not much clear, but their absence shows behavioral changes in mice (Wess, 2004).

The pharmacological response of mAChRs is mediated through the third intracellular loop of G protein. Receptor stimulation by agonist facilitates guanosine diphosphate (GDP)–guanosine triphosphate (GTP) exchange. This exchange helps in binding of γ-subunit present in G protein. Cleaving of γ-subunit occurs from α-subunit and β-subunit. Stimulation or inhibition of the activity of intracellular enzymes involved in the production of second messengers linked to the tissue response is due to GTP-bound γ-subunit. Activation of $M_1$, $M_3$, and $M_5$ occurs by the involvement of Gq of G-protein-stimulating PLC. Phosphoinositides like phosphatidylinositol 4,5-biphosphate get hydrolyzed by PLC to form $IP_3$ and DAG. Smooth muscle contraction and glandular secretion affect $Ca^{2+}$ release due to binding of $IP_3$ to receptors on intracellular sarcoplasmic reticulum store for $Ca^{2+}$. Protein phosphorylation associated with muscle contraction and influx of $Ca^{2+}$ is due to protein kinase C activated by DAG. $M_2$ and $M_4$ receptors involve Gi protein and adenylyl cyclase. Inhibition of adenylyl cyclase decreases levels of cyclic adenosine 3′,5′-monophosphate (cAMP) from ATP after cleavage of GTP-bound γ-subunit of the G protein. Stimulation of cAMP-dependent protein kinase after the formation of cAMP is involved in phosphorylation of many substrates responsible for tissue responses. In the heart,

$M_2$ receptor may be associated directly with ion channels through G protein without an intermediate second messenger (Caulfield and Birdsall, 1998; Hulme et al., 1990; Caulfield, 1993; Eglen et al., 1996). Properties of mAChRs are given in Table 1.1.

### 1.2.5 NICOTINIC RECEPTORS AND SIGNAL TRANSDUCTION

nAChRs are of two subtypes depending on their location. Nm subtypes of nAChRs are located in the NMJ. Plasma membranes of postganglionic cells in the autonomic ganglia contain Nn receptors belonging to a ligand-gated ion channel. Subunits of nAChRs form cation-selective channels. They are transmembrane polypeptides (Katzung et al., 2009).

### 1.3 CLASSIFICATION OF CHOLINERGIC AGONISTS

Based on their actions, cholinergic agonists are classified into three groups as direct acting, indirect acting, and reactivators of acetylcholinesterase (AChE). Classification of cholinergic agonists is given in Table 1.2.

### 1.4 MECHANISM OF ACTION OF CHOLINOMIMETIC DRUGS

Direct-acting cholinomimetics act by binding to and activating muscarinic and nicotinic receptors. ACh gets hydrolyzed of choline and acetic acid by an enzyme AChE. Indirect acting agents increase the endogenous concentration of ACh in the synaptic cleft and neuroeffector junctions by inhibition of AChE. The excess ACh, in turn, stimulates

cholinoceptors to induce increased responses (Brunton, 2011).

Cholinomimetics act mainly at sites where ACh is released to amplify its effects. Some ChE inhibitors also inhibit butyrylcholinesterase (BuChE) (pseudocholinesterase). BuChE is not important in the physiologic termination of indirect-acting cholinomimetic drugs. Hence, its inhibition is less important in the action of indirect-acting cholinomimetic drugs. Some quaternary ChE-inhibitors, for example, neostigmine, show a modest direct action as well by activating neuromuscular nicotinic cholinergic receptors along with blockade of ChE (Brunton, 2011).

### 1.5 PHARMACOKINETIC AND PHARMACODYNAMIC PROFILE OF DIRECTLY ACTING CHOLINERGIC AGONISTS

### 1.5.1 PHARMACOKINETICS

As choline esters are hydrophilic in nature, they are absorbed and distributed poorly into the CNS. Susceptibility of all choline esters to hydrolysis by ChE is different and is less active by oral route as they get hydrolyzed in the gastrointestinal (GI) tract (Brunton, 2011).

### 1.5.1.1 DISTRIBUTION AND FUNCTION

ChEs are of two types: AChE and BuChE with molecular structure resemblance and difference in distribution, specificity for substrate, and functions (Chatonnet and Lockridge, 1989). Normally, AChE and BuChE between them keep the plasma ACh at an undetectably low level, so ACh is strictly a neurotransmitter and not a hormone. The bound AChE at cholinergic

**TABLE 1.1**  Properties of Muscarinic Receptors.

| Receptor subtype | Term | Locations | Subtype of G-protein | Postreceptor mechanisms | Selective agonists | Selective antagonists |
|---|---|---|---|---|---|---|
| $M_1$ | Neural | Nerves | Gq/11 protein-linked | Inositol triphosphate–diacylglycerol cascade | McNA343, oxotremorine | Pirenzepine |
| $M_2$ | Cardiac | Smooth muscle, heart, nerves | Gi/0 protein-linked | cAMP production inhibition, potassium channel activation | – | Darifenacin, gallamine |
| $M_3$ | Smooth muscle/ glandular | Glands, endothelium, smooth muscle | Gq/11 protein-linked | Inositol triphosphate–diacylglycerol cascade | Cevimeline | Darifenacin |
| $M_4$ | – | CNS | Gi/0 protein-linked | cAMP production inhibition | – | Mamba toxin Muscarinic toxin 3 (MT3) |
| $M_5$ | – | CNS | Gq/11 protein-linked | Inositol triphosphate–diacylglycerol cascade | – | – |

cAMP, cyclic adenosine 3′,5′-monophosphate; CNS, central nervous system.

*Source*: Millar (2003) and Katzung et al. (2009).

**TABLE 1.2** Classification of Cholinergic Agonists.

| Class | Examples |
|---|---|
| A. Direct acting | |
| a. ACh and synthetic choline esters | Acetylcholine, bethanechol, carbachol, methacholine |
| b. Natural alkaloids | Pilocarpine, muscarine, arecoline, lobeline |
| c. Miscellaneous | Tremorine, oxotremorine |
| B. Indirect acting (anticholinesterases) | |
| a. Reversible anticholinesterases | |
| I. Carbamic acid derivatives | |
| i. Natural alkaloid | Physostigmine |
| ii. Quaternary compounds | Neostigmine, pyridostigmine, edrophonium |
| II. Acridine derivatives | Tacrine |
| III. Miscellaneous | Donepezil, rivastigmine, galantamine |
| b. Irreversible anticholinesterases | |
| I. Organophosphates | |
| i. Pesticides | Malathion, parathion |
| ii. War gases | Sarin, tabun, soman |
| iii. Miscellaneous | Isofluorophate, ecothiopate |
| II. Carbamate | Propoxur |
| C. Acetylcholinesterase reactivation | Pralidoxime |

*Source*: Sharma and Sharma (2017) and Harvey et al. (2011).

synapses hydrolyzes the released transmitter and terminate its action rapidly. Soluble AChE present in cholinergic nerve terminals regulates the free ACh concentration from which it may be secreted. The function of the secreted enzyme is not clear yet (Soreq and Seidman, 2001).

AChE and BuChE both belong to the class of serine hydrolases, which includes many proteases, such as trypsin. Two different regions are available on active site of AChE: an anionic site containing glutamate residue binding with choline moiety of ACh and a catalytic esteratic site. An acetylated enzyme and free choline are formed after the transfer of acetyl group of ACh to serine hydroxyl group. Serine acetyl group gets spontaneously hydrolyzed rapidly. The overall turnover number of AChE is extremely high. A single active site

hydrolyzes more than 10,000 molecules of ACh per second (Rang et al., 2011).

Difference between AChE and BuChE has been elaborated in Table 1.3 (Taylor et al., 2009; Massoulie et al., 1993).

### 1.5.2 PHARMACOLOGICAL ACTIONS

#### 1.5.2.1 CARDIOVASCULAR EFFECTS

#### 1.5.2.1.1 Heart

Directly acting cholinergic agonists show cardiac slowing. Sharp fall in arterial pressure is due to a decreased force of contraction of the atria decreasing cardiac output (CO) and nitric oxide-mediated generalized vasodilatation.

**TABLE 1.3**   Difference Between Acetylcholinesterase and Butyrylcholinesterase.

| Parameter | Acetylcholinesterase/ true-cholinesterase | Butyrylcholinesterase/ pseudocholinesterase |
|---|---|---|
| Distribution | Peripheral and central tissues, muscle and nerve, motor and sensory fibers, cholinergic and noncholinergic fibers | Skin, liver, brain, and GI smooth muscle; plasma (in soluble form) |
| Site of synthesis | Cholinergic neurons, muscle, and hematopoietic cells | Liver |
| Substrate specificity | Limited—quite specific for ACh and closely related esters such as methacholine | Broader—hydrolyzes synthetic substrate butyrylcholine more rapidly than ACh, as well as other esters, such as procaine, succinylcholine, and propanidid (a short-acting anesthetic agent) |
| Globular catalytic subunits, which constitute the soluble forms found | Cerebrospinal fluid | Plasma |

Ventricles have less parasympathetic innervation and low sensitivity to muscarinic agonists (Brunton, 2011).

Primary effects of ACh on the cardiovascular system include vasodilatation, negative inotropic (decrease in the force of contraction), chronotropic (heart rate), and dromotropic (conduction velocity in the AV node) effects (Rang et al., 2011).

Intravenous injection of small dose of ACh sows transient fall in blood pressure (BP) due to generalized vasodilatation mediated by vascular endothelial NO. This effect is usually accompanied by reflex tachycardia.

Bradycardia or atrioventricular (AV) nodal conduction block are direct effects of ACh on the heart and may be observed after considerably larger doses. Stimulation of mainly $M_3$ subtype of mAChRs is involved in generalized vasodilatation after exogenous administration of ACh (Khurana et al., 2004; Lamping et al., 2004). $M_3$ receptors are present on vascular endothelial cells, despite the apparent lack of cholinergic innervation. NO, endothelium-derived relaxing factor, production occurs after binding of agonist with

receptors activating Gq–PLC–$IP_3$ pathway, leading to $Ca^{2+}$–calmodulin-dependent activation of endothelial NO synthase (Moncada and Higgs, 1995). NO diffuses to adjacent vascular smooth muscle cells causing relaxation (Furchgott, 1999; Ignarro et al., 1999). Due to various pathophysiological conditions, if the endothelium is damaged, ACh acts predominantly on $M_3$ receptors located on vascular smooth muscle cells, causing vasoconstriction (Brunton, 2011).

### 1.5.2.1.2  *Blood Vessels*

Muscarinic agonists act on blood vessels and show generalized vasodilation due to availability of $M_3$ receptors on the endothelial lining of the vessel. They show fall in BP though they do not have a parasympathetic supply. Stimulation of $M_3$ receptors releases NO by the action of nitric oxide synthase on L-arginine. NO causes accumulation of cyclic guanosine monophosphate (cGMP) responsible for smooth muscle relaxation (Sneddon and Graham, 1992).

## 1.5.2.2 EFFECTS ON THE GASTROINTESTINAL TRACT

$M_1$ receptors are located in the parasympathetic ganglia of the gastric intramural plexus (Kromer and Eltze, 1991). Muscarinic ($M_3$) agonists increase gut motility, smooth muscle contraction, and secretions of gastric acid from parietal (oxyntic) cells of the stomach and digestive enzymes throughout the gut. Bethanechol, a nonselective agonist, increases intestinal motility and is used in the treatment of postoperative gastric distension and atony and in nonobstructive paralytic ileus (Goyal, 1989). Muscarinic agonists cause colicky pain due to increased peristaltic activity (Rang et al., 2011). Stimulation of vagal input to the GI tract increases tone, amplitude of contractions, and secretory activity of the stomach and intestine. $M_3$ receptors are mainly responsible for mediating the cholinergic control of GI motility (Matsui et al., 2002).

## 1.5.2.3 EFFECTS ON THE RESPIRATORY TRACT

Bronchoconstriction may occur due to a varied stimuli-causing reflex increase in parasympathetic actions. ACh causes bronchoconstriction and increases tracheobronchial secretions interfering with breathing (Rang et al., 2011). ACh stimulates chemoreceptors of the carotid and aortic bodies, primarily by $M_3$ receptors (Fisher et al., 2004).

## 1.5.2.4 EFFECTS ON THE URINARY TRACT

Detrusor muscle contraction, increased voiding pressure, and ureteral peristalsis are due to parasympathetic sacral innervations. Control of bladder contraction is facilitated by different mAChR subtypes, mainly $M_2$ receptors. According to studies with selective antagonists and $M_3$ knockout mice, $M_3$ receptor is involved in detrusor muscle contraction (Matsui et al., 2000).

## 1.5.2.5 EFFECTS ON THE GLANDULAR SECRETION

Stimulation of exocrine glands by ACh causes sweating, lacrimation, and salivation (Rang et al., 2011). $M_3$ are mainly associated with stimulation of lacrimal, nasopharyngeal, salivary, and sweat glands (Caulfield and Birdsall, 1998). $M_1$ receptor stimulation may also be involved in increased salivary secretion (Gautam et al., 2004).

## 1.5.2.6 EFFECTS ON THE EYE

The constrictor pupillae muscle runs circumferentially in the iris, whereas ciliary muscle adjusts the curvature of the lens. They both receive parasympathetic supply to the eye. Activation of mAChRs causes contraction of ciliary muscle pulling the ciliary body forward and inward. Tension on the suspensory ligament of the lens gets relaxed; lens bulges more and reduces its focal length, necessary for the eye to accommodate for near vision. The constrictor pupillae adjusts the pupil in response to changes in light intensity and also regulates the intraocular pressure (IOP). Normal IOP (10–15 mmHg above atmospheric) helps to keep the eye slightly distended. Aqueous humor gets secreted slowly and continuously by the cells of the epithelium covering the ciliary body. Aqueous humor is drained into the canal of Schlemm running around the eye close to the outer margin of the iris. One of the commonest preventable cause of blindness associated with glaucoma is abnormally raised IOP. In acute glaucoma, IOP increases due to obstruction to drainage of aqueous humor after dilation of the pupil. Folding of the iris tissue impedes the drainage angle. Muscarinic agonists activate constrictor pupillae muscle and lower the IOP. Normal individual show little effect. Increased tension in the ciliary muscle produced by these drugs helps in improving drainage by realigning the connective tissue trabeculae through which the canal of Schlemm passes (Rang et al., 2011).

### 1.5.2.7   EFFECTS ON THE CNS

Brain shows the presence of all mAChRs ($M_1$–$M_5$) (Volpicelli and Levey, 2004). They regulate perceptive function, motor control, hunger, nociception, and other processes (Wess et al., 2007). Muscarinic agonists crossing the blood–brain barrier (BBB) and acting on $M_1$ receptors produce significant central effects like tremor, hypothermia, increased locomotor activity, and improved cognition (Eglen et al., 1999).

### 1.5.3   CONTRAINDICATIONS

Most contraindications to the use of muscarinic agonists are predictable consequences of mAChR stimulation and include asthma, chronic obstructive pulmonary disease, urinary or GI obstruction, acid-peptic disease, and cardiovascular disease accompanied by bradycardia, hypotension, and hyperthyroidism. Atrial fibrillation may be precipitated by muscarinic agonists in hyperthyroid patients (Brunton, 2011).

### 1.5.4   ADVERSE EFFECTS

Diaphoresis, diarrhea, abdominal cramps, nausea/vomiting, and other GI side effects and

a sensation of tightness in the urinary bladder and presbyopia are observed. In hypotension, the severe reduction in coronary blood flow may be observed, especially if it is already compromised. These contraindications and adverse effects are generally of limited concern with topical administration for ophthalmic use (Brunton, 2011).

## 1.6   DIRECT-ACTING CHOLINERGIC AGONISTS

### 1.6.1   ACETYLCHOLINE

#### 1.6.1.1   CHOLINERGIC TRANSMISSION

During the synthesis of ACh, choline enters the neuron via carrier-mediated transport mechanism (Fig. 1.3). Cytosolic enzyme, CAT, present only in the cholinergic neuron is involved in the choline acetylation utilizing acetyl CoA as a source of the acetyl group.

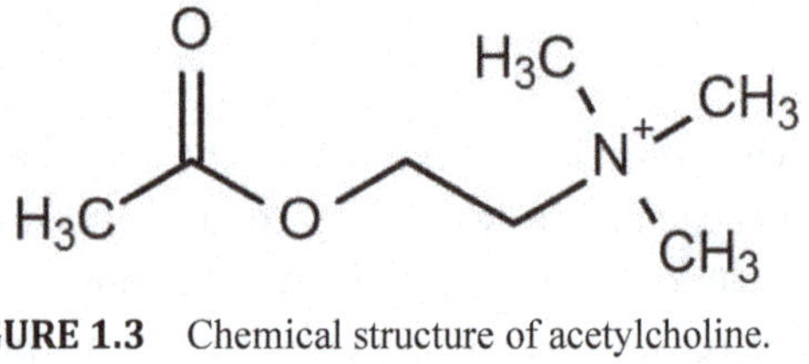

**FIGURE 1.3**   Chemical structure of acetylcholine.

ACh is packaged into synaptic vesicles at high concentration by carrier-mediated transport (Rang et al., 2011). ACh gets released by exocytosis involving $Ca^{2+}$. At the NMJ, single presynaptic nerve impulse releases 100–500 vesicles (Fig. 1.4). At "fast" cholinergic synapses, a presynaptic action potential produces only one postsynaptic action potential as ACh is hydrolyzed within about 1 ms by AChE. Transmission mediated by mAChRs is much slower in its time course. ACh functions as a modulator rather

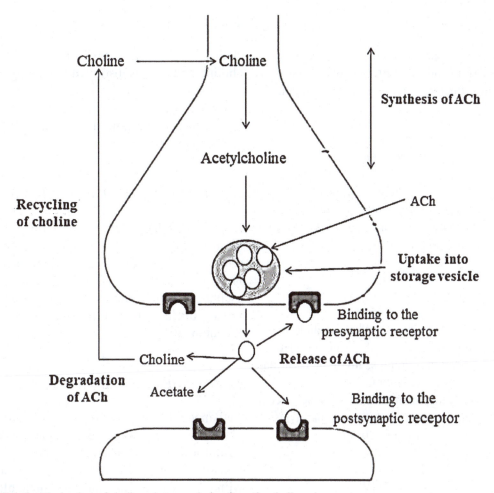

**FIGURE 1.4** Mechanism of cholinergic transmission from the cholinergic neuron.
*Source*: Adapted from Harvey et al. (2011).

than as a direct transmitter in many situations (Rang et al., 2011).

Positively charged quaternary ammonium group and ester group with a partial negative charge are the key features of ACh molecule required for its activity. Ester group is susceptible to rapid hydrolysis by ChE. Modifications of the choline ester structure reduce the susceptibility of the compound to hydrolysis by ChE. It also modifies the relative activity on mAChRs (Katzung et al., 2009).

### 1.6.1.2 PHARMACOKINETICS

ACh is very rapidly hydrolyzed. To attain necessary concentration for producing measurable actions, huge amounts of ACh must be infused intravenously. A large i.v. bolus injection has a short-term action of 5–20 s, whereas only local effects are produced after intramuscular and subcutaneous injections (Katzung et al., 2009).

### 1.6.1.3    MECHANISM OF ACTION

ACh and other related choline esters are agonists at muscarinic and nicotinic receptors, but they act more effectively on mAChRs (Katzung et al., 2009).

### 1.6.1.4    THERAPEUTIC USES

ACh is rarely given systemically. It is used topically and instilled into the eye for the induction of miosis during ophthalmologic surgery as a 1% solution (Brunton, 2011).

### 1.6.1.5    ADVERSE EFFECTS

mAChR-mediated adverse effects include bronchoconstriction, spasm of accommodation, flushing, abdominal cramps, involuntary urination, sweating, and salivation.

Nicotinic receptor-mediated adverse effects are CNS stimulation, twitching, and paralysis of skeletal muscles (Brunton, 2011).

### 1.6.1.6    CONTRAINDICATIONS

ACh is contraindicated in bronchial asthma, hyperthyroidism, peptic ulcer, myocardial infarction.

### 1.6.2    BETHANECHOL

### 1.6.2.1    PHARMACOKINETICS

It is less susceptible to hydrolysis as it is the fusion of carbachol and methacholine with selectivity for mAChRs. It is rarely used in clinical practice (Rang et al., 2011).

### 1.6.2.2    MECHANISM OF ACTION

Bethanechol is β-methyl analog of acetyl choline and an unsubstituted carbamoyl ester. It is strongly stereoselective for mAChRs showing almost 1000 times more potency with ($S$)-bethanechol than ($R$)-bethanechol (Katzung et al., 2009).

### 1.6.2.3    PHARMACOLOGICAL ACTIONS

Bethanechol has mainly muscarinic actions through $M_3$ receptors. It shows prominent effects on GI motility causing peristalsis and increased motility. It increases resting lower esophageal sphincter pressure. It also stimulates urinary bladder. It has little effect on the heart (Brunton, 2011).

### 1.6.2.4    THERAPEUTIC USES

Bethanechol is very occasionally used to stimulate GI motility or to assist bladder emptying (Brunton, 2011). Its use is restricted in conditions with the absence of organic obstruction with urinary retention and incomplete bladder emptying as in cases of postoperative urinary retention, diabetic autonomic neuropathy, and certain cases of chronic hypotonic, myogenic, or neurogenic bladder (Wein, 1991); catheterization can thus be avoided. When used chronically, 10–50 mg/day 3–4 times may be given orally. When given 1 h before or 2 h after the meal on an empty stomach, it minimizes nausea and vomiting. Bethanechol formerly was used to treat postoperative abdominal distention, gastric atony, gastroparesis, adynamic ileus, and gastroesophageal reflux (Brunton, 2011).

### 1.6.3   CARBACHOL (CARBAMYLCHOLINE)

#### 1.6.3.1   PHARMACOKINETICS

Carbachol is almost completely resistant to hydrolysis by ChE. It becomes distributed to areas of low blood flow due to long half-life (Brunton, 2011).

#### 1.6.3.2   MECHANISM OF ACTION

Carbachol is an unsubstituted carbamoyl ester. It retains significant nicotinic activity, particularly on autonomic ganglia (Brunton, 2011).

#### 1.6.3.3   THERAPEUTIC USES

Carbachol is used as experimental tools. It is used topically as a 0.01–3% solution and instilled into eye for the treatment of glaucoma and the induction of miosis during surgery (Brunton, 2011).

### 1.6.4   METHACHOLINE

#### 1.6.4.1   PHARMACOKINETICS

Increased resistance of methacholine to hydrolysis by ChE is due to the methyl group. It has greater duration and selectivity of action (Brunton, 2011).

#### 1.6.4.2   MECHANISM OF ACTION

Methacholine (acetyl-β-methylcholine), the β-methyl analog of ACh, is a synthetic choline ester. It shows predominant selectivity for muscarinic with only minor nicotinic actions. Muscarinic effect is useful in the cardiovascular system (Brunton, 2011).

#### 1.6.4.3   THERAPEUTIC USES

Methacholine is used as an experimental tool. It is administered by inhalation for the diagnosis of bronchial airway hyperactivity in patients who do not have clinically apparent asthma (Crapo et al., 2000). It is used as a powder that is diluted with 0.9% sodium chloride and administered via a nebulizer (Brunton, 2011).

#### 1.6.4.4   ADVERSE EFFECTS

Methacholine can cause bronchoconstriction and increased tracheobronchial secretions in all individuals (Brunton, 2011).

#### 1.6.4.5   CONTRAINDICATIONS/ PRECAUTIONS

Methacholine is contraindicated in asthma, severe airflow limitation, recent myocardial infarction or stroke, uncontrolled hypertension, or pregnancy. The response to methacholine also may be exaggerated or prolonged in patients taking β-adrenergic receptor antagonists. Emergency resuscitation equipment, oxygen, and medications to treat severe bronchospasm (e.g., $\beta_2$-adrenergic receptor agonists for inhalation) should be available during testing (Brunton, 2011).

### 1.6.5   PILOCARPINE

#### 1.6.5.1   PHARMACOKINETICS

Leaflets of South American shrubs of *Pilocarpus microphyllus* contain pilocarpine as the chief alkaloid. Being tertiary amine, pilocarpine readily gets absorbed from most sites of administration and crosses BBB. It mainly gets excreted through urine. Decreasing pH of the urine promotes its clearance (Katzung et al., 2009). Although the specific metabolic pathways have not been explained, pilocarpine clearance is decreased in patients with hepatic impairment, in whom doses may need to be reduced (Brunton, 2011).

### 1.6.5.2  MECHANISM OF ACTION

Pilocarpine has a dominant muscarinic action but is a partial agonist (Brunton, 2011).

### 1.6.5.3  PHARMACOLOGICAL ACTIONS

It shows some selectivity in stimulating secretion from various exocrine glands, such as sweat, salivary, lacrimal, and bronchial glands, and contracting iris smooth muscle. It has weak effects on GI smooth muscle and the heart (Rang et al., 2011). The sweat glands are sensitive to pilocarpine. After isolation of pilocarpine in 1875, shortly thereafter, Weber described its actions on the pupil and on the sweat and salivary glands (Brunton, 2011).

### 1.6.5.4  THERAPEUTIC USES

Currently, natural alkaloids like pilocarpine are used clinically as a sialagogue and miotic agent (Brunton, 2011). Pilocarpine can cross the conjunctival membrane. It is a stable compound and its actions last for about 1 day (Rang et al., 2011). Pilocarpine hydrochloride is used in the treatment of xerostomia due to head and neck radiation treatments or associated with Sjogren's syndrome (Porter et al., 2004; Wiseman and Faulds, 1995). Sjogren's syndrome is an autoimmune disorder occurring primarily in women with altered secretions of lacrimal

and salivary glands (Anaya and Talal, 1999). If salivary parenchyma maintains residual function, enhanced salivary secretion, ease of swallowing, and subjective improvement in hydration of the oral cavity are achieved. The usual dose is 5–10 mg three times daily; the dose should be lowered in patients with hepatic impairment. Pilocarpine is used topically in ophthalmology for the treatment of glaucoma and as a miotic agent. It is instilled in the eye as a 0.5–6% solution or may be delivered via an ocular insert (Brunton, 2011).

### 1.6.5.5  ADVERSE EFFECTS

Adverse effects are due to stimulation of cholinergic system. Sweating is the common side effect (Brunton, 2011).

### 1.6.6  MUSCARINE

Schmiedeberg in 1869 isolated alkaloid muscarine from the mushroom *Amanita muscaria*. Muscarine acts mainly at mAChR sites. The classification of these receptors derives from the actions of this alkaloid. Muscarine, a quaternary amine, is poorly absorbed following oral administration and does not cross the BBB easily. Even though these drugs resist hydrolysis, the choline esters are short-acting agents due to rapid elimination by the kidneys. Muscarine can still, however, be toxic when ingested and can even have CNS effects (Brunton, 2011).

### 1.6.7  ARECOLINE

It is the main alkaloid obtained from seeds of areca or betel nuts, that is, *Areca catechu*. It acts at nicotinic receptors. Being tertiary amine, it readily gets absorbed and crosses BBB. Natives of the Indian subcontinent and East Indies were consuming this red staining betel nut as a euphoric, a mixture containing nut, shell lime, and leaves of a climbing species of pepper, *Piper betle* (Brunton, 2011).

### 1.6.8 POISONING

Exaggeration of various parasympathetic effects was shown after poisoning from the ingestion of plants containing pilocarpine, muscarine, or arecoline, and they resemble those produced after ingestion of genus *Inocybe*. Atropine should be given parenterally in large doses to

cross BBB. Supportive measures for respiratory and cardiovascular systems and to counteract pulmonary edema may also be used (Brunton, 2011).

## 1.7 INDIRECT-ACTING CHOLINERGIC AGONIST (ANTICHOLINESTERASES)

ChE inhibitors affect peripheral and central cholinergic synapses. Most of the peripherally acting anti-ChE drugs inhibit AChE and BuChE about equally (Fig. 1.5). Centrally acting anti-ChEs are developed for the treatment of dementia (Rang et al., 2011).

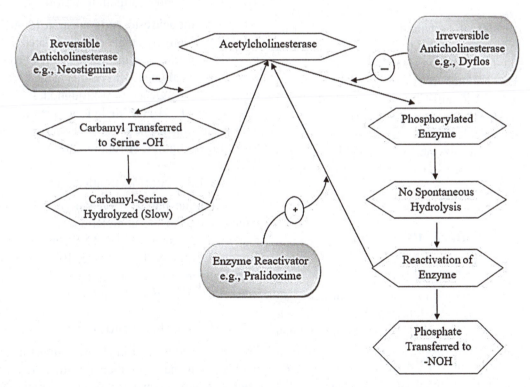

**FIGURE 1.5** Mechanism of action of indirect-acting cholinergic agonists. For reversible anti-ChE, for example, neostigmine, the recovery of activity by hydrolysis of the carbamylated enzyme takes many minutes. In case of irreversible anti-ChE, for example, dyflos, the reactivation of phosphorylated enzyme is accomplished by pralidoxime.

*Source*: Adapted from Rang et al. (2011).

### 1.7.1 REVERSIBLE ANTICHOLINESTERASES

At physiological pH, physostigmine and neostigmine possess positive charge, by serving as alternate substrates to ACh and generate carbamoylated enzyme by attacking the active center serine. Methylcarbamoyl AChE and dimethylcabamoyl AChE ($t_{1/2}$ for hydrolysis: 15–30 min) are more stable than acetyl enzyme (Sharma and Sharma, 2017).

### 1.7.1.1 CARBAMIC ACID DERIVATIVES

These are all carbamyl-possessing basic group binding with the anionic site. Like ACh, the transfer of carbamyl group to the serine hydroxyl group of the esteratic site occurs. The carbamylated enzymes get hydrolyzed slowly. It takes minutes for hydrolysis (Brunton et al., 2011). The anti-ChE drug gets hydrolyzed at a negligible rate compared with ACh. They are responsible for the slow recovery of the carbamylated enzyme, increasing the duration of action of these compounds (Brunton et al., 2011).

### 1.7.1.1.1 *Natural Alkaloid*

#### 1.7.1.1.1.1 *Physostigmine*

Physostigmine/eserine, a tertiary amine, is an alkaloid isolated by Jobst and Hesse in 1864. It was obtained from Calabar (ordeal) bean that was obtained from the dried, ripe seeds of a perennial plant of *Physostigma venenosum*. Calabar beans were used in witchcraft trials, in which guilt was judged by death from the poison, innocence by survival after ingestion of a bean. For the first time in 1877, physostigmine was used therapeutically that Laqueur used for

the treatment of glaucoma during his clinical practice (Karczmar, 1970; Holmstedt, 2000; Brunton, 2011).

#### 1.7.1.1.1.1 Pharmacokinetics

The absorption of physostigmine from the gastrointestinal tract, mucous membrane, and subcutaneous tissues are easy. Systemic effects may be observed after conjunctival instillation of solutions, if absorption from the nasal mucosa is not restricted. Hydrolytic cleavage of parenterally administered physostigmine occurs within 2–3 h by plasma esterases. Elimination by renal route is very less (Brunton, 2011).

#### 1.7.1.1.1.2 Mechanism of Action

It is a carbamic acid ester and a substrate for AChE forming a relatively stable carbamoylated intermediate with the enzyme which then becomes reversibly inactivated. This results in the potentiation of cholinergic activity throughout the body (Harvey et al., 2011). Methyl carbamate of an amine-substituted phenol is an essential moiety of the physostigmine (Brunton, 2011).

#### 1.7.1.1.1.3 Pharmacological Actions

Physostigmine has a broad variety of actions which is due to the stimulation of muscarinic and nicotinic sites of the ANS and nicotinic receptors of the NMJ. It crosses BBB to stimulate cholinergic receptors in the CNS. Its duration of action is about 2–4 h (Brunton, 2011).

#### 1.7.1.1.1.4 Therapeutic Uses

Physostigmine is used in atony of intestine and bladder as it increases their motility. It can be used to treat glaucoma as it produces miosis and spasm of accommodation as well as a lowering of IOP when instilled in the eye. It can be used to treat glaucoma, but pilocarpine is more effective. Also, it is used in the treatment of overdosage of drugs with anticholinergic drugs, such as atropine, phenothiazine, and tricyclic antidepressants (Brunton, 2011).

#### 1.7.1.1.1.5  Adverse Effects

In high doses, physostigmine affects CNS and may lead to convulsions. Bradycardia and fall in CO may also occur. Inhibition of AChE at NMJ leads to its deposition which causes paralysis. These effects are rarely seen in therapeutic doses (Harvey et al., 2011).

### 1.7.1.1.2  *Quaternary Compounds*

#### 1.7.1.1.2.1  *Neostigmine and Pyridostigmine*

##### 1.7.1.1.2.1.1  Pharmacokinetics

Larger doses of neostigmine and pyridostigmine are needed than by the parenteral route as absorption after oral administration is very poor. Both are destroyed by plasma esterases, $t_{1/2}$ is only 1–2 h. The parent compounds, as well as quaternary aromatic alcohols, are excreted via urine. Neostigmine is effective at 0.5–2-mg parenteral dose and the corresponding oral dose range between 15 and 30 mg or more (Cohan et al., 1976; Brunton, 2011).

##### 1.7.1.1.2.1.2  Mechanism of Action

They reversibly inhibit AChE similar to physostigmine. They have quaternary nitrogen. They are more polar and do not cross BBB. Their effect on skeletal muscle is greater than that of physostigmine. Before skeletal muscle gets paralyzed, they stimulate contractility (Brunton, 2011).

##### 1.7.1.1.2.1.3  Therapeutic Uses

Neostigmine stimulates the bladder and GIT. It is used as an antidote for tubocurarine and other competitive neuromuscular blockers and also in the symptomatic treatment of myasthenia gravis (Brunton, 2011). Pyridostigmine has slight (3–6 h) longer duration of action than neostigmine, it does not penetrate BBB and used in the chronic treatment of myasthenia gravis (Brunton, 2011).

##### 1.7.1.1.2.1.4  Adverse Effects

Neostigmine and pyridostigmine show adverse effects like flushing, salivation, nausea, abdominal pain, diarrhea, bronchial spasm, and decreased BP (Harvey et al., 2011).

#### 1.7.1.1.2.2  *Edrophonium*

##### 1.7.1.1.2.2.1  Pharmacokinetics

It is more rapidly absorbed. The activity of edrophonium is partial to synapses of the parasympathetic nervous system, has a reasonable attraction for AChE, and binds reversibly to the AChE active center. It is short-acting anti-ChE (duration of action: 10–20 min) with a limited volume of distribution and rapid renal elimination. Quaternary structure of edrophonium facilitates renal elimination (Brunton, 2011).

##### 1.7.1.1.2.2.2  Mechanism of Action

It is a synthetic, quaternary ammonium compound forming a readily reversible ionic bond by binding with anionic site of the enzyme (Sharma and Sharma, 2017).

##### 1.7.1.1.2.2.3  Therapeutic Uses

It is used mainly in the diagnosis of myasthenia gravis. Muscle weakness due to causes other than myasthenia gravis is not improved by an anti-ChE. Rapid increase in muscle strength is observed after i.v. injection (2 mg). An excess drug may provoke a cholinergic crisis and atropine may be used as an antidote (Harvey et al., 2011).

### 1.7.1.2 ACRIDINE DERIVATIVE

#### 1.7.1.2.1 Tacrine

Tacrine is more hydrophobic with a higher affinity for AChE. It inhibits the activity of AChE in the brain as it easily crosses the BBB. Partionining into lipid and higher affinity for AChE are responsible for a longer duration of action (Brunton, 2011).

### 1.7.1.3 MISCELLANEOUS

#### 1.7.1.3.1 Donepezil

It is a more selective AChE inhibitor used in Alzheimer's patients for the management of cognitive dysfunction associated with it. It is more hydrophobic, longer half-life, and administered once daily. It lacks the hepatotoxic effect of tacrine (Katzung et al., 2009).

### 1.7.1.4 THERAPEUTIC USES

The clinical uses of anti-ChEs are as follows:

- Neostigmine: At the end of an operation, it reverses the action of nondepolarizing neuromuscular blockers.
- Pyridostigmine or neostigmine: Treatment of myasthenia gravis.
- Edrophonium: A short-acting drug, given intravenously in the management of myasthenia gravis and to differentiate between muscle paralysis due to myasthenia gravis or cholinergic crisis at the motor end plate.
- Donepezil: In Alzheimer's disease.
- Ecothiopate: In glaucoma, as an eye drops (Katzung et al., 2009).

### 1.7.2 IRREVERSIBLE ANTICHOLINESTERASES

This group includes organophosphate (OP) compounds, such as war gases and pesticides, and carbamates.

### 1.7.2.1 MECHANISM OF ACTION

These compounds belong to pentavalent phosphorus group having fluoride as a labile group (in dyflos) or an organic group present in parathion and ecothiopate. This group gets released, by separating the serine hydroxyl molecule of the phosphorylated enzyme. Majority of the OP compounds were prepared for used as war gases, pesticides, and for clinical applications. They do not have cationic group except ecothiopate (contains quaternary nitrogen group binding with the anionic site). They interact only with the esteratic site of the enzyme (Katzung et al., 2009). The inactive phosphorylated enzyme is usually very stable. Significant hydrolysis does not occur with drugs such as dyflos and revival of enzyme activity based on the synthesis process of the new enzyme. The mechanism of action of ecothiopate is not firmly irreversible as the hydrolysis of drugs occurs slowly over the period of a few days (Katzung et al., 2009). As dyflos and parathion rapidly get absorbed from mucous membranes, steady skin, and insect cuticles, they are used as war gases or insecticides. They are nonpolar, volatile, and highly lipid soluble substances. They block other serine hydrolases (e.g., trypsin, thrombin) and may lack specificity—conferring to the quaternary group, although their pharmacological effects result mainly from ChE inhibition (Sharma and Sharma, 2017).

## 1.7.2.2 EFFECTS OF ANTICHOLINESTERASE DRUGS

### 1.7.2.2.1 Autonomic Cholinergic Synapses

Anti-ChE drugs mostly confirm improvement in ACh effects at parasympathetic postganglionic synapses by raising the secretions from glands such as bronchial, lacrimal, salivary, and GI. They are also responsible for increasing peristaltic property, constriction of bronchi, reduction in heart rate as well as BP, constriction of pupils, fixation of accommodation for near vision, and reduction in IOP (Sharma and Sharma, 2017). In large doses, first they activate and then inhibit autonomic ganglia resulting in complex effects related to ANS. Due to the deposition of ACh in plasma and body fluids, it produces a depolarization block. Neostigmine and pyridostigmine alter neuromuscular transmission greater than that produced by the autonomic system. Physostigmine and OP compounds show the exactly reverse pattern of action. Therapeutic usage of these compounds depends on this unfair selectivity. Acute anti-ChE poisoning (e.g., from contact with insecticides or war gases) produce a rigorous reduction in heart rate, reduction in BP, and complexity in respiration. These drugs may show fatal effects on the combination of a depolarizing neuromuscular block and central effects (Sharma and Sharma, 2017).

### 1.7.2.2.2 Neuromuscular Junction

Anti-ChEs prolong excitatory postsynaptic potential (EPP) due to repetitive firing in the muscle fiber, increasing twitch tension of a muscle stimulated via its motor nerve (Brunton, 2011). Usually, with each muscle fiber's stimulation, ACh produces one-action potential due to its fast hydrolysis, whereas in presence of AChE inhibitor, this is converted to a small coach of action potentials in the muscle fiber and hence greater tension developed. More significant action is observed when the transmission is blocked by tubocurarine; a competitive blocking agent and the transmission effects are dramatically restored by addition of anti-ChE. Blockade of a large proportion of the receptors increases the destruction of ACh molecules by an AChE molecule before getting an available receptor. But inhibition of AChE increases the chances of the ACh molecules to find a vacant receptor before being destroyed. This increases the EPP so that it reaches the threshold (Brunton, 2011). In myasthenia gravis, the transmission fails because there are too few ACh receptors, and ChE inhibition improves transmission just as it does in curarized muscle. In large doses, in case of poisoning, anti-ChEs initially cause twitching of muscles because spontaneous ACh release can give rise to EPPs that reach the firing threshold. Later, paralysis may occur due to depolarization block associated with the build-up of ACh in the plasma and tissue fluids (Brunton, 2011).

### 1.7.2.2.3 CNS

Physostigmine, a tertiary compound, and the organophosphorus, a non-polar compound, affect the brain region by easily crossing the BBB (Brunton, 2011). Activation of mAChRs is responsible for the entral effects, causing convulsions due to an initial excitation, followed by depression, which can cause unconsciousness and respiratory failure. These central effects are antagonized by atropine (Brunton, 2011).

### 1.7.2.3  ORGANOPHOSPHATE COMPOUNDS

#### 1.7.2.3.1  Malathion

For conferring the resistance to mammalians, malathion needs to change sulfur atom with oxygen in vivo. Detoxification occurs by hydrolysis of the carboxyl ester linkage by plasma carboxylesterases, whose activity indicates species resistance to malathion. Mammals and birds show a rapid detoxification reaction than in insects (Costa et al., 2003). Recently, malathion has been used in aerial spraying of relatively popular areas for control of citrus orchard-destructive Mediterranean fruitflies and mosquitoes that harbor and transmit viruses harmful to human beings, such as the West Nile encephalitis virus. Suicide attempts or deliberate poisoning is responsible for acute toxicity from malathion. The lethal dose of malathion in mammals is approximately 1 g/kg and skin–skin exposure leads to systemic absorption of a small fraction (<10%). In the management of pediculosis (lice) infestations, it is used as a topical preparation. Parathion and methyl parathion were widely used as insecticides because of their less volatility and good aqueous stability (Brunton, 2011).

#### 1.7.2.3.2  Isofluorophate

Isofluorophate [di-isopropyl flurophosphonate (DFP)] is low molecular weight, highly lipid soluble currently used insecticide. It creates almost irreversible inactivation of AChE and various esterases by alkyl phosphorylation. The DFP volatility characters facilitate inhalation and absorption through transdermal membrane and facilitate the entry into the CNS. After desulfuration, it forms dimethoxy or diethoxyphosphoryl enzyme (Brunton, 2011).

#### 1.7.2.3.3  Ecothiopate

It is a positively charged, nonvolatile, quaternary ammonium organophosphorus compound, useful clinically in ophthalmic conditions. It does not readily penetrate the skin (Brunton, 2011).

#### 1.7.2.3.4  Sarin, Tabun, and Soman

They are known as "nerve gases" and are the most powerful synthetic toxins known. In a preclinical study at nanogram doses, they show toxic effects. Dangerous used of these compounds occurred in warfare and terrorist attacks (Nozaki and Aikawa, 1995).

#### 1.7.2.3.5  Neurotoxicity

Accidental poisoning with many OPs like insecticides can produce a harsh type of demyelination in the peripheral nerve which further causes progressive weakness and loss of sensation. In 1931, contamination of fruit juice with an OP insecticide affected around 20,000 Americans (Brunton, 2011). The correct mechanism of the above reaction is ill understood, but it is considered to be a result from inhibition of an esterase (not ChE itself) precise to myelin. The activities of plasma and liver carboxylesterases (aliesterases) and plasma BuChE are blocked irreversibly by OPs (Lockridge and Masson, 2000); their scavenging capability for OPs can provide partial protection for inhibition of AChE in the nervous system. The carboxylesterases making malathion and other OPs less active or inactive by catalyzing the hydrolysis of carboxyl-ester linkages. Toxicity produced by simultaneous contact of two organophosphorus insecticides can show synergistic effects as carboxylesterases are also inhibited by OPs (Brunton, 2011).

### 1.7.2.4  TOXICITY

Potentially less harmful compounds have replaced the irreversible anti-ChEs for used at home and garden. Irreversible anti-ChEs are no active in inhibiting AChE in vitro; paraoxon is the dynamic metabolite (Brunton, 2011). Liver cytochrome oxidase enzymes carried out the phosphoryl oxygen for sulfur substitution, this reaction also occurs in insects, actually with high competence. Other insecticides, such as diazinon and chlorpyrifos, possess the phosphorothioate structure and are widely used for agriculture purpose, home, and in garden. The uses of such compounds are limited because of data of chronic toxicity in the new-born animal. They have been banned from indoor and outdoor residential use since 2005 (Brunton, 2011).

An increase in anti-ChE potency and duration of action can result from the linking of two quaternary ammonium moieties. Demecarium is the miotic agents. It consists of two neostigmine molecules connected by a series of 10 methylene groups, conferring additional stability to the interaction by associating with a negatively charged amino-side chain, Asn74, near the rim of the gorge. Carbamoylation inhibitors with high lipid solubilities (e.g., rivastigmine) readily cross the BBB and have longer durations of action. They are approved or in the clinical trial for the treatment of Alzheimer's disease (Cummings, 2004; Giacobini, 2000).

The carbamate insecticides used extensively as garden insecticides are carbaryl, aldicarb, and propoxur that inhibit ChE in a pattern similar to other carbamoylation inhibitors. The symptoms of poisoning closely resemble those of the OPs

(Baron, 1991; Ecobichon, 2000). Carbaryl has particularly low toxicity from dermal absorption. It is used topically for control of head lice in some countries. Not all carbamates in garden formulations are ChE inhibitors; the dithiocarbamates are fungicidal (Brunton, 2011).

## 1.8  CHOLINESTERASE REACTIVATION

Poisoning with OPs is very dangerous as spontaneous hydrolysis of phosphorylated ChE is extremely slow. Pralidoxime shows its action by reactivating the ChE enzyme by bringing the oxime group nearness with the phosphorylated esteratic site of the enzyme (Brunton, 2011). Oxime group is a strong nucleophile and attract the phosphate group missing from the serine hydroxyl group of the enzyme. Pralidoxime is effective in reactivating plasma ChE activity in a poisoned subject. It is used as an antidote in organophosphorus poisoning but the main disadvantage is that, within a short time, the phosphorylated enzyme undergoes a chemical alteration that renders it as no longer vulnerable to reactivation; to avoid this, pralidoxime is recommended to be given early for working better. Pralidoxime does not penetrate the brain, but compounds related to pralidoxime have been prepared to take care of central problems caused due to OP poisoning (Brunton, 2011).

## 1.9  ANTICHOLINESTERASE POISONING

### 1.9.1  SIGNS AND SYMPTOMS

#### 1.9.1.1  ACUTE INTOXICATION

Systemic effects, like ocular or respiratory, of anti-ChE acute intoxication appear within minutes after vapors or aerosols inhalation, while they are delayed after GI and percutaneous absorption. Ocular effects are ocular pain, miosis, conjunctival congestion, vision loss, ciliary contraction, and brow AChE. Symptoms related to respiratory system comprise chest tightness, whizzing respiration, and rhinorrhea and hyperemia due to combined effects of bronchoconstriction and increased bronchial secretion. GI symptoms include anorexia, nausea, vomiting, diarrhea, and abdominal cramp. Localized sweating and muscle fasciculations may be observed after percutaneous absorption of liquid (Brunton, 2011).

Signs and symptoms of muscarinic and nicotinic action are observed, CNS signs can be observed except with low lipid-soluble compounds (Costa, 2006). Toxic symptoms are due to lipid solubility which includes the strength of the OP–AChE linkage and phosphorylated enzyme aging (Brunton, 2011). Severe intoxication is visible by severe salivation, instinctive defecation and urination, sweating, lacrimation, erection of the penis, reduced heart rate, and hypotension (Brunton, 2011).

Fatigability, generalized weakness, involuntary twitchings, scattered fasciculations, severe weakness, and paralysis are due to nicotinic actions at NMJ. Paralysis of the respiratory muscles is the most serious effect (Xie et al., 2000). AChE appearance is lacking in skeletal muscle of mice, but in the brain and other organs innervated by ANS having a normal or near normal expression of AChE reproduce but have continuous tremors and severe compromise of skeletal muscle strength. According to these studies, the cholinergic system in the CNS adapts in development to chronically diminished hydrolytic capacity for AChE (Camp et al., 2008; Dobbertin et al., 2009).

### 1.9.1.2   BROAD SPECTRUM EFFECTS

Confusion, ataxia, inaudible speech, generalized seizures, no reflexes, Cheyne–Stokes respiration, unconsciousness, and central respiratory paralysis are some of the symptoms of acute AChE inhibition. Hypotension is due to an effect on vasomotor and other cardiovascular centers in the medulla oblongata (Brunton, 2011). Death after single-acute administration occurs from 5 min to about 24 h based on the nature of the agent, its dose, and route of exposure. Death occurs mainly because of the collapse of respiration associated with a secondary cardiovascular complication. Respiratory effects causing laryngospasm, increased salivary and tracheobronchial secretions, bronchoconstriction, cooperated voluntary control of the diaphragm and intercostal muscles, and central respiratory failure are because of peripheral muscarinic, nicotinic, and central actions (Brunton, 2011). Hypotension and cardiac arrhythmias are because of hypoxemia which is reversed by providing adequate ventilation to the lungs. Symptoms of the intermediate syndrome include delayed symptoms appearing after 1–4 days and with persistent low ChE and severe muscle weakness. Severe intoxication may be associated with delayed neurotoxicity (Lotti, 2002).

### 1.9.2   DIAGNOSIS AND TREATMENT

The diagnosis of acute anti-ChE intoxication is based on the history of its contact and specific symptoms. Measurement of ChE activities in erythrocytes and plasma helps the diagnosis in a suspected person with mild acute or chronic intoxication (Storm et al., 2000). The activities of ChE in the erythrocytes and plasma differ significantly in the normal people, whereas the activities almost suppressed beyond the normal value before the observation of symptoms. Sufficient doses of atropine significantly antagonize (mAChR) the activities such as rise in tracheobronchial secretion as well as salivary secretion, constriction of bronchi, reduction in heart rate, and to a reasonable level, peripheral ganglionic, and central actions. CNS effects of atropine are achieved after larger doses only. Pralidoxime, a ChE reactivator, reverses peripheral neuromuscular effects not reversed by atropine (Brunton, 2011).

Pralidoxime should be infused intravenously in a dose of 1–2 g over not <5 min in case of moderate or severe organophosphorus anti-ChE poisoning. In case if the weakness is not relieved, the dose of pralidoxime needs to be repeated. The initial treatment promises that the oxime group of drugs reaches the phosphorylated AChE, while the afterward still can be reactivated (Brunton, 2011). Most alkyl phosphates are highly soluble in lipid. Symptoms of toxicity may recur after initial treatment if extensive portioning between body fat has occurred. In such a situation, desulfuration is necessary for inhibition of AChE activity. It is always recommended to prolong the therapy atropine and pralidoxime for a week or longer for rigorous toxicities observed from the agents soluble in lipids (Brunton, 2011).

### 1.9.3 GENERAL SUPPORTIVE MEASURES

General supportive measures include the following:

- Exposure needs to be terminated by removal of the patient. If the atmosphere remains contaminated, a gas mask may be useful.
- Contaminated clothing to be removed and destructed. Contaminated skin or mucous membrane to be washed vigorously with water or gastric lavage may be used.
- Patient airway maintenance, including endobronchial aspiration. Artificial respiration may be provided.
- Diazepam may be used for alleviation of persistent convulsions (5–10 mg, intravenously).
- Treatment of shock (Brunton, 2011).

The doses of atropine should be administered in sufficient concentration so as to penetrate the BBB. Initially, 2–4 mg may be given intravenously, or 2 mg intramuscularly after every 5–10 min up to the symptoms of muscarinic agents disappear or the signs of atropine toxicity observed. The first time, the dose required may be more than 200 mg. A mild atropine block should be maintained until symptoms are evident (Brunton, 2011). The AChE reactivating agents are recommended as supplemental along with atropine therapy in the management of anti-ChE intoxication (Brunton, 2011).

### 1.9.4 DELAYED NEUROTOXICITY OF ORGANOPHOSPHORUS COMPOUNDS

Delayed neurotoxicity may be observed after some fluorine-containing organophosphorus anti-ChE agents, for example, DFP (Brunton, 2011). Muscle fatigue and twitching, mild sensory disturbances, ataxia, weakness, reduced tendon reflexes, and tenderness to palpitation are clinical symptoms of a severe polyneuropathy. The failing may develop to sagging paralysis and muscle slaying in severe cases. Improvement may not be complete even after several years as well (Brunton, 2011).

OP-induced delayed polyneuropathy is associated with a distinct esterase and neurotoxic esterase and is not dependent upon inhibition of ChEs (Johnson, 1993). Natural substrate and function of this enzyme are unclear, although it has specificity for hydrophobic esters (Glynn, 2000). After long-term exposure to OPs, myopathies resulting in generalized necrotic lesions and alters in end-plate cytostructure are observed in preclinical studies (De Bleecker et al., 1991).

## 1.10 CLINICAL USES OF CHOLINERGIC AGONISTS

### 1.10.1 GLAUCOMA

Glaucoma is a neurodegenerative condition; worldwide, it is the second-ranked principal problem of vision loss. Diagnosis of glaucoma is frequently delayed as it does not show any major symptoms up to its end stage of progress. It is a disorder of continuous optic neuropathies, which is observed by the degeneration of retinal ganglion cells and consequentially alters in the optic nerve head. Elevated IOP leads to loss of ganglion cells. Treatment usually involves the reduction of IOP and is initiated with ocular hypotensive drops. Sometimes, disease progression may be slowed by laser trabeculoplasty and surgery (Weinreb et al., 2014). Patients may experience gradual field loss and even lose sight completely if the disease remains untreated (Weinreb et al., 2014). Glaucoma may produce progressive neuropathy in the optic field of the eye, which is characterized by structural alterations in the optic nerve head as well as in optic disk causing functional alterations in patient's visual field (Fraser and Manvikar, 2010). The difference in two major types of glaucoma is given in Table 1.4.

**TABLE 1.4** Difference Between Open- and Close-Angle Glaucoma.

| Open-angle glaucoma | Angle-closure glaucoma |
|---|---|
| Observed in 90% cases | Rarely observed |
| It is due to slow obstruction of the drainage canals which elevates IOP | Develops very quickly |
| Angle between iris and cornea is large and unlock | Angle between iris and cornea is closed and thin angle |
| Symptoms and harm are difficult to observe | Symptoms and harm are easily observed |

IOP, intraocular pressure.

### 1.10.1.1 GLAUCOMA TREATMENT

The chronic form of open-angle glaucoma will be responded significantly to medical as well as pharmacological treatment. Primary drug treatment includes topical β-blockers (e.g., timolol) or prostaglandin analogs (e.g., latanoprost). Establishment of target IOP goal and initiation of first-line therapy is after through judgment of patient and after getting baseline IOP level. To create effectiveness and tolerability prior to starting therapy in left as well as right eye, treatment should continue a systematic approach and should start with a single drug in one eye only. But such approaches may be conservative when visual field loss has already set in (Fiscella et al., 2011; Weinreb et al., 2014).

### 1.10.1.1.1 Nonpharmacological Measures

Surgical intervention by iridectomy using a surgical laser is helpful in closed-angle glaucoma. Aqueous humor is allowed to drain freely through severely congested, hypertensive eye via an opening in the iris (Fiscella et al., 2011; Weinreb et al., 2014).

### 1.10.1.1.2 Pharmacological Measures

Drug therapy may also be employed in patients with acute closed-angle and simultaneous primary open-angle glaucoma and ophthalmic perioperative situation. For getting treatment success, it is necessary to have proper patient counseling concerning the proper and safe use of topical eye formulations (Fiscella et al., 2011; Weinreb et al., 2014).

The actual level of IOP determines the retinal ganglion cell death in glaucoma. Glaucoma treatment mainly involves decreasing the IOP using α-adrenergic agonists, β-blockers, carbonic anhydrase inhibitors, cholinergic agonists, and prostaglandin analogs. Appropriate drug as the first-line agent should be recommended in the patients having achievable glaucoma, such condition is with ocular hypertension (IOP > 22 mmHg) (Fiscella et al., 2011).

#### 1.10.1.1.2.1 Cholinergic Agonists

Pilocarpine and carbachol increase aqueous humor drainage through trabecular outflow route due to stimulation of mAChRs in meridional ciliary muscle fiber stimulation. Resistance to aqueous outflow decreases due to the contraction of smooth muscle, with a subsequent opening of the trabecular spaces (Fiscella et al., 2011; Brenner and Stevens, 2013). The adverse-effect profile and frequent dosing requirement, decreased the usefulness of these older type agents in the treatment of glaucoma. Pilocarpine was previously a primary drug which can achieve an equivalent reduction in IOP as that of β-blocking drugs (Fiscella et al., 2011; Brenner and Stevens, 2013).

### 1.10.2 MYASTHENIA GRAVIS

Myasthenia gravis is associated with irregular functioning of NMJ resulting in muscle weakness aggravating with muscle use and recovering with rest. Myasthenia may also be associated with the involvement of synaptic cleft or the presynaptic motor-nerve terminal. Most of the myasthenic disorders are due to the irregular functioning of the postsynaptic element of NMJ (Richman, 2015).

#### 1.10.2.1 TREATMENT

Treatment of Myasthenia gravis includes symptomatic, immunomodulatory (steroids and nonsteroidal agents) treatment or rapidly acting immune therapy and thymectomy (Gilhus and Verschuuren, 2015).

##### 1.10.2.1.1 Cholinesterase Inhibitors

The most common first treatment is pyridostigmine bromide with initial doses of 30–60 mg every 3–6 h, decreasing ACh breakdown by AChE inhibition. The maximum effectiveness is observed in the premature diseased state. Patients may develop tolerance after some time requiring an increase in dose. Single doses (120–180 mg) or greater hardly ever provide better benefit and lead to higher side effects (Kumar and Kaminski, 2011).

## 1.11 FUTURE OPPORTUNITIES AND CHALLENGES

As discussed above, cholinergic agonists are useful in many therapeutic conditions and as insecticides as well. But various side effects limit their use. Scientists have been working to develop new molecules to minimize or avoid these side effects. ACh, a direct acting agonist, has few clinical applications as it is rapidly hydrolyzed. Long-acting synthetic cholinergic agonists are preferred for clinical use (Ebert, 2013). Tacrine (Kromer and Eltze, 1991) is a ChE-blocking drug, and it was approved in 1993 by the FDA for the treatment of Alzheimer's disease. It has to be given four times a day due to short half-life of only 1.4–3.6 h. Donepezil, approved in 1996, has more selective mode of action, fewer side effects and long $t_{1/2}$ (70 h), and makes unit dose enough in a day (Mehta et al., 2012). Important advances are again required for the successful treatment of CNS problems such as Alzheimer's disease. Novel muscarinics provide leads for new classes of muscarinics. Many compounds are still in trials and which might be a novel substitutes for conventional cholinergic agonists in the near future. Safety and prolong the effectiveness of such compounds require sufficient time, in-depth clinical studies, and evaluation, but it can be a start for the development of safe cholinergic agonists.

## 1.12 CONCLUSION

In the above chapter, the authors have discussed the types of cholinergic agonists, their uses, mechanism of action, adverse effects, and poisoning. Although cholinergic agonists are widely used for mainly glaucoma, myasthenia gravis, and overdosage of parasympatholytics, ACh cannot be used therapeutically due to widespread action and easy hydrolysis. Irreversible anti-ChEs are also useful as insecticides worldwide. For this, various synthetic cholinergic agonists have been developed with good therapeutic usefulness.

## KEYWORDS

- **cholinomimetics**
- **acetylcholine**
- **Vagusstoff**
- **anticholinesterases**
- **glaucoma**

## REFERENCES

Anaya, J. M.; Talal, N. Sjogren's Syndrome Comes of Age. *Semin. Arthritis Rheum.* **1999**, *28*, 355–359.

Baron, R. L. Carbamate Insecticides. In *Handbook of Pesticide Toxicology*, Vol. 3; Hayes, W. J., Jr., Laws, E. R., Jr., Eds.; Academic Press: San Diego, CA, 1991.

Bonner, T. I.; Buckley, N. J.; Young, A. C.; Brann, M. R. Identification of a Family of Muscarinic *Acetylcholine* Receptor Genes. *Science* **1987**, *237*, 527.

Brenner, G. M. Stevens, C. W. *Pharmacology*, 4th ed.; Elsevier Saunders: Philadelphia, PA, 2013.

Broadley, K. J. *Autonomic Pharmacology*; Taylor & Francis: London, 1996.

Brunton, L. L.; Chabner, B. A.; Knollmann, B. C. *Goodman and Gillman: Pharmacological Basis of Therapeutics*, 12th ed.; McGraw-Hill, Medical Publishing Division: New York, 2011.

Caulfield, M. P. Muscarinic Receptors—Characterization, Coupling and Function. *Pharmacol. Ther.* **1993**, *58*, 319–379.

Caulfield, M. P.; Birdsall, N. J. International Union of Pharmacology, XVII. Classification of Muscarinic Acetylcholine Receptors. *Pharmacol. Rev.* **1998**, *50*, 279–290.

Chatonnet, A.; Lockridge, O. Comparison of Butyrylcholinesterase and Acetylcholinesterase. *Biochem. J.* **1989**, *260*, 625–634.

Costa, L. G. Current Issues in Organophosphate Toxicology. *Clin. Chim. Acta* **2006**, *366*, 1–13.

Costa, L. G.; Cole, T. B.; Furlong, C. E. Polymorphisms of Paroxonase and Their Significance in Clinical Toxicology of Organophosphates. *J. Toxicol. Clin. Toxicol.* **2003**, *41*, 37–45.

Crapo, R. O.; Casaburi, R.; Coates, A. L.; et al. Guidelines for Methacholine and Exercise Challenge Testing—1999. *Am. J. Respir. Crit. Care Med.* **2000**, *161*, 309–329.

Cummings, J. L. Alzheimer's Disease. *N. Engl. J. Med.* **2004,** *351,* 56–67.

De Bleecker, J.; Willems, J.; De Reuck, J.; Santens, P.; Lison, D. Histological and Histochemical Study of Paraoxon Myopathy in the Rat. *Acta Neurol. Belg.* **1991,** *91,* 255–270.

Ecobichon, D. J. Carbamates. In *Experimental and Clinical Neurotoxicology,* second ed.; Spencer, P. S., Schauburg, H. H., Eds.; Oxford University Press, New York, 2000.

Eglen, R. M.; Hegde, S. S., Watson, N. Muscarinic Receptor Subtypes and Smooth Muscle Function. *Pharmacol. Rev.* **1996,** *48,* 531–565.

Eglen, R. M.; Choppin, A., Dillon, M. P. *et al.* Muscarinic Receptor Ligands and Their Therapeutic Potential. *Curr. Opin. Chem. Biol.* **1999,** *3,* 426–432.

Fiscella, R. G.; Lesar, T. S.; Edward, D. P. Glaucoma. In *Pharmacotherapy: A Pathophysiological Approach,* eighth ed.; Dipiro, J. T., Talbert, R. L., Yee, G. C., et al., Eds.; McGraw-Hill Medical: New York, 2011.

Fisher, J. T.; Vincent, S. G.; Gomeza, J.; et al. Loss of Vagally Mediated Bradycardia and Bronchoconstriction in Mice Lacking M2 or M3 Muscarinic Acetylcholine Receptors. *FASEB J.* **2004,** *18,* 711–713.

Fraser, S. and Manvikar, S. Glaucoma-the pathophysiology and diagnosis. Hospital Pharmacist-London, **2005,** *12*(7), p.251.

Furchgott, R. F. Endothelium-Derived Relaxing Factor: Discovery, Early Studies, and Identification as Nitric Oxide. *Biosci. Rep.* **1999,** *19,* 235–251.

Gautam, D.; Heard, T. S.; Cui, Y.; et al. Cholinergic Stimulation of Salivary Secretion Studied with M1 and M3 Muscarinic Receptor Single- and Double-Knockout Mice. *Mol. Pharmacol.* **2004,** *66,* 260–267.

Giacobini, E. Cholinesterase Inhibitors: From the Calabar Bean to Alzheimer's Therapy. In *Cholinesterases and Cholinesterase Inhibitors*; Giacobini, E., Ed.; Martin Dunitz: London, 2000.

Gilhus, N. E.; Verschuuren, J. J. Myasthenia Gravis: Subgroup Classification and Therapeutic Strategies. *Lancet Neurol.* **2015,** *14,* 1023–1036.

Glynn, P. Neural Development and Neurodegeneration: Two Faces of Neuropathy Target Esterase. *Prog. Neurobiol.* **2000,** *61,* 61–74.

Goyal, R. K. Muscarinic Receptor Subtypes. Physiology and Clinical Implications. *N. Engl. J. Med.* **1989,** *321,* 1022–1029.

Goyal, R. K.; Rattan, S. Neurohumoral, Hormonal, and Drug Receptors for the Lower Esophageal Sphincter. *Gastroenterology* **1978,** *74,* 598–619.

Harvey, R. A.; Clark, M. A.; Finkel, R.; Rey, J. A.; Whalen, K. *Lippincott's Illustrated Reviews: Pharmacology,* fifth ed.; Wolters Kluwer: New York, 2011.

Heckmann, J. M.; Rawoot, A.; Bateman, K.; Renison, R.; Badri, M. A Single-Blinded Trial of Methotrexate versus Azathioprine as Steroid-Sparing Agents in Generalized Myasthenia Gravis. *BMC Neurol.* **2011,** *11,* 97.

Holmstedt, B. Cholinesterase Inhibitors: An Introduction. In *Cholinesterases and Cholinesterase Inhibitors*; Giacobini, E., Ed.; Martin Dunitz: London, 2000.

Hulme, E. C.; Birdsall, N. J. M.; Buckley, N. J. *Annu. Rev. Pharmacol. Toxicol.* **1990,** *30,* 633–673.

Johnson, M. K. Symposium Introduction: Retrospect and Prospects for Neuropathy Target Esterase (NTE) and the Delayed Polyneuropathy (OPIDP) Induced by Some Organophosphorus Esters. *Chem. Biol. Interact.* **1993,** *87,* 339–346.

Karczmar, A. G. History of the Research with Anticholinesterase Agents. In *Anticholinesterase Agents, vol. 1, International Encyclopedia of Pharmacology and Therapeutics, section 13*; Karczmar, A. G., Ed.; Pergamon Press, Oxford, 1970.

Katzung, B. G.; Masters, S. B.; Trevor, A. T. *Basic and Clinical Pharmacology,* 11th ed.; Lange Medical Publications: California, 2009.

Khurana, S.; Chacon, I.; Xie, G.; et al. Vasodilatory Effects of Cholinergic Agonists Are Greatly Diminished in Aorta from M3R–/– Mice. *Eur. J. Pharmacol.* **2004,** *493,* 127–132.

Kromer, W.; Eltze, M. J. Is Field (Vagal) Stimulation of Gastric Acid Secretion Mediated by M1 or non-M1 Muscarinic Receptors? A Methodical Problem Exemplified in the Mouse Stomach In Vitro. *Auton. Pharmacol.* **1991,** *11,* 337–342.

Kumar, V.; Kaminski, H. J. Treatment of Myasthenia Gravis. *Curr. Neurol. Neurosci. Rep.* **2011,** *11,* 89–96.

Lamping, K. G.; Wess, J.; Cui, Y.; et al. Muscarinic (M) Receptors in Coronary Circulation. *Arterioscler. Thromb. Vasc. Biol.* **2004,** *24,* 1253–1258.

Lockridge, O.; Masson, P. Pesticides and Susceptible Populations: People with Butyrylcholinesterase Genetic Variants May be at Risk. *Neurotoxicology* **2000,** *21,* 113–126.

Lotti, M. Low-Level Exposures to Organophosphorus Esters and Peripheral Nerve Function. *Musc. Nerve* **2002,** *25,* 492–504.

Massoulie, J., Pezzementi, L., Bon, S., Krejci, E.; Vallette, F. M. Molecular and Cellular Biology of Cholinesterases. *Progr. Neurobiol.* **1993,** *41,* 31–91.

Matsui, M.; Motomura, D.; Karasawa, H.; et al. Multiple Functional Defects in Peripheral Autonomic Organs in Mice Lacking Muscarinic Acetylcholine Receptor Gene for the M3 Subtype. *Proc. Natl. Acad. Sci. U.S.A.* **2000,** *97,* 9579–9584.

Millar, N. S. Assembly and Subunit Diversity of Nicotinic Acetylcholine Receptors. *Biochem. Soc. Trans.* **2003,** *31,* 869.

Moncada, S.; Higgs, E. A. Molecular Mechanisms and Therapeutic Strategies Related to Nitric Oxide. *FASEB J.* **1995,** *9,* 1319–1330.

Nozaki, H.; Aikawa, N. Sarin Poisoning in Tokyo Subway. *Lancet* **1995,** *346,* 1446–1447.

Porter, S. R.; Scully, C.; Hegarty, A. M. An Update of the Etiology and Management of Xerostomia. *Oral Surg. Oral Med. Oral Pathol. Oral Radiol. Endod.* **2004,** *97,* 28–46.

Rang, H. P.; Dale, M. M.; Ritter, J. M.; Flower, R. J.; Henderson, G. *Rang & Dale's Pharmacology,* seventh ed.; Churchill Livingstone: Edinburgh, 2011.

Richman, D. P. The Future of Research in Myasthenia. *JAMA Neurol.* **2015,** *72,* 812–814.

Sharma, H. L.; Sharma, K. K. *Principles of Pharmacology,* third ed.; Paras Medical Publisher: Hyderabad, 2017.

Sieb, J. P. Myasthenia Gravis: An Update for the Clinician. *Clin. Exp. Immunol.* **2014,** *175,* 408–418.

Sneddon, P.; Graham, A. J. Role of Nitric Oxide in the Autonomic Innervation of Smooth Muscle. *Auton. Pharmacol.* **1992,** *12,* 445–456.

Soreq, H.; Seidman, S. Acetylcholinesterase—New Roles for an Old Actor. *Nat. Rev. Neurosci.* **2001,** *2,* 294–302.

Storm, J. E.; Rozman, K. K.; Doull, J. Occupational Exposure Limits for 30 Organophosphate Pesticides Based on Inhibition of Red Blood Cell Acetylcholinesterase. *Toxicology* **2000,** *150,* 1–29.

Taylor, P.; Camp, S.; Radic, Z. Acetylcholinesterase; In *Encyclopedia*; Academic Press: Cambridge, MA, 2009.

Volpicelli, L. A.; Levey, A. I. Muscarinic Acetylcholine Receptor Subtypes in Cerebral Cortex and Hippocampus. *Progr. Brain Res.* **2004,** *145,* 59–66.

Weinreb, R. N.; Aung, T.; Medeiros, F. A. The Pathophysiology and Treatment of Glaucoma: A Review. *JAMA* **2014,** *311* (18), 1901–1911.

Wess, J. Muscarinic Acetylcholine Receptor Knockout Mice: Novel Phenotypes and Clinical Implications. *Annu. Rev. Pharmacol. Toxicol.* **2004,** *44,* 423–450.

Wess, J.; Eglen, R. M.; Gautam, D. Muscarinic Acetylcholine Receptors: Mutant Mice Provide New Insights for Drug Development. *Nat. Rev. Drug Discov.* **2007,** *6,* 721–733.

Wiseman, L. R.; Faulds, D. Oral Pilocarpine: A Review of Its Pharmacological Properties and Clinical Potential in Xerostomia. *Drugs* **1995,** *49,* 143–155.

# CHAPTER 2

# Cholinergic Antagonists

VISHAL S. GULECHA*, MANOJ S. MAHAJAN, AMAN UPAGANLAWAR, ABDULLA SHERIKAR, and CHANDRASHEKHAR UPASANI

*SNJB's SSDJ College of Pharmacy, Nashik, Maharashtra, India*

*Corresponding author. E-mail: vishalgulecha7@gmail.com*

## ABSTRACT

Cholinergic antagonists are drugs that bind to two principal cholinergic receptors, that is, muscarinic and nicotinic and prevent the effects of acetylcholine and other cholinergic agonists. Agents that are selective blockers of muscarinic receptors are known as antimuscarinic agents (more accurate) or parasympatholytics or anticholinergic agents. Neuromuscular-blocking agents, those mostly acting as nicotinic antagonist, interfere the motor transmission toward effectors, that is, skeletal muscles. Ganglionic blockers preferably act through nicotinic receptors associated with parasympathetic and sympathetic ganglia. These anticholinergics agents have the least clinical importance. The basic classes of cholinergic antagonists have distinct pharmacological properties. The muscarinic receptor antagonists are the most clinically important drug class, whereas the neuromuscular and ganglionic blockers are only used in specialized conditions. This chapter focused on the brief about the pharmacology of cholinergic antagonists.

## 2.1 INTRODUCTION

The term anticholinergic is used for the drugs having antimuscarinic as well as antinicotinic effects. But atropine and the related drugs dealt here lack any cognizable antinicotinic effects (Sharma and Sharma, 2017). Moreover, among quaternary ammonium antinicotinic drugs, those that block nicotinic ($N_N$) receptors of autonomic ganglia are called ganglion blockers, whereas those that block sympathetic transmission at somatic neuromuscular transmission ($N_M$ receptor) are called neuromuscular blocking drugs or skeletal muscle relaxants. Hence, atropine and allied group of drugs are preferably called antimuscarinic drugs because they inhibit and block the actions of acetylcholine (ACh) at muscarinic receptors with low affinity for nicotinic receptors (Brunton et al., 2011; Sharma and Sharma, 2017).

The well-known naturally occurring antimuscarinic agents are atropine and scopolamine. Synthetically derived antimuscarinic agents include homatropine and tropicamide; those possess specific selectivity for typical subtypes of muscarinic receptors and are short-acting agents than atropine. The semisynthetic derivatives have different distribution pattern in the body and duration of action than atropine. The drug pirenzepine, a synthetic derivative, shows partial selectivity (or acts as inverse agonist) for typical muscarinic receptor has gained advantages in the treatment of acid-peptic diseases. Tolterodine, oxybutynin, and so on are the other examples used in the urinary incontinence (UI). Atropine at high doses causes central nervous system (CNS) stimulation followed by depression, whereas quaternary ammonium antagonists poorly cross the blood–brain barrier (BBB), so they produce negligible CNS effects. Atropine at a dose of 0.5 mg depress salivary and bronchial secretion and sweating, but 1–2-mg dose results in dilatation pupil, inhibition of near vision accommodation, and increases heart rate due to inhibition of vagal effect. Still higher doses, such as 5 or 10 mg and above, lead to inhibition of parasympathetic control of urinary bladder causing inhibition of micturition. These higher doses also inhibit gastrointestinal tract and cause decreased tone and motility of gut. Therefore, atropine fails to show specific selectivity toward different muscarinic receptor subtype as the extent of various organs function is governed through parasympathetic tone and the involvement of intramural neurons and reflexes (Brunton et al., 2011).

Cholinergic antagonists form a main group for respiratory disorders. These agents were used anciently. In different parts of the world, especially several Asian countries have indigenous plants containing alkaloids possessing anticholinergic activities, as found and available in Gandevia (1975). Until the early part of the 20th century, these conventional herbal preparations were imported and used in Western medicine. The use of these agents declined partly from 1920 due to the discovery of adrenergic agents (Shrivastava, 2017; Gandevia, 1975).

*Atropa belladonna* (*A. belladonna*) is a perennial herbaceous plant belonging to the *Solanaceae* Juss family. Since ancient times, this medicinal plant has been well known for spasmolytic and mydriatic effect. Atropine, an alkaloid isolated from the plant, is used by physicians to examine patients' retina. Like atropine, scopolamine and hyoscyamine have some sedative effect and causes relaxation of smooth muscles. The high alkaloid content was observed in roots and leaves of *A. belladonna* which has promising spasmolytic activities (Genova, 1984). The belladonna alkaloid, atropine, is one of the earlier antimuscarinic compounds. The advancement in the knowledge of neurotransmission resulted in the effective use of these drugs. The type of antimuscarinic agents includes the naturally occurring alkaloids, such as atropine and hyoscine (which is also known as scopolamine); the semisynthetic derivatives, such as homatropine and ipratropium; and the synthetic congeners, such as darifenacin (Sharma and Sharma, 2017).

Various physiological studies were carried out to show bronchodilator activity of anticholinergic agents. Bronchodialation is controlled by the cholinergic (parasympathetic) division of the autonomic nervous system (ANS) (Widdicombe, 1979; Nadel, 1980). In this research, the drugs, such as ipratropium, were approved as a bronchodilator by inhalation route as it lacks systemic effects (Brunton et al., 2011). The objectives of the present chapter are to represent the pharmacological aspects of muscarinic antagonist and related drugs. Further, the therapeutic benefits of ganglionic blockers and neuromuscular blocking drugs are discussed.

## 2.2 CLASSIFICATION OF CHOLINERGIC ANTAGONISTS

Cholinergic antagonists are the agents or drugs which blocks the actions of ACh on cholinergic receptors (Barar, 2004). The brief classification of drugs used as cholinergic antagonist is given in Table 2.1.

**TABLE 2.1** Classification of Cholinergic Antagonist.

| Class | Examples |
|---|---|
| A. Antimuscarinic agents | Atropine, benztropine, cyclopentolate, darifenacin, fesoterodine, ipratropium, oxybutynin, scopolamine, solifenacin, tiotropium, tolterodine, trihexyphenidyl, tropicamide, trospium chloride |
| B. Neuromuscular blockers | Cisatracurium, pancuronium, rocuronium, succinylcholine, vecuronium |
| C. Ganglionic blockers | Nicotine, lobeline, epibatidine, succinylcholine, decamethonium, trimethaphan, hexamethonium, chlorisondamine |

## 2.3 ANTIMUSCARINIC AGENTS

### 2.3.1 CLASSIFICATION OF ANTIMUSCARINIC AGENTS

Classification of muscarinic antagonists is based on their selectivity toward specific muscarinic (M) receptors or their nonsensitivity toward the M receptors. The classes of M receptor and their subgroups are shown in Table 2.2.

### 2.3.2 NONSELECTIVE MUSCARINIC RECEPTOR ANTAGONIST

### 2.3.2.1 ATROPINE

The effects of ACh at muscarinic receptors are competitively antagonized by atropine. It was earlier used for peptic ulcer management. Currently, atropine is used mostly in resuscitation, anesthesia, and ophthalmological examinations. It is used as sulfate salt, which is a more soluble salt form of atropine. It is used as an antidote in the management of organophosphorus poisoning (Heath and Meredith, 1992). The other uses of atropine include minimizing and counteracting the adverse effects of pilocarpine or neostigmine during treatment of myasthenia gravis (Lullmann et al., 1982). Muscarine poisoning resulting from the ingestion of fungi *Clitocybe* and *Inocybe* species can also be treated by atropine (Hase et al., 1984). Atropin is cautiously used in hypoxic patients in which it may induce ventricular tachycardia or fibrillation. So, before giving atropine in hypoxic individuals, hypoxia should be overcome by clearing airways, oxygen, and through mechanical ventilation (Hase et al., 1984; Matthew and Lawson, 1970; Heath and Meredith, 1992).

**TABLE 2.2** Classification of Antimuscarinic Agents.

| Antimuscarinic agents | | | | |
|---|---|---|---|---|
| Nonselective antimuscarinics | Selective muscarinic antimuscarinics | | | |
| | $M_1$ selective | $M_2$ selective | $M_3$ selective | $M_4$ selective |
| Atropine, scopolamine | Pirenzipine, telenzepine, trihexylphenidyl | Methoctramine | *Para*-fluorohexahydro-sila-difenidol, 1,1-dimethyl-4-diphenylacetoxypiperidinium iodide (4-DAMP) | Himbacine |

### 2.3.2.1.1   Chemistry

Atropine is found with hyoscyamine in *A. belladonna*. Hyoscyamine readily hydrolyzes to atropine in aqueous alcohol. Therefore, atropine is a naturally occurring alkaloid in the form of a racemic mixture of D- and L-hyoscyamine in equal parts (Finar, 1965). It was found that upon hydrolysis, atropine gave (±)-tropic acid $(C_9H_{10}O_3)$ and tropine $(C_8H_{15}ON)$, which was shown to be alcohol. The presence of hydroxyl group in the structure of tropine makes it a saturated compound. On the other hand, dicarboxylic acid derivative of tropine, the tropinic acid contains the same number of carbon atoms as that of tropine (Rang and Dale, 2007; Shrivastava, 2017).

### 2.3.2.1.2   Pharmacokinetics

#### 2.3.2.1.2.1   Absorption

It is absorbed from gastrointestinal tract (GIT) mainly from the duodenum and jejunum. Maximum radioactivity was observed 1 h when given orally (Beermann et al., 1971; Dollery, 1991). Atropine injected intramuscularly shows a 38% increase in intestinal transit time (Hardison et al., 1979). In children, the oral administration of atropine at a dose of 0.03 mg/kg shows peak plasma concentrations at 90 min (Gervais et al., 1997; Ruggieri et al., 1985).

Fifteen minutes following rectal administration, atropine shows peak in plasma concentrations of 0.7 µg/L in comparison to 2.4 µg/L achieved following intramuscular (IM) administration (Bejersten et al., 1985). In children below 15 kg, the peak plasma concentration after rectal administration was seen lower, but it is clinically insignificant in older children (Bejersten et al., 1985). Atropine absorbed sublingually is of minor clinical significance as compared with IM or subcutaneous (SC) administration. It was also noted that sublingual absorption of atropine was variable and low in pregnancy (Kanto and Pihlajamaki, 1986; Rothrock et al., 1993). Atropine sulfate for inhalation in the form of pressurized metered-dose inhaler resulted in peak serum concentrations of 4.9 µg/L after administration of 1.7 mg of atropine (Shrivastava, 2017). When instilled, 1% atropine solution reached to peak plasma concentration within 8 min (Bermann et al., 1971; Hardman, 1996).

Atropine and its sulfate salt achieve peak plasma levels in about 30 min after IM injection (Bermann et al., 1971, Ruggieri et al., 1985; Gervais et al., 1997). Rate of absorption of atropine at a dose of 0.02 mg/kg in full-term pregnant women after IM or SC routes showed insignificant difference (Ali-Melkkila et al., 1993; Friedl et al., 1988). For optimal drug absorption by the endotracheal route, atropine should be administered at 2–2.5 times of the recommended intravenous (IV) dose (Lee and D'Alonzo, 1993; Barnes et al., 1987).

#### 2.3.2.1.2.2   Distribution

Atropine distributes rapidly following IV administration. It was observed that only 5% of administered atropine remained in the blood compartment after 5 min of IV injection (Brunton et al., 2006; Kanto and Klotz, 1988). Elimination kinetics of therapeutic doses of atropine is an example of a two-compartment model. The apparent volume of distribution $(aV_d)$ of atropine is around 1–1.7 L/kg. It shows clearance ranging from 5.9 to 6.8 mL/kg/min. The half-life $(t_{1/2})$ of atropine is found to be 2.6–4.3 h (Kanto et al., 1981; Aaltonen et al., 1984).

Atropine rapidly crosses the placenta but its distribution into the amniotic fluid is biased (Kanto et al., 1981, 1988; Kivalo and

Saarikoski, 1977). Atropine in small quantities is found to appear in breast milk but there is a scarcity of supporting data (Dollery, 1991). It was revealed that atropine appears to a lesser extent in cerebrospinal fluid (CSF) following IV administration (Virtanen et al., 1982; Kanto and Klotz, 1988). Local and systemic application of atropine shows slow and incomplete penetration into the eye (Morton, 1939; Friedl et al., 1988).

### 2.3.2.1.2.3 Metabolism

Primary metabolic site of atropine is liver where it undergoes microsomal oxidation, especially by monooxygenases. Chromatographic separation (HPLC) of urine showed five different metabolites such as atropine, noratropine, tropine, atropine-$N$-oxide and tropic acid (Van der Meer et al., 1983). Thus, the fraction of administered atropine is metabolized and the remainder is excreted unchanged in the urine (Van der Meer et al., 1986). Negligible amount of atropine is excreted in Biles (Hinderling et al., 1985). After IV injection, 57% injected fraction appears unchanged in the urine as atropine in addition to the presence of 29% of tropine. It is noted that hepatic and renal diseases may influence the pharmacokinetics of atropine (Hinderling et al., 1985).

### 2.3.2.1.2.4 Elimination

About 50–60% of atropine is excreted unchanged in the urine. The elimination half-life ($t_{1/2}$, 5–10 h) of atropine is longer in children (<2 years age) due to greater a$V_d$. Individuals with age 70 years and above show prolonged $t_{1/2}$ from 10 to 30 h due to decreased clearance (Aaltonen et al., 1984; Virtanen et al., 1982).

Age is an important determinant in the pharmacokinetics and pharmacodynamics of atropine; therefore, both children and adults have sensitivity for atropine (Berg et al., 1959; Smith et al., 1990). IV atropine administration in patients with Down's syndrome may induce abnormal cardioaccelerator response (Harris and Goodman, 1969). There is decreased susceptibility to actions of atropine seen with albinism patients (Prud'homme et al., 1999). The mechanisms for these differences are poorly understood.

### 2.3.2.1.3 Pharmacodynamics

### 2.3.2.1.3.1 Mechanism of Action

Atropine and related compounds act as a competitive antagonist of ACh or other muscarinic agonists at muscarinic ($M_1$–$M_3$) receptors (Sharma and Sharma, 2017). The antagonism is reversible (or surmountable) and therefore the blockade by smaller doses of atropine can be surmounted by increasing the local concentration of ACh or muscarinic agonists. When atropine binds to muscarinic receptors, it blocks all actions of ACh or muscarinic agonist (such as the release of inositol triphosphate from $M_1$ and $M_3$ activation and inhibition of adenylate cyclase from $M_2$ receptor activation). The muscarinic receptor agonists are more efficacious in antagonizing the effects of exogenously administered choline esters than antagonizing the responses obtained after parasympathetic postganglionic nerve stimulation. This is because, ACh during nerve stimulation is released so close to the muscarinic receptor that the atropine molecule, being bulky, cannot effectively intercept between ACh and the muscarinic receptor. Both atropine and scopolamine are nonselective antagonists and hence block all ($M_1$–$M_3$) muscarinic receptors as shown in Figure 2.1 (Sharma and Sharma, 2017; Barar 2004).

### 2.3.2.1.3.2    Pharmacological Actions

#### 2.3.2.1.3.2.1    Cardiovascular System

Sinoatrial (SA) nodal $M_2$ receptors are blocked by atropine leading to tachycardia. Slow parenteral administration of atropine causes paradoxical transient slowing of heart rate due to initial blockade of the presynaptic $M_1$ and $M_2$ receptors that facilitate the release of ACh on SA node (Sharma and Sharma, 2017). Atropine-induced tachycardia is mostly observed in teenagers because of high vagal tone and low pulse rate; however, infants and elders are not shown this tachycardia as they have low vagal tone and high pulse rate (Sharma and Sharma, 2017). The tachycardia along with increased conduction velocity facilitates excitation of ventricles through atrioventricular node that increases the ventricular rate in atrial flutter or atrial fibrillation patients. Due to this reason, atropine is advantageous in the emergencies in acute myocardial infarction (Schulte et al., 1991). Atropine fails to produce any effects on blood vessels, as blood vessels contain muscarinic receptors that are lack of parasympathetic innervations (Sharma and Sharma, 2017).

#### 2.3.2.1.3.2.2    Gastrointestinal Tract

Atropine like antimuscarinic agents reduces about 45–50% of basal gastric acid secretions. Also, these drugs cause reduction in salivary secretions causing dryness of mouth and difficulty in swallowing and talking. Antimuscarinics reduce the tone and motility of GIT, thereby prolongs gastric emptying time, closure of sphincters, and reduced peristaltic movements (Sharma and Sharma, 2017). The contraction of GIT mediated by activation of $M_3$ receptors by ACh can be antagonized and inhibited by muscarinic antagonists. Antimuscarinics show insignificant inhibitory actions against increased contractility and motility of GIT associated with parasympathetic stimulation (Sharma and Sharma, 2017).

#### 2.3.2.1.3.2.3    Urinary Bladder

Urinary bladder consists of detrusor muscle containing $M_3$ receptors. The contraction of detrusor muscle in response to bladder filling results in initiation of micturition reflex mediated through the action of ACh. Muscarinic antagonists block the response of ACh leading to urinary retention, hence useful in the management of UI (Milard, 2004). Muscarinic antagonists are therefore used in incontinence, nocturnal enuresis, and spastic paralysis. Stress incontinence is associated with weak skeletal muscles of external sphincter, which are insusceptible to the actions of antimuscarinics. Currently used antimuscarinic drugs, such as propantheline and emepronium, are quaternary ammonium compounds that have the inability to cross BBB but show peripheral unwanted action like xerostomia, blurred vision, and tachycardia arising as a result of the nonselective blockade of muscarinic receptors (Barar, 2004; Shrivastava, 2017).

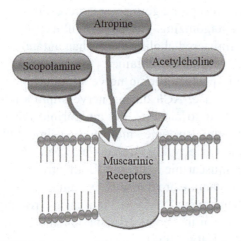

**FIGURE 2.1**   Competition of atropine and scopolamine with acetylcholine at muscarinic receptor.

Oxybutynin, a nonselective muscarinic blocker, is the drug of choice for UI. This agent has direct smooth muscle relaxant properties. The other drug, terodiline, possesses additional $Ca^{2+}$ channel-blocking activity that makes it important in the treatment of UI (Langtry and McTavish, 1990). A newer nonselective anti-muscarinic tolterodine has bladder selectivity over salivary secretions. Therefore, it is well tolerated and effective in patients with UI with reduced tendency to cause xerostomia (Chapple et al., 2004; Oyasu et al., 1994).

### 2.3.2.1.3.2.4   Airways

Airways innervated by parasympathetic nerves that release ACh through activation of M3 receptors induces bronchoconstriction along with increased mucus secretion without alteration in viscosity. In the treatment of asthma, antagonizing these effects by antimuscarinics are of limited utility but proved advantageous in acute exacerbations of asthma. Antimuscarinic agents, such as atropine, inhibit the nasal, oral, pharyngeal, and bronchial secretions, causing mucous membranes of respiratory tract to dry. They also cause reduction in mucous secretions from larynx, that is, laryngospasm during general anesthesia procedure (Sharma and Sharma, 2017; Shrivastava, 2017).

In case of patients with airway disease, conventional antimuscarinic drugs cause mucosal dryness and suppress mucociliary clearance. This result in the development of mucus plugs associated with airway obstruction and increases the incidence of infection in these patients.

Novel antimuscarinic congeners, such as ipratropium, tiotropium, and oxitropium, showed fewer effects on mucous secretion, therefore, have a lower risk of mucus plug formation, without inhibiting mucociliary movements. These agents, therefore, are advantageous over conventional antimuscarinics like atropine in blocking the indirect effects of inflammatory mediators of asthma (Sharma and Sharma, 2017).

Ipratropium, a muscarinic antagonist, is beneficial over atropine as it does not inhibit mucociliary movements (Gross, 1979). Selective blockade of $M_3$ together with $M_1$ receptors is useful in such conditions as $M_3$ receptors mediate bronchoconstriction and increased mucous secretion and $M_1$ receptors are associated with ganglionic transmission of the vagal reflex pathways. These actions may be counteracted by blocking presynaptic $M_2$ receptors of cholinergic nerve terminals that are autoinhibitory and facilitate ACh release (Alabaster, 1995). Tiotropium, a nonselective antimuscarinic used in chronic obstructive pulmonary disease (COPD), achieves $M_3$ selectivity (Maesen et al., 1993).

### 2.3.2.1.3.2.5   The Eye

Muscarinic antagonists are inhibitors of iris circular muscle contraction and also block the contraction of the ciliary body. The antimuscarinic actions such as pupillary dilation (mydriasis) and relaxation of the iris are mediated by $M_3$ receptors blockade. This may lead to raised intraocular pressure and aggravation of narrow-angle glaucoma in vulnerable individuals. Though antimuscarinics have ophthalmic uses, as in retinal examination by dilating the pupil, these agents may increase the risk of glaucoma and therefore contraindicated in patients at risk (Sharma and Sharma, 2017).

### 2.3.2.1.3.2.6   Prostate Gland

Human prostate glandular epithelium inner-vated by cholinergic nerves marked by the pres-ence of $M_1$ receptor is primarily responsible for secretions from the gland (Ruggieri et al., 1995; Sharma and Sharma, 2017). Muscarinic antagonists are generally contraindicated in patients with benign prostatic hypertrophies (BPH), but the use of $M_1$-selective muscarinic antagonists in the treatment of BPH is proved advantageous (Caine et al., 1975; Lau et al., 2000).

### 2.3.3 SELECTIVE MUSCARINIC RECEPTOR ANTAGONISTS

#### 2.3.3.1 PIRENZEPINE

A tricyclic antimuscarinic drug, pirenzepine, is an $M_1$-receptor selective agent (Caulfield and Birdsall, 1998). It has a higher affinity for $M_1$ receptors, specifically in the cerebral cortex and sympathetic ganglia. Pirenzepine shows minimal binding toward involuntary muscles (cardiac and smooth muscle) and glands at very low concentration. The drug is responsible for blocking the excitatory postsynaptic potential (EPSP) due to binding of ACh and other cholinergic agonists at ganglionic $M_1$ receptor thereby altering the functions of these $M_1$ receptors (Eglen et al., 1996; Caulfield and Birdsall, 1998). Pirenzepine, at low concentration, reduces the secretion of gastric acid and muscle spasm; therefore, it is used in the management of peptic ulcer. Pirenzepine also showed the contraction of lower esophageal sphincter due to blockade of ganglionic receptors (Wellstein and Pitschner, 1988). Adverse effects such as dry mouth, distorted visualization, and central muscarinic disturbances of other antimucarinics are observed to a lower extent with pirenzepine at therapeutic concentration. At a curative dose (100–150 mg/day), pirenzepine cure the duodenal and gastric ulcers as cimetidine and ranitidine which are $H_2$-receptor blockers (Carmine and Brogden, 1985; Tryba and Cook, 1997).

#### 2.3.3.2 TELENZEPINE

Telenzepine, a pirenzepine analog, possesses higher potency for $M_1$ receptor. Like pirenzepine, telenzepine is selective for $M_1$ receptor. It also binds to $M_3$ receptors without affecting their organizational structure (Pediani et al., 2017). This drug is mainly prescribed for acid-peptic disease in different parts of the world, but not in the United States (Brunton et al., 2011).

#### 2.3.3.3 TOLTERODINE

It is a newer antimuscarinic drug developed to treat overactive bladder in which there is an increased frequency of micturition and urgency and leakage. Tolterodine is equipotent with oxybutynin on muscarinic receptors of the urinary bladder, but it is shown to possess less affinity for salivary gland when compared to oxybutynin (Abrams, 1998). A 5-hydroxy-methyl metabolite of tolterodine, generated after metabolism by CYP2D6, possesses comparable activity to the parent drug (Hardman and Limbird, 1996).

#### 2.3.3.4 DARIFENACIN

Darifenacin is a drug meant for the treatment of UI. Its mechanism of action includes blocking of muscarinic $M_3$ receptor responsible for bladder muscle contractions, thereby decreases the urgency to urinate. About 98% of darifenacin is bound to plasma proteins mainly $\alpha_1$-acid–glycoprotein ($\alpha_1$-AGP). The drug has high a$V_d$, which is about 163 l (Steers, 2006). Darifenacin is extensively metabolized by cytochrome $P_{450}$ ($CYP_{450}$) enzymes, CYP2D6 and CYP3A4. The metabolic pattern of this drug has three steps including monohydroxylation in the dihydro-benzofuran ring, opening of previously formed dihydrobenzofuran ring, and at last pyrrolidine nitrogen undergoes *N*-dealkylation. Darifenacin is contraindicated in urinary retention, gastric retention, or uncontrolled narrow-angle glaucoma. The adverse effects of darifenacin are similar to other antimuscarinics and include dry mouth, dry eyes, constipation, dyspepsia, abdominal pain, nausea, diarrhea, and dizziness (Steers, 2006).

### 2.3.3.5  FESOTERODINE

It is a competitive antimuscarinic agent and a prodrug. It generates active metabolite known as 5-hydroxymethyl tolterodine, a hydrolytic product after metabolism by nonspecific esterases. Its bioavailability is about 52%, and it is bound to plasma protein mainly albumin and $\alpha_1$-AGP. The active metabolite is further biotransformed by hepatocytes by two cytochrome $P_{450}$ (CYP) enzymes, that is, CYP2D6 and CYP3A4. Most of the administered fraction (about 70%) of fesoterodine is found in urine. The drug is contraindicated in urinary and gastric retention, narrow-angle glaucoma, and so on. It is one of the drugs for the treatment of overactive bladder with UI, urgency, and frequency (Game et al., 2018).

### 2.3.3.6  SOLIFENACIN

Another competitive muscarinic receptor antagonist solifenacin shows peak plasma concentration within 3–8 h when administered orally. This drug has approximately 90% absolute bioavailability and about 98% is bound to plasma proteins, specifically to $\alpha_1$-AGP. Solifenacin is biotransformed by hepatocytes by CYP3A4 to four principal metabolites. Out of which, one is pharmacologically active

metabolite (4*R*-hydroxy solifenacin) and the remaining three are pharmacologically inactive metabolites. About 7% of the dose excreted unchanged in the urine. Solifenacin is contraindicated in urinary and gastric retention or uncontrolled narrow-angle glaucoma (Morales-Olivas, 2010).

### 2.3.4  THERAPEUTIC APPLICATIONS

### 2.3.4.1  PARKINSON'S DISEASE

Parkinson's disease (PD) characterized tremors and rigidity results from cholinergic overactivity in basal ganglia (Sharma and Sharma, 2017). The combination of antimuscarinic drugs with a dopaminergic drug is advantageous than using either drug alone. Most of the centrally acting antimuscarinics used in PD due to their ability to block muscarinic receptors in the striatum. These agents are used either alone in early stage or as an adjunct to levodopa in later stages of the disease. These drugs are also useful in relieving neuroleptic-induced extrapyramidal side effects. These drugs like benzhexol, procyclidine, benzatropine, and biperiden used in PD show adverse effects attributed to their anticholinergics effect that include dryness of mouth, blurred vision, confusion, reduced secretions, decreased motility as in urinary and gastric retention, hallucinations, and so on (Rang et al., 2003; Sharma and Sharma, 2017).

### 2.3.4.2  MOTION SICKNESS

Whenever the body is rotated or its equilibrium is distributed, the vestibular apparatus sends nauseating signals to the vomiting center. ACh serves as an excitatory neurotransmitter in such emesis of labyrinthine origin (Sharma and Sharma, 2017).

Scopolamine is the drug used for this purpose and is found to be more effective than the novel agents. It is administered as an injection or given orally or applied as a transdermal patch. Due to anticholinergics effects, scopolamine at therapeutic doses administered by any route may lead to sedation and dry mouth (Rang et al., 2003).

### 2.3.4.3  OPHTHALMOLOGIC DISORDERS

For carrying out an ophthalmoscopic examination, antimuscarinic agents are applied topically as eye drops or ointment, which proved advantageous (Sharma and Sharma, 2017). The shorter-acting antimuscarinics, such as tropicamide, eucatropine, or cyclopentolate, are preferred drugs used in adults and older children. Sometimes, greater efficacy of atropine is needed for children that at same time imposes more risk of antimuscarinic poisoning (McBrien et al., 2013). Antimuscarinics from tertiary amino group show enhanced penetration after conjunctival application. Preclinical evaluations indicate that a quaternary agent, glycopyrrolate which is analogous to atropine in onset and duration of action (McBrien et al., 2013).

Antimuscarinics should be avoided for mydriatic effects unless cycloplegia or prolonged action is required. α-Adrenoceptor agonists like phenylephrine is used for funduscopic examination to produce a short-lasting mydriasis. A second ophthalmologic use is in uveitis and iritis to prevent synechia (adhesion) formation. Homatropine like preparations are valuable and serves better for this purpose (McBrien et al., 2013).

### 2.3.4.4  RESPIRATORY DISORDERS

Atropine or scopolamine was used as a preanesthetic medication to prevent hazardous effects of ether that significantly increased airway secretions. This may lead to frequent episodes of laryngospasm, (McBrien et al., 2013). The preanaesthetic use of these agents became limited and declined after the availability of nonirritating agents, for example, halothane, enflurane, and so on. As glycopyrrolate has less adverse effects, it is preferred over scopolamine (Sharma and Sharma, 2017).

Ipratropium is used as an inhalational drug in the management of asthma and COPD. It is administered as aerosol due to advantages such as maximal concentration reaches the bronchial target tissue with reduced risk of systemic effects. Another antimuscarinic agent, tiotropium, has better bronchodilator action than ipratropium. Tiotropium is administered once in a daily basis. This drug is shown to reduce the severity COPD exacerbations. It is also used to increase exercise tolerance (Sharma and Sharma, 2017).

### 2.3.4.5  CARDIOVASCULAR DISORDERS

Antimuscarinics have limited use in cardiovascular disorders. It was observed that adequate doses of atropine can terminate reflux vagal cardiac slowing or asystole. In cases of second-degree heart block due to vagal activity as in digitalis toxicity, atropine may reduce the degree of heart attack (Sharma and Sharma, 2017). Administration of parenteral atropine or another similar muscarinic antagonist is beneficial in this situation. Patients with idiopathic dilated cardiomyopathy showed circulating autoantibodies that target the second extracellular loop of cardiac muscarinic receptors. Though the role of these antibodies in heart failure is unknown, the action of these antibodies is prevented by atropine (Field et al., 2010).

Intravenous administration of atropine can be done cautiously to treat sinus bradycardia accompanying myocardial infarction (Sharma and Sharma, 2017).

### 2.3.4.6 GASTROINTESTINAL DISORDERS

The main therapeutic use of antimuscarinics in GI disorders is to treat peptic ulcer. Drugs, such as propantheline and glycopyrrolate, were more preferred due to their inability to cross BBB evoking minimal unwanted effects including blurred vision, xerostomia, constipation, and urinary retention which limits their clinical utility (Sharma and Sharma, 2017). The development of selective $M_1$ blockers, such as pirenzepine and telenzepine, increases the utility of these drugs in peptic ulcer. These drugs were also effective in preventing the recurrence of duodenal ulcer (Srivastava, 2017).

Muscarinic antagonists are found to be effective in common traveler's diarrhea and other conditions associated with hypermotility. Combination of antimuscarinics with antidiarrheal opioids is an extremely effective therapy in diarrhea like condition. The classic combination of atropine with diphenoxylate is available under many names (e.g., Lomotil, Maidotril) in different dosage forms, like tablets and liquid orals. Belladonna alkaloids and the synthetic tertiary amine derivatives, such as dicyclomine, are very effective in reducing excessive salivation associated with metal poisoning (Srivastava, 2017).

### 2.3.4.7 URINARY DISORDERS

Urinary urgency induced by minor inflammatory bladder disorders can be treated by atropine and other muscarinic antagonists. Urinary bladder predominantly consists of $M_2$ and $M_3$ receptors mediating direct activation of contraction of detrusor muscles, especially by $M_3$ receptors subtype. Dicyclomine and oxybutynin are indicated in renal colic and urethral smooth muscle spasm (Sharma and Sharma, 2017; Barar, 2004). Trospium, a nonselective antagonist, is an alternative to oxybutynin. Trospium has better efficacy and reduced side effects than oxybutynin. Darifenacin and solifenacin are other approved agents with enhanced selectivity toward $M_3$ receptors. A directly acting smooth muscle relaxant, flavoxate, possesses anticholinergic and antispasmodic activities and is useful in urinary urge incontinence and for subrapubic pain in cystitis and urethritis (Srivastava, 2017; Sharma and Sharma, 2017).

### 2.3.4.8 SWEAT GLAND

Hyperhydrosis is sometimes treated with antimuscarinic drugs (like darifenacin, a selective $M_3$ receptor blocker), but these are rarely used because apocrine glands are involved in sweating (Sharma and Sharma, 2017).

### 2.3.5 ATROPIN/BELLADONNA POISONING

Use of organophosphorus insecticides, which are cholinesterase inhibitors, and ingestion of some wild mushrooms may lead to severe medical emergency resulting from excessive cholinergic activity that leads to intoxication. The symptoms of intoxications include severe dryness of mouth, and throat, wide pupillary dilation, dysphagia and thirst, tachycardia, redness of the skin, muscle incoordination, rise in body temperature, delirium, hallucinations, mania, apathy, stupor, coma, and finally respiratory collapse (Barar, 2004). The detailed side effects associated with anticholinergic drugs are given in Table 2.3.

**TABLE 2.3**   Spectrum of Anticholinergic Side Effects.

| Mild | Moderate | Severe |
|---|---|---|
| Dryness of mouth (modest) Moderately disturbing dry mouth/thirst | Moderately disturbing dry mouth/thirst; speech problems; reduced appetite | Difficulty in chewing, swallowing, speaking; impaired perception of taste and texture; mucosal damage, dental decay, malnutrition respiratory infection |
| Mild dilatation of pupils | Inability to accommodate; vision disturbances; dizziness; esophagitis; reduced gastric secretions, gastric emptying (atony); reduced peristalsis, constipation | Increased risk of accidents and falls, leading to decreased function; exac-erbation/precipitation of acute-angle closure glaucoma; fecal impaction (in patients with constipation); altered absorption of concomitant medications paralytic ileus, pseudo-obstruction |
| Urinary hesitancy | Increased heart rate | Urinary tract infections (in patients with urinary hesitancy); conduction disturbances, supraventricular tachyarrhythmia, exacerbation of angina, congestive heart failure |
| Decreased sweating | – | Thermoregulatory impairment leading to hyperthermia (heatstroke) |
| Drowsiness | Excitement | Profound restlessness and disorientation, agitation |
| Mild amnesia | Confusion | Hallucinations, delirium |
| Inability to concentrate | Memory impairment | Ataxia, muscle twitching, hyperreflexia, seizures; exacerbation of cognitive impairment (in patients with dementia) |

*Source:* Feinberg (1995).

### 2.3.5.1 DIAGNOSTIC TEST

Failure to produce typical muscarinic effects on the administration of cholinergic drugs such as methacholine (5 mg) or neostigmine (1 mg), by SC route easily makes the diagnosis apart from the clinical manifestation as mentioned above (Srivastava, 2017).

### 2.3.5.2 TREATMENT

#### 2.3.5.2.1 Antimuscarinic Therapy

Atropine is used for its effects on CNS and peripheral nervous system to treat organophosphate poisoning. Toxicity caused by parathion, an extremely potent agent, or by chemical warfare gases requires administration of high amount of atropine. To surmount the toxicity, IV administration of 1–2 mg of atropine sulfate for 5–15 min until signs of effect like xerostomia, reversal of miosis, and others appear. For full control of muscarinic excess, the dose of atropine must be repeated many times (e.g., 1 g of atropine per day for 1 month), as the short-term effects of the anticholinesterase agent may last for 24–48 h or longer (Chudoku, 2008; Sharma and Sharma, 2017).

#### 2.3.5.2.2 Cholinesterase Regenerators

Oxime group agents are also useful in the management of organophosphorus poisoning. These drugs govern hydrolysis of enzyme phosphorylated due to binding of organophosphorus. Hydrolytic action of these agents is capable of regenerating the active enzyme that allows dissociation from the complex. Pralidoxime (PAM) and diacetyl monoxime (DAM) are examples of oxime class. PAM is the highly efficient agent for the regeneration of cholinesterase found in neuromuscular junctions (NMJs) of skeletal muscle. PAM is unable to cross BBB; therefore, it is ineffective to overcome central manifestations of organophosphate poisoning (Sharma and Sharma, 2017).

DAM, on the other hand, crosses the BBB and showed regeneration of CNS cholinesterase in experimental animals. In severe poisoning, long-term treatment with PAM over several days is needed. Multiple-dose treatment with PAM is known to result in adverse effects, especially neuromuscular weakness. PAM is not used for the regeneration of enzyme resulting from carbamate inhibitors. Pretreatment with reversible enzyme inhibitors (pyridostigmine or physostigmine) to prevent binding of the irreversible organophosphate inhibitor is another approach in protection against excessive acetylcholinesterase (AChE) inhibition (Chudoku, 2008).

### 2.3.6 DRUG INTERACTIONS

Antimuscarinic agents show interaction with other categories of drugs, which are summarized in Table 2.4.

## 2.4 NEUROMUSCULAR BLOCKERS

Practice of anesthesia underwent revolution following the invention of neuromuscular blockers. Before the development of muscle relaxants, IV or inhalational anesthetic agents were used to induce and maintain anesthesia. Before the development of neuromuscular blockers, required muscle relaxation was induced by inhalation anesthesia that depresses respiratory or cardiac activities. With the development of muscle relaxants, anesthesia was redefined as a triad of narcosis, analgesia, and muscle relaxation. To produce these effects, specific drugs are used (Sharma and Sharma, 2017).

**TABLE 2.4**  Drug Interactions of Antimuscarinic Agents.

| Examples | Effects |
|---|---|
| Atropine or glycopyrrolate + alcohol | Impairment in attention and driving becomes hazardous |
| Antimuscarinic + antipsychotics | Produces additive effects and life-threatening interactions such as heat stroke in hot and humid climate, severe constipation, atropine like psychosis, dryness of mouth, etc. |
| Antimuscarinic + levodopa | Antimuscarinics may modestly reduce the rate and possibly the extent of levodopa absorption |
| Antimuscarinic + MAOIs | Increased antimuscarinic effects (synergistic effects) |
| Antimuscarinic + TCAs | Produces additive effects and atropine-like effects |
| Antimuscarinic + nitrates | Produces additive effects and atropine-like effects |
| Antimuscarinic + SSRIs | Delirium |

MAOIs, monoamine oxidase inhibitors; SSRIs, selective serotonin reuptake inhibitors; TCAs, tricyclic antidepressants.

*Source*: Baxter (2010).

The NMJ comprises the efferent nerve terminals, ACh as the neurotransmitter, and the muscle endplate postsynaptically (Lukasik, 1995). The action potential causes the release of ACh from the synaptic vesicles, which diffuses across the gap to the motor endplate (Lukasik, 1995; Barar, 2004). The motor endplate has specialized ligand-gated nicotinic ACh receptors, activation of which leads to depolarization in the postsynaptic membrane of striated muscle (Lukasik, 1995).

## 2.4.1 CLASSIFICATION OF NEUROMUSCULAR BLOCKERS

Neuromuscular blockers are classified into two groups based on the mode of action (Table 2.5), consisting of depolarizing blockers which depolarize the postsynaptic membrane to its refractory state (e.g., succinylcholine) and nondepolarizing blockers which competitively block the receptor site preventing the binding of ACh to its nicotinic receptor (e.g., D-tubocurarine) (Lukasik, 1995).

**TABLE 2.5**  Classification of Neuromuscular Blockers.

| Class | Examples |
|---|---|
| A. Depolarizing blockers | Succinylcholine, decamethonium, etc. |
| B. Nondepolarizing blockers | |
| a. Natural | D-Tubocurarine |
| b. Semisynthetic | Dimethyltubocurarine |
| c. Synthetic | Gallamine, pancuronium, atracurium, vecuronium, etc. |

*Source*: Barar (2004).

## 2.4.2 DEPOLARIZING NEUROMUSCULAR BLOCKERS

### 2.4.2.1 MECHANISMS OF ACTION

Succinylcholine structurally resembles ACh and thus binds and activates the ACh receptors. Succinylcholine and decamethonium are depolarizing agents that depolarize the postsynaptic membrane similar to ACh. This action produces prolong muscle fasciculation. After the initial muscle stimulation, constant depolarization results from continued binding to the nicotinic receptors (phase I). This leads to reduced receptor sensitivity and results in desensitization block that slow induce muscle paralysis (phase II). This dual block is a characteristic feature of depolarizers on all voluntary muscle. Phase II block is differently known as dual, biphasic, or antidepolarizing block and occurs clinically in response to a high dose or to prolong administration of either succinylcholine or decamethonium. It differs from phase I block which is slow in onset and similar to that produced by nondepolarizers like D-tubocurarine (Atherton, 1999; Barar, 2004; Donati and Bevan, 1996).

## 2.4.3 DEPOLARIZING NEUROMUSCULAR BLOCKING DRUGS

### 2.4.3.1 SUCCINYLCHOLINE

It is the only neuromuscular blocker that has ultrarapid onset of action with ultrashort duration. The effects of succinylcholine at NMJ are poorly understood. It depolarizes postsynaptic and extrajunctional receptors. It is well known that there is a reduction in receptor sensitivity if it remains in contact with agonist for a prolonged duration. This is the case with succinylcholine that stays at the endplate for prolonged duration developing desensitization after a short period of stimulation (Jonsson et al., 2006). It also inactivates $Na^+$ channels in the junctional and perijunctional areas during membrane depolarization preventing the propagation of the action potential (Jonsson et al., 2006). Succinylcholine injection frequently induces fasciculation, a type of muscle incoordination (Schreiber et al., 2005).

Plasma pseudocholinesterase is an enzyme that rapidly hydrolyzes succinylcholine. It has an elimination $t_{1/2}$ of <1 min in patients (Roy et al., 2002). As succinylcholine rapidly disappears from plasma, the maximal action is reached rapidly (Szalados et al., 1990).

### 2.4.3.2 SIDE EFFECTS

#### 2.4.3.2.1 Cardiovascular

Cardiovascular side effects of succinylcholine are mainly associated with increased catecholamine release. Succinylcholine-induced tachycardia is seen frequently. Use of succinylcholine in pediatric and adult patients may slow systole after the second dose. These effects can be overcome with atropine or glycopyrrolate (Lerman and Chinyanga, 1983).

#### 2.4.3.2.2 Fasciculation

Chances high degree of fasciculation, especially in muscular adults increases after rapid injection of succinylcholine. Though fasciculations are benign side effects of these agents, clinicians prefer to prevent such effects. A small dose about 0.05 mg/kg of nondepolarizing-type neuromuscular blockers like D-tubocurarine is administered prior to succinylcholine is effective in the prevention of fasciculation (Shrivastava, 2017). Rocuronium is the other alternative for this purpose.

#### 2.4.3.2.3 Muscle Pains

Generalized aches and pains are common with succinylcholine which is similar to the myalgia after 24–48 h of its administration in young and ambulatory patients (Wong and Chung, 2000).

#### 2.4.3.2.4 Intragastric Pressure

Intragastric pressure elevated by succinylcholine can be relieved by precurarization. Succinylcholine is shown to increase the lower esophageal sphincter pressure. Thus, succinylcholine is not involved in the risk of aspiration of gastric contents unless the lower esophageal sphincter is incompetent (Wong and Chung, 2000).

### 2.4.3.2.5   Intraocular and Intracranial Pressure

Intraocular pressure is increased by 5–15 mmHg with the administration of succinylcholine. The mechanism is not known, but it may be due to detachment of extraocular muscle, as in case of intraocular etiology. Pretreatment with curare compounds has little or no effect on this elevation. Due to this reason, succinylcholine is avoided in open-eye injuries (Vachon et al., 2003). Succinylcholine is also found to increase intracranial pressure.

### 2.4.3.2.6   Hyperkalemia

Serum potassium level is increased about 0.5 mEq/L by the use of succinylcholine. This hyperkalemia can be overcome by large doses of nondepolarizing blockers (Gronert, 2001). Patients with renal failure or those having pre-existing hyperkalemia do not show an increase in $K^+$ levels, but the absolute level might reach the toxic range. Cardiac arrest may be occasionally associated with severe hyperkalemia related reduced $K^+$ levels via a proliferation of extrajunctional receptors (Gronert, 2001; Thapa and Brull, 2000).

### 2.4.3.2.7   Abnormal Plasma Cholinesterase

Reduced plasma cholinesterase activity is seen in physiological and pathological conditions like pregnancy, liver disease, uremia, malnutrition, burns, and plasmapheresis. Also, some medications such as oral contraceptives

may decrease cholinesterase activity. This reduction in enzyme activity slightly increases the duration of action of succinylcholine (Davis et al., 1997).

## 2.4.4   NONDEPOLARIZING NEUROMUSCULAR BLOCKERS

### 2.4.4.1   MECHANISMS OF ACTION

These are the agents that competitively antagonize the action of ACh at the postsynaptic nicotinic receptor (Fig. 2.2). This is a dynamic binding that allows association and dissociation, that is, increased the concentration of ACh than the antagonist results in receptor occupation. With blockade by antagonist, endplate potential decreases gradually until it fails to reach the threshold to generate propagating action potential needed for muscle contraction. Under normal physiological conditions, more transmitter molecules than are needed to generate the endplate potential, evoking a greater than needed response. Simultaneously, only a fraction of the available receptors is used to generate the signal. Neuromuscular transmission, therefore, has a substantial margin of safety (Donati and Bevan, 1996).

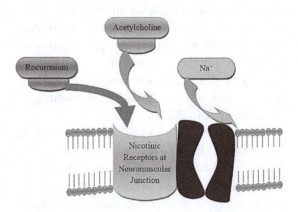

**FIGURE 2.2**   Mechanism of action of nondepolarizing competitive neuromuscular blockers.

Neuromuscular block, indicated by depression of the single twitch height, becomes evident only when 70–80% of the receptors are blocked by nondepolarizing neuromuscular blocking drugs (NMBDs). For a complete block, at least 92% of the receptor population must be occupied. Like the depolarizing drugs, nondepolarizing NMBDs also exhibit desensitization block. They bind tightly to desensitized receptors and can trap them in these states. This type of block is a noncompetitive block. There is a reduction in the margin of safety of transmission when more receptors are in the desensitized state. Many anesthetic medications increase the proportion of desensitized receptors, for example, inhalation anesthetics, thiopental, and local anesthetics (Donati, 2006).

### 2.4.4.2 PHARMACOKINETICS

Elimination $t_{1/2}$ of most of the NMBDs depends upon redistribution rather than elimination which is independent of their duration of action. However, understanding the pharmacokinetics of the drug provides an idea about the use of that agent in special situations such as long-term use, pathology of the excretory organs, and so

on (Kopman et al., 1999). There are different mechanisms explaining the variability in the duration of action as shown in Table 2.6.

### 2.4.5 NONDEPOLARIZING NEUROMUSCULAR BLOCKING DRUGS

Nondepolarizing NMBDs bind postsynaptic receptor to one of the α subunits of the receptor competitively (Naguib et al., 2002). An important characteristic of nondepolarizing NMBDs is identified in electromyography recordings that show fade in the response to high-frequency stimulation (>0.1 Hz) (Lee and Katz, 1977). Tetanic stimulation is followed by posttetanic facilitation, which is an increased response to any stimulation applied soon after the tetanus (Brull et al., 1991). Nondepolarizing blockade can be inhibited with edrophonium, neostigmine, or pyridostigmine. Succinylcholine, a depolarizing agent, can also antagonize such block provided that the blockade is intense and that the dose of succinylcholine is too small to produce a block of its own.

More than 50 nondepolarizing NMBDs have been introduced into clinical anesthesia. The first agent tested clinically was D-tubocurarine. It is obtained from the plant *Chondodendron*

**TABLE 2.6** Pharmacokinetic Properties of Nondepolarizing Neuromuscular Blockers.

| Properties | Notes |
|---|---|
| Onset of time | 2–7 min is longer than time to peak plasma concentrations (<1 min) due to drug transfer between plasma and neuromuscular junction and is represented quantitatively by a rate constant ($k$eo) This rate constant has $t_{1/2}$ of 5–10 min and is determined by factors like cardiac output, distance of the muscle from the heart, and blood flow to the muscle |
| Volume of distribution Elimination | 0.2–0.4 L/kg equal to ECF volume |
| Long duration | Elimination $t_{1/2}$ is 1–2 h that depends upon liver and/or kidney function |
| Intermediate duration | Intermediate elimination $t_{1/2}$, for example, atracurium and cisatracurium or have long elimination $t_{1/2}$, that is, 1–2 h, for example, vecuronium and rocuronium. Elimination depends upon redistribution rather than elimination |
| Short duration | Brief elimination $t_{1/2}$, for example, mivacurium |
| Ultrashort duration | Very brief elimination $t_{1/2}$, for example, succinylcholine |

ECF, Extracellular fluid.
*Source*: Kopman et al. (1999).

*tomentosum* in the crude form. This initial form requires purification and standardization to obtain D-tubocurarine. Nowadays, it has been completely replaced by more modern and safe synthetic analogs (Table 2.7).

### 2.4.6   DRUG INTERACTIONS

Interactions between NMBDs and anesthetics and other drugs have been identified. Although some of them are confirmed, many remained to be understood. Few of the clinically important interactions are listed in Table 2.8.

## 2.5   GANGLIONIC BLOCKERS

Ganglionic blockers are the drugs those competitively antagonize nicotinic ($N_N$) receptors of parasympathetic and sympathetic autonomic ganglia thereby inhibiting the action of ACh and other agonists. Some drugs belonging to the class are nonselective at both the autonomic ganglia therefore ineffective as neuromuscular blocker. Thus, these agents block the entire outflow of the ANS at the $N_N$ receptors (Sharma and Sharma, 2017; Barar, 2004).

Nondepolarizing blockers produce complex responses. The physiological actions of these agents are predictable and based on predominant tone of organ system, for example, arterioles have sympathetic predominant tone and application of nondepolarizing agent induces vasodilation. Also, many organ systems have parasympathetic predominant tone, for example, urinary bladder and GIT. Thus, the ganglionic blocker may produce antagonistic effects on these organ systems. The ganglion blocking drugs are used in

pharmacologic and physiologic research as they tend to block entire autonomic outflows. However, these agents lack selectivity; therefore, their use may lead to a broad range of undesirable effects with limited clinical usefulness (Pollard, 1994).

### 2.5.1   MECHANISM OF ACTION

Nicotine at large doses produces a two-phase prolonged blockade at ganglionic nicotinic receptors.

*Phase I*: It is a persistent depolarization of the autonomic ganglion. It results from the initial application of nicotine causing ganglionic depolarization generating an action potential which lasts for few seconds and later there is blockade of ganglionic transmission. There is a failure of antidromic stimuli-inducing action potential at this moment. Actually in phase I, the ganglia do not respond to any ganglionic stimulant, regardless of the type of receptor it activates. The main reason for the loss of electrical or receptor-mediated excitability during a period of maintained depolarization is that the voltage-sensitive $Na^+$ channel is inactivated and no longer opens in response to a brief depolarizing stimulus. During the latter part of phase I, all ganglionic stimulants other than nicotine, for example, histamine, angiotensin, bradykinin, and serotonin, become effective (Gurney and Rang, 1984).

*Phase II*: It takes place after Phase I. It is a postdepolarization phase that only allows blockade of the actions of nicotinic receptor agonists. Phase II starts several minutes after the action of nicotine. During this period, there is partial repolarization of the cell allows it to regain electrical excitability. It is found that desensitization of the receptor is the underlying factor for phase II block that

**TABLE 2.7** Properties of Individual Nondepolarizing Neuromuscular Blockers.

| Drug | Properties | Onset of action (min) | Duration of action (min) | Pharmacokinetics | Adverse drug reactions | Therapeutic uses |
|---|---|---|---|---|---|---|
| Tubocurarine | Competitive and longer acting | 6 | 80 | Slow metabolism; both renal (10%) and hepatic (45%) clearance; 30–50% protein bound | Hypotension due to the release of histamine; skin flushing | Precurarization to reduce fasciculations and muscle pains |
| Atracurium | Competitive and intermediate acting | 3 | 45 | Nonenzymatic degradation (Hofmann reaction); metabolism by tissue esterases; elimination half-life (22–25) min | Hypotension, tachycardia, and skin flushing; bronchospasm | Intubating in laryngoscopy |
| Cisatracurium | Competitive and intermediate acting | 2–8 | 45–90 | Nonenzymatic degradation (Hofmann reaction) and renal elimination; elimination half-life (22–25) min | Light headedness, flushing | To facilitate tracheal intubation |
| Doxacurium | Competitive and longer acting | 4–8 | 120 | Renal elimination; elimination half-life (1–2) h | Bronchospasm, whizzing, skeletal muscle weakness | To induce prolong anesthesia in cardiovascular surgery |
| Gantacurium | Competitive and ultrashort duration | 1–2 | 5–10 | Cysteine degradation and ester hydrolysis | Hypotension | Tracheal intubation |
| Mivacurium | Competitive and short acting | 2–3 | 15–21 | Hydrolysis by plasma cholinesterase | Hypotension, tachycardia, erythema, and flushing; bronchospasm is rare | In surgical procedures requiring brief muscle relaxation; to facilitate insertion of a laryngeal mask airway |
| Pancuronium | Competitive and longer acting | 3–4 | 85–100 | Hepatic and renal elimination; does not release histamine | Bronchospasm, flushing, salivation | To facilitate tracheal intubation |

*Source:* Martyn et al. (1992), Savarese et al. (1977), Fodale and Santamaria (2002), Basta et al. (1988), and Brunton et al. (2011).

**TABLE 2.8** Drug Interaction Between Neuromuscular Blocking Drugs.

| Interaction | Effects |
|---|---|
| NMBDs + cardiovascular agents | Reduce muscle contraction, decreased release of ACh presynaptically |
| NMBDs + antibiotics | Reduction of postsynaptic receptor sensitivity to ACh |
| | Decreased release of ACh presynaptically |
| | Impairment of ion channels |
| NMBDs + inhaled anesthetics | Reduce muscle contraction, reduction of postsynaptic receptor sensitivity to Ach |
| NMBDs + local anesthetics | Reduce muscle contraction, decreased release of ACh presynaptically |
| NMBDs + phenytoin | Augmentation of neuromuscular block |
| NMBDs + magnesium | Abolishes succinylcholine-induced fasciculation |

Ach, Acetylcholine.

*Source*: Adapted from Motamed and Donati (2002), Kopman et al. (2005), Suzuki et al. (2007), and Gopalakrishna et al. (2007).

leads to the failure of transmission (Gurney and Rang, 1984).

### 2.5.2 GANGLIONIC NEUROTRANSMISSION

Autonomic ganglia neurotransmission progression is more multifarious than a single neurotransmitter–receptor system. Activation of preganglionic nerve shows four distinct types of potential changes after intracellular recordings as indicated in Figures 2.3 and 2.4 (Adams et al., 1982; Ascher, 1979; Brunton et al., 2011; Prud'homme et al., 1999; Slavikova et al., 2003).

### 2.5.3 CLASSIFICATION OF GANGLION BLOCKERS

Ganglion blockers primarily act on the nicotinic receptor are categorized into two groups such as depolarizing ganglion blockers and nondepolarizing ganglion blockers (Table 2.9). The nondepolarizing ganglion blockers at first do not cause ganglionic stimulation or alteration in ganglionic potential but rather causes blocking of autonomic ganglia (e.g., hexamethonium and trimethaphan) by impairing the neurotransmission either by competitively inhibiting action of ACh at ganglionic nicotinic receptor or by inhibiting the channel. Trimethaphan is an example of the drug that competes with ACh at the NMJ, the action resembling curare drugs. Another agent, hexamethonium blocks the opened channel. This results in transient current flow as opened channel closes. Though trimethaphan and hexamethonium have different mechanisms of action, both the drugs block an initial EPSP is thereby inhibiting ganglionic transmission (Gurney and Rang, 1984).

**TABLE 2.9** Classification of Ganglion Blockers.

| Class | Example |
|---|---|
| A. Ganglion stimulants or depolarizing ganglion blockers | Nicotine, lobeline, epibatidine, succinylcholine, decamethonium, dimethylphenylpiperazinium, onium compounds (e.g., tetramethylammonium) |
| B. Competitive ganglionic blockers | Trimethaphan, hexamethonium |
| C. Noncompetitive ganglionic blockers | Chlorisondamine |

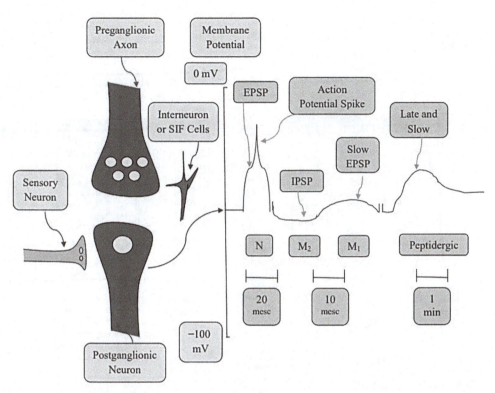

**FIGURE 2.3** Autonomic ganglion cells and the EPSP and IPSP recorded from the postganglionic cell body following stimulation of the preganglionic nerve fiber. SIF, Small intensely fluorescent; EPSP, excitatory postsynaptic potential; IPSP, inhibitory postsynaptic potential; M, muscarinic; N, nicotinic.

*Source*: Adams et al. (1982), Ascher (1979), Brunton et al. (2011), Prud'homme et al. (1999), and Slavikova et al. (2003).

### 2.5.4  GANGLION STIMULANTS

Nicotine and lobeline like natural alkaloids stimulate autonomic ganglia peripherally. Nicotine was first isolated in 1828, from tobacco leaves, *Nicotiana tabacum* by Posselt and Reiman. Later, it was evaluated pharmacologically by Orfila in 1943. Langley and Dickinson carried experiments on the superior cervical ganglion of rabbits discovered that nicotine act on ganglia rather than preganglionic or postganglionic nerve fibers. Lobeline which was isolated from *Lobelia inflate* is less potent than nicotine. Various synthetic compounds were developed during the late 19th and early 20th century,

showing distinguished effects at ganglionic receptor sites such as onium compounds like tetramehtylammonium (TMA) (Brunton et al., 2011).

### 2.5.5  NICOTINE

Nicotine is a naturally occurring colorless, liquid alkaloid from tobacco leaves. It is a volatile base with p$K$a of 8.5. Nicotine turns brown and smells like tobacco on exposure to air (Brunton et al., 2011; Wilson and Gisvold, 2004).

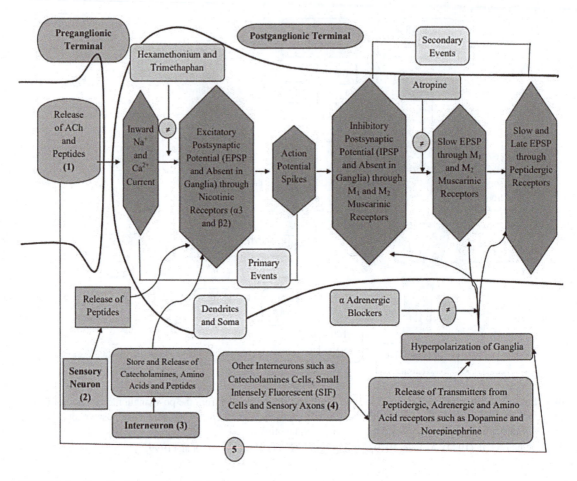

**FIGURE 2.4** Ganglionic neurotransmission.

*Source*: Adams et al. (1982), Ascher (1979), Brunton et al. (2011), Prud'homme et al. (1999), and Slavikova et al. (2003).

### 2.5.5.1 MECHANISM OF ACTION

Nicotine initially stimulates the ganglia showing ACh-like action with subsequent blockade of nicotinic receptors resulting from constant depolarization, as long-term use of nicotine cause cholinergic receptor desensitization and sustained blockade (Volle, 1980).

### 2.5.5.2 PHARMACOLOGICAL ACTIONS

#### 2.5.5.2.1 Peripheral Nervous System

As stated earlier, nicotine initially causes momentary excitation and subsequently more persistent depression at all autonomic ganglia. At low concentration, direct stimulation of ganglionic cells by nicotine facilitates impulse transmission. At larger doses, it initially stimulates ganglionic cells followed by obstruction of impulse transmission. The stimulation of ganglionic cells is concerned with depolarization and blockade of transmission is attributed to persistent depolarization leading to desensitization of nicotinic receptors at ganglionic cells. Nicotine produces a dual action at adrenal medulla. At lower doses, it evokes catecholamine discharge, whereas higher concentrations inhibit release. Like ACh, nicotine excites a number of sensory receptors, for example, mechanoreceptors in skin, mesentery, tongue, and stomach; chemoreceptors of carotid body, thermoceptors from skin and tongue; and nociceptors (Brunton et al., 2011).

### 2.5.5.2.2 *Central Nervous System*

Nicotine stimulates the CNS. At low concentration, nicotine shows weak analgesic effect, but at larger doses, it exerts a toxic effect and results in tremors that may precipitate convulsion. The excitation of respiration is a well-known action of nicotine. At larger doses, nicotine has direct actions on the medulla. Smaller doses are responsible for the enhancement in respiration reflex through chemoreceptors located in aorta carotid arteries. CNS stimulation followed by depression is seen at larger doses of nicotine that may lead to death from respiratory failure. It is a result of paralysis and blockade of muscles of respiration (MacDermott et al., 1999).

Nicotine is showed to stimulate chemoreceptor trigger zone (CTZ) leading to vomiting by central and peripheral mechanisms. It also activates vagal and spinal sensory nerves forming sensory input of the reflex pathways of vomiting. CNS releases excitatory amino acids and biogenic amines including dopamine to induce stimulatory actions of nicotine (MacDermott et al., 1999). There is a marked increase in the nicotinic receptors population after persistent contact with nicotine (Di Chiara, 2000; Stitzel et al., 2000).

### 2.5.5.2.3 *Cardiovascular System*

Stimulation of sympathetic ganglia and adrenal medulla by nicotine results in catecholamine discharge from adrenergic nerve endings. Nicotine also causes reflex stimulation and activation of chemoreceptors located on the aorta and carotid bodies producing vasoconstriction, tachycardia, and increase in blood pressure (Brunton et al., 2011).

### 2.5.5.2.4 *Gastrointestinal Tract*

Nicotine is showed to stimulate both parasympathetic ganglia and nerve endings resulting in increased tone and motor activity of the bowel. On first exposure, it induces nausea, vomiting, and occasionally diarrhea due to systemic absorption of nicotine (Brunton et al., 2011).

### 2.5.5.2.5 *Exocrine Glands*

Nicotine initially stimulates salivary and bronchial secretions followed by their inhibition (Brunton et al., 2011).

### 2.5.5.3 *PHARMACOKINETICS*

Nicotine is well and readily absorbed from respiratory tract, buccal cavity, and skin. Nicotine toxicity mainly arises due to its percutaneous absorption. As it is a comparatively strong base, it is incompletely absorbed from the stomach but absorption from intestine is more efficient. Tobacco chewing causes slow absorption of nicotine than smoking. Tobacco chewing, therefore, leads to prolonged duration of nicotine action that of inhaled during smoking. Cigarette smoke contains about 6–11 mg nicotine out of which 1–3 mg is going to systemic circulation. Three-fold raise in nicotine bioavailability is seen which depends on puffing intensity and smoking technique (Henningfield, 1995; Benowitz, 1998). To prevent a withdrawal or abstinence syndrome, different dosage forms of nicotine are available. It is administered orally, transdermal patch, nasal spray, and vapor inhaler. The gum and transdermal dosage form of nicotine are widely used to achieve a constant plasma concentration. There is 10 times increase in arterial blood concentrations than venous concentrations instantly following inhalation (Henningfield, 1995; Benowitz, 1999).

The liver is the principal metabolic site of nicotine. About 80–90% is metabolized by hepatocytes. Other organs like kidney and lung metabolize nicotine up to some extent involved in nicotine metabolism. Cotinine is the major metabolite of nicotine. Other minor metabolites found in lesser quantities include nicotine-1'-$N$-oxide and 3-hydroxycotinine and conjugated metabolites (Benowitz, 1998). There is similarity in the metabolic pattern of nicotine in smokers and nonsmokers. After inhalation and parenteral administration, the $t_{1/2}$ of nicotine is about 2 h. Nicotine and its metabolites are primarily excreted by renal mechanism in the urine. The rate of nicotine elimination gets reduced in alkaline urine. Nicotine is found to be excreted from milk. About 0.5 mg/L of nicotine is estimated in the milk of heavy-smoking women (Brunton et al., 2011).

### 2.5.5.4   ACUTE NICOTINE POISONING

Nicotine poisoning, especially in children, may result from an unintended intake of nicotine-containing insecticide sprays or chewing tobacco products. Severe lethal dose of nicotine in adults is about 60 mg of the base and tobacco smoke contains about 6–11 mg (1–2%) of nicotine. It appears that the absorption of nicotine from tobacco is delayed during tobacco chewing due to slow gastric emptying leading to vomiting and removal of tobacco from GIT. The symptoms of acute and severe nicotine poisoning appear quickly. It is characterized by nausea, increased salivary secretions, abdominal ache, vomiting, diarrhea, cold sweat, headache, vertigo, dizziness, impaired hearing, balance and vision, mental confusion, and weakness. These symptoms can be overcome by vomiting, gastric lavage, or by the administration of adsorbents like activated charcoal slurry through a tube in the left side of stomach and maintenance of respiration (Brunton et al. 2011).

### 2.5.6   GANGLION BLOCKERS

The drugs such as hexamethonium, trimethaphan, and mecamylamine cause blockade of autonomic ganglia without causing their stimulation. In 1913, Marshall had coined the term "nicotine-paralyzing" to describe the action of TMA on ganglia. Also, Acheson and Moe had explained in detail the correlation between effects of TMA on the cardiovascular system (CVS) and autonomic ganglia. Later, bisquaternary ammonium salts were developed by Baelow and coworkers. The drug-like hexamethonium consists of six methylene groups between the two quaternary nitrogen atoms with least neuromuscular and muscarinic blocking activities. Another drug, trimethylsulfonium, similar to quaternary and bisquaternary ammonium ions, possesses ganglionic blocking actions and is a hallmark in the development of sulfonium ganglionic blocking agents, for example, trimethaphan. In 1950, secondary amine compounds as mecamylamine were introduced into therapy for hypertension (Brunton et al., 2011).

### 2.5.6.1 MECHANISM OF ACTION

These agents block the postsynaptic actions of ACh without causing depolarization thereby preventing the transmission without initial stimulation (Elfvin et al., 1993; Fant et al., 1999).

### 2.5.6.2 PHARMACOLOGICAL ACTIONS

Ganglionic blockers result in blockade of both cholinergic and adrenergic divisions of ANS. The action of these drugs depends upon the dominant tone in the organ system. Table 2.10 summarizes pharmacological actions of ganglion blockers.

**TABLE 2.10** Pharmacological Effects of Ganglion Blockers.

| Site | Systems | Effects |
| --- | --- | --- |
| Arterioles | Sympathetic | Vasodilation; augmented peripheral blood flow; hypotension |
| Veins | Sympathetic | Peripheral pooling of blood; reduced venous return; reduced cardiac output |
| Ventricles | Sympathetic | Reduced contractile force |

**TABLE 2.10** *(Continued)*

| Site | Systems | Effects |
| --- | --- | --- |
| Sweat glands | | Anhidrosis, i.e., decreased secretion |
| Heart (myocardium, atrium, and SA node) | Parasympathetic | Tachycardia |
| Iris | Parasympathetic | Mydriasis |
| Ciliary muscle | Parasympathetic | Cycloplegia |
| GIT | Parasympathetic | Decreased tone and motility, constipation, reduced gastric and pancreatic secretions |
| Urinary bladder | Parasympathetic | Urinary retention |
| Salivary glands | Parasympathetic | Xerostomia, i.e., dry mouth |
| Reproductive tract | Sympathetic and parasympathetic | Decreased stimulation |

SA, Sinoatrial.

*Source*: Brunton et al. (2011), Elfvin et al. (1993), Fant et al. (1999), Lee and D'Alonzo (1993), and Sargent (1993).

### 2.5.6.3 PHARMACOKINETICS

Quaternary ammonium (hexamethonium) and sulfonium (trimethaphan) compounds are incompletely and erratically absorbed from GIT. Due to ionized nature, these drugs have reduced cell membranes penetration capacity. On the other hand, these agents reduce gastric emptying and peristalsis of the small intestine leading to retarded enteric absorption. It is to be noted that the absorption of mecamylamine is less unpredictable, but there exists a risk of frank paralytic ileus arising from a decreased bowel movement. Hexamethonium and trimethaphan are more concentrated in extracellular space and are primarily excreted unchanged by the kidney, whereas mecamylamine is slowly excreted through the kidney as it is more concentrated in liver and kidney (Brunton et al., 2011).

### 2.5.6.4   ADVERSE DRUG REACTIONS

Most adverse effects of ganglionic blockers are caused by excessive blockade of autonomic ganglia. The adverse drug reactions mostly associated with ganglion blockers are presented in Table 2.11.

**TABLE 2.11**   Outline of Adverse Drug Reactions of Ganglion Blockers.

| Mild adverse drug reactions | Severe adverse drug reactions |
|---|---|
| Visual disturbances, dry mouth, conjunctival redness, uncertain urination, decreased potency, subjective chilliness, constipation, rare diarrhea, abdominal distress, anorexia (loss of appetite), heartburn, nausea, bitter taste, and postural hypotension-induced fainting | Noticeable hypotension, constipation, syncope, paralytic ileus, urinary retention, and cycloplegia |

*Source*: Brunton et al. (2011).

### 2.5.6.5   THERAPEUTIC USES

Ganglionic blocking agents are used for the treatment of chronic hypertension but nowadays they are replaced by other superior agents. These drugs have utility in facilitating but newer alternatives like nitroprusside or other depressive sedatives are superior to produce controlled hypotension and to minimize hemorrhage and blood loss (Fukusaki et al., 1999). Currently, mecamylamine is used therapeutically in the United States. Trimethaphan, a short-acting drug, due to its direct vasodilating properties is used as antihypertensive agents particularly in patients with acute aortic aneurysm. It also produces prolonged neuromuscular blockade and may

potentiate the neuromuscular blocking action of tubocurarine (Brunton et al., 2011).

## 2.6   FUTURE OPPORTUNITIES AND CHALLENGES

Significant therapeutic benefits in a variety of physiological and pathophysiological are provided by anticholinergics. The development and understanding in the gene encoding, different types of muscarinic and nicotinic receptor subtypes has gained more advantage in discovery and development of novel antimuscarinic, neuromuscular-blocking agents, and ganglion-blocking agents with a wide range of favorable effects. This approach helps in minimizing the clinical toxicity associated with conventional anticholinergic drugs and targeting the specific cholinergic receptors for development of more beneficial and useful anticholinergic drugs for variety of ailments and in treatment of UI and irritable bowel diseases, such as peptic ulcer; respiratory disorders, such as COPD; and various neurodegenerative diseases, such as PD.

## 2.7   CONCLUSION

The present chapter focused largely on the pharmacology of cholinergic antagonists, neuromuscular-blocking agents, and ganglion-blocking drugs. All these medications are either directly or indirectly used for the treatment, management, and prophylaxis of the variety of ailments. Unlike cholinergic agonists, those have limited therapeutic utility; the cholinergic antagonists are advantageous in a range of clinical manifestations. Due to the inability of cholinergic antagonists in nicotinic receptors blockade, these agents do not produce effect at NMJ or autonomic ganglia.

## KEYWORDS

- **antimuscarinic poisoning**
- **anticholinergics**
- **antimuscarinic agents**
- **neuromuscular blockers**
- **ganglionic blockers**

## REFERENCES

Aaltonen, L.; Kanto, J.; Iisalo, E.; Pihlajamaki, K. Comparison of Radioreceptors Assay and Radioimmunoassay for Atropine Pharmacokinetic Application. *Eur. J. Clin. Pharmacol.* **1984,** *26*, 613–617.

Abrams, P.; Freeman, R.; Anderstrom, C.; Mattiasson, A. Tolterodine, a New Antimuscarinic Agent: As Effective but Better Tolerated than Oxybutynin in Patients with an Overactive Bladder. *Br. J. Urol.* **1998,** *81*, 801–810.

Adams, P. R.; Brown, D. A.; Constanti, A. Pharmacological Inhibition of the M-Current. *J. Physiol.* **1982,** *332*, 223–262.

Alabaster, V. A. Discovery and Development of Selective M3 Antagonists for Clinical Use. *Life Sci.* **1997,** *60*, 1053–1060.

Ali-Melkkila, T.; Kanto, J.; Lisalo, E. Pharmacokinetics and Related Pharmacodynamics of Anticholinergic Drugs. *Acta Anaesthesiol. Scand.* **1993,** *37*, 633–642.

American Heart Association. American Heart Association in Collaboration with the International Liaison Committee on Resuscitation Guidelines 2000 for Cardiopulmonary Resuscitation and Emergency Cardiovascular Care. Part 6, Section 5. Part 8, Section 2, Part 10. *Circulation* **2000,** *102* (Suppl. 1), I-112–128, 223–228, 291–342.

Ascher, P.; Large, W. A.; Rang, H. P. Studies on the Mechanism of Action of Acetylcholine Antagonists on Rat Parasympathetic Ganglion Cells. *J. Physiol.* **1979,** *295*, 139–170.

Barar, F. S. K. *Essential of Pharmacotherapeutics*; S. Chand & Company Ltd.: New Delhi, 2004.

Barnes, P. J.; Hansel, T. T. Prospects for New Drugs for Chronic Obstructive Pulmonary Disease. *Lancet* **2004,** *364*, 985–996.

Basta, S. J.; Savarese, J. J.; Ali, H. H.; et al. Clinical Pharmacology of Doxacurium Chloride. A New Long-Acting Nondepolarizing Muscle Relaxant. *Anesthesiology* **1988,** *69*, 478.

Baxter, K. *Stockley's Drug Interactions*; Pharmaceutical Press: London, 2010.

Beermann, B.; Hellstrom, K.; Rosen, A. The Gastrointestinal Absorption of Atropine in Man. *Clin Sci.* **1971,** *40*, 95–106.

Bejersten, A.; Olsson, G.; Palmer, L. The Influence of Body Weight on Plasma Concentration of Atropine after Rectal Administration in Children. *Acta Anaesthesiol. Scand.* **1985,** *29*, 782–784.

Benowitz, N. L. In *Nicotine Safety and Toxicity*; Benowitz, N. L., Ed.; Oxford University Press, New York, 1998; pp 3–28.

Berg, J. M.; Brandon, M. W; Kirman, B. H. Atropine in Mongolism. *Lancet* **1959,** *2*, 441.

Birdsall, N. J. M.; Buckley, N. J.; Caulfield, M. P.; et al. Muscarinic Acetylcholine Receptors. In *The IUPHAR Compendium of Receptor Characterization and Classification*. IUPHAR Media: London, 1998; pp 36–45.

Brull, S. J.; Connelly, N. R.; O'Connor, T. Z.; et al. Effect of Tetanus on Subsequent Neuromuscular Monitoring in Patients Receiving Vecuronium. *Anesthesiology* **1991,** *74*, 64.

Brunton, L. L.; Chabner, B. A.; Knollmann, B. C. *Goodman and Gillman: Pharmacological Basis of Therapeutics*, 12th ed.; McGraw-Hill, Medical Publishing Division: New York, 2011.

Caine, M.; Raz, S.; Zeigler, M. Adrenergic and Cholinergic Receptors in the Human Prostate. Prostatic Capsule and Bladder Neck. *Br. J. Urol.* **1975,** *47*, 193–202.

Carmine, A. A.; Brogden, R. N. Pirenzepine. A Review of Its Pharmacodynamic and Pharmacokinetic Properties and Therapeutic Efficacy in Peptic Ulcer Disease and Other Allied Diseases. *Drugs* **1985,** *30*, 85–126.

Caulfield, M. P.; Birdsall, N. J. International Union of Pharmacology, XVII. Classification of Muscarinic Acetylcholine Receptors. *Pharmacol. Rev.* **1998,** *50*, 279–290.

Chapple, R. R., Rechberger, T., Al-Shukri, S.; et al. Randomized, Double-Blind Placebo- and Tolterodine-Controlled Trial of the Once-Daily Antimuscarinic Agent Solifenacin in Patients with Symptomatic Overactive Bladder. *BJU Int.* **2004,** *93*, 303–310.

Chudoku, K. Antidose Therapy for Organophosphate Poisoning. *Clin. Pharmacol. Ther.* **2008,** *21*, 151–159.

Davis, L.; Britten, J. J.; Morgan, M. Cholinesterase: Its Significance in Anaesthetic Practice. *Anaesthesia* **1997,** *52*, 244.

Di Chiara, G. Behavioral Pharmacology and Neurobiology of Nicotine Reward and Dependence. In *Neuronal*

*Nicotinic Receptors*; Clementi, F., Fornasari, D., Gotti, C., Eds.; Springer-Verlag: Berlin, 2000; pp. 603–750.

Dollery, C. *Therapeutic Drugs*, Vol. 1; Churchill Livingstone: Edinburgh, 1991; pp A162–A167.

Donati, F. Dose Inflation When Using Precurarization. *Anesthesiology* **2006**, *105*, 222.

Donati, F.; Bevan, D. R. Postjunctional Mechanisms Involved in Neuromuscular Transmission. In *Neuromustular Transmission*; Booij, L. H. D., Ed.; BNJ Publishing Group: London, 1996; pp 28–44.

Eglen, R. M.; Hedge, S. S.; Watson, N. Muscarinic Receptor Subtypes and Smooth Muscle Function. *Pharmacol. Rev.* **1996**, *48*, 531–565.

Elfvin, L. G.; Lindh, B.; Hookfelt, T. The Chemical Neuroanatomy of Sympathetic Ganglia. *Annu. Rev. Neurosci.* **1993**, *16*, 471–507.

Fant, R. V.; Owen, C. C.; Henningfield, J. E. Nicotine Replacement Therapy. *Prim. Care* **1999**, *26*, 633–652.

Feinberg, M. The Problems of Anticholinergic Adverse Effects in Older Patients. *Drugs Aging* **1995**, *3*, 5–48.

Field, J. M.; Hazinski, M. F.; Sayre, M. R.; Chameides, L.; Schexnayder, S. M.; Hemphill, R.; et al. Part 1: Executive Summary: 2010 American Heart Association Guidelines for Cardiopulmonary Resuscitation and Emergency Cardiovascular Care. *Circulation* **2010**, *122*, 640–656.

Finar, I. L. *Organic Chemistry*, Vol. 2, 4th ed.; Longman: Harlow, 1965; pp 631–639.

Fodale, V.; Santamaria, L. B. Laudanosine, an Atracurium and Cisatracurium Metabolite. *Eur. J Anaesthesiol.* **2002**, *19*, 466.

Friedl, K. E.; Hannan, C. J.; Mader, T. H.; Patience, T. H.; Schadler, P. W. Effect of Eye Color on Heart Rate Response to Intramuscular Administration of Atropine. *J. Auton. Nerv. Syst.* **1988**, *24*, 51–56.

Fukusaki, M.; Miyako, M.; Hara, T. Effects of Controlled Hypotension with Sevoflurane Anesthesia on Hepatic Function of Surgical Patients. *Eur. J. Anaesthesiol.* **1999**, *16*, 111–116.

Game, X.; Peyronnet, B.; Cornu, J. Fesoterodine: Pharmacological Properties and Clinical Implications. *Eur. J. Pharmacol.* **2018**, *833*, 155–157.

Gandevia, B. Historical Review of the Use of Parasympatholytic Agents in the Treatment of Respiratory Disorders. *Postgrad. Med. J.* **1975**, *51*, 13–20.

Genova, E. *Atropa belladonna*. In *Red List of Bulgarian Vascular Plants*; Petrova, A., Vladimirov, V., Eds.; *Phytol. Balcan.* **2009**, *15*, 82.

Genova, E. *Atropa belladonna*. In *Red Data Book of the People's Republic of Bulgaria. Vol. I – Plants*; Velchev, V., Ed.; Publishing House Bulg. Acad. Sci.: Sofia, 1984; p 327.

Gervais, H. W.; Gindi, M.; Radermacher, P. R.; Volz-Zang, C.; Palm, D.; Duda, D.; Dick, W. F. Plasma Concentration Following Oral and Intramuscular Atropine in Children and Their Clinical Effects. *Paediatr. Anaesth.* **1997**, *7*, 13–18.

Gopalakrishna, M. D.; Krishna, H. M.; Shenoy, U. K. The Effect of Ephedrine on Intubating Conditions and Haemodynamics During Rapid Tracheal Intubation Using Propofol and Rocuronium. *Br. J. Anaesth.* **2007**, *99*, 191.

Gronert, G. A. Cardiac Arrest after Succinylcholine: Mortality Greater with Rhabdomyolysis than Receptor Upregulation. *Anesthesiology* **2001**, *94*, 523.

Gross, N. J. *Asthma: Basic Mechanisms and Clinical Management*, second ed.; Barnes, P. J., Hardison, W. G. Tomaszewski, N., Grundy S. M., Eds.; Effect of Acute Alterations in Small Bowel Transit Time upon the Biliary Excretion Rate of Bile Acids. *Gastroenterology* **1979**, *76*, 568–574.

Gurney, A. M.; Rang, H. P. The Channel-Blocking Action of Methonium Compounds on Rat Submandibular Ganglion Cells. *Br. J. Pharmacol.* **1984**, *82*, 623–642.

Hardison, W. G.; Tomaszewski, N.; Grundy, S. M. Effect of Acute Alterations in Small Bowel Transit Time upon the Biliary Excretion Rate of Bile Acids. *Gastroenterology* **1979**, *76*, 568–574.

Harris, W. S.; Goodman, R. M. Hyper-reactivity to Atropine in Down's Syndrome. *N. Engl. J. Med.* **1969**, *279*, 407–410.

Hase, N. K.; Shrinivasan, N. J.; Divekar, N. V.; Gore, A. G. Atropine Induced Ventricular Fibrillation in a Case of Diazinon Poisoning. *J. Assoc. Phys. India* **1984**, *32*, 536.

Heath, A. J. W.; Meredith, T. Atropine in the Management of Anticholinesterase Poisoning. In *Clinical and Experimental Toxicology of Organophosphates and Carbamates*; Ballantyne, B., Marrs, T. C., Eds.; Oxford: Butterworth, 1992; pp 543–554.

Henningfield, J. E. Nicotine Medications for Smoking Cessation. *N. Engl. J. Med.* **1995**, *333*, 1196–1203.

Hinderling, P. H.; Gundert-Remy, U.; Schmidlin, O. Integrated Pharmacokinetics and Pharmacodynamics of Atropine in Healthy Humans. I. Pharmacokinetics. *J. Pharm. Sci.* **1985**, *74*, 703–710.

Jonsson, M.; Dabrowski, M.; Gurley, D. A.; et al. Activation and Inhibition of Human Muscular and Neuronal

Nicotinic Acetylcholine Receptors by Succinylcholine. *Anesthesiology* **2006,** *104,* 724.

Kanto, J.; Klotz, U. Pharmacokinetic Implications for the Clinical Use of Atropine, Scopolamine, and Glycopyrrolate. *Acta Anaesthesiol. Scand.* **1988,** *32,* 69–78.

Kanto, J.; Pihlajamaki, K. Oropharyngeal Absorption of Atropine. *Int. J. Clin. Pharmacol. Ther. Toxicol.* **1986,** *24,* 627–629.

Kanto, J.; Virtanen, R.; Iisalo, E.; Maenpaa, K.; Liukko, P. Placental Transfer and Pharmacokinetics of Atropine after a Single Maternal Intravenous and Intramuscular Administration. *Acta Anaesthesiol. Scand.* **1981,** *25,* 85–88.

Kivalo, I.; Saarikoski, S. Placental Transmission of Atropine at Full Term Pregnancy. *Br. J. Anaesth.* **1977,** *49,* 1017–1021.

Kopman, A. F.; Chin, W. A.; Moe, J.; et al. The Effect of Nitrous Oxide on the Dose–Response Relationship of Rocuronium. *Anesth. Analg.* **2005,** *100,* 1343.

Kopman, A. F.; Klewicka, M. M.; Kopman, D. J.; et al. Molar Potency is Predictive of the Speed of Onset of Neuromuscular Block for Agents of Intermediate, Short, and Ultrashort Duration. *Anesthesiology* **1999,** *90,* 425.

Langtry, H. D.; McTavish, D. Terodiline: A Review of Its Pharmacological Properties, and Therapeutic Use in the Treatment of Urinary Incontinence. *Drugs* 1990, *40,* 748–761.

Lau, W. A. K.; Pennefather, J. N.; Mitchelson, F. *Proc. Aust. Soc. Clin. Exp. Pharmacol.* **2000,** *130,* 1013–1020.

Lee, C.; Katz, R. L. Fade of Neurally Evoked Compound Electromyogram During Neuromuscular Block by *d*-tubocurarine. *Anesth. Analg.* **1977,** *56,* 271.

Lee, E. W.; D'Alonzo, G. E. Cigarette Smoking, Nicotine Addiction and Its Pharmacological Treatment. *Arch. Intern. Med.* **1993,** *153,* 34–48.

Lemmens, H. J.; Brodsky, J. B. The Dose of Succinylcholine in Morbid Obesity. *Anesth. Analg.* **2006,** *102,* 438.

Lerman, J.; Chinyanga, H. M. The Heart Rate Response to Succinylcholine in Children: A Comparison of Atropine and Glycopyrrolate. *Can. Anaesth. Soc. J.* **1983,** *30,* 377.

Lukasik, V. M. Neuromuscular Blocking Drugs and the Critical Care Patient. *Vet. Emerg. Crit. Care* **1995,** *5,* 99.

Lullmann, H.; Schmaus, H.; Staemmler, E.; Ziegler, A. Comparison of Atropine and Dexetimide in Treatment of Intoxications by Selected Organophosphates. *Acta Pharmacol. Toxicol.* **1982,** *50,* 230–237.

MacDermott, A. B.; Role, L. W.; Siegelbaum, S. A. Presynaptic Ionotropic Receptors and the Control of Transmitter Release. *Annu. Rev. Neurosci.* **1999,** *22,* 443–485.

Maesen, F. P. V.; Smeets, J. J.; Costongs, M. A. L.; Cornelissen, P. J. G.; Wald, F. D. M.; Martin, T. R.; Kastor, J. A.; Kershbaum, K. L.; Engelman, K. The Effects of Atropine Administered with Standard Syringe and a Self-Injector Device. *Am. Heart J.* **1980,** *99,* 282–288.

Matthew, H.; Lawson, A. A. H. *Treatment of Common Acute Poisoning,* second ed.; Livingstone: Edinburgh, 1970.

Morales-Olivas, F. Solifenacin Pharmacology. Neurologic and Urodinamic Urology. *Arch. Esp. Urol.* **2010,** *63* (1), 43–52.

McBrien, N. A.; Stell, W. K.; Carr, B. How Does Atropine Exert Its Anti-Myopia Effects? *Ophthalm. Physiol. Opt.* **2013,** *33,* 373–378.

Minton, M. D.; Grosslight, K.; Stirt, J. A.; Bedford, R. F. Increases in Intracranial Pressure from Succinylcholine: Prevention by Prior Nondepolarizing Blockade. *Anesthesiology* **1986,** 65, 165.

Millard, R. J.; Moore, K.; Rencken, R.; Yalcin, I.; Bump, R. C. Duloxetine vs. Placebo in the Treatment of Stress Urinary Incontinence: A Four-Continent Randomized Clinical Trial. *BJU Int.* **2004,** *93,* 311–318.

Morton, H. G. Atropine Intoxication. Its Manifestation in Infants and Children. *J. Pediatr.* **1939,** *14,* 755–760.

Motamed, C.; Donati, F. Sevoflurane and Isoflurane, But Not Propofol, Decrease Mivacurium Requirements Over Time. *Can. J. Anaesth.* **2002,** *49,* 907.

Nadel, J. A. Autonomic Regulation of Airway Smooth Muscle. In *Physiology and Pharmacology of the Airways;* Nadel, J. A., Ed.; Marcel Dekker: New York, 1980, pp 217–257.

Naguib, M.; Flood, P.; McArdle, J. J. Advances in Neurobiology of the Neuromuscular Junction: Implications for the Anesthesiologist. *Anesthesiology* **2002,** *96,* 202.

Oyasu, H.; Yamamoto, T.; Sato, N.; Ozaki, R.; Mukai, T.; Ozaki, T.; Nishii, T.; Sato, H.; Parfitt, K. *Martindale: The Complete Drug Reference,* 32nd ed.; The Pharmaceutical Press: London, 1999; pp 455–457.

Pediani, J. D.; Ward, R. J.; Marsango, S.; Milligan, G. Spatial Intensity Distribution Analysis: Studies of G Protein-Coupled Receptor Oligomerisation. *Trends Pharmacol. Sci.* **2017,** *1465,* 30180–30183.

Pollard, B. J. Interactions Involving Relaxants. In *Applied Neuromuscular Pharmacology;* Pollard, B. J., Ed.; Oxford University Press: Oxford, England, 1994; pp 202–248.

Prud'homme, M.; Houdeau, E.; Serghini, R.; Tillet, Y.; Schemann, M.; Rousseau, J. Small Intensely Fluorescent Cells of the Rat Paracervical Ganglion. *Brain Res.* **1999,** *821,* 141–149.

Rang, H. P.; Dale; Ritter, J. M.; More. *Pharmacology*; Elsevier: Amsterdam, 2003; p 139.

Rothrock, S. G.; Green, S. M.; Schafermeyer, R. W.; Colucciello, S. A. *Ann. Emerg. Med.* **1993**, *22* (4), 751–753.

Roy, J. J.; Donati, F.; Boismenu, D.; Varin, F. Concentration-Effect Relation of Succinylcholine Chloride During Propofol Anesthesia. *Anesthesiology* **2002**, *97*, 1082.

Ruggieri, M. R.; Colton, M. D.; Wang, P.; Wang, J.; Smyth, R. J.; Pontari, M. A. Luthin, G. Saarnivaara, L.; Kautto, U. M.; Iisalo, E.; Pihlajamaki, K. Comparison of Pharmacokinetic and Pharmacodynamic Parameters Following Oral or Intramuscular Atropine in Children. Atropine Overdose in Two Small Children. *Acta Anaesthesiol. Scand.* **1985**, *29*, 529–536.

Sargent, P. B. The Diversity of Neuronal Nicotine Acetylcholine Receptors. *Annu. Rev. Neurosci.* **1993**, *16*, 403–443.

Savarese, J. J.; Ali, H. H.; Antonio, R. P. The Clinical Pharmacology of Metocurine: Dimethyltubocurarine Revisited. *Anesthesiology* **1977**, *47*, 277.

Schreiber, J. U.; Lysakowski, C.; Fuchs-Buder, T. Prevention of Succinylcholine-Induced Fasciculation and Myalgia: A Meta-analysis of Randomized Trials. *Anesthesiology* **2005**, *103*, 877.

Schulte, B.; Volz-Zang, C.; Mutschler, E.; Horne, C.; Palm, D.; Wellstein, A.; Pitschner, H. F.; Smith, D. S.; Orkin, F. K.; Gardner, S. M.; Zakeosian, G. Prolonged Sedation in the Elderly after Intraoperative Atropine Administration. *Anesthesiology* **1979**, *51*, 348–349.

Sharma, H. L.; Sharma, K. K. *Principles of Pharmacology*, 3rd ed.; Paras Medical Publisher: Hyderabad, 2017.

Slavikova, J.; Kuncova, J.; Reischig, J.; Dvorakova, M. Catecholaminergic Neurons in the Rat Intrinsic Cardiac Nervous System. *Neurochem. Res.* **2003**, *28*, 593–598.

Smith, C. E.; Saddler, J. M.; Bevan, J. C. Pretreatment with Non-Depolarizing Neuromuscular Blocking Agents and Suxamethonium-Induced Increases in Resting Jaw Tension in Children. *Br. J. Anaesth.* **1990**, *64*, 577.

Srivastava, S. K. *Complete Textbook of Medical Pharmacology*, Vol. I; Avichal Publication Company: New Delhi, 2017.

Steers, W. D. Darifenacin: Pharmacology and Clinical Usage. *Urol. Clin. N. Am.* **2006**, *33*, 475–482.

Stitzel, J. A.; Leonard, S. S.; Collins, A. C. Genetic Regulation of Nicotine-Related Behaviors and Brain

Nicotinic Receptors. In *Neuronal Nicotinic Receptors*; Clementi, F., Fornasari, D., Gotti, C., Eds.; Springer-Verlag: Berlin, 2000; pp. 563–586.

Suzuki, T.; Mizutani, H.; Ishikawa, K. Epidurally Administered Mepivacaine Delays Recovery of Train-of-Four Ratio from Vecuronium-Induced Neuromuscular Block. *Br. J. Anaesth.* **2007**, *99*, 721.

Szalados, J. E.; Donati, F.; Bevan, D. R. Effect of *d*-Tubocurarine Pretreatment on Succinylcholine Twitch Augmentation and Neuromuscular Blockade. *Anesth. Analg.* **1990**, *71*, 55.

Thapa, S.; Brull, S. J. Succinylcholine-Induced Hyperkalemia in Patients with Renal Failure: An Old Question Revisited. *Anesth. Analg.* **2000**, *91*, 237.

Tryba, M., and Cook, D. Current Guidelines on Stress Ulcer Prophylaxis. *Drugs* **1997**, *54*, 581–596.

Vachon, C. A.; Warner, D. O.; Bacon, D. R. Succinylcholine and the Open Globe. Tracing the Teaching. *Anesthesiology* **2003**, *99*, 220.

Van der Meer, M. J.; Hundt, H. K. L.; Muller, F. O. Inhibition of Atropine Metabolism by Organophosphate Pesticides. *Hum. Toxicol.* **1983**, *2*, 637–640.

Virtanen, R.; Kanto, J.; Iisalo, E.; Iisalo, E. U. M.; Salo, M.; Sjovall, S. Pharmacokinetic Studies on Atropine with Special Reference to Age. *Acta Anaesthesiol. Scand.* **1982**, *26*, 297–300.

Volle, R. L. Nicotinic Ganglion-Stimulating Agents. In *Pharmacology of Ganglionic Transmission*; Kharkevich, D. A., Ed.; Springer-Verlag: Berlin, 1980; pp 281–312.

Weiner, N. Atropine, Scopolamine, and Related Antimuscarinic Drugs. In *The Pharmacological Basis of Therapeutics*, seventh ed.; Gilman A. G., Goodman L. S., Rall T. W., Murad F., Eds.; MacMillan: New York, 1985; pp 130–138.

Wellstein, A.; Pitschner, H. F. Complex Dose-Response Curves of Atropine in Man Explained by Different Functions of M1- and M2 Cholinoceptors. *Naunyn Schmiedebergs Arch. Pharmacol.* **1988**, *338*, 19–27.

Widdicombe, J. G. The Parasympathetic Nervous System in Airways Disease. *Scand. J. Respir. Dis. Suppl.* **1979**, *103*, 38–43.

Wilson, C. O.; Gisvold, O. *Textbook of Organic Medicinal and Pharmaceutical Chemistry by Wilson and Gisvold*, 11th ed., J. Lippincot Co.: Philadelphia, PA, 2004.

Wong, S. F.; Chung, F. Succinylcholine-Associated Postoperative Myalgia. *Anaesthesia* **2000**, *55*, 144.

# CHAPTER 3

# Adrenergic Agonists

SEETHA HARILAL* and RAJESH K.

*Kerala University of Health Sciences, Kerala, India*

*Corresponding author. E-mail: seethaharilal1989@gmail.com*

## ABSTRACT

The amine neurotransmitters produced in the body such as dopamine, epinephrine, norepinephrine, and serotonin are important in the field of psychiatry as these neurotransmitters play an important role in regulating the mood, sleep, attention, and behavior. The neurotransmitters, epinephrine and norepinephrine act by binding centrally or peripherally to the adrenergic receptors; dopamine acts by binding to dopaminergic receptors and serotonin action is based on its binding to serotoninergic receptors. The chapter highlights this aspect by the discussion of neurotransmission of endogenous catecholamines and synthetic compounds with agonistic activity along with the compounds involved in the regulation of sympathetic system. The chapter discusses a wide range of drugs mimicking and modifying the sympathetic neurotransmission, mainly biogenic amine neurotransmitters, specifically, the catecholamines and synthetic non-catecholamines.

## 3.1 INTRODUCTION

Adrenergic pharmacology has its origin from the two words adrenergic and pharmacology.

The term "adrenergic" (derived from the English word "adrenaline" that means "adrenal" that indicate near kidney and a suffix "ine" meaning adrenal gland released hormone and a term "ergon" from Greek that means "work" and a suffix "ic") meaning "working on adrenaline or noradrenaline" (Stedman, 1918) and "Pharmacology" (originated from "pharmakon" the Greek that means "drugs" and "logos" meaning "study of") meaning study of drugs (Stedman, 1918). In short, adrenergic pharmacology is the learning and working of the endogenous or synthetic agents, noradrenaline, and adrenaline which are also known as norepinephrine and epinephrine, respectively. Adrenaline, an endogenous hormone and noradrenaline, a neurotransmitter that helps SNS/the sympathetic system functions. The SNS prepares our body to react to stressful conditions by producing "fight or flight response" to subjugate the challenge ahead. The fight response is produced by the activation of α receptors whereas the flight response, which is the reduced readiness to fight, is due to the β adrenoceptor stimulation or corticosterone secretion induced by noradrenaline (Haller et al., 1997).

So, to maintain body homeostasis to stressful conditions, certain vital functions in the body gets modulated within the areas like the smooth muscles, heart, adipose tissue, liver, salivary gland, and skeletal muscles due to neurotransmitter activity on the receptors and an aberration in the process can be dealt with the help of sympatho-mimetic compounds and the compounds which modify the neurotransmission regulation (Golan, 2012). Hence these compounds are clinically important in treating anxiety, depression, cognitive disorders, and also treating the diseases occurring in CVS such as hypertension; the respiratory system for instance asthma, nasal decongestion, and in ophthalmology (Brunton et al., 2011).

Sympathomimetic agents are agents who imitate the activity of endogenous catecholamines like adrenaline and noradrenaline on the adrenergic nervous system (Westfall, 2009). Sympathomimetic agents have other synonyms like adrenergic agonists, adrenergic amines. These work by directly binding to the adrenoceptors namely $\alpha$, $\beta$; or acting indirectly by an increase in the level of neurotransmitter noradrenaline by the action of drugs including cocaine, reserpine, or a mechanism involving both direct as well as indirect action (Brunton et al., 2011; Golan et al., 2012; Westfall, 2009). The $\alpha_2$ adrenoceptors present in the presynaptic neuron (called the autoreceptor) brings about a balancing feedback limiting the noradrenaline synthesis (Brunton et al., 2011; Golan et al., 2012). The chapter aims in giving an insight to the readers about various transmitters participating in the sympathetic system, its neurotransmission and the receptors onto which they are bound producing desired pharmacological effects and the various categories of drug producing adrenergic agonism thereby rendering it useful clinically.

## 3.2   ADRENERGIC PATHWAY

It deals with the neurotransmission of the endogenous catecholamines like noradrenaline, and dopamine; the sympathetic neurotransmitters and adrenaline, the major hormone released by the adrenal medulla (Tortora et al., 2008). Except for sweat glands and few blood vessels, noradrenaline is released from all the postganglionic sympathetic nerve endings. Dopamine action is predominant in the brain, some peripheral neurons, renal, and mesenteric vessels. The synthesis of dopamine in dopaminergic neuron (in CNS) stops further synthesis of catecholamine (Golan, 2012).

### 3.2.1   ADRENERGIC NEUROTRANSMISSION

#### 3.2.1.1   CATECHOLAMINES

These are compounds with a catechol nucleus (benzene ring containing two hydroxy groups in the third and fourth position) and an amine containing side chain attached to it (Katzung et al., 2014) The natural or endogenous catecholamines are adrenaline, noradrenaline, and dopamine (Brunton et al., 2011). The synthetic catecholamines available are isoprenaline (Rang et al., 2007), dobutamine, dopexamine, dipivefrine.

The endogenous catecholamines like adrenaline, noradrenaline act on $\alpha$ and $\beta$ receptors and produce sympathomimetic or agonistic action (Tortora et al., 2008; Golan, 2012).

Adrenaline also known as epinephrine has a predominant action on $\beta_1$ and $\beta_2$ receptors while weak agonistic action on $\beta_3$ receptors (Udaykumar, 2016; Katzung et al., 2014). Due to its action on $\beta_1$ adrenoceptors, there is an elevation in the systolic BP because of increased inotropy (myocardial contraction) and the output of the heart (may be due to increased renin secretion); while action on $\beta_2$ receptors produces lowered diastolic BP and the peripheral resistance because of the blood vessel dilation. Dilation of smooth muscles of the bronchi due to $\beta_2$ receptors makes it useful clinically in treating asthmatic attacks. Increase in plasma glucose level is observed due to stimulation of the $\beta_2$ receptors within liver and the skeletal muscles (Emorine et al., 1989; Gibson et al., 2006).

Noradrenaline, also known as norepineph-rine, has predominant action on $\alpha_1$, $\alpha_2$ receptors ($\alpha_2$ receptor having inhibitory action), and $\beta_1$ while weak agonistic action on $\beta_2$ leading to increased diastolic and systolic BP and total PR (Lohse et al., 2003; Bremner et al., 2001; Heubach et al., 2002).

Dopamine acts predominantly on $D_1$ dopa-minergic receptors in renal, mesenteric, and coronary vascular beds leading to vasodilation. It also has agonistic action on $\beta_1$ receptors causing positive inotropic effect and at high doses causes vasoconstriction due to its action on $\alpha_1$ receptors (Stevens et al., 2008; Sibley et al., 1993; Katzung et al., 2014).

Isoprenaline, also called isoproterenol, is a semisynthetic sympathomimetic agent has agonistic action on $\beta$ receptors with small affinity for $\alpha$ receptors. Activation of $\beta_1$ recep-tors can cause cardiac stimulation which raises the myocardial contractility and also the rate of the heart leading to an increase in the systolic BP. $\beta_2$ receptor activation causes vasodilation thereby lowering diastolic BP. Also, relaxation of the smooth muscles in the bronchi is observed because of $\beta_2$ receptor stimulation. So it can be used for treating refractory atrioventricular block, refractory bradycardia, and asthma (Stevens et al., 2008; Katzung et al., 2014).

Dobutamine has agonistic action in $\alpha$ and $\beta$ receptors. Stimulation of cardiac $\beta_1$ receptor causes both an increase in myocardial contrac-tion and heart rate leading to increased cardiac output while activation of $\beta_2$ receptor in smooth muscles causes relaxation (Golan, 2012; Uday-kumar, 2016; Stevens et al., 2008). Dopexamine

has agonistic action in $D_1$, $D_2$ as well as $\beta_3$ receptors. It can be used for managing sepsis, heart failure, and shock. When a high dose is used, tachycardia and hypotension may occur (Golan, 2012; Udaykumar, 2016; Stevens et al., 2008; Katzung et al., 2014).

## 3.2.2 ADRENERGIC NEUROTRANSMISSION PATHWAY

Adrenergic neurotransmission involves a series of events which are important as endogenous catecholamines and agents having its action on the adrenergic system or those modulating the neurotransmission are employed in treating diseases of the heart, respiratory tract, psycho-logical, or other cognitive disorders (Golan, 2012; Tripathi, 2014). Steps involved in the neurotransmission of the adrenergic system are as follows:

### 3.2.2.1 SYNTHESIS OF CATECHOLAMINES

Tyrosine, an essential amino acid, is a precursor of noradrenaline (Fig. 3.1) and is synthesized from phenylalanine, an amino acid obtained from food. Phenylalanine gets oxidized in liver with phenylalanine hydroxylase as enzyme to yield tyrosine which gets taken up by the adren-ergic neurons from circulation via an aromatic acid transporter. In the adrenergic neurons, tyrosine hydroxylase aided by tetrahydrobiop-terin hydroxylates tyrosine in the cytoplasm to L-dihydroxyphenylalanine (L- DOPA) (Purves et al., 2004; Wevers et al., 1999). Deficiency of tyrosine hydroxylase causes a high depletion in catecholamine synthesis resulting in neurolog-ical defects accompanied by a defect in motor coordination and other extrapyramidal clinical features like Parkinson's disease (Wevers et al., 1999). L-DOPA undergoes decarboxylation in the presence of L-DOPA decarboxylase with pyridoxal phosphate as a cofactor, thereby forming dopamine. A VNT namely VMAT causes dopamine transfer to the synapses (Njus et al., 1986) followed by a conversion to noradrenaline with the assistance by ascorbic acid with dopamine β-hydroxylase as the enzyme. Noradrenaline then moves to the adrenal medulla then undergoing methylation and leads to the synthesis of adrenaline with the aid of phenylethanolamine N-methyl transferase which is invigorated by glucocorticoids (Sharara Chami et al., 2010; Viskupic et al., 1994).

### 3.2.2.2   STORAGE OF CATECHOLAMINES

Noradrenaline is stored in storage vesicles and then released by exocytosis triggered by calcium ions when subjected to action potential (Sudhof, 2004).

### 3.2.2.3   RELEASE OF CATECHOLAMINES

The emergence of an action potential causes $Ca^{2+}$ influx into the axoplasm thereby releasing the neurotransmitter noradrenaline by exocytosis. Noradrenaline release can be produced by vesicular calcium sensor synaptotagmin 1, maintained by its RIM1α and 2α binding, which are two active zone proteins. The released noradrenaline can either bind to postsynaptic receptors like $α_1$, $β_1$, $β_2$, $β_3$, and $α_2$ presynaptic receptors thereby producing certain pharmacological effects or subjected to reuptake into the vesicles of the presynaptic neurons (Katz, 1969; Sudhof, 2004)

### 3.2.2.4   REUPTAKE OF CATECHOLAMINES

The response produced by the noradrenaline liberated from nerve terminals can be stopped by any of the following reuptake mechanisms:

#### 3.2.2.4.1   Neuronal/Axonal Uptake

Reuptake of noradrenaline into the neuronal cytoplasm occurs with the help of $Na^+$/$Cl^-$-dependent transporters, norepinephrine transporter (NET) which concentrates noradrenaline in the neuronal cytoplasm in the presence of $Na^+$ ions. This further reduces the binding of noradrenaline to postsynaptic neurons producing pharmacological responses (Golan, 2012).

#### 3.2.2.4.2   Vesicular Uptake

The noradrenaline that is uptaken into the axonal cytoplasm is partly destroyed by MAO and partly subjected to transportation into the synaptic vesicles for storage with the help of VMAT-2 by the interchanging of $H^+$ ions (Njus et al., 1986).

#### 3.2.2.4.3   Extraneuronal Uptake

Reuptake of adrenaline at a higher rate compared to that of noradrenaline in non-neuronal peripheral tissues (such as cardiac muscles, intestinal and vascular smooth muscles) is mediated by extraneuronal amine transporter (ENT). Adrenaline is not stored but rapidly destroyed by MAO and catechol-O-methyltransferase (COMT).

### 3.2.2.5   METABOLISM OF CATECHOLAMINES

Catecholamines are metabolized in the body by two enzymes MAO and COMT. MAO, a mitochondrial enzyme metabolizes noradrenaline that leaks out from axonal cytoplasm leading to the formation of 3, 4-dihydroxy mandelic acid (DOMA), then metabolized by COMT to 3-methoxy 4-hydroxy mandelic acid (VMA), which is the major metabolite excreted in urine (Golan, 2012; Wevers et al., 1999).

COMT, an enzyme in the cytoplasm, metabolizes norepinephrine that moves out into systemic circulation in the form of normetanephrine which in presence of MAO gets converted to MOPGAL which in turn undergoes dehydrogenation to form VMA (Fig. 3.1) (Wevers et al., 1999).

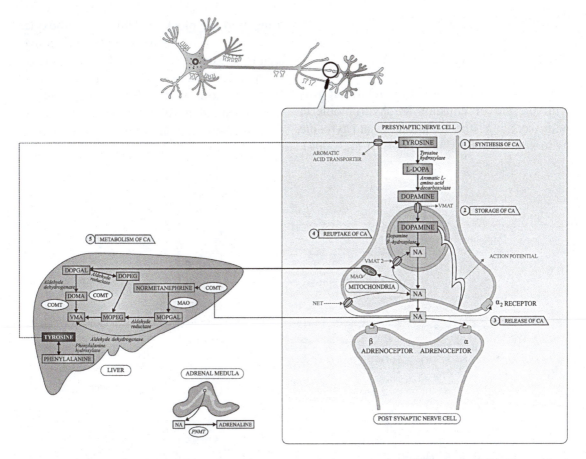

**FIGURE 3.1 (See color insert.)** The adrenergic neurotransmission mechanism: 1. Synthesis of CA: The synthesis of endogenous CA like NA, dopamine produced from tyrosine which crosses the blood–brain barrier via a carrier. 2. Storage of catecholamines: Dopamine, as well as noradrenaline synthesized, is translocated into synaptic vesicles via VMAT. 3. Release of CA: When subjected to action potential noradrenaline is liberated from the vesicle by exocytosis. In the cells of adrenal medulla, noradrenaline returns to the cytoplasm and gets converted into adrenaline with the aid of PNMT. Then Adrenaline is accumulated within the vesicle. 4. Reuptake of catecholamines: Reuptake of noradrenaline occurs by axonal uptake via NET and vesicular uptake via VMAT. 5. Metabolism of catecholamines are metabolized with the aid of the enzymes MAO and COMT. CA, catecholamines; DOPA, dihydroxyphenylalanine; NA, noradrenaline; VMAT, vesicular monoamine transporter; MAO, monoamine oxidase; NET, norepinephrine transporter; PNMT, phenylethanolamine N-methyl transferase; DOPGAL, 3,4-didydroxyphenylclycoaldehyde; DOMA, 3,4-dihydroxy mandelic acid; VMA, 3-methoxy 4-hydroxy mandelic acid; MOPEG, methoxy-4-hydroxyphenylethyl-glycol; DOPEG, 3,4-dihydroxyphenylethylene glycol; COMT, catechol-O-methyl transferase.

*Source*: Adapted from Tripathi (2014), Golan (2012), and Wevers et al. (1999).

### 3.2.3 DRUGS MODIFYING VARIOUS STEPS IN THE ADRENERGIC NEUROTRANSMISSION PATHWAY

The effect of drugs that modify adrenergic neurotransmission is as follows:

### 3.2.3.1 SYNTHESIS OF NORADRENALINE

The tyrosine hydroxylation to DOPA happens with the aid of cofactor tetrahydrobiopterin with oxygen as the co-substrate and tyrosine hydroxylase as enzyme (Purves et al., 2004).

α-methyl-p-tyrosine blocks tyrosine hydroxylase thereby blocking the tyrosine conversion to dihydroxyphenylalanine, leading to a depletion of noradrenaline. DOPA which is produced from tyrosine is decarboxylated to form dopamine with the aid of L-DOPA decarboxylase. α-methyldopa blocks L-DOPA decarboxylase needed for the dihydroxyphenylalanine conversion to dopamine leading to the formation of a methylated noradrenaline which may act like false neurotransmitter. Figure 3.2 condenses the various steps involved in catecholamine neurotransmission and also the drugs affecting them.

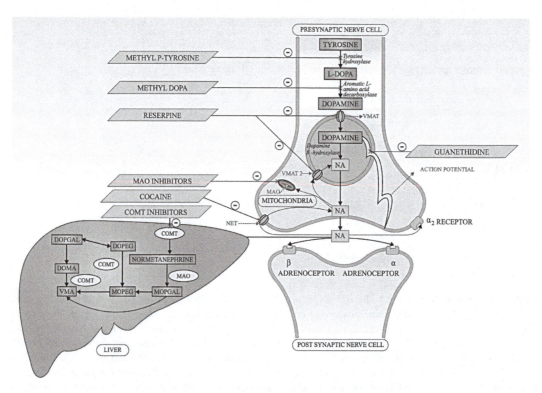

**FIGURE 3.2   (See color insert.)** Drugs modifying adrenergic neurotransmission. DOPA, dihydroxyphenylalanine; NA, noradrenaline; VMAT, vesicular monoamine transporter; MAO, monoamine oxidase; NET, norepinephrine transporter; DOPGAL, 3,4-didydroxyphenylclycoaldehyde; DOMA, 3,4-dihydroxy mandelic acid; VMA, 3-methoxy 4-hydroxy mandelic acid; MOPEG, methoxy-4-hydroxyphenylethyl-glycol; DOPEG, 3,4-dihydroxyphenylethylene glycol; COMT, catechol-O-methyl transferase.

*Source*: Adapted from Tripathi (2014) and Golan (2012).

### 3.2.3.2   STORAGE OF NORADRENALINE

VMAT, a VNT, prevents leakage and deterioration of the neurotransmitters thereby maintaining a check in its concentration within the cytoplasm. These are of two types VMAT-1 and VMAT-2. VMAT-1 is located in the peripheral organs particularly the cells of the paracrine and endocrine glands whereas VMAT-2 is located in the vesicular transporter located in CNS, histaminergic cells of adrenal medulla, blood cells, and stomach (Mahata et al., 1993; Peter et al., 1995; Weihe et al., 1994; Erickson et al., 1996). Reserpine blocks both VMAT-1 as well as VMAT-2 leading to reduced uptake of dopamine as well as noradrenaline into the synaptic vesicles leading to its degradation by MAO thereby inhibiting adrenergic neurotransmission at the synapses. Apart from reserpine, tetrabenazine and methamphetamine also have competitive blocking action on VMAT-2, which is more sensitive compared to VMAT-1 (Erickson et al., 1996; Peter et al., 1994). The enhancement of storage of neurotransmitters like dopamine and noradrenaline can be a good remedy in the discipline of psychiatry and in behavioral, locomotory defective disorders like parkinsonism.

### 3.2.3.3   NORADRENALINE RELEASE

Noradrenaline is released from vesicles and performs their pharmacological action by its binding to the respective adrenoceptors like α, β adrenoceptors. Drugs like guanethidine and bretylium inhibit the impulse coupled noradrenaline release from synaptic vesicle by exocytosis (Brunton et al., 2011).

### 3.2.3.4   REUPTAKE OF NORADRENALINE AND ADRENALINE

According to Tanda et al. (1997), in the region of prefrontal cortex, intravenous administration of cocaine and amphetamine showed a marked increase in the extracellular dopamine and noradrenaline by blocking noradrenaline carrier. Cocaine binds to NET thereby blocking the reuptake of noradrenaline into the axonal cytoplasm thereby potentiating adrenergic neurotransmission at the synapses (Giros et al., 1992; Giros et al., 1993).

Indirect sympathomimetics such as tyramine and ephedrine diffuse through NET and produces sympathomimetic responses. Guanethidine also displaces vesicular noradrenaline leading to an initial sympathomimetic response followed by noradrenaline depletion. Corticosterone blocks ENT thereby inhibiting adrenaline uptake into cells of other tissues (Golan, 2012).

### 3.2.3.5   CATECHOLAMINES METABOLISM

Inhibitors of MAO such as nialamide, tranylcypromine, moclobemide potentiate noradrenaline whereas MAO- B inhibitors such as selegiline potentiate dopamine in the brain (Golan, 2012; Wevers et al., 1999; Brunton et al., 2011). Inhibitors of COMT like tolcapone, entacapone potentiate both noradrenaline and dopamine (Fig. 3.2).

### 3.3   ADRENERGIC RECEPTORS

These receptors come under the superfamily of GPCR. It functions by activating or inhibiting the signaling mediators, $G_s$, $G_q$, $G_i$, and $G_0$ which then activates or inhibits the inositol triphosphate ($IP_3$)/diacylglycerol (DAG) or cAMP, which are the second messengers. Adrenergic receptors are classified into α, β adrenoceptors. α adrenergic receptors are further classified into $α_1$, $α_2$, β adrenergic receptors can be further classified into $β_1$, $β_2$, $β_3$ adrenergic receptors (Insel, 1996; Brodde et al., 1999).

### 3.3.1 α ADRENERGIC RECEPTORS

α adrenergic receptors are subclassified as $α_1$, $α_2$ adrenergic receptors. $α_1$ receptors are further classified into $α_{1A, 1B, 1a}$ and $α_2$ are classified as $α_{2A}$, $α_{2B}$, and $α_{2C}$ (Insel, 1996). Tables 3.1 and 3.2 give an outline of the location and functioning

of α adrenoceptors in different areas of the body system and its signaling pathway.

$α_1$ adrenergic receptors (Fig. 3.3) can be seen predominantly distributed in the smooth muscles (of the blood vessels, genitourinary system, and intestine) (Walden et al., 1997), liver, heart, and the salivary gland. These are brought about by stimulating $G_q$ or $G_0$ signaling mediators. The enzyme phospholipase C is activated by $G_q$ thereby activating the second messengers such as $IP_3$/DAG. $α_{1A}$ receptor works by activating calcium ion channels whereas $α_{1B}$ receptor acts by exerting its activity on $IP_3$ which potentiates influx of $Ca^{2+}$ ions mediating vascular and genitourinary smooth muscle contractions which are probably by activating $Ca^{2+}$-troponin complex causing increased myocardial contraction.

**TABLE 3.1**    Characteristics of $α_1$ Adrenergic Receptors.

| $α_1$ Adrenoceptors | | |
|---|---|---|
| **Location and function** | Smooth muscles in the blood vessels | Contraction |
| | Genitourinary smooth muscle | Contraction |
| | Intestinal smooth muscle | Relaxation |
| | Heart | Positive inotropy and excitability |
| | Liver | Glycogenolysis and gluconeogenesis |
| | Salivary gland | $K^+$ release |
| **Signal transduction mechanism** | 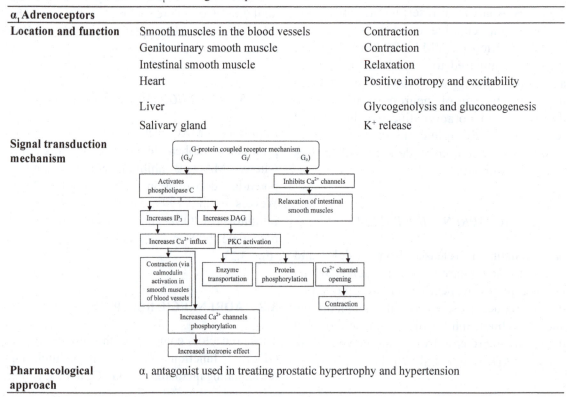 | |
| **Pharmacological approach** | $α_1$ antagonist used in treating prostatic hypertrophy and hypertension | |

PKC, protein kinase C; $IP_3$, inositol triphosphate; DAG, diacylglycerol.

**TABLE 3.2** Characteristics of $\alpha_2$ Adrenoceptors.

| $\alpha_2$ Adrenoceptors | | |
|---|---|---|
| **Location and function** | Pancreatic β cells | Reduced insulin secretion |
| | Nerve endings | Reduced noradrenaline release |
| | Platelets | Aggregation |
| | Blood vessels | Constriction/dilatation |

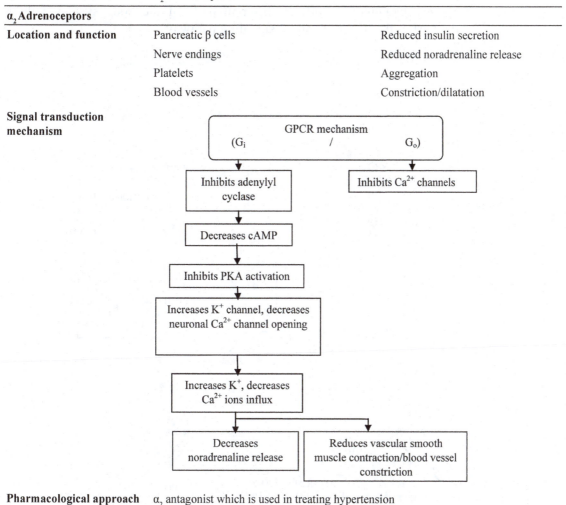

**Signal transduction mechanism**

| **Pharmacological approach** | $\alpha_2$ antagonist which is used in treating hypertension |

PKA, protein kinase A; cAMP, cyclic adenosine monophosphate.

DAG produces activation of the inactive Protein Kinase C. The probable mechanism for relaxation of intestinal smooth muscles and $K^+$ release from salivary glands may be due to $G_0$ signaling mediator stimulation. $\alpha_{1a}$ is the recombinant of receptor $\alpha_{1A}$. In some blood vessels, the vasoconstriction response is brought about by $\alpha_{1B}$ adrenoceptor which is involved in the heart and blood pressure functioning. The $\alpha_{1B}$ as well $\alpha_{1a}$ adrenoceptors both play a vital function in memory and the nociceptive responses (Han et al., 1990; Muramatsu, 1995; Tanoue et al., 2002).

### 3.3.2 β ADRENERGIC RECEPTORS

β adrenergic receptors are further subclassified into $\beta_1$, $\beta_2$, $\beta_3$ adrenergic receptors. Studies on recombinant β adrenoceptors revealed a novel inactive binding domain in $\beta_1$ adrenoceptor identified as $\beta_4$ adrenoceptor expressed in the fat and heart tissues (Granneman, 2001). Tables 3.3–3.5 summarize the location and functioning of $\beta_1$, $\beta_2$, and $\beta_3$ receptors in different parts of the body system with its signaling pathway.

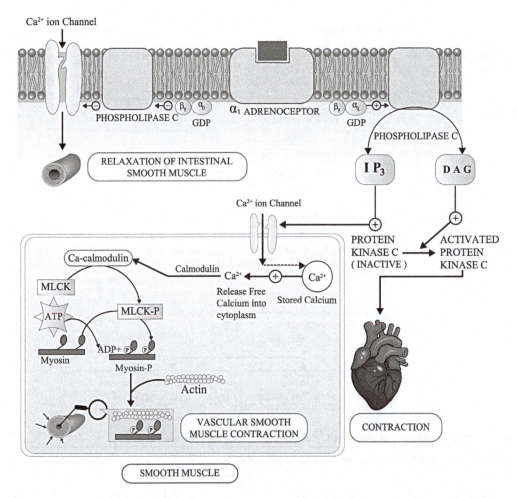

**FIGURE 3.3 (See color insert.)** Signal transduction and pharmacological action of the neurotransmitters, noradrenaline, adrenaline, and other exogenous sympathetic drugs on $\alpha_1$ adrenergic receptors. GDP, guanosine 5'-diphosphate; IP$_3$, inositol triphosphate; DAG, diacylglycerol; Ca$^{2+}$, calcium ions; ATP, adenosine triphosphate; ADP, adenosine diphosphate; MLCK, myosin light-chain kinase.

*Source*: Adapted from Katzung et al. (2014) and Rang et al. (2007). $\alpha_2$ adrenergic receptors (Fig. 3.4) are predominantly distributed in pancreatic $\beta$ cells, adrenergic nerve endings, platelets, and blood vessels and response of all $\alpha_{2A}$, $\alpha_{2B}$, and $\alpha_{2C}$ are mediated by G$_o$ or G$_i$ signaling mediators. $\alpha_2$ autoreceptors in the presynaptic neurons mediate reduced norepinephrine release by stimulating G$_o$ having reduced $\alpha$ action inhibiting the enzyme adenylyl cyclase thereby decreasing cAMP, hence reducing Ca$^{2+}$ influx and inhibiting exocytosis of the transmitter. Activation of the signaling mediator G$_i$ in the postsynaptic neuron results in inhibition of adenylyl cyclase thereby decreasing cAMP. Hence inactive protein kinase A cannot be activated which might be the probable process for the platelet aggregation and reduced insulin secretion (Bylund, 1992; Michelotti et al., 2000).

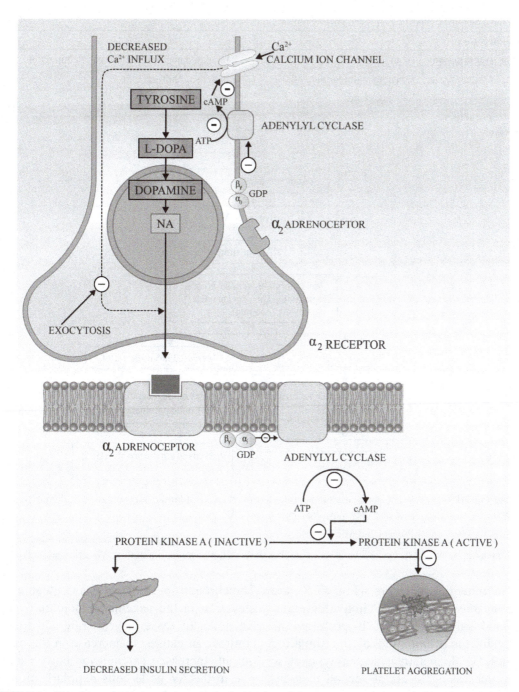

**FIGURE 3.4**  Signal transduction and pharmacological action of the neurotransmitters, noradrenaline, adrenaline, and other exogenous sympathetic drugs on $\alpha_2$ adrenergic receptors. DOPA, dihydroxyphenylalanine; NA, noradrenaline; cAMP, cyclic adenosine monophosphate; ATP, adenosine triphosphate; GDP, guanosine 5'-diphosphate.

*Source*: Adapted from Katzung et al. (2014) and Rang et al. (2007).

**TABLE 3.3**    Characteristics of $\beta_1$ Adrenergic Receptors.

| $\beta_1$ Adrenoceptors | | |
|---|---|---|
| **Location and function** | Heart | Increased heart rate and force of contraction |
| | Renal juxtaglomerular cells | Increases renin secretion |
| | Salivary gland | Amylase secretion |
| **Signal transduction mechanism** | 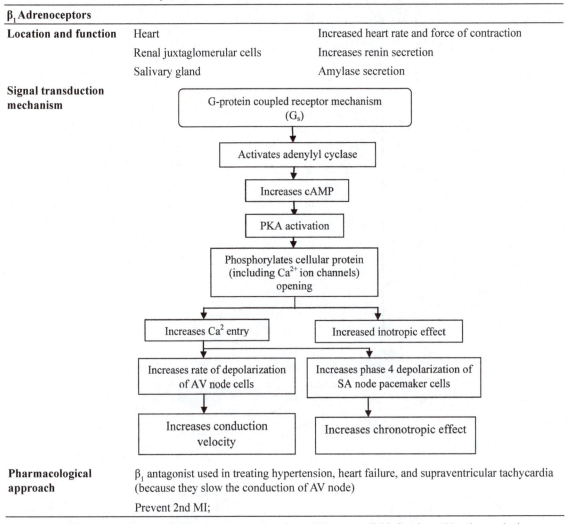 | |
| **Pharmacological approach** | $\beta_1$ antagonist used in treating hypertension, heart failure, and supraventricular tachycardia (because they slow the conduction of AV node) | |
| | Prevent 2nd MI; | |

PKA, protein kinase A; cAMP, cyclic adenosine monophosphate; MI, myocardial infarction; AV, atrioventricular.

$\beta_1$ adrenergic receptors (Fig. 3.5) are found in the heart, renal juxtaglomerular cells, and salivary glands. It performs its activity by the stimulation of the signaling mediator, $G_s$ which activates adenylyl cyclase thereby increasing levels of second messengers such as cAMP. This produces activation of protein kinase including ion channels phosphorylation like $Ca^{2+}$ channels present in sarcolemma resulting in positive inotropy. The chronotropic activity is an outcome of an elevation in the pacemaker potential rate of SA node. Increase in heart rate and force of contraction causes an increase in the cardiac output. Increased calcium ion entry causes a rise in the AV node cells depolarization rate giving rise to increased conduction velocity. $G_s$ stimulation also increases renin and amylase secretion in salivary glands and renal juxtaglomerular cells, respectively.

**TABLE 3.4** Characteristics of $\beta_2$ Adrenergic Receptors.

| $\beta_2$ Adrenoceptors | | |
|---|---|---|
| **Location and function** | Smooth muscle | Relaxation/dilation |
| | Blood vessels | Dilate |
| | Bronchi | Dilate |
| | GIT | Relax |
| | Uterus | Relax |
| | Bladder detrusor | Relax |
| | Seminal tract | Relax |
| | Ciliary muscle | Relax |
| | Heart (minor function only) | Increased HR<br>Increased force of contraction |
| | Skeletal muscle | Tremor<br>Increased muscle mass and speed of contraction<br>Glycogenolysis |
| | Liver | Glycogenolysis |
| | Nerve terminals | Rise in adrenergic neurotransmitter release;<br>Inhibition of histamine release |

**Signal transduction mechanism**

| G-protein coupled receptor mechanism |
|---|

| $G_s$ independent activation of $K^+$ channels | Activates adenylyl cyclase |
|---|---|
| Rise in $K^+$ efflux | Increases cAMP |
| Smooth muscle cell (bronchial) hyperpolarization | PKA activation |
| Opposes depolarization necessary to elicit contraction | Phosphorylates intracellular protein |
| | Glycogen phosphorylase activation and glycogen catabolism |
| | Rise in plasma glucose (liver skeletal muscles) |

**Pharmacological approach**   $\beta_2$ agonists are used in treating asthma

GIT, gastrointestinal tract; HR, heart rate; PKA, protein kinase A; cAMP, cyclic adenosine monophosphate.

**TABLE 3.5**   Characteristics of $\beta_3$ Adrenoceptors.

| $\beta_3$ Adrenoceptors | | |
|---|---|---|
| **Location and function** | Adipose tissue | Lipolysis<br>Thermogenesis |
| | Skeletal muscle | Thermogenesis |
| **Signal transduction mechanism** | | |

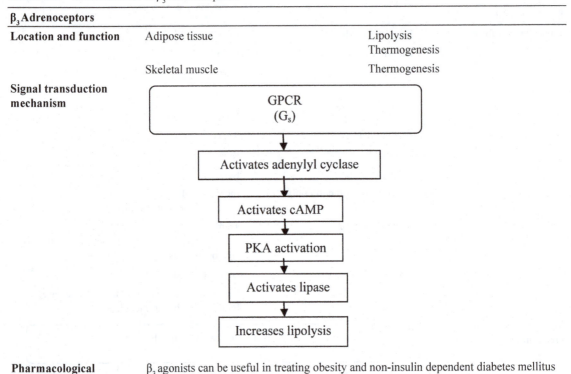

| **Pharmacological approach** | $\beta_3$ agonists can be useful in treating obesity and non-insulin dependent diabetes mellitus | |

PKA, protein kinase A; cAMP, cyclic adenosine monophosphate.

$\beta_2$ adrenergic receptors (Fig. 3.6) are present in the smooth muscles (of blood vessels, bronchi, GIT, uterus, bladder detrusor, ciliary muscle, seminal tract), heart, liver, skeletal muscle, and nerve endings which respond by stimulation of $G_s$ signaling mediator activating adenylyl cyclase thereby increasing cAMP levels (Lohse, 2003). This event stimulates contractile protein (for example, myosin light chain kinase) phosphorylation which produces a reduced affinity for the calcium–calmodulin complex resulting in smooth muscle relaxation. Also increased $K^+$ efflux contributes to bronchial smooth muscle relaxation which may be due to the result of $G_i$ signaling or $G_s$ independent $K^+$ ion channel activation regulated by the exchange of $N^+/H^+$ an outcome of $\beta_2$ adrenoceptor interaction with NHERF (Hall et al., 1998). A minor action of $\beta_2$ receptors on the heart can be the result of $G_s$ activation thereby activating ion channel (such as calcium ion channel) phosphorylation producing an increased myocardial contraction. In humans, $\beta_1$, $\beta_2$ adrenoceptors activation via $G_s$ signaling pathway promote phosphorylation in cardiomyocytes thereby forming $Ca^{2+}$-troponin complex causing myocardial contractility (Lohse, 2003; Bartel et al., 2003; Molenaar et al., 2000). Due to the $G_s$ signaling activation protein kinase A activation takes place in the hepatocytes resulting in the stimulation of phosphorylase kinase which stimulates conversion of glycogen to glucose; while

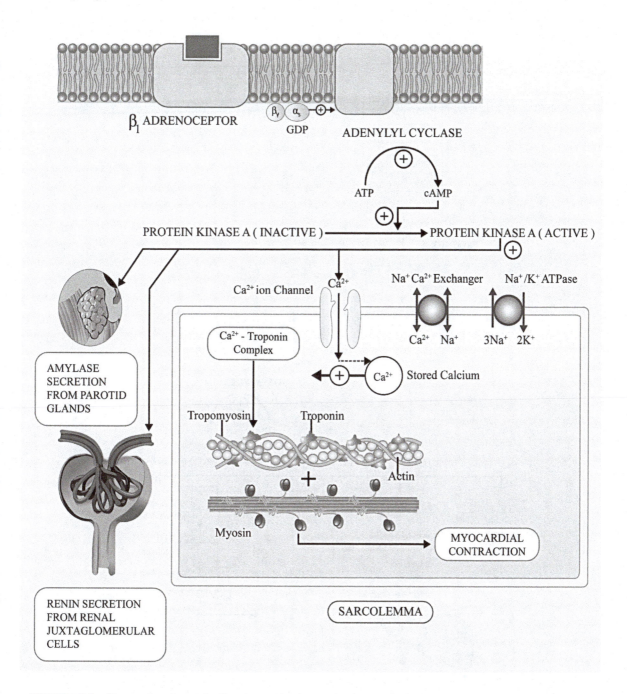

**FIGURE 3.5 (See color insert.)** Signal transduction and pharmacological action of the neurotransmitters, noradrenaline, adrenaline, and other exogenous sympathetic drugs on $\beta_1$ adrenergic receptors. cAMP, cyclic adenosine monophosphate; ATP, adenosine triphosphate; GDP, guanosine 5′-diphosphate; $Ca^{2+}$, calcium ions; $Na^+$ $Ca^{2+}$ exchanger, sodium–calcium exchanger or sodium–potassium pump; $Na^+$ $K^+$ ATPase, sodium–potassium adenosine triphosphatase.

*Source*: Adapted from Katzung et al. (2014) and Rang et al. (2007).

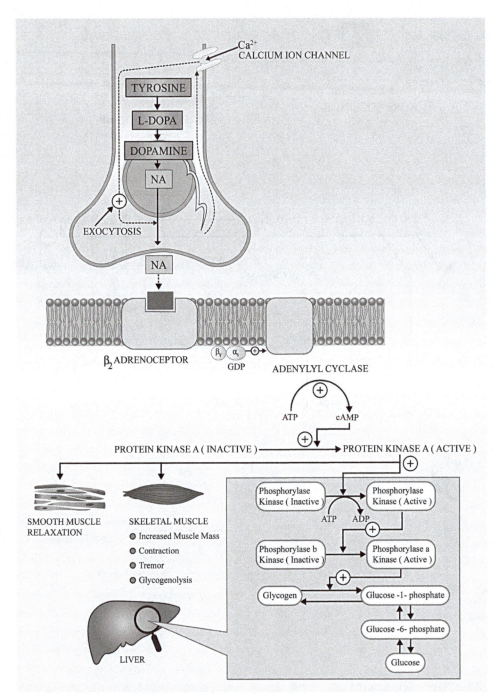

**FIGURE 3.6**   Signal transduction and pharmacological action of neurotransmitters such as noradrenaline, adrenaline, and other exogenous sympathetic drugs on $\beta_2$ adrenoceptors. DOPA, dihydroxyphenylalanine; NA, noradrenaline; cAMP, cyclic adenosine monophosphate; ATP, adenosine triphosphate; GDP, guanosine 5′-diphosphate.

*Source*: Adapted from Katzung et al. (2014) and Rang et al. (2007).

stimulating glycogenolysis in the skeletal muscles. Also, the probable mechanism behind increased noradrenaline release can be stimulating $G_s$ in presynaptic neuron activating adenylyl cyclase enzyme thereby increasing the cAMP level, hence increasing influx of $Ca^{2+}$ and potentiating exocytosis of the neurotransmitter.

$\beta_3$ adrenergic receptors (Fig. 3.7) are present in skeletal muscle and adipose tissue bringing about an activation of $G_s$ signaling pathway, activating adenylyl cyclase, thereby activating cAMP levels which is sufficient for protein kinase A activation. This is succeeded by the mobilization of fat from the WAT to BAT where lipase activation causes triglyceride oxidation to yield fatty acid within the brown adipose tissue fueling thermogenesis within the striated skeletal muscles. In humans, the BAT level decreases with age progression and its stimulation by $\beta_3$ sympathomimetics can possibly be a good remedy for type 2 DM and obesity. According to Weyer et al. (1999), WAT contains dormant adipose cells which if subjected to $\beta_3$ adrenergic activation exert thermogenic activity but again this topic demands investigation. Also uncoupling the

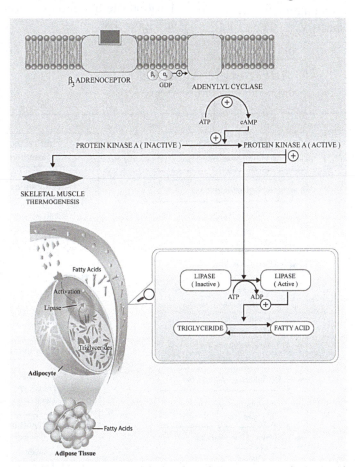

**FIGURE 3.7** Signal transduction and pharmacological action of the neurotransmitters, noradrenaline, adrenaline, and other exogenous sympathetic drugs on $\beta_3$ adrenergic receptors. cAMP, cyclic adenosine monophosphate; ATP, adenosine triphosphate; GDP, guanosine 5'-diphosphate.

*Source*: Adapted from Katzung et al. (2014) and Rang et al. (2007).

protein expression in skeletal muscles, WAT and BAT contribute to thermogenesis. Anti-diabetic activity can be possible by the result of glucose disposal but studies on rodents showed increased insulin secretion and also non-oxidative disposal of glucose improved the volume of insulin-mediated glucose (Gold-stein, 1997; Weyer et al., 1999; Arch et al., 1984; Yen, 1984; Yen et al., 1984; Wilbraham et al., 1987; Holloway et al., 1992; Largis et al., 1994; Yoshida et al., 1994; Yoshida et al., 1994). $\beta_4$ adrenoceptors act via G-protein coupled receptor mechanism showing a mixed agonistic as well as antagonistic action on BAT and heart respectively. The agonistic

action in adipocytes enhances thermogenesis whereas having a blocking action on heart rate (Granneman, 2001; Ahlquist, 1948).

## 3.4 CLASSIFICATION OF ADRENERGIC AGONISTS

These are agents or drugs mimicking the cellular or pharmacological responses occurring in the adrenergic system by interacting with the adren-ergic receptors (Katzung et al., 2014). It can be divided based on the selectivity as well as the location/site of action of receptors as illustrated in Figure 3.8.

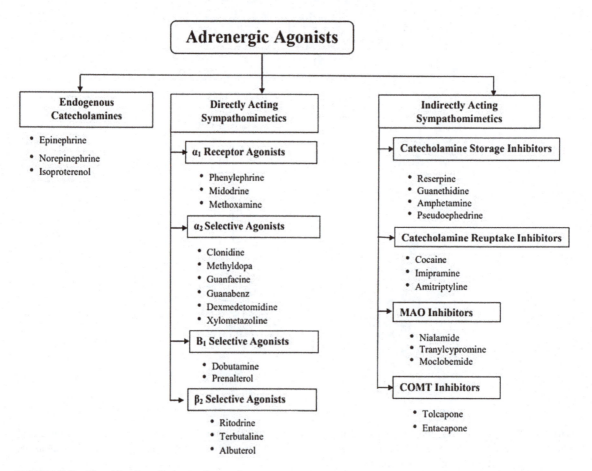

**FIGURE 3.8**  Classification of adrenergic agonists.

## 3.5 ADRENERGIC AGONISTS

### 3.5.1 ENDOGENOUS CATECHOLAMINES

#### 3.5.1.1 ADRENALINE

It acts by binding to α and β adrenoceptors activating or inhibiting G-protein coupled receptor. Cardiovascular effects include cardiac arrhythmia and tachycardia; effects on CNS include cerebral hemorrhage; endocrinal effects include hyperglycemic conditions and the other effects include are tissue ischemia and necrosis. It can be used as a local anesthetic, in treating an acute asthmatic attack, anaphylaxis, cardiac arrest as well as ventricular fibrillation. It is also used for the reduction of bleeding during surgery (Breuer et al., 2013; Simons et al., 2010).

#### 3.5.1.1.1 *Pharmacological Actions*

##### 3.5.1.1.1.1 *Heart*

Both α and β receptors can be seen in heart tissue out of which β receptors are more in number. Adrenaline acts mainly on $\beta_1$ subtype of β adrenoceptor with some minor cardiac activity expressed by $\beta_2$ producing some blood vessel dilation resulting in the reduction of peripheral resistance as well as diastolic BP. Binding to $\beta_1$ produces an increase in heart rate thereby causing a change in rhythm of the heart. A rise in myocardial contraction produces increased systolic BP (Lohse et al., 2003). Adrenaline maintains the diastole by raising the depolarization of SA nodal cells, hence increase heart rate corresponding to shortened AP and RP. Adrenaline improves the conduction velocity in AV node, Purkinje fibers and atrioventricular fibers and enhancement of automaticity can be a leading cause for arrhythmia. The action of adrenaline on $\alpha_1$ adrenoceptors improves the myocardial contraction and refractory period (Bremner et al., 2001; Heubach et al., 2002).

##### 3.5.1.1.1.2 *Blood Vessels*

Adrenaline interaction with $\beta_2$ receptors causes blood vessel dilation in the skeletal muscles as well as the coronary vessels. Blood vessels present in the mucous membrane and in the skin constrict because of its interaction with $\alpha_1$ receptors. It results in reduced peripheral resistance as well as diastolic blood pressure (Han et al., 1990; Muramatsu et al., 1995).

##### 3.5.1.1.1.3 *Blood Pressure*

There is an increase in the systolic BP due to positive inotropy ($\beta_1$ effect) and a reduction in diastolic BP ($\beta_2$ effect). Noradrenaline increases the systolic ($\beta_1$ effect) and the diastolic BP (Lohse et al., 2003; Hieble et al., 1995).

##### 3.5.1.1.1.4 *Respiratory Effects*

Adrenaline and isoprenaline are strong bronchodilators while noradrenaline exhibit weak action. This is because of its activity on $\beta_2$ adrenoceptors inhibiting antigen-induced inflammatory mediators (e.g., chemokines, cytokines, prostanoids) release and bronchial smooth muscle relaxation proving its benefit in asthmatic conditions. Adrenaline also causes decongestion of mucosa of upper airway by directly acting upon the respiratory center which is not clinically used. Intravenous dose of epinephrine causes apnoea whereas toxic doses cause pulmonary edema (Nadel et al., 1986; Rhoden et al., 1988).

##### 3.5.1.1.1.5 *Eye*

Adrenaline produces mydriasis and increased aqueous humor outflow because of its activity on $\alpha_1$ receptor producing contraction of radial iris muscles and $\beta_2$ receptor enhancing ciliary secretion (Tripathi, 2014). Adrenaline also acts on the β receptor and produce an increase in the endogenous eicosanoid, prostaglandin E (PGE) levels then resulting in ocular hypotension in rabbits (Naveh et al., 2000).

### 3.5.1.1.1.6  GIT

Adrenaline acts directly causing smooth muscle dilation by its interaction with the $\beta_2$ receptors and indirectly relaxes gut by decreasing the release of acetylcholine from the cholinergic nerve terminals because of its action on $\alpha_2$ adrenoceptors seen in the ganglionic cells thereby decreasing electrolyte and water secretion to lumen of intestine (Cellek et al., 2007; Quinson et al., 2001; Schemann et al., 2010).

### 3.5.1.1.1.7  Bladder

Adrenaline interacts with $\beta_2$ receptors relaxing the detrusor muscle and $\alpha_1$ receptors contracting bladder trigone muscles thereby contracting the sphincter muscles leading to urinary retention (Goepel et al., 1997).

### 3.5.1.1.1.8  Genitalia

Depending upon the species, hormonal, and gestational conditions the effect of catecholamines on uterus varies. Adrenaline interacts with $\alpha_1$ receptors leading to a contraction in non-pregnant uterus and $\beta_2$ receptors leading to relaxation during pregnancy; while noradrenaline causes uterine contraction during pregnancy. Uterine stimulation occurs due to an interaction of adrenaline or related substances with α receptors while uterine inhibition may occur when the agents combine with the β receptors (Miller, 1967).

Noradrenaline produces the activation of $\alpha_1$ receptors of smooth muscles resulting in the contraction of the prostate in males which leads to ejaculation. The amount of evidence available now, including both the in vivo as well as in

vitro, shows that a flaccid penile state is partly due to release of noradrenaline from sympathetic nerves corporal contraction of smooth muscle subsequent to stimulation of homogeneous post-synaptic α-adrenoceptor population (Adaikan et al., 1981; Hedlund et al., 1985; Tejada et al., 1989; Tejada et al., 1989). Noradrenaline and phenylephrine-induced contractions in isolated HET are resultant of predominantly $\alpha_1$ activation and not $\alpha_2$ adrenoceptors (Christ et al., 1990).

### 3.5.1.1.1.9  Splenic Capsule

Interaction with α receptors increases RBC in circulation. Stress including exercise increases pooling of the lymph nodes from the blood to the efferent lymphatics a resultant of bone marrow stimulation of spleen contraction because of interaction with α receptors (Lucia et al., 1937; Bierman et al., 1952; Tripathi, 2014).

### 3.5.1.1.1.10  Skeletal Muscle

Interaction with $\beta_2$ receptors stimulate the adenylyl cyclase thereby increasing cAMP levels leading to the activation of protein kinase A producing increased muscle mass, contraction, tremor, glycogenolysis. Interaction with $\beta_3$ receptors produces thermogenesis (Emorine et al., 1989).

### 3.5.1.1.1.11  CNS

Excess adrenaline release causes metabolic dysfunction by promoting glucose synthesis. This can disturb the circadian rhythm of the body causing excess glucose to be utilized by the brain exposing to psychosomatic diseases such as stress and anxiety as a result of its stimulatory action and may cause secondary effects on cardiovascular systems and other systems. Sympathomimetic like ephedrine, amphetamine shows increased alertness, insomnia, mood elevation as they cross the blood–brain barrier

(Bennun, 2014; Rektor et al., 2003; Richter, 1960).

### 3.5.1.1.1.12  Metabolic

Adrenaline via $G_i$ signaling pathway suppresses extracellular signal-regulated protein kinases 1 and 2 (ERK1/2) present in pancreatic beta cells because of its activity on $\alpha_2$ receptors, then causing a reduced insulin release and also stimulating an insulin-independent production of glucose (Gibson et al., 2006).

### 3.5.1.1.2  Pharmacokinetics

Rapid absorption is observed followed by IM injection. It cannot be administered orally. Being hydrophilic the drug does not bypass the BBB but crosses placenta; therefore, is seen in the breast milk. Metabolism occurs rapidly by COMT and MAO because of catechol ring and the occurrence of rapid phase I (oxidation) as well as phase II (conjugation in GIT mucosa and liver) reaction. Excretion occurs through urine (Brunton et al., 2011; Florey, 2008; Rataboli, 2010).

### 3.5.1.2  NORADRENALINE

Its difference from the epinephrine is the absence of substitution of a methyl within amino group. It is a major neurotransmitter secreted by the sympathetic postganglionic nerves. In the adrenal medulla, 10–20% of catecholamine present is noradrenaline (Tortora et al., 2008).

### 3.5.1.2.1  Pharmacodynamics and Pharmacokinetics

Noradrenaline has a direct agonist action on the effector cells. It is an α agonist with less potency when compared to adrenaline. It acts by binding on to $\alpha_1$ and $\beta_1$ receptors thereby activating G-Protein Coupled receptor cascade. It stimulates $\beta_1$ receptor and has little agonist activity on $\beta_2$ receptor (Katzung et al., 2014). Noradrenaline produces decreased or unchanged output of the heart, raised diastolic, systolic and the pulse pressure, raised TPR and raised stroke volume. There is a rise in the coronary flow because of dilation and a rise in the blood pressure. But small doses do not induce vasodilator effects. The metabolic activity of noradrenaline is comparable with those induced by adrenaline, but is not much significant in humans. Noradrenaline does not produce effects during administration via oral and subcutaneous route and is inactivated by those enzymes which act on adrenaline. A small quantity is excreted through the urine. Noradrenaline produces undesired effects comparable to adrenaline like tachycardia and other arrhythmias; endocrinal effects are hyperglycemia and the other effects include tissue ischemia and necrosis. High dose produces hypertension. So, blood pressure determination has to be done frequently during systemic administration. Noradrenaline can be used in intensive care to raise the BP and also can be used in shock (Brunton et al., 2011). Noradrenaline interaction with $\beta_2$ receptor mediated by activating adenylyl cyclase activated phosphorylase cascade in the liver resulting in glycogenolysis and gluconeogenesis. Increased plasma glucose and reduced insulin secretion (due to $\alpha_2$ receptor interaction) precipitate diabetes mellitus. $\beta_3$ adrenergic receptor interaction activates lipase thereby converting TG into fatty acid within adipose tissue and occurrence of thermogenesis within skeletal muscles. It is employed in treating shock and hypotension (Weyer et al., 1999; Tripathi, 2014). It has poor absorption following oral and SC administration. It penetrates well into the CNS because of an absence of catechol in the structure. Metabolism is rapid with the help of COMT and MAO. The half-life is 2 min. It is excreted rapidly through urine (Brunton et al., 2011; Rataboli, 2010).

### 3.5.1.3 PHARMACODYNAMICS AND PHARMACOKINETICS OF VARIOUS ENDOGENOUS CATECHOLAMINES

#### 3.5.1.3.1 Isoproterenol

Isoproterenol is also known as isoprenaline and shows its sympathomimetic nature by binding specifically to the $\beta_1$ and $\beta_2$ adrenergic receptors. Due to the bronchodilation and contraction of heart muscles effects, it can be used in treating refractory atrioventricular block, refractory bradycardia, and asthma (Crawford et al., 2010). According to O'Neill et al., increase in consciousness was reported on administration of isoproterenol infusion in atrial fibrillation condition performing catheter ablation. Effects on the heart such as cardiac arrhythmias and tachycardia are reported. Isoproterenol interacts with propranolol, nortriptyline, amitriptyline, and MAO inhibitors. Adverse events include tachycardia, headache, palpitation, skin flushing, tremor, nausea, dizziness, weakness, and sweating. According to Van et al., human mononuclear cells release from spleen on administration of isoproterenol through intravenous route (Brunton et al., 2011; Rataboli, 2010; Florey, 2008; Van et al., 1990; O'Neill et al., 2012). It is absorbed readily following parenteral as well as aerosol administration. Drug absorption following oral administration is unreliable and it undergoes large scale first-pass metabolism. The drug is 14–18% plasma protein bound and is absorbed readily after inhalation and via the intramuscular route.

The drug penetrates well into the CNS where it undergoes metabolism by COMT and not by the MAO. Metabolism occurs to form o-methylated form. Conjugation happens at the wall of gut. Excretion of the metabolites takes place through the bile and urine. About 90% of drug undergoes excretion within a time period of 24 h in the urine. The elimination half-life is 2.5–5 min (O'Neill et al., 2012; Kadar et al., 1974).

#### 3.5.1.3.2 Dopamine

Dopamine, a neurotransmitter present in CNS is also known to be the precursor of the catecholamines, adrenaline, and noradrenaline. Epithelial cells synthesize dopamine which locally produces diuretic effects in the region of proximal tubule. Dopamine undergoes metabolism by COMT and MAO, so the oral administration cannot be effective. Dopamine acts on many receptors; at high levels, dopamine acts via $\beta_1$ receptors showing an elevation in the contraction on myocardium, whereas at reduced concentration dopamine interact with vascular $D_1$ receptors mainly at renal and coronary vessels. Vasodilation happens due to $D_1$ receptor activation. Glomerular filtration rate, renal blood flow, and sodium excretion increases. Dopamine helps the noradrenaline release and it also plays a role in the cardiac activity (Stevens et al., 2008; Sibley et al., 1993; Katzung et al., 2014). Dopamine increases pulse and systolic pressure then causes a small rise in diastolic BP. Total peripheral resistance remains unchanged. High concentration of dopamine also stimulates $\alpha_1$ receptors causing vasoconstriction. Dopamine does not induce central effects because of the difficulty to pass through the blood–brain barrier. Overdose of dopamine produces undesired action because of excess sympathomimetic activity. Headache, nausea, vomiting, hypertension, arrhythmia, and tachycardia may happen while dopamine administration. Dose has to be reviewed in patients consuming MAO inhibitors and tricyclic antidepressants. Dopamine is used for treating congestive heart failure, septic, and cardiogenic shock. Dopamine produces a short-term improvement in cardiac and renal function in people having chronic cardiac disease and kidney failure. Dopamine is administered through intravenous route to one large vein. It has a short duration of activity, therefore to control its intensity of activity, the administration rate is adjusted (Brunton et al., 2011; Crooks, 1978).

## 3.5.2 DIRECTLY ACTING SYMPATHOMIMETICS

### 3.5.2.1 $\alpha_1$ RECEPTOR AGONISTS

#### 3.5.2.1.1 Phenylephrine

Phenylephrine is an $\alpha_1$ selective receptor agonist used as a mydriatic and nasal decongestant in various ophthalmic and nasal formulations. The pharmacological activities produced are identical to methoxamine and produce arterial vasoconstriction (Brunton et al., 2011). Phenylephrine activates $\alpha_1$ receptors which result in contraction of the smooth muscles producing vasoconstriction; increased BP and vascular resistance. Ocular administration causes iris dilator muscle to constrict resulting in pupil dilation (mydriasis). Nasal decongestion, ocular decongestion, mydriasis during an ophthalmological examination, maintenance of BP during surgery, treating drug-induced, and neurogenic shock (Katzung et al., 2014). It is adequately absorbed after oral or topical administration. It is administered by oral, topical and parenteral routes. It is partly metabolized by MAO in

intestine and liver (Brunton et al., 2011; Stevens et al., 2008; Florey, 2008).

#### 3.5.2.1.2 Midodrine

Midodrine is used as a prodrug with $\alpha_1$ receptor agonist activity. Active metabolite formed is desglymidodrine. Its duration of activity is about 4–6 h and the half-life is about 3 h (Katzung et al., 2014). Midodrine activates $\alpha_1$ receptors causing vasoconstriction resulting in a rise in the diastolic and systolic blood pressure while standing, sitting, and in supine positions, hence used for treating postural hypotension, hypotension caused by infections in infants, psychotropic agents induced hypotension, hypotension in persons having renal dialysis, and autonomic insufficiency (Golan, 2012; Brunton et al., 2011). It is absorbed rapidly after administering orally. The peak concentration is achieved approximately 1 h following the administration. Metabolism occurs in the liver and in other tissues (Brunton et al., 2011; Stevens et al., 2008; Tripathi, 2014).

#### 3.5.2.1.3 Methoxamine

Methoxamine activates $\alpha_1$ adrenoreceptors thereby causing vagally mediated bradycardia and vasoconstriction resulting in a prolonged rise in the blood pressure. Its use is in the treatment of hypotension occurring during surgery. This drug has contraindication in CAD and severe hypertension and can be used in caution after parenteral injection of ergot alkaloids, with congestive heart failure and hyperthyroidism (Brown et al., 1966, Florey, 2008). Methoxamine is administered by intravenous route and also by intramuscular route. Methoxamine intravenous administration gives its presser effect within 1–2 min whereas onset of activity of the intramuscular injection occurs within 15–20 min. Duration of activity of methoxamine via intravenous route is about 60 min while comparing to 90 min via the intramuscular route. Methoxamine undergoes dealkylation to form a metabolite O-dealkylated metabolite, 2-hydroxymethoxamine (Florey, 2008).

### 3.5.2.2    $\alpha_2$ SELECTIVE AGONISTS

#### 3.5.2.2.1    Clonidine

Clonidine acts on peripheral regions of $\alpha_1$ adrenoceptors and can be employed as a topical nasal decongestant and also as a vasoconstrictor. While given orally it produces a fall in the BP and influences the rate of the heart. Parenteral administration shows an acute hypertensive effect. Clonidine activates $\alpha_2$ receptor lowering the rate of the heart and produces a dilation of capacitance vessels causing a decrease in PVR resulting in lowered sympathetic effusion from CNS producing a decreased cardiac output resulting in decreased blood pressure. The adverse events include bradycardia, sedation, and hypotension. It is also formulated as a transdermal patch. Clonidine is majorly used for treating hypertension. It is also employed for treating addicted patients to withdraw alcohol, tobacco, and narcotics. It improves the signs occurring in postural hypotension. It has several off labels uses in attention deficit hyperactivity disorder (ADHD), mania, Tourette's syndrome, psychosis, post-hepatic neuralgia, ulcerative colitis, and restless legs syndrome (RLS). It can be used as sympatholytic in treating nasal decongestion and signs of drug withdrawal (Brunton et al., 2011; Brittain, 2008; Golan, 2012). After administering orally, the drug shows 100% bioavailability. It has a half-life of about 6–24 h. The drug shows a maximum response at approximately 1–3 h. The volume of distribution is 2–4 L/kg and is about 20–40% protein bound. Being lipophilic in nature it crosses the BBB then acts in the CNS. Metabolism of clonidine takes place in the liver by entering enterohepatic circulation. About 65% of the drug moves out of the body through the urine where 32% remains unchanged and 22% excreted through feces. The half-life is 8–12 h (Rataboli, 2010; Brittain, 2008)

#### 3.5.2.2.2    Methyldopa

It is used as an antihypertensive which acts centrally. It acts via $\alpha_2$ receptors. The hypotensive mechanism is comparable to clonidine. Methyldopa is metabolized to $\alpha$-methylnorepinephrine which is discharged by an adrenergic nerve terminal stimulates $\alpha_2$ receptor lowering the heart rate producing a decrease in the sympathetic effusion from CNS producing a decrease in the cardiac output leading to lowered blood pressure. It can be used for treating hypertension in pregnancy, sedation, persistent mental lassitude, and impaired mental concentration, nightmares, mental depression, vertigo, extrapyramidal signs can occur but are infrequent, elevated prolactin secretion in men and women (Brunton et al., 2011; Rataboli, 2010; Golan, 2012). It is orally absorbed and attains its maximal pharmacological activity in 4–6 h and persists for 24 h. It crosses the blood–brain barrier and being a prodrug converts to active metabolite to perform its action. It also crosses the placenta and can be present in breast milk. Metabolism occurs to form inactive methyldopa-O-sulfate formed because of the first pass effect occurring in the intestine and in the liver (Brunton et al., 2011; Stevens et al., 2008; Rataboli, 2010).

### 3.5.2.2.3 Guanfacine

It is more selective when compared to clonidine and is an $\alpha_2$ receptor agonist. It produces a hypotensive effect and also employed in treating ADHD in children. It shows good absorption after administering orally and the half-life range is about 12–24 h. It can cause withdrawal symptoms when stopped suddenly. Guanfacine centrally stimulates $\alpha$ adrenoceptor. It has antihypertensive property but is used rarely (Brittain, 2008; Brunton et al., 2011).

### 3.5.2.2.4 Guanabenz

It is an agonist of the $\alpha_2$ receptor which acts centrally and lowers the BP. It acts similar to clonidine. The half-life range is about 4–6 h. Metabolism takes place in the liver. The adverse events are comparable to clonidine. Guanabenz centrally stimulates $\alpha$ adrenoceptor, also have antihypertensive activity but rarely used (Brunton et al., 2011; Florey, 2008)

### 3.5.2.2.5 Dexmedetomidine

Dexmedetomidine centrally stimulates $\alpha_2$ adrenoceptor and used in sedation of subjects initially subjected to intubation and mechanical ventilation during treatment in ICU and reduces the requirement for opioids in pain control (Gertler et al., 2001; Rataboli, 2010).

### 3.5.2.2.6 Xylometazoline

Xylometazoline stimulates $\alpha_2$ adrenoceptor resulting in constriction of the nasal mucosa. It has CNS-related side effects such as anxiety, dizzy, tremor, and other effects such as rebound congestion on overuse. Topical products may cause local adverse effects like burning, stinging, and dryness. It causes a rise in the heart rate, palpitations, blood pressure so, contraindicated in cardiac disease. The drug is not used in the case of patients treated with MAOIs. It is mainly used as a Topical decongestant (Brunton et al., 2011; Rataboli, 2010; Florey, 2008).

### 3.5.2.2.7 Oxymetazoline

Oxymetazoline stimulates $\alpha_2$ adrenoceptor resulting in constriction of the nasal mucosa. In a large dose, it may cause hypotension. It is mainly used topically as a decongestant and to constrict the vascular smooth muscle for relieving ophthalmic hyperemia (Brunton et al., 2011; Rataboli, 2010).

### 3.5.2.3 $\beta_1$ SELECTIVE AGONISTS

### 3.5.2.3.1 Dobutamine

Dobutamine interacts with $\beta$ and $\alpha$ receptors and was earlier thought being a selective $\beta_1$ receptor agonist. It is similar in structure to dopamine. Both enantiomeric forms exist as a racemic mixture. The levo (−) isomer is an $\alpha_1$ receptor agonist while the dextro (+) isomer is an antagonist of $\alpha_1$ receptors. Both isomers have agonist effects on beta receptors although dextro isomer is more potent. Dobutamine produces more inotropic effects binding to the $\alpha_1$ receptors than the chronotropic effects on the heart. Cardiac output rises and then there is a change in the peripheral resistance. In some of the patients, there might be a significant increase in the heart rate and blood pressure and long-term effect is unclear. There might be a chance to develop tolerance. Dobutamine is used for a short-term period for treating decompensation of heart which happens in congestive heart failure or MI or after heart surgery. The onset of activity is fast, within 1–10 min and has a 2-min half-life (Brunton et al., 2011; Golan, 2012; Stevens et al., 2008). It penetrates well into CNS because of the absence of catechol in structure. Metabolism happens in the tissues and in the liver. The plasma half-life of the drug is approximately 2 min and excreted via urine (Rataboli, 2010; Florey, 2008).

### 3.5.2.3.2  *Prenalterol*

It is less effective in $\beta_2$ receptor activation hence categorized as a partial $\beta_1$ selective agonist which raises the output of the heart with fewer reflex to tachycardia compared to selective $\beta$ agonist which reported hemodynamic effects on long-term use of prenalterol in congestive heart patients (Brunton et al., 2011; Lambertz et al., 1984; Rataboli, 2010).

### 3.5.2.4  *$\beta_2$ SELECTIVE AGONISTS*

### 3.5.2.4.1  *Ritodrine*

Ritodrine functions as a $\beta_2$ receptor selective agonist. After administering intravenously, about half the amount of drug excretes unchanged. Following administration orally, the absorption is fast but incomplete at about 30%. It is mainly employed as a uterine relaxant for stopping premature labor and can prolong pregnancy. Ritodrine binds selectively to $\beta_2$ adrenergic receptor stimulating adenylyl cyclase thereby increasing cAMP level resulting in relaxation of uterine and bronchial smooth muscles. Adverse events occur because of excess $\beta$ receptor activation. The patients

with heart disease might be at risk because of the effects of the drug on the heart. Tremor is one common adverse event and there is a chance to develop anxiety and restlessness. Drug tolerance also occurs. tachycardia is another common adverse event because of $\beta_1$ receptor stimulation and CNS related effects, for example, muscle tremor, nervousness and so on are caused by stimulation of mainly the $\beta$ adrenergic receptors seen in skeletal muscle, heart, and the CNS. There is a rare possibility of arrhythmia or myocardial ischemia while people suffering from CAD are at risk. Patients taking MAO inhibitors (MAOIs) can have an increase in cardiovascular adverse events. Parenteral administration produces an increase in glucose concentration of lactate and fatty acids and potassium concentration decrease in plasma. The drug is used for treating premature labor and asthma (Brunton et al., 2011; Rataboli, 2010).

### 3.5.2.4.2  *Terbutaline*

Terbutaline selectively binds to $\beta_2$ adrenergic receptor stimulating adenylyl cyclase thereby increasing cAMP level resulting in relaxation of uterine and bronchial smooth muscles. Cardiovascular effects for example tachycardia and the CNS effects including muscle tremor, nervousness have been reported. It is used treatment of asthma and premature labor (Brunton et al., 2011; Rataboli, 2010). It can be used for long-term for treating COPD, bronchospasm and parenterally for status asthmaticus (Florey, 2008). The non-catecholamine structure of terbutaline makes it better efficacious in administering orally. 60% of terbutaline is absorbed following administration via the oral route. The onset of activity starts about in 1 to 2 h. The activity lasts up to 6 h with a bioavailability of 50%. Being hydrophilic in nature it cannot cross the blood–brain barrier, mainly crosses the placenta and enters breast milk. The drug is 14–25% protein bound compared to erythrocyte with a plasma level ratio of 2.0:2.6. Its peak bronchodilation is 10 min. Its non-catecholamine structure prevents its metabolism by COMT or MAO hence glucuronic or sulfuric acid conjugation takes place. Metabolism occurs in the gut wall and in the liver. The half-life is 3–4 h and 60% excretion happens unchanged through urine following administering orally and 2% following administering parenterally (Florey, 2008).

### 3.5.2.4.3   Albuterol

The drug is a $\beta_2$ receptor selective agonist. The therapeutic use and pharmacological action look similar to terbutaline. It is mainly used in the management of bronchospasm and administered via oral route or as an inhalation. The effect occurs at about 15 min, then lasting up to 3 to 4 h. Albuterol selectively binds to the $\beta_2$ receptor, then stimulating the adenylyl cyclase thereby increasing cAMP level resulting in the relaxation of bronchial and uterine smooth muscles. It produces side effects like tachycardia, muscle tremor, and nervousness. It is employed for treating premature labor and asthma (Brunton et al., 2011; Rang et al., 2007). It is formulated as inhalation and produces a fast onset of activity within 5–15 min, lasting for 3–6 h. It undergoes first pass effect in the gut wall and in the liver but does not happen in the lungs. Approximately 30% of the unchanged drug along with the metabolites are rapidly excreted and has half-life 4–6 h (Brunton et al., 2011; Rataboli, 2010).

### 3.5.2.4.4   Other Short-acting $\beta_2$ Adrenergic Agonists

Other short-acting $\beta_2$ adrenergic agonists are levalbuterol, bitolterol, fenoterol, isoetharine, metaproterenol, and procaterol. Levalbuterol is a dextro isomer of albuterol, which is a racemic mixture. It is $\beta_2$ selective and formulated and used as an inhalation. Bitolterol is a $\beta_2$ receptor selective agonist and a prodrug of the active colterol. After inhalation, duration of activity is 3–6 h. Fenoterol, a $\beta_2$ receptor selective agonist, has duration of activity of 4–6 h, the effect of the drug starts immediately after inhalation. It binds to $\beta_1$ receptors producing cardiac effects. It is discontinued from the market. Isoetharine is a $\beta_2$ receptor selective agonist. MAO does not affect the drug and the metabolism is by COMT. It is formulated as an inhalation and is used in treating bronchoconstriction. Metaproterenol, a $\beta_2$ receptor selective agonist belonging to the category of resorcinol bronchodilators is used long-termly for treating COPD, asthma, and bronchospasm. After administering orally, approximately 40% drug absorption takes place as the active form. It is usually resistant to COMT. It is not much $\beta_2$ selective in comparison to terbutaline, so it may cause cardiac stimulation. While taking orally, the drug has a slow onset of action and the effect lasts up to 3 to 4 h. Inhalation produces an immediate effect. It may be used via oral route or by inhalation. Pirbuterol is a $\beta_2$ receptor selective agonist and it is almost similar to albuterol. It is administered as an inhalation and effect lasts up to 3–4 h. Procaterol is a $\beta_2$ receptor selective agonist and is administered as an inhalation. It has an immediate onset of activity and a 5-h duration of action (Brunton et al., 2011; Barisione et al., 2010).

### 3.5.2.4.5   Other Long-acting β₂ Receptor Selective Agonists

Other long-acting $\beta_2$ receptor selective agonists are salmeterol, formoterol, arformoterol, carmoterol, and indacaterol. Salmeterol is a $\beta_2$ selective receptor agonist, highly specific and showing long duration of activity which is more than 12 h. It is mainly used for treating COPD. It is metabolized by CYP3A4. The onset of activity is slow. Formoterol $\beta_2$ receptor selective agonist with quick action while used as an inhalation and a duration of action which is long and lasts for 12 h. The drug is used in treating bronchospasm, asthma, and obstructive pulmonary disease. Arformoterol is a long-acting $\beta_2$ receptor selective agonist. The drug is used long-term for treating COPD and bronchoconstriction. Metabolism is by the enzymes CYP2C19 and CYP2D6. The adverse events include insomnia, tachycardia, reduction in plasma potassium level and a rise in the level of plasma glucose. Carmoterol, a $\beta_2$ selective adrenergic agonist is having greater selectivity. The drug shows a fast onset of activity and the duration of activity is very long which lasts more than 24 h. It is a bronchodilator used in treating asthma and COPD. Indacaterol is a $\beta_2$ adrenergic receptor selective agonist with a rapid onset and long duration in activity. The drug is used in treating COPD and asthma (Brunton et al., 2011; Barisione et al., 2010).

### 3.5.3   INDIRECTLY ACTING ADRENERGIC RECEPTOR AGONISTS

### 3.5.3.1   CATECHOLAMINE STORAGE INHIBITORS

### 3.5.3.1.1   Reserpine

It is a $\beta_2$ receptor selective agonist. The therapeutic use and pharmacological activity are similar to terbutaline. It is used for the management of bronchospasm and administered orally or by inhalation. The effect occurs within 15 min and lasts for 3 to 4 h (Florey, 2008). The absorption is rapid after oral administration. Peak plasma concentration reaches in approximately 2 h. Metabolism happens in the liver and about 90% of this drug gets excreted in the form of metabolites. The half-life of the drug is 50–386 h (Brunton et al., 2011).

### 3.5.3.1.2   Amphetamine

Amphetamine produces a stimulation of the CNS and also the peripheral $\beta$ and $\alpha$ receptors. The drug is highly potent and a strong stimulant. The effect lasts up to many hours. It increases diastolic and systolic BP and also produces its effect on smooth muscles. It decreases fatigue, sleep, elevates the mood, and produces euphoria (Brunton et al., 2011). It reduces the appetite, increase respiration rate, and is a minor analgesic. Adverse effects include dizziness, tremor, restlessness, irritability, weakness, talkativeness, hallucinations, delirium, and anxiety following the central stimulation, depression, and fatigue. Cardiac effects include arrhythmias, angina, and palpitation. It can produce nausea, abdominal cramps, vomiting, anorexia, and diarrhea. The drug can be abused and so will show dependence. It can be used in treating ADHD and narcolepsy. Amphetamine can enter into sympathetic nerve terminals and can displace the reserved catecholamine transmitter. They might antagonize the NET thereby producing a rise in the norepinephrine release resulting in vasoconstriction, cardiac stimulation, and increased BP. It penetrates well into the CNS due to absence of catechol in structure gets rapidly excreted through urine (Brunton et al., 2011; Golan, 2012; Tanda et al., 1997).

### 3.5.3.1.3 Pseudoephedrine

It is a nasal decongestant with a half-life of 5–8 h. It undergoes phase I biotransformation which include para hydroxylation, N-demethylation, and oxidative deamination (Brunton et al., 2011; Florey, 2008). According to Arafa et al. (2008), pseudoephedrine and a combination of pseudoephedrine and imipramine can be useful as a method of treatment of retrograde ejaculation in diabetic patients (Arafa et al., 2008).

### 3.5.3.2 CATECHOLAMINE REUPTAKE INHIBITORS

### 3.5.3.2.1 Cocaine

Cocaine blocks the neuronal reuptake of norepinephrine at the central and peripheral synapses by inhibiting NET thereby potentiating neurotransmission at adrenergic synapses producing vasoconstriction and cardiac stimulation resulting in elevation of blood pressure. It has cardiovascular side effects which include accelerating coronary atherosclerosis, tachycardia, and effects in the CNS such as seizure, CNS depression/excitation, anxiety. It is contraindicated with ester-type local anesthetics, parabens, or PABA, as it might produce hypersensitivity reactions and also in ophthalmologic anesthesia as it can produce sloughing of the corneal epithelium. It should be used carefully in patients having cardiovascular disease. Avoid beta-blocker use in patients treated for cardiovascular complications induced by cocaine abuse. It is also used as mucosal and ophthalmic local anesthetic and in diagnosing Horner's syndrome pupil (Florey, 2008; Giros et al., 1992; Giros et al., 1993; Golan, 2012; Tanda et al., 1997). The Cocaine can be absorbed from the mucosa

and is bound highly to the protein and the onset of activity is 1 min with a 30-min duration of action. Metabolism of cocaine is by hepatic enzymes in humans and plasma esterases in many animals. The half-life is 75 min and some cocaine is excreted through the urine unchanged (Brunton et al., 2011; Rataboli, 2010; Florey, 2008).

### 3.5.3.2.2 Imipramine

It comes under the category of tricyclic antidepressants and is absorbed well after administration orally. The bioavailability is 30–77% in the case of oral administration. The plasma half-life is 6–25 h and 56–95% is protein bound. Metabolism takes place in liver yielding the metabolite, desipramine. Imipramine excretion is by urine, feces, as well as breast milk. It is employed in treating depression (Florey, 2008). Imipramine is also used with pseudoephedrine and imipramine for treating the diabetic patients for retrograde ejaculation (Arafa et al., 2008). It is also employed in treating panic attacks, attention deficient hyperkinetic disorder. It has an enuretic effect but is not preferred. Side effects include sedation, tremor, insomnia, orthostatic hypotension, conduction defects, arrhythmias, aggravation of psychosis, withdrawal syndrome, seizure, weight gain, loss of libido (Katzung et al., 2014).

### 3.5.3.3 MAO INHIBITORS

### 3.5.3.3.1 Tranylcypromine

Tranylcypromine resembles dextroamphetamine, a weak MAO inhibitor. It is absorbed readily after an oral dose and is metabolized extensively. Both the in vivo and in vitro experiments reveal tranylcypromine as one potent MAO inhibitor. It shows a long duration of activity (Britain, 2008; Katzung et al., 2014).

### 3.5.3.4   COMT INHIBITORS

#### 3.5.3.4.1   Entacapone

Entacapone is selective and also a reversible inhibitor of COMT. It is used for treating Parkinson's disease. The absorption is rapid after administering orally. Metabolism is by isomerization and glucuronidation. It is majorly excreted as metabolites through the bile while minor elimination is through urine. Common adverse events include dyskinesia and nausea (Mosby, 2005; Munchau et al., 2000).

## 3.6   FUTURE OPPORTUNITIES AND CHALLENGES

Detailed studies and understanding of adrenergic pharmacology including adrenoceptor subtypes are significant in developing newer drugs. Recently, development of many useful drugs has taken place and many more drugs are in various phases of development process as an outcome of these investigations. Such studies on various receptor subtypes include studies on $\alpha_1$ receptor in which diverse studies and knowledge of $G_q$ signaling pathway suggest novelty of $\alpha_1$ agonist in treating heart failure (O'Connell et al., 2014). Development of dabuzalgron, an $\alpha_{1A}$ subtype specific agonist administered orally for treating urinary incontinency, and further studies on cardiotoxicity models demonstrated its cardioprotective action indicating a scope for the drug in near future for treating cardiac failure (Beak et al., 2017). Agonist-selective signaling and structure-function studies on $\alpha_1$ adrenoceptors can help in developing drugs with selective therapeutics and decreased side effects (Perez, 2007). Considering $\alpha_2$ receptors, dexmedetomidine, an $\alpha_2$ agonist known to lend a significant contribution to the area of anesthetic pharmacology. Further investigations are needed in establishing the exact potentiality of these molecules (Eisenach, 1993; Kallgren, 1995; Mark et al., 1995). Currently, apart from its analgesic, sedation, muscular relaxation, and anxiolytic (reported in animals) properties $\alpha_2$ adrenergic agonists also exhibits certain collateral effects in CVS, respiratory and GIT. In order to minimize adverse effects, it is co-administered in low doses along with opioid analgesics and benzodiazepines (Carroll et al., 1999; Crassous et al., 2007; Gross et al., 2001; Guimaraes et al., 2001; Maze et al., 1991; Paddleford et al., 1999; Scheinin et al., 1989; Tranquilli et al., 1993). Similar to $\alpha_1$ adrenoceptors, further study on $\alpha_2$ adrenoceptors subtypes is the need of the hour as this can result in developing subtype specific-selective therapeutic agents (Crassous et al., 2007).

Regarding $\beta_2$ adrenoceptors, $\beta_2$ agonists which are long-acting in nature are administered together with inhalational corticosteroids which are given two times daily for treating asthma and COPD (Cazzola et al., 2011; Sears et al., 2005; Tamm et al., 2012; Tashkin et al., 2010). Recent researches have resulted in developing ultra-long-acting $\beta_2$ agonists (used once in a day) such as indacaterol, olodaterol, and vilanterol trifenatate. Pharmacogenetic studies of $\beta_2$ adrenoceptors helped in predicting genetic profile based asthmatic treatment by $\beta$ agonists (Ortega, 2015). Likewise, the occurrence of $\beta_3$ adrenoceptors in human adipocytes (both WAT and BAT) and the striated skeletal muscles make way for developing selective $\beta_3$ agonist which can prove a promising remedy for obesity and insulin sensitive diabetes mellites (Arch, 2002;

Roland, 2013). Mirabegron, a $\beta_3$ adrenoceptor agonist is clinically demonstrated to be effectual against OAB because of $G_s$ protein stimulation producing a rise in cAMP level. Long-term exposure of GPCR to adrenergic agonists results in desensitization and is relevant in many medical conditions including OAB (Igawa et al., 2013; Nantel et al., 1993; Roland, 2013; Sacco et al., 2012; Scott et al., 2013). Further agonist-induced desensitization studies on HEK and CHO cells showed a targeted desensitization probably because of cAMP-dependent or independent mechanism hence require detailed attention in future (Roland, 2013; Scott et al., 2013).

Endothelial dysfunction in the region of pulmonary as well as extrapulmonary circulation seems to be challenging for development of new molecules as these are associated with diseases like asthma and cardiovascular diseases (Wanner et al., 2010). So, further investigations are needed to restore endothelial-dependent vasodilation.

## 3.7 CONCLUSION

The chapter discusses the neurotransmission of adrenergic system and the endogenous catecholamines which play a role in the process of neurotransmission and various drugs which alter the management of neurotransmission of the adrenergic system. Many drugs are now clinically used for treating and managing several diseases including asthma, heart failure, COPD, and in ophthalmology. More drugs are currently developed with several drug candidates in initial stages of the development process and many more will be developed as time moves on and the understandings, as well as, the studies go deeper. Pharmacogenetics and personalized treatment are gaining popularity and several

studies are moving on in these directions. All the above developments make a promising future for adrenergic agonists.

## KEYWORDS

- **adrenaline**
- **endogenous catecholamines**
- **adrenoceptors**
- **noradrenaline**
- **sympathomimetics**

## REFERENCES

Adaikan, P. G.; Karim, S. M. Adrenoceptors in the Human Penis. *J. Auton. Pharmacol.* **1981,** *1*, 199–203.

Ahlquist, R. P. A Study of the Adrenotropic Receptors. *Am. J. Physiol- Legacy. Content.* **1948,** *153* (3), 586–600.

Arafa, M.; El, T. O. Medical Treatment of Retrograde Ejaculation in Diabetic Patients: A Hope for Spontaneous Pregnancy. *J. Sex. Med.* **2008,** *5* (1), 194–198.

Arch, J. R. S. B3-Adrenoceptor Agonists: Potential, Pitfalls and Progress. *Eur. J. Pharmacol.* **2002,** *440* (2–3), 99–107.

Arch, J. R. S.; Ainsworth, A. T.; Ellis, R. D. M. Treatment of Obesity with Thermogenic Adrenoceptor Agonists, Studies on BRL 26830A in Rodents. *Int. J. Obes.* **1984,** *8* (Supp 1), 1–11.

Barisione, G.; Baroffio, M.; Crimi, E.; Brusasco, V. Beta-Adrenergic Agonists. *Pharmaceuticals* **2010,** *3* (4), 1016–1044.

Bartel, S.; Krause, E. G.; Wallukat, G.; Karczewski, P. New Insights into B2-Adrenoceptor Signaling in the Adult Rat Heart. *Cardiovasc. Res.* **2003,** *57*, 694–703.

Beak, J. Y.; Huang, W.; Parker, J. S.; Hicks, S. T.; Patterson, C.; Simpson, P. C.; Ma, A.; Jin, J.; Jensen, B. C. An Oral Selective Alpha-1A Adrenergic Receptor Agonist Prevents Doxorubicin Cardiotoxicity. *JACC. Basic. Transl. Sci.* **2017,** *2* (1), 39–53.

Bennun, A. The Psychosomatic Separation of Noradrenaline at Brain and Adrenaline at Blood is a Homeostatic Lame Axis Allowing that the Fight-or-Flight Stressing of

Metabolic Controls Could Turn on Related Diseases. *Int. J. Med. Bio. Front.* **2014,** *20,* 59–101.

Bierman, H. R.; Kelly, K. H.; Cordes, F. L.; Byron, R. L. Jr.; Polhemus, J. A.; Rappoport, S. The Release of Leucocytes and Platelets from the Pulmonary Circulation by Adrenaline. *Blood* **1952,** *7,* 683–692.

Bremner, J. B.; Griffith, R.; Coban, B. Ligand Design for $A_1$ Adrenoceptors. *Curr. Med. Chem.* **2001,** *8,* 607– 620.

Breuer, C.; Wachall, B.; Gerbeth, K.; Abdel-Tawab, M.; Fuhr, U. Pharmacokinetics and Pharmacodynamics of Moist Inhalation Epinephrine Using a Mobile Inhaler. *Eur. J. Clin. Pharmacol.* **2013,** *69,* 1303–1310.

Brittain, H. G., Ed. *Analytical Profile of Drug Substances and Excipients*; Academic Press: New Jersey, 2008; Vol. 25, pp 529–530.

Brittain, H. G., Ed. *Analytical Profile of Drug Substances and Excipients*; Academic Press: New Jersey, 2008; Vol. 21, pp 133–136.

Brittain, H. G., Ed. *Analytical Profile of Drug Substances and Excipients*; Academic Press: New Jersey, 2008; Vol. 21, pp 160–162.

Brodde, O. E.; Michel, M. C. Adrenergic and Muscarinic Receptors in the Human Heart. *Pharmacol. Rev.* **1999,** *51,* 651–690.

Brown, R. S.; Carey, J. S; Mohr, P. A.; Shoemaker, W. C. Comparative Evaluation of Sympathomimetic Amines in Clinical Shock. *Circulation* **1966,** *34* (2), 260–271.

Brunton, L., Ed. *Goodman and Gilman's the Pharmacological Basis of Therapeutics*, 12th Ed.; Mc Graw Hill: New York, 2011; pp 277–304.

Bylund, D. B. Subtypes of A1 and A2 Adrenergic Receptors. *FASEB J.* **1992,** *6* (3), 832–839.

Carroll, G. L. Analgesic and Pain. *Vet. Clin. N. Am-small* **1999,** *29* (3), 701–717.

Cazzola, M.; Calzetta, L.; Matera, M. G. $B_2$-Adrenoceptor Agonists: Current and Future Direction. *Brit. J. Pharmacol.* **2011,** *163* (1), 4–17.

Cellek, S.; Thangiah, R.; Bassil, A. K.; Campbel, C. A.; Gray, K. M.; Stretton, J. L; Lalude, O.; Vivekanandan, S.; Wheeldon, A.; Winchester, W. J.; Sanger, G. J. Demonstration of Functional Neuronal B3-Adrenoceptors Within the Enteric Nervous System. *Gastroenterology* **2007,** *133* (1), 175–183.

Christ, G. J.; Maayani, S.; Valcic, M.; Melman, A. Pharmacological Studies of Human Erectile Tissue: Characteristics of Spontaneous Contractions and Alterations in A-Adrenoceptor Responsiveness with Age and Disease in Isolated Tissues. *Br. J. Pharmacol.* **1990,** *101* (2), 375–381.

Contopoulos-Ioannidis, D. G.; Kouri, I.; Ioannidis, J. P. Pharmacogenetics of the Response to B2 Agonist Drugs: A Systematic Overview of the Field. *Pharmacogenomics* **2007,** *8* (8), 933–58.

Crassous, P. A.; Denis, C.; Paris, H.; Senard, J. M. Interest of $A_2$-Adrenergic Agonists and Antagonists in Clinical Practice: Background, Facts and Perspectives. *Curr. Top. Med. Chem.* **2007,** *7* (2), 187–194.

Crawford, T.; Chugh, A.; Good, E.; Yoshida, K.; Jongnarangsin, K.; Ebinger, M.; Pelosi, Jr. F.; Bogun, F.; Morady, F.; Oral, H. Clinical Value of Noninducibility by High-Dose Isoproterenol Versus Rapid Atrial Pacing After Catheter Ablation of Paroxysmal Atrial Fibrillation. *J. Cardiovasc. Electr.* **2010,** *21* (1), 13–20.

Crooks, P. A, Breakefield, X. O, Sulens, C. H.; Castiglione, C. M.; Coward, J. K. Extensive Conjugation of Dopamine (3, 4-Dihydroxyphenethylamine) Metabolites in Cultured Human Skin Fibroblasts and Rat Hepatoma Cells. *Biochem. J.* **1978,** *176* (1), 187.

Eisenach, J. C. Overview: First International Symposium on $A_2$-Adrenergic Mechanisms of Spinal Anesthesia. *Region. Anesth.* **1993,** *18,* I–VI.

Emorine, L. J; Marullo, S.; Briend-Sutren, M. M; Patey, G.; Tate, K.; Delavier-Klutchko, C.; Strosberg, A. D. Molecular Characterization of the Human $B_3$-Adrenergic Receptor. *Science* **1989,** *245* (4922), 1118–1121.

Erickson, J. D.; Schafer, M. K; Bonner, T. I.; Eiden, L. E.; Weihe, E. Distinct Pharmacological Properties and Distribution in Neurons and Endocrine Cells of Two Isoforms of the Human Vesicular Monoamine Transporter. *Proc. Natl. Acad. Sci.* **1996,** *93* (10), 5166–5171.

Florey, K., Ed. *Analytical Profile of Drug Substances and Excipients*; Academic Press: New Jersey, 2008; Vol 7, p 193.

Florey, K., Ed. *Analytical Profile of Drug Substances and Excipients*; Academic Press: New Jersey, 2008; Vol 12, pp 392–407.

Florey, K., Ed. *Analytical Profile of Drug Substances and Excipients*; Academic Press: New Jersey, 2008; Vol 3, pp 493–494.

Florey, K., Ed. *Analytical Profile of Drug Substances and Excipients*; Academic Press: New Jersey, 2008; Vol 15, pp 331– 332.

Florey, K., Ed. *Analytical Profile of Drug Substances and Excipients*; Academic Press: New Jersey, 2008; Vol 14, pp 150.

Florey, K., Ed. *Analytical Profile of Drug Substances and Excipients*; Academic Press: New Jersey, 2008; Vol 8, pp 151–153.

Florey, K., Ed. *Analytical Profile of Drug Substances and Excipients*; Academic Press: New Jersey, 2008; Vol 19, pp 616–619.

Florey, K., Ed. *Analytical Profile of Drug Substances and Excipients*; Academic Press: New Jersey, 2008; Vol 13, p 737.

Florey, K., Ed. *Analytical Profile of Drug Substances and Excipients*; Academic Press: New Jersey, 2008; Vol 8, p 489.

Florey, K., Ed. *Analytical Profile of Drug Substances and Excipients*; Academic Press: New Jersey, 2008; Vol 15, p 151.

Florey, K., Ed. *Analytical Profile of Drug Substances and Excipients*; Academic Press: New Jersey, 2008; Vol 4, p 64.

Gertler, R.; Brown, H. C; Mitchell, D. H; Silvius, E. N. Dexmedetomidine: A Novel Sedative-Analgesic Agent. *Taylor Francis* **2001,** *14* (1), 13–21.

Gibson, T. B; Lawrence, M. C; Gibson, C. J; Vanderbilt, C. A; Mcglynn, K.; Arnette, D.; Chen, W.; Collins, J.; Naziruddin, B.; Levy, M. F.; Ehrlich, B. E. Inhibition of Glucose-Stimulated Activation of Extracellular Signal-Regulated Protein Kinases 1 and 2 by Epinephrine in Pancreatic B-Cells. Diabetes **2006,** *55* (4), 1066–1073.

Giros, B. R.; Mestikawy, S. A; Godinot, N. A; Zheng, K. E; Han, H. O.; Yang-Feng, T. E.; Caron, M. G. Cloning, Pharmacological Characterization, And Chromosome Assignment of the Human Dopamine Transporter. *Mol. Pharmacol.* **1992,** *42* (3), 383–390.

Giros, B.; Caron, M. G. Molecular Characterization of the Dopamine Transporter. *Trends Pharmacol. Sci.* **1993,** *14*, 43–49.

Goepel, M.; Michel, M. C.; Rubben, H.; Wittmann, A. Comparison of Adrenoceptor Subtype Expression in Porcine and Human Bladder and Prostate. *URO Res.* **1997,** *25* (3), 199–206.

Golan, D. E., Ed. *Principles of Pharmacology the Pathophysiologic Basis of Drug Therapy*, 3rd ed.; Lippincott Williams & Wilkins: Philadelphia, 2012; pp 132–145.

Goldstein, D. S. Catecholamines in the Periphery. *Adv. Pharmacol.* **1997,** *42*, 529–539.

Granneman, J. G. The Putative B4-Adrenergic Receptor is a Novel State of the B1-Adrenergic Receptor. *Am. J. Physiol. Endocrinol. Metab.* **2001,** *280* (2), E199–E202.

Gross, M. E.; Booth, N. H. Tranquilizers, Alpha-2 Adrenergic Agonist and Related Agents. *J. Vet. Pharmacol. Ther.* **2001,** *41* (56), 43–66.

Guimaraes, S.; Moura, D. Vascular Adrenoceptors: An Update. *Pharmacol Rev.* **2001,** *53* (2), 319– 356.

Hall, R. A.; Premont, R. T.; Chow, C. W.; Blitzer, J. T.; Pitcher, J. A.; Claing, A.; Stoffel, R. H.; Barak, L. S.; Shenolikar, S.; Weinman, E. J.; Grinstein, S. The B 2-Adrenergic Receptor Interacts with the $Na^+/H^+$-Exchanger Regulatory Factor to Control $Na^+/H^+$ Exchange. *Nature* **1998,** *392* (6676), 626.

Haller, J.; Makara, G. B.; Kruk, M. R. Catecholaminergic Involvement in the Control of Aggression: Hormones, the Peripheral Sympathetic, and Central Noradrenergic Systems. *Neurosci. Biobehav. R.* **1997,** *22*, 85–97.

Han, C.; Li, J.; Minneman, K. P. Subtypes of A1-Adrenoceptors in Rat Blood Vessels. *Eur. J. Pharmacol.* **1990,** *190* (1–2), 97–104.

Hedlund, H.; Andersson, K. E. Comparison of the Responses to Drugs Acting on Adrenoceptors and Muscarinic Receptors in Human Isolated Corpus Cavernosum and Cavernous Artery. *J. Auton. Pharmacol.* **1985,** *5* (1), 81–88.

Heubach, J. F.; Rau, T.; Eschenhagen, T.; Ravens, U.; Kaumann, A. J. Physiological Antagonism Between Ventricular $B_1$-Adrenoceptors and $A_1$-Adrenoceptors but no Evidence for $B_2$-and $B_3$-Adrenoceptor Function in Murine Heart. *Brit. J. Pharmacol.* **2002,** *136* (2), 217– 229.

Hieble, J. P.; Bylund, D. B; David, E. C.; Douglas, C. E.; Langer, S. Z.; Lefkowitz, R. J.; Minneman, K. P.; Ruffolo, R. R. International Union of Pharmacology X. Recommendation for Nomenclature of A1 Adrenoceptors: Consensus Update. *Pharmacol. Rev.* **1995,** *47* (2), 267–270.

Holloway, B. R.; Howe, R.; Rao, B. S.; Stribling, D. ICI D7114: A Novel Selective Adrenoceptor Agonist of Brown Fat and Thermogenesis. *Am. J. Clin. Nutr.* **1992,** *55* (1), 262S–264S.

Igawa, Y.; Michel, M. C. Pharmacological Profile of B3-Adrenoceptor Agonists in Clinical Development of the Treatment of Overactive Bladder Syndrome. *N-S Arch. Pharmacol.* **2013,** *386*, 177–183.

Insel, P. A. Adrenergic Receptors—Evolving Concepts and Clinical Implications. *N. Engl. J. Med.* **1996,** *334* (9), 580–585.

Kadar, D.; Tang, H. Y.; Conn, A. W. Isoproterenol Metabolism in Children After Intravenous Administration. *Clin. Pharmacol. Therap.* **1974,** *16* (5), 789–795.

Kallgren. M. A. A2-Adrenergic Agonists. *J. Arthroplasty* **1995,** *9* (1), 93–103.

Katz, B. *The Release of Neural Transmitter Substances*; Liverpool University Press, 1969; pp 5–39.

Katzung, B. G.; Trevor, A. T. *Basic and Clinical Pharmacology*, 13th Edition; Mcgraw-Hill Education: New York, 2014.

Lambertz, H. E.; Meyer, J. O.; Erbel, R. Long-term Hemodynamic Effects of Prenalterol in Patients with Severe Congestive Heart Failure. *Circulation* **1984,** *69* (2), 298–305.

Largis, E. E.; Burns, M. G.; Muenkel, H. A.; Dolan, J. A.; Claus, T. H. Antidiabetic and Antiobesity Effects of a Highly Selective B3-Adrenoceptor Agonist (CL 316,243). *Drug. Dev. Res.* **1994,** *32* (2), 69–76.

Lohse, M. J.; Engelhardt, S.; Eschenhagen, T. What is the Role of B-Adrenergic Signaling in Heart Failure? *Circ. Res.* **2003,** *93* (10), 896–906.

Lucia, S. P.; Leonard, M. E.; Falconer, E. H. The Effect of the Subcutaneous Injection of Adrenalin on the Leucocyte Count of Splenectomized Patients and of Patients with Certain Diseases of the Hematopoietic and Lymphatic Systems. *Am. J. Med. Sci.* **1937,** *194* (1), 35–43.

Mahata, S. K.; Mahata, M.; Fischer-Colbrie, R.; Winkler, H. Vesicle Monoamine Transporters 1 and 2: Differential Distribution and Regulation of their mRNAs in Chromaffin and Ganglion Cells of Rat Adrenal Medulla. *Neurosci. Lett.* **1993,** *156* (1–2), 70–72.

Maze, M.; Tranquilli, W. Alpha-2 Adrenoceptor Agonists: Defining The Role in Clinical Anesthesia. *Anesthesiology* **1991,** *74,* 581–605.

Michelotti, G. A.; Price, D. T.; Schwinn, D. A. A1-Adrenergic Receptor Regulation: Basic Science and Clinical Implications. *Pharmacol. Ther.* **2000,** *88* (3), 281–309.

Miller, J. W. Adrenergic Receptors in the Myometrium. *Ann. N.Y. Acad. Sci.* **1967,** *139* (1), 788–798.

Molenaar, P.; Bartel, S.; Cochrane, A.; Vetter, D.; Jalali, H.; Pohlner, P.; Burrell, K.; Karczewski, P.; Krause, E. G.; Kaumann, A. Both B2-And B1-Adrenergic Receptors Mediate Hastened Relaxation and Phosphorylation of Phospholamban and Troponin I in Ventricular Myocardium of Fallot Infants, Consistent with Selective Coupling of B2-Adrenergic Receptors to Gs-Protein. *Circulation* **2000,** *102* (15), 1814–1821.

Mosby. *Mosby's Drug Consult*, 15th ed.; Mosby's Inc: St. Louis Missouri, 2005.

Munchau, A.; Bhatia, K. P. Pharmacological Treatment of Parkinson's Disease. *Postgrad. Med. J.* **2000,** *76* (900), 602–610.

Muramatsu, I. Functional Subclassification of Vascular $A_1$-Adrenoceptors. *Pharmacol. Commun.* **1995,** *6,* 23–28.

Nadel, J. A; Barnes, P. J; Holtzman, M. J. *Comprehensive Physiology*; American Physiological Society: USA, 1986; Vol 3, pp 693–702.

Nantel, F.; Bonin, H.; Emorine, L. J.; Zilberfarb, V.; Strosberg, A. D.; Bouvier, M. The Human B3-Adrenergic Receptor is Resistant to Short Term Agonist-Promoted Desensitization. *Mol. Pharmacol.* **1993,** *43,* 548–555.

Naveh, N.; Kaplan-Messas, A.; Marshall, J. Mechanism Related to Reduction of Intraocular Pressure by Melanocortins in Rabbits. *Brit. J. Ophthalmol.* **2000,** *84* (12), 1411–1414.

Njus, D.; Kelley, P. M.; Harnadek, G. J. Bioenergetics of Secretory Vesicles. *Biochim. Biophys. Acta.* **1986,** *853* (3–4), 237–265.

O' Neill, D. K.; Aizer, A.; Linton, P.; Bloom, M.; Rose, E.; Chinitz, L. Isoproterenol Infusion Increases Level of Consciousness During Catheter Ablation of Atrial Fibrillation. *J. Interv. Card. Electr.* **2012,** *34* (2), 137–142.

O'Connell, T. D.; Jensen, B. C.; Baker, A. J.; Simpson, P. C. Cardiac Alpha1-Adrenergic Receptors: Novel Aspects of Expression, Signaling Mechanisms, Physiologic Function, and Clinical Importance. *Pharmacol. Rev.* **2014,** *66* (1), 308–333.

Ortega, V. E. Predictive Genetic Profiles for B-Agonist Therapy in Asthma. A Future Under Construction. *Am. J. Resp. Crit. Care.* **2015,** *191* (5), 494–496.

Paddleford, R. R.; Harvey, R. C. Alpha 2 Agonists And Antagonists. *Vet. Clin. N. Am-small* **1999,** *29* (3), 737–745.

Perez, D. M. Structure-Function of A1-Adrenergic Receptors. *Biochem. Pharmacol.* **2007,** *73* (8), 1051–1062.

Peter, D.; Jimenez, J.; Liu, Y. O.; Kim, J. U.; Edwards, R. H. The Chromaffin Granule and Synaptic Vesicle Amine Transporters Differ in Substrate Recognition and Sensitivity to Inhibitors. *J. Biol. Chem.* **1994,** *269* (10), 7231–7237.

Peter, D.; Liu, Y. O.; Sternini, C.; De, G. R.; Brecha, N.; Edwards, R. H. Differential Expression of Two Vesicular Monoamine Transporters. *J. Neurosci.* **1995,** *15* (9), 6179–6188.

Purves, D.; Augustine, J. G.; Fitzpatrick, D.; Hall, W. C.; Lamantia, A. S.; Mcnamara, J. O.; Williams, S. M. *Neuroscience*, 3rd ed.; Sinauer Associates, Inc. Publishers: Sutherland, 2004; pp 147–150.

Quinson, N.; Robbins, H. L.; Clark, M. J.; Furness, J. B. Locations and Innervation of Cell Bodies of Sympathetic Neurons Projecting to the Gastrointestinal Tract in the Rat. *Arch. Histol. Cytol.* **2001,** *64* (3), 281–294.

Rang, H. P.; Dale, M. M.; Ritter, J. M.; Flower, R. J.; Henderson, G. *Rang & Dale's Pharmacology*, 7th ed.; Churchill Livingstone: Edinbergh, 2011.

Rataboli, P. V. *Clinical Pharmacology and Rational Therapeutics*, 2nd ed.; Ane Books Pvt. Ltd: India, 2010; pp 51–54.

Rektor, I.; Kanovsky, P.; Bares, M.; Brazdil, M.; Streitova, H.; Klajblova, H.; Kuba, R.; Daniel, P. A SEEG Study of ERP in Motor and Premotor Cortices and in the Basal Ganglia. *Clin. Neurophysiol.* **2003**, *114* (3), 463–471.

Rhoden, K. J.; Meldrum, L. A.; Barnes, P. J. Inhibition of Cholinergic Neurotransmission in Human Airways by B2-Adrenoceptors. *J. Appl. Physiol.* **1988**, *65* (2), 700–705.

Richter, C. P. Biological Clocks in Medicine And Psychiatry: Shock-Phase Hypothesis. *Proc. Natl. Acad. Sci.* **1960**, *46* (11), 1506–1530.

Robert, J. L.; Lee, E. L.; Chhabirani, M.; Marc, G. C. The B-Adrenergic Receptor and Adenylate Cyclase. *BBA-Rev. Biomembr.* **1976**, *457* (1), 1–39.

Roland, S. A Door Opener for Future Research: Agonist-Induced B$_3$-Adrenoceptor Desensitization in HEK Cells but not CHO Cells. *N-S Arch. Pharmacol.* **2013**, *386*, 841–842.

Scheinin, M.; Macdonald, E. An Introduction to the Pharmacology of A$_2$- Adrenoceptors in the Central Nervous System. *Acta Vet. Scand.* **1989**, *85*, 11–19.

Schemann, M.; Hafsi, N.; Michel, K.; Kober, O. I; Wollmann, J.; Li, Q.; Zeller, F.; Langer, R.; Lee, K.; Cellek, S. The B3-Adrenoceptor Agonist GW427353 (Solabegron) Decreases Excitability of Human Enteric Neurons via Release of Somatostatin. *Gastroenterology* **2010**, *138* (1), 266– 274.

Scott, J. D.; Dessauer, C. W.; Tasken, K. Creating Order from Chaos: Cellular Regulation by Kinase Anchoring. *Ann. Rev. Pharmacol. Toxicol.* **2013**, *53*, 187–210.

Sears, M. R.; Lotvall, J. Past, Present and Future-B$_2$-Adrenoceptor Agonists in Asthma Management. *Resp. Med.* **2005**, *99* (2), 152–170.

Sharara-Chami, R. I; Joachim, M.; Pacak, K.; Majzoub, J. A. Glucocorticoid Treatment—Effect on Adrenal Medullary Catecholamine Production. *Shock* **2010**, *33* (2), 213.

Sibley, D. R.; Monsma, Jr. F. J.; Shen, Y. Molecular Neurobiology of Dopaminergic Receptors. *Int. Rev. Neurobiol.* **1993**, *35*, 391–415.

Simons, K. J.; Simons, F. E. Epinephrine and its use in Anaphylaxis: Current Issues. *Curr. Opin. Allergy. Clin. Immunol.* **2010**, *10* (4), 354–361.

Stedman. *Stedman's Medical Dictionary*, 3rd ed.; Lippincott Williams & Wilkins: Philadelphia, 1918; pp 21.

Stedman. *Stedman's Medical Dictionary*, 3rd ed.; Lippincott Williams & Wilkins: Philadelphia, 1918; pp 327.

Stedman. *Stedman's Medical Dictionary*, 3rd ed.; Lippincott Williams & Wilkins: Philadelphia, 1918; pp 752.

Stevens, C. W.; Brenner, G. M. *Pharmacology*, 2nd ed.; Saunders: Philadelphia, 2008; pp 71–80.

Sudhof, T. C. The Synaptic Vesicle Cycle. *Ann. Rev. Neurosci.* **2004**, *27*, 509–547.

Tamm, M.; Richards, D. H.; Beghe, B.; Fabbri, L. Inhaled Corticosteroid and Long-Acting B$_2$- Agonist Pharmacological Profiles: Effective Asthma Therapy in Practice. *Resp. Med.* **2012**, *106*, S9–S19.

Tanda, G.; Pontieri, F. E.; Frau, R.; Chiara, G. Contribution of Blockade of the Noradrenaline Carrier to the Increase of Extracellular Dopamine in the Rat Prefrontal Cortex by Amphetamine and Cocaine. *Eur. J. Neurosci.* **1997**, *9* (10), 2077–2085.

Tanoue, A.; Koshimizu, T. A.; Tsujimoto, G. Transgenic Studies of A1-Adrenergic Receptor Subtype Function. *Life. Sci.* **2002**, *71* (19), 2207–2215.

Tashkin, D. P.; Fabbri, L. M. Long-Acting Beta-Agonists in the Management of Chronic Obstructive Pulmonary Disease: Current and Future Agents. *Resp. Res.* **2010**, *11* (1), 149.

Tejada, S. D.; Goldstein, I.; Azadzoi, K.; Krane, R. J.; Cohen, R. A. Impaired Neurogenic and Endothelium Mediated Relaxation of Penile Smooth Muscle from Diabetic Men with Impotence. *N. Engl. J. Med.* **1989**, *320* (16), 1025–1030.

Tejada, S. D.; Kim, N.; Lagan, I.; Krane, R. J.; Goldstein, I. Regulation of Adrenergic Activity in Penile Corpus Cavernosum. *J. Urol.* **1989**, *142* (4), 1117–1121.

Thompson, M. D.; Burnham, W. M.; Cole, D. E. The G Protein-Coupled Receptors: Pharmacogenetics and Disease. *Cr. Rev. Clin. Lab Sci.* **2005**, *42* (4), 311–389.

Tortora, G. J.; Derrickson, B. H. *Principles of Anatomy and Physiology*, 13th Ed.; John Wiley & Sons: USA, 2008; pp 596.

Tranquilli, W. J.; Maze, M. Clinical Pharmacology and Uses of A$_2$-Adrenergic Agonist in Veterinary Anesthesia. *Anesth. Pharmacol. Rev.* **1993**, *1*, 293–309.

Tripathi, K. D. *Essentials of Medical Pharmacology*, 7th ed.; Jaypee Brothers Medical Publishers (P) Ltd: New Delhi, 2014; pp 124–139.

Udaykumar, P. *Pharmacology Companion*, 1st Ed.; CBS Publishers and Distributors Pvt. Ltd: India, 2016; pp 16–33.

Van, T. L. J.; Michel, M. C.; Grosse-Wilde, H.; Happel, M.; Eigler, F. W.; Soliman, A.; Brodde, O. E. Catecholamines Increase Lymphocyte B2-Adrenergic Receptors via a B2-Adrenergic, Spleen-Dependent Process. *Am. J. Physiol.* **1990**, *258*, 191–202.

Viskupic, E.; Kvetnansky, R.; Sabban, E. L.; Fukuhara, K.; Weise, V. K.; Kopin, I. J.; Schwartz, J. P. Increase in Rat Adrenal Phenylethanolamine N-Methyltransferase mRNA Level Caused by Immobilization Stress Depends on Intact Pituitary-Adrenocortical Axis. *J. Neurochem.* **1994**, *63* (3), 808–814.

Walden, P. D.; Durkin, M. M.; Lepor, H.; Wetzel, J. M.; Gluchowski, C.; Gustafson, E. L. Localization of mRNA and Receptor Binding Sites for the Alpha Sub-1a-Adrenoceptor Subtype in the Rat, Monkey and Human Urinary Bladder and Prostate. *J. Urol.* **1997**, *157* (3), 1032–1038.

Wanner, A.; Mendes, E. S. Airway Endothelial Dysfunction in Asthma and Chronic Obstructive Pulmonary Disease: A Challenge for Future Research. *Am. J. Resp. Crit. Care* **2010**, *182* (11), 1344–1351.

Weihe, E.; Schafer, M. K.; Erickson, J. D.; Eiden, L. E. Localization of Vesicular Monoamine Transporter Isoforms (VMAT1 and VMAT2) to Endocrine Cells and Neurons in Rat. *J. Mol. Neurosci.* **1994**, *5* (3), 149–164.

Westfall, T. C. *Reference Module in Neuroscience and Biobehavioral Psychology: Encyclopedia of Neuroscience*; Academic Press: USA, 2009; pp 685–695.

Wevers, R. A.; De, R. A. J. F.; Brautigam, C.; Geurtz, B.; Van Den, H. L. P.; Steenbergen-Spanjers, G. C.; Smeitink, J. A.; Hoffmann, G. F.; Gabreels, F. J. A Review of Biochemical and Molecular Genetic Aspects of Tyrosine Hydroxylase Deficiency Including a Novel Mutation (291delc). *J. Inherit. Metab. Dis.* **1999**, *22* (4), 364–373.

Weyer, C.; Gautier, J. F.; Danforth, E. Development of $B_3$- Adrenoceptor Agonists for the Treatment of Obesity and Diabetes: An Update. *Diabetes Metab.* **1999**, *25* (1), 11–21.

Wilbraham, J. M.; Wooton, C. L.; Martin, D. A.; Holloway, B. R. B3- Adrenergic Agonist, Zeneca ZD2079, Improves Glucose Homeostasis in Insulin-Resistant Rodents. *Diabetes* **1987**, *46* (Suppl 1), A212.

Yen, T. T. The Antiobesity and Metabolic Activities of LY79771 in Obese and Normal Mice. *Int. J. Obesity* **1984**, *8* (1), 69–78.

Yen, T. T.; Fuller, R. W.; Hemrick, L. S. K.; Dininger, N. B. The Antiobesity and Metabolic Activities of LY104119 in Obese and Normal Mice. *Int. J. Obes.* **1984**, *8* (1), 69–87.

Yoshida, T.; Sakane, N.; Wakabajashi, Y.; Umekawa, T.; Kondo, M. Antiobesity and Anti-diabetic Effects of CL 316243, a Highly Specific Beta 3 Adrenoceptor Agonist, in Yellow KK Mice. *Life. Sci.* **1994**, *54* (7), 491–498.

Yoshida, T.; Sakane, N.; Wakabajashi, Y.; Umekawa, T.; Kondo, M. Antiobesity Effect of CL 316243, A Highly Specific Beta 3 Adrenoceptor Agonist, with Monosodium-L-Glutamate-Induced Obesity. *Eur. J. Endocrinol.* **1994**, *131* (1), 97–102.

# CHAPTER 4

# Adrenergic Antagonists

SEETHA HARILAL* and RAJESH K.

*Kerala University of Health Sciences, Kerala, India*

*Corresponding author. E-mail: seethaharilal1989@gmail.com*

## ABSTRACT

Adrenergic antagonists are agents capable of antagonizing the adrenergic neurotransmitters like noradrenaline and adrenaline competitively or noncompetitively bound to α, β adrenoceptors. Hence these have a wide range of application clinically. In spite of having a wide range of pharmacological action in areas of CNS, locomotion, ophthalmology, and metabolism, these are of importance in the case of diseases associated with CVS as adrenergic neurotransmitters play a role in vasoconstriction, increased heart rate, and contractility effects of the heart due to interaction with α and β adrenoceptors. Additionally, activity on smooth muscle renders it effective against BPH. Other off-label uses include regulation of anxiety, glaucoma, in improving HDL, an adjuvant in hypothyroidism, and migraine prophylaxis. This chapter highlights this aspect by the discussion of neurotransmission blockade of the sympathetic system by some synthetic agents with antagonistic action. The chapter covers the area of drugs blocking sympathetic system via adrenoceptors α, β either individualized or by combination. In addition to this, few drugs with specific binding to the receptor subtypes are also mentioned.

## 4.1 INTRODUCTION

Noradrenaline (NA) and adrenaline are endogenous agents that help in functioning of SNS which prepares our body respond to stressful conditions by producing "fight or flight response" to overcome challenges ahead (Haller et al., 1997). Keeping in mind the end goal to keep up homeostasis of the body to distressing conditions, modulation of certain vital functioning happens inside the body in various regions, for example, adipose tissue, heart, non-striated smooth muscles, striated skeletal muscles, salivary glands, and the liver. This causes an aberration in the typical physiology which can be managed by using sympathomimetic or sympatholytic agents (Golan et al., 2012). The term "sympatholytics" has its root from the words "sympathetic" indicating sympathetic system, "lysis" in Greek meaning destruction. Hence, sympatholytics are agents causing destruction or blockade of impulses of sympathetic system from postganglionic neurons to the effector organs (Miller-Keane et al., 2003). Synonyms of sympatholytics include adrenergic antagonists, adrenergic blockers, antiadrenergic agents, and sympathoplegics (Farlex, 2017). These are agents binding primarily to adrenoceptors found in postsynaptic neurons resulting in a hindrance to the downstream signalling pathways of adrenoceptors (Brunton et al., 2011; Golan et al., 2012).

Inhibition of adrenoceptors signalling pathways can occur by diverse mechanisms like non-competitive and competitive adrenergic blockers bound to the adrenoceptors in postsynaptic level; by drugs modulating the regulation of catecholamines like catecholamine synthesis blockade by methyltyrosine; vesicular blockade by reserpine which prevents release of noradrenaline; drugs for instances bretylium, guanethidine and 6-hydroxydopamine that act specifically on the sympathetic nerve terminal causing an amassing of neurotransmitter within sympathetic neurones or $\alpha_2$ adrenoceptor agonists acting on $\alpha_2$ adrenoceptor presynaptically and inhibiting postsynaptic activation of adrenoceptors and downstream signalling pathway (Brock et al., 1988). Hence, these medications are clinically important in the therapy of anxiety disorders like GAD, posttraumatic

stress disorder (PTSD) and panic disorders (Tyrer, 1992; Kaplan, 1998); in cardiovascular diseases like hypertension; respiratory disorders like asthma, nasal decongestion; ophthalmology for glaucoma; BPH and so on (Brunton et al., 2011). The chapter aims in giving an insight to the readers about various sympatholytic agents affecting the sympathetic system, its action on neurotransmission and explain about specific receptor subtypes bound drugs showing desired pharmacological action thereby rendering it useful clinically.

## 4.2 CLASSIFICATION OF ADRENERGIC ANTAGONISTS

Adrenergic antagonists can be categorized in view of their selectivity to adrenoceptors; site and the duration of activity. Figure 4.1 categorizes and describes these drugs on the basis of its site of activity and receptor selectivity.

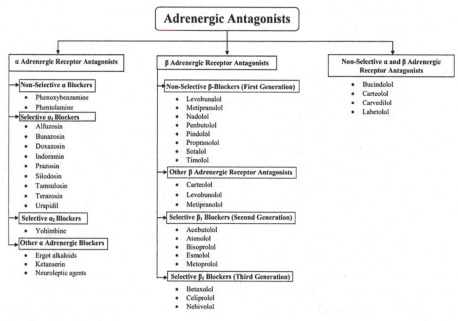

**FIGURE 4.1**   Classification of adrenergic antagonists.

## 4.3 ADRENERGIC ANTAGONISTS

Adrenergic antagonists are agents that block the activity of endogenous catecholamines and sympathomimetic agents acting on the adrenoceptors. Many are antagonists with a competitive nature and have selective affinities for various adrenoceptors like α adrenoceptors and β adrenoceptors (Brunton et al., 2011; Katzung, 2009).

### 4.3.1 α ADRENOCEPTOR ANTAGONIST

Numerous activities of the catecholamines of the body are through α adrenergic receptors. $\alpha_1$ receptor mediates the constriction of venous, arterial, and visceral smooth muscles. The $\alpha_2$ receptors are associated with increasing the vagal tone, suppressing sympathetic output, regulating metabolic effects thereby decreasing secretion of insulin and inhibit lipolysis (Brunton et al., 2011). They are likewise associated with acetylcholine and norepinephrine release from the nerve terminals, in promoting aggregation of platelets, and in contraction of many veins and arteries. The α receptor adrenergic antagonists produces an extended pharmacological activity based on the selective affinity toward various α adrenoreceptors like $\alpha_1$ and $\alpha_2$ (Golan et al., 2012). It produces effects primarily on the CVS and furthermore has activities on the periphery and CNS. Some recent drugs are even selective to various subtype of a particular receptor like $\alpha_{1A}$ and $\alpha_{1B}$ (Katzung, 2009).

### 4.3.1.1 PHARMACODYNAMICS AND PHARMACOKINETICS

$\alpha_1$ receptor antagonists obstruct the $\alpha_1$ adrenoceptors thereby hindering endogenous catecholamines mediated vasoconstriction. Vasodilation happens in both the arterial and venous supply. The blood pressure diminishes as an outcome of the abatement in PR (Golan et al., 2012). Positive inotropy can be observed. $\alpha_1$ receptor obstruction can relieve BPH symptoms for instance, resistance to urinal output by relaxing the smooth muscle. The $\alpha_2$ adrenoceptors play an imperative function both in the central and peripheral SNS activity regulation. Stimulation of presynaptic $\alpha_2$ adrenoceptor inhibits norepinephrine and other co-transmitters release from spn endings (Andersson, 1996; Brunton et al., 2011). In CNS, $\alpha_2$ adrenoceptor stimulation in pontomedullary region produces a hindrance in SNS action and lowering of BP. Many drugs like clonidine demonstrate its activity on the above mentioned receptors. Antagonists for example yohimbine thus increase the sympathetic activity via increased discharge of norepinephrine from the terminal part of the nerve reducing the $\beta_1$ and $\alpha_1$ adrenoceptor activation (Brunton et al., 2011).

### 4.3.1.2 NON-SELECTIVE α BLOCKERS

These are agents that act by blocking both $\alpha_1$ and $\alpha_2$ adrenoceptors competitively or non-competitively. The non-selective α adrenergic blockers are phenoxybenzamine and phentolamine (Brunton et al., 2011; Katzung, 2009; Tripathi, 2014). Table 4.1 depicts the mechanism, pharmacological effects, and clinical responses of non-selective α blockers.

#### 4.3.1.2.1 Phenoxybenzamine

It is an α-adrenoceptor antagonist with a nonselective nature. Being a haloalkyl amine compound, it produces irreversible antagonism.

Phenoxybenzamine undergoes a spontaneous chemical transformation to yield an intermediate active electrophilic carbonium ion (Brunton et al., 2011; Katzung, 2009). This electrophilic carbonium ion makes a stable covalent bond with the α receptors causing a non-competitive antagonism thus blocks GPCR mechanism by blocking $G_{\alpha q}$ protein. Blockade of $G_{\alpha q}$ happens to reduce $IP_3$ level which inhibits accessibility of free calcium required for the activation of protein kinase necessary for vasoconstriction (Fig. 4.2). Subsequently cardiac output tends to elevate producing an abatement of peripheral resistance. Moreover, relaxation of smooth

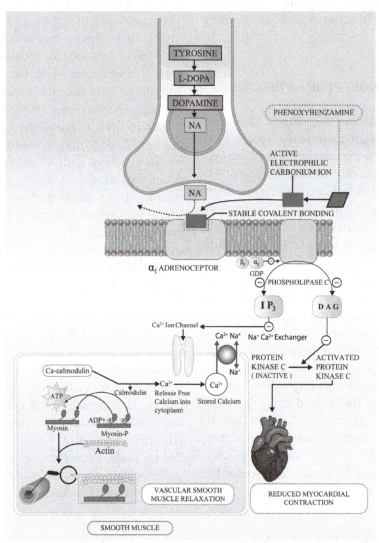

**FIGURE 4.2**   Mechanism of action of phenoxybenzamine on smooth muscles. DOPA, dihydroxyphenylalanine; NA, noradrenaline; GDP, guanosine 5′-diphosphate; $IP_3$, inositol triphosphate; DAG, diacylglycerol; $Ca^{2+}$, calcium ions; ATP, adenosine triphosphate; ADP, adenosine diphosphate; $Na^+$–$Ca^{2+}$ Exchanger, sodium–calcium exchanger or sodium–potassium pump.

*Source*: Adapted from Tripathi (2014), Golan et al. (2012), and Stevens and Brenner (2008).

**TABLE 4.1**  Pharmacodynamic and Pharmacokinetic Profile of Non-selective α Blockers.

| Non-selective α blocker | Pharmacodynamics | | | Pharmacokinetics |
|---|---|---|---|---|
| | Pharmacological effects | Mechanism of action | Adverse effects | Clinical responses | Pharmacokinetics |
| Phenoxybenzamine | Increases cardiac output; decrease PR; relaxes smooth muscles in the region of/near neck, prostate, and urinary bladder | Irreversible covalent binding to α receptors and action via $G_{\alpha q}$ receptor mechanism | Postural hypotension; reflex tachycardia; cardiac arrhythmias | BPH; causalgia; episodes of HTN in pheochromocytoma; secondary shock | Route of administration: PO; incomplete absorption; Duration of action: 3 to 4 days; Excretion: Urine |
| Phentolamine | Elevates CO; decrease PR and BP; vasodilation | Competitive pharmacological antagonism at α adrenoceptors | Dizziness; headache; nasal congestion; tachycardia | Episodes of HTN phaeochromocytoma; erectile dysfunction | Route of administration: IV, IM, SC; Duration of action: 10–15 min (IV), 3–4 h (IM); Metabolism: Liver; Excretion: Urine |

CO, cardiac output; PR, peripheral resistance; IM, intramuscular; BPH, benign prostatic hyperplasia; BP, blood pressure; SC, subcutaneous; HTN, hypertension; PO, per oral; IV: intravenous.

muscles in the region of neck, prostate, and urinary bladder. Hence used in treating BPH (Caine et al., 1981; Caine et al., 1978; Majid et al., 1971). It also finds its usage in treating secondary shock and pheochromocytoma (Russell et al., 1998). According to Ghostine et al. (1984), it can be used in treating causalgia. It can cause postural hypotension along with reflex tachycardia which can produce cardiac arrhythmias. Reports demonstrates incomplete absorption when this drug is administered via oral route with a gradual onset of action. The impact of the medication lasts for about 3–4 days and is eliminated via urine in a 24-h time limit (Brunton et al., 2011; Seideman, 1982).

### 4.3.1.2.2 Phentolamine

It is an α-adrenoceptor nonselective antagonist which produces competitive antagonism. Phentolamine, a derivative of imidazoline has a similar/comparable cardiovascular activity to that of phenoxybenzamine. It causes vasodilation by competitively blocking both $\alpha_1$, $\alpha_2$ adrenoceptors with an abatement in PR and systemic BP (Majid et al., 1971; Russell et al., 1998; Juenemann et al., 1986). Activation of reflex sympathetic nerve elevates CO as outlined in Figure 4.3. It is utilized for managing short-term hypertension and bowel pseudo obstruction in pheochromocytoma patients. Moreover, it is utilized to counter anesthesia effect by antagonism of α receptor mediated vasoconstriction brought about by sympathomimetics usually administered along local anesthetics. A main adverse event observed is hypotension. Other effects include cardiac arrhythmia, ischemic cardiac events MI, and tachycardia. The drug administration is via parenteral route (Juenemann et al., 1986). Activity is instant on administration through intravenous route and about 15–20

min through intramuscular (IM) or subcutaneous (SC) route of administration. Its effect endures for about 10–15 min on intravenous administration and through intramuscular it takes 3–4 h. Biotransformation happens in the liver and elimination through urine (Brunton et al., 2011; Seideman, 1982).

### 4.3.1.3   SELECTIVE $\alpha_1$ BLOCKERS

These are agents that act by selectively blocking $\alpha_1$ adrenoceptors competitively or non-competitively. The non-selective α adrenergic blockers are alfuzosin, bunazosin, doxazosin, indoramin, prazosin, silodosin, tamsulosin, terazosin, urapidil (Brunton et al., 2011; Katzung, 2009). Table 4.2 describes the mechanism of action, pharmacological effects, and clinical responses of selective $\alpha_1$ blockers.

### 4.3.1.3.1   Alfuzosin

Alfuzosin (Fig. 4.4) has selective antagonistic action on $\alpha_1$ adrenoceptor having analogous selectivity to all the $\alpha_1$ receptor subtypes. Alfuzosin, being a quinazoline derivative performs its activity on $\alpha_1$ adrenoceptor by competitive binding there by alleviating the prostatic smooth muscles, rendering it useful in treating BPH but not preferred in treating hypertension (Brunton et al., 2011; Katzung, 2007; Gugger, 2011; Wilde et al., 1993). It shows 64% bioavailability succeeding oral administration. Literature reports the protein binding of alfuzosin to be 90% accompanied by a $V_d$ of 2.5 L/kg. It is contraindicated by CYP3A4 inhibitors like clarithromycin, ketoconazole, ritonavir and so on due to CYP3A4 metabolism. Only 11% of alfuzosin is eliminated unaltered in urine; about 75–91% of metabolites remain inert and are eliminated in feces. It has a $t_{1/2}$ of 3–5 h

(Wilde et al., 1993). Most commonly observed adverse effects were postural hypotension, tachycardia, dizziness, and headache (Jardin A et al., 1991).

### 4.3.1.3.2 Bunazosin

Bunazosin, an $\alpha_1$ adrenoceptor antagonist and a member of the quinazoline class is used mainly in treating hypertension (Hara et al., 2006). Additionally, it finds its use as hypotensive

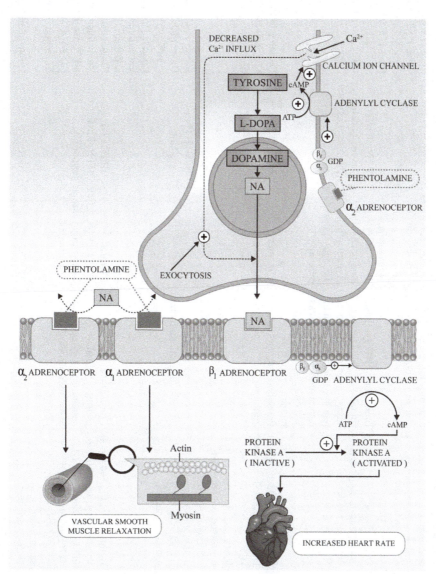

**FIGURE 4.3** Mechanism of action of phentolamine on heart and smooth muscles. DOPA, dihydroxyphenylalanine; NA, noradrenaline; GDP, guanosine 5'-diphosphate; ATP, adenosine triphosphate; cAMP, cyclic adenosine monophosphate; $Ca^{2+}$, calcium ions.

*Source*: Tripathi (2014), Golan et al. (2012), and Stevens and Brenner (2008).

**TABLE 4.2** Pharmacodynamic and Pharmacokinetic Profile of Selective $\alpha_1$ Blockers.

| Selective $\alpha_1$ blocker | Pharmacodynamics | | | | Pharmacokinetics |
|---|---|---|---|---|---|
| | Pharmacological effects | Mechanism of action | Adverse effects | Clinical responses | |
| Alfuzosin | Prostatic smooth muscles relaxation | Competitive pharmacological antagonism at $\alpha_1$ receptors | Dizziness; headache; postural hypotension on treatment initiation; tachycardia | BPH | Route of administration: PO; 64% bioavailability; Protein binding is 90%; $V_d$: 2.5 L/kg; $t_{1/2}$: 3 to 4 h; Excretion: Urine (11% unchanged) and feces (75–91% metabolites) |
| Bunazosin | Direct ocular neuroprotective effect; improves the ocular circulation | Inhibition of NO synthase (in rabbits); Na$^+$ channel antagonistic activity (in rats); rise in blood velocity in the retinal and choroidal areas and optic nerve head (in humans) | Blepharitis conjunctival; hyperaemia; headache; throbbing sensation | HTN; in retinal vascular occlusive diseases; glaucoma | Bioavailability: 81%; Metabolism: Liver; Excretion: Urine |
| Doxazosin | Vasodilation effect; drop in BP; decreased PR; increase HDL; decrease LDL, TC levels; beneficial in insulin resistance and impaired glucose metabolism | Phosphodiesterases inhibition | Dizziness; fatigue; headache; hypotension | BPH; HTN | Duration of drug activity: 36 h; Metabolism: Liver; $t_{1/2}$: 20 h; Excretion: Feces |
| Indoramin | Symptomatic relief of bronchoconstriction in asthmatic patients | Competitive pharmacological antagonism at $\alpha_1$ receptors | — | BPH; HTN; migraine | Route of administration: PO; Bioavailability: <30%; 90% protein bound; $V_d$: 2.5 L/kg; $t_{1/2}$: 3–4 h; Excretion: Urine (11% unchanged) and feces (75–91% metabolites) |
| Prazosin | Reduction in PR and blood flow to heart; vasodilation; fall in BP; decrease LDL, TG; elevation of HDL; sympathetic outflow suppression in CNS | Competitive pharmacological antagonism at $\alpha_1$ receptors; phosphodiesterases inhibition | Postural hypotension; reflex tachycardia; cardiac arrhythmias | BPH; CHF; HTN; Raynaud's syndrome | Route of administration: PO; Bioavailability: 50–70 %; 90% protein bound (AGP); Metabolism: Liver; $t_{1/2}$: 3 h; Excretion: Kidney |
| Silodosin | Hypogastric nerve stimulation-induced increase in intraurethral pressure | Selective antagonism of $\alpha_{1A}$ subtype receptors | Dizziness; orthostatic hypotension; retrograde ejaculation | BPH | Route of administration: PO; Rapid absorption; Bioavailability: 32%; 95.6% protein bound (AAG); Metabolism: Glucuronide conjugation; $t_{1/2}$: 3 h; Excretion: Kidney |

**TABLE 4.2** *(Continued)*

| Selective $\alpha_1$ blocker | Pharmacodynamics | | | Clinical responses | Pharmacokinetics |
|---|---|---|---|---|---|
| | Pharmacological effects | Mechanism of action | Adverse effects | | |
| Tamsulosin | $\alpha_{1A}$ antagonism relaxes prostatic smooth muscles | Selective antagonism of $\alpha_{1A}$ and $\alpha_{1D}$ subtype receptors | Abnormal ejaculation; decreased libido and impotence | BPH | Route of administration: PO; Metabolism: Liver; $t_{1/2}$: 5–10 h |
| Terazosin | Induce apoptosis in the prostate smooth muscle cells | Antagonism of $\alpha_{1A}$, $\alpha_{1B}$, and $\alpha_{1D}$ subtype receptors | Asthenia; dizziness; hypotension; impotence; rhinitis; somnolence | BPH; HTN | Duration of action: >18 h; Bioavailability: 90%; $t_{1/2}$: 12 h |
| Urapidil | Higher doses having antiarrhythmic properties | Antagonism of $\alpha_1$ and agonism of 5-HT$_{1A}$ receptors | Dizziness; headache; hypotension; fatigue; nausea; palpitations | BPH; HTN | Bioavailability: 72%; 75–80 % protein bound (AAG); $V_d$: 0.41 to 0.77 L/kg; Metabolism: Liver; $t_{1/2}$: 3 h; Excretion: Urine |

AGP, $\alpha$1-acid glycoprotein; NO, nitric oxide; PR, peripheral resistance; IM, intramuscular; BP, blood pressure; SC, subcutaneous; HTN, hypertension; BPH, benign prostatic hyperplasia; PO, per oral; $V_d$, volume of distribution; IV, intravenous; $t_{1/2}$, half-life; HDL, high density lipoprotein; CHF, congestive heart failure; LDL, low density lipoprotein; TC, total cholesterol.

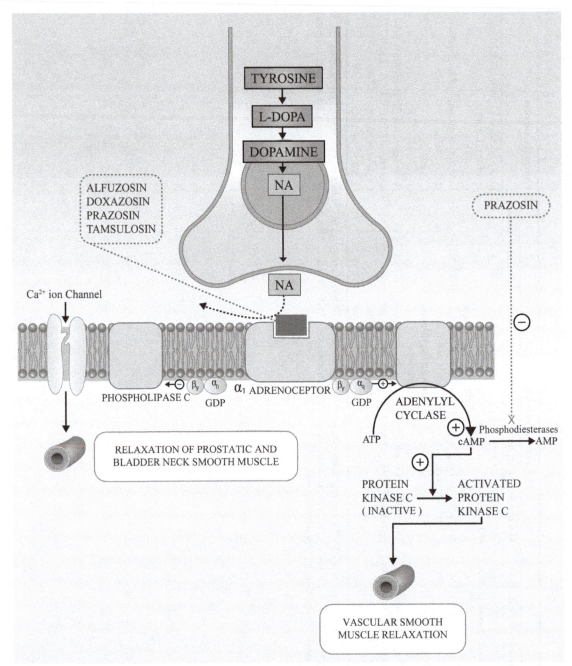

**FIGURE 4.4**  Mechanism of action of selective $\alpha_1$ adrenergic antagonists. Mechanism of action of alfuzosin, doxazosin, prazosin, and tamsulosin on non striated smooth muscles. DOPA, dihydroxyphenylalanine; NA, noradrenaline; GDP, guanosine 5′-diphosphate; ATP, adenosine triphosphate; cAMP, cyclic adenosine monophosphate; AMP, adenosine monophosphate; $Ca^{2+}$, calcium ions.

*Source*: Adapted from Tripathi (2014), Golan et al. (2012), and Stevens and Brenner (2008).

medicament in the field of ophthalmology, hence therapeutically used for treating ischemic retinal diseases like retinal vascular occlusive diseases and glaucoma associated with ocular circulation disturbances as it produces a direct neuroprotective effect and improves the ocular circulation. Instillation of bunazosin in rabbit eyes has showed an apparent improvement in optic nerve head blood flow impairment and perseverance of visual evoked potentials (VEP) implicit time probably by inhibiting NO synthase; in rats through $Na^+$ channel action, glutamate induced neuronal death was reduced; in humans, an elevation in the velocity of blood in the retinal and choroidal areas and optic nerve head were observed (Hara et al., 2006).

### 4.3.1.3.3 Doxazosin

It is structurally similar to prazosin with highly selective $\alpha_1$ receptor antagonism. It affects the subtypes namely $\alpha_{1D}$, $\alpha_{1B}$, and $\alpha_{1A}$. Similar to prazosin it acts by obstructing phosphodiesterases thereby activating the protein kinase bringing about an abatement in the tone of smooth muscles present in blood vessels causing vasodilation effect and fall in BP, resultant of declined peripheral resistance. Similar/analogous activity in tone of prostatic has been reported as a consequence of $\alpha_1$-adrenoceptor blockade relieving obstruction of bladder outflow (Fulton et al., 1995). Studies report that doxazosin (Fig. 4.4) produce increase in HDL level and decrease the LDL and total cholesterol levels; beneficial in insulin resistance and impaired glucose metabolism (Fulton et al., 1995; Grimm et al., 1996). The bioavailability and biotransformation in doxazosin is analogous to prazosin but has lengthy duration of activity extending as far as 36 h. The $t_{1/2}$ is about 20 h with majority metabolites excreted/elimination

via feces (Fulton et al., 1995; Li et al., 2015). Similar to terazosin it produces apoptosis of $\alpha$ adrenoceptors in the smooth muscles of prostate, hence utilized for treating urinary tract issues connected with BPH and hypertension. Adverse events are fatigue, dizziness, hypotension, and headache (Dutkiewicz, 1997; Lepor, 1995; Gugger, 2011).

### 4.3.1.3.4 Indoramin

It is an $\alpha_1$ selective antagonist and is competitive in nature. It causes a reduction in BP and utilized for treating migraine, BPH, and hypertension. Studies report that indoramin causes symptomatic alleviation of bronchoconstriction most commonly observed in the asthmatics (Holmes et al., 1986; Lewis et al., 1973; Stott et al., 1991). After an oral dose, bioavailability is below 30% due to hepatic metabolism. The drug elimination is via urine. After administration via intravenous route, the $V_d$ observed was 7.4 L/kg when the dose of 0.14 mg/kg, 15 mg were given. 80–90% of the drug is protein bound with a half-life of 5 h. Primary adverse effects observed are dry mouth, sedation, and ejaculation problems (Draffan et al., 1976; Holmes et al., 1986; Volans et al., 1982).

### 4.3.1.3.5 Prazosin

Prazosin, a piperazinyl quinazoline compound is the prototype of selective $\alpha_1$ blockers. It blocks $\alpha_1$ receptors in veins and arterioles producing an abatement in PVR/PR and flow of blood to heart (Hanon et al., 2000; Jaillon, 1980; Stanaszek et al., 1983). It is more $\alpha_1$ receptor selective with little to no $\alpha_2$ receptor blockade effect and has replaced the non-selective $\alpha$ receptor agonist drugs such as phentolamine and phenoxybenzamine. The $\alpha_1$

selectivity is 1000 times greater than $\alpha_2$ receptors (Brunton et al., 2011). It affects the $\alpha_{1A,}$ $\alpha_{1B}$ and $\alpha_{1D}$ subtypes. It is an effective blocker of cyclic nucleotide phosphodiesterases that brings about an elevation in cAMP level within smooth muscles succeeding a stimulation of protein kinase C resulting in vasodilation and diminished BP (Fig. 4.4; Brunton et al., 2011; Stanaszek et al., 1983). Hence, utilized for treating hypertension, CHF, and decrease baroreflex functioning in hypertensives. It decreases LDL, triglycerides and elevates the HDL concentration. It also suppresses the sympathetic outflow in CNS, hence efficacious in PTSD and nightmares (Gugger, 2011; Menkes et al., 1981). After an oral dose, it has a good absorption and bioavailability is about 50–70%. Plasma concentration is attained after 1–3 h of oral administration. It is 90% bound to protein particularly $\alpha_1$-acid glycoprotein (Hobbs et al., 1978; Seideman, 1982; Stanaszek et al., 1983). It is subjected to hepatic biotransformation and is eliminated by the aid of kidneys. It has approximately a half-life of 3 h and drug action ranges for 10 h. The off label use is treating BPH (Brunton et al., 2011; Craig et al., 1991; Katzung, 2007; Stanaszek et al., 1983).

### 4.3.1.3.6  Silodosin

Silodosin is an $\alpha_{1A}$ subtype selective adrenoceptor antagonist utilized for treating BPH accompanied by little impact on BP. Rapid absorption with 32% bioavailability can be observed succeeding oral administration. It is 95.6% bound to protein specifically with AGP. Metabolism is by several pathways and the main metabolites produced are alcohol dehydrogenase (ADH/ALDH) via UDP-glucuronosyltransferase. Adverse effects include dizziness, orthostatic hypotension and the main adverse effect is retrograde ejaculation (Gugger, 2011;

Matsubara et al., 2006; Michel, 2010; Montorsi, 2010; Rossi et al., 2010).

### 4.3.1.3.7  Tamsulosin

Tamsulosin (Fig. 4.4) is an $\alpha_{1A}$ and $\alpha_{1D}$ subtype selective adrenoceptor antagonist. It finds application in treating BPH. Due to a little impact on BP, it is not a drug of choice for treating hypertension. Absorbed well following oral administration and the half-life ranges from 5 to 10 h. Biotransformation occurs by CYP enzymes. Commonly observed adverse effects are abnormal ejaculation and additionally other adverse events related to sexual function so far reported are declined/deteriorated libido and impotence (Chapple et al., 1996; Chapple et al., 2002; Gugger, 2011; Hofner, 1998).

### 4.3.1.3.8  Terazosin

It is a structural analog of prazosin with high selectivity for $\alpha_1$ receptors but having less potency. It affects the $\alpha_{1A}$, $\alpha_{1B}$, and $\alpha_{1D}$ subtypes. Terazosin is more aqueous soluble than prazosin with the bioavailability more prominent than 90%. It has a 12-h period half-life with an increased length of activity which is more prominent 18 h. It is primarily used to treat BPH and hypertension. It develops prostatic smooth muscles apoptosis, hence utilized for treating urinary tract issues related to BPH. Adverse events observed are hypotension, dizziness, asthenia, somnolence, impotence, and rhinitis (Fulton et al., 1995; Schwinn et al., 2004).

### 4.3.1.3.9  Urapidil

It is an $\alpha_1$ selective adrenergic receptor antagonist with weak $\alpha_2$-agonistic and 5-HT$_{1A}$-agonistic

**TABLE 4.3** Pharmacodynamic and Pharmacokinetic Profile of Selective $\alpha_2$ Blocker.

| Selective $\alpha_2$ blocker | Pharmacodynamics | | | Pharmacokinetics |
|---|---|---|---|---|
| | **Pharmacological effects** | **Mechanism of action** | **Adverse effects** | **Clinical responses** | |
| Yohimbine | Increase central sympathetic outflow along with noradrenaline release peripherally, thus raises BP and HR; CNS effects like excitation, tremor, antidiuresis, nausea, vomiting; Genital congestion | Competitive pharmacological antagonism at $\alpha_2$ adrenoceptors | Anxiety might occur | Aphrodisiac; erectile dysfunction; hypotension | Route of administration: PO; Rapidly absorbed and eliminated; $t_{1/2}$: 0.25–2.5 h; Excretion: Urine |

HR, heart rate; BP, blood pressure; PO, per oral; $t_{1/2}$, half-life.

actions. It possesses a distinct structure while compared to prazosin. It produces hypotension and clinically applied for treating hypertension and BPH. At large dosages antiarrhythmic properties are seen. It also shows CNS effects by competitively blocking 5 $HT_{1a}$. Commonly observed adverse events are headache, dizziness, fatigue, nausea, and palpitations, which subsides on cessation of drug. It has 72% bioavailability and $V_d$ is 0.41–0.77 L/kg and it penetrates the BBB. It is 75–80% bound to protein with a half-life of 3 h. It is subjected to first pass and 50–70% is eliminated via urine (Kirsten et al., 1988; Langtry et al., 1989).

### 4.3.1.4   SELECTIVE $\alpha_2$ BLOCKERS

These are agents that act by selectively blocking $\alpha_2$ adrenoceptors competitively or non-competitively. The selective $\alpha_2$ adrenergic blocker available is yohimbine (Brunton et al., 2011; Katzung, 2009; Tripathi, 2014). Table 4.3 describes the mechanism of action, pharmacological effects, and clinical responses of selective $\alpha_2$ blockers.

#### 4.3.1.4.1   Yohimbine

Yohimbine is an $\alpha_2$ receptor competitive antagonist. It is obtained from Rauwolfia roots, cortex of *Coryanthe yohimbe* tree, and the bark of *Pausinystalia yohimbe* (Feuerstein et al., 1985; Hedner et al., 1992; Morales, 2000; Steinegger et al., 1988). Yohimbine, an indolealkylamine alkaloid acts centrally producing effects like excitation, tremor, antidiuresis, nausea, vomiting; raises the blood pressure and the rate of the heart due to increased centrally mediated sympathetic outflow and noradrenaline release peripherally; and genital congestion, hence utilized as aphrodisiac. It can produce tremors

and stimulates the motor activity. It has an antagonistic action toward/against serotonin. It is mainly employed in treating sexual dysfunction chiefly in males (Morales, 2000; Rowland et al., 1997). The drug shows rapid absorption and elimination. The bioavailability in the case of oral administration shows great variability, which ranges from 7% to 87% and the mean value 33%. This incomplete oral bioavailability may be because of incomplete gastrointestinal absorption or because of first-pass effect. The distribution is rapid and half-life ranges from 0.25 to 2.5 h. About 0.5–1% of yohimbine is excreted unchanged through urine which indicates drug elimination by hepatic clearance (Owen et al., 1987; Hedner et al., 1992; Guthrie et al., 1990).

### 4.3.1.5   OTHER A ADRENERGIC BLOCKERS

These include ergot alkaloids, ketanserin, and neuroleptic agents. Ergot alkaloids were the firstly discovered adrenergic antagonists and possess complex pharmacological properties (Brunton et al., 2011; Katzung, 2009; Tripathi, 2014). It acts on $\alpha$ receptors, dopaminergic receptors, and 5HT receptors to a varying extent. It is clinically used in treating migraine and postpartum hemorrhage. Ketanserin, an $\alpha_1$ adrenergic antagonist selectively antagonizes $5HT_{2A}$ receptors and can be utilized effectively against hypertension because of its $\alpha$ adrenergic blockade. Additionally, it improves Raynaud's disease (Brunton et al., 2011). Several neuroleptic drugs like chlorpromazine, haloperidol, butyrophenone, phenothiazine and so on inspite of being used as antagonists of dopaminergic $D_2$ receptors antagonizes the $\alpha$ receptor. Adverse events observed are stuffed nasal passage, reduction in BP, and ejaculation inhibition (Brunton et al., 2011).

### 4.3.2 β ADRENOCEPTOR ANTAGONIST

β adrenoceptor antagonistic action depends on their relative affinity toward $\beta_1$, $\beta_2$ adrenoceptors, their action on α receptor, lipid solubility, ability to cause vasodilation, and their pharmacokinetic parameters (Frishman et al., 2011). All these characteristics are useful in the appropriate drug selection for the patient. β antagonists can be categorized as non-selective, selective, and those with additional cardiovascular activity in connection to β adrenoceptor antagonism. Some β blockers partially activate the β adrenoceptor without catecholamines presence. So, these partial agonists possess a slight sympathomimetic activity. For example, acebutolol, pindolol. Some other β antagonists possess a feature like inverse agonism nature. Many β antagonists like propranolol, pindolol, acebutolol, metoprolol, carvedilol, and so on possess a membrane stabilizing or local anesthetic action similar to lidocaine. Some β antagonists like bucindolol, carvedilol, and labetalol blocks $\alpha_1$ along with β receptors. Many other β antagonists along with these also cause vasodilator activity (Brubacher, 2015; Brunton et al., 2011). The mechanism by which β antagonists acts and its action on different organs is demonstrated in Figure 4.5.

### 4.3.2.1 PHARMACODYNAMICS AND PHARMACOKINETICS

#### 4.3.2.1.1 Cardiovascular System

Adrenergic β antagonist produces major therapeutic actions on the CVS. They cause a decrease in myocardial contractility and the rate of the heart while at rest, the effect seen is low; but during exercise or stress or when the SNS is active the antagonist effect is profound (Brunton et al., 2011; Katzung, 2009). β antagonist decreases cardiac output during short term administration. The peripheral resistance rises and the BP is maintained due to vascular $\beta_2$ blockade and raised activity of the sympathetic system which cause vascular α receptor activation (Brunton et al., 2011). But while using long term, the TPR comes back to the initial stage (Mimran and Ducailar, 1988) or may decrease in patients with hypertension (Manin't Veld et al., 1988). In β antagonists with $\alpha_1$ blockade activity or with direct vasodilator activity, CO is preserved but a higher fall in PR is observed, for example, labetalol, bucindolol, and carvedilol. β antagonists affects significantly the cardiac rhythm and automaticity due to inhibition of $\beta_1$, $\beta_2$ adrenoceptors (Altschuld and Bilman, 2000; Brodde and Michel, 1999).

#### 4.3.2.1.2 Antihypertensive Effects

Adrenergic β antagonists lower the blood pressure in hypertension without lowering the normal BP. The correct mechanism of the effect is not understood clearly (Brunton et al., 2011). The β antagonists blocks $\beta_1$ receptor mediated renin release from the JG apparatus, but its relationship with the lowering of the BP is not known even though some researchers found the antihypertensive activity of drug, propranolol profound in patients with increased renin plasma level, but β antagonists can be also useful for people with low to normal renin concentration in the plasma (Brunton et al., 2011; Katzung, 2009). In some β receptor antagonists, reduction of the BP might be also due to additional effects like peripheral vasodilation, $\beta_2$ receptor activation, nitric oxide production, $\alpha_1$ receptor blockade opening of potassium channels, calcium entry blockade and antioxidant property (Frishman et al., 2011).

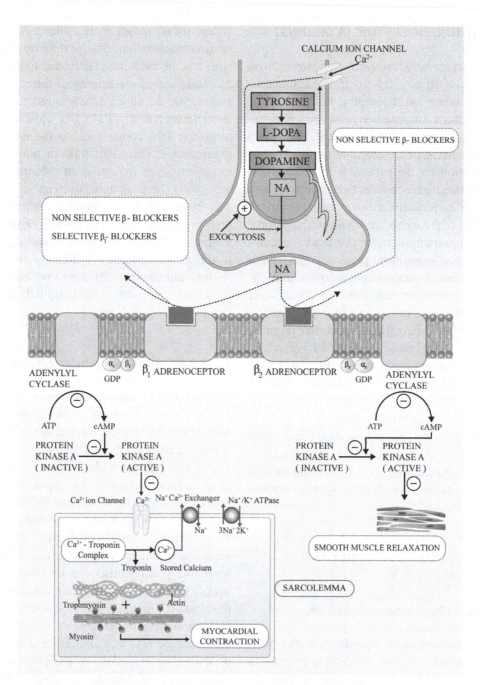

**FIGURE 4.5** Mechanism of action of β adrenergic antagonists. DOPA, dihydroxyphenylalanine; NA, noradrenaline; GDP, guanosine 5′-diphosphate; ATP, adenosine triphosphate; cAMP, cyclic adenosine monophosphate; $Ca^{2+}$, calcium ions; $Na^+$ $Ca^{2+}$ Exchanger, sodium–calcium exchanger or sodium–potassium pump; $Na^+$ $K^+$ ATPase, sodium–potassium adenosine triphosphatase.

*Source*: Adapted from Tripathi (2014), Golan et al. (2012), and Stevens and Brenner (2008).

### 4.3.2.1.3   Respiratory System

Non-selective adrenergic β antagonists produces blockade of β receptor located in the nonstriated smooth muscle of bronchi which has little effects in normal people; but it causes severe bronchoconstriction in patients having COPD (Brubacher, 2015; Brunton et al., 2011; Javeed et al., 1996). Selective $\beta_1$ adrenergic antagonists with ISA does not cause such bronchoconstriction but should be used carefully in bronchospastic patients (Brubacher, 2015; Brunton et al., 2011; Javeed et al., 1996).

### 4.3.2.1.4   Metabolic Effects

Adrenergic β receptor antagonists alter the metabolism of lipids and carbohydrates. Non-selective β blockers block glycogenolysis and can delay recovery from insulin-dependent hypoglycemia in diabetes mellitus. They also influence the counter regulatory action of catecholamine occurring in hypoglycemia by reducing the symptoms like tachycardia, nervousness and tremor. It must be used carefully in patients having frequent hypoglycemic reactions and labile diabetes. Selective $\beta_1$ antagonists are normally preferred in above mentioned hypoglycemic condition as they are less likely to delay the recovery (DiBari et al., 2003). β adrenergic antagonists can cause lowered free fatty acid release from the adipose tissue by antagonism as the β receptors mediates the release of free fatty acids into the blood which phones which forms an important energy source for the exercising muscles. Non-selective adrenergic β antagonists lower HDL cholesterol and increases triglycerides and LDL cholesterol. $\beta_1$ selective antagonists make better the lipid profile of the serum in case of dyslipidemic patients. The β antagonists having vasodilating property increases insulin sensitivity contrary to clinically utilized β blockers of non-selective nature which decreases the insulin sensitivity. They also show cardioprotective effects. β antagonist may also block tremor induced by catecholamines. They may also block degranulation of mast cell inhibition by catecholamines (Brubacher, 2015; Brunton et al., 2011; Lacey et al., 1991).

### 4.3.2.1.5   Adverse Effects

The frequently seen adverse events arise because of the β receptor blockade.

#### 4.3.2.1.5.1   Central Nervous System

In CNS, the adverse reactions of adrenergic β antagonists include fatigue, depression, and sleep disturbances including insomnia as well as nightmares (Brunton et al., 2011; Frishman et al., 2011).

#### 4.3.2.1.5.2   Respiratory System

Major adverse events are blockade of $\beta_2$ receptor located in the nonstriated smooth muscles of bronchi and is relevant in producing bronchodilation in patients having bronchospastic disease. So, a blockade can cause life threatening resistance to airflow in those patients. $\beta_1$ selective drugs with ISA are not probable to cause such a bronchospasm (Brunton et al., 2011; Frishman et al., 2011).

#### 4.3.2.1.5.3   Cardiovascular System

In patients having impaired myocardial functions, vital support for the cardiac functions is provided by SNS, hence considering susceptible patients, β adrenergic antagonists can produce congestive cardiac failure. Blockade of β receptors can also produce bradycardia (Brubacher, 2015; Brunton et al., 2011; Javeed et al., 1996).

### 4.3.2.1.5.4    Overdose

Manifestations of poisoning is based on the pharmacology of drug ingested such as $\beta_1$ selectivity, membrane stabilizing property, and ISA. Bradycardia, hypotension, widened QRS complex, prolonged AV conduction times are routinely seen in the dosage. Hypoglycemia, bronchospasm, seizures, and depression may occur (Brunton et al., 2011). Bradycardia may be managed with atropine. Isoproterenol is employed in treating hypotension. Glucagon produces elevated rate and contraction of heart and is helpful during overdose of $\beta$ adrenergic antagonist (Brunton et al., 2011).

### 4.3.2.1.6    Drug Interactions

Both pharmacokinetic and pharmacodynamic drug interactions are observed in case of $\beta$ adrenergic receptor antagonists. $\beta$ agonists, $\alpha$ blockers, and sympathomimetics interact with $\beta$ antagonists (Brunton et., 2011; Frishman et al., 2011). Colestipol, cholestyramine, and ammonium salts can reduce the $\beta$ blocker absorption. Drugs like rifampin, phenobarbital, and phenytoin and habits such as smoking stimulate hepatic enzymes thereby reducing the plasma levels of many $\beta$ antagonists metabolized extensively in liver. Drugs like hydralazine and cimetidine produce an increase in the bioavailability of various $\beta$ antagonist drugs like metoprolol, propranolol by its action on the liver blood flow (Brunton et al., 2011; Frishman et al., 2011). Pharmacodynamic interactions include calcium blockers and $\beta$ antagonists producing additive effects on the cardiac conducting system. $\beta$ blockers produce synergistic action with other antihypertensive agents on the BP. $\beta$ antagonists induced antihypertensive effect can be opposed by NSAIDs and indomethacin (Brunton et., 2011; Frishman et al., 2011).

### 4.3.2.1.7    Therapeutic Uses

$\beta$ antagonists are extensively employed in treating coronary heart diseases, angina, hypertension, and congestive cardiac failure. They are also employed in treating myocardial infarction, supraventricular and ventricular arrhythmias, in treating HOCM, palpitation, syncope and in managing pheochromocytoma, acute dissecting aortic aneurysm (Poirier et al., 2012; Toda, 2003). In ophthalmology, its use is in treating glaucoma by lowering aqueous humor production. They can also be used as adjuvant in managing hyperthyroidism, migraine prophylaxis, acute panic symptoms, essential tremor, akathisia, and in the prevention of variceal bleeding in portal hypertension. Competitive beta receptor adrenergic antagonists are extensively employed in treating hypertension, ischemic cardiac disease, congestive heart failure, and in cardiac arrhythmias (Brubacher, 2015; Brunton et al., 2011 Poirier et al., 2012; Toda, 2003).

### 4.3.2.2    NON-SELECTIVE B-BLOCKERS (FIRST GENERATION)

These are agents that act by blocking $\beta$ adrenoceptors competitively or non-competitively. The non-selective $\beta$ adrenergic blockers are levobunolol, metipranolol, nadolol, penbutolol, pindolol, propranolol, sotalol, timolol (Brunton et al., 2011; Katzung, 2009; Tripathi, 2014). Table 4.4 describes the mechanism of action, pharmacological effects, and clinical responses of non-selective $\beta$ blockers.

**TABLE 4.4**  Pharmacodynamic and Pharmacokinetic Profile of Non-selective β Blockers.

| Selective β₁ blocker | Pharmacodynamics | | | Clinical responses | Pharmacokinetics |
|---|---|---|---|---|---|
| | Pharmacological effects | Mechanism of action | Adverse effects | | |
| Levobunolol | Reducing intraocular pressure | Antagonism at $\beta_1$, $\beta_2$ receptors | Atrioventricular block; bradycardia; bronchospasm; decreased libido; mask hypoglycemic symptoms; sedation | Glaucoma | Topical application; Prolonged activity compared to timolol |
| Nadolol | Decreases cardiac output, AV node conduction and oxygen ($O_2$) demand; decreases BP | Antagonism at $\beta_1$ and $\beta_2$ receptors | Atrioventricular block; bradycardia; bronchospasm; decreased libido; mask symptoms of hypoglycemia Sedation. | Angina pectoris; HTN; indicated in arrhythmias; migraine headache | Excretion: Urine |
| Penbutolol | Decreases cardiac output, AV node conduction and $O_2$ demand; decreases BP | Antagonism at $\beta_1$, $\beta_2$ adrenoceptors | Atrioventricular block; bradycardia; bronchospasm; decreased libido; mask symptoms of hypoglycemia; sedation | Angina pectoris | Route of administration: PO; $t_{\frac{1}{2}}$: 4.5 h |
| Pindolol | Decreases cardiac output, AV Node conduction and $O_2$ demand; decreases BP; MSA | Antagonism at $\beta_1$, $\beta_2$ adrenoceptors with ISA and MSA | Bronchospasm but less compared to propranolol | HTN | Route of drug administration: PO, IV; 57% protein bound; $V_d$: 136 l; $t_{\frac{1}{2}}$: 3.6 (oral) and 3.1 (IV) h |
| Propranolol | Decreases cardiac output, AV Node conduction and $O_2$ demand; decreases BP; MSA | Competitive pharmacological antagonism at β receptors with MSA | Atrioventricular block; bradycardia; bronchospasm; decreased libido; mask symptoms of hypoglycemia; sedation | Angina pectoris; arrhythmia prophylaxis; CHF; HTN; prophylaxis of MI; pheochromocytoma | Route of administration: PO; well absorbed; Bioavailability: Low; Distribution: > 90% protein bound; Excretion: Urine |
| Sotalol | Addition K⁺ blocking | Antagonism at $\beta_1$ and $\beta_2$ receptors | | Antiarrhythmic | Excretion: Urine. |
| Timolol | Reducing intraocular pressure | Antagonism at $\beta_1$ and $\beta_2$ receptors | Atrioventricular block; bradycardia; bronchospasm; decreased libido; mask symptoms of hypoglycemia; sedation | Glaucoma | Metabolism: Liver |

AV, atrioventricular; HTN, hypertension; PR, peripheral resistance; PO, per oral; $t_{\frac{1}{2}}$, half-life; MSA, membrane stabilizing activity; BP, blood pressure; IV, intravenous; $V_d$, volume of distribution; CHF, congestive heart failure; MI, myocardial infarction.

### 4.3.2.2.1 *Nadolol*

It is an adrenergic β receptor non-selective antagonist which shows equal affinity toward β₁ and β₂ receptors. It has a long duration of activity and it does not show ISA and membrane stabilizing effect. The half-life is about 20 h (Brunton et al., 2011; Florey, 2008). It is employed in treating angina pectoris, arrhythmia, and hypertension. Off-label indications are in the management of migraine, variceal bleeding in portal hypotension, and parkinsonian tremors. It soluble in water and it has a bioavailability of 35%. CNS activity is low compared to lipid soluble antagonists. Metabolism is not extensive and the excretion is almost intact in urine (Brunton et al., 2011; Florey, 2008).

### 4.3.2.2.2 *Pindolol*

It is an adrenergic β receptor non-selective antagonist. It possesses ISA and shows a low MSA. It is mainly used in treating angina pectoris and hypertension. It blocks exercise induced increase in the heart rate and the output of the heart. After an oral dose it is absorbed completely and shows a good bioavailability. Almost 50% of the metabolism happens in liver and excretion occurs via the urine. The drug has a half-life of about 4 h (Brubacher, 2015; Brunton et al., 2011; Pritchard et al., 1971).

### 4.3.2.2.3 *Propranolol*

It is an adrenergic non-selective competitive β receptor antagonist showing equal affinity to β₁ and β₂ receptors. The drug does not induce α receptor blockade and it lacks ISA. It is usually a prototype to which all other β antagonists are

compared (Brunton et al., 2011). It is mainly used in treating angina and hypertension and is also helpful in ventricular and supraventricular arrhythmias, tachycardias, prematurely occurring contractions of the ventricles, myocardial infarction, digitalis induced tachyarrhythmias, pheochromocytoma, essential tremor, and migraine. Off label use is for parkinsonian tremors, antipsychotic induced akathisia, variceal hemorrhage, a complication of portal hypertension, and in GAD. It is intravenously administered for managing life threatening arrhythmias (Brunton et al., 2011; Katzung, 2009). After an oral dose it is fully absorbed and it is highly lipophilic. Metabolism happens in the liver and after first-pass metabolism only 25% reaches the blood circulation. It has a $V_d$ of 4 L/kg. It enters the CNS penetrating the BBB. About 90% drug in the circulation is plasma protein bound. Excretion occurs via urine. The half-life of the drug is 4 h (Brubacher, 2015; Brunton et al., 2011).

### 4.3.2.2.4 *Sotalol*

It is an adrenergic nonselective β adrenergic blocker which is hydrophilic in nature. The drug does not show ISA or membrane-stabilizing activity. Apart from its β blockade property it has activity on cardiac ion channels which helps in prolonging action potential and because of this it is used in the treatment of arrhythmia. Studies report a typical hemodynamic activity of sotalol; reduced cardiac output and the rate of the heart with not much change in the BP and the stroke volume, respectively. After an oral dose the drug has a bioavailability of 100% as the first-pass metabolism is not significant. It reaches the peak plasma concentration in about 2–3 h. The $V_d$ is 1.3 L/kg. The drug shows negligible plasma protein binding. Excretion occurs renally with about 75% excretion occurring in 72 h. Half-life range is about 7–18 h (Antonaccio et al., 1990; Brunton et al., 2011; Taboulet et al., 1993).

### 4.3.2.2.5 Timolol

It is an adrenergic non-selective potent β antagonist. It shows no membrane stabilizing and ISA. It is employed in the treatment of congestive heart failure, hypertension, myocardial infarction, and in managing migraine (Brunton et al., 2011; Florey, 2008; Mayama et al., 2013). In ophthalmology, its use is in treating open angle–glaucoma and in intraocular hypertension. It blocks β receptors present in the ciliary epithelium and reduces aqueous humor generation (Chiou et al., 1993; Florey, 2008; Mayama et al., 2013; Van et al., 1990). It shows good absorption from the GIT. The drug attains a peak concentration in the plasma in about 1–2.4 h. The bioavailability is 61–75% due to first-pass metabolism. While used as ophthalmic drops, absorption is fast and the intraocular pressure is lowered within about 3 h. Metabolism happens in liver with the help of CYP2D6 enzyme. After an oral dose first-pass metabolism happens to the drug and its half-life is about 4 h. It can also be useful for patients suffering from coronary cardiac disease (Florey, 2008; Brunton et al., 2011). After an oral dose of timolol the absorption is fast and complete. The drug reaches maximum plasma concentration within 1–2.4 h. The bioavailability is 61–75% due to first-pass metabolism. When used as ophthalmic drops, absorption is rapid and the intraocular pressure is decreased within 3 h. The half-life is approximately 2.5 h. The drug is majorly excreted in the urine (Florey, 2008).

### 4.3.2.2.6 Other β Adrenergic Receptor Antagonists

These include carteolol, levobunolol, metipranolol, and timolol which are utilized in ophthalmology for treating glaucoma and elevated BP in the eye (Brunton et al., 2011; Katzung, 2009; Tripathi, 2014).

#### 4.3.2.2.6.1 Carteolol

Carteolol is a β receptor non-selective antagonist and has intrinsic sympathomimetic activity (Janczewski et al., 1988; Mayama et al., 2013; Veld et al., 1982). Carteolol produces a vasodilating effect, which is higher when compared to β-blockers without ISA such as timolol. The actual mechanism of ISA within the body is not much understood.

#### 4.3.2.2.6.2 Levobunolol

Levobunolol is an adrenergic β receptor non-selective antagonist. After administration it changes to a metabolite, dihydrobunolol (DHB) with same potency as levobunolol (Di et al., 1977; Mayama et al., 2013). In the retina and choroid, the diffusion of DHB is less because of its polarity and possesses a decreased risk for vasoconstriction (Acheampong et al., 1995; Dong et al., 2007; Mayama et al., 2013). There is no major variation in the blood flow parameters following an administration of a single dose of levobunolol. The rate of blood flow in the retinal vein increase slightly and in retrobulbar arteries, blood flow was unchanged (Altan-Yaycioglu et al., 2001; Bloom et al., 1997; Leung et al., 1997; Mayama et al., 2013; Schmetterer et al., 1997). Several studies report an increase in pulsatile ocular blood flow following a single drop of levobunolol 0.5% or with levobunolol treatment daily twice for a week (Bosem et al., 1992; Mayama et al., 2013; Morsman et al., 1995).

**TABLE 4.5** Pharmacodynamic and Pharmacokinetic Profile of Selective $\beta_1$ Blockers (Second Generation).

| Selective $\beta_1$ blocker | Pharmacodynamics | | | | Pharmacokinetics |
|---|---|---|---|---|---|
| | Pharmacological effects | Mechanism of action | Adverse effects | Clinical responses | |
| Acebutolol | Decreases cardiac output, AV node conduction, and O$_2$ demand; decreases BP; ISA; MSA | Antagonism at $\beta_1$ receptors with ISA and MSA | Acebutolol has minimal metabolic effects | In bradyarrhythmia or PVD patients | Metabolism: Rapid metabolism to diacetyl; Oral bioavailability: 40%; $t\frac{1}{2}$: 8–12 h; Excretion: Kidney |
| Atenolol | Pure $\beta_1$ Blockers; decreases cardiac output, AV Node conduction and O$_2$ demand; decreases BP | Antagonism at $\beta_1$ receptors | Atrioventricular block; bradycardia; decreased libido; mask symptoms of hypoglycemia; sedation | Angina pectoris; acute MI; HTN | Oral absorption: Prolonged activity; Oral bioavailability: 50%; Metabolism: Liver |
| Bisoprolol | Increases TG; reduction in HDL | Antagonism at $\beta_1$ receptors | Dizziness; headache; tiredness | Angina, HTN | Bioavailability: 90%; Prolonged activity; 30% bound to protein; Metabolism: Liver; $t\frac{1}{2}$: 10–11 h; Excretion: Kidney |
| Esmolol | Pure $\beta_1$ blockers; decreases cardiac output, AV node conduction and O$_2$ demand; decreases BP | Antagonism at $\beta_1$ receptors | Atrioventricular block; bradycardia; decreased libido; mask symptoms of hypoglycemia | early mi; htn; sedation; terminate symptoms occurring in anesthesia such as atrial fibrillation episodes, arrhythmia, tachycardia | Route of administration: IV infusion; Metabolism: Hydrolysis; $t\frac{1}{2}$: 3 to 4 min |
| Metoprolol | Bronchoconstriction; bradycardia; MSA; vasoconstriction (reflex) | Antagonism at $\beta_1$ receptors with MSA | Atrioventricular block; bradycardia; decreased libido; mask symptoms of hypoglycemia; sedation | Angina; arrhythmia prophylaxis; atrial flutter; HTN | Route of administration: PO; Oral bioavailability: 40%; Metabolism: 90% of drug subjected to hepatic metabolism prior to excretion; $t\frac{1}{2}$: 3–4 h; can enter CNS |

AV, atrioventricular; HTN, hypertension; $t\frac{1}{2}$, half-life; MSA, membrane stabilizing activity; BP, blood pressure; IV, intravenous; MI, myocardial infarction; ISA, intrinsic sympathomimetic activity; TG, triglyceride; HDL, high density lipoprotein; PVD, peripheral vascular disease.

### 4.3.2.2.6.3 Nipradilol

Nipradilol is a β receptor non-selective blocker having a weak $\alpha_1$ blocking activity. It is used as a topical application for glaucoma therapy in Japan (Kanno et al., 2000; Mayama et al., 2013). It also produces NO-releasing activity similar to nitroglycerin and acts as a vasodilator showing highly potent activity (Sugiyama et al., 2001; Mayama et al., 2013). Systemic administration of nipradilol produces a vasodilating effect similar to nitroglycerin or nifedipine (Araki et al., 1992; Mayama et al., 2013).

### 4.3.2.3 SELECTIVE $\beta_1$ BLOCKERS (SECOND GENERATION)

These are agents that act by selectively blocking $\beta_1$ adrenoceptors competitively or non-competitively. The selective $\beta_1$ adrenergic blockers are acebutolol, atenolol, bisoprolol, esmolol, metoprolol (Brunton et al., 2011; Katzung, 2009; Tripathi, 2014). Table 4.5 describes the mechanism of action, pharmacological effects, and clinical responses of selective $\beta_1$ blockers (second generation).

### 4.3.2.3.1 Acebutolol

It is an antagonist of $\beta_1$ receptor. It shows intrinsic sympathomimetic and also some membrane stabilizing activity. It is employed in treating hypertension, ventricular, and atrial cardiac arrhythmia, Smith Magnus syndrome, and myocardial infarction (Brunton et al., 2011). It is rapidly and well absorbed after an oral dose and forms diacetolyl, an active metabolite after extensive first-pass metabolism. It produces majority of the activity of the drug. The drug has a bioavailability of 35–50%. It has a half-life of 3 h. The metabolite half-life is 8–12 h.

The drug gets excreted through urine. The drug has a systemic availability is between 35% and 45%. The drug is about 11–19% plasma protein bound. Diacetolyl, the active metabolite has a potency comparable to acebutolol. Acebutolol and diacetolyl is excreted renally and it ranges from 25% to 45% after administration through oral route and 40–60% in the case of intravenous dose. It is lipophilic, so it can easily cross the BBB (Florey, 2008).

### 4.3.2.3.2 Atenolol

It is an antagonist of $\beta_1$ receptor and do not have membrane stabilizing activity and also the ISA (Brunton et al., 2011; Florey, 2008). It is hydrophilic so penetrate the BBB slightly. It can be used in the treatment of angina pectoris, hypertension, arrhythmias, coronary heart disease, and also in cardiac complications after a myocardial infarction. After an oral dose a little amount of drug is metabolized, which is about 10% of the dose. The bioavailability is 50%. The drug is excreted almost unchanged through urine. The drug shows low hydrophilicity and shows poor plasma protein binding which is lower than 5%. The excretion through urine is about 40% and in parenteral dosage, urinary excretion is about 75–100%. About 40–50% of the drug after an oral dose is excreted in the feces (Brunton et al., 2011; McDevitt, 1987).

### 4.3.2.3.3 Bisoprolol

It acts as a highly specific $\beta_1$ receptor antagonist. The drug does not show MSA and ISA. The drug is highly $\beta_1$ selective compared to metoprolol, atenolol and so on but not as nebivolol (Brunton et al., 2011). It is mainly used in treating hypertension which can be helpful in chronic heart

**TABLE 4.6** Pharmacodynamic and Pharmacokinetic Profile of Selective $\beta_1$ Blockers (Third Generation).

| Selective $\beta_1$ blocker | Pharmacodynamics | | | Clinical responses | Pharmacokinetics |
| --- | --- | --- | --- | --- | --- |
| | Pharmacological effects | Mechanism of action | Adverse effects | | |
| Betaxolol | Slight MSA | Selective antagonism at $\beta_1$ receptors without ISA. | Atrioventricular block; bradycardia; decreased libido; mask symptoms of hypoglycemia; sedation | Angina pectoris; glaucoma; HTN; in ophthalmology, it lessened aqueous humor production bringing forth a reduction in intraocular pressure | Route of administration: PO, IV; $t_{1/2}$: 4–5 h; Excretion: Feces |
| Celiprolol | ISA; weak bronchodilating and vasodilating effect | $\beta_2$ adrenoceptors agonism; peripheral $\alpha_2$ adrenoceptor antagonism; oxidative stress inhibition | Fewer side effects of the peripheral vascular system | Angina pectoris; HTN | |
| Nebivolol | Antioxidant properties and endothelial NO mediated vasodilation effects. It decreases the BP by dropping the PVR and elevates the stroke volume preserving the CO maintaining the systemic flow to the organs. It does not affect serum lipids but can increase insulin sensitivity | Selective antagonism at $\beta_1$ receptors | Fatigue; headache | HTN | Route of administration: PO; $t_{1/2}$: 10 h |

CO: cardiac output, HTN, hypertension; $t_{1/2}$, half-life; MSA, membrane stabilizing activity; IV, intravenous; ISA, intrinsic sympathomimetic activity; NO, nitric oxide.

failure, arrhythmias, and IHD. The adverse events include bradycardia, dizziness, hypotension, and fatigue. After an oral dose, the drug shows a bioavailability of 90% and metabolism occurs in liver; elimination occurs via renal route. The drug has a half-life of about 11–17 h (Brunton et al., 2011; Katzung, 2009).

### 4.3.2.3.4  Esmolol

It is an adrenergic $\beta_1$ receptor selective antagonist. It shows rapid onset and a short duration of activity. The ISA is little and it does not possess MSA (Brubacher, 2015; Brunton et al., 2011; Reynolds, 1986). It is administered intravenously for producing a short period $\beta$ blockade and in seriously ill patients. It can be used in treating supraventricular tachycardia, treating or preventing tachycardia during surgery. It can be used as an antiarrhythmic agent. It has a half-life of about 8 min. The $V_d$ is 2 L/kg. It is excreted through urine (Brunton et al., 2011).

### 4.3.2.3.5  Metoprolol

It is an adrenergic $\beta_1$ selective receptor antagonist without having ISA and MSA (Brunton et al., 2011; Florey, 2008). It is employed in the treatment of essential hypertension, tachycardia, heart failure, angina pectoris, vasovagal syncope, secondarily in the prevention of myocardial infarction, migraine treatment, and hyperthyroidism. It can be administered orally and used intravenously as metoprolol tartrate. It is helpful in chronic heart failure. After an oral dose, the drug has 40% bioavailability because of first-pass effect. Extensive metabolism occurs in the liver by the enzyme CYP2D6. The drug excretion takes place via

the urine after biotransformation and 85% drug excreted are metabolites. The drug has a half-life of about 4 h (Brunton et al., 2011; Florey, 2008).

### 4.3.2.4  SELECTIVE $\beta_1$ BLOCKERS (THIRD GENERATION)

These are agents that act by selectively blocking $\beta_1$ adrenoceptors with or without intrinsic sympathomimetic activity. The selective $\beta_1$ adrenergic blockers (third generation) are betaxolol, celiprolol, nebivolol (Brunton et al., 2011; Katzung, 2009; Tripathi, 2014). Table 4.6 describes the mechanism of action, pharmacological effects, and clinical responses of selective $\beta_1$ blockers (third generation).

### 4.3.2.4.1  Betaxolol

It is a selective antagonist of $\beta_1$ receptor and it does not show intrinsic sympathomimetic action. It shows a slight membrane stabilizing activity (Brunton et al., 2011; Mayama et al., 2013). It is employed in treating hypertension, angina and glaucoma. In ophthalmology, it lowers the production of the aqueous humor therefore reduces the intraocular pressure. It is well absorbed and shows a high bioavailability. The drug has a half-life of about 14–22 h. Drug elimination is partly by hepatic and renal route (Brunton et al., 2011; McDevitt, 1987).

### 4.3.2.4.2  Celiprolol

It is an antagonist of the adrenergic $\beta$ receptor and is cardioselective. It does not show MSA. It produces weak bronchodilating and vasodilating action due to $\beta_2$ agonism and produces a relaxation action on smooth muscle (Milne et al., 1991; Florey, 2008). It also blocks peripheral $\alpha_2$ receptors, promotes NO production, and inhibits oxidative stress. At $\beta_2$ receptor it produces intrinsic sympathomimetic action. It lowers the BP and the rate of the heart. The adverse events include headache, fatigue, dizziness, and swollen ankles. The drug absorption is fast from gastrointestinal tract after an oral dose and there is no first pass metabolism. It is employed in treating angina and hypertension (Brunton et al., 2011; Florey, 2008). The drug has a bioavailability of 30–70% and reaches a maximum plasma concentration at 2–4 h. About 25% of given dose of the drug is protein bound. The drug is not affected by first-pass effect and excretion is majorly via feces with about 11% through the urine. For parenteral administered drug 50% excretion happens via urine and 31% through feces.

### 4.3.2.4.3  *Nebivolol*

It is a $\beta_1$ selective adrenergic receptor antagonist with antioxidant properties and endothelial NO mediated vasodilation effects (Bakris et al., 2006; McNeely et al., 1999). It decreases PVR and shows a reduction in blood pressure. Elevation of stroke volume preserved the output of the heart and maintaining the flow of blood to organs. The drug does not alter the serum lipids but can produce a raise in the insulin sensitivity. It is employed in treating hypertension. Adverse events reported include fatigue, dizziness, paraesthesias, headache, and are comparable to the other $\beta$ blockers having vasodilator property (Bakris et al., 2006; McNeely et al., 1999).

## 4.3.3  NON-SELECTIVE $\alpha$ AND $\beta$ ADRENERGIC RECEPTOR ANTAGONISTS (THIRD GENERATION)

These drugs possess vasodilator effect other than $\beta$ receptor blockade. These effects may happen because of $\alpha_1$ receptor blockade, a rise in the NO levels, $\beta_2$ agonism, potassium channel opening, calcium entry, or antioxidant action as mentioned in Figure 4.6. Drugs coming under this class are bucindolol, carteolol, carvedilol, and labetalol (Brunton et al., 2011; Katzung, 2009; Tripathi, 2014; Toda, 2003). Table 4.7 describes the mechanism of action, pharmacological effects, and clinical responses of $\alpha$ and $\beta$ adrenergic receptor antagonist.

### 4.3.3.1  BUCINDOLOL

It is an adrenergic $\beta$ receptor non-selective antagonist which possesses intrinsic sympathomimetic activity. It produces weak adrenergic $\alpha_1$ antagonistic action. It is helpful in the patients with congestive cardiac failure. Following oral administration maximum plasma level reaches in about 2 h (Hershberger et al., 1990; Pollock et al., 1990; Woodley et al., 1987). The drug is 87% protein bound. It undergoes metabolism in the liver. The drug shows a half-life about 8 h. The major metabolite produced is 5-hydroxybucindolol (Meredith et al., 1985).

### 4.3.3.2  CARVEDILOL

It acts as an antagonist of adrenergic $\beta$ receptor and it has membrane stabilizing action but does not have ISA. It produces blockade of $\alpha_1$, $\beta_2$, and $\beta_1$ receptors. Additionally, it shows calcium channel antagonistic activity,

**TABLE 4.7** Pharmacodynamic and Pharmacokinetic Profile of Non-Selective α and β Adrenergic Receptor Antagonists (Third Generation).

| Non-selective α and β adrenergic receptor antagonists | Pharmacodynamics | | | Clinical responses | Pharmacokinetics |
|---|---|---|---|---|---|
| | Pharmacological effects | Mechanism of action | Adverse effects | | |
| Bucindolol | Mild vasodilatory property; decrease plasma renin activity and plasma norepinephrine | Selective antagonism at β receptors with ISA; weak adrenergic antagonism at $\alpha_1$ receptors | Bradycardia; diarrhea, dizziness; hyperglycemia; intermittent claudication | CHF | Route of administration: PO; Metabolism: Liver, major metabolite is, 5-hydroxybucindolol. |
| Carteolol | Non-selective β-adrenoceptor antagonist with partial agonistic action and no local anaesthetic activity; Reduction in raised intraocular pressure | Non-selective antagonism at β adrenoceptors with strong to moderate ISA | Atrioventricular block, bradycardia, bronchospasm, decreased libido, sedation, masks hypoglycemic symptoms | Glaucoma | Route of drug administration: PO; $t_{1/2}$: 5–5.4 h; Major metabolite: 8-hydroxy-carteolol Excretion: urine |
| Carvedilol | Anti-inflammatory and antioxidant effects; cardioprotective effects | Antagonism at $\beta_1$, $\beta_2$, and $\alpha_1$ receptor blockade with MSA | Failure of ejaculation; postural hypotension; rashes | CHF; treatment of HTN and in LVD after MI | Oral bioavailability: 30% |
| Labetalol | ISA; MSA; vasodilating activity; decreases BP and HR; anti-inflammatory and antioxidant effects | Competitive antagonism at both $\alpha_1$ and β receptors | Failure of ejaculation; liver damage; postural hypotension; rashes | Treat chronic HTN; intravenously for emergency hypertensive conditions; CHF | Absorption completion occurs from gut; Bioavailability: 20–40%; Metabolism: Liver; Excretion: Urine |

ISA, intrinsic sympathomimetic activity; BP, blood pressure; MSA, membrane stabilizing activity; HR, heart rate; HTN, hypertension; CHF, congestive heart failure; LVD, left ventricular dysfunction; PO, per oral; $t_{1/2}$, half-life.

anti-inflammatory, antiapoptotic, antiproliferative, and antiarrhythmic activity and antioxidant effects. It is mainly used for treating congestive heart failure and has cardioprotective effects. It can be used in treating hypertension and can be used for dysfunction of left ventricles occurring in succession to myocardial infarction (Van Tassell et al., 2008; Weir et al., 2005). Following oral administration, the drug shows rapid absorption and reaches the maximum concentration in the plasma at

about 1–2 h. The drug is about 95% protein bound and it undergoes metabolism in liver with the help of enzymes CYP2C9 and CYP2D6. The drug has a half-life of 7–10 h (Weir et al., 2005).

### 4.3.3.3    LABETALOL

It is a competitive antagonist of the β and the α1 receptor. The pharmacology is somewhat

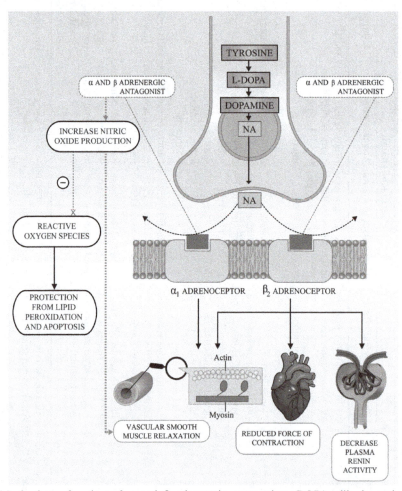

**FIGURE 4.6**    Mechanism of action of α and β adrenergic antagonists. DOPA, dihydroxyphenylalanine; NA, noradrenaline.

*Source*: Adapted from Tripathi (2014), Golan et al. (2012), and Stevens and Brenner (2008).

complicated and its properties include blockade of the $\beta_1$ and $\beta_2$ receptors, $\alpha_1$ receptors blockade, partial agonism of the $\beta_2$ receptors, suppression of norepinephrine uptake by the neurons (Brittain et al., 1976). Orally, it can be used in treating chronic hypertension and intravenously for emergency hypertensive conditions. It can show some vasodilator activity. The drug action is rapid, about 2–5 min following IV administration then lasting for 2–4 h. Bioavailability after administration via oral route is 20–40%. The drug half-life is 8 h (Martin et al., 1976).

## 4.4 FUTURE OPPORTUNITIES AND CHALLENGES

The study of adrenergic pharmacology, structure, and activity of endogenous catecholamines brought about development of various structurally and functionally similar drugs that mimic as well as block the sympathetic system (Golan et al., 2012). These agents are of diverse pharmacological profile. Elucidating the mechanism and functioning of these agents has provided us with a new insight which proceeded to development of novel molecules with ameliorated/better pharmacokinetic as well as pharmacodynamic profile. In spite of these developments, limited data is accessible on the pharmacokinetic and pharmacodynamics profile of several newer drugs. Research is currently going on in several fields like pharmacogenetic study of adrenergic receptors (Thompson et al 2005; Contopoulos et al., 2007) and further investigation is needed to perceive the pharmacological profile of such drugs. Pharmacogenetic studies of adrenergic system have brought a new insight in treating heart diseases, myasthenia gravis, asthma (Wanner et al., 2010).

## 4.5 CONCLUSION

This chapter discusses the agents capable of blocking the sympathetic system by competitively or non-competitively by binding to the adrenergic receptors such as $\alpha$ and $\beta$ receptors. Interaction of sympatholytics with adrenergic receptors based on their affinity to specific receptors such as $\alpha$ or $\beta$ receptors or their selectivity to specific subtypes of adrenergic receptors determines the pharmacological and therapeutic actions. Detailed knowledge of adrenergic receptors binding and several case studies are the key to understand the pharmacological and therapeutic activity of these categories of drugs to explore newer molecules with better pharmacokinetic profile, therapeutic uses, and fewer side effects.

## KEYWORDS

- BPH
- catecholamines
- epinephrine
- norepinephrine
- HDL
- sympatholytics

## REFERENCES

Acheampong, A. A.; Breau, A.; Shackleton, M.; Luo, W.; Lam, S.; Tang-Liu, D. D. Comparison of Concentration–Time Profiles of Levobunolol and Timolol in Anterior and Posterior Ocular Tissues of Albino Rabbits. *J. Ocul. Pharmacol. Ther.* **1995,** *11,* 489–502.

Altan-Yaycioglu, R.; Turker, G.; Akdol, S.; Acunas, G.; Izgi, B. The Effects of Beta-Blockers on Ocular Blood Fow in Patients with Primary Open Angle Glaucoma: A Color Doppler Imaging Study. *Eur. J. Ophthalmol.* **2001,** *11,* 37–46.

Andersson, K. E. Prostatic and Extraprostatic Alpha-Adrenoceptors-Contributions to the Lower Urinary Tract Symptoms in Benign Prostatic Hyperplasia. *Scand. J. Urol. Nephrol.* **1996**, *179* (Suppl), 105–111.

Antonaccio, M. J.; Gomoll, A. Pharmacology, Pharmacodynamics and Pharmacokinetics of Sotalol. *Am. J. Cardiol.* **1990**, *65* (2), 12–21.

Araki, H.; Itoh, M.; Nishi, K. Effects of Nipradilol on the Microvascular Tone of Rat Mesentery: Comparison with Other Beta Blockers and Vasodilators. *Arch. Int. Pharmacodyn. Ther.* **1992**, *318*, 47–54.

Arrizabalaga, P.; Montoliu, J.; Martinez, A. V.; Andreu, L.; López, J. P.; Revert, L. Increase in Serum Potassium Caused by $B_2$ Adrenergic Blockade in Terminal Renal Failure: Absence of Mediation by Insulin or Aldosterone. *Proc. Eur. Dial. Transplant. Assoc.* **1983**, *20*, 572–576.

Bakris, G. L.; Hart, P.; Ritz, E. Beta Blockers in the Management of Chronic Kidney Disease. *Kidney Int.* **2006**, *70* (11), 1905–1913.

Beta-Blocker Evaluation of Survival Trial Investigators. A Trial of the Beta-Blocker Bucindolol in Patients with Advanced Chronic Heart Failure. *N. Engl. J. Med.* **2001**, *344*, 1659–1667.

Black, J. W.; Duncan, W. A.; Shanks, R. G. Comparison of Some Properties of Pronethalol and Propranolol. *Br. J. Pharmacol. Chemother.* **1965**, *25*, 577–591.

Bloom, A. H.; Grunwald, J. E.; Dupont, J. C. Effect of One Week of Levobunolol Hcl 0.5% on the Human Retinal Circulation. *Curr. Eye Res.* **1997**, *16*, 191–196.

Bosem, M. E.; Lusky, M.; Weinreb, R. N. Short-Term Effects of Levobunolol on Ocular Pulsatile Fow. *Am. J. Ophthalmol.* **1992**, *114*, 280–286.

Boskabady, M. H.; Snashall, P. D. Bronchial Responsiveness to B-Adrenergic Stimulation and Enhanced B-Blockade in Asthma. *Respirology* **2000**, *5* (2), 111–118.

Bristow, M. R. What Type of Beta-Blocker Should be Used to Treat Chronic Heart Failure? *Circulation* **2000**, *102*, 484–486.

Brittain, H. G., Ed. *Analytical Profile Of Drug Substances And Excipients*; Academic Press: New Jersey, 2008; Vol. 21, pp 529–533.

Brittain, R. T.; Levy, G. P. A Review of the Animal Pharmacology of Labetalol, a Combined Alpha-and Beta-Adrenoceptor-Blocking Drug. *Brit. J. Clin. Pharmaco.* **1976**, *3* (4 Suppl. 3), 681–684.

Brock, J. A.; Cunnane, T. C. Studies on the Mode of Action of Bretylium and Guanethidine in Post-Ganglionic Sympathetic Nerve Fibres. *NS Arch. Pharmacol.* **1988**, *338* (5), 504–509.

Brubacher, J. R. *B-Adrenergic Antagonists*: *Goldfrank's Toxicologic Emergencies*, 10th ed.; Mcgraw-Hill Education: New York, 2015; pp 856–869.

Brunton, L., Ed. *Goodman and Gilman's the Pharmacological Basis of Therapeutics*, 12th Ed.; Mc Graw Hill: New York, 2011; pp 277–304.

Caine, M.; Perlberg, S.; Meretyk, S. A Placebo Controlled Double Blind Study of the Effect of Phenoxybenzamine in Benign Prostatic Obstruction. *Br. J. Urol.* **1978**, *50*, 551–554.

Chapple, C.; Andersson, K. E. Tamsulosin: An Overview. *World J. Urol.* **2002**, *19* (6), 397–404.

Chapple, C. R.; Wyndaele, J. J.; Nordling, J.; Boeminghaus, F.; Ypma, A. F.; Abrams, P. Tamsulosin, The First Prostate-Selective Alpha 1A-Adrenoceptor Antagonist. A Meta-Analysis of Two Randomized, Placebo-Controlled, Multicentre Studies in Patients with Benign Prostatic Obstruction (Symptomatic BPH). *Eur. Urol.* **1996**, *29*, 155–167.

Chiou, G. C.; Chen, Y. J. Effects of Antiglaucoma Drugs on Ocular Blood Flow in Ocular Hypertensive Rabbits. *J. Ocul. Pharmacol.* **1993**, *9*, 13–24.

Di, C. F. J.; Leinweber, F. J.; Szpiech, J. M.; Davidson, I. W. Metabolism of L-Bunolol. *Clin. Pharmacol. Ther.* **1977**, *22*, 858–863.

Dong, Y.; Ishikawa, H.; Wu, Y.; Yoshitomi, T. Vasodilatory Mechanism of Levobunolol on Vascular Smooth Muscle Cells. *Exp. Eye. Res.* **2007**, *84*, 1039–1046.

Draffan, G. H.; Lewis, P. J.; Firmin, J. L.; Jordan, T. W.; Dollery, C. T. Pharmacokinetics of Indoramin in Man. *Brit. J. Clin. Pharmaco.* **1976**, *3* (3), 489–495.

Epstein, M. O. Jr.; Hollenberg, N. K. B-Blockers and the Kidney: Implications for Renal Function and Renin Release. *Physiologist* **1985**, *28*, 53–63.

Farlex. *Farlex Partner Medical Dictionary*; Farlex Inc: USA, 2017.

Farmer, J. B.; Kennedy, I.; Levy, G. P.; Marshall, R. J. Pharmacology of AH 5158; A Drug Which Blocks Both A-and B-Adrenoceptors. *Brit. J. Pharmacol.* **1972**, *45* (4), 660–675.

Feuerstein, T. J.; Hertting, G.; Jackish, R. Endogenous Adrenaline as Modulator of Hipocampal Serotonin Release. Dual Effect of Yohimbine, Rauwolscine and Corynanthine as A-Adrenoceptor Antagonists and 5-HT-Receptor Agonists. *NS Arch. Pharmacol.* **1985**, *329*, 216–221.

Florey, K., Ed. *Analytical Profile of Drug Substances and Excipients*; Academic Press: New Jersey, 2008; Vol. 20, pp 241–301.

Florey, K. *Analytical Profile of Drug Substances and Excipients*; Academic Press: New Jersey, 2008; Vol. 9, pp 456–482.

Florey, K., Ed. *Analytical Profile of Drug Substances and Excipients*; Academic Press: New Jersey, 2008; Vol. 19, pp 21–26.

Florey, K., Ed. *Analytical Profile of Drug Substances and Excipients*; Academic Press: New Jersey, 2008; Vol. 16, pp 645–681.

Florey, K., Ed. *Analytical Profile of Drug Substances and Excipients*; Academic Press: New Jersey, 2008; Vol. 13, pp 1–23.

Florey, K., Ed. *Analytical Profile of Drug Substances and Excipients*; Academic Press: New Jersey, 2008; Vol. 12, pp 326–343.

Fonarow, G. C. Role of Carvedilol Controlled-Release in Cardiovascular Disease. *Exp. Rev. Cardiovasc. Ther.* **2009,** *7* (5), 483–498.

Fonseca, V. A. Effects of B-Blockers on Glucose and Lipid Metabolism. *Curr. Med. Res. Opin.* **2010,** *26*, 615–629.

Frishman, W. H.; Saunders, E. Beta-Adrenergic Blockers. *J. Clin. Hypertens.* **2011,** *13*, 649–653.

Fulton, B.; Wagstaff, A. J.; Sorkin, E. M. Doxazosin: An Update of its Clinical Pharmacology and Therapeutic Applications in Hypertension and Benign Prostatic Hyperplasia. *Drugs* **1995,** *49*, 295–320.

Ghostine, S. Y.; Comair, Y. G.; Turner, D. M.; Kassell, N. F.; Azar, C. G. Phenoxybenzamine in the Treatment of Causalgia: Report of 40 Cases. *J. Neurosurg.* **1984,** *60* (6), 1263–1268.

Golan, D. E., Ed. *Principles of Pharmacology the Pathophysiologic Basis of Drug Therapy*, 3rd Ed.; Lippincott Williams & Wilkins: Philadelphia, 2012; pp 132–145.

Grimm, R. H.; Flack, J. M.; Grandits, G. A.; Elmer, P. J.; Neaton, J. D.; Cutler, J. A.; Lewis, C.; Mcdonald, R.; Schoenberger, J.; Stamler, J. Long-Term Effects on Plasma Lipids of Diet and Drugs to Treat Hypertension. *JAMA* **1996,** *275* (20), 1549–1556.

Gugger, J. J. Antipsychotic Pharmacotherapy and Orthostatic Hypotension. *CNS Drugs* **2011,** *25* (8), 659–671.

Guthrie, S. K.; Hariharan, M.; Grunhaus, L. J. Yohimbine Bioavailability in Humans. *Eur. J. Clin. Pharmacol.* **1990,** *39* (4), 409–411.

Hague, C.; Chen, Z.; Uberti, M.; Minneman, K. P. $A_1$ Adrenergic Receptor Subtypes: Non-Identical Triplets with Different Dancing Partners? *Life Sci.* **2003,** *74*, 411–418.

Haller, J.; Makara, G. B.; Kruk, M. R. Catecholaminergic Involvement in the Control of Aggression: Hormones,

the Peripheral Sympathetic, and Central Noradrenergic Systems. *Neurosci. Biobehav. Res.* **1997,** *22*, 85–97.

Hanon, O.; Giacomino, A.; Troy, S.; Bernaud, C.; Girerd, X.; Weber, S. Efficacy of and Tolerance to Prolonged Release Prazosin in Patients with Hypertension and Non-Insulin Dependent Diabetes. *Ann. Cardiol. Angeiol.* **2000,** *49*, 390–396.

Hara, H.; Ichikawa, M.; Oku, H.; Shimazawa, M.; Araie, M. Bunazosin, A Selective $A_1$-Adrenoceptor Antagonist, as an Anti-Glaucoma Drug: Effects on Ocular Circulation and Retinal Neuronal Damage. *Cardiovasc. Drug. Rev.* **2006,** *23*, 43–56.

Hedner, T.; Edgar, B.; Edvinsson, L.; Hedner, J.; Persson, B.; Pettersson, A. Yohimbine Pharmacokinetics and Interaction with the Sympathetic Nervous System in Normal Volunteers. *Eur. J. Clin. Pharmacol.* **1992,** *43* (6), 651–656.

Hedner, T.; Edgar, B.; Edvinsson, L.; Hedner, J.; Persson, B.; Pettersson, A. Yohimbine Pharmacokinetics and Interaction with the Sympathetic Nervous System in Normal Volunteers. *Eur. J. Clin. Pharmacol.* **1992,** *43* (6), 651–656.

Hershberger, R. E.; Wynn, J. R.; Sundberg, L.; Bristow, M. R. Mechanism of Action of Bucindolol in Human Ventricular Myocardium. *J. Cardiovasc. Pharm.* **1990,** *15* (6), 959–967.

Hobbs, D. C.; Twomey, T. M.; Palmer, R. F. Pharmacokinetics of Prazosin in Man. *J. Clin. Pharmacol.* **1978,** *18* (8–9), 402–406.

Hofner, K. Tamsulosin: Effect on Sexual Function in Patients with LUTS Suggestive of BPO (Symptomatic BPH). *Eur. Urol.* **1998,** *33* (Suppl. 1), 129.

Holmes, B.; Sorkin, E. M. Indoramin. *Drugs* **1986,** *31* (6), 467–499.

Hugh, C. H. Jr.; Talmage, D. E. *Pharmacology and Physiology for Anesthesia*, 2nd Ed.; Elsevier: Philadelphia, 2013; pp 218–234.

Jaillon, P. Clinical Pharmacokinetics of Prazosin. *Clin. Pharmacokinet.* **1980,** *5* (4), 365–76.

Janczewski, P.; Boulanger, C.; Iqbal, A.; Vanhoutte, P. M. Endothelium-Dependent Effects of Carteolol. *J. Pharmacol. Exp. Ther.* **1988,** *247*, 590–595.

Javeed, N.; Javeed, H.; Javeed, S.; Moussa, G.; Wong, P.; Rezai, F. Refractory Anaphylactoid Shock Potentiated by Betablockers. *Cathet. Cardiovasc. Diagn.* **1996,** *39*, 383–384.

Juenemann, K. P.; Lue, T. F.; Fournier, G. R.; Tanagho, E. A. Hemodynamics of Papaverine-and Phentolamine-Induced Penile Erection. *J. Urol.* **1986,** *136* (1), 158–61.

Kanno, M.; Araie, M.; Koibuchi, H.; Masuda, K. Effects of Topical Nipradilol, A Beta Blocking Agent with

Alpha Blocking and Nitroglycerin-Like Activities, on Intraocular Pressure and Aqueous Dynamics in Humans. *Br. J. Ophthalmol.* **2000**, *84*, 293–299.

Kaplan, H. I.; Sadock, B. *Kaplan and Sadock's Synopsis of Psychiatry,* 8th Ed.; Lippincott Williams & Wilkins: Baltimore, 1998.

Katzung, B. G., Ed. *Basic & Clinical Pharmacology*, 13th Ed.; Mc Graw Hill: Singapore, 2015; pp 121–140.

Keating, G. M.; Jarvis, B. Carvedilol, A Review of its Use in Chronic Congestive Heart Failure. *Drugs* **2003**, *63*, 1697–1741.

Khilnani, G.; Khilnani, A. K. Inverse Agonism and its Therapeutic Significance. *Ind. J. Pharmacol.* **2011**, *43*, 492–501.

Kirsten, R.; Nelson, K.; Steinijans, V. W.; Zech, K.; Haerlin, R. Clinical Pharmacokinetics of Urapidil. *Clin. Pharmacokinet.* **1988**, *14* (3), 129–140.

Kozlovski, V. I.; Lomnicka, M.; Chlopicki, S. Nebivovol and Carvedilol Induce NO-Dependent Coronary Vasodilatation that is Unlikely to be Mediated by Extracellular ATP in the Isolated Guinea Pig Heart. *Pharmacol. Rep.* **2006**, *58* (Suppl), 103–110.

Lacey, R. J.; Berrow, N. S.; Scarpello, J. H.; Morgan, N. G. Selective Stimulation of Glucagon Secretion by B2-Adrenoceptors in Isolated Islets of Langerhans of the Rat. *Br. J. Pharmacol.* **1991**, *103* (3), 1824–1828.

Langtry, H. D.; Mammen, G. J.; Sorkin, E. M. Urapidil. *Drugs* **1989**, *38* (6), 900–940.

Leung, M.; Grunwald, J. E. Short-Term Effects of Topical Levobunolol on the Human Retinal Circulation. *Eye* **1997**, *11* (3), 371.

Lewis, P. J.; George, C. F.; Dollery, C. T. Clinical Evaluation of Indoramin, A New Antihypertensive Agent. *Eur. J. Pharmacol.* **1973**, *6* (4), 211–216.

Li, Q.; Kong, D.; Du, Q.; Zhao, J.; Zhen, Y.; Li, T.; Ren, L. Enantioselective Pharmacokinetics of Doxazosin and Pharmacokinetic Interaction Between the Isomers in Rats. *Chirality* **2015**, *27* (10), 738–744.

Maack, C.; Cremers, B.; Flesch, M.; Hoper, A.; Sudkamp, M.; Bohm, M. Different Intrinsic Activities of Bucindolol, Carvedilol and Metoprolol in Human Failing Myocardium. *Br. J. Pharmacol.* **2000**, *130*, 1131–1139.

Maccarthy, E. P.; Bloomfield, S. S. Labetalol: A Review of its Pharmacology, Pharmacokinetics, Clinical Uses and Adverse Effects. *Pharmacotherapy* **1983**, *3* (4), 193–217.

Majid, P. A.; Sharma, B.; Taylor, S. H. Phentolamine for Vasodilator Treatment of Severe Heart-Failure. *Lancet* **1971**, *298* (7727), 719–724.

Mangrella, M.; Rossi, F.; Fici, F. Pharmacology of Nebivolol. *Pharmacol Res.* **1998**, *38*, 419–431.

Martin, L. E.; Hopkins, R.; Bland, R. Metabolism of Labetalol by Animals and Man. *Brit. J. Clin. Pharmacol.* **1976**, *3* (4 Suppl. 3), 695–710.

Matsubara, Y.; Kanazawa, T.; Kojima, Y.; Abe, Y.; Kobayashi, K.; Kanbe, H.; Harada, H.; Momose, Y.; Terakado, S.; Adachi, Y.; Midgley, I. Pharmacokinetics and Disposition of Silodosin (KMD-3213). *Yakugaku. Zasshi.* **2006**, *126*, 237–245.

Mayama, C.; Araie, M. Effects of Antiglaucoma Drugs on Blood Flow of Optic Nerve Heads and Related Structures. *Jpn. J. Ophthalmol.* **2013**, *57* (2), 133–149.

Mayama, C.; Araie, M. Effects of Antiglaucoma Drugs on Blood Flow of Optic Nerve Heads and Related Structures. *Jpn. J. Ophthalmol.* **2013**, *57* (2), 133–149.

Mayama, C.; Araie, M. Effects of Antiglaucoma Drugs on Blood Flow of Optic Nerve Heads and Related Structures. *Jpn. J. Ophthalmol.* **2013**, *57* (2), 133–149.

Mcdevitt, D. G. Comparison of Pharmacokinetic Properties of Beta-Adrenoceptor Blocking Drugs. *Eur. Heart. J.* **1987**, *8* (Suppl M), 9–14.

Mcneely, W.; Goa, K. L. Nebivolol in the Management of Essential Hypertension: A Review. *Drugs* **1999**, *57*, 633–651.

Menkes, D. B.; Baraban, J. M.; Aghajanian, G. K.; Prazosin Selectively Antagonizes Neuronal Responses Mediated by $A_1$-Adrenoceptors in Brain. *NS Arch. Pharmacol.* **1981**, *317* (3), 273–275.

Meredith, P. A.; Kelman, A. W.; Mcsharry, D. R.; Vincent, J.; Reid, J. L. The Pharmocokinetics of Bucindolol and its Major Metabolite in Essential Hypertension. *Xenobiotica* **1985**, *15* (11), 979–985.

Michel, M. C. The Pharmacological Profile of the $A_{1a}$-Adrenoceptor Antagonist Silodosin. *Eur. Urol. Suppl.* **2010**, *9* (4), 486–490.

Miller-Keane, O. T.; O'Toole, M. T. *Miller-Keane Encyclopedia and Dictionary of Medicine, Nursing, and Allied Health,* 7th Ed.; Saunders: Philadelphia, 2003.

Milne, R. J.; Buckley, M. M. Celiprolol: An Updated Review of its Pharmacodynamic and Pharmacokinetic Properties. *Drugs* **1991**, *41*, 941–969.

Montorsi, F. Profile of Silodosin. *Eur. Urol. Suppl.* **2010**, *9* (4), 491–495.

Morales, A. Yohimbine in Erectile Dysfunction: The Facts. *Int. J. Impot. Res.* **2000**, *12* (S1), S70.

Morsman, C. D.; Bosem, M. E.; Lusky, M.; Weinreb, R. N. The Effect of Topical Beta-Adrenoceptor Blocking Agents on Pulsatile Ocular Blood Flow. *Eye* **1995**, *9*, (Pt 3), 344–347.

Nagaraja, S. I.; Liu, S. X.; Eichberg, J.; Bond, R. A. Treatment with Inverse Agonists Enhances Baseline Atrial Contractility in Transgenic Mice with Chronic B₂-Adrenoceptor Activation. *Br. J. Pharmacol.* **1999,** *127,* 1099–1104.

Nagarathnam, D.; Wetzel, J. M.; Miao, S. W.; Marzabadi, M. R.; Chiu, G.; Wong, W. C.; Hong, X. F. J.; Forray, C.; Branchek, T. A.; Heydom, W. E.; Chang, R. S.; Broten, T.; Schorn, T. W.; Gluchowski, C. Design and Synthesis of Novel A_{1a} Adrenoceptor-Selective Dihydropyridine Antagonists for the Treatment of Benign Prostatic Hyperplasia. *J. Med. Chem.* **1998,** *41,* 5320–5333.

Owen, J. A.; Nakatsu, S. L.; Fenemore, J.; Condra, M.; Surridge, D. H.; Morales, A. The Pharmacokinetics of Yohimbine in Man. *Eur. J. Clin. Pharmacol.* **1987,** *32* (6), 577–582.

Pedersen, M. E, Cockcroft, J. R. The Vasodilatory Beta-Blockers. *Curr. Hypertens. Rep.* **2007,** *9,* 269–277.

Poirier, L.; Lacourciere, Y. The Evolving Role of B-Adrenergic Receptor Blockers in Managing Hypertension. *Can. J. Cardiol.* **2012,** *28* (3), 334–340.

Pollock, S. G.; Lystash, J.; Tedesco, C.; Craddock, G.; Smucker, M. L. Usefulness of Bucindolol in Congestive Heart Failure. *Am. J. Cardiol.* **1990,** *66* (5), 603–607.

Prichard, B. N.; Cruickshank, J. M.; Graham, B. R. B-Adrenergic Blocking Drugs in the Treatment of Hypertension. *Blood Press.* **2001,** *10* (5–6), 366–386.

Pritchard, B. N.; Thorpe, P. Pindolol in Hypertension. *Med. J. Aust.* **1971,** *58,* 1242.

Rath, G.; Balligand, J. L.; Dessy, C. Vasodilatory Mechanisms of Beta Receptor Blockade. *Curr. Hypertens. Res.* **2012,** *14,* 310–317.

Reynolds, R. D.; Gorczynski, R. J.; Quon, C. Y. Pharmacology and Pharmacokinetics of Esmolol. *J. Clin. Pharmacol.* **1986,** *26* (Suppl. A), A3–A14.

Rosei, E. A.; Rizzoni, D. Metabolic Profile of Nebivolol, A B-Adrenoceptor Antagonist with Unique Characteristics. *Drugs* **2007,** *67* (8), 1097–1107.

Rossi, M.; Roumeguere, T. Silodosin in the Treatment of Benign Prostatic Hyperplasia. *Drug Des. Dev. Ther.* **2010,** *4,* 291.

Rowland, D. L; Kallan, K.; Slob, A. K. Yohimbine, Erectile Capacity, and Sexual Response in Men. *Arch. Sex. Behav.* **1997,** *26* (1), 49–62.

Ruffolo, R. R. Jr.; Hieble, J. P. Alpha-Adrenoceptors. *Pharmacol. Ther.* **1994,** *61,* 1–64.

Russell, W.; Metcalfe, I.; Tonkin, A.; Frewin, D. The Preoperative Management of Phaeochromocytoma. *Anaesth. Intens. Care* **1998,** *26* (2), 196–200.

Schmetterer, L.; Strenn, K.; Findl, O.; Breiteneder, H.; Graselli, U.; Agneter, E.; Eichler, H. G.; Wolzt, M. Effects of Antiglaucoma Drugs on Ocular Hemodynamics in Healthy Volunteers. *Clin. Pharmacol. Ther.* **1997,** *61,* 583–595.

Schwinn, D. A.; Price, D. T.; Narayanm, P. A_1-Adrenoceptor Subtype Selectivity and Lower Urinary Tract Symptoms. *Mayo. Clin. Proc.* **2004,** *79* (11), 1423–1434.

Seideman, P. Pharmacokinetic and Dynamic Aspects of A–Adrenoceptor Blockade in Hypertension. *J. Intern. Med.* **1982,** *212* (S665), 61–66.

Singh, B. N.; Whitlock, R. M.; Comber, R. H.; Williams, F. H.; Harris, E. A. Effects of Cardioselective B-Adrenoceptor Blockade on Specific Airways Resistance in Normal Subjects and in Patients with Bronchial Asthma. *Clin. Pharmacol. Ther.* **1976,** *19,* 493–501.

Sponer, G.; Feuerstein, G. Z. The Adrenergic Pharmacology of Carvedilol. *Heart. Fail. Rev.* **1999,** *4* (1), 21–28.

Stanaszek, W. F.; Kellerman, D.; Brogden, R. N.; Romankiewicz, J. A. Prazosin Update. *Drugs* **1983,** *25* (4), 339–84.

Steinegger, E.; Hansel, R. *Lehrbuch Der Pharmakognosie Und Phytopharmazie*; Springer: Berlin, 1988; pp 542–546.

Stott, M. A.; Abrams, P. Indoramin in the Treatment of Prostatic Bladder Outflow Obstruction. *BJU Int.* **1991,** *67* (5), 499–501.

Sugiyama, T.; Kida, T.; Mizuno, K.; Kojima, S.; Ikeda, T. Involvement of Nitric Oxide in the Ocular Hypotensive Action of Nipradilol. *Curr. Eye Res.* **2001,** *23,* 346–351.

Taboulet, P.; Cariou, A.; Berdeaux, A.; Bismuth, C. Pathophysiology and Management of Self-Poisoning with Beta-Blockers. *J. Toxicol. Clin. Toxicol.* **1993,** *31,* 531–551.

Toda, N. Vasodilating Beta-Adrenoceptor Blockers as Cardiovascular Therapeutics. *Pharmacol. Ther.* **2003,** *100,* 215–234.

Tuttle, R. R.; Mills, J. Dobutamine: Development of a New Catecholamine to Selectively Increase Cardiac Contractility. *Circ. Res.* **1975,** *36,* 185–196.

Tyrer, P. Anxiolytics not Acting at the Benzodiazepine Receptor: Beta Blockers. *Prog. Neuro-Psychoph.* **1992,** *16* (1), 17–26.

Van, B. E. M.; Bacon, D. R.; Fahrenbach, W. H. Ciliary Vasoconstriction After Topical Adrenergic Drugs. *Am. J. Ophthalmol.* **1990,** *109,* 511–517.

Van, T. B.W.; Rondina, M. T.; Huggins, F.; Gilbert, E. M.; Munger, M. A. Carvedilol Increases Blood Pressure Response to Phenylephrine Infusion in Heart Failure Subjects with Systolic Dysfunction: Evidence of Improved Vascular A1-Adrenoreceptor Signal Transduction. *Am. Heart. J.* **2008,** *156* (2), 315–321.

Veld, M. I.; Schalekamp, M. A. How Intrinsic Sympathomimetic Activity Modulates the Haemodynamic Responses to Beta-Adrenoceptor Antagonists. A Clue to the Nature of their Antihypertensive Mechanism. *Br. J. Clin. Pharmacol.* **1982,** *13*, 245S–57S.

Volans, G. N.; Jeffereys, D.; Latham, A. N.; Frost, T. Pharmacokinetics of Oral Indoramin. *Curr. Med. Res. Opin.* **1982,** *8* (1), 51–53.

Weir. R. A.; Dargie, H. J. Carvedilol in Chronic Heart Failure: Past, Present and Future. *Future Cardiol.* **2005,** *1* (6), 723–734.

Wilde, M. I.; Fitton, A.; Mctavish, D. Alfuzosin. *Drugs* **1993,** *45* (3), 410–429.

Williams, M. E.; Gervino, E. V.; Rosa, R. M.; Landsberg, L.; Young, J. B.; Silva, P. Catecholamine Modulation of Rapid Potassium Shifts During Exercise. *N Engl. J. Med.* **1985,** *312*, 823–827.

Woodley, S. L.; Gilbert, E. M.; Anderson, J. L.; O'connell, J. B.; Deitchman, D.; Yanowitz, F. G.; Mealey, P. C.; Volkman, K.; Renlund, D. G.; Menlove, R. Beta-Blockade with Bucindolol in Heart Failure Caused by Ischemic Versus Idiopathic Dilated Cardiomyopathy. *Circulation* **1991,** *84* (6), 2426–2441.

Zhong, H.; Minneman, K. P. $A_1$ Adrenoceptor Subtypes. *Eur. J. Pharmacol.* **1999,** *375*, 261–276.

# PART II

## Drugs Affecting the
## Central Nervous System

# CHAPTER 5

# Pharmacological Management of Alzheimer's Disease

RAKESH KUMAR, RAJAN KUMAR, ABHINAV ANAND, NEHA SHARMA, and
NAVNEET KHURANA*

*Lovely Professional University, Punjab, India*

*Corresponding author. E-mail: navi.pharmacist@gmail.com*

## ABSTRACT

Alzheimer's disease (AD) is a neurodegenerative disorder, which is characterized by severe progressive loss of neurons in the central nervous system. AD is the basic type of dementia among the geriatrics that slowly decreases memory and alters the thought process, reduces understanding the simple tasks of the daily routine, and with passing time, it becomes more problematic. A neuronal destruction occurs in the AD with less number of viable neuron cells. As it is progressive in nature, it ultimately leads to shrinkage of the brain. The exact pathology of the AD is still unknown but it involves the accumulation of the amyloid-$\beta$ (A$\beta$) and neurofibrillary tangles (NFTs) in the hippocampus and cortex region of the brain that both A$\beta$ and NFTs lead degeneration and alteration function of the neurons. Various medicines are recommended for the treatment of the AD, but these medicines provide only symptomatic relief and are not able to treat the underline pathological cause. In the present chapter, various drugs which are used for the management of AD with their mechanism and side effects along with the drugs which are under investigation are discussed.

## 5.1 INTRODUCTION

The word dementia, obtained from the Latin word "de" and "mente" means "out (de) of one's mind (mente)." Dementia is a common term which is characterized by deterioration of cognitive function, involving loss of memory, impairment of language and disorientation as well as a degree of mental ability in the patient to interfere daily life (Boller and Forbes, 1998). Alzheimer's disease (AD) is the general type of dementia. AD accounts about 80% cases that are more than other all type of dementia such as 25% vascular dementia, 15% Lewy body dementia and frontotemporal (FT) dementia (Boller and Forbes, 1998; Alzheimer's Association, 2015).

AD was described first time by the neuropathologist and the German psychologist Alois Alzheimer in 1906. At the beginning of the 21st century, the highly pervasive dementia type in people older than 60 years was acknowledged (World Health Organization, 2012). In 2016, worldwide people suffering from dementia were estimated to be more than 47.5 million. By 2030, it is being estimated globally up to 75.6 million people with dementia (World Health

Organization, 2016). This is a loss of neuron disorder with a progressive decline in memory and cognitive function in the brain that usually appears in late adulthood. In this disease, the continuous neuronal damage occurs in the hippocampus and cerebral cortex area of the brain which leads to reduce the mass of the brain (Uddin and Amran, 2018; Perl, 2010).

AD is among the five most basic cause of mortality in the United States (Berríos-Torres et al., 2017). As it is a disorder of above the 60 years age, it appears rarely at 40 years and 50 years age. As per clinical and epidemiology studies, about two lakh people of age 65 years are suffering from this disorder. By 2050, as per estimate, a new case of the AD is supposed to develop at every 33 s (Alzheimer's Association, 2014; Prince et al., 2016). The main objective is to provide information regarding the etio-pathogenesis and available drug therapy of the AD. Along with FDA (Food and Drug Administration) approved drugs, some drugs that are investigational drugs and recent drugs which are used as the symptomatic treatment of the AD along with the mechanism of action, pharmacokinetics, their adverse effects and future opportunities are discussed in this chapter.

## 5.2 PATHOPHYSIOLOGY OF ALZHEIMER'S DISEASE

The extracellular senile plaques composed mainly of amyloid β (Aβ) and neurofibrillary tangles (NFTs) composed of tau protein are represented as the major pathological hallmarks of the AD. The relatively less prominent familial form of the AD which may present in the early years of one's life is the consequence of a mutation in any of three genes, namely *APP*, *PS-1*, or *PS-2*. The sporadic variant generally appears after 65 years of age that oversees for a large majority of the cases. Mostly, it occurs from an association of both the genetic and the environmental factors (Aprahamian et al., 2013).

The only definite risk factors for the sporadic are the E4 allele of apolipoprotein E and senescence. The plaques are constituted chiefly of the neurotoxic peptide. The Aβ is the by-product of the cleavage of a large precursor protein (APP) by enzymes (β-secretase and γ-secretase) (Allen et al., 2011). However, the Aβ is invisible in case of APP, which is cleaved by the enzyme α-secretase. NFTs mainly comprise of the tau protein that is a microtubule-associated protein. The function of tau protein is to facilitate the neuronal transport system by binding with microtubules in cells. In case of the AD, tau is uncoupled microtubules and aggregates into tangles by inhibiting transport and resulting in microtubule disassembly. It also depends on the phosphorylation of tau(Allen et al., 2011; Ravi et al., 2018). In a hypothesis, it has been reported that the mitochondrial dysfunction leads to amyloidosis, alter the kinase and tau phosphatase enzyme, and cell cycle re-entry (Moreira, 2018; Shoshan-Barmatz et al., 2018).

## 5.3 ETIOLOGY OF ALZHEIMER'S DISEASE

The strategy behind the formation of the AD is not yet explained properly. But, the certain factors are such as variation in tau phosphorylation, oxidative stress, calcium metabolism alteration, neuroinflammation, protein processing and atypical energy metabolism. It is the accumulation of Aβ observed to be an essential pathological mark of the AD (Munoz and Feldman, 2000). The cause of the AD is based on aging, genetic factors, environmental factors, drug-induced factors, mitochondrial dysfunction, vascular factors, infectious agents, and decline of the immune system. This disorder has been proposed as a multi-factorial disorder in which external and internal factors act together to ameliorate the rate of normal aging "allostatic load." The progressive neurodegeneration and blood vessels result in the formation of Aβ and tau protein cluster which spread via cell to cell, a transfer from the medial temporal lobe to affect association areas of the brain and then the primary sensory areas. The heterogeneity characteristic of the AD is due to the differentiation in the routes (Armstrong, 2013).

## 5.4 STAGES AND SYMPTOMS OF ALZHEIMER'S DISEASE

The stages and symptoms of AD progress are variables that get worse with time. Changes in the brain starts years before the appearance of AD symptoms. This time period, which lasts for a few years, is called as a preclinical AD. The AD occurs slowly in the form of three stages, namely mild AD (early-stage), moderate AD (intermediate-stage) and severe AD (late-stage) (Alzheimer's Association, 2017). In Table 5.1, the symptoms are associated with each stage.

## 5.5 DIAGNOSIS OF ALZHEIMER'S DISEASE

Diagnosis involves knowing some questions from the patient and their friends, colleague, and relatives. The patient is diagnosed by scanning and imaging techniques such as positron emission tomography (PET), magnetic resonance imaging (MRI), and computerized tomography (CT) scan of the temporal lobe of brain. This can be carried out depending upon the need, severity of symptoms and the gravity of the situation (Armstrong, 2013). In fact, as per the presence of other severe diseases in geriatric people, several cases of the AD and other types of dementia fall under the line. The new technologies come in research studies on AD that is placing toward diagnosis at the earlier stages when symptoms are mild or even absent. Although the drugs developing for AD has demonstrated a monstrous challenge, while according to the experts to treat the AD at early stages will stand with the best opportunity (Munoz and Feldman, 2000). In Table 5.2, the very common diagnostic procedures along with their purposes are shown.

**TABLE 5.1** The Stages and Symptoms of Alzheimer's Disease (Alzheimer's Association, 2017).

| Stage | Symptoms |
| --- | --- |
| Preclinical AD | The brain alteration occurs with no symptom. |
| Mild cognitive impairment due to AD | That is considered as a decrease in the memory and decline thinking skills, but no harm of function. |
| Mild AD | Unable to recognize the objects and not able to perform normal task. |
| Moderate AD | Increased risk of diverging, behavioural and personality changes, forgetting history events, moody, confused, the sleep patterns change. |
| Severe AD | Decrease of awareness about current experiences, fail to judge, increased speaking problem, facing trouble during walking, sitting, and engulfing, trouble for the patient who needs round the clock help for daily activities. |

**TABLE 5.2**  Diagnostic Procedure of Alzheimer's Disease.

| Procedure | Purposes |
|---|---|
| Medical history | Gives an image of cognitive and behavioral alterations over time. This helps doctors to search that the cognitive decrement is due to medication. |
| Family history | To know about the disease history of the family members and individuals at higher risk of forming the AD. |
| Physical and neurological exam | To screen the physical medical monitoring such as blood pressure, hearing, vision, sensation, reflexes, muscle strength, and to find an abnormality that reflects causes of cognitive loss other than the AD. |
| Lab tests (glucose, hormones, vitamin deficiencies, electrolytes, etc.) | To diagnose metabolic disorders, this can guide to cognitive symptoms after lab tests. |
| Tests for cognitive, functional, and psychological status | To assess daily routine functions, to check depression, that can cause issues of memory, assess cognitive symptoms and other types of dementia which can manifest in various ways. |
| MRI or CT scan | Look for brain structural lesions, atrophy and stroke. |
| PET scan | PET may be used to diagnose the certain forms of dementia including AD. |
| Cerebrospinal fluid analysis | Cerebrospinal fluid analysis is used to know the free-floating proteins in brain that form Aβ plaques and NFTs of AD. |

## 5.6  ANTI-ALZHEIMER'S DRUGS

The treatment approach includes use of cholinergic agents with the expectation of enhancing brain acetylcholine levels. Cholinesterase inhibitors block the enzyme-induced metabolism of intra-synaptic acetylcholine. The barrier of developing muscarinic agonist includes finding a medicine with specific M1 subtype agonistic activity which is not affecting other muscarinic receptor subtypes, thus preventing adverse events associated with these receptors (Schneider, 2013). The drugs for the treatment of AD are represented in Figure 5.1.

### 5.6.1  APPROVED DRUGS

#### 5.6.1.1  CHOLINESTERASE INHIBITORS

##### 5.6.1.1.1  Donepezil

Donepezil is an extended-performing reversible acetylcholinesterase inhibitor. It shows selectivity for cholinesterase enzyme within the brain. Once a day administration, donepezil has protracted long time inhibition of cholinesterase enzyme. Donepezil is certified for the symptomatic remedy of dementia in mild to the moderately excessive AD (Jones, 2002). The preliminary recommended 5 mg dose of donepezil should be given in the evening once a day for as a minimum 1 month to maintain the therapeutic response. Then after, the dose may be enhanced to 10 mg once daily. The two-phase III clinical trials demonstrated proof of efficacy to get FDA approval. The 6 and 12 month's duration randomized trials were conducted in both chronic impaired and nursing home patients. The barrier of developing muscarinic agonist includes finding a medicine with specific M1 subtype agonistic activity which is not affecting other muscarinic receptor subtypes, thus preventing adverse events associated with these receptors (Jones, 2002).

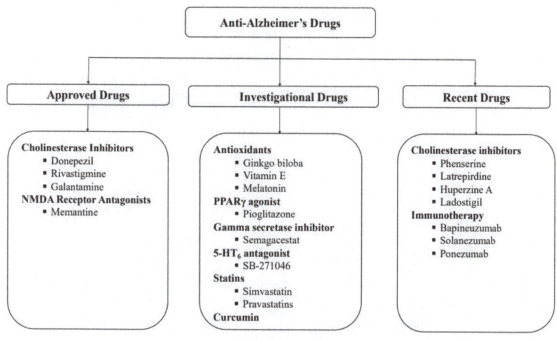

**FIGURE 5.1**   Classification of drugs used for Alzheimer's disease.

#### 5.6.1.1.1.1   Pharmacokinetics

The relative oral bioavailability of donepezil is about 100% with 3–4 h peak concentration. Donepezil is metabolized largely in the liver and eliminated in unchanged form in the urine. It has a long elimination half-life up to 70 h, and steady-state occurs in approximately 2 weeks. 10 mg per day dose was maintained for the patients with intermediate to chronic AD and might benefit from an enhanced dose (Perry et al., 1982; Davies and Maloney, 1976).

#### 5.6.1.1.1.2   Mechanism of Action

Donepezil is an acetylcholinesterase (AChE) inhibitor (Fig. 5.2) with dose-dependent activity. Donepezil binds to inhibit the hydrolysis of acetylcholine that reversibly inactivates cholinesterase by enhancing acetylcholine concentration at cholinergic synapses. The actual mode of action of donepezil in the AD patient is not fully understood (Perry et al., 1982). The symptoms of the AD are related to the cholinergic deficit, mainly in the brain (cerebral cortex region) and other area of the brain. The AD implicates a substantial decline in the elements of the cholinergic system (Davies and Maloney, 1976).

#### 5.6.1.1.1.3   Adverse Effects

During a clinical trial, the general side effects were observed, such as nausea, vomiting, and diarrhoea. Loss of appetite, difficulty in sleeping, and muscle cramp were also observed as per marketed brand Aricept® (Davies and Maloney, 1976; Perry et al., 1982).

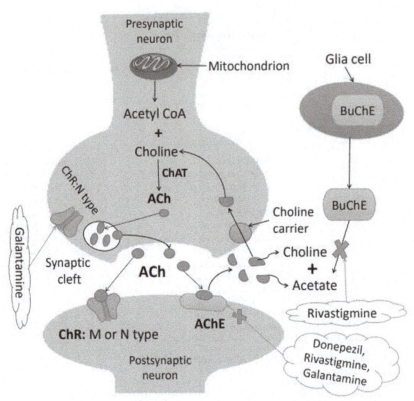

**FIGURE 5.2** Mechanism action of donepezil, rivastigmine, and galantamine inhibit AChE enzyme, rivastigmine also inhibits the BuChE enzyme and galantamine potentiate the ChR. ChR, N type channel; ChAT, choline acetyltransferase; ACh, acetylcholine; ChR, cholinergic receptor; AChE, acetylcholinesterase; BuChE, butyrylcholinesterase; M, muscarinic type receptor; N, nicotinic type receptor.

### 5.6.1.1.2  *Rivastigmine*

Rivastigmine is used for mild to the intermediate stage of the AD treatment. The pseudo-irreversible mechanism that results in prolonged inhibition of AChE. Rivastigmine, primarily 1.5 mg twice daily dose, is assigned treatment for the AD patient, which would be initiated by the experienced physician in diagnosis (Winblad et al., 2007). This therapy is given to regularly monitor patient of the AD. If it is tolerated well, the dose can be enhanced with a maintenance dose of 3–6 mg twice daily. In two published clinical trials, it was showing efficacy, and doses were titrated weekly over 7 weeks with one of two dosage ranges, 1–4 mg per day or 6 mg per day and 12 mg per day. The low dose of rivastigmine was not accepted, possibly contributing to less tolerability and considerably high adverse effects. The clinical study was conducted on 1195 patients with a moderately severe AD and based on a placebo-controlled study differentiating among the rivastigmine dose of 17.4, 9.5, and 6 mg administered orally twice daily with mini-mental state examination scores of 10–20 over 6 months. All formulations were demonstrated to have efficacy; however, fewer side effects obtained with the patch formulations (Winblad et al., 2007).

### 5.6.1.1.2.1  Pharmacokinetics

Rivastigmine is administered orally and has 40% plasma protein binding and well absorbed at dose 3 mg 2 times a day. The elimination half-life is less than 2 h, and enzyme inhibition lasts about 9 h. This drug is unmetabolized by the hepatic. Rivastigmine is slowly hydrolysed and then excreted by the renal. The peak of plasma concentration is 1 h and the peak of cerebrospinal fluid concentration at 1.4–3.8 h. It is extrahepatic metabolism, which is not likely to have significant pharmacokinetic interactions (Cummings et al., 2007; Jann et al., 2002).

### 5.6.1.1.2.2  Mechanism of Action

Rivastigmine is selective for AChE and BChE (Fig. 5.2) within the brain. But, the precise mechanism of rivastigmine is unknown (Winblad et al., 2007).

### 5.6.1.1.2.3  Adverse Effects

The common side effects were observed during clinical studies such as loss of weight, loss of appetite, nausea, and vomiting (Winblad et al., 2007; Cummings et al., 2007; Jann et al., 2002).

### 5.6.1.1.3  Galantamine

Galantamine is used in cognitive decline in mild to the moderate AD. Galantamine acts on nicotinic receptors as an allosteric modulator. It is possibly increasing cholinergic transmission by presynaptic nicotinic stimulation (Birks, 2012). At the 8 and 16 mg dose twice daily, the efficacy of galantamine was demonstrated with some side effects at the decreased dose in at least four randomized trials for 3–6 months. As per Cochrane review, it was concluded that galantamine showed positive effects for 3–6 months study with no additional changes at the overdose of 16 mg per day (Birks, 2012; Andrea et al., 2013).

### 5.6.1.1.3.1  Pharmacokinetics

Galantamine has approximately 90% bioavailability with a peak of plasma concentration approximately 1 h and the half-life of about 7 h. Galantamine undergo hepatic metabolism, and excreted by renal system. Galantamine has a minimum effect on the metabolism with other drugs. However, the drug paroxetine and ketoconazole are the inhibitors of CYP2D6 or 3A4 that may decrease clearance; enhance the bioavailability and plasma levels of galantamine (Birks, 2012; Andrea et al., 2013).

### 5.6.1.1.3.2  Mechanism of Action

Galantamine is a reversible competitive inhibitor of enzyme AChE (Fig. 5.2). The competitive inhibitors are potentially less active within brain areas having high acetylcholine level and more active in other brain areas (Andrea et al., 2013).

### 5.6.1.1.3.3  Adverse Effects

Nausea, vomiting, diarrhoea, weight loss and anorexia are the most common side effects due to cholinesterase inhibitors, which are cholinergic mediated. The early cholinergic effects are very common at the primarily dosing. Decreasing the dose temporarily may decline the reemergence of acute cholinergic adverse effects. But, several patients may tend to become tolerant to the adverse effects (Andrea et al., 2013).

### 5.6.1.2  NMDA RECEPTOR ANTAGONISTS

### 5.6.1.2.1  Memantine

In late 2003, memantine was approved by the FDA for intermediate to severe AD. Memantine is usually prescribed to ameliorate the memory, attention, language, and the ability to perform common tasks. It was the first drug for the AD that is approved in the United States as the NMDA receptor antagonist. Memantine is particularly given to treat moderate-to-severe AD. In 2005, the FDA declined the approval of memantine for the mild AD. Memantine is characterized as a moderate-affinity and uncompetitive NMDA receptor antagonist. The rationale for the use of memantine is that it may protect against overstimulation of NMDA receptors. It may also occur in the AD as well as consequent glutamate and calcium-mediated neurotoxicity. For the approval, it was shown the positive outcomes during 6 months of placebo-controlled clinical trials. In one clinical trial, cholinesterase inhibitors were not permitted, but in a second clinical trial, donepezil was received by all patients for at least 6 months duration before being randomized to memantine or the placebo. In third moderate to severe AD clinical trial, it was not shown the significant effects for memantine; however, the trials were similar to the design of cholinesterase inhibitors trials that included only patients with mini-mental state examination (Mount and Downton, 2006; McShane et al., 2006).

### 5.6.1.2.1.1 Pharmacokinetics

Memantine has a better rate of absorption to approach 100% bioavailability. The maximum plasma protein binding concentration is 3–7 h, with 60–80 h elimination half-life. It has no food interaction and is broadly distributed throughout the body. Memantine has minimum metabolism that occurs in the liver and excreted unchanged in the urine by kidneys (Mount and Downton, 2006; McShane et al., 2006).

### 5.6.1.2.1.2 Mechanism of Action

Memantine is known as NMDA receptor antagonist (Fig. 5.3). But, the therapeutic mode of

action is not known properly. Memantine did not show apparent pharmacological activity. At the higher level of glutamate triggers the receptor and potentiates the ion channel opening. It is hypothesized that the memantine inhibits NMDA receptor and prevents calcium influx, depolarization, and high activation of the neuron. The glutamatergic activity of memantine could be seen. In 6 months of the clinical trial, it was not observed to indicate the short-term and symptomatic effect (Mount and Downton, 2006; McShane et al., 2006).

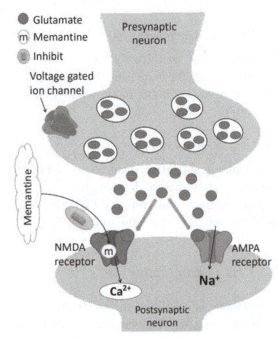

**FIGURE 5.3** NMDA receptor antagonist effect of memantine. NMDA, *N*-methyl-D-aspartate; AMPA, α-amino-3-hydroxy-5-methyl-4-isoxazolepropionic acid.

### 5.6.1.2.1.3 Adverse Effects

The common side effects are included as confusion, diarrhoea, headache, dizziness, occasional hallucinations, and somnolence. Memantine may decrease the cholinergic effects of donepezil. It is observed to have no interaction with cholinesterase inhibitors (Mount and Downton, 2006; McShane et al., 2006).

### 5.6.2 INVESTIGATIONAL DRUGS

#### 5.6.2.1 ANTIOXIDANTS

##### 5.6.2.1.1 Ginkgo biloba

*Ginkgo biloba* belongs to family Ginkgoaceae, a member of the gymnosperms. The rest plants of the Ginkgoaceae family are now extinct and found as fossils. *Ginkgo biloba* is a small tree that originated in China but now planted worldwide as an ornamental plant. It is also cultivated in Korea, France, and the United States. The major chemical constituents present in *Ginkgo biloba* as the mixture of flavonoids and terpenoids. The dried leaves of *Ginkgo biloba* contain ginkgolides (A, B, C, J, and M), bilobalide and 0.1–0.25% terpene, and lactones. 30–40% of the mixture is bilobalide whereas ginkgolide A is present in maximum amount and accounts about 30% of the mixture. The nature of ginkgolides is diterpenoid while bilobalide is sesquiterpenoid (Paul, 2002).

##### 5.6.2.1.1.1 Pharmacokinetics

After oral intake, the intestinal microflora hydrolyzes the flavonoid glycoside into sugar moiety and aglycone, which further get absorbed. In human, a minor part of the flavonoid is absorbed without hydrolysis. The major metabolic pathway is glucuronidation in the intestine (Hakk and Larsen, 2002; Hollman et al., 1997; Mauri et al., 1999).

##### 5.6.2.1.1.2 Mechanism of Action

The mode of action includes antioxidant effect, radical scavenging, alteration of the membrane fluidity and inhibition of the glucocorticoid synthesis and platelet activating factors (Diamond and Bailey, 2013; Pereira et al., 2013).

##### 5.6.2.1.1.3 Medicinal Uses

The extract of the leaves of *Ginkgo biloba* has a long history, and in today's world, it is the most commonly used herbal medicinal product in Europe and the United States. The extract of this plant is basically used in cerebral vascular disease and senile dementia. It is reported to provide positive results in the treatment of the AD. It is also reported to improve the cerebral blood circulation. The bilobalide is not able to control the platelet activating factors, but still, it provides a neuroprotective effect. EGb761® is the standardized extract of the plant. The extract is available in a suitable form such as film-coated tablets, oral liquids, and parenteral solutions (Heinonen and Gaus, 2015).

##### 5.6.2.1.1.4 Adverse Effects

Mostly it is considered that the natural herbs don't have any side effects, but it is not true. As they all have the chemicals, and different chemicals have different kind of action with the targets. So, *Ginkgo biloba* also reported having a certain adverse effect such as gastrointestinal upset, headache, dizziness. In the animal study, it is reported to have carcinogenic at high dose level (Diamond and Bailey, 2013; Ulbricht et al., 2008).

#### 5.6.2.1.2 Vitamin E

Vitamin E is a fat-soluble vitamin that is chemically known as tocopherol. Vitamin E was discovered by Evans and Bishop in

1922. There are total eight forms of vitamin E, namely, alfa, beta, gamma, and delta form of tocopherol and tocotrienol. It is present in fatty food material, and it can get stored in the fat tissues of the body. So, taking vitamin E on daily basis is not required. The best source of vitamin E is vegetable oils (Rizvi et al., 2014).

### 5.6.2.1.2.1    Pharmacokinetics

Plasma concentration of vitamin E directly proportions to its absorption. As it is fat soluble, so it is directly absorbed from a diet in the intestinal lumen. Then it gets incorporated into chylomicrons and distributed in the body (Boccardi et al., 2016; Yap et al., 2001).

### 5.6.2.1.2.2    Medicinal Uses

Vitamin E is highly beneficial for the body as it has the excellent antioxidant properties, helping to maintain the integrity of the cell membrane, regulate platelets aggregation and also control the activity of protein kinase C (Burton et al., 1983; Howard et al., 2011). Now, it is reported that tocopherol can be helpful as a remedy of the AD. As in the pathology of AD, oxidative stress is the main cause, and vitamin E has an excellent antioxidant, that is free radical scavenging activity. So it will help to cure the AD and other diseases also which are related to oxidative stress (Boccardi et al., 2016; Rizvi et al., 2014).

### 5.6.2.1.3    Melatonin

Melatonin is an indole hormone. It is produced by the pineal gland and a small amount is also produced from the gastrointestinal tract. This compound involves various kinds of biological activities of the body. The sources of melatonin are human milk, banana, beets, cucumbers, and tomatoes (Boccardi et al., 2016; Yap et al., 2001).

### 5.6.2.1.3.1    Medicinal Uses

The main physiological role of melatonin is to control the circadian rhythm. The role of melatonin in some other disease conditions is also reported as it is helpful in the treatment of breast cancer, fibrocystic breast disease, and colon cancer. It is also reported that melatonin modifies the immunity, stress response, and also helps to reduce the aging effect. The antioxidant effect of melatonin is well established. As it was reported, that administration of the melatonin enhances the expression of antioxidant enzyme superoxide dismutase and glutathione peroxidase. In the animal study, it was found that melatonin helps to reduce the symptoms of the AD, as it reduces the plasma homocysteine and lipid level, which is due to its antioxidant property. It was also found that the level of melatonin was reduced in cerebrospinal fluid of the patients of the AD (Baydas et al., 2003; Malhotra et al., 2012).

### 5.6.2.1.3.2    Adverse Effects

Melatonin has very low toxicity in animals and humans. But, the headache, insomnia, rash, stomach upset and nightmares are the common side effect of melatonin (Boccardi et al., 2016; Yap et al., 2001).

### 5.6.2.2    PPARγ AGONIST

### 5.6.2.2.1    Pioglitazone

Pioglitazone is a compound of the category thiazolidinediones (TZDs). It is a specific agonist of receptor nuclear peroxisome-proliferator-activated receptor γ (PPARγ) (Galimberti and Scarpini, 2017).

### 5.6.2.2.1.1    Pharmacokinetics

Pioglitazone is shown to have a better absorption rate with an absolute bioavailability of 83%.

Approximately at 1.5 h, pioglitazone is observed to reach a maximum concentration and metabolized by the liver cytochrome P450 (CYP). But, only 15–30% of pioglitazone is recovered after oral administration from the urine. The negligible elimination of pioglitazone occurs through kidneys. The drug is excreted initially as metabolites and their conjugates. The most of the oral dose is excreted into the bile either unchanged or as metabolites and eliminated in the feces (Harrington et al., 2011).

### 5.6.2.2.1.2 Therapeutic Uses

The main use of pioglitazone is as a remedy for type 2 diabetes. It acts via receptor nuclear PPARγ. It attaches and forms a heterodimer with retinoid X receptor (RXR), which further leads to the alteration of the transcription of the gene and proteins related to the glucose and lipid metabolism (Application, 2013). Now it was reported that PPARγ is also involved in the neuroinflammation of the AD. Rosiglitazone is one of first PPARγ compound which was tested in the AD and entered in phase III of the clinical trials but later abandoned. Later on, pioglitazone was reported to reduce the glial cell inflammation and Aβ level in transgenic mice (Harrington et al., 2011).

### 5.6.2.3   GAMMA-SECRETASE INHIBITOR

#### 5.6.2.3.1   Semagacestat

Semagacestat is azepine class of gamma-secretase inhibitor and is the only one form of gamma-secretase which entered into clinical trials for the treatment of the AD (Olson and Albright, 2008).

##### 5.6.2.3.1.1   Pharmacokinetics

It is quickly absorbed and has a maximum concentration in plasma between 0.5–1 h (York et al., 2010).

##### 5.6.2.3.1.2   Therapeutic Uses

Semagacestat is developed to reduce the Aβ peptide production by inhibiting intramembranous cleavage of APP. Initially, it was found to be a beneficial compound in the AD, but later on, it was found that it aggravates the cognitive dysfunction and produces certain other side effects (Tagami et al., 2017).

### 5.6.2.4   5-HT$_6$ ANTAGONIST

#### 5.6.2.4.1   SB-271046

It is developed as a potential cognition enhancer and 5-HT6 receptor antagonist. Blockade of the 5-HT6 receptor is found to improve the cognitive function. Cognitive dysfunction is recognized as a key symptom of the AD; SB-271046 can be useful as therapeutic potential in the treatment of the AD. In various research articles, it has been found to improve cognitive function. Still, there is a lack of facts availability regarding SB-271046 (Upton et al., 2008).

### 5.6.2.5   STATINS

#### 5.6.2.5.1   Simvastatin and Pravastatin

HMG-CoA reductase inhibitors (commonly known as statins) have been widely prescribed groups of drugs in the world since their introduction to the market (Ramkumar et al., 2016).

##### 5.6.2.5.1.1   Pharmacokinetics

Statins have low bioavailability due to high first-pass metabolism except for pravastatin, which is an active metabolite. Mostly statins vary in the extent of metabolism, active and inactive metabolites—the therapeutic activity of pravastatin based on the kinetic profile of both parent compound and active metabolite. Pravastatin has the lowest plasma protein binding profile. The metabolism and excretion occur in liver and kidneys, respectively. Atorvastatin showed to have the longest terminal half-life (Flint-Garcia et al., 2003).

### 5.6.2.5.1.2 Therapeutic Uses

Statins are basically used to lower the increased lipid profile. The statins have a therapeutic effect by inhibiting HMG-CoA that is a rate-limiting step in cholesterol biosynthesis. Statin therapy was observed to be effective in lowering low-density lipoprotein cholesterol levels 20–50%, as well as lowering triglyceride levels 10–20% and causing a possible rise in serum high-density lipoprotein cholesterol levels (5–10%). It is reported to reduce the level of cholesterol, which can be supposed to have significant of statins in the treatment of the AD (Taylor et al., 2014).

### 5.6.2.5.1.3 Adverse Effects

Mostly statins are associated with certain side effects like musculoskeletal, hepatic and renal dysfunction, malignancy (Ramkumar et al., 2016).

### 5.6.2.6 CURCUMIN

Curcumin is a polyphenolic compound, obtained from turmeric. Turmeric is a spice that has received much interest in the medical and scientific field worldwide. Turmeric is a rhizomatous herbaceous perennial plant (*Curcuma longa*) of the ginger family, Zingiberaceae (Hewlings and Kalman, 2017).

### 5.6.2.6.1 Pharmacokinetics

Curcumin was reported to have low systemic availability after oral administration due to high first-pass metabolism and metabolism (glucuronidation and sulfation of curcumin) in the intestine (Hewlings and Kalman, 2017).

### 5.6.2.6.2 Medicinal Uses

Curcumin is reported to have various pharmacological properties. It provides benefits for inflammation, pain and metabolic disturbance. In many studies, the curcumin has been reported as an antioxidant having a positive effect in various kinds of diseases with oxidative stress. The curcumin can be of beneficial therapeutic potential in the treatment of the AD (Hewlings and Kalman, 2017).

### 5.6.3 RECENT DRUGS

### 5.6.3.1 CHOLINESTERASE INHIBITORS

### 5.6.3.1.1 Phenserine

Phenserine is a phenyl carbamate derivative, well known as the backbone of physostigmine which is a natural product obtained from the calabar bean of West Africa. Phenserine was synthesized and prepared as a crystalline compound, selected from preclinical evaluation among the various novel carbamates (Greig et al., 2005).

### 5.6.3.1.1.1 Pharmacokinetics

As per Phase II of clinical studies, phenserine showed the safety and pharmacological profile at the dose (5–10 mg b.i.d) in patients with the AD. It produced significant amelioration in the cognitive function. Phenserine showed a low pharmacokinetic half-life, decreased dosing frequency, reduced drug exposure in the body and decline the dependence on drug metabolism in the elderly (Greig et al., 2005).

#### 5.6.3.1.1.2 Mechanism of Action

Phenserine has two modes of action. At first, it showed a selective, non-competitive inhibitor of AChE, and in second, it showed the depletion secretion of Aβ (Greig et al., 2005).

#### 5.6.3.1.1.3 Therapeutic Uses

Phenserine is used as therapy for the AD by ameliorating cognitive, behavioral, and functional impairments in patients with the AD (Greig et al., 2005).

#### 5.6.3.1.1.4 Adverse Effects

During Phase II of clinical studies, the basic adverse events such as nausea and vomiting were observed (Greig et al., 2005).

### 5.6.3.1.2 Latrepirdine

Dimebon is a trade name of latrepirdine which is also known as dimebolin. In 1983 it was used as antihistamine drug in Russia. But as per clinical research, latrepirdine improved cognitive function in AD patients while destroying the psychopathic symptoms. Dimebon is administered orally to show cholinesterase inhibition and NMDA-antagonist activities. It has been predicted the brain cell death inhibition in the animal model of the AD. Due to a non-reactive result in a human trial, it couldn't show the positive response in AD patients. That is why dimebon was unlicensed for neurodegenerative conditions (Doody et al., 2008; Jones, 2010).

### 5.6.3.1.3 Huperzine A

Huperzine A is a phytochemical from Chinese herbs, novel lycopodium alkaloid that is isolated from herb *Huperzia serrata*. In China, traditionally, it has been used as strains, confusion swelling, and schizophrenia since centuries, but now it is used for AD and vascular dementia (Qian and Ke, 2014).

#### 5.6.3.1.3.1 Pharmacokinetics

Huperzine A has greater oral bioavailability, better penetration through the blood–brain barrier (BBB) and prolonged inhibition of AChE. In pharmacokinetic studies, huperzine A was absorbed rapidly in rats, canines, and healthy human volunteers. It has been manifested to have better distribution and elimination at a moderate rate administration (Zhang et al., 2008).

#### 5.6.3.1.3.2 Mechanism of Action

Huperzine A is a reversible inhibitor of acetylcholinesterase and improves cognitive deficits of animal models in the AD. Huperzine A also has cell protection against Aβ protein, $H_2O_2$ and glutamate. It prevents cell against ischemia, staurosporine-induced cytotoxicity and apoptosis. Such protective effects protect mitochondria, regulate the expression of apoptotic proteins Bcl-2, Bax, P53, caspase-3, attenuates oxidative stress, and upregulates nerve growth factor. Huperzine A interferes with APP metabolism. It shows antagonizing effects for NMDA receptors and potassium currents that can also provide as neuroprotection (Wang et al., 2006).

#### 5.6.3.1.3.3 Adverse Effects

Huperzine A showed the usual adverse effect of nausea during the clinical study, but apart from this, it has no serious adverse effect during the study (Rafii et al., 2011).

### 5.6.3.1.4    Ladostigil (TV3326)

Ladostigil has both neuroprotective effects cholinesterase inhibitory and monoamine oxidase MAO-A and MAO-B activities in one molecule. Ladostigil has observed about 25–40% cholinesterase inhibitor in rats. When ladostigil was administered orally at the doses range from 12–35 mg per kg it antagonizes scopolamine-induced spatial memory impairments. Ladostigil has successfully completed a preclinical study and entered in clinical trial phase II. The researchers hypothesized that an increase in the level of synaptic AChE induced by ladostigil might be one of the mode of action involved in cognitive function improvement. However, salivation, diarrhoea, and muscle weakness are the symptoms arise due to access cholinergic stimulation in the periphery until the high dose of 139 mg per kg was administered orally, which inhibited 50–60% cholinesterase. Such a wide therapeutic ratio was probably observed due to the relatively enhanced small rate in cholinesterase inhibition with increasing drug doses over the ranges of 17–69 mg per kg. This increase in cholinesterase inhibition with oral administration of ladostigil at dose 17–69 mg per kg was markedly lower and compared to subcutaneous injection of ladostigil at the doses range of 8.6–17 mg/kg; 40% increase in cholinesterase inhibition. Ladostigil was observed to inhibit both AChE and BuChE in an in vitro study. In which, the inhibitory effect was 100 times more potent against AChE than BuChE (Weinstock et al., 2000).

### 5.6.3.1.4.1    Mechanism of Action

Ladostigil showed both cholinesterase inhibition and MAO-A and MAO-B activities in a single molecule (Weinreb et al., 2012). The APP can be processed two alternatives ways, one is mutually exclusive posttranslational pathways, and another is the amyloidogenic pathway. The APP is cleaved sequentially by β-secretase and γ-secretase the N-terminal and C-terminal of the Aβ domain respectively to release Aβ. Similarly, in the non-amyloidogenic pathway, APP is cleaved by α-secretase and γ-secretase within the Aβ sequence, and form of a soluble APP, hence, it is precluding the formation of the Aβ (Vetrivel and Thinakaran, 2006).

### 5.6.3.2    IMMUNOTHERAPY

The monoclonal antibodies are administered intravenously in the AD against Aβ peptide, to ameliorate cognitive function and the slight improvement in functional scores (Salloway et al., 2014).

### 5.6.3.2.1    Bapineuzumab

Bapineuzumab, a humanized monoclonal antibody and investigated by Elan/Johnson & Johnson, can have the potential therapeutic effect for the remedy of the AD. It was found to acknowledge the N-terminal five residues of Aβ peptide in a helical conformation. This is stabilized by internal hydrogen bonds involving the first three amino acids. It is speculated to cause changes in the underlying AD neuropathology (Miles et al., 2013). After getting an active immunization with it, the AD patients were found to have positive with the reduction of plaques, but 6% subjects developed aseptic meningitis, and hence the clinical trial was stopped (Woodhouse et al., 2007). Bapineuzumab is in clinical development for the remedy of the AD (Salloway et al., 2014).

### 5.6.3.2.2 Solanezumab

Solanezumab is an anti-Aβ monoclonal antibody and neuroprotector that is under the clinical trial for the remedy of the AD (McCartney, 2015). Solanezumab is investigated by Eli Lilly. In a clinical study of 24 months old PDAPP, transgenic mice observed the antibody 'm266' with a rapid normalization of the cognitive deficits (Dodart et al., 2002).

### 5.6.3.2.3 Ponezumab (PF4360365)

Ponezumab is an anti-Aβ monoclonal antibody, investigated by Pfizer. Ponezumab was in phase II of the clinical trial for the remedy of the AD and the cerebral amyloid angiopathy, but it has been discontinued in 2011 and 2016, respectively (Salloway et al., 2014; Dodart et al., 2002).

### 5.7 FUTURE OPPORTUNITIES AND CHALLENGES

The AD is a very general and economically burdening disease today. From many decades, pharmacotherapeutics of the AD have seen tremendous research inputs. According to the limitations posed by the multifactorial pathogenetic pathway for the AD, no disease-modifying therapeutic approach has come out till date. The advances in the pharmacotherapy of the AD is having the threat of pursuit the wrong pathology and therefore the symptomatic treatment is once again coming to the fore.

Cholinergic agents and NMDA glutamate antagonist are being profoundly used for management of symptoms of the AD. There are several plant products which have demonstrated promising results in managing the cognitive abilities and memory in patients suffering from the AD. However, these plant products should be further explored to isolate individual phytoconstituents. The isolated compounds can then be subjected to further modifications to obtain a new and promising drug candidate for the AD.

The worldwide researchers are also focusing extensively on developing drugs against Aβ plaques and tau tangles, but these approaches have not yet delivered significantly considerable results. There is a strong need to consider the multifactorial approaches for treating AD. Targeting only one etiological factor is not sufficient. For coming up, with an effective treatment of the AD, the drug candidates should be designed in many ways that can act on multiple biological targets at once.

### 5.8 CONCLUSION

In this chapter, the various drug treatments for the AD are discussed, including its etiology, pathophysiology and diagnosis. But in a nutshell, we can say yes that is not proper treatment as the available drug therapy of the AD provides only symptomatic relief. So, it is necessary for the people worldwide to get proper treatment by developing a new drug therapy which can treat the underline pathological conditions of AD.

### KEYWORDS

- **cognition**
- **dementia**
- **amyloid β**
- **neurofibrillary tangles**
- **Alzheimer's disease**

## REFERENCES

Armstrong, A. R. Review Article: What Causes AD? *Folia Neuropathol* **2013**, *3*, 169–188.

Allen, S. J.; Watson, J. J.; Dawbarn, D. The Neurotrophins and Their Role in Alzheimer's Disease. *Curr. Neuropharmacol.* **2011**, *9*, 559–573.

Alzheimer's Association. AD Facts and Figures. *Alzheimer's & Dementia* **2014**, *10*, 1–80.

Alzheimer's Association. AD Facts and Figures. *Alzheimer's & Dementia* **2015**, *11*, 332–384.

Tepzz Z. T. An application. *European Patent* **2013**, *19*, 1–8.

Aprahamian, I.; Stella, F.; Forlenza, O. V. New Treatment Strategies for AD: Is There a Hope? *Indian J. Med. Res.* **2013**, *138*, 449–460.

Baydas, G.; Kutlu, S.; Naziroglu, M.; Canpolat, S.; Sandal, S.; Ozcan, M.; Kelestimur, H. Inhibitory Effects of Melatonin on Neural Lipid Peroxidation Induced by Intracerebroventricularly Administered Homocysteine. *J. Pineal Res.* **2003**, *34*, 36–39.

Berríos-Torres, S. I.; Umscheid, C. A.; Bratzler, D. W.; Leas, B.; Stone, E. C.; Kelz, R. R. et al. Centers for Disease Control and Prevention Guideline for the Prevention of Surgical Site Infection. *JAMA Surg.* **2017**, *152*, 784.

Birks, J. Cholinesterase Inhibitors for Alzheimer's Disease: Review. *Cochrane Database Syst Rev.* **2012**, *5*, 1–51.

Boccardi, V.; Baroni, M.; Mangialasche, F.; Mecocci, P. Vitamin E Family: Role in the Pathogenesis and Treatment of AD. *Alzheimer's Dement.* **2016**, *2*, 182–191.

Boller, F.; Forbes, M. M. History of Dementia and Dementia in History: An Overview. *J. Neurol. Sci.* **1998**, *158*, 125–133.

Burton, G. W.; Joyce, A.; Ingold, K. U. Is Vitamin E, the Only Lipid-soluble, Chain-breaking Antioxidant in Human Blood Plasma and Erythrocyte Membranes? *Arch. Biochem. Biophys.* **1983**, *221*, 281–290.

Cummings, J.; Lefèvre, G.; Small, G.; Appel-Dingemanse, S. Pharmacokinetic Rationale for the Rivastigmine Patch. *Neurology* **2007**, *69*, 4 Suppl., 1, S10–S13.

Davies, P.; Maloney, A. J. F. Selective Loss of Central Cholinergic Neurons in AD. *The Lancet* **1976**, *308*, 1403.

Diamond, B. J.; Bailey, M. R. *Ginkgo biloba*. Indications, Mechanisms, and Safety. *Psychiatr. Clin. North. Am.* **2013**, *36*, 73–83.

Dodart, J. C.; Bales, K. R.; Gannon, K. S.; Greene, S. J.; DeMattos, R. B.; Mathis, C. et al. Immunization Reverses Memory Deficits Without Reducing Brain Aβ Burden in Alzheimer's Disease Model. *Nat. Neurosci.* **2002**, *5*, 452–457.

Doody, R. S.; Gavrilova, S. I.; Sano, M.; Thomas, R. G.; Aisen, P. S.; Bachurin, S. O. et al. Effect of Dimebon on Cognition, Activities of Daily Living, Behaviour, and Global Function in Patients with Mild-to-moderate AD: A Randomised, Double-blind, Placebo-controlled Study. *The Lancet* **2008**, *372*, 207–215.

Flint-Garcia, S. A.; Thornsberry, J. M.; Buckler, E. S. Structure of Linkage Disequilibrium in Plants. *Annu. Rev. Plant. Biol.* **2003**, *54*, 357–374.

Galimberti, D.; Scarpini, E. Pioglitazone for the Treatment of AD. *Expert Opin. Investig. Drugs.* **2017**, *26*, 97–101.

Greig, N. H.; Sambamurti, K.; Yu, Q.; Brossi, A.; Bruinsma, G. B.; Lahiri, D. K. An Overview of Phenserine Tartrate, A Novel Acetylcholinesterase Inhibitor for the Treatment of AD. *Curr. Alzheimer Res.* **2005**, *2*, 281–290.

Hakk, H.; Larsen, G.; Klasson, W. E. Tissue disposition, excretion and metabolism of 2,2′,4,4′,5-pentabromodiphenyl ether (BDE-99) in the male Sprague-Dawley rat. *Xenobiotica.* **2002**, *32*(5), 369–382.

Harrington, M. J.; Razghandi, K.; Ditsch, F.; Guiducci, L.; Rueggeberg, M.; Dunlop, J. W. C. et al. Origami-like Unfolding of Hydro-actuated Ice Plant Seed Capsules. *Nat. Commun.* **2011**, *2*, 337.

Heinonen, T.; Gaus, W. Cross Matching Observations on Toxicological and Clinical Data for the Assessment of Tolerability and Safety of *Ginkgo biloba* Leaf Extract. *Toxicology* **2015**, *327*, 95–115.

Hewlings, S.; Kalman, D. Curcumin: A Review of Its' Effects on Human Health. *Foods* **2017**, *6*, 92.

Hollman, P. C. H.; Van Trijp, J. M. P.; Buysman, M. N. C. P.; Martijn, M. S.; Mengelers, M. J. B.; De Vries, J. H. M. et al. Relative Bioavailability of the Antioxidant Flavonoid Quercetin from Various Foods in Man. *FEBS Letters* **1997**, *418*, 152–156.

Howard, A. C.; McNeil, A. K.; McNeil, P. L. Promotion of Plasma Membrane Repair by Vitamin E. *Nat. Commun.* **2011**, *2*, 1–8.

Jones, R. W. Drug Treatment of AD. *Rev. Clin. Gerontol.* **2002**, *12*, 165–173.

Jones, R. W. Dimebon Disappointment. *Alzheimers. Res. Ther.* **2010**, *2*, 3–5.

Jaan, M. W.; Shirley, K. L.; Small, G. W. Clinical Pharmacokinetics and Pharmacodynamics of Cholinesterase Inhibitors. *Clin. Pharmacokinet.* **2002**, *41*, 719–739.

Malhotra, S.; Sawhney, G.; Pandhi, P. The Therapeutic Potential of Melatonin: A Review of the Science. *Med. Gen. Med.* **2012**, *6*, 1–10.

Mauri, P. L.; Iemoli, L.; Gardana, C.; Riso, P.; Simonetti, P.; Porrini, M. et al. Liquid Chromatography/electrospray Ionization Mass Spectrometric Characterization of Flavonol Glycosides in Tomato Extracts and Human Plasma. *Rapid Commun. Mass Spectrom.* **1999**, *13*, 924–931.

McCartney, M. Margaret McCartney: The "Breakthrough" Drug that's not been Shown to Help in AD. *BMJ (Clinical Research Ed.).* **2015**, *351*, h4064.

McShane, R.; Areosa Sastre, A.; Minakaran, N. Memantine for Dementia. *Cochrane Database Syst Rev.* **2006**, 1.

Miles, L. A.; Crespi, G. A. N.; Doughty, L.; Parker, M. W. Bapineuzumab Captures the N-terminus of the AD Amyloid-beta Peptide in a Helical Conformation. *Sci. Rep.* **2013**, *3*, 1–6.

Moreira, P. I. Sweet Mitochondria: A Shortcut to AD. *J. Alzheimers Dis.* **2018**, *62*, 1391–1401.

Mount, C.; Downton, C. Alzheimer disease: Progress or Profit? *Nat. Med.* **2006**, *12*, 780–784.

Munoz, D. G.; Feldman, H. Causes of AD. *Can. Med. Assoc. J.* **2000**, *162*, 65–72.

Olson, R. E.; Albright, C. F. Recent Progress in the Medicinal Chemistry of Gamma-secretase Inhibitors. *Curr. Top Med. Chem.* **2008**, *8*, 17–33.

Paul, M. D. A Biosynthetic Approach. *Pharm Sci.* **2002**, *471496405*.

Pereira, E.; Barros, L.; Ferreira, I. C. F. R. Chemical Characterization of *Ginkgo biloba* L. and Antioxidant Properties of Its Extracts and Dietary Supplements. *Ind Crops Prod.* **2013**, *51*, 244–248.

Perl, D. P. Neuropathology of Alzheimer's Disease. *Mt Sinai J. Med.* **2010**, *77*, 32–42.

Perry, E. E.; Perry, R. H. The Cholinergic System in AD. *Trends Neurosci.* **1982**, *5*, 261–262.

Prince, M.; Comas-Herrera, A.; Knapp, M.; Guerchet, M.; Karagiannidou, M. World Alzheimer Report: Improving Healthcare for People Living with Dementia. Coverage, Quality and Costs Now and in the Future. *AD Intern.* **2016**, 1–140.

Qian, Z. M.; Ke, Y. Huperzine A: Is It An Affective Disease-modifying Drug for AD? *Front Aging Neurosci.* **2014**, *6*, 1–6.

Rafii, M. S.; Walsh, S.; Little, J. T.; Behan, K.; Reynolds, B.; Ward, C. et al. A Phase II Trial of Huperzine A in Mild to Moderate Alzheimer Disease. *Neurology* **2011**, *76*, 1389–1394.

Ramkumar, S.; Raghunath, A.; Raghunath, S. Statin Therapy: Review of Safety and Potential Side Effects. *Acta. Cardiol. Sin.* **2016**, *32*, 631–639.

Ravi, S. K.; Ramesh, B. N.; Mundugaru, R.; Vincent, B. Multiple Pharmacological Activities of *Caesalpinia crista* Against Aluminium-induced Neurodegeneration in Rrats: Relevance for AD. *Environ. Toxicol. Pharmacol.* **2018**, *58*, 202–211.

Rizvi, S.; Raza, S. T.; Ahmed, F.; Ahmad, A.; Abbas, S.; Mahdi, F. The Role of Vitamin E in Human Health and Some Diseases. *Sultan Qaboos Univ. Med. J.* **2014**, *14*, 157–165.

Salloway, S.; Sperling, R.; Fox, N. C.; Blennow, K.; Klunk, W.; Raskind, M. et al. Two Phase III Trials of Bapineuzumab in Mild-to-Moderate AD. *N. Engl. J. Med.* **2014**, *370*, 322–333.

Schneider, L. Alzheimer Disease Pharmacologic Treatment and Treatment Research. *Continuum* **2013**, *19*, 339–357.

Shoshan-Barmatz, V.; Nahon-Crystal, E.; Shteinfer-Kuzmine, A.; Gupta, R. VDAC1, Mitochondrial Dysfunction, and AD. *Pharmacol. Res.* **2018**, 1–45.

Tabgled Bank Studios. A Journalist's Guide to Alzheimer's Disease and Drug Development **2016**, 1–20.

Tagami, S.; Attwater, J.; Holliger, P. Simple Peptides Derived from the Ribosomal Core Potentiate RNA Polymerase Ribozyme Function. *Nat. Chem.* **2017**, *9*, 325–332.

Taylor, M. J.; McNicholas, C.; Nicolay, C.; Darzi, A.; Bell, D.; Reed, J. E. Systematic Review of the Application of the Plan-do-study-act Method to Improve Quality in Healthcare. *BMJ Qual. Saf.* **2014**, *23*, 290–298.

Tricco, A. C.; Soobiah, C.; Berliner, S.; Ho, J. M.; Ng, C. H.; Ashoor, H. M.; Straus, S.E. Efficacy and safety of cognitive enhancers for patients with mild cognitive impairment: a systematic review and meta-analysis. *Can. Med. Assoc. J.* **2013**, *185*(16), 1393–1401.

Ulbricht, C.; Chao, W.; Costa, D.; Rusie-Seamon, E.; Weissner, W.; Woods, J. Clinical Evidence of Herb-Drug Interactions: A Systematic Review by the Natural Standard Research Collaboration. *Curr. Drug. Metab.* **2008**, *9*, 1063–1120.

Upton, N.; Chuang, T. T.; Hunter, A. J.; Virley, D. J. 5-HT6 Receptor Antagonists as Novel Cognitive Enhancing Agents for AD. *Neurotherapeutics* **2008**, *5*, 458–469.

Uddin, M. S.; Amran, M. S. *Handbook of Research on Critical Examinations of Neurodegenerative Disorders,* 1st ed.; USA: IGI Global, 2018.

Vetrivel, K. S.; Thinakaran, G. Amyloidogenic Processing of -amyloid Precursor Protein in Intracellular Compartments. *Neurology* **2006**, *66*, S69–S73.

Wang, R.; Yan, H.; Tang, X. C. Progress in Studies of Huperzine A, A Natural Cholinesterase Inhibitor from Chinese Herbal Medicine. *Acta. Pharmacol. Sin.* **2006**, *27*, 1–26.

Weinreb, O.; Amit, T.; Bar-Am, O.; B. H. Youdim, M. Ladostigil: A Novel Multimodal Neuroprotective Drug with Cholinesterase and Brain-Selective Monoamine Oxidase Inhibitory Activities for Alzheimers Disease Treatment. *Curr.* Drug Targets **2012,** *13,* 483–494.

Weinstock, M.; Bejar, C.; Wang, R. H.; Poltyrev, T.; Gross, A.; Finberg, J. P. et al. TV3326, A Novel Neuroprotective Drug with Cholinesterase and Monoamine Oxidase Inhibitory Activities for the Treatment of AD. *J. Neural. Transm. Suppl.* **2000,** *60,* 157–169.

Winblad, B.; Grossberg, G.; Frolich, L.; Farlow, M.; Zechner, S.; Nagel, J. et al. IDEAL: A 6-month, Double-blind, Placebo-controlled Study of the First Skin Patch for Alzheimer Disease. *Neurology* **2007,** *69,* S14–S22.

Woodhouse, A.; Dickson, T. C.; Vickers, J. C. Vaccination Strategies for AD: A New Hope? *Drugs & Aging* **2007,** *24,* 107–119.

World Health Organization. Dementia: A Public Health Priority. *Dementia* **2012,** *112.*

World Health Organization. World Health Statistics Monitoring Health for the SDGs. *WHO* **2016,** *1,* 121.

Yap, S. P.; Yuen, K. H.; Wong, J. W. Pharmacokinetics and Bioavailability of α-, γ- and δ-tocotrienols Under Different Food Status. *J. Pharm. Pharmacol.* **2001,** *53,* 67–71.

Yi, C.; Li, R.; Wolbeck, J.; Xu, X.; Nilsson, M.; Aires, L.; et al. Climate Control of Terrestrial Carbon Exchange Across Biomes and Continents. *Environ. Res. Lett.* **2010,** *5,* 10.

Zhang, H. Y.; Zheng, C. Y.; Yan, H.; Wang, Z. F.; Tang, L. L.; Gao, X. et al. Potential Therapeutic Targets of Huperzine A for AD and Vascular Dementia. *Chem.-Biol. Interact.* **2008,** *175,* 396–402.

# CHAPTER 6

# Pharmacological Management of Parkinson's Disease

NEWMAN OSAFO[1*], SAMUEL OBENG[2], DAVID D. OBIRI[1],
ODURO K. YEBOAH[1], and LESLIE B. ESSEL[1,3]

*1Kwame Nkrumah University of Science and Technology,
Kumasi, Ghana*

*2Virginia Commonwealth University, Virginia, USA*

*3University of Missouri, Missouri, USA*

*\*Corresponding author. E-mail: nosafo.pharm@knust.edu.gh*

## ABSTRACT

Parkinson's disease (PD), a chronic progressive neurodegenerative disorder, is associated with the loss of dopaminergic neurons in the substantia nigra pars compacta resulting in reduced dopamine levels in the striatum. The available treatments target either dopaminergic or nondopaminergic pathways of motor activity control, and are for relieving symptoms rather than altering disease progression. Till date, levodopa remains the most effective drug used in the management of PD. It restores the dopamine balance, hence improves motor symptoms of PD. Within the scope of this chapter, we will review the pharmacotherapy of PD with dopamine replacement therapies, dopamine agonist, anticholinergics, and the monoamine oxidase B inhibitors. The mechanisms of action, pharmacokinetics, and pharmacodynamics of these drug groups are discussed in detail. Also, a brief review of the various neurotransmitters, which play significant roles in motor control and hence the pathogenesis of PD is offered in order to understand the basis of the pharmacotherapy.

## 6.1 INTRODUCTION

Parkinson's disease (PD) is a neurodegenerative condition that is progressive in its course, resulting from the death of dompaminergic cells in the substantia nigra. Motor disorders namely bradykinesia, muscle rigidity, and resting tremor, characterizes PD clinically (Hung and Schwarzschild, 2014; NICE, 2017). While PD is known as a condition marked by motor symptoms, nonmotor symptoms form vital parts of this syndrome. The motor symptoms can be worrying, since they greatly affect day to day activities, though they are often not realized by health-care experts (Shulman et al., 2001, 2002). Nonmotor presentations also occur and may include autonomic dysfunction, sleep disorders, anxiety, and fatigue (Zesiewicz et al., 2010). Autonomic dysfunction in PD may manifest as orthostatic hypotension, sexual dysfunction, gastrointestinal disorders, and urinary incontinence. Excessive daytime somnolence, occasional limb movements, rapid-eye movement (REM) sleep behavior disorder, restless leg syndrome, and insomnia represents sleep disorders that may occur in PD (Zesiewicz et al., 2010).

Principally, the symptoms of PD arises as a consequence of the selective death of dopamine-producing nerve cells in the substantia nigra pars compacta. The death of these nerve cells culminates in decrease in dopamine concentrations in the corpus striatum (Hung and Schwarzschild, 2014). Dopaminergic impairment leads to changes in basal glutamatergic synaptic transmission of striatum and plasticity in the efferent cells of the striatum, the spiny projection neurons (Bagetta et al., 2012). A large number of PD incidence are irregular, with no single identifiable causative factor. However, a multifactorial etiology is believed to play a part in the gradual process of the death of the cells in the nigral region. This includes mitochondrial dysfunction and increased concentrations of reactive oxygen species (most important pathologic mechanisms) (Greenamyre et al., 2004), head trauma, exposure to environmental toxins, and neuroinflammation (Blandini, 2013; Jafari et al., 2013). Cytokines and neurotoxins released by the reactive inflammatory cells, distract dopamine-producing neurons projecting in the substantia nigra and corpus striatum (Blandini, 2013). Strategies that employ the replacement of dopamine appears to be the backbone of management for motor symptoms that occur in cases of PD. In spite of its positive impact on motor function, dopaminergic therapy has noteworthy drawbacks (Lang and Obeso, 2004). Typical drawbacks include the commencement of motor fluctuations (in which periods of relative immobility and mobility alternate), as well

as dyskinesias that occur involuntarily. In addition, motor symptoms, which includes postural instability and freezing of gait, and nonmotor symptoms associated with advanced phases of PD mostly are refractory to dopaminergic therapy (Braak et al., 2003). Deterioration of nondopamine-producing neurons in other neural areas is probably the cause of these symptoms (Braak et al., 2003; Hung and Schwarzschild, 2014).

To better appreciate the utilization of pharmacologic agents that target nondopaminergic and dopaminergic systems, it is important to have a short review of the influence that the various neurotransmitter systems have as part of regulation of motor function. As far as the classical basal ganglia organization model is concerned, the corpus striatum receives excitatory glutamatergic inputs from the cerebral cortex under the control of dopamine through the nigrostriatal pathway. Dopamine directly causes a modulation of these inputs. The dopaminergic mediated modulation occurs via a stimulatory effect on a subgroup of striatal neurons which contain substance P and γ-aminobutyric acid (GABA). An indirect pathway also exists where modulation of these glutamatergic inputs by dopamine, occurs via the inhibition of different subgroups of neurons that express both encephalin and GABA. The effects on these pathways are mediated via the interaction of dopamine with its cognate dopamine-1 ($D_1$) and dopamine-2 ($D_2$) receptors, respectively, which are very much expressed in the corpus striatum (Hung and Schwarzschild, 2014).

The direct pathway has to do with the sending directly, GABAergic signals, to the output nucelei of the globus pallidus pars interna (GPi), the basal ganglia and substantia nigra pars reticulata (SNr), with subsequent conduction of GABAergic fibers to the ventral thalamic nuclei. Striatofugal neuronal axons, however, forms

GABAergic synapses. These carry GABAergic projections to the subthalamic nucleus (STN), which then utilizes glutamate to cause a modulation of basal ganglia output from the GPi/SNr. From this, one could infer that dopamine acts as a regulatory molecule in the activity of the basal ganglia via the balancing of contrasting effects on the indirect and direct pathways. As such the loss or gain of striatal dopamine, as in PD or following therapy with dopaminergic agents correspondingly, disturbs this state of balance, leading to respective hypokinetic (parkinsonian), or hyperkinetic (dyskinetic) states (Hung and Schwarzschild, 2014). Evidence from recent studies puts forward that the circuitry of the basal ganglia may be much more complicated than it was thought previously (Cui et al., 2013), and the classical model may account for just some of the phenomenology that occur with PD (Hung and Schwarzschild, 2014).

Glutamate receptors are highly expressed in the corpus striatum and also occur at a great density in the brain, not like dopaminergic receptors, which are extremely concentrated in the basal ganglia. There is also an extensive expression of GABA in the central nervous system (CNS). Having a principal function in the circuitry of the basal ganglia, these neurotransmitter systems are hypothetically essential therapeutic targets in PD and the management of associated symptoms. Nevertheless, due to the absence of regional specificity, there is an enhanced risk of side effects from the actions of these neurotransmitters on other brain regions, posing a potential challenge to the use of these agents in PD (Hung and Schwarzschild, 2014).

Another neurotransmitter that is caught up in the regulation of the function of the basal ganglia is adenosine. This purine nucleoside has a modulatory effects on neural synaptic function, with its action being mediated by the 4 G-protein-coupled receptor subtypes $A_1$,

$A_{2A}$, $A_{2B}$, and $A_3$. Much attention has however, been given to the $A_{2A}$ receptor as it represents a possible treatment target due to its heavy presence in the corpus striatum (Hickey and Stacy, 2012; Schwarzschild et al., 2005). Also, serotonergic system alterations play a role in the etiology of PD (Huot et al., 2011). Of the 14 serotonin (5-HT) receptor subtypes (Nichols and Nichols, 2008), several subtypes, including the $5$-$HT_{1A}$ and $5$-$HT_{2C}$ receptor subtypes, are found in neurons of the striatum. In addition to the corpus striatum, inputs by the serotonergic system from the raphe nuclei create extensive networks in other regions of the brain, such as the substantia nigra, the globus pallidus, thalamus, the STN, and cortex (Hung and Schwarzschild, 2014).

A body of proof from neuropathological findings implicated degeneration of neurons in PD as not being limited to the basal ganglia and dopamine containing neurons. Per the Braak staging system (Braak et al., 2003), inclusion bodies that contain α-synuclein occur in caudal brainstem nuclei (stage 1) with the involvement of the substantia nigra (stage 3). During stage 2, serotonin containing neurons that are located in the raphe nucleus become affected, so do the locus coeruleus derived noradrenaline producing projection neurons. At advanced phases of PD progression, there are extension of the neurodegeneration to cholinergic neurons in the neocortex and pedunculopontine tegmental nucleus. The mutiliplicity of affected neurotransmitter systems leads to the development of several symptoms. It is imperative for clinicians and their patients to note that, these symptoms may only be responsive to adjunctive nondopamine therapies (Hung and Schwarzschild, 2014). This chapter discusses a detailed pathogenesis of PD and the possible CNS mediators that contribute to the development of the neurodegenerative disorder. Again, the chapter particularizes on PD pharmacotherapy and the pharmacodynamics and pharmacokinetic profiles of these agents employed in managing the disorder, as well as its signs.

## 6.2   ANTI-PARKINSON'S DRUGS

The motor symptoms of PD are principally treated by employing the dopamine replacement therapies, and nearly 50 years of using levodopa (LD), that is, the precursor compound of dopamine, it is still the most effective treatment. Nonetheless, even though it is beneficial in modulating motor function, the use of dopamine therapy has major drawbacks. This has made it important to develop various therapeutic methods focusing on nondopaminergic pathways (Lang and Obeso, 2004).

### 6.2.1   *DOPAMINE REPLACEMENT THERAPY*

#### 6.2.1.1   *LEVODOPA*

Levodopa (LD) has the chemical name (-)-L-α-amino-β-(3,4-dihydroxy-benzene) propanoic acid (Fig. 6.1). It was realized from initial studies to reversed reserpine-induced parkinsonian syndrome in animals. This was proceeded by the mapping of the regional distribution of dopamine in human brain, which eventually resulted in the important observation that deficiency of dopamine in the nigrostriatum results in the condition of PD (Ehringer et al., 1960; Khor and Hsu, 2007). Long after its introduction into the therapy of PD, levodopa stays the gold standard of management for this incapacitating neurodegenerative disorder (Deleu et al., 2002). The control of the motor symptoms when managing PD result from the effects of dopamine, which is the active

metabolite of levodopa. However, it (dopamine) is limited by its inability to cross the blood–brain barrier (BBB). Because LD has good BBB penetrating ability, it serves as a prodrug of dopamine. Decarboxylation of levodopa to dopamine occurs quickly in the extracerebral tissues, particularly in the gastrointestinal tract, after giving the drug via the oral route. Only a little amount of a administered dose of LD is therefore able to reach CNS (Khor and Hsu, 2007). The result of this is reduced biosynthesis in the striatum, as well as the storage of dopamine from the exogenous levodopa. There is also the subsensitization of postsynaptic dopamine receptors, a pharmacodynamic mechanism that invokes motor complications like motor response variations (end-of-dose akinesia or wearing-off phenomenon, "early-morning" dystonia, freezing, "on–off" fluctuations, and dyskinesias) and neuropsychiatric manifestations upon chronic treatment (Deleu et al., 2002). Henceforth, LD is mostly given with an agent that inhibits the decarboxylase enzyme like benserazide or carbidopa (CD) (Fig. 6.1) to prevent its peripheral decarboxylation (Khor and Hsu, 2007).

Currently, LD is marketed in different formulations, such as the orally disintegrated tablet (ODT), the controlled-release tablet (CR), immediate-release tablet (IR), and IR CD-LD in combination with a fixed 200 mg dose of entacapone. The entacapone is added to inhibit catechol-O-methyltransferase (COMT)-mediated LD metabolism (Descombes et al., 2001; Ghika et al., 1997).

### 6.2.1.1.1   *Pharmacokinetics*

#### 6.2.1.1.1.1   *Absorption*

Absorption of levodopa (LD) occurs through the saturable L-neutral amino acid transport

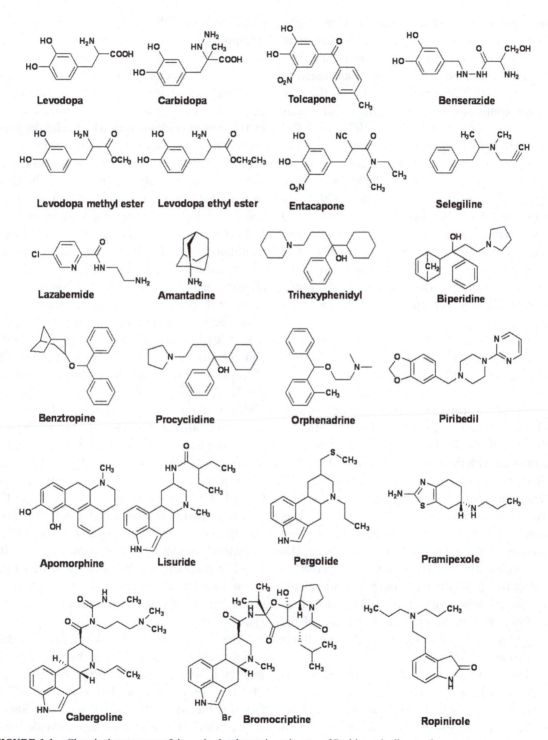

**FIGURE 6.1** Chemical structures of drugs in the dopaminergic arm of Parkinson's disease therapy.

mechanism for large amino acids. Conversely, there is a more than proportional elevation of the plasma concentration of LD when the dose of LD is increased. This has been attributed to the main metabolic pathway of LD in the GIT which comprises saturable amino acid decarboxylation (Muenter and Tyce, 1971; Nutt and Fellman, 1984; Sasahara et al., 1980). When LD is given orally, there is almost complete absorption, with just 2% occuring in the feces (Morgan et al., 1971; Peason and Bianchine, 1970; Sasahara et al., 1980). Yet, the amount of the oral LD dose reaching the systemic circulation complete when no peripheral decarboxylase inhibitor, CD is added is just about 30%. The addition of the latter culminating in a two- to three-fold upsurge in bioavailability (Bianchine et al., 1972; Kuruma et al., 1972).

In current practice of medicine, LD is mostly given as the fixed-combination formulation with a peripheral decarboxylase inhibitor. Either bensarazide [(+)-D, L-seryl-(2,3,4-trihydroxybenzyl)-hydrazine] or CD (1-α-methyldopahydrazine) is the preferred choice with the former having 10-fold potency compared with the latter (Da Prada et al., 1987). Neither of the two, at the usual doses, have BBB penetrating power. Kaakkola et al. (1985) also report that when the ratio of levodopda to CD is increased from 1:10 to 1:5, rather, increases the tolerance to the adverse effects of levodopa. The coadminstration of levodopa with CD gives an almost two-fold rise in plasma elimination half-life of levodopa (Nutt et al., 1985), as well as a 75% reduction in total the daily dose of the drug that is needed to yield the clinical effects (Cedarbaum, 1987).

The stomach's capacity for absorption is quite limited, yet, it is a very vital area of where the decarboxylation of levodopa can take place. There is about 80–90% of the absorption of levodopa occurring in the proximal area of the small intestine. This is mediated via an active saturable carrier system for larger neutral amino acids (e.g., phenylalanine, histidine, leucine, tyrosine, isoleucine, threonine, valine, and tryptophan) (Wade et al., 1973). The gastric emptying rate is the key determining element in the nature of levodopa which also in itself can change the pattern of gastric emptying in healthy volunteers. This may provide an explanation to the irregular absorption of the drug from the GIT (Robertson et al., 1990). Food can cause a delay in gastric emptying (especially fat). This is noteworthy since there could be a resultant rise in the time of maximum plasma concentration ($T_{max}$) of levodopa by three-fold (Baruzzi et al., 1987). Thus, levodopa should be given at minimum, 30 min before food. Moreover, decreased gastric pH, simultaneous use of antimuscarinic drugs, tricyclic antidepressants, and ferrous sulfate delay and/or decrease the absorption absorption of levodopa (Pfeiffer, 1996). N-soluble fiber diets (Astarloa et al., 1992) as food rich in cisapride, a benzamide derivative, all seem to increase the absorption of levodopa and hence, motor function (Neira et al., 1995). Under fasting conditions in healthy geriatric subjects, the LD absorption from an IR CD–LD formulation is quick, with 60% of the dose absorbed in 30 min, with the completion of absorption 2–3 h (Lovenberg and Victor, 1974). In general, coadministration of entacapone with CD–LD products results in an increased total levodopa AUC by elevating the concentrations at the later time points after dosing (Khor and Hsu, 2007).

The transportation of LD via the L-neutral amino acid transport system, has resulted in increased studies into exploring the influence of dietary protein on the clinical outlook of LD (Karstaedt et al., 1991; Nutt et al., 1984; Simon et al., 2004). These studies have revealed that the clinical outcome of LD is decreased by a daily diet which contains protein in excess of 1.6 g/kg or a single protein load of about 28 g (Carter et al., 1989; Mena and Cotzias, 1975). Protein-containing diet seems to reduce the oral bioavailabity of LD (Baruzzi et al., 1987; Robbertson et al., 1991). The reduced response to LD however, appears not to correspond to concentrations of LD in the plasma (Carter et al., 1989). The findings from a study that employed positron emission tomography identified the diminished effect is due to the reduction in uptake of LD into the brain. This is possibly due to competition resulting from elevated concentrations of amino acids in plasma (Leenders et al., 1986) as patients experience respite from parkinsonian symptoms when the amino acids levels reduce in the presence of increased plasma levodopa concentrations (Alexander et al., 1994).

## 6.2.1.1.1.2 Distribution

The transport of LD occurs across the BBB by a stereospecific and saturable facilitated process via the large neutral amino acid (LNAA) transport carrier system. When an intravenous dose of 50 mg LD was given alone, the volume of distribution (Vd) at steady state in young healthy volunteers was approximated to be 70% higher compared to that recorded in elderly volunteers (Robertson et al., 1989). It was also observed that there was approximately 50% increase in the Vd in young healthy study participants than that in elderly study participants when an intravenous LD dose was given together with CD. The binding of Levodopa to plasma proteins is low (Hinterberger and Andrews, 1972; Rizzo et al., 1996). CD on the other hand does not penetrate the BBB even at elevated doses (Lotti and Porter, 1970; Porter, 1973). However, similar to levodopa, CD is does not considerably bound to human plasma proteins (Vickers et al., 1974).

## 6.2.1.1.1.3 Elimination

The conversion of L-tyrosine to L-3,4-dihyrdo-phenylalanine (L-dopa) by tyrosine hydroxylase takes place in noradrenaline-producing neurons of the sympathetic ganglia, in catecholaminergic neurons in the brain, as well as in norepinephrine- and epinerphrine-containing cells of the adrenal medulla (Deleu et al., 2002). There are four different pathways involved in the metabolism of levodopa. These are decarboxylation by aromatic L-amino acid decarboxylase (AAAD), transamination by tyrosine aminotransferase, oxidation by tyrosinase or other oxidants and 3-O-methylation by COMT. The ubiquitous, nonspecific enzyme, AAAD is widely distributed in the gastrointestinal tract, liver, lungs, adrenal, spleen, heart, kidneys, brain, and its capillaries (Porter et al., 1973). Decarboxylation by AAAD is the principal metabolic pathway for LD and unlike tyrosine hydroxylase, intensity of dopaminergic transmission does not affect its regulation in any way (Lovenberg and Victor, 1974).

When LD is given orally, about 95% of the dose is decarboxylated peripherally, with only about 1% entering the brain (Bianchine et al., 1972; Dingemanse, 2000). Dopamine is the primary product of the decarboxylation process. Further metabolism may occur, typically the formation of homovanillic acid (HVA); 3,4-dihydroxyphenyl acetic acid (DOPAC); and to a minor degree, noradrenaline by dopamine β-hydroxylase, and vanillinemandelic acid (Khor and Hsu, 2007). However, HVA can also be obtained from an alternative pathway (3-O-methyldopa [3-OMD] pathway) (Deleu et al., 2002).

The COMT pathway product 3-OMD has a 15-h plasma half-life and is a poor substrate for AAAD. 3-OMD appears to compete with LD for transport into the brain via the LNAA transport carrier system. Even though oral challenges with 3-OMD decrease the clinical effects of LD, 3-OMD is not as potent as phenylalanine as far as competition with LD for transport into the brain. The contribution by 3-OMD to the total concentration of LNAAs competing with LD is therefore small, and as such not an important determinant of clinical effects to LD (Nutt et al., 1997). Figure 6.2 summarizes the various pathways implicated in the metabolism of levodopa.

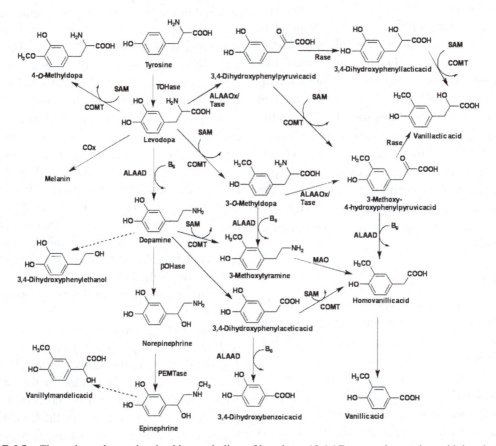

**FIGURE 6.2**    The major pathways involved in metabolism of levodopa. ALAAD, aromatic L-amino acid decarboxylase; ALAAOx, aromatic L -amino acid oxidase; βOHase, β-hydroxylase; COMT, catechol-O-methyltransferase; COx, catechol oxidase; MAO, monoamine oxidase; PEMTase, phenylethanolmethyltransaminase; Rase, reductase; SAM, *S*-adenosyl- L -methionine; Tase, transaminase; TOHase, tyrosine hydroxylase.

When LD is given via the oral route, about 90% of the given dose is recovered in urine after 48 h. Coadministration with CD reduces the recovered dose to about 60%, probably because a large portion of the LD dose is biotransformed to the long half-life metabolite, 3-OMD, during coadministration with CD (Bianchine et al., 1972). There is an age-dependent plasma clearance of LD following intravenous administration in the presence of CD (Yeh et al., 1989). The plasma half-life of LD when given as monotherapy is about 50 min, and increases to ~90 min, when coadministered with CD (Khor and Hus, 2007).

Following several dosages, the pharmacokinetic profile of CD and LD stabilizes and no sudden nonlinear accumulation occurs. LD pharmacokinetic parameters are however, comparable to treatment-naïve, stable, and fluctuating parkinsonian patients (Nutt et al., 1992), irrespective of their disease states (Contin et al., 2001).

### 6.2.1.1.2  *Pharmacodynamics*

So far, four types of responses to levodopa have been realized, that is, the long-duration response (LDR), short-duration response (SDR), dyskinesias and negative response (Nutt and Holford, 1996). The SDR, measured in minutes to hours, is positively correlates with the pharmacokinetics of levodopa and is what causes erratic vacillations in motor performance. LDR to long-term levodopa therapy is measured in days and its detection is via the gradual surge in motor dysfunction over a number of days after withdrawing of the drug from PD patients. The negative effect is the deteriorating of motor activity, whose time course ranges from minutes to an hour and mostly manisfests as the SDR of levodopa or apomorphine disappears. LD-associated dyskinesias can occur in varying forms and are easily encountered complications of long-term levodopa therapy. These instabilities with motor function are superimposed on a diurnal motor disparity, varying from normal circadian motor rhythms (Frankel et al., 1990).

It is proposed that, the intensified Parkinsonian symptoms observed following LD therapy is caused by the inhibitory influence of low concentration of dopamine on striatal dopamine receptors with opposed action, rather than those receptors stimulated by higher concentrations to produce the SDR (Paalzow and Paalzow, 1986). In advanced PD, this is a unique response possibly due to the prominence of SDR, and may happen when the plasma levodopa concentration reaches the locality of minimum drug effect. The approach of giving frequent small doses can add to the incidence of erratic variabilities in clinical response imposed by pharmacokinetic factors.

Levodopa-associated dyskinesias are connected to its SDR. "Off" dystonia occurs when plasma concentrations of levodopa are minimum and the SDR fades. "Peak-dose" dystonia on the other hand is observed for the period of the SDR while "biphasic" dyskinesia is observed as the SDR surfaces and also when it disappears. While progression of the disease results in a significant reduction in the duration of motor response, that of dyskinesias remains almost unaltered (Mouradian et al., 1989). At the outset, clinical response to levodopa lasts longer than the duration of dyskinesia. The clinical response, however, gradually parallels the dyskinesia profile.

### 6.2.1.2  *DOPAMINE AGONISTS*

Structurally, dopamine agonists can be classified into ergoline derivatives which are the ergot alkaloids (bromocriptine, pergolide, lisuride, and cabergoline), nonergoline derivatives (pramipexole, piribedil, and ropinirole), and the aporphines (apomorphine). Most of these agonists possess an ethanolamine moiety incorporated within their chemical structures (Fig. 6.1) (Deleu et al., 2002). Dopamine agonists act by their ability to stimulate directly, dopaminergic receptors. Currently, there exist at least five distinct dopamine receptors ($D_1$ to $D_5$), but are usually classified as $D_1$-like ($D_1$, $D_5$) and $D_2$-like ($D_2$, $D_3$, $D_4$) due to the lack of a selective agonist for each distinct receptor (Vallone et al., 2000). Principally, adenylate cyclase-associated

$D_1$ receptors are located on intrastriatal neurons. $D_2$ receptors, however, are present mainly at axons of the descending corticostriatal tract and are not associated with adenylate cyclase, and may even inhibit its activity. These two afore-mentioned receptors are believed to be postsyn-aptic while $D_3$ receptors are considered partially pre- and postsynaptic (Deleu et al., 2002).

### 6.2.1.2.1 Pharmacokinetics

When orally administered, bromocriptine is rapidly absorbed but with a low systemic bioavailability, due to its extensive first-pass metabolism (Friis et al., 1979a). Metabolites of bromocriptine, which are pharmacologi-cally inactive, are excreted mainly (94–98%) in the bile and feces and only 3–6% of the drug given orally is excreted unchanged in the urine. Oxidation of the proline residue as part of the tricyclic peptide moiety results in the forma-tion of hydroxylates. These hydroxylates are then subjected to glucuronidation. Bromocrip-tine has rapid entry into the CNS with about 8% of the drug crossing the BBB at approxi-mately 0.3 h after administration (Friis et al., 1979b).

Following absorption after an oral adminis-tration, pergolide undergoes extensive hepatic metabolism. The drug partially inhibits cyto-chrome P450 (CYP) 3A4 and strongly inhibits CYP2D6 (Wynalda and Wienkers, 1997). Atleast 10 inactive pergolide metabolites have been identified to be excreted in urine and feces. Within 48 h, 55% of the drug is excreted in urine and the remainder in the feces (Ruben et al., 1981).

Apomorphine administered orally is associ-ated with poor absorbed with a bioavailability of less than 4% (Gancher et al., 1991). It therefore requires 10–20 times the parenteral dose to achieve similar therapeutic response (Lees et al., 1989). Concomitant administration with vitamin C further decreases the relative bioavailability of sublingual apomorphine (Gancher et al., 1991). The drug is very lipo-philic. The metabolic pathways of apomorphine include glucuronidation, sulfate conjugation and autooxidation with the latter being the main metabolic route. No active metabolites of apomorphine however, have been identified. Its total body clearance exceeds the hepatic blood flow, implying extrahepatic clearance. However, just about 4% of the administered dose have been detected unchanged in urine (Deleu et al., 2002).

Pramipexole, however, is readily absorbed with distinct pharmacokinetic profiles for men and women, respectively. The AUC for each dose level being 35–43% greater in women. This is mostly because, oral clearance decreases 24–27% in women. Unlike pergolide, the prami-pexole undergoes minimal hepatic biotransfor-mation and is excreted virtually unchanged in the urine by renal tubular secretion (Bennett and Piercey, 1999). Similarly, ropinirole absorp-tion after oral administration is high (94%). However, unlike pramipexole, ropinirole is metabolized extensively by the hepatic micro-somal enzyme system (CYP1A2, CYP3A4, and CYP2D6) (Wynalda and Wienkers, 1997). The *N*-despropyl metabolite of ropinirole is further metabolized to 7-hydroxy and carboxylic acid derivatives. These metabolites subsequently undergo glucuronidation and are excreted in the urine. The 7-hydroxy-ropinirole metabolite is the only metabolite that possesses significant in vivo agonistic activity on dopaminergic recep-tors (Deleu et al., 2002).

Cabergoline is well absorbed when orally administration. The drug undergoes extensive hydrolysis in the liver leading to the breaking of the acylurea bond of the urea moiety and is excreted in the bile and feces. Its extremely long half-life is advantageous, as it alleviates the problem of "off" reactions and a once daily administration is favored (Andreotti et al., 1995).

Piribedil extensively undergoes first-pass metabolism, resulting in poor oral bioavailability. Its metabolic pathways, primarily demethylation, *p*-hydroxylation, and *N*-oxidation, give rise to numerous metabolites, with 1-(3,4-dihydroxybenzyl) 4-(2-pyrimidinyl)-piperazine, being the pharmacologically active one (Sarati et al., 1991). Jenner et al. (1973) identified that, the metabolites are renally excreted with no amount of the parent compound excreted unchanged in the urine.

### 6.2.1.2.2  *Pharmacodynamics*

Presynaptic $D_2$ receptor stimulation is thought to result in neuroprotective effects while anti-Parkinsonian activity is closely associated with stimulation of the postsynaptic $D_2$ receptor. It's been proposed that, optimal therapeutic response to dopamine agonist requires stimulation of both $D_1$ and $D_2$ receptors. Sure thing, the response to dopamine agonist activity on $D_2$ receptors is enhanced by prior stimulation of $D_1$ receptors (Robertson and Robertson, 1986). Stimulation of these $D_1$ receptors, however, has been linked to the development of dyskinesia (Bedard et al., 1999). Most of the dopamine agonists currently utilized in the treatment of PD are $D_2$ agonists with or without $D_1$ receptor activity. Ropinirole and pramipexole are dopamine agonists that show some selective agonistic activity at $D_3$ receptors (Lange, 1998).

Research into the clinical relevance of dopamine agonists, either as monotherapy or as adjuvant therapy to levodopa, continues to attract interest. In particular is their potential to delay the occurrence of late motor symptoms induced by levodopa. Dopamine agonists such as bromocriptine, ropinirole, pramipexole, and pergolide have been shown, in a number of studies, to act as free radical scavengers against hydroxyl radicals and nitric oxide radicals (Lange et al., 1994; Zou et al., 1999). The antioxidant effect of dopamine agonist may be exerted through $D_2$-autoreceptor stimulation, which decreases dopamine formation and metabolism (Grünblatt et al., 1999).

### 6.2.1.3  *PERGOLIDE*

Pergolide is a synthetic $D_2$ receptor-activating ergoline moiety which is also a partial $D_1$ receptor agonist with affinity for $D_3$ receptors (Langtry and Clissold, 1990). In advanced disease, the drug, used as an adjunct to levodopa, reduces motor fluctuations and improves motor function (Olanow et al., 1994).

The most significant adverse effects seen with pergolide are dyskinesias (30%), hallucinations (10–20%), disturbances of the gastrointestinal system (10%), insomnia (10%), somnolence, mild and transient bradycardia, and rarely hepatic injury (Langtry and Clissold, 1990). Coadministration with domperidone may substantially improve tolerability (Barone et al., 1999).

### 6.2.1.4  *APOMORPHINE*

Structurally, apomorphine is a derivative of the opioid morphine. Apomorphine has short-lasting stimulatory effect on $D_1$ and $D_2$ receptors. At low doses, it inhibits dopamine turnover by acting on presynaptic autoreceptors. The drug, as an adjunct to levodopa, is used in rescue therapy for severe and levodopa-resistant "off" periods during treatment with levodopa. Acute challenge with apomorphine has been employed in the testing of dopaminergic responsiveness in PD. The clinical response to apomorphine may therefore predict the effect of chronic levodopa therapy in 90% of PD patients (Lees, 1993).

The most significant adverse effects seen with apomorphine therapy are nausea and vomiting (almost 73% of patients), which can be attenuated by administration of domperidone before apomorphine therapy. Neuropsychiatric adverse effects, however, are associated with long-term use, but with a lower incidence than with other dopamine agonists (Frankel et al., 1990). Long term use have been associated with significant increase, about 67%, in the mean daily duration of dyskinesia compared with placebo (Ostergaard et al., 1995).

### 6.2.1.5 PRAMIPEXOLE

Pramipexole, a synthetic aminobenzothiazole derivative, has selective actions on $D_2$-like receptors, particularly with agonist activity on $D_3$ receptors (Piercey, 1998). It is believed that pramipexole's selective stimulation of $D_3$ receptors is responsible for its antiparkinsonian activity. By stimulating presynaptic dopamine autoreceptors, pramipexole decreases the synthesis and turnover of endogenous dopamine, thus decreases oxidative stress and associated neuronal degeneration (Deleu et al., 2002).

The adverse effect profile of pramipexole is associated with stimulation of peripheral and central dopamine receptors. The development of tolerance may be avoided by slow titration of pramipexole or the addition of domperidone in sensitive patients. Because the drug is a nonergoline its activity is devoid of the rare but nuisant ergot-related adverse effects associated with bromocriptine and pergolide therapy. One disturbing effect is the sudden onset of sleep

which may precipitate road and other forms of accidents. Patients beginning treatment with pramipexole should be counseled on the risk and the need to exercise extreme caution when driving or operating heavy machinery (Frucht et al., 1999).

### 6.2.1.6 CABERGOLINE

Cabergoline is a dopamine agonist with high affinity for the $D_2$-like receptors. This ergot derivative binds to $D_2$ and $D_3$ receptors, but possesses decreased affinity for $D_1$ receptors (Fariello, 1998). The binding of cabergoline to its receptors is long lasting, up to 72 h. It significantly increases the threshold for dyskinesia and further delays the incidence of serious motor complications (Rinne et al., 1998).

Nausea, vomiting, gastritis, and dyspepsia represents the main adverse effects of cabergoline, occuring in a third of patients. Other significant adverse effects include dizziness, hypotension, and peripheral edema. The incidence of constrictive pericarditis and pleuropulmonary disease with cabergoline therapy remain rare (Hutton et al., 1996; Rinne et al., 1998).

Therapy with dopamine-receptor agonists is associated with unwanted effects, such as impulse control disorders, including pathological gambling, binge eating, and hypersexuality. Pramipexole is the most implicated agent in pathological gambling. Its induction of pathological gambling is believed to be as a result of disproportionate stimulation of dopamine $D_3$ receptors, primarily localized to the limbic system (Dodd et al., 2005).

### 6.2.1.7 AMANTADINE

Initially, amantadine was indicated for the prophylaxis for influenza A, but emerging evidence found it to also posses antiparkinsonian activity (Schwab et al., 1969). Clinically, amantadine is used in monotherapy or as add-on to levodopa/PDI or dopamine receptor agonists in early and late stage PD (Deleu et al., 2002).

### 6.2.1.7.1  Pharmacokinetics

Amantadine has slow and variable absorption if administered orally. It's been reported that, amantadine achieves steady-state plasma concentrations often within 4–7 days in both healthy volunteers and PD patients. It is extensively bound to tissues with a lesser percentage, as free form, in circulation (Aoki and Sitar, 1988). Amantadine's apparent Vd is inversely related to the dose and is partly responsible for a notable dose-dependent logarithmic increase in plasma concentration. Dosage regimens that results in plasma concentrations below 1–1.5 mg/L is advised to reduce the risk of toxicity (Horadam et al., 1981). Only 5–15% of the drug is metabolized via acetylation (Aoki and Sitar, 1988). The remaining fraction (90%) is almost entirely excreted by the kidney, particularly by renal tubular secretion. As such, reduction in dosage is ideal in people of advanced age as well as in individuals with comprimized renal function (Deleu et al., 2002). Amantadine therapy is marked by hallucinations, insomnia, nightmares, livedo reticularis, and swelling of the ankles (Münchau and Bhatia, 2000).

### 6.2.1.7.2  Mechanism of Action

There is no known established mechanism of amantadine in the treatment of PD, as well as drug-associated extrapyramidal systems. However, several mechanisms have been proposed, the most likely being the enhancement of dopaminergic transmission and the mild inhibition of muscarinic receptors by amantadine (Bailey and Stone, 1975; Kulisevsky and Tolosa, 1990). Amantadine may enhance dopaminergic transmission by enhancing dopamine synthesis, and blocking presynaptic reuptake of dopamine and noradrenaline (Oertel and Quinn, 1996). The drug predominantly improves impaired voluntary movement, as well as rigor. Moreover, amantadine at the therapeutic low micromolar concentration, noncompetitively inhibits the $N$-methyl-D-aspartate (NMDA) receptor-associated exocytosis of acetylcholine from rat neostriatum in vitro (Stoof et al., 1992). This means that, likewise to centrally-acting muscarinic receptor antagonists, the neuroprotective NMDA antagonism may contribute to the positive outcome of amantadine therapy (Deleu et al., 2002). It is thought that, amantadine's NMDA antagonistic properties may be of importance in the management of dyskinesia, as viable sign of glutamatergic projections to the basal ganglia has been demonstrated to be relevant to the progression of dyskinesias (Chase et al., 1996). Stimulation of cerebral dopa decarboxylase has been reported by Deep et al. (1996) to contribute to its therapeutic benefits in PD.

### 6.2.2  ANTICHOLINERGICS

Although they are used as adjuncts in the management of PD, anticholinergic drugs, notwithstanding, are relevant to PD management since they offer different action path that attenuate some of the worrying symptoms of the disease, predominantly, reflex resting tremor. Anticholinergics are employed as monotherapy during early stages but as adjunctive therapy administered with L-dopa at advanced stages of the disease (Olanow and Koller, 1998; Sweeney, 1995). They are as such potentially advantageous in delaying the need for LD treatment when used early in the disease progression. Anticholinergics used in the management of PD makes it possible to reduce the dose of LD which would have otherwise been given in advanced stage of the disease hence prolonging the use of LD (Bassi et al., 1986). In practice, these drugs are also relevant as they help attenuate extrapyramidal symptoms seen with treatment with antipsychotics (Brocks, 1999).

### 6.2.2.1 PHARMACOKINETICS

#### 6.2.2.1.1 Absorption

All of the anticholinergic drugs but benztropine used in PD (ACPs), are given as the hydrochloride derivative of its salt. Benztropine is administered in the form of its mesylate salt. The absorption of ACPs, with the exception of benztropine mesylate, is comparatively faster after oral administration of the solid formulation with a time to maximal plasma concentrations ($T_{max}$) mostly less than 2.5 h. The variations in the $T_{max}$ observed with orally administered ACPs have been realized to be dependent on the administered dose of the drug (Brocks, 1999).

#### 6.2.2.1.2 Distribution

For each of the ACP which is given intravenously, a Vd in excess of total body water has been reported. In the postdistributive phase, the plasma concentrations of these drugs decrease and accounts for the prolonged $t_{1/2}$ values. For drugs like procyclidine, relatively low Vd and low plasma clearance have been identified to account for the relatively elevated plasma levels of the drug when compared with other ACPs (Whiteman et al., 1985).

Tissue distribution of ACPs has been extensively reported in animal experimental models. The tissue:plasma partition of these agents are shown to be large. It has been shown that there is rapid uptake of biperiden, orphenadrine, and ethopropazine into targeted site of drug action. It has been demonstrated that biperiden achieves maximal brain levels in 3–10 min of iv infusion of the drug (Yokogawa et al., 1992). Intravenous bolus doses of ethopropazine in rats, achieved maximal concentrations in brain tissues in less than 30 min (Maboudian-Esfahani and Brocks, 1999). Radiolabeled orphenadrine and tofenacine, the principal $N$-demethylated metabolite of ophenadrine, achieved maximum levels in rat brain in less than 15 min and by 60 min of intraperitoneal administration (Roozemond et al., 1968). The high concentrations and short $T_{max}$ of the drugs are attributed to their high lipophilic nature. It has been proposed that, the presence of sphingomyelin and phosphatidyl choline in the brain (Lehninger et al., 1993) increases the tissue uptake of ACPs (Brocks, 1999; Ishizaki et al., 1998).

The extent of plasma protein binding of ACP is poorly understood in humans. However, in a rat model, ethopropazine has been identified to extensively bind (>95%) nonlinearly to plasma proteins (Maboudian-Esfahani and Brocks, 1999a,b). Orphenadrine and biperiden on the other hand are approximately 90% bound to plasma proteins (Ellison, 1972; Nakashima et al., 1987, 1993).

#### 6.2.2.1.3 Elimination

There is wide-ranging metabolism of ACPS in humans and animals. These agents contain an aliphatic tertiary amine moeity that is susceptible to $N$-dealkylation and $N$-oxidation (He et al., 1995; Hespe, 1965). The alicyclic groups of procyclidine and trihexyphenidyl have been established to experience hydroxylation (Nation et al., 1978; Peame et al., 1982).

After oral administration in rats, benztropine is metabolized into eight phase I metabolites and four other glucuronide conjugates that undergo biliary and urinary excretion (He et al., 1995). The clearance of the ACPs is low relative to hepatic blood flow hence, there is decrease hepatic first-pass effect associated when given orally (Brocks, 1999). Plasma elimination of trihexyphenidyl following prolong treatment course follows a two-compartmental model (Deleu et al., 2002). The drug is mainly metabolized via hydroxylation of the alicyclic moeity and is excreted 76% renally (Nation et al., 1978).

In addition to *N*-demethylation, orphenadrine is also metabolized via deamination and conjugation with 60% excretion via the renal system within 72 h. Only 8% of the drug is excreted unchaged (Guo et al., 1997). The drug interacts with CYP2B6 and CYP2D6, and cause up to 45–57% and 80–90% reduction in microsomal

enzymes, respectively (Ellison, 1972). CYP1A2, CYP2A6, CYP3A4, and CYP2C19 marker activities are however decreased to a lesser extent. Only about a percentage of biperiden is excreted unmetabolized in urine with a chunk of the drug metabolized (Grimaldi et al., 1986).

### 6.2.2.2 MECHANISM OF ACTION

It has been hypothesized that in PD, there is a discordance in the dopaminergic and cholinergic neurological pathways. Researchers believe that cholinergic inhibitors can correct this imbalance during the early stages of PD. The cholinergic antagonists achieve via a decreased extent of neostriatal acetylcholine-mediated neurotransmission (Olanow and Koller, 1998; Sweeney, 1995). Although the mechanism of action of ACP is most probably attributed to their potential blockade muscarinic receptors in the striatum (Fig. 6.3), studies have identified that many of these agents potently inhibit the presynaptic carrier-mediated dopamine

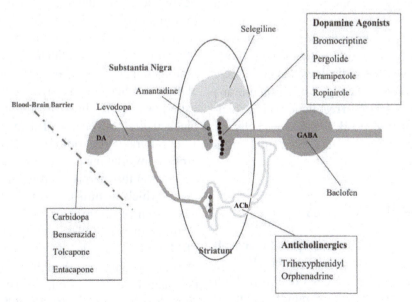

**FIGURE 6.3** Site of action of drugs used in Parkinson's disease.

transport process, where dopamine molecule inflowing to the nerve terminal through the carrier is complemented by two or more sodium ions and one chloride ion (Krueger, 1990). Also, ACPs have potent agonist action at the noradrenergic synapse with some having NMDA receptor antagonist action. This is established to reverse akinesia and/or augment the effects of LD in animal models of PD. ACPs are beneficial for managing tremor-predominant disease phenotypes. Procyclidine seems to have the most potent NMDA antagonistic activity amongst the ACP.

### 6.2.3   *MONOAMINE OXIDASE B INHIBITORS*

No drug has been shown, beyond doubt, to arrest neurodegeneration in PD. As the disease advances, a growing number of dopamine-producing neurons in the substantia nigra die. The resulting decrease in dopaminergic neuronal mass subsequently fuels dopaminergic turnover of the remaining neurons with an upsurge in hydrogen peroxide production, an outcome of dopamine metabolism in the MAO-B pathway. Hydrogen peroxide in turn initiates the production of toxic hydroxyl radicals causing oxidative stress with resultant damage to cell membrane (Chiueh et al., 1994). Treatment with selegiline, an inhibitor MAO-B, is therefore believed to be neuroprotective in the substantia nigra (Marsden, 1990).

#### 6.2.3.1   *PHARMACOKINETICS*

Selegiline has rapid GIT absorption and well distributed in tissues, such as the brain. The drug rapidly penetrates into the CNS due to its high lipid-solubility. Selegiline is rapidly metabolized in hepatocytes by microsomal cytochrome

P450 system to L-*N*-desmethyl-selegiline, L-methamphetamine and L-amphetamine, with the latter accounting for most of the metabolite pool. All these metabolites can be identified in human serum, cerebrospinal fluid, and urine (Mahmood, 1997).

The clearance of selegiline is many times higher than liver blood flow, which indicates the involvement of extrahepatic elimination. Addition of selegiline to LD may enable a reduction in L-dopa dose by 10–15%, sometimes up to 30% (Oertel and Quinn, 1996). The addition often can also reduce mild LD response fluctuations. Monotherapy in de novo patients delay the need for additional treatment by approximately a year (Parkinson's Study Group, 1998). Side effects of LD including dyskinesias, orthostatic hypotension and psychiatric problems are, however, likely increased by selegiline (Churchyard et al., 1997).

#### 6.2.3.2   *MECHANISM OF ACTION*

Selegiline slows the metabolism of dopamine to dihydroxyphenylacetic acid and hydrogen peroxide via selective and irreversible inhibition of intra- and extracellular MAO-B. It also inhibits dopamine's reuptake from the synaptic cleft (Münchau and Bhatia, 2000). In animal experiments selegiline, by inhibiting MAO-B, blocked the conversion of 1-methyl-4-phenyl-1,2,3,6-tetrahydropyridine (MPTP) to *N*-methyl-4-phenylpyridinium (MPP$^+$), an exogenous neurotoxin that harms neurons in the substantia nigra, which in turn prevented MPTP-induced Parkinsonism in primates (Fig. 6.4). It was therefore theorized that antagonizing the activity of MAO-B might also dawdle the development of PD in humans, though via a dissimilar mechanism to MPTP-induced parkinsonism in animals (Münchau and Bhatia, 2000).

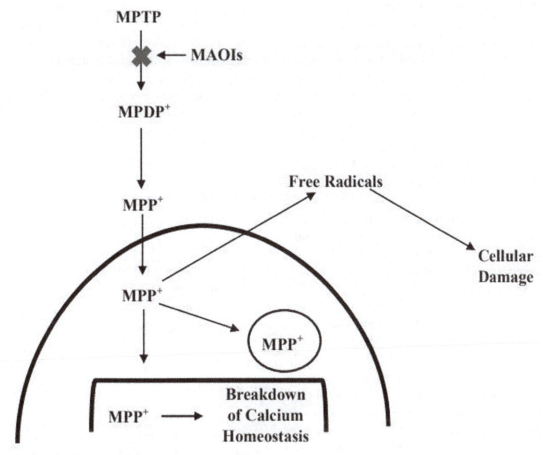

**FIGURE 6.4**   Mechanism of action of monoamine oxidase B inhibitors in management of Parkinson's disease. MPDP$^+$, 1-methyl-4-phenyl-2,3-dihydropyridinium; MPP$^+$, *N*-methyl-4-phenylpyridinium; MPTP, 1-methyl-4-phenyl-1,2,3,6-tetrahydropyridine.

Selegiline can elicit symptomatic effects which is realized in the augmented dopamine presence or its amphetamine metabolites. Nevertheless, the exact mechanism for increasing dopaminergic neurotransmission has not been fully elucidated. This is because MAO-B in the striatum is largely present in glia. In addition, selegiline possesses trophic effects, free of inhibition of MAO-B (Tatton and Greenwood, 1991). Other MAO-B inhibitors include rasagiline.

### 6.3   FUTURE OPPORTUNITIES AND CHALLENGES

Some studies have established the safety of gene therapy in the management of PD (Witt and Marks, 2011). One such study focused on improving dopamine availability to the striatum by employing a much more continues delivery via gene transfer to target protein expression. Also the same study targeted reducing STN activity with local induction of GABA expression via gene transfer. Gene transfer was also employed in protecting, as well as reinstating nigrostriatal neuronal role with trophic factor expression (Lewitt et al., 2011). Hopefully, these gene therapy progress into effective pharmacotherapy.

With the understanding of the pathophysiology and modes of treatment being a significant progress in modern medicine, there continuous to be growing research into identifying much more effective management and/or treatment options for PD. Building on the MPTP primate model of PD, development of fetal mesencephalic and other tissue implant which may have the ability to reverse parkinsonism although it is still in the experimental stage (Kopin, 1993). There is however, optimism surrounding these techniques which could also be of importance in other neuropsychiatric disorders.

The research into agents that retard the development of PD has obtained significant consideration with some success although further research on them are ongoing. A number of nondopaminergic drugs, such as $\alpha_2$-adrenergic antagonists are also under various stages of development (Schapira, 2007). The $\alpha_2$-adrenoceptor antagonists, idazoxan and fipamezole, decrease levodopa-induced dyskinesia and lengthen the duration of action of levodopa in MPTP-managed primates (Savola et al., 2003).

## 6.4  CONCLUSION

This chapter has dealt extensively with drugs used for the management of PD. The dopamine replacement therapies target dopaminergic pathways and are the mainstay drugs used, whiles the other drug groups, such as the anticholinergics affect the nondopaminergic

pathway and are used as adjuncts in the management of PD. Levodopa remains the gold-standard in the management, but it is associated with dyskinesia. Although the current treatments only offer symptomatic relief, they do not affect disease progression. This necessitates the development of other therapeutic approaches that may affect the natural history of PD.

## KEYWORDS

- bradykinesia
- neurodegeneration
- dopamine
- levodopa
- Parkinson's disease

## REFERENCES

Alexander, G. M.; Schwartzman, R. J.; Grothusen, J. R.; Gordon, S. W. Effect of Plasma Levels of Large Neutral Amino Acids and Degree of Parkinsonism on the Blood-to-Brain Transport of Levodopa In Naïve and MPTP Parkinsonian Monkeys. *Neurology* **1994**, *44*, 1491–1499.

Andreotti, A. C.; Pianezzola, E.; Persiani, S.; Pacciarini, M. A.; Strolin Benedetti, M.; Pontiroli, A. E. Pharmacokinetics, Pharmacodynamics, and Tolerability of Cabergoline, a Prolactin-Lowering Drug, After Administration of Increasing Oral Doses (0.5, 1.0, and 1.5 milligrams) in Healthy Male Volunteers. *J. Clin. Endocrinol. Metab.* **1995**, *80*, 841–845.

Aoki, F. Y.; Sitar, D. S. Clinical Pharmacokinetics of Amantadine Hydrochloride. *Clin. Pharmacokinet.* **1988**, *14*, 35–51.

Astarloa, R.; Mena, M. A.; Sanchez, V.; de la Vega, L.; de Yébenes, J. G. Clinical and Pharmacokinetic Effects of a Diet Rich in Insoluble Fiber on Parkinson's Disease. *Clin. Neuropharmacol.* **1992**, *15*, 375–380.

Bagetta, V.; Sgobio, C.; Pendolino, V.; et al. Rebalance of Striatal NMDA/AMPA Receptor Ratio Underlies the Reduced Emergence of Dyskinesia During D$_2$-Like Dopamine Agonist Treatment in Experimental Parkinson's Disease. *J. Neurosci.* **2012,** *32,* 17921–17931.

Bailey, E. V.; Stone, P. W. The Mechanism of Action of Amantadine in Parkinsonism: a Review. *Arch. Int. Pharmacodyn. Ther.* **1975,** *216,* 246–260.

Barone, P.; Bravi, D.; Bermejo-Pareja. F.; Marconi, R.; Kulisevsky, J.; Malagù, S.; Weiser, R.; Rost, N. Pergolide Monotherapy in the Treatment of Early PD: A Randomized, Controlled Study. Pergolide Monotherapy Study Group. *Neurology* **1999,** *53,* 573–579.

Baruzzi, A.; Contin, M.; Riva, R.; Procaccianti, G.; Albani, F.; Tonello, C.; Zoni, E.; Matinelli, P. Influence of Meal Ingestion Time on Pharmacokinetics of Orally Administered Levodopa in Parkinsonian Patients. *Clin. Neuropharmacol.* **1987,** *10,* 527–537.

Bassi, S.; Albizzati, M. G.; Calloni, E.; Sbacchi, M.; Frattola, L. Treatment of Parkinson's Disease with Orphenadrine Alone and in Combination With L-dopa. *Br. J. Clin. Pract.* **1986,** *40,* 273–275.

Bedard, P. J.; Blanchet, P. J.; Levesque, D.; et al. Pathophysiology of L-Dopa-Induced Dyskinesias. *Mov. Disord.* **1999,** *14*(1), 4–8.

Bennett Jr., J. P.; Piercey, M. F. Pramipexole: A New Dopamine Agonist for the Treatment of Parkinson's Disease. *J. Neurol. Sci.* **1999,** *163,* 25–31.

Bianchine, J. R.; Messiha, F. S.; Hsu, T. H. Peripheral Aromatic L-Amino Acids Decarboxylase Inhibitor in Parkinsonism. II. Effect on Metabolism of L-2- 14 C-Dopa. *Clin. Pharmacol. Ther.* **1972,** *13,* 584–594.

Blandini, F. Neural and Immune Mechanisms in the Pathogenesis of Parkinson's Disease. *J. Neuroimmune Pharmacol.* **2013,** *8,* 189–201.

Braak, H.; Del Tredici, K.; Rüb, U.; de Vos, R. A.; Jansen Steur, E. N.; Braak, E. Staging of Brain Pathology Related to Sporadic Parkinson's Disease. *Neurobiol. Aging.* **2003,** *24,* 197–211.

Brocks, D. R. Anticholinergic drugs used in Parkinson's disease: An Overlooked Class of Drugs From a Pharmacokinetic Perspective. *J. Pharm. Pharmaceut. Sci.* **1999,** *2* (2), 39–46.

Carter, J. H.; Nutt, J. G.; Woodward, W. R.; Hatcher, L. F.; Trotman, T. L. Amount and Distribution of Dietary Protein Affects Clinical Response to Levodopa in Parkinson's Disease. *Neurology* **1989,** *39,* 552–556.

Cedarbaum, J. M. Clinical Pharmacokinetics of Anti-Parkinsonian Drugs. *Clin. Pharmacokinet.* **1987,** *13,* 141–178.

Chase, T. N.; Engber, T. M.; Mouradian, M. M. Contribution of Dopaminergic and Glutamatergic Mechanisms to the Pathogenesis of Motor Response Complications in Parkinson's Disease. *Adv. Neurol.* **1996,** *69,* 497–501.

Chiueh, C. C.; Wu, R. M.; Mohanakumar, K. P.; Sternberger, L. M.; Krishna, G.; Obata, T.; Murphy, D. L. *In vivo* Generation of Hydroxyl Radicals and MPTP-Induced Dopaminergic Toxicity in the Basal Ganglia. *Ann. N.Y. Acad. Sci.* **1994,** *738,* 25–36.

Churchyard, A.; Mathias, C. J.; Boonkonchuen, P.; Lees, A. J. Autonomic Effects of Selegeline: Possible Cardio-Vascular Toxicity in Parkinson's Disease. *J. Neurol. Neurosurg. Psychiatry.* **1997,** *63,* 228–234.

Contin, M.; Riva, R.; Martinelli, P.; Albani, F.; Avoni, P.; Baruzzi, A. Levodopa Therapy Monitoring in Patients with Parkinson Disease: A Kinetic-Dynamic Approach. *Ther. Drug Monit.* **2001,** *23,* 621–629.

Cui, G.; Jun, S. B.; Jin, X.; Pham, M. D.; Vogel, S. S.; Lovinger, D. M.; Costa, R. M. Concurrent Activation of Striatal Direct and Indirect Pathways During Action Initiation. *Nature* **2013,** *494,* 238–242.

Da Prada, M.; Kettler, R.; Zürcher, G.; Schaffner, R.; Haefely W. E. Inhibition of Decarboxylase and Levels of Dopa and 3-O-Methyldopa: A Comparative Study of Benserazide Versus Carbidopa in Rodents and of Madopar Standard Versus Madopar HBS in Volunteers. *Eur. Neurol.* **1987,** *27*(1), 9–20.

Deep, P.; Dagher, A.; Sadikot, A.; Gjedde, A.; Cumming, P. Stimulation of Dopa Decarboxylase Activity in Striatum of Healthy Human Brain Secondary to NMDA Receptor Antagonism with a Low Dose of Amantadine. *Synapse* **1999,** *34,* 313–318.

Deleu, D.; Northway, M. G.; Hanssens, Y. Clinical Pharmacokinetic and Pharmacodynamic Properties of Drugs Used in the Treatment of Parkinson's Disease. *Clin. Pharmacokinet.* **2002,** *41* (4), 261–309.

Descombes, S.; Bonnet, A.; Gasser, U.; Thalamas, C.; Dingemanse, J.; Arnulf, I.; Bareille, M. P.; Agid, Y.; Rascol, O. Dual-Release Formulation, a Novel Principle in L-Dopa Treatment of Parkinson's Disease. *Neurology.* **2001,** *56,* 1239–1242.

Dingemansex, J. Issues Important for Rational COMT Inhibition. *Neurology* **2000,** *55,* 24–32.

Ehringer, H.; Hornykiewicz, O. Verteilung Von Noradrenalin Und Dopamin (3-Hydroxytyramin) Im Gehirn Des Menschen Und Ihr Verhalten Bei Erkrankungen Der Extrapyramidalen Systems. *Wien. Klin. Wochenschr.* **1960,** *38,* 1236–1239.

Ellison, T. Metabolic Studies of 3H-Orphenadrine Citrate in the Rat, Dog and Rhesus Monkey. *Arch. Int. Pharmacodyn. Ther.* **1972,** *195,* 213–230.

Fariello, R. G. Pharmacodynamic and Pharmacokinetic Features of Cabergoline: Rationale for Use in Parkinson's Disease. *Drugs* **1998,** *55*(1), 10–16.

Frankel, J. P.; Pirtosek, Z.; Kempster, P. A.; Bovingdon, M.; Webster, R.; Lees, A. J.; Stern, G. M. Diurnal Differences in Response to Oral Levodopa. *J. Neurol. Neurosurg. Psychiatry.* **1990,** *53,* 948–950.

Friis, M. L.; Gron, U.; Larsen, N. E.; Pakkenberg, H.; Hvidberg, E. F. Pharmacokinetics of Bromocriptine During Continuous Oral Treatment of Parkinson's Disease. *Eur. J. Clin. Pharmacol.* **1979,** *15,* 275–280.

Friis, M. L.; Paulson, O. B.; Hertz, M. M. Transfer of Bromocriptine Across the Blood–Brain Barrier in Man. *Acta Neurol. Scand.* **1979,** *59,* 88–95.

Frucht, S.; Rogers, J. D.; Greene, P. E.; Gordon, M. F.; Fahn, S. Falling Asleep at the Wheel: Motor Vehicle Mishaps in Persons Taking Pramipexole and Ropinirole. *Neurology* **1999,** *52,* 1908–1910.

Gancher, S. T.; Nutt, J. G.; Woodward, W. R. Absorption of Apomorphine by Various Routes in Parkinsonism. *Mov. Disord.* **1991,** *6,* 212–216.

Ghika, J.; Gachoud, J.; Gasser, U.; Study Group L-DD-R. Clinical Efficacy and Tolerability of a New Levodopa/Benserazide Dualrelease Formulation in Parkinsonian Patients. L-Dopa Dual-Release Study Group. *Clin. Neuropharmacol.* **1997,** *20,* 130–139.

Greenamyre, J. T.; Hastings, T. G. Biomedicine. Parkinson's-Divergent Causes, Convergent Mechanisms. *Science.* **2004,** *304,* 1120–1122.

Grimaldi, R.; Perucca, E.; Ruberto, G.; Gelmi, C.; Trimarchi, F.; Hollmann, M.; Crema, A. Pharmacokinetic and Pharmacodynamic Studies Following the Intravenous and Oral Administration of the Antiparkinsonian Drug Biperiden to Normal Subjects. *Eur. J. Clin. Pharmacol.* **1986,** *29,* 735–737.

Grünblatt, E.; Mandel, S.; Berkuzki, T.; Youdim, MB. H. Apomorphine Protects Against MPTP-Induced Neurotoxicity in Mice. *Mov. Disord.* **1999,** *14,* 612–618.

Guo, Z.; Raeissi, S.; White, R. B.; Stevens, J. C. Orphenadrine and Methimazole Inhibit Multiple Cytochrome P450 Enzymes in Human Liver Microsomes. *Drug Metab. Dispos.* **1997,** *25,* 390–393.

He, H.; McKay, G.; Midha, K. K. Phase I and II Metabolites of Benztropine in Rat Urine and Bile. *Xenobiotica* **1995,** *25,* 857–872.

Hespe, W.; De Roos, A. M.; Nauta, W. T. H. Investigation into the Metabolic Fate of Orphenadrine Hydrochloride

After Oral Administration to Male Rats. *Arch. Int. Pharmacodyn. Ther.* **1965,** *156,* 180–200.

Hickey, P.; Stacy, M. Adenosine A$_{2A}$ Antagonists in Parkinson's Disease: What's Next? *Curr. Neurol. Neurosci. Rep.* **2012,** *12,* 376–385.

Hinterberger, H.; Andrews, C. J. Catecholamine Metabolism During Oral Administration of Levodopa. *Arch. Neurol.* **1972,** *26,* 245–252.

Horadam, V. W.; Sharp, J. G.; Smilack, J. D.; McAnalley, B. H.; Garriott, J. C.; Stephens, M. K.; Prati, R. C.; Brater, D. C. Pharmacokinetics of Amantadine Hydrochloride in Subjects With Normal and Impaired Renal Function. *Ann. Intern. Med.* **1981,** *94,* 454–458.

Hung, A. Y.; Schwarzschild, M. A. Treatment of Parkinson's Disease: What's in the Non-Dopaminergic Pipeline? *Neurotherapeutics* **2014,** *11,* 34–46.

Huot, P.; Fox, S. H.; Brotchie, J. M. The Serotonergic System in Parkinson's Disease. *Prog. Neurobiol.* **2011,** *95,* 163–212.

Hutton, J. T.; Koller, W. C.; Ahlskog, J. E.; et al. Multicenter, Placebo Controlled Trial of Cabergoline Taken Once Daily in the Treatment of Parkinson's Disease. *Neurology* **1996,** *46,* 1062–1065.

Ishizaki, J.; Yokogawa, K.; Nakashima, E.; Ohkuma, S.; Ichimura, F. Influence of Ammonium Chloride on the Tissue Distribution of Anticholinergic Drugs in Rats. *J. Pharm. Pharmacol.* **1998,** *50,* 761–766.

Jafari, S.; Etminan, M.; Aminzadeh, F.; Samii, A. Head Injury and Risk of Parkinson Disease: A Systematic Review and Meta-Analysis. *Mov. Disord.* **2013,** *28,* 1222–1229.

Jenner, P.; Taylor, A. R.; Campbell, D. B. Preliminary Investigation of the Metabolism of Piribedil (ET 495); A New Central Dopaminergic Agonist and Potential Anti-Parkinson Agent. *J. Pharm. Pharmacol.* **1973,** *25,* 749–750.

Kaakkola, S.; Mannisto, P. T.; Nissinen, E.; Vuorela, A.; Mäntylä, R. The Effect of an Increased Ratio of Carbidopa to Levodopa on the Pharmacokinetics of Levodopa. *Acta. Neurol. Scand.* **1985,** *72,* 385–391.

Karstaedt, P. J.; Pincus, J. H.; Coughlin, S. S. Standard and Controlled Release Levodopa/Carbidopa in Patients with Fluctuating Parkinson's Disease on a Protein Redistribution Diet. A Preliminary Report. *Arch. Neurol.* **1991,** *48,* 402–405.

Khor, S. P. & Hsu A. (2007). The Pharmacokinetics and Pharmacodynamics of Levodopa in the Treatment of Parkinson's Disease. *Curr. Clin. Pharmacol.* **2007,** *2,* 234–243.

Kopin, I. J. Parkinson's Disease: Past, Present, and Future. *Neuropsychopharmacology* **1993,** *9,* 1–12.

Krueger, B. K. Kinetics and Block of Dopamine Uptake in Synaptosomes From Rat Caudate Nucleus. *J. Neurochem.* **1990,** *55,* 260–267.

Kulisevsky, J.; Tolosa, E. Amantadine in Parkinson's Disease. In *Therapy of Parkinson's Disease*; Koller, W. C., Paulson, G. W., Eds.; Marcel-Dekker: New York, 1990; pp 143–160.

Kuruma, I.; Bartholini, G.; Tissot, R.; Fletscher, A. Comparative Investigation of Inhibitors of Extracerebral Dopa Decarboxylase in Man and Rats. *J. Pharm. Pharmacol.* **1972,** *24,* 289–294.

Lang, A. E.; Obeso, J. A. Challenges in Parkinson's Disease: Restoration of the Nigrostriatal Dopamine System is not Enough. *Lancet Neurol.* **2004,** *3,*309–316.

Lange, K. W. Clinical pharmacology of Dopamine Agonists in Parkinson's Disease. *Drugs Aging.* **1998,** *13,* 381–389.

Lange, K. W.; Rausch, W. D.; Gsell, W.; Naumann, M.; Oestreicher, E.; Riederer, P. Neuroprotection by Dopamine Agonists. *J. Neural Transm.* **1994,** *43,* 183–201.

Langtry, H. D.; Clissold, S. P. Pergolide: A Review of its Pharmacological Properties and Therapeutic Potential in Parkinson's Disease. *Drugs* 1990, *39,* 491–506.

Leenders, K. L.; Poewe, W. H.; Palmer, A. J.; Brenton, D. P.; Frackowiak, R. S. Inhibition of L-[18F]Fluorodopa Uptake into Human Brain by Amino Acids Demonstrated by Positron Emission Tomography. *Ann. Neurol.* **1986,** *20,* 258–262.

Lees, A. J.; Montastruc, J. L.; Turjanski, N.; Rascol, O.; Kleedorfer, B.; Peyro Saint-Paul, H.; Stern, G. M.; Rascol, A. Sublingual Apomorphine and Parkinson's Disease [Letter]. *J. Neurol. Neurosurg. Psychiatry.* **1989,** *52,* 1440

Lehninger, A. L.; Nelson, D. L.; Cox, M. M. Biological Membranes and Transport. In *Principles of Biochemistry*, 2nd ed.; Worth Publishers: New York; 1993, pp. 269–297.

Lewitt, P. A., Rezai, A. R., Leehey, M. A., et al. AAV2-GAD Gene Therapy for Advanced Parkinson's Disease: A Double-Blind, Sham-Surgery Controlled, Randomised Trial. *Lancet Neurol.* **2011,** *10* (4), 309–319.

Lotti, V. J.; Porter, C. C. Potentiation and Inhbition of Some Central Actions of L(−)-Dopa by Decarboxylase Inhibitors. *J. Pharmacol. Exp. Ther.* **1970,** *172,* 406–15.

Maboudian-Esfahani, M.; Brocks, D. R. Disposition of Ethopropazine Enantiomers in the Rat: Tissue Distribution and Plasma Protein Binding. *J. Pharm. Pharm. Sci.* **1999,** *2,* 23–29.

Maboudian-Esfahani, M.; Brocks, D. R. Pharmacokinetics of Ethopropazine in the Rat After Oral and Intravenous Administration. *Biopharm. Drug Dispos.* **1999,** *20,* 159–163.

Mahmood, I. Clinical Pharmacokinetics and Pharmacodynamics of Selegiline: An Update. *Clin. Pharmacokinet.* **1997,** *33,* 91–102.

Marsden, C. D. Parkinson's Disease. *Lancet* **1990,** *335,* 948–952.

Mena, I.; Cotzias, G. C. Protein Intake and Treatment of Parkinson's Disease with Levodopa. *N. Engl. J. Med.* **1975,** *292,* 181–184.

Morgan, J. P.; Bianchine, J. R.; Spiegel, H. E.; Rivera-Calimlim, L.; Hersey, R. M. Metabolism of Levodopa in Patients with Parkinson's Disease. Radioactive and Fluorometric Assays. *Arch. Neurol.* **1971,** *25,* 39–44.

Mouradian, M. M.; Heuser, I. J.; Baronti, F.; Fabbrini, G.; Juncos, J. L.; Chase, T. N. Pathogenesis of Dyskinesias in Parkinson's Disease. *Ann. Neurol.* **1989,** *25,* 523–526.

Muenter, M. D.; Tyce, G. M. L-Dopa Therapy of Parkinson's Disease: Plasma L-Dopa Concentration, Therapeutic Response, and Side Effects. *Mayo Clin. Proc.* **1971,** *46,* 231–239.

Münchau, A.; Bhatia, K. P. Pharmacological Treatment of Parkinson's Disease. *Postgrad. Med. J.* **2000,** *76,* 602–610.

Nakashima, E.; Ishizaki, J.; Takeda, M.; Matsushita, R.; Yokogawa, K.; Ichimura, F. Pharmacokinetics of Anticholinergic Drugs and Brain Muscarinic Receptor Alterations in Streptozotocin Diabetic Rats. *Biopharm. Drug Dispos.* **1993,** *14,* 673–684.

Nakashima, E.; Yokogawa, K.; Ichimura, F.; Hashimoto, T.; Yamana, T.; Tsuji, A. Effects of Fasting on Biperiden Pharmacokinetics in the Rat. *J. Pharm. Sci.* **1987,** *76,* 10–13.

Nation, R. L.; Triggs, E. J.; Vine, J. Metabolism and Urinary Excretion of Benzhexol in Humans. *Xenobiotica* **1978,** *8,* 165–169

National Institute for Health and Clinical Services. Parkinson's Disease in Over 20s: Diagnosis and Management, 2006.

Neira, W. D.; Sanchez, V.; Mena, M. A.; de Yebenes, J. G. The Effects of Cisapride on L-Dopa Levels and Clinical Response in Parkinson's Disease. *Mov. Disord.* **1995,** *10,* 66–70.

Nichols, D. E.; Nichols, C. D. Serotonin Receptors. *Chem. Rev.* **2008,** *108,* 1614–1641.

Nutt, J. G.; Carter, J. H.; Lea, E. S.; Woodward, W. R. Motor Fluctuations During Continuous Levodopa Infusions in Patients with Parkinson's Disease. *Mov. Disord.* **1997,** *12,* 285–292.

Nutt, J. G.; Fellman, J. H. Pharmacokinetics of Levodopa. *Clin. Neuropharmacol.* **1984,** *7,* 35–49.

Nutt, J. G.; Holford, NH. G. The Response to Levodopa in Parkinson's Disease: Imposing Pharmacological Law and Order. *Ann. Neurol.* **1996,** *39,* 561–573.

Nutt, J. G.; Woodward, W. R.; Anderson, J. L. The Effect of Carbidopa on the Pharmacokinetics of Intravenously Administered Levodopa: The Mechanism of Action in the Treatment of Parkinsonism. *Ann. Neurol.* **1985,** *18,* 537–543.

Nutt, J. G.; Woodward, W. R.; Carter, J. H.; Gancher, S. T. Effect of Longterm Therapy on the Pharmacodynamics of Levodopa. Relation to On-Off Phenomenon. *Arch. Neurol.* **1992,** *49,* 1123–1130.

Nutt, J. G.; Woodward, W. R.; Hammerstad, J. P.; Carter, J. H.; Anderson, J. L. The "On-Off" Phenomenon in Parkinson's Disease: Relation to Levodopa Absorption and Transport. *N. Engl. J. Med.* **1984,** *310,* 483–488.

Oertel, W. H.; Quinn, N. P. Parkinsonism. In *Neurological Disorders. Course and Treatment*; Brandt, T., Caplan, L. R., Dichgans, J.; et al., Eds.; Academic Press: San Diego; 1996, pp. 715–772.

Olanow, C. W.; Fahn, S.; Muenter, M.; et al. A Multicenter Doubleblind Placebo-Controlled Trial of Pergolide as an Adjunct to Sinemet in Parkinson's Disease. *Mov. Disord.* **1994,** *9,* 40–47.

Olanow, C. W.; Koller, W. C. An Algorithm (Decision Tree) for the Management of Parkinson's Disease. *Neurology* **1998,** *50* (3), S1–S57.

Ostergaard, L.; Werdelin, L.; Odin, P.; et al. Pen Injected Apomorphine Against off Phenomena in Late Parkinson's Disease: A Double Blind, Placebo Controlled Study. *J. Neurol. Neurosurg. Psychiatry.* **1995,** *58,* 681–687.

Paalzow, G. H. M.; Paalzow, L. K. L-Dopa: How it May Exacerbate Parkinsonian Symptoms. *Trends Pharmacol. Sci.* **1986,** *9,* 15–19.

Paeme, G.; Van Bossuyt, H.; Vercruysse, A. Phenobarbital Induction of Procyclidine Metabolism: In Vitro Study. *Eur. J. Drug Metab. Pharmacokinet.* **1982,** *7,* 229–231.

Parkinson Study Group. Mortality in DATATOP: A Multicenter Trial in Early Parkinson's Disease. *Ann. Neurol.* **1998,** *43,* 318–325.

Peaston, M. J.; Bianchine, J. R. Metabolic Studies and Clinical Observations During L-Dopa Treatment of Parkinson's Disease. *BMJ* **1970,** *1,* 400–403.

Pfeiffer, R. F. Antiparkinsonian Agents: Drug Interactions of Clinical Significance. *Drug Safety.* **1996,** *14,* 343–354.

Piercey, M. F. (1998). Pharmacology of Pramipexole, a Dopamine $D_3$- Preferring Agonist Useful in Treating Parkinson's Disease. *Clin. Neuropharmacol.* **1998,** *21,* 141–151.

Porter, B. A.; Rosenfield, R. L.; Lawrence, A. M. The Levodopa Test of Growth Hormone Reserve in Children. *Am. J. Dis. Child.* **1973,** *126,* 589–592.

Porter, C. C. Inhibitors of Aromatic Amino Acid Decarboxylase—Their Biochemistry, Advances in Neurology, Raven Press: New York. **1973,** *2,* 37–58.

Reuter, I.; Harder, S.; Engelhardt, M.; Baas, H. The Effect of Exercise on Pharmacokinetics and Pharmacodynamics of Levodopa. *Mov. Disord.* **2000,** *15,* 862–868.

Rinne, U. K.; Bracco, F.; Chouza, C.; Dupont, E.; Gershanik, O.; Marti Masso, J. F.; Montastruc, J. L.; Marsden, C. D. Early Treatment of Parkinson's Disease With Cabergoline Delays the Onset of Motor Complications: Results of a Double-Blind Levodopa Controlled Trial. The PKDS009 Study Group. *Drugs* **1998,** *55* (1), 23–30.

Rizzo, V.; Memmi, M.; Moratti, R.; Melzi d'Eril, G.; Perucca, E. Concentrations of L-Dopa in Plasma and Plasma Ultrafiltrates. *J. Pharm. Biomed. Anal.* **1996,** *14,* 1043–1046.

Robertson, D. R. C.; Renwick, A. G.; Wood, N. D.; Cross, N.; Macklin, B. S.; Fleming, J. S.; Waller, D. G.; George, C. F. The Influence of Levodopa on Gastric Emptying in Man. *Br. J. Clin. Pharmacol.* **1990,** *29,* 47–53.

Robertson, D. R.; Higginson, I.; Macklin, B. S.; Renwick, A. G.; Waller, D. G.; George, C. F. The Influence of Protein Containing Meals on the Pharmacokinetics of Levodopa in Healthy Volunteers. *Br. J. Clin. Pharmacol.* **1991,** *31,* 413–417.

Robertson, G. S.; Robertson, H. A. Synergistic Effects of $D_1$ and $D_2$ Dopamine Agonists on Turning Behaviour in Rats. *Brain Res.* **1986,** *384,* 387–390.

Roozemond, R. C.; Hespe, W.; Nauta, W. T. The Concentrations of Orphenadrine and Its *N*-Demethylated Derivatives in Rat Brain, After Intraperitoneal Administration of Orphenadrine and Tofenacine. *Int. J. Neuropharmacol.* **1968,** *7,* 293–300.

Rubin, A.; Lemberger, L.; Dhahir, P. Physiologic Disposition of Pergolide. *Clin. Pharmacol. Ther.* **1981,** *30,* 258–265.

Sarati, S.; Guiso, G.; Spinelli, R.; Caccia, S. Determination of Piribedil and its Basic Metabolites in Plasma by High-Performance Liquid Chromatography. *J. Chromatogr.* **1991,** *563,* 323–332.

Sasahara, K.; Nitanai, T.; Habara, T.; Morioka, T.; Nakajima, E. Dosage form Design for Improvement of Bioavailability of Levodopa III: Influence of Dose on Pharmacokinetic Behavior of Levodopa in Dogs and Parkinsonian Patients. *J. Pharm. Sci.* **1980,** *69,* 1374–1378.

Savola, J. M.; Hill, M.; Engstrom, M.; et al. Fipamezole (JP-1730) is a Potent Alpha2 Adrenergic Receptor Antagonist That Reduces Levodopa-Induced Dyskinesia in the MPTP-Lesioned Primate Model of Parkinson's Disease. *Mov. Disord.* **2003**, *18*, 872–883.

Schapira, A. H. V. Future Directions in the Treatment of Parkinson's Disease. *Mov. Disord.* **2007**, *22* (17), S385–S391.

Schwab, R. S.; England, A. C.; Poskanzer, D. C.; Poskanzer, D. C. Amantadine in the Treatment of Parkinson's Disease. *JAMA* **1969**, *208*, 1168–1170.

Schwarzschild, M. A.; Agnati, L.; Fuxe, K.; Chen, J. F.; Morelli, M. Targeting Adenosine A2A Receptors in Parkinson's Disease. *Trends Neurosci.* **2006**, *29*, 647–654.

Shulman, L. M.; Taback, R. L.; Bean, J.; Weiner, W. J. Comorbidity of the Nonmotor Symptoms of Parkinson's Disease. *Mov. Disord.* **2001**, *16*, 507–510.

Shulman, L. M.; Taback, R. L.; Rabinstein, A. A.; Weiner, W. J. Non-Recognition of Depression and Other Non-Motor Symptoms in Parkinson's Disease. *Parkinsonism Relat. Disord.* **2002**, *8*, 193–197.

Simon, N.; Gantcheva, R.; Bruguerolle, B.; Viallet, F. The Effects of a Normal Protein Diet on Levodopa Plasma Kinetics in Advanced Parkinson's Disease. *Parkinsonism Relat. Disord.* **2004**, *10*, 137–142.

Stoof, J. C.; Booij, J.; Drukarch, B.; Wolters, E. C. The Anti-Parkinsonian Drug Amantadine Inhibits the *N*-Methyl-D-Aspartic Acid-Evoked Release of Acetylcholine from Rat Neostriatum in a Non-Competitive Way. *Eur. J. Pharmacol.* **1992**, *213*, 439–443.

Sweeney, P. J. Parkinson's Disease: Managing Symptoms and Preserving Function. *Geriatrics* **1995**, *50*, 24–31.

Tatton, W. G.; Greenwood, C. E. Rescue of Dying Neurons: A New Action for Deprenyl in MPTP Parkinsonism. *J. Neurosci. Res.* **1991**, *30*, 666–672.

Vallone, D.; Picetti, R.; Borrelli, E. Structure and Function of Dopamine Receptors. *Neurosci. Biobehav. Rev.* **2000**, *24*, 125–132.

Vickers, S.; Stuart, E. K.; Bianchine, J. R.; Hucker, H. B.; Jaffe, M. E.; Rhodes, R. E.; Vandenheuvel, W. J. Metabolism of Carbidopa (1-(-)-Alpha-Hydrazino-3,4-Dihydroxy-alpha-Methylhydrocinnamic Acid Monohydrate), an Aromatic Amino Acid Decarboxylase Inhibitor, in the Rat, Rhesus Monkey, and Man. *Drug Metab. Dispos.* **1974**, *2*, 9–22.

Wade, D. N.; Mearrick, P. T.; Morris, J. L. Active Transport of L-Dopa in the Intestine. *Nature* **1973**, *242*, 463–465.

Whiteman, P. D.; Fowles, AS. E.; Hamilton, M. J.; Peck, A. W.; Bye, A.; Webster, A. Pharmacokinetics of Procyclidine in Man. *Eur. J. Clin. Pharmacol.* **1985**, *28*, 79–83.

Witt, J., Marks, W. J. Jr. An Update on Gene Therapy in Parkinson's Disease. *Curr. Neurol. Neurosci. Rep.* **2011**, *11* (4), 362–370.

Wynalda, M. A.; Wienkers, L. C. Assessment of Potential Interactions Between Dopamine Receptor Agonists and Various Human Cytochrome P450 Enzymes Using a Simple In Vitro Inhibition Screen. *Drug Metab. Dispos.* **1997**, *25*, 1211–1214.

Yeh, K. C.; August, T. F.; Bush, D. F.; Lasseter, K. C.; Musson, D. G.; Schwartz, S.; Smith, M. E.; Titus, D. C. Pharmacokinetics and Bioavailability of Sinemet CR: A Summary of Human Studies. *Neurology* **1989**, *39*, 25–38.

Yokogawa, K.; Nakashima, E.; Ichimura J.; Hasegawa, M.; Kido, H.; Ichimura, F. Brain Regional Pharmacokinetics of Biperiden in Rats. *Biopharm. Drug Dispos.* **1992**, *13*, 131–140.

Zesiewicz, T. A.; Sullivan, K. L.; Arnulf, I.; Chaudhuri, K. R.; Morgan, J. C.; Gronseth, G. S.; Miyasaki, J.; Iverson, D. J.; Weiner, W. J. Practice Parameter: Treatment of Nonmotor Symptoms of Parkinson Disease: Report of the Quality Standards Subcommittee of the American Academy of Neurology. *Neurology* **2010**, *74*, 924–931.

Zou, L.; Jankovic, J.; Rowe, D. B.; Xie, W.; Appel, S. H.; Le, W. Neuroprotection by Pramipexole Against Dopamine- and Levodopa-Induced Cytotoxicity. *Life Sci.* **1999**, *64*, 1275–1285.

# CHAPTER 7

# Pharmacological Management of Huntington's Disease

SONIA SHARMA[1*], SUSHANT SHARMA[2], and SHALLINA GUPTA[1]

[1]*Guru Nanak Dev University Amritsar, Punjab, India*

[2]*University of KwaZulu Natal, Durban, South Africa*

*Corresponding author. E-mail: soniasharma.bot@gmail.com*

## ABSTRACT

Huntington's disease (HD) is a neurodegenerative disease characterized by cognitive dysfunction, abnormal involuntary movements, psychiatric disease, and behavioral disturbance. As there is no accurate treatment to delay the progression of HD, symptomatic treatment may be beneficial in HD patients as it may affect motor function, safety, and quality of life of patients. HD is caused by a cysteine adenosine-guanine (CAG) trinucleotide repeats in exon 1 of the huntingtin (*htt*) gene located at the fourth chromosome, 4p16.9. The present chapter has concentrated on the pharmacological treatment or therapies used for the management of HD. Symptomatic treatment of HD involve use of anti-glutamatergic agents, neuroleptics, γ-aminobutyric acid (GABA) agonists, acetylcholinesterase inhibitors, antidepressants, dopamine depleting agents, and potential neuroprotective agents. The aim of the present chapter is to provide current knowledge of therapeutic agents or drugs for treatment of different symptoms, mainly on the pharmacotherapy of non-motor symptoms of HD.

## 7.1 INTRODUCTION

Huntington's disease (HD) is a progressive neurodegenerative disease, first explained in 1872 by George Huntington, an American physician, after he examined and analyzed many affected individuals (Neylan, 2003). The symptoms of HD can begin at any age but the mean age of symptom onset is 40 years. Life expectancy after finding is around 20 years, for the period of which the patient may become fully affected. This disease found in all racial groups but very common in people of Northern Europe (Harper, 1992). HD is a familial disease, transmitted from parents to child through mutation in gene and results in accumulation of toxic protein in the brain. HD is diagnosed based on clinical research in the family history and always be confirmed with genetic testing (Huntington's Disease Collaborative Research Group, 1993). Genetic testing is applicable to family members at high risk, but only clinicians should perform the testing and counseling. Testing include two categories that is, diagnostic and predictive testing.

Diagnostic testing (DT) is carried out in a patient with symptoms of HD to confirm the disease, whereas the predictive testing (PT) is carried out in a person showing no symptoms of HD but has family history. A positive predictive test indicates that person will develop HD in the future. But PT does not tell that when person will have the disease or what the symptoms will be. PT is performed only in some specialist genetic centers following international guidelines (Craufurd and Harris, 1989; International Huntington Association, 1994). These include pretest, posttest, emotional, and practical counseling.

Many areas of the brain degenerate, involving the glutamate, neurotransmitters dopamine, and GABA. HD is a characterized by abnormal involuntary movements, cognitive dysfunction, psychiatric disease, and behavioral disturbance. In HD, the normal small projecting neurons are mainly affected and inhibited the function. So as a result, long axons of neurons terminate in substantia nigra and use GABA as a neurotransmitter. Therefore, the level of GABA in HD patients remains very low (Perry et al., 1973). It is the member of a group called 'trinucleotide-repeat' neurodegenerative disease, concerned with the expansion of cysteine adenosine-guanine (CAG) sequence in the huntingtin (*htt*) gene. A normal *htt* gene has fewer than 36 CAG repeats, whereas in abnormal gene, it has 36 or more repeats. Abnormal genes with CAG repeat of lengths between 36 and 39 show low penetrance, which mean that some people will develop HD and some will not, but repeats of 40 or more will always cause HD (Langbehn et al., 2004). The appearance of disease depends upon the number of CAG sequence in a gene (Kumar et al., 2003). In HD patients, dopaminergic nigral neurons remain intact and level of dopamine remain higher

than normal (Spokes, 1980). As HD is known as dopamine-predominant disease, so therefore anti-dopaminergic drugs are effective against HD disease. Tetrabenazine (TBZ) is anti-dopaminergic more effective drug for reducing chorea, but it also has a high risk of adverse effects. Recently some newer neuroleptic drugs, such as aripiprazole and olanzapine, with adequate efficacy and less adverse effect profile than older drugs for treating chorea and psychosis is available in the market. As such there are no current treatments of HD, but symptomatic therapies, counseling, and education can be very effective tool for doctors to use with patients affected by HD. The purpose of this chapter is to give an overview of HD and its genetic biology. This chapter explains different drugs used for the treatment of this disease on the basis of various types of characters because it is important to consider the symptom first while choosing the medicated procedure and future opportunities and challenges to be faced.

## 7.2  GENETIC BIOLOGY OF THE HUNTINGTON'S DISEASE

HD is autosomal dominant inheritance disease, detected in about 5 in 100,000 individuals worldwide (Clarke, 2005; Kumar et al., 2015). HD is caused by a CAG trinucleotide repeats in exon 1 of the huntingtin (*htt*) gene located at 4p16.9. Age of onset of this disease varies from childhood to eighties, whereas peak age of onset is in the forties and fifties and progresses more quickly in women (Karadima et al., 2012; Zielonka et al., 2013, 2015). The *htt* is known to be involved in various cellular functions such as internal cell signaling, maintenance of cyclic adenosine monophosphate response element binding protein, and preventing neuronal toxicity (Godin et al., 2010; Nucifora et al., 2001). The *htt* gene is highly expressed in the brain and testes (Li et al., 1993; Strong et al., 1993). It has been co-localized with various other organelles such as endoplasmic reticulum, nucleus, endosomes, and Golgi apparatus (DiFiglia et al., 2007; Hoffner et al., 2002). This disease is characterized by diffuse loss of neurons, involve in the striatum and cortex. In the striatum, the spiny neurons contain enkephalin and GABA is affected initially in the disease and target primary neurons. Disease occurs when the number of CAG repeats increased from 36, and repeats of 40 or more results in full penetrance of the disease. At particular level, the number of CAG repeats gets inversely proportional to the age of a disease (Tabrizi et al., 2009). The main symptom of HD is chorea while other motor symptoms include gait disturbances, dystonia, tics, myoclonus, dysarthria problems, and saccadic eye movements (Haskins and Harrison, 2000; Walker, 2007; Wider and Luthi-Carter, 2006). Other common behavioral and cognitive symptoms include dementia, depression, irritability, psychosis, apathyaggression, impaired executive function, and personality changes (Haskins and Harrison, 2000; Wider and Luthi-Carter, 2006). Chorea tends to diminish with disease progression as dystonia emerge and parkinsonism. Death of HD patients mainly occurs by various other secondary causes such as chronic illness, pneumonia, and even suicide (Ross and Tabrizi, 2011; Walker, 2007; Wider and Luthi-Carter, 2006).

As there is no treatment to delay the onset or progression of HD, symptomatic treatment of chorea may be beneficial in HD patients as it may affect motor function, safety and quality of life (Andrich et al., 2007; Frank et al., 2004; Verhagen et al., 2002). Researchers also consider treatment for the dystonia and various other non-motor and movement disorders of HD. Glutamate, dopamine and GABA are known to be the most affected neurotransmitters in HD (Feigin et al., 2007; Glass et al., 2000; Paulsen, 2009). There are number of theories on the pathogenesis of HD, and likely multiple individual mechanism occurring at once such as excitotoxicity, mitochondrial dysfunction, transcription dysregulation, and so on (Cepeda et al., 2003; Gunawardena and Goldstein, 2005; Li and Li, 2004; Ravikumar et al., 2002).

## 7.3 THERAPEUTICS

At this time, there is no neuroprotective, curative, or disease modifying treatment for HD (Adam and Jankoric, 2008). The pharmacotherapy currently used for the treatment designed to treat the symptomatology of HD patients such as to control motor sedative, behavioral symptoms, neuroprotective agent, and cognitive enhances to improve the quality of HD patients (Handey et al., 2006a,b). In early stages of disease, chorea may not interfere with the lifestyle, so may not need any treatment, but intervention is necessary when symptoms start affecting lifestyle such as eating, walking, writing, and so on. Drugs used to treat the cognitive and behavioral symptoms include antidepressants such as SSRIs (selective serotonin reuptake inhibitors), neuroleptics and acetylcholinesterase inhibitors, used to treat both motor and psychotic problem (Adam and Jankovic, 2008; Frank, 2014). Till now, seven compounds have been tested in HD patient at different stages of disease. These compounds are in phase I, that is, cysteamine, ethyl-EPA, memantine and in phase II, that is, coenzyme Q10 and creatine (Table 7.1). Here we discuss, the drugs used as HD therapeutics focusing on their benefits and side effects.

**TABLE 7.1**　Potential Pathway in the Pipeline Targeting Pathogenesis of Huntington's Disease.

| Therapies | Pathways | Neuroprotective compounds |
|---|---|---|
| Mitochondrial dysfunction | Inhibition of complex II of mitochondria ETC, that is, malonate, 3-nitropropionic acid cause neuronal loss; bioenergetic defects | Coenzyme Q10, cysteamine, creatine, cysteamine |
| Excitotoxicity | Cortical afferents release glutamate and glutamate agonist result in neuron death | Riluzole, memantine, tetrabenazine |
| Transcriptional dysregulation | m*htt* fragments and associated transcription factors; dysregulation of transcription factors involved in protein aggregation | HDACi 4b, sodium phenylbutyrate |
| *htt* proteolysis | Impairment of ubiquitin-proteosome pathway result in storage of m*htt* | Minocycline, caspase 6 inhibitors |
| *htt* storage and clearance | Implication of autophagic mechanism | C2-8, trehalose |
| Defects in axonal transport and synaptic dysfunction | Disruption of axonal transport | Fetal cells, embryonic stem cell |
| Neuronal aggregation | m*htt;* truncated N-terminal mutant; chaperones | Artificial peptides and intrabodies, antisense oligo-nucleotide, iRNA |

## 7.4　ANTI-HUNTINGTON'S DRUGS

Various drugs and surgical methods have been evaluated in this disease to suppress chorea such as dopamine antagonists, acetylcholinesterase inhibitors, glutamate antagonists, fetal cell transplantation, dopamine antagonists and so on (Armstrong and Miyasaki, 2012; Bagchi, 1983; Paleacu, 2007; Pidgeon and Rickards, 2013). Figure 7.1 shows the potential drugs used against the various symptoms of HD. Various other therapies such as alternative and complementary therapies, cognitive interventions, adjunctive therapy, and behavioral plans also play an important role in showing HD symptoms. So therefore, there is a need to consider the symptom while choosing the medicated procedure.

## 7.4.1　*CHOREA*

In literature, several reports are there to summarize the treatment of chorea (Adam and Jankovic, 2008; Bonelli and Hofmann, 2004; Frank and Jankovic, 2010; Handley et al., 2006; Imarisio et al., 2008; Jankovic, 2009; Nakamura and Aminoff, 2007; Phillips et al., 2008; Roze et al., 2008). But there is a lack of evidence for the long term treatment strategies in HD (Bonelli and Wenning, 2006). For the treatment of chorea, various agents, therapies were commonly used such as antidopaminergic agents, antipsychotic agents, N-methyl-D-aspartic acid receptor antagonists, and omega-3 fatty acid (Fig. 7.1). Recently the American Academy of Neurology Guidelines Publication recommended TBZ, riluzole, or amantadine for the treatment of chorea (Armstrong and Miyasaki, 2012; Table 7.2).

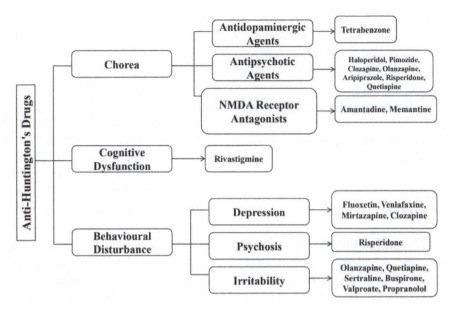

**FIGURE 7.1** Outline of drugs used for the different symptoms for HD patients.

### 7.4.1.1 ANTIDOPAMINERGIC AGENTS

#### 7.4.1.1.1 Tetrabenazine

The efficacy of tetrabenazine (TBZ) as a drug used against chorea was demonstrated by the Huntington Study Group in a double blind, placebo controlled study (Huntington Study Group, 2006). TBZ is an inhibitor of dopamine pathway, alter the motor deficits and degrade cell loss in test organism, confirming that the dopamine pathway plays an essential role in HD pathogenesis (HDCRG, 2006; Tang et al., 2007). Frank et al. (2008) examined 30 HD patients, who had been taking this drug for last 2 months and observed that this treatment resulted in reduction of 5.0 units of chorea.

#### 7.4.1.1.2 Cannabinoid

Cannabinoid compounds showed promising neuroprotective effects in cellular and animal models (Sagredo et al., 2012). These compounds have been proposed for attenuating involuntary movements in HD patients (Fernandez-Ruiz, 2009; Lastres-Becker et al., 2003). The endo-cannabinoid signaling system play an important role in the control of neuronal homeostasis in various chronic and acute neurodegenerative diseases (Fernandez-Ruiz et al., 2005, 2007, 2010; Marsicano et al., 2003; Mechoulam and Shohami, 2007).

### 7.4.1.2 ANTIPSYCHOTIC AGENTS

#### 7.4.1.2.1 Haloperidol

Girotti et al. (1984) examined the effect of haloperidol on 18 HD patients and observed that this drug reduced abnormal involuntary movement. In another study, Koller and Trimble (1985) investigated effect of haloperidol drugs on 13 HD patients. At a dose of 2–80 mg of haloperidol, chorea decreased in all patients. The main caution is advised to a patient when prescribed with QT-prolongation condition. The increased concentration of the drug has been reported when it was given with other drugs used as inhibitors or substrate of CYP2D6 or CYP3A4 isoenzymes. Intake of haloperidol with carbamazepine or rifampin has been associated with a reduction of haloperidol level in plasma.

### 7.4.1.2.2   Pimozide

Girotti et al. (1984) studied the effect of pimozide on 18 chorea affected patients of HD. Therapy include combination of pimozide with haloperidol and TBZ drugs is more effective against chorea. The additive effect on QT-prolongation may be increased when administered pimozide with tricyclic antidepressants, phenothiazines or anti-arrhythmic agents. Pimozide drug also show many side effects such as neuroleptic malignant syndrome, sedation or parkinsonism, and so on.

### 7.4.1.2.3   Clozapine

Van Vught et al. (1997) investigated 33 chorea patients treated with clozapine drug. They observed −4.0 score of UHDRS chorea as compared to −0.3 score in the placebo group. Various other studies also showed beneficial effects of clozapine on chorea HD (Bonuccelli et al., 1994; Vallette et al., 2001). In another study by Colosimo et al. (1995) found clozapine ineffective against chorea. Co-administration of drugs used to induce cytochrome 450 enzymes may reduce the level of clozapine in the plasma. A low dose of clozapine drug should be considered when combined with paroxetine and fluvoxamine. Some side effects

of this drug are also known such as seizures, agranulocytosis, syncope, and even death.

### 7.4.1.2.4   Olanzapine

Several recent reports suggest that olanzapine drug is also effective for the treatment of chorea or motor symptoms of HD. In another study Bonelli et al. (2002) reported significant reduction in chorea 13.4±6.8 to 6.9±5.0 after 14 days of treatment. Other drugs such as omeprazole, carbamazepine or rifampin, known to induce glucoronyl transferase enzyme or CYP1A2 may reduce the level of this drug in plasma. Some side effects of this drug are nausea, weight gain, asthenia, fatigue, and so on.

### 7.4.1.2.5   Aripiprazole

Brusa et al. (2009) studied the effect of aripiprazole on six chorea patients affected with HD. Dose of this drug results in reduction of UHDRS chorea. Ciammola et al. (2009) examined a great improvement in patients have been taking drug dose of 7.5–15 mg. When quinidine and ketoconazole are given along with aripiprazole, should result in the reduction to one half of its normal dose, whereas if aripiprazole will be coadministered with carbamazepine, dose of former should be doubled. This drug also showed some side effects such as vomiting, headache, anxiety, tremor, insomnia, and so on.

### 7.4.1.2.6   Risperidone

Risperidone showed significant improvement in choreoathetoid involuntary movement at a dose of 6 mg in patient affected by HD (Parsa et al., 1997). Dallacchio et al. (1999) examined four HD patients and observed similar effects

**TABLE 7.2** Drugs Used in the Treatment of Huntington's Disease with Their Contraindications and Side Effects.

| Drug | Dose | Contraindications | Drug interaction | Side effects |
|---|---|---|---|---|
| Tetrabenazine | 25.5–100 mg | Depression, suicidality, hepatic impairment | Haloperidol, MAO inhibitors, metoclopramid, chlorpromazine, CYP2D6 inhibitors, | Depression, parkinsonism, drowsiness, fatigue, anxiety, gastrointestinal distress |
| Haloperidol | 2–80 mg | CNS depression, Parkinson's syndrome, bradycardia, anticoagulants | Alcohol, QT prolongation-inducing drugs, CYP2D6, and CYP3A4 isoenzymes, | Hypo and hypertension, Tachycardia, insomnia, agitation, depression, lactation, impotence, blurred vision, urinary retention, tardive syndrome |
| Pimozide | 1 mg | Patient with congenital long QT syndrome, CNS depression, Tourette's disorder | Grape fruit | Constipation, parkinsonism, neuroleptic malignant syndrome, depression |
| Clozapine | 25 mg | Uncontrolled epilepsy, myeloproliferative disorder, paralytic ileus | Liver cytochrome p450 enzymes, polycyclic aromatic hydrocarbon | Hypertension, syncope, sweating, tremor, headache, sedation |
| Olanzapine | 2.5 mg | Parkinson's syndrome, breast-feeding, prostatic hypertrophy, hepatic dysfunction, epilepsy, dementia | Fluvoxamine, Carbamazepine, dopamine agonists, CYP4501A2-inhibitors, alcohol, sedative | Sedation, nausea, dry mouth, asthenia, increased appetite, extrapyramidal symptoms |
| Ziprasidone | 20 mg | Patients with QT prolongation, heart failure, acute myocardial infarction | Metoclopramide, alcohol, antihistamine, drugs for sleep and anxiety | Dizziness, asthenia, weight gain |
| Aripiprazole | 2 mg | Increase in carbamazepine could cause low blood level; ketoconazole, quinidine and paroxetine cause increase blood level | QT prolongation-inducing drugs, CYP3A4 and CYP2D6 isoenzymes | Vomiting, nausea, headache, akathisia, tremor, constipation |
| Risperidone | 0.5 mg | Parkinson's syndrome, hepatic and renal dysfunction, diabetes, dementia, prolactin, cardiovascular diseases | Phenothiazine, liver enzymes including drugs, tricyclic antidepressive, drugs inducing QT prolongation, antihypertensive | Increased appetite, insomnia, fatigue, cough, anxiety, nausea, tremor |
| Quetiapine | 25 mg | Cerebrovascular diseases, QT prolongation | HIV protease inhibitors, azol antifungals, CYP450-3A4 inhibitors, erythromycin | Dizziness, constipation, dyspepsia, hypotension, |
| Amantadine | 100 mg | Depression, psychosis, seizure disorders, renal dysfunction, hypotension | Drugs with anticholinergic properties | Insomnia, ataxia, headache, anorexia, diarrhea, fatigue, agitation, hallucinations |
| Rivastigmine | 3 mg | — | Cytochrome p450 enzymes | Nausea, fatigue, insomnia, vomiting, confusion, headache |

**TABLE 7.2**   *(Continued)*

| Drug | Dose | Contraindications | Drug interaction | Side effects |
|---|---|---|---|---|
| Donepezil | 5 mg | – | Anti-cholinergic drugs, liver enzymes | Diarrhea, headache, agitation, vomiting |
| Fluoxetine | 20 mg | Contra-indicated with pimozide, MAOIS, thioridazine | Aspirin, rosuvastatin, duloxentine, aripiprazole, pregabalin, metoprolol | Headache, anxiety, dizziness, asthenia, tremor, nausea |
| Venlafaxine | 37.5 mg | SSRI or SNRI treatment | CYP3A3/4-inhibitors, MAO inhibitors, Warfarin | Anxiety, dizziness, constipation, tremor, blurred vision |
| Mirtazapine | 30 mg | Hepatic and renal dysfunction, breast-feeding, hypotension, epilepsy, diabetes, mania, suicidal behavior | CYP3A4-inhibitors, and inductors, sedative, cimetidine, warfarin, serotonergic drugs | Weight gain, constipation, somnolence |
| Sertraline | 100 mg | Pimozide, MAOIs | Salmeterol, aspirin, pregabalin, albuterol, metoprolol, esomeprazole | Dizziness, fatigues, diarrhea, nausea, insomnia |
| Tiapride | 200–400 mg | Renal failure, QT prolongation, breast-feeding, pregnancy | L-DOPA | Tardive, chorea, dyskinesia |
| Clonazepam | 20 g | Ataxia, alcohol, renal dysfunction | Lithium, antipsychotics, hypnotics | Sleep disorders, chorea, anxiety, epilepsy |
| Citalopram (SSRI) | 40 g | Bleeding anomalies, QT prolongation, diabetes mellitus, renal insufficiency | Anticoagulants, desipramin, cimetidine, neuroleptics, metoprolol | Depression |
| Buspirone (serotonin-agonist) | 15 g | Liver and renal dysfunction, metabolic acidosis diabetes | Haloperidol, CYP3A4, rifampicin, MAO inhibitors, carbamazepine, itraconazole nefazodone | Aggression, anxiety |
| Riluzole | 200 mg | Elevated liver enzymes, liver dysfunction | CYPIA2 inhibitors; CYPIA2 inductors | Chorea |
| Valproic acid | 60 g | Carnitine-palmitoyltransferase type II deficiency, hepatic and pancreatic dysfunction | Cytochrome P450 inducers, plasma levels decrease | Myoclonus, epilepsy, mood stabilizer |
| Levetiracetam | 500 mg | Suicidal tendency, hepatic and renal dysfunction, hemodialysis, | Trazodone, baclofen, duloxetine | Epilepsy |

at 6 mg dose of risperidone. Combination of this drug with other enzyme inducing drugs such as phenobarbital, phenytoin and rifampin, and so on resulted in decreased concentration of risperidone in plasma (Videnovic, 2013). On the other hand fluoxetine, ranitidine, paroxetine, and cimetidine and so on drugs are known to increase its concentration in plasma. This drug also showed some side effects such as dizziness, somnolence, drooling, akathisia, headache, dry mouth, insomnia, rhinorrhea, anxiety, and so on.

### 7.4.1.2.7 Quetiapine

Quetiapine drug is known to reduce choreiform movement in HD patients (Bonelli and Niederwresar, 2002). The use of this drug should be prevented in combination with drugs used to increase QT interval. High doses of quetiapine are recommended for the patients already taking phenytoin, quetiapine, rifampin, barbiturates, or glucocorticoids. Side effects of this drug include constipation, weight gain, hypotension, dizziness, somnolence, and so on.

### 7.4.1.3 N-METHYL-D-ASPARTIC ACID RECEPTOR ANTAGONISTS

### 7.4.1.3.1 Amantadine

Vertagen et al. (2002) examined 24 patients of HD and observed significant reduction of chorea after 4 weeks of treatment with 400 mg/day oral dose of amantadine. Intravenous treatment of amantadine drug showed significant improvement in dyskinesia score in nine patients (Lucetti et al., 2003). O'Suilleabhain and Dewey (2003) observed that 300 mg/day dose of amantadine had no impact on HD chorea compared to placebo on 19 patients. Combination of amantadine with quinine reduces the renal clearance by 30%. Some side effects shown by this drug are dizziness, anorexia, livedo reticularis, agitation, diarrhea, and so on.

### 7.4.1.3.2 Memantine

Memantine hold a promising effect by reducing striatal cell death, slow progression, and cognitive improvement in HD patients (Cankurtran et al., 2006; Lee et al., 2006). Ondo et al. (2007) examined nine patients suffering from HD disease. On giving the treatment of memantine at 20 mg/day to patients, they observed significant improvement in chorea score. Very few side effects of this drug are known such as dizziness, headache, confusion, hypertension, and so on.

### 7.4.1.3.3 Remacemide

Remacemide is a non-competitive antagonist receptor of N-methyl-D-aspartate (NMDA) had promising effect in mice affected by HD (Ferrante et al., 2002; Schilling et al., 2001). Kieburtz et al. (1996) observed effect of 200 mg/day dose of this drug on the HD patients and found improvement in chorea level (Kremer et al., 1999). HSG (2001) examined 347 patients and looked into the positive effect of remacemide on decline in HD.

### 7.4.1.3.4 Riluzole

Riluzole is glutamate inhibitor in the CNS was neuroprotective in HD patients and upregulate the level of brain-derived neurotrophic factor (BDNF) and glial derived neurotrophic level, known neuroprotective factors (Mary et al., 1997; Palfi et al., 1997). Huntington Study Group (2003) investigated the dosage effects of this drug on 63 patients and found great reduction in the level of chorea after 8 weeks of treatment at dose 200 mg/day. Later in another study by Landwehrmeyer et al., (2007) got contrasting result to the previous study. They examined 537 patients of HD, taking 50 mg/day dose of riluzole, but they observed no beneficial effect of drug on these patients.

### 7.4.1.3.5   Pridopidine

Pridopidine is a new drug used to reduce involuntary movement in HD but its mechanism of action is not known till now (Dyhring et al., 2010). This drug is also known as "dopaminergic stabilizer" from the observation of animal studies where it increased locomotor activity (Ponten et al., 2010; Rung et al., 2005, 2008; Seeman et al., 2009). Sahlholm et al. (2013) suggested that pridopidine acts by sigma-1 receptors and this negate the "dopaminergic stabilizer" hypothesis.

### 7.4.2   COGNITIVE DYSFUNCTION

#### 7.4.2.1   RIVASTIGMINE

De Tommaso et al. (2007) observed an effect of rivastigmine on 11 patients of HD. The team observed that 3–6 mg dose of this drug results in the improvement of cognitive function after 24 months of treatment. The effect of drug was measured by mini-mental state examination (MMSE). This drug also shows some side effects such as nausea, diarrhea, dizziness, agitation, and so on.

### 7.4.3   BEHAVIORAL DISTURBANCE

#### 7.4.3.1   DEPRESSION

##### 7.4.3.1.1   Fluoxetine

The significant result of fluoxetine was observed in clinical trial of 30 patients affected with HD and Hamilton Depression Inventory (Combo et al., 1997). This drug is contraindicated with pimozide, thioridazine, and Monoamine oxidase inhibitor (MAOIs). Fluoxetine drug is known to inhibit the activity of CYP2D6. Side effects of fluoxetine are headache, dizziness, anxiety, asthenia, tremor, and so on.

##### 7.4.3.1.2   Venlafaxine

Holl et al. (2010) examined 26 HD patients and observed the effect of venlafaxine drug on them. The team found significant improvement in the depression. Only one among five patients developed side effects of this drug like irritability and nausea. This drug is not taken along with serotonergic drugs, that is, SSRI. This drug also has some side effects such as nausea, asthenia, anorexia, vomiting, somnolence, dizziness, anxiety, tremor, abnormal orgasm/ejaculation, and so on.

##### 7.4.3.1.3   Mirtazapine

This drug is used to treat depression and suicidal ideations in HD patients at the dose of 65–90 mg/day (Bonelli, 2003). This drug is not safe to use along MAOIs. Dose of drug varies for the patients as dose of mirtazapine increased for the patients taking carbamazepine and phenytoin whereas its concentration has decreased when concomitant with therapy includes ketoconazole or cimetidine and so on. Side effects of this drug include weight gain, constipation, dizziness, somnolence, and so on.

### 7.4.3.1.4  Clozapine

175 mg dose of clozapine drug has been reported for treating the depression in HD patients (Sajatovic et al., 1991). This drug is contraindicated in patients with some disorders like epilepsy, myeloproliferative disorder, agranulocytosis (clozapine induced), or granulocytopenia. Clozapine drug should be used with precaution when taken along with drugs/therapy known to prolong level of the QT or cytochrome P450 enzyme. A low dose of this drug is required when combined with fluvoxamine or paroxetine. Main side effects of this drug include tachycardia, syncope, constipation, nausea, drowsiness, myocarditis, and even death.

### 7.4.3.2  PSYCHOSIS

### 7.4.3.2.1  Risperidone

The risperidone drug is used to treat psychotic symptoms observed in HD patients (Erdemoglu and Boratav, 2002; Madhusoodanan and Brenner, 1998). Drugs used as enzyme inducers such as rifampin, phenytoin, and so on and carbamazepine reduce plasma concentration of risperidone. Whereas two drugs, that is, paroxetine and fluoxetine increase its level in plasma. Main side effects of this drug are drooling, akathisia, nasal congestion, dizziness, insomnia, and so on.

### 7.4.3.3  IRRITABILITY

### 7.4.3.3.1  Olanzapine

Squitieri et al. (2001) examined 11 patients of HD and observed significant improvement in the behavioral symptoms such as anxiety, depression, and irritability. Later on Paleacu and the team (2002) also observed improvement in Unified *Huntington's Disease* Rating Scale (UHDRS). Drugs related with glucuronyl transferase enzymes and CYP1A2 have been known to reduce olanzapine plasma level. This drug is also known to induce many side effects such as extrapyramidal, akathisia, fatigue, nausea, weight gain, and sedation, and so on.

### 7.4.3.3.2  Quetiapine

150–300 mg dose of quetiapine drug is very effective against behavioral disorders such as agitation, irritability, and insomnia in five HD patients (Alpay and Koroshetz, 2006). The use of this drug should be avoided with drugs used to increase QT interval. High dose of quetiapine is required in patients taking hepatic enzyme inducers, phenytoin, or quetiapine. Very few side effects are known such as dyspepsia, somnolence, constipation, weight gain, and so on.

### 7.4.3.3.3  Sertraline

Sertraline drug is also used against behavioral disturbance such as aggressiveness or irritability. Caution is advised when sertraline is coadministered with warfarin, linezolid, tramadol, lithium, and so on. Main side effects of this drug are dizziness, fatigues, nausea, insomnia, and so on (Alpay and Koroshetz, 2006).

### 7.4.3.3.4  Buspirone

Three studies documented the management of aggressive and irritability behavior by buspirone at the dose range 20–60 mg/day (Bhandary and Masand, 1997; Byrne et al., 1994; Findling,

1993). Caution is advised during coadministration with MAOIs, CYP3A4, phenytoin, carbamazepine, dexamethasone, and so on. Dizziness, headache, nausea, and lightheadedness are the main side effects of buspirone drug.

### 7.4.3.3.5  Valproate

Valproate drug is used as a mood stabilizer in relieving aggression and agitation in patients affected with HD (Grove et al., 2000). This drug contra-indicated in patients affected with urea cycle disorder or hepatic disease. Main side effects of this drug include hepatotoxicity, asthenia, vomiting, blurred vision, insomnia, teratogenicity, and so on.

### 7.4.3.3.6  Propranolol

Propranolol drug is known to be effective in controlling aggressive behavior in HD patients at dose of 32–240 mg (Stewart, 1987; Stewart et al., 1987). The substrate or inhibitors CYP2D6, CYP2C19, and CYP1A2 are known to increase blood levels of propranolol whereas its level decreases by co-administration with phenytoin, ethanol, phenobarbital, and rifampin. This drug also causes some side effects such as insomnia, nausea, fatigue, irritability, and so on.

### 7.5  FUTURE OPPORTUNITIES AND CHALLENGES

Mainly there are three factors contribute to the challenges in selecting drug targets and developing new treatments for HD. First one is that the cellular and pathological pathways targeted for the development of HD therapy are still not clear. These include role of *mhtt* gene in mitochondrial dysfunction, high mitochondrial immobility in the brain neurons, mitochondrial dynamics, breakdown of mitochondrial biogenesis, and so on (Knott et al., 2008). It is not clear that which of the molecular event and pathways play an important role in the pathology? And how many of these should be blocked by suppressing genes to prevent HD. What type of practices or treatment is required to test the potential possibilities? Proteomic studies showed that *htt* gene interact with hundreds of proteins and which interaction is deregulated in HD is not known (Li and Li, 2004; Shirasaki et al., 2012). It is also not clear that HD is a result of gain of function or loss of function of protein. Second challenge is lack of animal models to analyze therapeutic effect for HD because of the difference in number of polyQ repeats in HD patients and also different size of mutated gene (Costa et al., 2010). This is also difficult to assess issue related to age dependency, slow increase of disease in patients in animal models. Third challenge is in developing various trials in a rare chronic neurodegenerative disease. These are limited number of patients, number of trial for same patient, and geographical and ecological challenges for patient and so on (Munoz-Sanjuan and Bates, 2011). The main future research on treatment of HD is to diagnose various laboratory methods used for monitoring the progression of disease before the onset of symptoms. HD is a result of mutation in a single *htt* gene, so therefore expected to be easier to elucidate the clinical trial, pathology, and identification of pathway involved in progression of this disease. But still research in the past 20 years showed that the challenge is far from being completed. As the research is still ongoing; therefore, it is hoped that the final effect would become more perceptible in the near future.

## 7.6 CONCLUSION

The therapeutic advances used in the management of HD have been discussed in the chapter. Most focus of these therapies is on the development of protective strategies, essential for delaying the onset and slowing the development of HD. Till now, seven compounds have been tested in patients at different stages of the HD disease. These compounds are in phase I (cysteamine, minocycline, ethyl-EPA, memantine) and in phase II (coenzyme Q10, creatine). TBZ is very common drug available in several countries for the treatment of chorea. Many discrepancies draw attention to the difficulty in predicting the value of new drugs in humans based on animal models of HD. As there is intensive research into HD and recent findings seem effective and promising therapeutic strategies. At this stage, it is really difficult to suggest that one treatment is better than the other as many of them have not been examined in humans so as to examine their effect. Therefore, general care, current treatment, and knowledge of onset of symptoms are crucial for people affected with HD.

## KEYWORDS

- **chorea**
- **dopamine antagonists**
- **huntingtin**
- **Huntington's disease**
- **dopamine-predominant disease**

## REFERENCES

Adam, O. R.; Jankovic, J. Symptomatic Teatment of Huntington Disease. *Neurotherapeutics* **2008**, *5*, 181–197.

Alpay, M.; Koroshetz, W. J. Quetiapine in the Treatment of Behavioral Disturbances in Patients with Huntington's Disease. *Psychosomatics* **2006**, *47*, 70–72.

Andrich, J.; Saft, C.; Ostholt, N.; Müller, T. Complex Movement Behavior and Progression of Huntington's Disease. *Neurosci. Lett.* **2007**, *416*, 272–274.

Armstrong, M. J.; Miyasaki, J. M. American Academy of Neurology. Evidence-based Guideline: Pharmacologic Treatment of Chorea in Huntington disease: Report of the Guideline Development Subcommittee of the American Academy of Neurology. *Neurology* **2012**, *79*, 597–603.

Bagchi, S. P. Differential Interactions of Phencyclidine With Tetrabenazine and Reserpine Affecting Intraneuronal Dopamine. *Biochem. Pharmacol.* **1983**, *32*, 2851–2856.

Bhandary, A. N.; Masand, P. S. Buspirone in the Management of Disruptive Behaviors due to Huntington's Disease and Other Neurological Disorders. *Psychosomatics* **1997**, *38*, 389–391.

Bonelli, R. M.; Hofmann, P. A Review of the Treatment Options for Huntington's Disease. *Expert Opin. Pharmacother.* **2004**, *5*, 767–776.

Bonelli, R. M.; Niederwieser, G. Quetiapine in Huntington's Disease: A First Case Report. *J. Neurol.* **2002**, *249*, 1114–1115.

Bonelli, R. M.; Wenning, G. K. Pharmacological Management of Huntington's Disease: An Evidence-Based Review. *Curr. Pharm. Des.* **2006**, *12*, 2701–2720.

Bonelli, R. M. Mirtazapine in Suicidal Huntington's Disease. *Ann. Pharmacother.* **2003**, *37*, 452.

Bonuccelli, U.; Ceravolo, R.; Maremmani, C.; et al. Clozapine in Huntington's Chorea. *Neurology* **1994**, *44*, 821–823.

Brusa, L.; Orlacchio, A.; Moschella, V.; et al. Treatment of the Symptoms of Huntington's Disease: Preliminary Results Comparing Aripiprazole And Tetrabenazine. *Mov. Disord.* **2009**, 24, 126–129.

Byrne, A.; Martin, W.; Hnatko, G. Beneficial Effects of Buspirone Therapy in Huntington's Disease. *Am. J. Psychiatry.* **1994**, *151*, 1097.

Cankurtaran, E. S.; Ozalp, E.; Soygur, H.; et al. Clinical Experience with Risperidone and Memantine in the Treatment of Huntington's Disease. *J. Natl. Med. Assoc.* **2006**, *98*, 1353–1355.

Cepeda, C.; Hurst, R. S.; Calvert, C. R.; et al. Transient and Progressive Electrophysiological Alterations in the Corticostriatal Pathway in a Mouse Model of Huntington's Disease. *J. Neurosci.* **2003**, *23*, 961–969.

Ciammola, A.; Sassone, J.; Colciago, C.; et al. Aripiprazole in the Treatment of Huntington's Disease: A Case Series. *Neuropsychiatr. Dis. Treat.* **2009**, *5*, 1–4.

Clarke, C. R. A. Neurological Disease, In *Clinical Medicine;* Kumar, P., Clark, M., Eds.; Kumar And Clark., 6th ed.; W.B. Saunders: Edinburgh And New York, 2005; pp 1173–1271.

Colosimo, C.; Cassetta, E.; Bentivoglio, A. R.; Albanese, A. Clozapine in Huntington's Disease. *Neurology* **1995,** *45,* 1023–1024.

Como, P. G.; Rubin, A. J.; O'Brien, C. F.; et al. A Controlled Trial of Fluoxetine in Nondepressed Patients with Huntington's Disease. *Mov. Disord.* **1997,** *12,* 397–401.

Costa, V.; Giacomello, M.; Hudec, R.; et al. Mitochondrial Fission and Cristae Disruption Increase the Response of Cell Models of Huntington's Disease to Apoptotic Stimuli. *EMBO Mol. Med.* **2010,** *2,* 490–503.

Craufurd, D.; Harris, R. Predictive Testing for Huntington's Disease. *Br. Med. J.* **1989,** *298,* 892.

Dallocchio, C.; Buffa, C.; Tinelli, C.; Mazzarello, P. Effectiveness of Risperidone in Huntington Chorea Patients. *J. Clin. Psychopharmacol.* **1999,** *19,* 101–103.

De Tommaso, M.; Difruscolo, O.; Sciruicchio, V.; et al. Two Years' Follow-Up of Rivastigmine Treatment in Huntington Disease. *Clin. Neuropharmacol.* **2007,** *30,* 43–46.

Difiglia, M.; Sena-Esteves, M.; Chase, K.; Sapp, E.; Pfister, E.; Sass, M.; et al. Therapeutic Silencing of Mutant Huntingtin with Sirna Attenuates Striatal and Cortical Neuropathology and Behavioral Deficits. *Proc. Natl. Acad. Sci. U S A.* **2007,** *104,* 17204–17209.

Dyhring, T.; Nielsen, E. O.; Sonesson, C.; Pettersson, F.; Karlsson, J.; Svensson, P.; et al. The Dopaminergic Stabilizers Pridopidine (ACR16) and (-)-OSU6162 Display Dopamine D(2) Receptor Antagonism and Fast Receptor Dissociation Properties. *Eur. J. Pharmacol.* **2010,** *628,* 19–26.

Erdemoglu, A. K.; Boratav, C. Risperidone in Chorea and Psychosis of Huntington's Disease. *Eur. J. Neurol.* **2002,** *9,* 182–183.

Feigin, A.; Tang, C.; Ma, Y.; et al. Thalamic Metabolism and Symptom Onset in Preclinical Huntington's Disease. *Brain* **2007,** *130,* 2858–2867.

Fernández-Ruiz, J.; González, S.; Romero, J.; Ramos, J. A. Cannabinoids in Neurodegeneration and Neuroprotection. In *Cannabinoids As Therapeutics (MDT);* Mechoulam, R., Ed.; Birkhäuser Verlag: Switzerland, 2005; pp 79–109.

Fernández-Ruiz, J.; Romero, J.; Velasco, G.; Tolón, R. M.; Ramos, J. A.; Guzmán, M. Cannabinoid CB₂ Receptor: A New Target for the Control of Neural Cell Survival? *Trends Pharmacol. Sci.* **2007,** *28,* 39–45.

Fernández-Ruiz, J. The Endocannabinoid System as a Target for the Treatment of Motor Dysfunction. *Br. J. Pharmacol.* **2009,** *156,* 1029–1040.

Fernández-Ruiz, J.; García, C.; Sagredo, O.; Gómez-Ruiz, M.; De Lago, E. The Endocannabinoid System as a Target for the Treatment of Neuronal Damage. *Exp. Opin. Ther. Targets* **2010,** *14,* 387–404.

Ferrante, R. J.; Andreassen, O. A.; Dedeoglu, A.; et al. Therapeutic Effects of Coenzyme Q10 and Remacemide in Transgenic Mouse Models of Huntington's Disease. *J. Neurosci.* **2002,** *22,* 1592–1599.

Findling, R. L. Treatment of Aggression in Juvenile-Onset Huntington's Disease with Buspirone. *Psychosomatics* **1993,** *34,* 460–461.

Frank, S.; Marshall, F.; Plumb, S.; et al. Functional Decline due to Chorea in Huntington's Disease. *Neurology* **2004,** *62,* 204.

Frank, S.; Ondo, W.; Fabn, S.; Hunter, C.; Oakes, D.; Plumb, S.; et Al. A Study of Chorea After Tetrabenazine Withdrawal in Patients with Huntington's Disease. *Clin. Neuropharmacol.* **2008,** *31,* 127–133.

Frank, S. Treatment of Huntington's Disease. *Neurotherapeutics* **2014,** *11,* 153–160.

Frank, S. A.; Jankovic, J. Advances in the Pharmacological Management of Huntington's Disease. *Drugs* **2010,** *70,* 561–571.

Girotti, F.; Carella, F.; Scigliano, G.; et al. Effect of Neuroleptic Treatment on Involuntary Movements and Motor Performances in Huntington's Disease. *J. Neurol. Neurosurg. Psychiatry* **1984,** *47,* 848–852.

Glass, M.; Dragunow, M.; Faull, R. L. The Pattern of Neurodegeneration in Huntington's Disease: A Comparative Study of Cannabinoid, Dopamine, Adenosine and GABA(A) Receptor Alterations in the Human Basal Ganglia in Huntington's Disease. *Neuroscience* **2000,** *97,* 505–519.

Godin, J. D.; Colombo, K.; Molina-Calavita, M.; Keryer, G.; Zala, D.; Charrin, B. C.; et al. Huntingtin is Required for Mitotic Spindle Orientation and Mammalian Neurogenesis. *Neuron* **2010,** *67,* 392–406.

Grove, V. E.; Jr. Quintanilla, J.; Devaney, G. T. Improvement of Huntington's Disease with Olanzapine and Valproate. *N. Engl. J. Med.* **2000,** *343,* 973–974.

Gunawardena, S.; Goldstein, L. S. Polyglutamine Diseases and Transport Problems: Deadly Traffic Jams on Neuronal Highways. *Arch. Neurol.* **2005,** *62,* 46–51.

Handley, O. J.; Naji, J. J.; Dunnett, S. B.; et al. Pharmaceutical, Cellular And Genetic Therapies for Huntington's Disease. *Clin. Sci.* **2006,** *110,* 73–88.

Harper, P. S. The Epidemiology of Huntington's Disease. *Hum. Genet.* **1992**, *89*, 365–376.

Haskins, B. A.; Harrison, M. B. Huntington's Disease. *Curr. Treat. Option N.* **2000**, *2*, 243–262.

HDRG. A Novel Gene Containing a Trinucleotide Repeat That is Expanded and Unstable on Huntington's Disease Chromosomes. The Huntington's Disease Collaborative Research Group. *Cell* **1993**, *72*, 971–983.

Hoffner, G.; Kahlem, P.; Djian, P. Perinuclear Localization of Huntingtin as a Consequence of its Binding to Microtubules Through an Interaction with Beta-Tubulin: Relevance to Huntington's Disease. *J. Cell Sci.* **2002**, *115*, 941–948.

Holl, A. K.; Wilkinson, L.; Painold, A.; et al. Combating Depression in Huntington's Disease: Effective Antidepressive Treatment with Venlafaxine XR. *Int. Clin. Psychopharmacol.* **2010**, *25*, 46–50.

Huntington Study Group. A Randomized, Placebo-Controlled Trial of Coenzyme Q10 And Remacemide in Huntington's Disease. *Neurology* **2001**, *57*, 397–404.

Huntington Study Group. Dosage Effects if Riluzole in Huntington's Disease: A Multicentre Placebo-Controlled Study. *Neurology* **2003**, *61*, 1551–1556.

Huntington Study Group. Tetrabenazine as Antichorea Therapy in Huntington Disease: A Randomized Controlled Trial. *Neurology* **2006**, *66*, 366–372.

Imarisio, S.; Carmichael, J.; Korolchuk, V.; et al. Huntington's Disease: From Pathology and Genetics to Potential Therapies. *Biochem. J.* **2008**, *412*, 191–209.

International Huntington Association, World Federation of Neurology Research Group on Huntington's Chorea. Guidelines for the Molecular Genetics Predictive Test in Huntington's Disease. International Huntington Association (IHA) and the World Federation of Neurology (WFN) Research Group on Huntington's Chorea. *Neurology* **1994**, *44*, 1533–1536.

Jankovic, J.; Treatment of Hyperkinetic Movement Disorders. *Lancet Neurol.* **2009**, *8*, 844–856.

Karadima, G.; Dimovasili, C.; Koutsis, G.; Vassilopoulos, D.; Panas, M. Age at Onset in Huntington's Disease: Replication Study on the Association of HAP1. *Park. Relat. Disord.* **2012**, *18*, 1027–1028.

Kieburtz, K.; Feigin, A.; Mcdermott, M.; Como, P.; Adwender, D.; Zimmerman, C. et al. A Controlled Trial of Remacemide Hydrochloride in Huntington's Disease. *Mov. Disord.* **1996**, *11*, 273–277.

Knott, A. B.; Perkins, G.; Schwarzenbacher, R.; Bossy-Wetzel, E. Mitochondrial Fragmentation in Neurodegeneration. *Nat. Rev. Neurosci.* **2008**, *9*, 505–518.

Koller, W. C.; Trimble, J. The Gait Abnormality of Huntington's Disease. *Neurology* **1985**, *35*, 1450–1454.

Kremer, B.; Clark, C. M.; Almqvist, E. W.; Raymond, L. A.; Graf, P.; Jacova, C.; et al. Influence of Lamotrigine on Progression of Early Huntington's Disease: A Randomized Clinical Trial. *Neurology* **1999**, *53*, 1000–1011.

Kumar, A.; Singh, S. K.; Kumar, V.; Kumar, D.; Agarwal, S.; Rana, M. K. Huntington's Disease: An Update of Therapeutic Strategies. *Gene* **2015**, *556*, 91–97.

Landwehrmeyer, G. B.; Dubois, B.; De Yebenes, J. G.; Kremer, B.; Gaus, W.; Kraus, P. H.; et al. Riluzole in Huntington's Disease: A 3-year, Randomized Controlled Study. *Ann. Neurol.* **2007**, *62*, 262–272.

Langbehn, D. R.; Brinkman, R. R.; Falush, D.; Paulsen, J. S.; Hayden, M. R. A New Model for Prediction of the Age of Onset and Penetrance for Huntington's Disease Based on CAG Length. *Clin. Genet.* **2004**, *65*, 267–277.

Lastres-Becker, I.; De Miguel, R.; De Petrocellis, L.; Makriyannis, A.; Di Marzo, V.; Fernández-Ruiz, J. Compounds Acting at the Endocannabinoid And/or Endovanilloid Systems Reduce Hyperkinesias in a Rat Model of Huntington's Disease. *J. Neurochem.* **2003**, *84*, 1097–109.

Lee, S. T.; Chu, K.; Park, J. E.; et al. Memantine Reduces Striatal Cell Death with Decreasing Calpain Level in 3-Nitropropionic Model of Huntington's Disease. *Brain Res.* **2006**, *1118*, 199–207.

Li, S. H.; Schilling, G.; Young, W. S.; Li, X. J.; Margolis, R. L.; Stine, O. C.; et al. Huntington's Disease Gene (IT15) is Widely Expressed in Human and Rat Tissues. *Neuron* **1993**, *11*, 985–993.

Li, S. H.; Li, X. J. Huntingtin-Protein Interactions And the Pathogenesis of Huntington's Disease. *Trends Genet.* **2004**, *20*, 146–154.

Lucetti, C.; Del Dotto, P.; Gambaccini, G.; et al. IV Amantadine Improves Chorea in Huntington's Disease: An Acute Randomized, Controlled Study. *Neurology* **2003**, *60*, 1995–1997.

Madhusoodanan, S.; Brenner, R. Use of Risperidone in Psychosis Associated with Huntington's Disease. *Am. J. Geriatr. Psychiatry* **1998**, *6*, 347–349.

Marsicano, G.; Goodenough, S.; Monory, K.; Hermann, H.; Eder, M.; Cannich, A.; et al. CB1 Cannabinoid Receptors and On-demand Defense Against Excitotoxicity. *Science* **2003**, *302*, 84–88.

Mary, V.; Wahl, F.; Stutzmann, J. M. Effect of Riluzole on Quinolinate Induced Neuronal Damage in Rats: Comparison with Blockers of Glutamatergic Neurotransmission. *Neurosci. Lett.* **1995**, *201*, 92–96.

Mechoulam, R.; Shohami, E. Endocannabinoids and Traumatic Brain Injury. *Mol. Neurobiol.* **2007,** *36,* 68–74.

Munoz-Sanjuan, I.; Bates, G. P. The Importance of Integrating Basic and Clinical Research Toward the Development of New Therapies for Huntington Disease. *J. Clin. Invest.* **2011,** *121,* 476–483.

Nakamura, K.; Aminoff, M. J. Huntington's Disease: Clinical Characteristics, Pathogenesis and Therapies. *Drugs Today* **2007,** *43,* 97–116.

Neylan, T. C. Neurodegenerative Disorders: George Huntington's Description of Hereditary Chorea. *J. Neuropsychiatry Clin. Neurosci.* **2003,** *15,* 108.

Nucifora, F. C.; Jr, Sasaki, M.; Peters, M. F.; et al. Interference by Huntingtin and Atrophin-1 with Cbp-Mediated Transcription Leading to Cellular Toxicity. *Science* **2001,** *291,* 2423–2428.

O'Suilleabhain, P.; Dewey, R. B. A Randomised Trial of Amantadine in Huntington Disease, *Arch. Neurol.* **2003,** *60,* 996–998.

Ondo, W. G.; Mejia, N. I.; Hunter, C. B. A Pilot Study of the Clinical Efficacy and Safety of Memantine for Huntington's Disease. *Parkinsonism Relat. Disord.* **2007,** *13,* 453–454.

Paleacu, D.; Anca, M.; Giladi, N. Olanzapine in Huntington's Disease. *Acta Neurol. Scand.* **2002,** *105,* 441–444.

Paleacu, D. Tetrabenazine in the Treatment of Huntington's Disease. *Neuropsychiatr. Dis. Treat.* **2007,** *3,* 545–551.

Palfi, S.; Riche, D.; Brouillet, E.; et al. Riluzole Reduces Incidence of Abnormal Movements but not Striatal Cell Death in a Primate Model of Progressive Striatal Degeneration. *Exp. Neurol.* **1997,** *146,* 135–141.

Parsa, M. A.; Szigethy, E.; Voci, J. M.; Meltzer, H. Y. Risperidone in Treatment of Choreoathetosis of Huntington's Disease. *J. Clin. Psychopharmacol.* **1997,** *17,* 134–135.

Paulsen, J. S. Functional Imaging in Huntington's Disease. *Exp. Neurol.* **2009,** *216,* 272–277.

Perry, T. L.; Hansen, S.; Kloster, M. Huntington's Chorea. Deficiency of Gammaaminobutyric Acid in Brain. *N. Engl. J. Med.* **1973,** *288,* 337–342.

Phillips, W.; Shannon, K. M.; Barker, R. A. The Current Clinical Management of Huntington's Disease. *Mov. Disord.* **2008,** *23,* 1491–1504.

Pidgeon, C.; Rickards, H. The Pathophysiology and Pharmacological Treatment of Huntington Disease. *Behav. Neurol.* **2013,** *26,* 245–253.

Ponten, H.; Kullingsjo, J.; Lagerkvist, S.; Martin, P.; Pettersson, F.; Sonesson, C.; et al. In Vivo Pharmacology of the Dopaminergic Stabilizer Pridopidine. *Eur. J. Pharmacol.* **2010,** *644,* 88–95.

Ravikumar, B.; Duden, R.; Rubinsztein, D. C. Aggregate-prone Proteins with Polyglutamine and Polyalanine Expansions are Degraded by Autophagy. *Hum. Mol. Genet.* **2002,** *11,* 1107–1117.

Ross, C. A.; Tabrizi, S. J. Huntington's Disease: From Molecular Pathogenesis to Clinical Treatment. *Lancet Neurol.* **2011,** *10,* 83–98.

Roze, E.; Saudou, F.; Caboche, J. Pathophysiology of Huntington's Disease: From Huntingtin Functions to Potential Treatments. *Curr. Opin. Neurol.* **2008,** *21,* 497–503.

Rung, J. P.; Carlsson, A.; Markinhuhta, K. R.; Carlsson, M. L. The Dopaminergic Stabilizers (-)-OSU6162 and ACR16 Reverse (Ð)-MK-801-Induced Social Withdrawal in Rats. *Prog. Neuro. Psychopharmacol. Biol. Psychiatry* **2005,** *29,* 833–839.

Rung, J. P.; Rung, E.; Helgeson, L.; Johansson, A. M.; Svensson, K.; Carlsson, A.; et al. Effects of (-)-OSU6162 and ACR16 on Motor Activity in Rats, Indicating a Unique Mechanism of Dopaminergic Stabilization. *J. Neural. Transm.* **2008,** *115,* 899–908.

Sagredo, O.; Pazos, M. R.; Valdeolivas, S.; Fernandez-Ruiz, J. Cannabinoids: Novel Medicines for the Treatment of Huntington's Disease. *Recent Pat. CNS Drug Discov.* **2012,** *7,* 1–8.

Sahlholm, K.; Arhem, P.; Fuxe, K.; Marcellino, D. The Dopamine Stabilizers ACR16 and (-)-OSU6162 Display Nanomolar Affinities at the Sigma-1 Receptor. *Mol. Psychiatry* **2013,** *18,* 12–14.

Sajatovic, M.; Verbanac, P.; Ramirez, L. F.; Meltzer, H. Y. Clozapine Treatment of Psychiatric Symptoms Resistant to Neuroleptic Treatment in Patients with Huntington's Chorea. *Neurology* **1991,** *41,* 156.

Schilling, G.; Coonfield, M. L.; Ross, C. A.; et al. Coenzyme Q10 and Remacemide Hydrochloride Ameliorate Motor Deficits in a Huntington's Disease Transgenic Mouse Model. *Neurosci. Lett.* **2001,** *315,* 149–153.

Seeman, P.; Tokita, K.; Matsumoto, M.; Matsuo, A.; Sasamata, M.; Miyata, K. The Dopaminergic Stabilizer ASP2314/ACR16 Selectively Interacts with D2 (High) Receptors. *Synapse* **2009,** *63,* 930–934.

Shirasaki, D. I.; Greiner, E. R.; Al-Ramahi, I.; Gray, M.; et al. Network Organization of the Huntingtin Proteomic Interactome in Mammalian Brain. *Neuron* **2012,** *75,* 41– 57.

Spokes, E. G. Neurochemical Alterations in Huntington's Chorea: A Study of Postmortem Brain Tissue. *Brain* **1980,** *103,* 179–210.

Squitieri, F.; Cannella, M.; Piorcellini, A.; et Al. Short-Term Effects of Olanzapine in Huntington Disease. *Neuropsychiatry Neuropsychol. Behav. Neurol.* **2001**, *14*, 69–72.

Stewart, J. T.; Mounts, M. L.; Jr. Clark, R. L. Aggressive Behavior in Huntington's Disease: Treatment with Propranolol. *J. Clin. Psychiatry* **1987**, *48*, 106–108.

Stewart, J. T. Paradoxical Aggressive Effect of Propranolol in a Patient with Huntington's Disease. *J. Clin. Psychiatry* **1987**, *48*, 385–386.

Strong, T. V.; Tagle, D. A.; Valdes, J. M.; Elmer, L. W.; Boehm, K.; Swaroop, M.; et al. Widespread Expression of the Human and Rat Huntington's Disease Gene in Brain and Nonneural Tissues. *Nat. Genet.* **1993**, *5*, 259–265.

Tabrizi, S. J.; Langbehn, D. R.; Leavitt, B. R.; Roos, R. A. C.; Durr, A.; Crauford, D.; et al. Biological and Clinical Manifestations of Huntington's Disease in the Longitudinal TRACK-HD Study: Cross Sectional Analysis of Baseline Data. *Lancet Neurol.* **2009**, *8*, 791–801.

Tang, T. S.; Chen, X.; Liu, J.; et al. Dopaminergic Signalling and Striatal Neurodegeneration in Huntington's Disease. *J. Neurosci.* **2007**, *27*, 7899–7910.

The Huntington's Disease Collaborative Research Group. A Novel Gene Containing A Trinucleotide Repeat That is Expanded and Unstable on Huntington's Disease Chromosomes. The Huntington's Disease Collaborative Research Group. *Cell* **1993**, *72*, 971–983.

The HDCRG. Tetrabenazine as Antichorea Therapy in Huntington Disease: A Randomized Controlled Trial. *Neurology* **2006**, *66*, 366–372.

Vallette, N.; Gosselin, O.; Kahn, J. P. Efficacy of Clozapine in the Course of Huntington Chorea: Apropos of a Clinical Case. *Encephale* **2001**, *27*, 169–171.

Van Vugt, J. P.; Siesling, S.; Vergeer, M.; et al. Clozapine Versus Placebo in Huntington's Disease: A Double Blind Randomised Comparative Study. *J. Neurol. Neurosurg. Psychiatry* **1997**, *63*, 35–39.

Verhagen, M. L.; Morris, M. J.; Farmer, C.; et al. Huntington's Disease: A Randomized, Controlled Trial Using the NMDA-Antagonist Amantadine. *Neurology* **2002**, *59*, 694–699.

Videnovic, A. Treatment of Huntington Disease. *Curr. Treat. Options Neurol.* **2013**, *15* (4), 424–438.

Walker, F. O. Huntington's Disease. *Lancet* **2007**, *369*, 218–228.

Wider, C.; Luthi-Carter, R. Huntington's Disease: Clinical and Aetiological Aspects. *Schweiz. Arch. Neurol. Psychiatr.* **2006**, *157*, 378–383.

Zielonka, D.; Marinus, J.; Roos, R. A.; De Michele, G.; Di Donato, S.; Putter, H.; et al. The Influence of Gender on Phenotype and Disease Progression in Patients with Huntington's Disease. *Park. Relat. Relat. Disord.* **2013**, *19*, 192–197.

Zielonka, D.; Mielcarek, M.; Landwehrmeyer, B. G. Update on Huntington's Disease: Advances in Care and Emerging Therapeutic Options. *Parkinsonism Relat. Disord.* **2015**, *21*, 169–178.

# CHAPTER 8

# Pharmacological Management of Amyotrophic Lateral Sclerosis

SHALINI MANI*, CHAHAT KUBBA, TANYA SHARMA, and
MANISHA SINGH

*Jaypee Institute of Information Technology, Uttar Pradesh, India*

*Corresponding author. E-mail: mani.shalini@gmail.com*

## ABSTRACT

Amyotrophic lateral sclerosis (ALS) is a motor neuron disease, belonging to a wider category of neurodegenerative disorders, which primarily lead to death of motor neurons that initiate and serve as a vital link for communication between the brain and the voluntary muscles. Thus, their degeneration may lead to paralysis state and long-term disability. Till date, no effective treatment is known to exist for this disease phenotype. However, many other drugs have been investigated and therapies are currently under clinical development, which may offer hope in the near future. This chapter reviews various modes of therapies and treatments which are presently being implemented for ALS treatment. To manage the diseased condition, different therapies such as physical, occupational, breathing, speech, and nutritional support therapies are discussed. The authors have further highlighted the use of various common drugs such as riluzole and edaravone and their distinct mechanism of action. Along with discussing the clinical trials and their developments, authors summarize the treatment development in this motor neuron disease and discuss the strengths and shortcomings of past as well as upcoming clinical trials.

## 8.1 INTRODUCTION

Amyotrophic lateral sclerosis (ALS) is one of the many diseases that cause neurodegeneration and progressive deterioration of motor neurons, whose predominance have been observed in upper portions of cortex, lower brain stem in particular and spinal cord (Ripps et al., 1995). The selective death of motor neurons causes paralysis of voluntary muscles, which begins focally in the limb or bulbar muscle but disseminates gradually throughout the body (Brown and Chalabi, 2017). ALS is the most frequent and life threatening neurodegenerative disorder of midlife that marks its onset in the late 60s (Turner et al., 2011) with nearly 10% of ALS cases being familial and the remaining 90% being sporadic, occurring without a family history (Driss and Siddique, 2011). Epidemiological studies conducted in 1990s across Europe and North America found that the crude prevalence rate of ALS is between 1.5 and 2.7 per 100,000 population/year but the statistics are fairly uniform globally (Logroscino et al., 2008). Numerous studies in general have described that gender differences play a crucial role in making an individual susceptible to developing ALS; the reported data indicated an inclination of males toward higher incidence rates that females (McCombe et al., 2011).

Extensive research has elucidated some of the cellular processes causing the ALS pathogenesis. The disease is distinguished by abnormalities in mitochondrial morphology and biochemistry, slowed anterograde or retrograde axonal transport, accumulation of neuro-filaments, intra-cytoplasmic inclusions, decreased levels of neurotrophic factors, generation of free radicals, excitotoxicity, and inflammation (Leigh and Wijesekera, 2009). Apart from the above mentioned attributes, majority of the researchers hypothesize a complex interplay of both genetic and exogenous environmental factors leading to death of motor neuron. In nearly 20% of the cases having autosomal dominant familial ALS and 2% of patients suffering from sporadic ALS, mutations in the copper–zinc superoxide dismutase (*SOD1*) gene has been reported (Rosen et al., 1993). Other genes leading to familial motor neuron degeneration include angiogenin (Greenway et al., 2006), aalsin (*ALS2)* (Yang et al., 2001), vesicle associated membrane protein (*VAPB, ALS8)* (Nishimura et al., 2004), senataxin (*ALS4)* (Chen et al., 2004), and p150 subunit of dynactin (*DCTN1)* (Münch et al., 2004). Amongst the environmental factors or elements, it has been witnessed that pesticides and metal exposure pose that maximum risk for onset of ALS (Bozzoni et al., 2016). Metals with a potential relevance for ALS are uranium, aluminum, zinc, arsenic, cadmium, copper, vanadium, and cobalt. Traces of these metals have been found to exist in the cerebrospinal fluid (CSF) of ALS patients in significantly elevated concentration (Ingre et al., 2015). Both genetic and environmental factors are not mutually exclusive and are most likely to complement each other.

Despite of the prevailing knowledge about molecular pathways underlying the motor neuron degeneration, the definite cause of ALS cannot be explicitly stated and still remains to be unspecified. Typical manifestations of ALS are muscle stiffness and spasticity due to the failure of corticospinal motor neurons (Handy et al., 2011). Many symptoms such as foot drop, difficulty in walking, compromised hand dexterity, and shoulder weakness are seen in the early stage of ALS (Gordon, 2011). However, as the disease progresses, affecting the lower motor neurons, excessive electrical irritability is initiated leading to spontaneous muscle fasciculation and loss of synaptic connectivity with the target muscles, which then atrophy (Kiernan et al., 2011). Patients gradually become anarthric and show signs of diminished cognition (Hoerth et al., 2003).

The life expectancy associated with ALS is nearly 3 years soon after the emergence of symptoms. However, the longevity may extend from a few months to decades, as observed in approximately 5% of the patients (Chio et al., 2009). The data collected by the researchers clearly indicates the fact that patients with symptoms onset before 40 years of age have longer survival rates (Ferrante et al., 1997; Aguila et al., 2003). As the age of onset increases, the survival time decrease (Testa et al., 1989).

Although a rare disease, its unvarying lethality necessitates the need of scientific indulgence to augment our basic knowledge about the pathways that progressively leads to lethality as well as designing novel therapies to prevent its outset. It has only been 10 years or so that this deadly disease has been publicly recognized and the awareness has increased among all strata of the society; various diagnostic techniques have emerged, which try to achieve the diagnosis in a timelier fashion, and overall care is better. Few drugs have also been designed based on this knowledge, though without manifestation of complete cure. Thus a real therapeutic drug which can cure the devastating condition remains elusive. Unraveling the ALS genome (Ravtis and Traynor, 2008), better drug delivery (Mazibuko et al., 2015) and the application of pharmacogenomics (Paratore et al., 2012), as well as stem cell therapy (Lunn et al., 2014) are some future avenues that can bring success to the treatment of ALS. This chapter emphasizes on the various modes of therapies and treatments which have been so far implemented for ALS treatment. Authors further discuss the current therapeutic advances in the fields of ALS research. It has been followed by the current developments in ALS therapy and its future applications.

## 8.2 ANTI-AMYOTROPHIC LATERAL SCLEROSIS DRUGS

Discovery about the putative causes of ALS in the late 1980s opened new therapeutic avenues. The early experimental findings reported a systemic defect in the metabolism of glutamate (Caroscio and Platakis, 1987), deficiency of glutamate dehydrogenase (GDH) (Hugon et al., 1989), and a defective glutamate transport (Rothstein et al., 1992) as some of the prominent causes of ALS. All these studies are in accordance with the hypothesis that accumulation of glutamate, the most common excitatory neurotransmitter in the central nervous system (CNS) of all vertebrates, at the synapses in

the toxic range of the concentration, leads to neuronal cell death.

Thereupon, clinical trials are primarily being oriented toward finding out or exploring the usefulness of compounds which could potentially detain the rate of progression of the degenerative process associated with ALS. Branched-chain amino acids such as valine, isoleucine, and so on, that could modulate the glutamatergic system were proposed as possible treatments but the results failed to get ratified because a substantial amount of variability has been observed in each individual patient due to which, foreseeing the evolution of the disease becomes problematic (Plaitakis et al., 1988; Testa et al., 1989). A drug named lamotrigine underwent a double blind placebo controlled trial but failed to show significant benefits in ALS patients (Andrew et al., 1993). Another drug gabapentin provided beneficial effects yet required a well-controlled clinical trial to critically assess its effect before recommending it as a potential drug for the treatment of ALS (Welty et al., 1995). Till date only two drugs, riluzole and edaravone are known to be FDA approved and recommended for ALS treatments.

### 8.2.1 RILUZOLE

After so many trials with different drugs, a placebo-controlled trial was conducted which was randomized in nature. Riluzole, the presumed blocker of glutamate release, demonstrated a significantly positive effect with respect to the survival of the participants and thus in 1994, it was established as the first disease-modifying treatment for ALS (Bensimon et al., 1994). However, the discovery was subjected to many questions in view of the disproportionate benefit it offered to the patients; the drug increased survival rate in one group of participants with ALS but failed to manifest the same result in another group of patients experiencing similar conditions (Rowland, 1994).

To address the uncertainty, a scientific research was again carried out, demonstrating a statistically significant prolongation in patients' survival rate who received intermediate to high dose of riluzole thereby confirming that riluzole is well tolerated; riluzole dose of 100 mg dose had the maximum benefit than risk (Lacomblez et al., 1996). These results overlapped with the findings of Bensimon et al. (1994), who suggested that every day 50-mg tablets containing 100 mg of may ameliorate the survival rate (Bensimon et al., 1994). The drug received an approval for marketing and utilization by FDA (Food and Drug Administration) in 1995 with a recommended dose of 50 mg every 12 h. Another placebo-controlled trial of riluzole treatment conducted in 959 patients with ALS reanalyzed the beneficial effects of ALS and successfully displayed that the individuals receiving the drug treatment stayed in a better health condition, that too for a significantly longer duration of time (Riviere et al., 1998). Similarly, the data collected by another group of researchers suggested that survival rate among the patients which primarily exhibited bulbar dysfunction was enhanced by active and aggressive management (Traynor et al., 2003). In spite of the proposed benefits, a number of concerns about the therapeutic effects of riluzole still persist. A relatively high cost of the drug, adverse effects and moderate prolongation of survival, being few months on an average, continues to be the central consideration (Zocolella et al., 2007). Moreover, trials conducted in ALS patients aged over 75 years and having an advanced stage disease observed no significant deviation from the parallel treatment group thereby highlighting that the produce of riluzole is less likely to occur in advance state (Bensimon et al., 2002).

Considering the above implications, recommendation of treatment (drug oriented) to the diseased individuals should be done with some restrictions. In 1997, a practice advisory was issues by the Quality Standards Subcommittee of the American Academy of Neurology which suggested that riluzole may prolong survival in patients with definite or probable ALS having symptoms for less than 5 years and forced vital capacity (FVC) >60% with no signs of tracheotomy. According to their expert opinion, a potential benefit could be seen in those who are suspected to have ALS based upon the symptoms present for more than 5 years (Neurology, 1997).

More than two decades since its discovery, riluzole still remains the only approved drug by FDA to cure the fatality of the disease. During this period, understanding the biologic consequence accountable for the therapeutic action of riluzole in ALS has vastly increased our understanding of the pathways that are followed during the disease progression, along with the mechanism responsible in slowing the progression of this disease.

### 8.2.1.1   PHARMACOKINETICS

The basic one compartment pharmacokinetic model of riluzole shows 51.4% inter-individual variability in plasma clearance values while the intra-individual variability being 28%. Its clearance is not dependent on any of the factors such as dosage, duration of treatment (only up to 10 months), age, kidney function, gender and smoking. In non-smoking male patients, clearance is observed to be 51.4 L/h; the values being 32% lesser in women and 36% lower in non-smoker patients (Bruno et al., 1997).

## 8.2.1.2   MECHANISM OF ACTION

In the mammalian neurons, glutamic acid is the key neurotransmitter responsible for excitation in the CNS (Meldrum, 2000). When depolarization occurs, the release of glutamate is observed from the glutamatergic nerve terminals; by crossing the synaptic cleft it acts on postsynaptic target molecules. Following its action, glutamic acid is efficiently removed from the synapse by glutamate uptake systems in glia and nerve terminals (Auger et al., 2000). However, in ALS, elevated concentrations of glutamic acid can gather in the synaptic cleft causing excitotoxic conditions and prolonged depolarization thereafter (Bosch et al., 2006). Riluzole drug acts by blocking the excitotoxic process by targeting various processes such as inhibition of glutamic acid release and inhibition of voltage-dependent sodium channels or voltage-dependent calcium channels on nerve endings and cell bodies. All processes work in synergy to contribute to its powerful neuroprotective action (Doble, 1996). The potential sites of action of riluzole have been depicted in Figure 8.1.

### 8.2.1.2.1   Decreased Neuronal Firing

Plethora of observations enforcing the delineating effect in the rhythmic firing of numerous types of neurons has been evident from the action of riluzole. Experimental observations in various model systems indicate that riluzole, in response to sustained current injection, reduced repetitive neuronal firing without interfering single action potential triggered by temporary current injection. The following Table 8.1 depicts the various concentration of riluzole used in different organisms, which were effective in lowering down the neuronal firing.

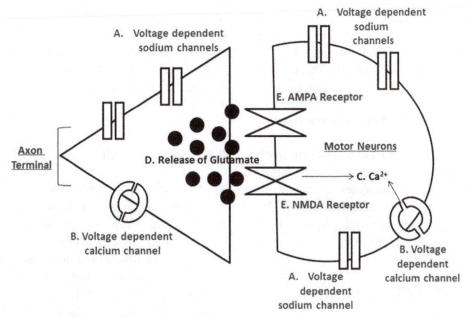

**FIGURE 8.1**   The potential sites of action of riluzole. A. Targeting the voltage dependent sodium channels. B. Targeting the voltage dependent calcium channels. C. Activating the intracellular calcium ion buffering process. D. Inhibiting the release of glutamic acid. E. Blocking the NMDA or AMDA receptors.

**TABLE 8.1**   The Effect of Riluzole on Different Model Systems.

| Concentration (µM) | Experimental model | Effects | References |
|---|---|---|---|
| 3 | Rat cortical neurons | Riluzole reduced the tonic firing | Siniscalchi et al., 1997 |
| 0.1–100 | Rat cortical neurons | Effective in inhibiting persistent sodium current | Urbani et al., 2000 |
| ≤5 | Rat brainstem neurons | Reduced membrane resonance, post-inhibitory rebound, sub-threshold Oscillations, and bursting | Wu et al., 2004 |
| 5–10 | Mouse spinal neurons | Failure of repetitive firing | Kuo et al., 2006 |
| 5–500 | Mouse model with mutated protein copper, zinc superoxide dismutase (Gly[93] Ala, G93A) | Rapid decay of $Na^+$ currents | Zona et al., 2006 |
| 5 | Rat ventral horn neurons | Blocked 69.5% of the TTX-sensitive $Na_p$ current | Theiss et al., 2007 |
| ≤5 | Mouse spinal inter neurons | Blocked the sodium current gradually and suppressed both the amplitude and frequency of voltage oscillations | Conhaim et al., 2007 |

In some cases, riluzole was also effective in decreasing the duration, amplitude, and frequency of action potential in growth-hormone-secreting pituitary ($GH_3$) cells, block the voltage-gated $Na^+$ currents and inhibit $Ca^{2+}$ signaling (Charles and Parrazal, 2003).

### 8.2.1.2.2   Effect on Neurotransmission

It is believed that riluzole exerts its neuroprotective and anticonvulsant action by decreasing the release of excitatory amino acid glutamate or by blocking the glutamate receptors. The results of various studies supporting this hypothesis have been summarized in Table 8.2.

### 8.2.1.3   ADVERSE EFFECTS

Data generated in the clinical trials revealed that patients treated with riluzole had more frequent nausea occurrences in comparison to the placebo group (Lacomblez et al., 1996). Tends of asthenia were also observed in some patients (Bensimon et al., 2002). Patients also developed an elevated enhanced concentration of alanine transferase (serum) in comparison to the control or placebo group of patients (Lacomblez et al., 1996).

### 8.2.2   EDARAVONE

Since its launch in 1996, riluzole, in spite of having an unsatisfactory effect, remained the only approved drug available in the market for the treatment of ALS patients until recently, when a newly discovered drugedaravone, sold under the brand name radicava, was received an approval in May 2017 by FDA after 20 years for the treatment of this fatal disease. Unlike riluzole, edaravone was approved due to its efficacy to make the patients' symptoms

**TABLE 8.2** The Effect of Riluzole on Neurotransmission.

| Experimental model | Concentration | Results | References |
|---|---|---|---|
| Caudate nucleus of the cat | $10^{-5}$ M | Reduction in the spontaneous efflux of glutamate; maximum effect appearing after 40 min. Prolonged application enhanced the neuronal pool of glutamate, which was released under potassium depolarization. | Cheramy et al., 1992 |
| Hippocampal slices of 30- to 50-day-old female Sprague-Dawley rats | 10–30 µM | Riluzole inhibits the $K^+$-evoked release of endogenous excitatory amino acid glutamate from CA1 of hippocampal area. | Martin et al., 1993 |
| Pyramidal neurons in rat hippocampal brain slices | 5 µM | Riluzole blocks glutamate transmission via depression in presynaptic conductance in glutamatergic nerve fibers. | MacIver et al., 1996 |
| Male Sprague-Dawley rats | 4 mg/kg (Dose of 1 mg/kg and 2 mg/kg had no effect) | When administered alone, at a concentration of 4 mg/kg riluzole acted as a NMDA receptor antagonist. However, its co-administration with NMDA receptor ligands at a concentration of 30 mg/kg showed that riluzole indirectly acted upon glutamate receptors. | Kretschmer et al., 1998 |
| Mouse neocortex | 100 mM ($IC_{50}$ = 19.5 mM) | Glutamate release was diminished by riluzole upto 77%. | Jehle et al., 2000 |
| Rat cerebrocortical synaptosomes | 1 µM | Riluzole reduces the P/Q-type calcium channels mediated calcium influx and thsui blocks the release of glutamate. | Wang et al., 2004 |
| Male long Evans rat | 6 and 12 mg/kg | Riluzole prominently reduces the basal levels as well as and formalin-triggered elevations in the spinal glutamate concentration. | Coderre et al., 2007 |
| Neonatal rat hypoglossal motoneurons | 10 µM | Riluzole decreased glutamatergic transmission by inhibiting PKC, which regulates presynaptic NMDA receptors. | Lamanauskas and Nistri, 2008 |

deteriorate more slowly rather than extending survival (Kalin et al., 2017).

### 8.2.2.1 PHARMACOKINETICS

The route of administration of edaravone is intravenous therapy (IV) infusion. According to the data generated by clinical trials, the maximum concentration of edaravone is achieved by the end of administration of the drug. Trends also showed a non-proportional increase in area under concentration time curve (AUC) versus maximum plasma concentration ($C_{max}$). Even with multiple dosages, no accumulation is observed. 92% of the injected drug binds mainly to albumin and its half-life is 4.5–6 h. It is metabolized to sulfate conjugates and excreted in the urine in the form of glucuronide conjugate form (Mitsubishi Tanabe Pharma America, 2017).

### 8.2.2.2   MECHANISM OF ACTION

It has been evident from various scientific studies that oxidative stress could be the important mechanisms which can either directly affect the cells or aggravate other mechanisms that cause motor neuron deterioration observed in ALS (Ferrante et al., 1997; D'Amico et al., 2013). Edaravone scavenges these free radicals and relieves the destructive effects of oxidative stress. Edaravone can exist in two forms and both of them react differently with the radicals, exhibiting different mechanism of action. Single electron transfer mechanism has been proposed to contribute to both the anionic form, the overall reactivity toward •OH radical being (~44%), closely followed by the formation of radical adduct (~40%). On the other hand, formation of radical adduct was established to be the key mechanism contributing greater than 98%, no matter what the polarity of the surroundings was. However, the anionic form was suggested to be 8.6× reactive than its neutral form (Galano and Prezgonzlez, 2011). Apart from the hydroxyl radicals, edaravone can also scavenge other free radical including lipid peroxyl radical (LOO•) and peroxynitrite (ONOO$^-$). Edaravone has successfully demonstrated its antioxidative effects against water-soluble peroxyl radicals such as vitamin E (lipid soluble peroxyl radical) and vitamin C (Yamamoto et al., 1996) and is also effective in scavenging free radicals, including LOO•, ONOO$^-$ along with the other reactive oxygen species (ROS) acting through its electron donating properties (ABE et al., 2004; Kamogawa and Sueishi, 2014; Fujisawa and Yamamoto, 2016). Edaravone also exhibits

protective effects against oxidative neurotoxicity produced by peroxynitrite and activated microglia (Banno et al., 2005), HT22 neuronal cells and primary rat astrocytes glia (Lee et al., 2010), and oligodendrocyte precursor cells of rat (Miyamoto et al., 2013).

The safety and effectiveness of edaravone were investigated in an open trial phase II design consisting of 20 subjects, which were followed by the administration of edaravone for 2 weeks and a subsequent 2-week observation period; the 4-weeks cycle was repeated six times. The results revealed diminished levels of 3NT (oxidative stress marker) in CSF after the first cycle of treatment with 60 mg edaravone/day and the concentration being significantly low in majority of the patients after the sixth round of treatment. From the data of the trial, it was concluded that edaravone progressively delayed the motor disturbances and was safe to use (Kimura and Yoshino, 2006). The FDA approval of edaravone was based on this randomized, 6-month, placebo-controlled trial conducted on Japanese patients with the recommended dosage of edaravone of 60 mg administered as an intravenous infusion over a 60-min period.

Though regular administration of edaravone makes the patient's symptoms deteriorate at a slower pace, the drug still does not addressees the primary concern of preventing the degeneration of the motor neurons. Moreover, the patients would have to bear a heavy cost of around $1087 per infusion and multiple infusions costing them approximately $146,000 per year as per the commentary published in *Pharmaceutical Technology* in May 2017.

### 8.2.2.3   ADVERSE EFFECT

A history of hypersensitive reactions to edaravone such as multi-form, redness, erythema, anaphylaxis, decreased blood pressure, and dyspnea has been observed in some patients. The post marketing data from outside the US also reported sulfite allergic reactions. Approximately 10% of the patients treated with edaravone also experienced headache, dermatitis, eczema, brusing, and disturbances in gait (Mitsubishi Tanabe Pharma America, 2017).

## 8.3 MANAGEMENT OF AMYOTROPHIC LATERAL SCLEROSIS

Out of the several drugs that underwent clinical trials, riluzole and edaravone are the only ones that have received approval so far by FDA for treatment or curbing the advancement of ALS. Considering the fatality, the disease poses and the middling effect these drugs, the treatment cannot be restricted to riluzole and edaravone alone and necessitates multidisciplinary approach to manage the condition. Symptomatic treatments have therefore gained importance in containment of chief ramifications of the disease such as depression, pain, spasticity, sleep disorders, hypersialhorroea, constipation, and reflux. In one of the study (population based) conducted from 1996 to 2000, authors have depicted the importance of multidisciplinary clinic in better prognosis as compared to the patients attending a general neurology clinic with the data indicating a 29.7% decrease in the mortality rate (Traynor et al., 2003). With respect to the multidisciplinary approach in order to improve the prognosis of ALS, the different approaches toward the management of ALS have also been studied.

### 8.3.1 NUTRITIONAL MANAGEMENT

One of the major issues in managing ALS is denutrition and weight loss, serving as a fundamental pointer of prognosis and increasing the death toll rate by seven fold if the body mass index reaches below 18.5 kg/m$^2$ (Desport et al., 1999; Marin et al., 2011). The consumption of calories and protein in ALS was first investigated by (Slowie et al., 1983) using the 24-h dietary recall method. Their results indicated that out of the 20 randomly selected ambulatory ALS patients, 70% of them had a lower calorie intake than the recommended daily allowance. This alteration in the nutritional status can be attributed to a multitude of factors. A highly recognized one amongst them is dysphagia; as the disease advances, patient's efficacy of swallowing gets impaired which increases the susceptibility to developing malnutrition (Clave et al., 2006). Another significant factor in hypermetabolism (Desport et al., 2005). Higher disbursement of energy than intake profoundly impairs the energy balance due to which rapid decrease in the body fat and body mass index occurs (Dupuis et al., 2011). Other factors such as dyspnea (Holm et al., 2013), cognitive impairment, fatigue and difficulty in self-feeding may also contribute to malnutrition (Daniel, 2013).

Various alternative methods can be adopted to maintain nutritional intake orally. The foremost step in nutritional care is to alter the diet content; food should be chewable and rich in calorie content (Taft and Tanenbaum, 1990). Special techniques of swallowing can be taught by an occupational therapist or a physiotherapist (Heffernan et al., 2004). Nutritional supplements can be taken to alter food consistency (Miller et al., 2009). Enteral feeding using percutaneous endoscopic gastrostomy (PEG) can also serve as substitute path for administrating medication and fluid thereby stabilizing the body weight (Mazzini et al., 1995).

### 8.3.2 SPEECH MANAGEMENT

Speech impairment is commonly seen in the ALS affected patients (Kent et al., 1991). The decline of speech intelligibility in ALS can be attributed to articulatory impairments, which is characterized by reduction in the velocity of jaw and lip movement, and resonatory impairments marked by increase in the nasal airflow (Rong et al., 2016). In order to facilitate and maintain communication, speech therapy is necessitated. Occupational therapists or physiotherapists can be useful in teaching the patients with adaptive techniques that can enhance the clarity of the speech. Strategies such as partner support, speech supplementation and other non-verbal cues can aid the speech process (Hanson et al., 2011). Additionally, exploring alternate methods of communication such as pen and paper, charts, alphabet board, picture boards and so on can be useful (Andersen et al., 2007). Other measures include avoidance of background noises such as that emanating from a radio, television, or multiple people talking simultaneously (Borasio and Mitchell, 2007). Table 8.3 summarizes some of the communication strategies to augment the speech process.

**TABLE 8.3** Communication Strategies to Augment the Speech Process.

| Communication strategies | |
|---|---|
| Speaking strategies | • Speaking at a slower rate; |
| | • Use of keywords; |
| | • Breathing after every 3 to 4 words; |
| | • Voice resting (15–20 min); |
| | • Not overusing the muscles and allowing; them to recuperate |
| Conversation strategies | • Avoid talking in noisy areas; |
| | • Interpretation by the partner; |
| | • Confirmation; |
| | • Use of topic cues |

**TABLE 8.3** *(Continued)*

| Communication strategies | |
|---|---|
| Low-tech devices | • Alphabet board; |
| | • Communication chart; |
| | • Pen and paper |
| Non-verbal cues | • Body gestures; |
| | • Facial expression; |
| | • Eye contact |

### 8.3.3 RESPIRATORY MANAGEMENT

One of the key requirements for controlling the ALS advancement is managing the respiratory function. Respiratory involvement is constantly seen during the advancement of the disease and it has a direct impact on survival. A significant prolongation in survival rate and an improvement in the qualitative aspect of life have been observed in the people suffering from ALS, with the use of non-invasive ventilation (NIV); NIV evidently palliates symptomatic outcomes of the disease in these patients (Bourke et al., 2006; Radunovic et al., 2013). As the disease progresses, the medical care becomes more complicated. To manage the respiratory failure, the disease demands an interdisciplinary care team, consisting of a respiratory specialist to monitor the pulmonary function every 3 months as per the recommendation of ALS Practice Parameter (Miller et al., 1999). Therapies can also be directed to clear the respiratory secretions, which pool in the upper respiratory tract of individuals and are very common in ALS. Therapies can also be adopted to prevent atalectasis, and improve gas exchange or relieving the symptoms of dyspenea (Gelanis, 2001). Little changes in the diet and lifestyle can also be of great advantage (Gelanis, 2001).

### 8.3.4 PHYSICAL MANAGEMENT

Throughout the span of this incurable degenerative disease, the quality of life deteriorates. So the primary objective of incorporating physical therapy into the treatment of ALS is to prevent the deteriorating effect ALS has on an individual's quality of life and provide real benefits in terms of spasticity. Various studies have demonstrated that focusing on the physical symptomatic aspect of ALS can be useful to the ALS suffering population (Bohannon, 1983; Ashworth et al., 2012). A common intervention to gain physical independence is the use of assistive devices based on the extremity of the condition and energy requirement. For instance, a walker with wheels can be used since it requires lesser expenditure of energy to advance (Foley et al., 1996), or adaptivity of an object's characteristics like the length of the handle can enhance motor performance in patients that show limited use of hand (Fuller et al., 1997). Those patients who suffer spasticity, stretching, and ROM exercise can be useful. A study conducted, though not specific to ALS condition, shows that spasticity could be decreased by a range of passive motion done on a regular basis (Skold et al., 2000). To ameliorate pain, modalities such as massage heat and transcutaneous electrical nerve stimulation (TENS) can be used (Walling, 1999).

### 8.3.5 OCCUPATIONAL MANAGEMENT

Often, the patients appoint an occupational therapist, who serves the responsibility of evaluating the daily performance of the patients and monitor their functional tasks which include personal care, activities involving work and mobility, and pinpointing what the patient exactly requires. During the early stages of the disease, an occupational therapist can use splints and orthotics to reduce pain in neck and extremities, optimizing the range of motion at home and implement the strategies to manage pain. Whereas, the goal at the later stages of the disease should be to assess positioning and maintain the integrity of the skin, optimize safety of the patient and employ some modifications in the environment to suit the diseased (Radomski and Trombly, 2014). Moreover, the therapist can aid in providing instructions to the patients on the best possible use of body parts that are still functional, and introduce special aids like utensils that are easier to hold for patients with dysfunction their hands (Fuller and Trombly, 1997).

## 8.4 CURRENT TRENDS IN EXPERIMENTAL THERAPEUTICS FOR AMYOTROPHIC LATERAL SCLEROSIS

Though some potential drugs and different approaches for the management of ALS have been designed and proven, however all these approaches separately or combined are not enough for the complete prognosis of the disease. Hence there is a need to develop alternative novel therapies for ALS treatment. In the recent past few years, certain new therapeutics with claimed efficacy in CT scans of ALS patients, have emerged. This section discusses the efforts made toward the development of new ALS therapy, from the perspective of therapy targets such as axon guidance signaling pathways, RNA targeted therapeutics, anti-glutamatergic, anti-oxidative methods. It further highlights the outcomes of major clinical trials conducted currently.

### 8.4.1 THERAPIES TARGETING AXON GUIDANCE SIGNALING PATHWAYS

Neurons lengthen their axons in excess of extensive distances during developing connections with other synaptic neurons. Proteins involved in guiding the axon may control the progression of axon guidance. These proteins can play a function in attracting or repelling axons thus guide them to a definite area or stopping them from rising into unsuitable regions, correspondingly. Recently several report sustain the testing theory that abnormal expression or function of axon guiding proteins such as ephrins, semaphorins, slits, and netrins usually implicated in curating and maintaining circuits of motor neuron may bring on pathological variations in circuits of motor neuron (Hollis, 2015). Interestingly, pathological variations occurring in nerve terminals and motor axons are found to lead degeneration of motor neurons and associated clinical abnormalities (Fischer et al., 2004). It can be inferred from this discovery that the disease development may begin at the nerve endings and ultimately grow toward the body of the neuronal cell. Numerous diverse molecules responsible for guiding the axons changed expressions in patients of ALS. Thus single-nucleotide polymorphisms (SNPs) in genes expressing proteins for axon guidance may important for diagnosis of ALS (Lesnick et al., 2008). After realizing the importance of these proteins, recently the cell replacement strategies have been designed for corrections of the degenerating motor system of these patients (Silva and Yu, 2008).

In the therapy based upon cell-replacement, transplanting the motor neurons which are derived from stem-cell is measured to be a probable move to treat ALS (Christou et al., 2007). These motor neurons may extend to axons which eventually form neuromuscular junctions (Thonhoff et al., 2009). These transplanted neurons may face the challenge due to inhibitory action of myelin associated proteins such as Nogo-A, myelin associated glycoprotein (MAG) and oligodendrocyte-myelin glycoprotein (Omgp). Inhibition may also be due to inhibitory proteins specific for the neurite growth. A winning transplantation outcome came in to lime light a decade back, in which rat spinal cord was injected with motor neurons along with dibutyryl cyclic adenosine monophosphate (dbcAMP)(Deshpande et al., 2006). The systemic administration of phosphodiesterase 4 inhibitor was also performed to obstruct the repulsive outcome of myelin (Deshpande et al., 2006). These studies highlighted that the alteration of the proteins responsible for regulation of growth and guidance of axon is important to re-establish appropriate functioning of motor neurons subsequent to therapy by cell replacement.

Furthermore, some of the proteins guiding the axons may pretend as effectual curative target as well. For example, upregulation of Nogo-Λ (a myelin-associated neurite outgrowth inhibitor) (Lee and Zheng, 2012) and Sema3A (inhibitor of axonal outgrowth and vital for normal neuronal pattern development (Dent et al., 2010) has been linked with ALS and either inhibiting or down regulating both may repair the circuits of motor neuron circuitry and arrest deterioration of these neuron. A fascinating move toward the treatment could be the employment of antagonists to thwart these axons in response to increased concentration of Sema3A and/ or Nogo-A. As developing such inhibitors may be a potential candidate for inhibiting the Sema3A and Nogo-A mediated function in ALS and interestingly some preliminary studies using different antibodies, have also been conducted (Buchli et al., 2007; GrandPre et al., 2002). In 2014, Bros-Facer et al. performed a study regarding the treatment of ALS using an antibody specific for Nogo-A in the SOD1G93A mouse model. The results showed that treatment with anti-Nogo-A antibody drastically enhances neuromuscular function in the SOD1G93A mouse model of ALS during the early stage of the disease. It is further suggested that pharmacological blockage of Nogo-A may be a disease-curing approach in case of ALS (Bros-Facer et al., 2014).

## 8.4.2 RNA-TARGETED THERAPEUTICS

In case of ALS, both the upper and lower motor neurons undergo for progressive degeneration. It leads to a range of disorders that lead to front temporal dementia at the other end (Philips and Robberecht, 2013). Although many of the sporadic cases lack the knowledge of a known genetic cause, a large number of the familial forms of ALS have now been known to bear a genetic background which are associated by means of some mutation in genes such as *SOD1*, *FUS, C9ORF72,* and *TARDBP* (Renton et al., 2013). The mutations reported in these proteins are known to trigger their aggregate formation which can further exhibit the toxic effect that leads to cell-specific deterioration (Ferraiuolo et al., 2011). Till date, the suitable drugs which can considerably diminish this gain-of-toxic function of these proteins are yet to be discovered. However, targeting the synthesis of such toxic protein by various specific approaches such by targeting the RNA can probably attain this goal. RNA-targeted Therapeutic approaches based upon targeting the RNA, which involve small interfering RNA (siRNA) as well as antisense oligonucleotides (ASOs) are in the developmental process for genetic forms of ALS (Evers et al., 2015).

RNA-targeted therapies consist of ASOs as well as minute interfering RNA are being applied and studied for RNA based therapeutics of ALS. Both of these employ Watson–Crick base pairing-targeted antisense mechanisms that lead to deterioration or inactivation of a specific mRNA (Burnett and Rossi, 2012). These different pathways are notable by the enzymes responsible for degradation of the RNA, by the various processes employed to use siRNA/ASOs therapeutically and by their sites of action. The use of ASOs external to the nervous system has resulted in the US Food and Drug Administration-approved drug for familial hypercholesterolemia as well as many other numerous illnesses. As per current studies, the prevalent distribution of ASOs all through the brain as well as the spinal cord while delivering to the CSF (DeVos et al., 2013) has greatly strengthened the guarantee of employing ASOs for the therapy of neurological disorders. An over expression of mutant human *SOD1* can be seen in the murine ALS G93A model which suffer through the progressive kind of motor neuron disease. This is a proven model providing a distinct opportunity for RNAi studies. The notion of RNAi for diminishing mutant *SOD1* gained substantiation *in vitro* with flourishing decline in mutant human *SOD1* in mutant *SOD1 transfected* cell lines, in the existence of siRNA against *SOD1* (Ding et al., 2003). Employment of RNAi in opposition to a downstream path which might include a comparable effect of retarding the neurodegenerative process is a substitute move toward reduction of the known abnormal mRNA directly involves. This approach can be used in cases where the mutation causing the disease is unidentified, which is very common in a large number of patients with spontaneous ALS. An ASO produced in opposition to miR-155 has been then shown to lengthen survival nearly by 7% and lengthen disease duration nearly by 38%, when injected intra-ventricularly following disease onset (Koval et al., 2013). Comparable slowing down of disease in SOD1G93A mice and furtherance of endurance was achieved by a second group inhibiting miR-155 (Butovsky et al., 2014). Thus therapies based upon RNA-targeting exhibit vast guarantee for disorders of neurodegeneration and elevate the expectation for suitable therapy to retard the phenotype of ALS.

### 8.4.3   TARGETING MOTOR NEURON DEATH PATHWAYS BY BLOCKING GLUTAMATE-LINKED EXCITOTOXICITY

Glutamate communicates with an assortment of definite receptor and system for transportation to create a functional synapse during the excitatory action in the CNS. Excitotoxicity is a phenomenon caused due to excessive stimulation of glutamate receptors shown in both acute as well as chronic neurodegenerative diseases. Extreme and deregulated activation of glutamate receptors is the primary cause of excitotoxicity. When these receptors are exposed to high or steadily increasing concentrations of glutamate for lengthened periods of time, the cells expressing these receptors begin to die (Choi, 1994). In physiologic circumstances, glutamate level are retained at nanomolar concentration range (Herman and Jahr, 2007), which is insufficient to cause high-affinity glutamate receptor activation. However, glutamate concentration can rise upto millimolar amounts during synaptic discharge events (Beato and Scimemi, 2009). $Ca^{2+}$-permeable receptors primarily cultivate excitotoxicity. Incursion of $Ca^{2+}$ is buffered by the endoplasmic reticulum (ER) and the mitochondria are responsible for the moderation of $Ca^{2+}$, and disturbance of intracellular compartmentalization of $Ca^{2+}$ or

its surplus can lead to cell death (Bonda et al., 2011).

Lastly, $Ca^{2+}$ can cause direct activation of catabolic enzymes, like proteases, phospho-lipases, as well as nucleases that lead to cell death and damage to and tissue damage. After the failure of the mitochondrial buffering system, the cell becomes highly susceptible to mitochondria-mediated cell death, ROS, manu-facture, and electron-chain dysfunction (Emerit et al., 2004).

Treatment of ALS using anti-excitotoxins is still in active research phase and has shown a promising clinical outcome (Venkova-Hristova et al., 2017). The benzothiazole compound R-pramipexole came into view as a drug from research on Parkinson's disease that can save dopaminergic neurons from glutamate excito-toxicity. The neuroprotective effects observed to be triggered downstream from the glutamate-gated receptors and further intended to perform action in the mitochondria (Izumi et al., 2007). Thus, potential neuroprotective compounds are hypothesized to have a considerable clinical benefit in ALS patients. Unsurprisingly, thera-peutic approach that belongs to this class has recruited numerous patients in their respective phase 3 trials. Clinical trials with TCH346 (Miller et al., 2007) and olesoxime (Lenglet et al., 2014) have recruited more than 500 patients each, with dexpramipexole (Cudkowicz et al., 2011) and xaliproden (Lacomblezetal, 2004) recruiting more than 1000 and 2000 patients, respectively. Dexpramipexole falls under the category of compounds termed as benzothia-zoles, which shows a broad range of biological activity. As per studies conducted in preclinical models, dexpramipexole has demonstrated its neuroprotective activity in CNS to be dependent on the function of mitochondrial (Alavian et al., 2012), including the improvement in surviv-ability of SOD1G93A mouse model (Bennett et al., 2014). However, this compound was not capable to exhibit its therapeutic effects in ALS (Cudkowiczetal., 2013).

### 8.4.4 TARGETING NEUROINFLAMMATION BY SUPPRESSING GLIAL ACTIVATION AND TNF-α PRODUCTION

ALS cannot be categorized as an autoimmune disorder, but processes of neuroinflammation induced by microglia and astrocytes seem to play essential roles in pathology of ALS (Turner et al., 2004). It is evident from these studies that in ALS, motor neurons, astrocytes, and distinct immune cells communicate with each other which lead to activation at the sites where injury of neurons have taken place. Motor neurons produce neurotoxic signaling which is shown to stimulate cells to generate ROS and pro-inflammatory cytokines, inducing stress of motor neuron, damage to cells and triggering a chain of cell death (Boillée et al., 2006). Microglia, astrocytes, and T lymphocytes are some of the cell populations that are involved in this neuro-inflammatory reaction (Philips and Robberecht, 2011). Microglia are important players in all the damages and disorders of brain thus can exhibit harmful or favorable effects based on their intrinsic characteristics, their communication with the microenvironment, and the existence of pathogens (Harry and Kraft, 2012). Microglia's behavior is strongly connected to the action of astrocytes and T lymphocytes. Astrocytes do not belong to the immune system but are of neuroectodermal origin and may contribute in the immune response, particularly in pathological environment involving damage to neurons (Rizzo et al., 2013). In ALS, astrocytes attain toxic characteristics and consequently add to motor neuron deterioration (Di Giorgio et al., 2008). T lymphocyte appears to add to an endogenous neuron protection action in ALS by escalating the defensive ability of microglia and restricting their toxic responses (Hooten et al., 2015).

The special effects of cells derived from bone marrow on replacing microglia cell have been researched extensively in ALS rodent models and have shown promising results. Bone marrow stem cells, once transplanted, are able to produce mature microglia in CNS that lead to an enhancement in neuroprotective functions. These observations are quite similar to the outcomes of a clinical research based on hematopoietic stem cells intravenously injected into irradiated sporadic ALS patients. In that study, transplanted hematopoietic stem cells infiltrated areas marked by damaged cells as well as neuroinflammation, attaining an immunomodulatory cellular phenotype, although, ALS patients were not observed to be clinically benefitted (Appel et al., 2008). Lately, it was established that neuronal stem cell (NSC) transplantation could improve pathological processes in the mutant SOD1$^{G93A}$ mouse model of ALS. These cells were shown to release trophic factors, conserve neuro-muscular function, and decrease astrogliosis and inflammation (Teng et al., 2012) A phase I clinical trial began at Emory University in 2010 to evaluate the safety and admissibility of intraspinal administrations of fetal human spinal cord-derived NSCs in around 18 ALS patients (Glass et al., 2012). Initial reports from this trial stated that all patients responded well to the treatment without any lasting complications related to the surgical procedure or the stem cell transplantation. Also, no evidence of progression of the disease was found after injecting stem cells. However, a significantly larger study is required to be performed to finally ascertain these effects (Glass et al., 2012).

## 8.4.5   TARGETING PROTEIN AGGREGATION AND OXIDATIVE STRESS—COMMON CELL BIOLOGICAL LINKS

All of the aforementioned targets-excitotoxicity, protein cytoskeletal dysfunction neuroinflammation and glial activation, are not mutually exclusive but probably communicate with each other rather dynamically and may be mechanically cohesive either through involvement of a common origin in protein accumulation disorders or phenomenon of enhanced level of oxidative stress (Poon et al., 2005). For example, chief proteins including the tubulin-binding/stabilizing protein TCTP (translationally controlled tumor protein) and mutant *SOD1* are both exceptionally carbonylated in the SOD1$^{G93A}$ spinal cord at later stages of disease (Poon et al., 2005). The protein deterioration system component ubiquitin carboxy-terminal hydrolase-L1 and alpha/beta-crystallinis are also evidently hypercarbonylated in this model animal (Poon et al., 2005). Therefore, expression of mutant *SOD1* impacts the stability of protein by the process of enhanced carbonylation of protein; the oxidative impact might aggravate the pathways responsible for protein deterioration. The oxidative stress to cytoskeletal proteins might reduce the stress handling ability of the cells. The aggregates of post translationally modified proteins may enhance the activation of neuroinflammatory microglial cells.

The biomarkers which are dependent on oxidative stress are the most significant step toward ascertaining oxidative damage or stress. The concentration and impact of oxidative stress can be estimated by determining the different biomarkers of oxidative/nitrosative stress induced damage in tissues as well as biological fluids (Beckman and Ischiropoulos, 2003). The oxidized DNA, lipids, and proteins are few of the thoroughly studied biomarkers.

Bogdanov et al. found elevated levels of 8OH2'dG which is a known marker for oxidative stress induced damage of DNA in different samples like plasma, urine, and CSF at a particular time point from ALS patients, other neurological diseases, or no known disorders. In the plasma and CSF the level of 8OH2'dG levels is found to be increased with age which further provides the providing further evidence for a role of oxidative damage in normal aging (Bogdanov et al., 2000).

## 8.4.6   OTHER THERAPIES

There are a large number of compounds which are being studied for therapy of ALS and currently are at different phases of trials. First, in the history of the clinical trials of ALS, lithium (Li) is a unique case. It has been found that Li may have possible neuroprotective properties (Forlenza et al., 2014). In an atypical report, which included publications of results of Li treatment done in SOD1$^{G93A}$ mice and also the results of a pilot human trial, the findings involved notable progression of endurance in patients (Fornai et al., 2008). Another trial on human which involved randomization of sixteen ALS patients into lithium plus riluzole mediated treatment division versus 28 patients in the control division (only riluzole), no deaths was observed during the treatment division versus eight deaths in the riluzole division after 15-month treatment period.

Secondly, edaravone is a free radical scavenger used intravenously which gets rid of lipid peroxide and hydroxyl radicals and might accordingly be categorized as an anti-oxidative agent. As mentioned earlier, the action of ROS is believed to add to pathology of ALS.

Therapeutic properties of edaravone were established in a wobbler mice bearing signs and symptoms similar to ALS, previous to the beginning of clinical upgrading (Kimura and Yoshino, 2006). The first two trails had been unsuccessful. On the premise of the outcomes of the most effective phase three trial, edaravone acquired advertising and marketing authorization solely in Japan in 2015.

Thirdly, tirasemtiv activates the complex of troponin, present in fast-twitch skeletal muscle fibers to presence of calcium (Hwee et al., 2014). A demonstrable mechanism of action is advancement in the muscle tightening as a reaction to decreased neural participation thus significantly diminishing muscle weariness (Shefner et al., 2012). A good number of trials on human were performed in ALS patients because the wasting in muscle is a sign of ALS (Shefner et al., 2013). The, recruiting total 605 patients was a failure on the main endpoint (Shefner et al., 2016). As the research had shown the statistical importance on main two secondary endpoints, thus these researchers offer the probability to effectively initiate further Phase 3 study.

Fourthly, masitinib is an exceptionally discerning inhibitor for tyrosine kinase which was formerly shown to avert neuroinflammation of CNS in numerous models of neurodegenerative disorders which involves ALS SOD1$^{G93A}$ rats (Piette et al., 2011; Trias et al., 2016). Masitinib is presently in Phase 3 development in ALS sufferers as an addition to riluzole mediating treatment ecruiting 382 patients divided into three treatments (trial ID# NCT02588677). It will be of benefit to assess probable clinical ability of edaravone and masitinib. The current status of some of the compounds which have been used for ALS therapy is mentioned in Table 8.4.

**TABLE 8.4** Summary of a Few Important Compounds Which Have Been Used for the Therapy of Amyotrophic Lateral Sclerosis and the Current Status of Their Clinical Trials.

| Compound | Therapy | Phases of clinical development | Outcome | References |
|---|---|---|---|---|
| Riluzole | Anti-glutamatergic | 3 | Mixed | Lacomblez et al., 1996 |
| Celecoxib | Anti-inflammatory | 2–3 | Failure | Cudkowicz et al., 2006 |
| Minocycline | Anti-inflammatory | 1–2; 3 | Failure | Gordon et al., 2007 |
| Memantine | Anti-glutamatergic | 2; 3 | Failure | de Carvalho et al., 2010 |
| Masitinib | CSF1R inhibition | 2–3 | Positive | Cadot et al., 2011 |
| Dexpramipexole | Neuroprotective | 2; 3 | Failure | Cudkowicz et al., 2011; Cudkowicz et al., 2013 |
| Lithium | Others | 2; 2–3; 3 | Failure | Verstraete et al., 2012 Aggarwal et al., 2010 |
| Creatine | Anti-oxidative | 2; 2–3; 3 | Failure | Pastula et al., 2012 |
| Ceftriaxone | Anti-glutamatergic | 1–2, 3 | Failure | Berry et al., 2013 |
| Edaravone | Anti-oxidative | 2; 3 | Mixed | Abe et al., 2014 |
| Olesoxime | Neuroprotective | 2–3 | Failure | Lenglet et al., 2014 |
| Erythropoietin | Anti-inflammatory | 2;3 | Failure | Lauria et al., 2015 |
| Tirasemtiv | others | 2 | Failure | Shefner et al., 2016 |

## 8.5   FUTURE OPPORTUNITIES AND CHALLENGES

As there is lack in understanding the pathophysiology of ALS, hence different studies showed promising outcomes when tested experimentally in animal model systems but failed translation to the human subjects. The translational failures could be a result of various reasons, such as:

1.  Inability of many drugs to cross blood–brain barrier.
2.  Animal models reacted positively to only those drugs which were administered before the commencement of the disease and not in the later stages.
3.  The small population of ALS and its pathogenesis.

There are many effective targets which can be involved in the progression of ALS however if the targets are not simultaneously acted upon, the therapeutic strategy can turn out to be useless as the pathological approach revolves around all the barriers of therapy that the researchers create. As the pathophysiology of ALS is multifactorial, hence instead of one or two drugs, a combinatorial therapy is required. Thus a rational decision has to be taken while selecting patient targeted therapies; multiple and distinctly acting drugs or a "cocktail" of may be significant to attempt in recent future in both ALS model systems and human subjects.

Further to effectively treat ALS, a multidisciplinary approach is needed, due to the gathering of inter-related pathological stressors used in ALS. Clinicians who explains this phenotype, epidemiologists from diverse backgrounds, and lab scientists/researchers with an expertise in biology of oxidative stress geneticists, and experts in genomic imprinting, biostatistics, and clinical psychologists must participate in these kind of research, as their combined expertise is important to discover the etiology of this complicated disease.

## 8.6   CONCLUSION

ALS remains a vulnerable motor neuron disease, foremost to the loss of distinct group of neuronal cells and has an average age of survival between 2 and 5 years. At various multidisciplinary clinics, the symptomatic management of ALS is being offered as an important current treatment strategy for the individual patient. As per studies, riluzole is the main important treatment till today that significantly reduces the progressiveness of the disease, the advancement of more effective riluzole analogs must therefore gain importance as an issue in the near future. Thus, if combined via prospective of research, such assumption ought to shape the premise for an integrated principle of such disorder and pave the way for drug discovery projects directed in the near future. To conclude, efforts concentrated toward establishment of ALS therapy have witnessed cycles of success and failures. In spite of that, a continual or consistent scientific approach may result in new milestones for therapy of ALS.

## KEYWORDS

- **amyotrophic lateral sclerosis**
- **motor neuron**
- **riluzole**
- **superoxide dismutase 1**

# REFERENCES

Abe, S.; Kirima, K.; Tsuchiya, K.; Okamoto, M.; Hasegawa, T.; Houchi, H.; Yoshizumi, M.; Tamaki, T. The Reaction Rate of Edaravone (3-Methyl-1-Phenyl-2-Pyrazolin-5-One (MCI-186)) with Hydroxyl Radical. *Chem. Pharm. Bull.* **2004**, *52*, 186–191.

Abe, K.; Itoyama, Y.; Sobue, G.; Tsuji, S.; Aoki, M.; Doyu, M.; Hamada, C.; Kondo, K.; Yoneoka, T.; Akimoto, M. et al. Confirmatory Double-Blind, Parallel-Group, Placebo-Controlled Study of Efficacy and Safety of Edaravone (MCI-186) in Amyotrophic Lateral Sclerosis Patients. *Amyotroph Lateral Scler Frontotemporal Degener.* **2014**, *15*, 610–617.

Aggarwal, S.; Zinman, L.; Simpson, E.; McKinley, J.; Jackson, K.; Pinto, H.; Kaufman, P.; Conwit, R.; Schoenfeld, D.; Shefner, J. et al. Safety and Efficacy of Lithium in Combination with Riluzole for Treatment of Amyotrophic Lateral Sclerosis: A Randomised, Double-Blind, Placebo-Controlled Trial. *Lancet Neurol.* **2010**, *9*, 481–488.

Aguila, M.; Longstreth, W.; McGuire, V.; Koepsell, T.; van Belle, G. Prognosis in Amyotrophic Lateral Sclerosis: A Population-Based Study. *Neurology* **2003**, *60*, 813–819.

Alavian, K.; Dworetzky, S.; Bonanni, L.; Zhang, P.; Sacchetti, S.; Mariggio, M.; Onofrj, M.; Thomas, A.; Li, H.; Mangold, J. et al. Effects of Dexpramipexole on Brain Mitochondrial Conductances And Cellular Bioenergetic Efficiency. *Brain Res.* **2012**, *1446*, 1–11.

Andrew, E.; Stewart H.; Schulzer M.; Cameron D. Anti-Glutamate Therapy in Amyotrophic Lateral Sclerosis: A Trial Using Lamotrigine. *Can. J. Neurol. Sci.* **1993**, *20*, 297–301.

Andersen, P.; Borasio, G.; Dengler, R.; Hardiman, O.; Kollewe, K.; Leigh, P.; Pradat, P.; Silani, V.; Tomik, B. Good Practice in the Management of Amyotrophic Lateral Sclerosis: Clinical Guidelines. An Evidence-Based Review With Good Practice Points. EALSC Working Group. *Amyotrophic Lateral Scler.* **2007**, *8*, 195–213.

Appel, S.; Engelhardt, J.; Henkel, J.; Siklos, L.; Beers, D.; Yen, A.; Simpson, E.; Luo, Y.; Carrum, G.; Heslop, H. et al. Hematopoietic Stem Cell Transplantation in Patients with Sporadic Amyotrophic Lateral Sclerosis. *Neurology* **2008**, *71*, 1326–1334.

Ashworth, N. L.; Satkunam, L. E.; Deforge, D. Treatment for Spasticity in Amyotrophic Lateral Sclerosis/Motor Neuron Disease. *Cochrane Database Syst. Rev.* **2012**, *2*, 1–18.

Auger, C.; Attwell, D. Fast Removal of Synaptic Glutamate by Postsynaptic Transporters. *Neuron* **2000**, *28*, 547–558.

Banno, M.; Mizuno, T.; Kato, H.; Zhang, G.; Kawanokuchi, J.; Wang, J.; Kuno, R.; Jin, S.; Takeuchi, H.; Suzumura, A. The Radical Scavenger Edaravone Prevents Oxidative Neurotoxicity Induced by Peroxynitrite and Activated Microglia. *Neuropharmacology* **2005**, *48*, 283–290.

Beltran-Parrazal, L.; Charles, A. Riluzole Inhibits Spontaneous $Ca^{2+}$ Signaling in Neuroendocrine Cells by Activation of K+ Channels and Inhibition of Na+ Channels. *Br. J. Pharmacol.* **2003**, *140*, 881–888.

Bennett, E.; Mead, R.; Azzouz, M.; Shaw, P.; Grierson, A. Early Detection of Motor Dysfunction in the SOD1G93A Mouse Model of Amyotrophic Lateral Sclerosis (ALS) Using Home Cage Running Wheels. *PLoS One* **2014**, *9*, e107918.

Bensimon, G.; Lacomblez, L.; Delumeau, J.; Bejuit, R.; Truffinet, P.; Meininger, V. A Study of Riluzole in the Treatment of Advanced Stage or Elderly Patients with Amyotrophic Lateral Sclerosis. *J. Neurol.* **2002**, *249*, 609–615.

Berry, J.; Shefner, J.; Conwit, R.; Schoenfeld, D.; Keroack, M.; Felsenstein, D.; Krivickas, L.; David, W.; Vriesendorp, F.; Pestronk, A. et al. Design and Initial Results of a Multi-Phase Randomized Trial of Ceftriaxone in Amyotrophic Lateral Sclerosis. *PLoS One* **2013**, *8*, e61177.

Bogdanov, M.; Brown, R.; Matson, W.; Smart, R.; Hayden, D.; O'Donnell, H.; Flint Beal, M.; Cudkowicz, M. Increased Oxidative Damage to DNA in ALS Patients. *Free Radic. Biol. Med.* **2000**, *29*, 652–658.

Bohannon, R. Results of Resistance Exercise on a Patient with Amyotrophic Lateral Sclerosis. *Phys Ther.* **1983**, *63*, 965–968.

Boillée, S.; Vande Velde, C.; Cleveland, D. ALS: A Disease of Motor Neurons and their Nonneuronal Neighbors. *Neuron* **2006**, *52*, 39–59.

Boillee, S.; Yamanaka, K.; Lobsiger, C.; Copeland, N.; Jenkins, N.; Kassiotis, G.; Kollias, G.; Cleveland, D. Onset and Progression in Inherited ALS Determined by Motor Neurons and Microglia. *Science* **2006**, *312*, 1389–1392.

Bonda J, D.; A. Smith, M.; Perry, G.; Lee, H.; Wang, X.; Zhu, X. The Mitochondrial Dynamics of Alzheimers Disease and Parkinsons Disease Offer Important Opportunities for Therapeutic Intervention. *Curr. Pharm. Des.* **2011**, *17*, 3374–3380.

Bourke, S.; Tomlinson, M.; Williams, T.; Bullock, R.; Shaw, P.; Gibson, G. Effects of Non-Invasive Ventilation on Survival and Quality of Life in Patients with Amyotrophic Lateral Sclerosis: A Randomised Controlled Trial. *Lancet Neurol.* **2006,** *5,* 140–147.

Bozzoni V.; Pansarasa O.; Diamanti L.; Nosari G.; Cereda C.; Ceroni M. Amyotrophic Lateral Sclerosis and Environmental Factors. *Funct. Neurol.* **2016,** *31,* 7–19.

Bros-Facer, V.; Krull, D.; Taylor, A.; Dick, J.; Bates, S.; Cleveland, M.; Prinjha, R.; Greensmith, L. Treatment with an Antibody Directed Against Nogo-A Delays Disease Progression in the SOD1G93A Mouse Model of Amyotrophic Lateral Sclerosis. *Hum. Mol. Genet.* **2014,** *23,* 4187–4200.

Brown, R. H.; M. D.; Chalabi, A. A. Amyotrophic Lateral Sclerosis. *N. Engl. J. Med.* **2017,** *377,* 162–72.

Buchli, A.; Rouiller, E.; Mueller, R.; Dietz, V.; Schwab, M. Repair of the Injured Spinal Cord. *Neurodegener Dis.* **2007,** *4,* 51–56.

Burnett, J.; Rossi, J. RNA-Based Therapeutics: Current Progress and Future Prospects. *ACS Chem. Biol.* **2012,** *19,* 60–71.

Butovsky, O.; Jedrychowski, M.; Cialic, R.; Krasemann, S.; Murugaiyan, G.; Fanek, Z.; Greco, D.; Wu, P.; Doykan, C.; Kiner, O. et al. Targeting Mir-155 Restores Abnormal Microglia And Attenuates Disease In SOD1 Mice. *Ann. Neurol.* **2014,** *77,* 75–99.

Bruno, R.; Vivier, N.; Montay, G.; Liboux, A.; Powe, L.; Delumeau, J.; Rhodes, G. *Clin. Pharmacol. Therap.* **1997,** *62,* 518–526.

Cadot, P.; Hensel, P.; Bensignor, E.; Hadjaje, C.; Marignac, G.; Beco, L.; Fontaine, J.; Jamet, J.; Georgescu, G.; Campbell, K. et al. Masitinib Decreases Signs of Canine Atopic Dermatitis: A Multicentre, Randomized, Double-Blind, Placebo-Controlled Phase 3 Trial. *Vet Dermatol.* **2011,** *22,* 554–564.

Cady, J.; Allred, P.; Bali, T.; Pestronk, A.; Goate, A.; Miller, T.; Mitra, R.; Ravits, J.; Harms, M.; Baloh, R. Amyotrophic Lateral Sclerosis Onset is Influenced by the Burden of Rare Variants in Known Amyotrophic Lateral Sclerosis Genes. *Ann. Neurol.* **2014,** *77,* 100–113.

Campanari, M.; García-Ayllón, M.; Ciura, S.; Sáez-Valero, J.; Kabashi, E. Neuromuscular Junction Impairment in Amyotrophic Lateral Sclerosis: Reassessing the Role of Acetylcholinesterase. *Front. Mol. Neurosci.* **2016,** *9.*

Chen, Y.; Bennett, C.; Huynh, H.; Blair, I.; Puls, I.; Irobi, J.; Dierick, I.; Abel, A.; Kennerson, M.; Rabin, B. et al. DNA/RNA Helicase Gene Mutations in a Form of Juvenile Amyotrophic Lateral Sclerosis (ALS4). *Am. J. Hum. Genet.* **2004,** *74,* 1128–1135.

Chéramy, A.; Barbeito, L.; Godeheu, G.; Glowinski, J. Riluzole Inhibits the Release of Glutamate in the Caudate Nucleus of the Cat in vivo. *Neurosci. Lett.* **1992,** *147,* 209–212.

Chio, A.; Logroscino, G.; Hardiman, O.; Swingler, R.; Mitchell, D.; Beghi, E.; Traynor, B. Prognostic Factors In ALS: A Critical Review. *Amyotroph Lateral Scler.* **2009,** *10,* 310–323.

Choi, D. Glutamate Receptors and the Induction of Excitotoxic Neuronal Death. *Prog. Brain Res.* **1994,** *100,* 47–51.

Christou, Y.; Moore, H.; Shaw, P.; Monk, P. Embryonic Stem Cells And Prospects for their use in Regenerative Medicine Approaches to Motor Neurone Disease. *Neuropathol. Appl. Neurobiol.* **2007,** *33,* 485–498.

Clavé, P.; De Kraa, M.; Arreola, V.; Girvent, M.; Farré, R.; Palomera, E.; Serra-Prat, M. The Effect of Bolus Viscosity on Swallowing Function in Neurogenic Dysphagia. *Aliment Pharmacol Ther.* **2006,** *24,* 1385–1394.

Coderre, T.; Kumar, N.; Lefebvre, C.; Yu, J. A Comparison of the Glutamate Release Inhibition and Anti-Allodynic Effects of Gabapentin, Lamotrigine, and Riluzole in a Model of Neuropathic Pain. *J. Neurochem.* **2006,** *100,* 1289–1299.

Correction for Fornai et al., Lithium Delays Progression of Amyotrophic Lateral Sclerosis. *Proc. Natl. Acad. Sci.* **2008,** *105,* 16404‑16407.

Cudkowicz, M.; Bozik, M.; Ingersoll, E.; Miller, R.; Mitsumoto, H.; Shefner, J.; Moore, D.; Schoenfeld, D.; Mather, J.; Archibald, D. et al. The Effects of Dexpramipexole (KNS-760704) in Individuals with Amyotrophic Lateral Sclerosis. *Nat. Med.* **2011,** *17,* 1652–1656.

Cudkowicz, M.; Shefner, J.; Schoenfeld, D.; Zhang, H.; Andreasson, K.; Rothstein, J.; Drachman, D. Trial of Celecoxib in Amyotrophic Lateral Sclerosis. *Ann. Neurol.* **2006,** *60,* 22–31.

Cudkowicz, M.; van den Berg, L.; Shefner, J.; Mitsumoto, H.; Mora, J.; Ludolph, A.; Hardiman, O.; Bozik, M.; Ingersoll, E.; Archibald, D. et al. Dexpramipexole Versus Placebo for Patients with Amyotrophic Lateral Sclerosis (EMPOWER): A Randomised, Double-Blind, Phase 3 Trial. *Lancet Neurol.* **2013,** *12,* 1059–1067.

D'Amico, E.; Factor-Litvak, P.; Santella, R.; Mitsumoto, H. Clinical Perspective on Oxidative Stress in Sporadic Amyotrophic Lateral Sclerosis. *Free Radic. Biol. Med.* **2013,** *65,* 509–527.

de Carvalho, M.; Pinto, S.; Costa, J.; Evangelista, T.; Ohana, B.; Pinto, A. A Randomized, Placebo-Controlled Trial of Memantine for Functional Disability in Amyotrophic

Lateral Sclerosis. *Amyotroph Lateral Scler.* **2010**, *11*, 456–460.

Dent, E.; Gupton, S.; Gertler, F. The Growth Cone Cytoskeleton in Axon Outgrowth and Guidance. *Cold Spring Harb. Perspect Biol.* **2010**, *3*, a001800–a001800.

Deshpande, D.; Kim, Y.; Martinez, T.; Carmen, J.; Dike, S.; Shats, I.; Rubin, L.; Drummond, J.; Krishnan, C.; Hoke, A. et al. Recovery From Paralysis in Adult Rats Using Embryonic Stem Cells. *Ann. Neurol.* **2006**, *60*, 32–44.

Desport, J.; Torny, F.; Lacoste, M.; Preux, P.; Couratier, P. Hypermetabolism in ALS: Correlations with Clinical and Paraclinical Parameters. *Neurodegener Dis.* **2005**, *2*, 202–207.

Desport, J.; Preux, P.; Truong, T.; Vallat, J.; Sautereau, D.; Couratier, P. Nutritional Status is a Prognostic Factor for Survival in ALS Patients. *Neurology* **1999**, *53*, 1059–1059.

DeVos, S.; Goncharoff, D.; Chen, G.; Kebodeaux, C.; Yamada, K.; Stewart, F.; Schuler, D.; Maloney, S.; Wozniak, D.; Rigo, F. et al. Antisense Reduction of Tau in Adult Mice Protects Against Seizures. *J. Neurosci.* **2013**, *33*, 12887–12897.

Di Giorgio, F.; Boulting, G.; Bobrowicz, S.; Eggan, K. Human Embryonic Stem Cell-Derived Motor Neurons are Sensitive to the Toxic Effect of Glial Cells Carrying an ALS-Causing Mutation. *Cell Stem Cell* **2008**, *3*, 637–648.

Ding, H.; Schwarz, D.; Keene, A.; Affar, E.; Fenton, L.; Xia, X.; Shi, Y.; Zamore, P.; Xu, Z. Selective Silencing by Rnai of a Dominant Allele that Causes Amyotrophic Lateral Sclerosis. *Aging Cell* **2003**, *2*, 209–217.

Doble, A. The Pharmacology And Mechanism of Action of Riluzole. *Neurology* **1996**, *47*, 233S–241S.

Emerit, J.; Edeas, M.; Bricaire, F. Neurodegenerative Diseases And Oxidative Stress. *Biomed. Pharmacother.* **2004**, *58*, 39–46.

Evers, M.; Toonen, L.; van Roon-Mom, W. Antisense Oligonucleotides in Therapy for Neurodegenerative Disorders. *Adv. Drug Deliv. Rev.* **2015**, *87*, 90–103.

Ferrante, R.; Browne, S.; Shinobu, L.; Bowling, A.; Baik, M.; MacGarvey, U.; Kowall, N.; Brown, R.; Beal, M. Evidence of Increased Oxidative Damage in Both Sporadic and Familial Amyotrophic Lateral Sclerosis. *J. Neurochem.* **1997**, *69*, 2064–2074.

Ferraiuolo, L.; Kirby, J.; Grierson, A.; Sendtner, M.; Shaw, P. Molecular Pathways of Motor Neuron Injury in Amyotrophic Lateral Sclerosis. *Nat. Rev. Neurol.* **2011**, *7*, 616–630.

Fischer, L.; Culver, D.; Tennant, P.; Davis, A.; Wang, M.; Castellano-Sanchez, A.; Khan, J.; Polak, M.; Glass, J. Amyotrophic Lateral Sclerosis is a Distal Axonopathy: Evidence in Mice and Man. *Exp. Neurol.* **2004**, *185*, 232–240.

Foley, M.; Prax, B.; Crowell, R.; Boone, T. Effects of Assistive Devices on Cardiorespiratory Demands in Older Adults. *Phys Ther.* **1996**, *76*, 1313–1319.

Forlenza, O.; De-Paula, V.; Diniz, B. Neuroprotective Effects Of Lithium: Implications for the Treatment of Alzheimer's Disease and Related Neurodegenerative Disorders. *ACS Chem. Neurosci.* **2014**, *5*, 443–450.

Fujisawa, A.; Yamamoto, Y. Edaravone, A Potent Free Radical Scavenger, Reacts with Peroxynitrite to Produce Predominantly 4-NO-Edaravone. *Redox Rep.* **2016**, *21*, 98–103.

Fuller, Y.; Trombly, C. Effects of Object Characteristics on Female Grasp Patterns. *Am. J. Occup. Ther.* **1997**, *51*, 481–487.

Gelanis, D. Respiratory Failure or Impairment in Amyotrophic Lateral Sclerosis. *Curr. Treat. Options Neurol.* **2001**, *3*, 133–138.

Glass, J.; Boulis, N.; Johe, K.; Rutkove, S.; Federici, T.; Polak, M.; Kelly, C.; Feldman, E. Lumbar Intraspinal Injection of Neural Stem Cells in Patients with Amyotrophic Lateral Sclerosis: Results of a Phase I Trial in 12 Patients. *Stem Cells* **2012**, *30*, 1144–1151.

Gordon, P. Amyotrophic Lateral Sclerosis. *CNS Drugs* **2011**, *25*, 1–15.

Gordon, P.; Moore, D.; Miller, R.; Florence, J.; Verheijde, J.; Doorish, C.; Hilton, J.; Spitalny, G.; MacArthur, R.; Mitsumoto, H. et al. Efficacy of Minocycline in Patients with Amyotrophic Lateral Sclerosis: A Phase III Randomised Trial. *Lancet Neurol.* **2007**, *6*, 1045–1053.

GrandPré, T.; Li, S.; Strittmatter, S. Nogo-66 Receptor Antagonist Peptide Promotes Axonal Regeneration. *Nature* **2002**, *417*, 547–551.

Greenway, M.; Andersen, P.; Russ, C.; Ennis, S.; Cashman, S.; Donaghy, C.; Patterson, V.; Swingler, R.; Kieran, D.; Prehn, J. et al. ANG Mutations Segregate with Familial and 'Sporadic' Amyotrophic Lateral Sclerosis. *Nat Genet.* **2006**, *38*, 411–413.

Greenwood, D. Nutrition Management of Amyotrophic Lateral Sclerosis. *Nutr. Clin. Pract.* **2013**, *28*, 392–399.

Handy, C.; Krudy, C.; Boulis, N.; Federici, T. Pain in Amyotrophic Lateral Sclerosis: A Neglected Aspect of Disease. *Neurol. Res. Int.* **2011**, *2011*, 1–8.

Hanson, E. K.; Yorkston, K. M.; Britton, D. Dysarthria in Amyotrophic Lateral Sclerosis: A Systematic Review of Characteristics, Speech Treatment and Augmentative and Alternative Communication Options. *J. Med. Speech Lang. Pathol.* **2011**, *19*, 12–30.

Harry, G.; Kraft, A. Microglia in the Developing Brain: A Potential Target with Lifetime Effects. *NeuroToxicology* **2012**, *33*, 191–206.

Heffernan, C.; Jenkinson, C.; Holmes, T.; Feder, G.; Kupfer, R.; Leigh, P.; McGowan, S.; Rio, A.; Sidhu, P. Nutritional Management in MND/ALS Patients: An Evidence Based Review. *Amyotroph Lateral Scler Other Motor Neuron Disord.* **2004**, *5*, 72–83.

Herman, M.; Jahr, C. Extracellular Glutamate Concentration in Hippocampal Slice. *J. Neurosci.* **2007**, *27*, 9736–9741.

Hollis, E. Axon Guidance Molecules and Neural Circuit Remodeling after Spinal Cord Injury. *Neurotherapeutics* **2015**, *13*, 360–369.

Holm, T.; Maier, A.; Wicks, P.; Lang, D.; Linke, P.; Münch, C.; Steinfurth, L.; Meyer, R.; Meyer, T. Severe Loss of Appetite in Amyotrophic Lateral Sclerosis Patients: Online Self-Assessment Study. *Interact. J. Med. Res.* **2013**, *2*, e8.

Hooten, K.; Beers, D.; Zhao, W.; Appel, S. Protective and Toxic Neuroinflammation in Amyotrophic Lateral Sclerosis. *Neurotherapeutics* **2015**, *12*, 364–375.

Hugon, J.; Tabaraud, F.; Rigaud, M.; Vallat, J.; Dumas, M. Glutamate Dehydrogenase and Aspartate Aminotransferase in Leukocytes of Patients with Motor Neuron Disease. *Neurology* **1989**, *39*, 956–956.

Hwee, D.; Kennedy, A.; Ryans, J.; Russell, A.; Jia, Z.; Hinken, A.; Morgans, D.; Malik, F.; Jasper, J. Fast Skeletal Muscle Troponin Activator Tirasemtiv Increases Muscle Function and Performance in the B6SJL-SOD1G93A ALS Mouse Model. *PLoS One* **2014**, *9*, e96921.

Ingre C.; Roos P.M.; Piehl F.; Kamel F.; Fang F. Risk Factors for Amyotrophic Lateral Sclerosis. *Clin. Epidemiol.* **2015**, *7*, 181–193.

Ischiropoulos, H.; Beckman, J. Oxidative Stress and Nitration in Neurodegeneration: Cause, Effect, or Association? *J. Clin. Investig.* **2003**, *111*, 163–169.

Izumi, Y.; Sawada, H.; Yamamoto, N.; Kume, T.; Katsuki, H.; Shimohama, S.; Akaike, A. Novel Neuroprotective Mechanisms of Pramipexole, an Anti-Parkinson Drug, Against Endogenous Dopamine-Mediated Excitotoxicity. *Eur. J. Pharmacol.* **2007**, *557*, 132–140.

Izumi, Y.; Sawada, H.; Yamamoto, N.; Kume, T.; Katsuki, H.; Shimohama, S.; Akaike, A. Novel Neuroprotective Mechanisms of Pramipexole, an Anti-Parkinson Drug, Against Endogenous Dopamine-Mediated Excitotoxicity. *Eur. J. Pharmacol.* **2007**, *557*, 132–140.

Jehle, T.; Bauer, J.; Blauth, E.; Hummel, A.; Darstein, M.; Freiman, T.; Feuerstein, T. Effects of Riluzole on

Electrically Evoked Neurotransmitter Release. *Br. J. Pharmacol.* **2000**, *130*, 1227–1234.

Kalin, A.; Medina-Paraiso, E.; Ishizaki, K.; Kim, A.; Zhang, Y.; Saita, T.; Wasaki, M. A Safety Analysis of Edaravone (MCI-186) During the First Six Cycles (24 Weeks) of Amyotrophic Lateral Sclerosis (ALS) Therapy From the Double-Blind Period in Three Randomized, Placebo-Controlled Studies. *Amyotroph Lat. Scler. Frontotemp. Degener* **2017**, *18*, 71–79.

Kamogawa, E.; Sueishi, Y. A Multiple Free-Radical Scavenging (MULTIS) Study on the Antioxidant Capacity of a Neuroprotective Drug, Edaravone as Compared with Uric Acid, Glutathione, and Trolox. *Bioorg. Med. Lett.* **2014**, *24*, 1376–1379.

Kent, R.; Sufit, R.; Rosenbek, J.; Kent, J.; Weismer, G.; Martin, R.; Brooks, B. Speech Deterioration in Amyotrophic Lateral Sclerosis. *J. Speech Lang. Hear Res.* **1991**, *34*, 1269.

Koval, E.; Shaner, C.; Zhang, P.; du Maine, X.; Fischer, K.; Tay, J.; Chau, B.; Wu, G.; Miller, T. Method for Widespread Microrna-155 Inhibition Prolongs Survival in ALS-Model Mice. *Hum. Mol. Genet.* **2013**, *22*, 4127–4135.

Kiernan, M.; Vucic, S.; Cheah, B.; Turner, M.; Eisen, A.; Hardiman, O.; Burrell, J.; Zoing, M. Amyotrophic Lateral Sclerosis. *Lancet* **2011**, *377*, 942–955.

Kretschmer, B.; Kratzer, U.; Schmidt, W. Riluzole, A Glutamate Release Inhibitor, and Motor Behavior. *Naunyn Schmiedebergs Arch. Pharmacol.* **1998**, *358*, 181–190.

Kuo, J.; Lee, R.; Zhang, L.; Heckman, C. Essential Role of the Persistent Sodium Current in Spike Initiation During Slowly Rising Inputs in Mouse Spinal Neurones. *J. Physiol.* **2006**, *574*, 819–834.

Lacomblez, L.; Bensimon, G.; Meininger, V.; Leigh, P.; Guillet, P. Dose-Ranging Study of Riluzole in Amyotrophic Lateral Sclerosis. *Lancet* **1996**, *347*, 1425–1431.

Lamanauskas, N.; Nistri, A. Riluzole Blocks Persistent Na+ and Ca2+ Currents and Modulates Release of Glutamate via Presynaptic NMDA Receptors on Neonatal Rat Hypoglossal Motoneuronsin Vitro. *Eur. J. Neurosci.* **2008**, *27*, 2501–2514.

Lacomblez, L.; Bensimon, G.; Douillet, P.; Doppler, V.; Salachas, F.; Meininger, V. Xaliproden in Amyotrophic Lateral Sclerosis: Early Clinical Trials. *Amyotroph. Lateral Scler. Other Motor Neuron Disord.* **2004**, *5*, 99–106.

Lacomblez, L.; Bensimon, G.; Meininger, V.; Leigh, P.; Guillet, P. Dose-Ranging Study of Riluzole in

Amyotrophic Lateral Sclerosis. *The Lancet* **1996,** *347,* 1425–1431.

Lauria, G.; Dalla Bella, E.; Antonini, G.; Borghero, G.; Capasso, M.; Caponnetto, C.; Chiò, A.; Corbo, M.; Eleopra, R.; Fazio, R. et al. Erythropoietin in Amyotrophic Lateral Sclerosis: A Multicentre, Randomised, Double Blind, Placebo Controlled, Phase III Study. *J. Neurol. Neurosurg. Psychiatry* **2015,** *86,* 879–886.

Lee, J.; Zheng, B. Role of Myelin-Associated Inhibitors in Axonal Repair After Spinal Cord Injury. *Exp. Neurol.* **2012,** *235,* 33–42.

Lenglet, T.; Lacomblez, L.; Abitbol, J.; Ludolph, A.; Mora, J.; Robberecht, W.; Shaw, P.; Pruss, R.; Cuvier, V.; Meininger, V. A Phase II–III Trial of Olesoxime in Subjects with Amyotrophic Lateral Sclerosis. *Eur. J. Neurol.* **2014,** *21,* 529–536.

Lesnick, T.; Sorenson, E.; Ahlskog, J.; Henley, J.; Shehadeh, L.; Papapetropoulos, S.; Maraganore, D. Beyond Parkinson Disease: Amyotrophic Lateral Sclerosis and the Axon Guidance Pathway. *PLoS One* **2008,** *3,* e1449.

Lewis, C.; Suzuki, M. Therapeutic Applications of Mesenchymal Stem Cells for Amyotrophic Lateral Sclerosis. *Stem Cell Res. Ther.* **2014,** *5,* 32.

Lee, B.; Egi, Y.; van Leyen, K.; Lo, E.; Arai, K. Edaravone, A Free Radical Scavenger, Protects Components of the Neurovascular Unit Against Oxidative Stress in vitro. *Brain Res.* **2010,** *1307,* 22–27.

Logroscino, G.; Traynor, B.; Hardiman, O.; Chio', A.; Couratier, P.; Mitchell, J.; Swingler, R.; Beghi, E. Descriptive Epidemiology of Amyotrophic Lateral Sclerosis: New evidence and unsolved issue. *J. Neurol. Neurosurg. Psychiatry* **2008,** *79,* 6–11.

Lomen-Hoerth, C.; Murphy, J.; Langmore, S.; Kramer, J.; Olney, R.; Miller, B. Are Amyotrophic Lateral Sclerosis Patients Cognitively Normal? *Neurology* **2003,** *60,* 1094–1097.

Lunn, J.; Sakowski, S.; Feldman, E. Concise Review: Stem Cell Therapies for Amyotrophic Lateral Sclerosis: Recent Advances and Prospects for the Future. *Stem Cells* **2014,** *32,* 1099–1109.

MacIver, M.; Amagasu, S.; Mikulec, A.; Monroe, F. Riluzole Anesthesia: Use-Dependent Block of Presynaptic Glutamate Fibers. *Anesthesiology* **1996,** *85,* 626–634.

Marin, B.; Desport, J.; Kajeu, P.; Jesus, P.; Nicolaud, B.; Nicol, M.; Preux, P.; Couratier, P. Alteration of Nutritional Status at Diagnosis is a Prognostic Factor for Survival of Amyotrophic Lateral Sclerosis Patients. *J. Neurol. Neurosurg. Psychiatry.* **2011,** *82,* 628–634.

Martin, D.; Thompson, M.; Nadler, J. The Neuroprotective Agent Riluzole Inhibits Release of Glutamate and Aspartate from Slices of Hippocampal Area CA1. *Eur. J. Pharmacol.* **1993,** *250,* 473–476.

Mazibuko, Z.; Choonara, Y.; Kumar, P.; Du Toit, L.; Modi, G.; Naidoo, D.; Pillay, V. A Review of the Potential Role of Nano-Enabled Drug Delivery Technologies in Amyotrophic Lateral Sclerosis: Lessons Learned from Other Neurodegenerative Disorders. *J. Pharm. Sci.* **2015,** *104,* 1213–1229.

Mazzini, L.; Corrà, T.; Zaccala, M.; Mora, G.; Del Piano, M.; Galante, M. Percutaneous Endoscopic Gastrostomy and Enteral Nutrition in Amyotrophic Lateral Sclerosis. *J. Neurol.* **1995,** *242,* 695–698.

McCombe, P.; Henderson, R. Effects of Gender in Amyotrophic Lateral Sclerosis. *Gend. Med.* **2010,** *7,* 557–570.

McKerracher, L.; Winton, M. Nogo on the Go. *Neuron* **2002,** *36,* 345–348.

Meldrum B. S. Glutamate as a Neurotransmitter in the Brain: Review of Physiology and Pathology. *J. Nutr.* **2000,** *130,* 1007S–10015S.

Miller, R.; Rosenberg, J.; Gelinas, D.; Mitsumoto, H.; Newman, D.; Sufit, R.; Borasio, G.; Bradley, W.; Bromberg, M.; Brooks, B. et al. Practice Parameter: The Care of the Patient with Amyotrophic Lateral Sclerosis (an Evidence-Based Review). *Muscle Nerve* **1999,** *22,* 1104–1118.

Miller, R.; Jackson, C.; Kasarskis, E.; England, J.; Forshew, D.; Johnston, W.; Kalra, S.; Katz, J.; Mitsumoto, H.; Rosenfeld, J. et al. Practice Parameter Update: The Care of the Patient with Amyotrophic Lateral Sclerosis: Drug, Nutritional, and Respiratory Therapies (an Evidence-Based Review): Report of the Quality Standards Subcommittee of the American Academy of Neurology. *Neurology* **2009,** *73,* 1218–1226.

Miller, R.; Bradley, W.; Cudkowicz, M.; Hubble, J.; Meininger, V.; Mitsumoto, H.; Moore, D.; Pohlmann, H.; Sauer, D.; Silani, V. et al. Phase II/III Randomized Trial of TCH346 in Patients with ALS. *Neurology* **2007,** *69,* 776–784.

Mitchell, J.; Borasio, G. Amyotrophic Lateral Sclerosis. *Lancet* **2007,** *369,* 2031–2041.

Mitsubishi Tanabe Pharma America. *Radicava (edaravone) prescribing information;* 2017.

Miyamoto, N.; Maki, T.; Pham, L.; Hayakawa, K.; Seo, J.; Mandeville, E.; Mandeville, J.; Kim, K.; Lo, E.; Arai, K. Oxidative Stress Interferes with White Matter Renewal after Prolonged Cerebral Hypoperfusion in Mice. *Stroke* **2013**, *44*, 3516–3521.

Munch, C.; Sedlmeier, R.; Meyer, T.; Homberg, V.; Sperfeld, A.; Kurt, A.; Prudlo, J.; Peraus, G.; Hanemann, C.; Stumm, G. et al. Point Mutations of the P150 Subunit of Dynactin (DCTN1) Gene in ALS. *Neurology* **2004**, *63*, 724–726.

New Evidence and Unsolved Issues. *J. Neurol. Neurosurg. Psychiatry* **2008**, *79*, 6–11.

Nishimura, A.; Mitne-Neto, M.; Silva, H.; Richieri-Costa, A.; Middleton, S.; Cascio, D.; Kok, F.; Oliveira, J.; Gillingwater, T.; Webb, J. et al. A Mutation in the Vesicle-Trafficking Protein VAPB Causes Late-Onset Spinal Muscular Atrophy and Amyotrophic Lateral Sclerosis. *Am. J. Hum. Genet.* **2004**, *75*, 822–831.

Paratore, S.; Pezzino, S.; Cavallaro, S. Identification of Pharmacological Targets in Amyotrophic Lateral Sclerosis Through Genomic Analysis of Deregulated Genes and Pathways. *Curr. Genomics* **2012**, *13*, 321–333.

Pastula, D.; Moore, D.; Bedlack, R. Creatine for Amyotrophic Lateral Sclerosis/Motor Neuron Disease. *Cochrane Database Syst. Rev.* **2012**.

Pérez-González, A.; Galano, A. OH Radical Scavenging Activity of Edaravone: Mechanism and Kinetics. *J. Phys. Chem. A* **2011**, *115*, 1306–1314.

Petrov, D.; Mansfield, C.; Moussy, A.; Hermine, O. ALS Clinical Trials Review: 20 Years of Failure. Are we any Closer to Registering a New Treatment? *Front. Aging Neurosci.* **2017**, *9*, 68.

Philips, T.; Robberecht, W. Neuroinflammation in Amyotrophic Lateral Sclerosis: Role of Glial Activation in Motor Neuron Disease. *Lancet Neurol.* **2011**, *10*, 253–263.

Piette, F.; Belmin, J.; Vincent, H.; Schmidt, N.; Pariel, S.; Verny, M.; Marquis, C.; Mely, J.; Hugonot-Diener, L.; Kinet, J. et al. Masitinib as an Adjunct Therapy for Mild-to-Moderate Alzheimer's Disease: A Randomised, Placebo-Controlled Phase 2 Trial. *Alzheimers Res. Ther.* **2011**, *3*, 16.

Plaitakis, A.; Caroscio, J. Abnormal Glutamate Metabolism in Amyotrophic Lateral Sclerosis. *Ann. Neurol.* **1987**, *22*, 575–579.

Plaitakis, A.; Mandeli, J.; Smith, J.; Yahr, M. Pilot Trial of Branched-Chain Aminoacids in Amyotrophic Lateral Sclerosis. *Lancet* **1988**, *331*, 1015–1018.

Poon, H.; Hensley, K.; Thongboonkerd, V.; Merchant, M.; Lynn, B.; Pierce, W.; Klein, J.; Calabrese, V.; Butterfield, D. Redox Proteomics Analysis of Oxidatively Modified Proteins in G93A-SOD1 Transgenic Mice—A Model of Familial Amyotrophic Lateral Sclerosis. *Free Radic. Biol. Med.* **2005**, *39*, 453–462.

Practice Advisory on the Treatment of Amyotrophic Lateral Sclerosis with Riluzole: Report of the Quality Standards Subcommittee of the American Academy Of Neurology. *Neurology* **1997**, *49*, 657–659.

Radomski, M.; Trombly, C. *Occupational Therapy For Physical Dysfunction*; Wolters Kluwer Health: Philadelphia, 2014.

Radunovic, A.; Annane, D.; Rafiq, M.; Brassington, R.; Mustfa, N. Mechanical Ventilation for Amyotrophic Lateral Sclerosis/Motor Neuron Disease. *Cochrane Database Syst. Rev.* **2013**.

Ravits, J.; Traynor, B. Current And Future Directions in Genomics of Amyotrophic Lateral Sclerosis. *Phys. Med. Rehabil. Clin. N Am.* **2008**, *19*, 461–477.

Reddy, L.; Miller, T. RNA-Targeted Therapeutics for ALS. *Neurotherapeutics* **2015**, *12*, 424–427.

Renton, A.; Chiò, A.; Traynor, B. State of Play in Amyotrophic Lateral Sclerosis Genetics. *Nat. Neurosci.* **2013**, *17*, 17–23.

Ripps, M. E.; Huntley, G. W.; Hof, P. R.; Morrison, J. H.; Gordon, J. W. Transgenic Mice Expressing an Altered Murine Superoxide Dismutase Gene Provide an Animal Model of Amyotrophic Lateral Sclerosis. *Proc. Natl. Acad. Sci.* **1995**, *92*, 689–693.

Riviere, M.; Meininger, V.; Zeisser, P.; Munsat, T. An Analysis of Extended Survival in Patients with Amyotrophic Lateral Sclerosis Treated with Riluzole. *Arch Neurol.* **1998**, *55*, 526.

Rizzo, F.; Riboldi, G.; Salani, S.; Nizzardo, M.; Simone, C.; Corti, S.; Hedlund, E. Cellular Therapy to Target Neuroinflammation in Amyotrophic Lateral Sclerosis. *Cell. Mol. Life Sci.* **2013**, *71*, 999–1015.

Robberecht, W.; Philips, T. The Changing Scene of Amyotrophic Lateral Sclerosis. *Nat. Rev. Neurosci.* **2013**, *14*, 248–264.

Rong, P.; Yunusova, Y.; Wang, J.; Zinman, L.; Pattee, G.; Berry, J.; Perry, B.; Green, J. Predicting Speech Intelligibility Decline in Amyotrophic Lateral Sclerosis Based on the Deterioration of Individual Speech Subsystems. *PLoS One* **2016**, *11*, e0154971.

Rosen, D.; Siddique, T.; Patterson, D.; Figlewicz, D.; Sapp, P.; Hentati, A.; Donaldson, D.; Goto, J.; O'Regan, J.; Deng, H. et al. Mutations in Cu/Zn Superoxide Dismutase Gene are Associated with Familial Amyotrophic Lateral Sclerosis. *Nature* **1993**, *362*, 59–62.

Rothstein, J.; Martin, L.; Kuncl, R. Decreased Glutamate Transport by the Brain and Spinal Cord in Amyotrophic Lateral Sclerosis. *N. Engl. J. Med.* **1992**, *326*, 1464–1468.

Rowland, L. P. Amyotrophic Lateral Sclerosis. *Curr. Opin. Neurol.* **1994**, *7*, 310–315.

Scimemi, A.; Beato, M. Determining the Neurotransmitter Concentration Profile at Active Synapses. *Mol. Neurobiol.* **2009**, *40*, 289–306.

Shefner, J.; Cedarbaum, J.; Cudkowicz, M.; Maragakis, N.; Lee, J.; Jones, D.; Watson, M.; Mahoney, K.; Chen, M.; Saikali, K. et al. Safety, Tolerability And Pharmacodynamics of a Skeletal Muscle Activator in Amyotrophic Lateral Sclerosis. *Amyotroph Lateral Scler.* **2012**, *13*, 430–438.

Shefner, J.; Watson, M.; Meng, L.; Wolff, A. A Study to Evaluate Safety and Tolerability of Repeated Doses of Tirasemtiv in Patients with Amyotrophic Lateral Sclerosis. *Amyotroph. Lateral Scler. Frontotemporal Degener.* **2013**, *14*, 574–581.

Shefner, J.; Wolff, A.; Meng, L.; Bian, A.; Lee, J.; Barragan, D.; Andrews, J. A Randomized, Placebo-Controlled, Double-Blind Phase Iib Trial Evaluating the Safety and Efficacy of Tirasemtiv in Patients with Amyotrophic Lateral Sclerosis. *Amyotroph. Lateral Scler. Frontotemporal Degener.* **2016**, *17*, 426–435.

Siddique, T.; Driss S. A. Familial Amyotrophic Lateral Sclerosis, A Historical Perspective. *Acta Myologica* **2011**, *30*, 117–120.

Siniscalchi, A.; Bonci, A.; Mercuri, N.; Bernardi, G. Effects of Riluzole on Rat Cortical Neurones: Anin Vitroelectrophysiological Study. *Br. J. Pharmacol.* **1997**, *120*, 225–230.

Skold, C. Spasticity in Spinal Cord Injury: Self- and Clinically Rated Intrinsic Fluctuations and Intervention-Induced Changes. *Arch. Phys. Med. Rehabil.* **2000**, *81*, 144–149.

Slowie, L. A.; Paige, M. S.; Antel, J. P. Nutritional Considerations in the Management of Patients with Amyotrophic Lateral Sclerosis (ALS). *J. Am. Diet Assoc.* **1983**, *83*, 44–47.

Suzuki, M.; McHugh, J.; Tork, C.; Shelley, B.; Klein, S.; Aebischer, P.; Svendsen, C. GDNF Secreting Human Neural Progenitor Cells Protect Dying Motor Neurons, but not their Projection to Muscle, in a Rat Model of Familial ALS. *PLoS One* **2007**, *2*, e689.

Tanenbaum, B.; Taft, J. Maintaining Good Nutrition with ALS. *ALSA* **1990**.

Teng, Y.; Benn, S.; Kalkanis, S.; Shefner, J.; Onario, R.; Cheng, B.; Lachyankar, M.; Marconi, M.; Li, J.; Yu, D. et al. Multimodal Actions of Neural Stem Cells in a Mouse Model of ALS: A Meta-Analysis. *Sci. Transl. Med.* **2012**, *4*, 165ra164.

Testa, D.; Caraceni, T.; Fetoni, V. Branched-Chain Amino Acids in the Treatment of Amyotrophic Lateral Sclerosis. *J. Neurol.* **1989**, *236*, 445–447.

Theiss, R.; Kuo, J.; Heckman, C. Persistent Inward Currents in Rat Ventral Horn Neurones. *J. Physiol.* **2007**, *580*, 507–522.

Thonhoff, J.; Ojeda, L.; Wu, P. Stem Cell-Derived Motor Neurons: Applications and Challenges in Amyotrophic Lateral Sclerosis. *Curr. Stem Cell Res. The.* **2009**, *4*, 178–199.

Trias, E.; Ibarburu, S.; Barreto-Núñez, R.; Babdor, J.; Maciel, T.; Guillo, M.; Gros, L.; Dubreuil, P.; Díaz-Amarilla, P.; Cassina, P. et al. Post-Paralysis Tyrosine Kinase Inhibition with Masitinib Abrogates Neuroinflammation and Slows Disease Progression in Inherited Amyotrophic Lateral Sclerosis. *J. Neuroinflamm.* **2016**, *13*, 1–12.

Traynor, B.; Alexander M.; Corr B.; Frost E.; Hardiman O. Effect of a Multidisciplinary Amyotrophic Lateral Sclerosis (ALS) Clinic on ALS Survival: A Population Based Study, 1996–2000. *J. Neurol. Neurosurg. Psychiatry* **2003**, *74*, 1258–1261.

Turner, M.; Barnwell, J.; Al-Chalabi, A.; Eisen, A. Young-Onset Amyotrophic Lateral Sclerosis: Historical and Other Observations. *Brain* **2012**, *135*, 2883–2891.

Turner, M.; Cagnin, A.; Turkheimer, F.; Miller, C.; Shaw, C.; Brooks, D.; Leigh, P.; Banati, R. Evidence of Widespread Cerebral Microglial Activation in Amyotrophic Lateral Sclerosis: An [11C](R)-PK11195 Positron Emission Tomography Study. *Neurobiol. Dis.* **2004**, *15*, 601–609.

Urbani, A.; Belluzzi, O. Riluzole Inhibits the Persistent Sodium Current in Mammalian CNS Neurons. *Eur. J. Neurosci.* **2000**, *12*, 3567–3574.

Van Den Bosch, L.; Van Damme, P.; Bogaert, E.; Robberecht, W. The Role of Excitotoxicity in the Pathogenesis of Amyotrophic Lateral Sclerosis. *Biochimica et Biophysica Acta.* **2006**, *1762*, 1068–1082.

Venkova-Hristova, K.; Christov, A.; Kamaluddin, Z.; Kobalka, P.; Hensley, K. Progress in Therapy Development for Amyotrophic Lateral Sclerosis. *Neurol. Res. Int.* **2012**, *2012*.

Verstraete, E.; Veldink, J.; Huisman, M.; Draak, T.; Uijtendaal, E.; van der Kooi, A.; Schelhaas, H.; de Visser, M.; van der Tweel, I.; van den Berg, L. Lithium Lacks Effect on Survival in Amyotrophic Lateral Sclerosis: A Phase Iib Randomised Sequential Trial. *J. Neurol. Neurosurg. Psychiatry* **2012**, *83*, 557–564.

Walling, A. D. Amyotrophic Lateral Sclerosis: Lou Gehrig's Disease. *Am. Fam Phys.* **1999**, *59*, 1489–1496.

Wang, S.; Wang, K.; Wang, W. Mechanisms Underlying the Riluzole Inhibition of Glutamate Release From Rat Cerebral Cortex Nerve Terminals (Synaptosomes). *Neuroscience* **2004**, *125*, 191–201.

Welty, D.; Schielke, G.; Rothstein, J. Potential Treatment of Amyotrophic Lateral Sclerosis With Gabapentin: A Hypothesis. *Ann. Pharmacother.* **1995**, *29*, 1164–1167.

Wijesekera, L.; Leigh, P. Amyotrophic Lateral Sclerosis. *Orphanet J. Rare Dis.* **2009**, *4*, 3.

Wu, N.; Enomoto, A.; Tanaka, S.; Hsiao, C.; Nykamp, D.; Izhikevich, E.; Chandler, S. Persistent Sodium Currents in Mesencephalic V Neurons Participate in Burst Generation and Control of Membrane Excitability. *J. Neurophysiol.* **2005**, *93*, 2710–2722.

Yamamoto, Y.; Kuwahara, T.; Watanabe, K.; Watanabe, K. Antioxidant Activity of 3-Methyl-1-Phenyl-2-Pyrazolin-5-One. *Redox Rep.* **1996**, *2*, 333–338.

Yang, Y.; Hentati A.; Deng, H. X.; Dabbagh, O.; Sasaki, T.; Hirano, M.; Hung, W. Y.; Ouahchi, K.; Yan, J.; Azim, A. C.; Cole, N.; Gascon, G.; Yagmour, A.; Hamida, M. B.; Vance, M. P.; Hentati, F.; Siddique, T. The Gene Encoding Alsin, a Protein with Three Guanine-nucleotide Exchange Factor Domains, is Mutated in a Form of Recessive Amyotrophic Lateral Sclerosis. *Nat. Genet.* **2001**, *29*, 160–165.

Yoshino, H.; Kimura, A. Investigation of the Therapeutic Effects of Edaravone, a Free Radical Scavenger, on Amyotrophic Lateral Sclerosis (Phase II Study). *Amyotroph. Lateral Scler.* **2006**, *7*, 247–251.

Yu, D.; Silva, G. Stem Cell Sources and Therapeutic Approaches for Central Nervous System and Neural Retinal Disorders. *Neurosurg. Focus* **2008**, *24*, E11.

Zoccolella, S.; Beghi, E.; Palagano, G.; Fraddosio, A.; Guerra, V.; Samarelli, V.; Lepore, V.; Simone, I.; Lamberti, P.; Serlenga, L. et al. Riluzole and Amyotrophic Lateral Sclerosis Survival: A Population-Based Study in Southern Italy. *Eur. J. Neurol.* **2007**, *14*, 262–268.

Ziskind-Conhaim, L.; Wu, L.; Wiesner, E. Persistent Sodium Current Contributes to Induced Voltage Oscillations in Locomotor-Related Hb9 Interneurons in the Mouse Spinal Cord. *J. Neurophysiol.* **2008**, *100*, 2254–2264.

# CHAPTER 9

# Antianxiety Drugs

VAIBHAV WALIA and MUNISH GARG*

*Maharshi Dayanand University, Haryana, India*

*Corresponding author. E-mail: mgarg2006@gmail.com*

## ABSTRACT

Anxiety is a neurobehavioral state characterized by autonomic, neuroendocrine, and behavioral changes that serve to facilitate coping mechanism in an organism in response to potential or actual stressor. Anxiety when start to interfere with the normal activities of individual is considered as pathological anxiety. The exact pathophysiology of the anxiety is not known till today; however, it is considered as the disorder characterized by the imbalance in the neurotransmission of the various neurochemicals in the brain. On the basis of these neurochemicals, various pharmacotherapeutic agents have been developed for the treatment of anxiety and related disorders. However, each class of the pharmacotherapeutic agents suffers from its own limitations. Thus, despite the therapeutic advantages of these pharmacotherapeutic agents for the treatment of anxiety, adverse effects imposed by these agents made the health care professionals and the researchers to look for the alternatives which are safe and free from the life-threatening risk. In the present chapter, authors describe anxiety, anxiety disorders, neurochemicals implicated in the anxiety, pharmacotherapeutic agents and the possible alternatives used for the treatment of anxiety.

## 9.1 INTRODUCTION

Anxiety is a neurobehavioral state characterized by autonomic, neuroendocrine, and behavioral changes that serve to facilitate coping mechanism in an organism in response to potential or actual stressor (Steimer, 2002). Behavioral and cognitive adjustment of an organism in response to stress is known as coping. Coping can be of two types: active coping (use by the organism when the escape from the threat is possible) and passive coping (use by the organism when the escape from the threat is impossible). It is suggested that the organism develop the anxiety related behavior to cope up with the stressful situation or stressor (Boonen, 2010; Folkman and Greer, 2000; Gongora, 2000; Stanton et al., 2007). Anxiety is a defense mechanism and the normal that increase the awareness and responsiveness in an organism to deal with novel situations. However, when it becomes so excessive that it starts to interrupt the normal activities of an organism then it is pathological anxiety (APA, 2000; 2013). Pathological anxiety led to emotional state characterized by the consistent apprehension and vaguely defined negative events (Nutt et al., 2008). The difference between the normal and pathological anxiety is very difficult to make due to the thin boundary between these two forms of anxiety (Ganellen et al., 1986; Starcevic et al., 2012; Uhlenhuth et al., 2002). Further anxiety-related behavior occurs more frequently in the elderly patients (Brenes et al., 2008). Further anxiety has been found to be responsible for the cognitive deficits in the elderly patients (DeLuca et al., 2005; Palmer et al., 2007; Sinoff and Werner, 2003) and increases the chances of the late-life depression and relapse in younger population (Andreescu et al., 2007; Dombrovski et al., 2007; Flint and Rifat, 1997; Hettema et al., 2006; Steffens and McQuoid, 2005). Anxiety is often experienced by the females greater than the males (Samuelsson et al., 2005). Anxiety and depression are the two distinct conditions that occur together frequently (Brown et al., 1996) and is therefore often get misdiagnosed and thus patients do not get adequate pharmacotherapy. Anxiety increases disability (Brenes et al., 2005) and the risk of mortality (Brenes et al., 2007; Martens et al., 2010; Tully et al., 2008; van Hout et al., 2004). Anxiety without depression is common than depression without anxiety (Mineka et al., 1998; Morilak and Frazer, 2004). Further chronic anxiety often led to the emergence of depression (Schapira et al., 1972). Up to 90% of the patients suffering from the depression suffer from anxiety (Regier et al., 1998) and the presence of anxiety in depression patients increase the risk of relapse (Flint and Rifat, 1997). Both anxiety and depression are characterized by the elevated level of glucocorticoids (Airan et al., 2007; Gould et al., 1998; Grippo et al., 2005; McEwen et al., 1999; Popa et al., 2008; Sheline, 1996). Further both anxiety and depression is characterized by the increased levels of the glucocorticoids, further glucocorticoids have been used widely to induce anxiety and depression in rodent models (APA, 1994; Ardayfio and Kim, 2006; Gourley et al.,

2008; Murray et al., 2008; Zhao et al., 2008). In the present chapter, authors describe different types of the anxiety disorders, neurochemicals implicated in the anxiety, and pharmacotherapy of the anxiety.

## 9.2 CLASSIFICATION OF ANXIETY DISORDERS

According to WHO, approximately 3.6% of the individuals suffer from the anxiety and the prevalence is more in females than in males (WHO, 2017). Anxiety disorders are generally characterized by the increased levels and the incidence of anxiety and fear. According to DSM-IV, anxiety disorders can be of various types (APA, 2013) and they include:

### 9.2.1 GENERALIZED ANXIETY DISORDER

Generalized anxiety disorder (GAD) is accompanied by the increased irritability, apprehension, and worry. Thus the patients with GAD are considered as "lifetime worriers."

### 9.2.2 PANIC DISORDER

Panic Disorder (PD) is characterized by the short events of anxiety, fear, and discomfort accompanied by the increase in heart rate, chest pain, rapid breathing, and sweating.

### 9.2.3 SOCIAL ANXIETY DISORDER

Social anxiety disorder (SAD) is accompanied by the anxiety attacks in an individual in response to the social contexts and situations such that patients fear to interact with unfamiliar things and the people.

### 9.2.4 *OBSESSIVE-COMPULSIVE DISORDER*

Obsessive-compulsive disorder (OCD) is accompanied by repetitive intrinsic anxiogenic thoughts which further results in the marked anxiety (obsessions) and stereotyped behaviors or acts (compulsions) aimed to reduce the distress of the obsessions.

### 9.2.5 *POST-TRAUMATIC STRESS DISORDER*

Post-traumatic stress disorder (PTSD) is characterized by the prior incidence of intense trauma which results in the anxiety response in the patients.

## 9.3 NEUROCHEMICAL APPROACHES FOR THE TREATMENT OF ANXIETY

Disturbances in the level and neurotransmission of serotonin (5-HT), dopamine (DA) and norepinephrine (NE) are implicated in the regulation of anxiety (Charney, 2003; Goddard et al., 2010). Neurochemicals implicated in the anxiety related behavior is shown in Figure 9.1.

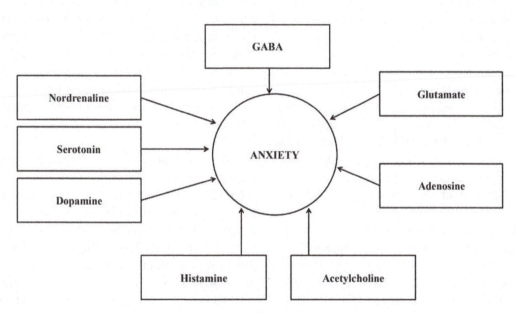

**FIGURE 9.1**  Neurochemicals implicated in the regulation of anxiety.

Serotonergic neurons are mainly confined to the raphe nuclei (Kocsis et al., 2006) and known to regulate the mood and behavior (Heninger and Charney, 1988). Some of the studies suggested that increasing the serotonergic transmission reduces the anxiety while some suggested that increasing the serotonergic transmission increases the anxiety.

Dopamine and dopaminergic pathways in the brain affect the anxiety-related behavior in many ways (de la Mora et al., 2010) and the blockade of the dopamine (D2) receptors is responsible for the anxiolysis (Lorenz et al., 2010). Thus it is suggested that increasing the dopaminergic signaling exerts anxiolytic effect (Ascher et al., 1995; Bystritsky et al., 2008).

Noradrenergic neurons of the locus coeruleus (Charney and Heninger, 1985) also regulate the anxiety-related behavior (Morilak et al., 2005) and emotional learning and processing (Pitman and Delahanty, 2005).

Glutamate is an excitatory neurotransmitter that modulates anxiety related states (Carobrez et al., 2001; O' Connor et al., 2010). Further, the agonism at the N-methyl-D-aspartate (NMDA) receptor is responsible for the anxiogenic response (Myers et al., 2011) while the antagonism at NMDA receptor exerts anxiolytic effect (Feusner et al., 2009; Minkeviciene et al., 2008; Otto et al., 2007).

Gamma amino butyric acid (GABA) is the main neurotransmitter that is known to regulate the anxiety-related behavior and the potentiation of GABAergic neurotransmission in the brain mediated anxiolytic effect (Mohler, 2011). However, the GABAergic modulators often imposed the risk of abuse tolerance and potentially fatal withdrawal symptoms (Roy-Byrne et al., 1993).

Acetylcholine (ACh) is a fast-acting neurotransmitter at ganglion (Changeux, 2010) known to regulate the mood and behavior (Janowsky et al., 1972; Risch et al., 1981). ACh signals through muscarinic receptors (mAChRs) and nicotinic receptors (nAChRs; Picciotto et al., 2000; Wess, 2003a). $M_2/M_4$ acts as an inhibitory autoreceptors (Douglas et al., 2002; Raiteri et al., 1984) and their expression reduces the release of glutamate (Gil et al 1997; Higley et al., 2009). Stimulation of M1/M5 receptors stimulates the dopamine release from striatum (Zhang et al., 2002). nAChRs

regulates the release of glutamate, GABA, dopamine, ACh, NA, and 5-HT (McGehee et al., 1995; Wonnacott, 1997). nAChRs occur as homomeric or heteromeric assemblies of α- and β-subunits (Picciotto et al., 2000). Nicotine at higher dose exerts anxiogenic effects (File et al., 1998, 2000) while at lower dose exerts anxiolytic effects (Tucci et al., 2003; Smythe et al., 1998).

Adenosine in brain acts through two types of receptor, that is, $A_1$ and $A_2$ receptors. $A_1$ receptors are present in highest concentration in cerebrum, hippocampus, and thalamus (Fastbom et al., 1987; Goodman and Synder, 1982) while $A_{2A}$ receptors are present in high concentration in corpus striatum, nucleus accumbens (NAc), and tuberculum olfactorium (Svenningsson et al., 1999). It has been reported that the pre-synaptic $A_1$ receptors function through the inhibition of neurotransmitters release (Svenningsson et al., 1999). Blockade of $A_{2A}$ receptors is responsible for the stimulatory effects of caffeine (Svenningsson et al., 1999; Ferre et al., 1997).

Histamine also regulates anxiety and the related behavior (Serafim et al., 2013) and alters the behavioral, motor, neuroendocrine, and sympathetic response accompanied by the behavioral changes (Brown et al., 2001; Gray and McNaughton, 2003; Miklos and Kovacs, 2003; Westerink et al., 2002) and the antagonism of histamine functions at H1 receptors is responsible for the anxiolytic effect (Ito, 2000).

## 9.4 CLASSIFICATION OF ANTIANXIETY DRUGS

Drugs used in the treatment of anxiety include (shown in Fig. 9.2)

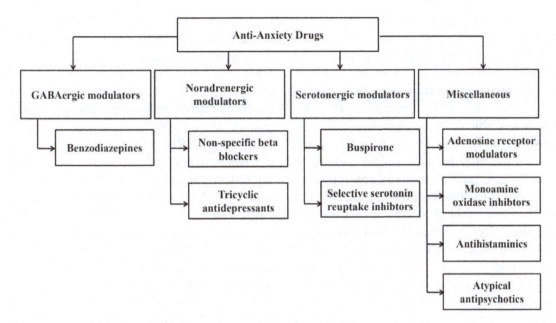

**FIGURE 9.2** Classification of antianxiety drugs.

## 9.5 ANTIANXIETY DRUGS

### 9.5.1 GABAERGIC MODULATORS

#### 9.5.1.1 BENZODIAZEPINES

Benzodiazepines (BZs) are the highly effective antianxiety, with shorter onset time, better tolerability, and acceptability. These qualities have made these drugs superior (Cascade and Kalali, 2008; Chouinard, 2004; Rosenbaum, 2005). In the CNS, GABA is responsible for the reduction of neuronal excitability and produces calming effect (Fox et al., 2011). GABA-BZ-Cl⁻ channel complex is a ligand-gated chloride-selective ion channel complex consisting of GABA and BZ binding sites (refer to Fig. 9.3). The BZ binding site is located at the interface between α and γ subunits. It is suggested that the stimulation of the GABA-BZ-Cl channel complex led to the hyperpolarization and inhibition of the neurons

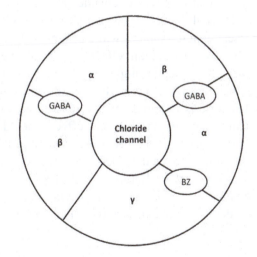

**FIGURE 9.3** GABA and BZ binding site on GABA-BZ-chloride channel complex.

(Kelly et al., 2002). Further 60% of GABA$_A$ receptors contain α1 subunit responsible for amnesia and anticonvulsant effects (Crestani et al., 2001; Mattila-Evenden et al., 2001). BZ receptors (based on α subunit isoforms) can be further classified into two types: $BZ_1$ and $BZ_2$ receptors. $BZ_1$ receptors contain α1 isoform and mediate sedative effects (Kaufmann, 2003; Sieghart, 1994) while $BZ_2$ receptors contain α2 isoform and mediate anxiolytic and myorelaxant effects (Sieghart, 1994; Crestani et al., 2001; Kaufmann, 2003). It is suggested that lipid solubility of the BZs affects their therapeutic profiles also such that more is lipid solubility of these drugs higher is the rate of absorption and faster onset of action occurs (Fox et al., 2011) but greater is the incidence of amnesia associated with these drugs (Fox et al., 2011). Diazepam belongs to the category of BZs, interacts with BZ-binding site on GABA-BZ-Cl⁻receptors and exerts anxiolytic effect (Crestani et al., 2001; Fox et al., 2011). Diazepam enhances the binding of GABA on its binding site (Wieland et al., 1992; Amin et al., 1997) resulting in the enhancement of chloride channel conductance (Rowlet et al., 2005). Diazepam at lower doses interacts with α2-subunit while at higher doses interacts with α1-subunit effect (van Rijnsoever et al., 2004) suggesting that lower dose is responsible for the anxiolytic effect while the higher dose is required for myorelaxation and anticonvulsant effect. Adverse effects include cognitive impairment, sedation, psychomotor impairment, and fetal defects (Fontaine et al., 1984; Curran, 1986; Bergman et al., 1992). BZs carry abuse potential, associated with physical dependence and withdrawal (O'Brien, 2005).

transporters (Stark et al., 1985). It is suggested that the increased extracellular levels of 5-HT upon SSRIs treatment led to the activation of $5-HT_{1A}$ receptors (Sprouse et al., 2001; Celada et al., 2004) responsible for the reduction of the firing rate of serotonergic neurons initially (Savitz et al., 2009). However, upon the continuous SSRIs treatment, these receptors get desensitized and the firing rate of the serotonergic neurons increases neurons (Blier et al., 1987; Hamon et al., 1991; Li et al., 1997, 1996; Raap et al., 1999). However, there is no desensitization or downregulation of postsynaptic $5-HT_{1A}$ receptors (De Montigny and Blier, 1991; Hamon et al., 1991). It has been reported that the SSRIs decrease the density of $5-HT_{1A}$ receptors (Klimek et al., 1994) while increasing the density of 5-HT uptake sites (Maj and Moryl, 1993) in brain. The increased levels of the 5-HT may contribute to 5-HT syndrome which presents one or combination of mental changes, autonomic hyperactivity, neuromuscular dysfunction (Lane and Baldwin, 1997) and sleeping problems (Antai-Otong, 2004), greater fragmentation of sleep (DeMartinis and Winokur, 2007; Popa et al., 2006; Thase, 1999), and reduced total sleep time (Thase, 1999). Adverse effects include sexual dysfunction, headache, nausea (Rossi et al., 2004), tremor, tachycardia, nausea, vomiting (Borys et al., 1992), and increased bleeding risk (Schelleman et al., 2011).

### 9.5.2   *SEROTONERGIC MODULATORS*

#### 9.5.2.1   *SELECTIVE SEROTONIN REUPTAKE INHIBITORS*

Selective serotonin reuptake inhibitors (SSRIs) inhibit the reuptake of 5-HT back inside the neurons by inhibiting the activity of 5-HT

#### 9.5.2.2   *BUSPIRONE*

$5-HT_{1A}$ is heteroreceptor and expressed widely in the amygdala, hippocampus, and the prefrontal cortex of brain (Albert et al., 1990) and modulates anxiety and the related behavior (Savitz et al., 2009; Meltzer et al., 2012; Donaldson et al., 2013). $5-HT_{1A}$ receptor is an autoreceptor coupled with $G_i$ and $G_o$ proteins and the related pathways (Lanfumey and Hamon, 2004; Barnes and Sharp, 1999) and the stimulation of these receptors inhibit the activity of these neurons (Celada et al., 2004). Buspirone shows partial agonism at the $5\text{-HT}_{1A}$ receptors and is used in the treatment of behavioral disturbances in dementia including agitation (Salzman, 2001; Apter and Allen, 1997; Holzer et al., 1995; Stanislav et al., 1994). Adverse effects include headache, dizziness, nervousness, and lightheadedness (Jann, 1988).

### 9.5.3 NORADRENERGIC MODULATORS

#### 9.5.3.1 NON-SPECIFIC BETA BLOCKERS

Propranolol shows antagonism at β receptors and is widely used for the treatment of anxiety (Tyrer and Lader, 1974). Further the blockade of β receptors by the propranolol reduces the consolidation of emotional memory (Mills and Dimsdale, 1991; Pitman and Delahanty, 2005; Stein et al., 2007), reduced the response of amygdala to several emotional stimuli (Hurlemann et al., 2010; Van Stegeren et al., 2005) and counteracts the physiological responses of anxiety (Davis et al., 1993; Mills and Dimsdale, 1991).

#### 9.5.3.2 TRICYCLIC ANTIDEPRESSANTS

Tricyclic antidepressants (TCAs) inhibit the reuptake of both NA and 5-HT and are used in the treatment of anxiety disorders (Mavissakalian, 2003; Schmitt et al., 2005). Further the highly serotonergic TCA, that is, clomipramine has been found to exert beneficial effect in OCD (Katz et al., 1990). TCAs absorbed rapidly following the administration by role route, bind to the plasma albumin strongly and to the extravascular tissues responsible for the large volume of distribution of these drugs (Dawson, 2004; Brunton et al., 2006). TCAs are mainly metabolized by the CYP450 enzymes and excreted in urine. The plasma concentration for therapeutic effect ranges from 50 to 300 ng/mL. When the plasma concentrations are greater than 450 ng/mL cognitive or behavioral toxicity develop and when the concentrations rise above 1000 ng/mL major toxicity and death can occur (Dawson, 2004; Schulz and Schmoldt, 2003). Adverse effects include dryness of mouth, blurred vision, postural hypotension, headache, dizziness, constipation, cardiac arrhythmia, and urinary retention.

### 9.5.4 MISCELLANEOUS DRUGS

Different categories of drugs that lies under the miscellaneous category are shown in Table 9.1.

**TABLE 9.1** Various Antianxiety Drugs and Their Adverse Effects.

| Class | Examples | Adverse effects |
|---|---|---|
| Adenosine modulators | Caffeine | Reduced sleep, increase BP, heart attack, and reduced fetal growth. |
| MAO inhibitors | Moclobemide and selegiline | The use might results in the increased BP when consumed with the milk products and thus imposed dietary restrictions in the prescribed patients. |
| Antihistaminics | Promethazine | Drowsiness, sedation, somnolence, fatigue, headache, impairs cognitive function, memory, and psychomotor performance (Simon and Simon, 2008). |
| Atypical antipsychotics | Aripiprazole | Weight gain, hyperlipidemia, and cardiac dysfunction. |

### 9.5.4.1   ADENOSINE RECEPTOR MODULATOR—CAFFEINE

Caffeine competitively blocks adenosine receptors (Holtzman et al., 1991; Ribeiro and Sebastião, 2010) but at lower doses shows antagonism at $A_{2A}$ receptor only (Svenningsson et al., 1997) and reduced the risk of the development of Parkinson's disease (Rogers et al., 2013). Caffeine consumption modulates the cognition, perception, alertness, wakefulness (Greenberg et al., 2007; Leathwood and Pollet, 1982–1983), the anxiety related behavior (Baldwin and File, 1989; Loke et al., 1985; Rusted, 1999), and also the memory related functions (Cunha and Agostinho, 2010). Lower dose of caffeine stimulates the locomotor activity (Rogers et al., 2010) while higher concentrations induce anxiety (Kaplan et al., 1997; Rhoads et al., 2011; Sawyer et al., 1982). Caffeine at higher doses shows antagonism at the adenosine receptors, inhibition of the phosphodiesterases, and activation of the β1-receptors responsible for the positive inotropic and chronotropic effects (Hartley et al., 2004; Savoca et al., 2005; Rudolph and Knudsen, 2010). However, beside this, caffeine inhibits the reuptake of the calcium and stimulates the release of the calcium from sarcoplasmic reticulum (Endo, 1977; Supinski et al., 1984). This rise in the intracellular calcium is responsible for the activation of endothelial nitric oxide synthase (eNOS) known to synthesize nitric oxide, (Goodman and Synder, 1982) responsible for the vasodilatation (Mahmud and Feely, 2001), suggesting that the high caffeine consumption is responsible for the marked hypotension secondary to vasodilation (Hatano et al., 1995). Caffeine inhibits the enzyme phosphodiesterase (Umemura et al., 2006) responsible for the hydrolysis of cAMP, which further results in the increased intracellular level of the cAMP and cyclic guanosine monophosphate (cGMP)

both of which are known to affect the cardiac contractility and impose the risk of the ventricular arrhythmias (Klatsky et al., 2011). Caffeine consumption shows reinforcement in both human and animals (Griffiths and Mumford, 1995; Wright et al., 2013). Adverse effects include nervousness, unstable BP, increased urinary frequency, insomnia, gastroesophageal reflux and so on (Juliano et al., 2011).

### 9.5.4.2   MONOAMINE OXIDASE INHIBITORS

Monoamine oxidase inhibitors (MAOIs) inhibit the enzyme MAO and inhibit the breakdown of catecholamines in the body and are used in the treatment of the resistant anxiety disorders (Bakish et al., 1995). Hypertensive reactions occur when the patients prescribed with these drugs consume milk products as they are the rich source of tyramine.

### 9.5.4.3   ANTIHISTAMINICS

Hydroxyzine is an antihistaminic drug that exerts anxiolytic effect by inhibiting histamine $H_1$ receptor (Bystritsky, 2006; Guaiana et al., 2007) and are used as the adjunct therapy for the treatment of anxiety. Adverse effects include dizziness, lethargy, sedation, somnolence, headache, fatigue, cognitive, and memory impairment (Simon and Simon, 2008).

### 9.5.4.4   ATYPICAL ANTIPSYCHOTICS

Atypical antipsychotics (AAPs) bind to $D_2$ receptors and shows antagonism at $D_2$ and histamine receptors (Kapur et al., 2000) and mediates anxiolysis (Blier, 2005). Aripiprazole exerts anxiolysis by antagonism at $5\text{-}HT_{2A}$

receptor and partial agonism at 5-HT$_{1A}$ receptor (Burris et al., 2002). AAPs, for example, olanzapine counteract stress-induced anxiety-like behavior (Locchi et al., 2008). Adverse effects include rise in blood glucose and lipids, weight gain, sexual dysfunction, and cardiovascular dysfunction (Ucok and Gaebel, 2008).

## 9.6 ALTERNATIVES TO MEDICATIONS

Prescription of most of the psychotherapeutic agents is often accompanied by the prescription of various vitamins or other supplements. Recently, it has been reported that these accompanied vitamins often exert the therapeutic benefits themselves. For example, vitamin-b6 has been shown to modulate the synthesis of GABA and 5-HT neurotransmitters and has shown benefits in the treatment of the various neurodegenerative and neuropsychiatric disorders (De Souza et al., 2000; Lerner et al., 2002) and is also known to exert positive effect on the mental state (McCarty, 2000).

Magnesium (Mg) is known to modulate the hypothalamic pituitary adrenal axis (HPAA; Held et al., 2002; Murck et al., 1998) and inhibits the activity of NMDA receptors implicated in anxiety-related behavior (Coan and Collingridge, 1985; Johnson and Shekhar, 2006). Further, the exposure to stressors also influences the level of Mg (Mocci et al., 2001; Takase et al., 2004). It has been reported that the reduction in the Mg level is characterized by the elevated level of anxiety (Laarakker et al., 2011; Pyndt et al., 2015; Sartori et al., 2012) and Mg supplementation reduces the anxiety-related behavior (Iezhitsa et al., 2011).

Vitamin C affects the mood and the behavioral state of individuals (Koizumi et al., 2016). Vitamin C supplementation is known to reduce the anxiety (de Oliveira et al., 2015) and provides neuroprotection against the reactive oxygen species (ROS). Vitamin C also modulates the activity of GABA and glutamate receptors and prevents the NMDA mediated excitotoxic damage to the neurons (Kocot et al., 2017).

Vitamin D is an important regulator of gene expression and implicated in the various disorders. It has been reported that the vitamin D supplementation exerts therapeutic benefits in the seasonal affective disorders (Huang et al., 2014). Vitamin D supplementation is considered as the cost-effective way to treat and prevent the development of mood disorders (Li et al., 2013).

## 9.7 FUTURE OPPORTUNITIES AND CHALLENGES

The exact pathophysiology of anxiety and related disorder is not known completely until today and without the complete knowledge of the disorder, it is impossible to treat the disorder. Additional, anxiety often co-exists with the other conditions such as depression and there is very thin boundary between these two, and thus diagnosed wrongly, and the patients remain untreated. Also, the wrong diagnosis led to the wrong prescription of the drugs, which sometimes may worsen the conditions of the patients. Therefore, the future research should be based on the understanding of the anxiety and an appropriate biomarker for the anxiety should be developed. Further, the research should aim to explore the newer pharmacotherapeutic targets for the better pharmacotherapy of the anxiety patients.

## 9.8   CONCLUSION

Antianxiety drugs used for the treatment of anxiety have their own limitations and merits. Different categories of the anxiolytic drugs are available in the market and each of the categories has its own limitations. The use of BZDs results in the motor impairment, mental confusion, amnesia and so on while the use of SSRIs imposed the risk of akathisia and suicidal ideations and attempts. Therefore, care should be taken while selecting the anxiolytic drugs. Also, the vitamins and minerals are frequently prescribed with the anxiolytic drugs and these vitamins play a key role in the synthesis of the various neurotransmitters and thus can be used alone in the treatment of anxiety and related disorders.

## KEYWORDS

- anxiety
- benzodiazepines
- dopamine
- noradrenaline
- pharmacotherapeutic agents

## REFERENCES

Airan, R. D.; Meltzer, L. A.; Roy, M.; Gong, Y.; Chen, H.; Deisseroth, K. High-speed Imaging Reveals Neurophysiological Links to Behavior in an Animal Model of Depression. *Science* **2007**, *317*, 819–823.

Albert, P. R.; Zhou, Q. Y.; Van Tol, H. H.; Bunzow, J. R.; Civelli, O. Cloning, Functional Expression, and mRNA Tissue Distribution of the Rat 5-hydroxytryptamine1A Receptor Gene. *J. Biol. Chem.* **1990**, *265*, 5825–5832.

American Psychiatric Association. *Diagnostic and Statistical Manual of Mental disorders-IV-TR*; APA: Washington, DC, 2000.

American Psychiatric Association. *Diagnostic and Statistic Manual of Mental Disorder*; APA, 1994; pp 124–129.

American Psychiatric Association. *Diagnostic and Statistical Manual of Mental Disorders (DSM-5®)*; American Psychiatric Pub, 2013.

Amin, J.; Brooks-Kayal, A.; Weiss, D. S. Two Tyrosine Residues on the α Subunit are Crucial for Benzodiazepine Binding and Allosteric Modulation of γ-aminobutyric Acid A Receptors. *Mol. Pharmacol.* **1997**, *51*, 833–841.

Andreescu, C.; Lenze, E. J.; Dew, M. A.; Begley, A. E.; Mulsant, B. H.; Dombrovski, A. Y.; Pollock, B. G.; Stack, J.; Miller, M. D.; Reynolds, C. F. Effect of Comorbid Anxiety on Treatment Response and Relapse Risk in Late-Life Depression: Controlled Study. *Br. J. Psychiatry* **2007**, *190*, 344–349.

Antai-Otong, D. Antidepressant-induced Insomnia: Treatment Options. *Perspect. Psychiatr. Care* **2004**, *40*, 29–33.

Apter, J. T.; Allen, L. A. Buspirone: Future Directions. *J. Clin. Psychopharmacol.* **1997**, *19*, 86–93.

Ardayfio, P.; Kim, K. S. Anxiogenic-like Effect of Chronic Corticosterone in the Light-dark Emergence Task in Mice. *Behav. Neurosci.* **2006**, *120*, 249.

Ascher, J. A.; Cole, J. O.; Colin, J. N.; Feighner, J. P.; Ferris, R. M.; Fibiger, H. C.; Richelson, E. Bupropion: A Review of its Mechanism of Antidepressant Activity. *J. Clin. Psychiatry* **1995**.

Bakish, D.; Hooper, C. L.; West, D. L.; Miller, C.; Blanchard, A.; Bashir, F. Moclobemide and Specific Serotonin Re-uptake Inhibitor Combination Treatment of Resistant Anxiety and Depressive Disorders. *Human Psychopharmacol. Clin. Exp.* **1995**, *10*, 105–109.

Baldwin, H. A.; File, S. E. Caffeine-induced Anxiogenesis: The Role of Adenosine, Benzodiazepine and Noradrenergic Receptors. *Pharmacol. Biochem. Behav.* **1989**, *32*, 181–186.

Barnes, N. M.; Sharp, T. A Review of Central 5-HT Receptors and Their Function. *Neuropharmacol* **1999**, *38*, 1083–1152.

Bergman, U.; Wiholm, B. E.; Rosa, F.; Baum, C.; Faich, G. A. Effects of Exposure to Benzodiazepine During Fetal Life. *Lancet* **1992**, *340*, 694–696.

Blier, P. A Typical Antipsychotics for Mood and Anxiety Disorders: Safe and Effective Adjuncts? *J. Psychiatr. Neurosci.* **2005**, *30*, 232–233.

Blier, P.; Chaput, Y. Modifications of the Serotonin System by Antidepressant Treatments: Implications for the Therapeutic Response in Major Depression. *J. Clin. Psychopharmacol.* **1987,** *7,* 24S–35S.

Boonen, A. Toward a Better Understanding of the Role of Psychological Variables in Arthritis Outcome Research. *Arthritis Res. Ther.***2010,** *12,* 106.

Borys, D. J.; Setzer, S. C.; Ling, L. J.; Reisdorf, J. J.; Day, L. C.; Krenzelok, E. P. Acute Fluoxetine Overdose: A Report of 234 Cases. *Am. J. Emerg. Med.* **1992,** *10,* 115–120.

Brenes, G. A.; Guralnik, J. M.; Williamson, J. D.; Fried, L. P.; Simpson, C.; Simonsick, E. M.; Penninx, B. W. The Influence of Anxiety on the Progression of Disability. *J. Am. Ger. Soc.* **2005,** *53,* 34–39.

Brenes, G. A.; Kritchevsky, S. B.; Mehta, K. M.; Yaffe, K.; Simonsick, E. M.; Ayonayon, H. N.; Penninx, B. W. Scared to Death: Results from the Health, Aging, and Body Composition study. *Am. J. Ger. Psychiatry* **2007,** *15,* 262–265.

Brenes, G. A.; Penninx, B. W.; Judd, P. H.; Rockwell, E.; Sewell, D. D.; Wetherell, J. L. Anxiety, Depression and Disability Across the Lifespan. *Aging Mental Health* **2008,** *12,* 158–163.

Brown, C.; Schulberg, H. C.; Madonia, M. J.; Shear, M. K.; Houck, P. R. Treatment Outcomes for Primary Care Patients with Major Depression and Lifetime Anxiety Disorders. *Am. J. Psychiatry* **1996,** *153,* 1293.

Brunton, L. L.; Lazo, J. S.; Parker, K. L. *Goodman and Gilman's The Pharmacological Basis of Therapeutics*; McGraw-Hill: New York, 2006.

Burris, K. D.; Molski, T. F.; Xu, C.; Ryan, E.; Tottori, K.; Kikuchi, T.; Molinoff, P. B. Aripiprazole, a Novel Antipsychotic, is a High-affinity Partial Agonist at Human Dopamine D2 Receptors. *J. Pharmacol. Exp. Therap.* **2002,** *302,* 381–389.

Bystritsky, A.; Kerwin, L.; Feusner, J. D.; Vapnik, T. A Pilot Controlled Trial of Bupropion XL Versus Escitalopram in Generalized Anxiety Disorder. *Psychopharmacol. Bull.* **2008,** *41,* 46–51.

Bystritsky, A. Treatment-resistant Anxiety Disorders. *Mol. Psychiatry.* **2006,** *11,* 805.

Carobrez, A. P.; Teixeira, K. V.; Graeff, F. G. Modulation of Defensive Behavior by Periaqueductal Gray NMDA/Glycine-B Receptor. *Neurosci. Biobehav. Rev.* **2001,** 25, 697–709.

Cascade, E.; Kalali, A. H. Use of Benzodiazepines in the Treatment of Anxiety. *Psychiatry (Edgmont)* **2008,** *5,* 21.

Celada, P.; Puig, M. V.; Amargós-Bosch, M.; Adell, A.; Artigas, F. The Therapeutic Role of 5-HT1A and 5-HT2A

Receptors in Depression. *J. Psychiatr. Neurosci.* **2004,** *29,* 252.

Changeux, J. P. Allosteric Receptors: From Electric Organ to Cognition. *Annu. Rev. Pharmacol. Toxicol.* **2010,** *50,* 1–38.

Charney, D. S. Neuroanatomical Circuits Modulating Fear and Anxiety Behaviors. *Acta. Psychiatr. Scandinavica* **2003,** *108,* 38–50.

Charney, D. S.; Heninger, G. R. Noradrenergic Function and the Mechanism of Action of Antianxiety Treatment: I. The Effect of Long-term Alprazolam Treatment. *Arch. Gen. Psychiatry* **1985,** *42,* 458–467.

Chouinard, G. Issues in the Clinical use of Benzodiazepines: Potency, Withdrawal, and Rebound. *J. Clin. Psychiatry* **2004,** *65,* 7–12.

Coan, E. J.; Collingridge, G. L. Magnesium Ions Block an *N*-methyl-D-aspartate Receptor-mediated Component of Synaptic Transmission in Rat Hippocampus. *Neurosci. Lett.* **1985,** *53,* 21–26.

Crestani, F.; Löw, K.; Keist, R.; Mandelli, M. J.; Möhler, H.; Rudolph, U. Molecular Targets for the Myorelaxant Action of Diazepam. *Mol. Pharmacol.* **2001,** *59,* 442–445.

Cunha, R. A.; Agostinho, P. M. Chronic Caffeine Consumption Prevents Memory Disturbance in Different Animal Models of Memory Decline. *J. Alzheimer. Dis.* **2010,** *20,* S95–S116.

Curran, H. V. Tranquillising Memories: A Review of the Effects of Benzodiazepines on Human Memory. *Biol. Psychol.* **1986,** *23,* 179–213.

Davis, M.; Falls, W. A.; Campeau, S.; Kim, M. Fear-potentiated Startle: A Neural and Pharmacological Analysis. *Behav. Brain. Res.* **1993,** *58,* 175–198.

Dawson, A. H. The Tricyclic Antidepressants. *Med. Toxicol.* Lippincott Williams and Wilkins: Baltimore **2004,** 851–861.

de la Mora, M. P.; Gallegos-Cari, A.; Arizmendi-García, Y.; Marcellino, D.; Fuxe, K. Role of Dopamine Receptor Mechanisms in the Amygdaloid Modulation of Fear and Anxiety: Structural and Functional Analysis. *Prog. Neurobiol.* **2010,** *90,* 198–216.

De Montigny, C.; Blier, P. Enhancement of 5-HT Synaptic Efficacy by Antidepressant Treatments: A Tenable Unitary Theory. *Excerpta. Medica. Int. Cong. Ser.* **1991,** *968,* 301–304.

de Oliveira, I. J.; de Souza, V. V.; Motta, V.; Da-Silva, S. L. Effects of Oral Vitamin C Supplementation on Anxiety in Students: A Double-Blind, Randomized, Placebo-Controlled Trial. *Pak. J. Biol. Sci.* **2015,** *18,* 11–18.

De Souza, M. C.; Walker, A. F.; Robinson, P. A.; Bolland, K. A Synergistic Effect of a Daily Supplement for 1 month of 200 mg Magnesium plus 50 mg Vitamin B6 for the Relief of Anxiety-related Premenstrual Symptoms: A Randomized, Double-blind, Crossover Study. *J. Women's Health Gender-Based Med.* **2000,** *9,* 131–139.

DeLuca, A. K.; Lenze, E. J.; Mulsant, B. H.; Butters, M. A.; Karp, J. F.; Dew, M. A.; Reynolds, C. F. Comorbid anxiety Disorder in Late Life Depression: Association with Memory Decline over Four Years. *Int. J. Ger. Psychiatry* **2005,** *20,* 848–854.

DeMartinis, N. A.; Winokur, A. Effects of Psychiatric Medications on Sleep and Sleep Disorders. *C.N.S. Neurol. Disord. Drug. Target* **2007,** *6,* 17–29.

Dombrovski, A. Y.; Mulsant, B. H.; Houck, P. R.; Mazumdar, S.; Lenze, E. J.; Andreescu, C.; Reynolds, C. F. Residual Symptoms and Recurrence During Maintenance Treatment of Late-life Depression. *J. Affect. Disord.* **2007,** *103,* 77–82.

Donaldson, Z. R.; Nautiyal, K. M.; Ahmari, S. E.; Hen, R. Genetic Approaches for Understanding the Role of Serotonin Receptors in Mood and Behavior. *Curr. Opin. Neurobiol.* **2013,** *23,* 399–406.

Douglas, C. L.; Baghdoyan, H. A.; Lydic, R. Postsynaptic Muscarinic M1 Receptors Activate Prefrontal Cortical EEG of C57BL/6J Mouse. *J. Neurophysiol.* **2002,** *88,* 3003–3009.

Endo, M. Calcium Release from the Sarcoplasmic Reticulum. *Physiol. Rev.* **1977,** *57,* 71–108.

Fastbom, J.; Pazos, A.; Palacios, J. M. The Distribution of Adenosine A1 Receptors and 5'-Nucleotidase in the Brain of Some Commonly Used Experimental Animals. *Neuroscience* **1987,** *22,* 813–826.

Ferré, S.; Fuxe, K.; Fredholm, B. B.; Morelli, M.; Popoli, P. Adenosine–dopamine Receptor–receptor Interactions as an Integrative Mechanism in the Basal Ganglia. *Trend. Neurosci.* **1997,** *20,* 482–487.

Feusner, J. D.; Kerwin, L.; Saxena, S.; Bystritsky, A. Differential Efficacy of Memantine for Obsessive-Compulsive Disorder vs. Generalized Anxiety Disorder: An Open-Label Trial. *Psychopharmacol. Bull.* **2009,** *42,* 81–93.

File, S. E.; Gonzalez, L. E.; Andrews, N. Endogenous Acetylcholine in the Dorsal Hippocampus Reduces Anxiety Through Actions on Nicotinic and Muscarinic Receptors. *Behav. Neurosci.* **1998,** *112,* 352.

File, S. E.; Kenny, P. J.; Cheeta, S. The Role of the Dorsal Hippocampal Serotonergic and Cholinergic Systems in the Modulation of Anxiety. *Pharmacol. Biochem. Behav.* **2000,** *66,* 65–72.

Flint, A. J.; Rifat, S. L. Two-year Outcome of Elderly Patients with Anxious Depression. *Psychiatr. Res.* **1997,** *66,* 23–31.

Folkman, S.; Greer, S. Promoting Psychological Well-Being in the Face of Serious Illness: When Theory, Research and Practice Inform Each Other. *Psycho-oncol.* **2000,** *9,* 11–19.

Fontaine, R.; Chouinard, G.; Annable, L. Rebound Anxiety in Anxious Patients After Abrupt Withdrawal of Benzodiazepine Treatment. *Am. J. Psychiatry* **1984,** *141,* 848–852.

Fox, C.; Liu, H.; Kaye, A. D.; Manchikanti, L.; Trescot, A. M.; Christo, P. J.; Falco, F. J. R. Clinical Aspects of Pain Medicine and Interventional Pain Management: A Comprehensive Review. Paducah, KY: ASIP Publishing. *Antianxiety Agents.* **2011,** 543–552.

Ganellen, R. J.; Matuzas, W.; Uhlenhuth, E. H.; Glass, R.; Easton, C. R. Panic Disorder, Agoraphobia, and Anxiety-relevant Cognitive Style. *J. Affect. Disord.* **1986,** *11,* 219–225.

Gil, Z.; Connors, B. W.; Amitai, Y. Differential Regulation of Neocortical Synapses by Neuromodulators and Activity. *Neuron* **1997,** *19,* 679–686.

Goddard, A. W.; Ball, S. G.; Martinez, J.; Robinson, M. J.; Yang, C. R.; Russell, J. M.; Shekhar, A. Current Perspectives of the Roles of the Central Norepinephrine System in Anxiety and Depression. *Depress. Anx.* **2010,** *27,* 339–350.

Gongora, E. A. *Coping to Problems and the Role of Control: A Ethnopsciological Point of View Inside Ecosystem With Tradition [dissertation] [in Spanish];* Universidad Nacional Autonoma de Mexico: Mexico City (Mexico), 2000; p 271.

Goodman, R. R.; Synder, S. H. Autoradiographic Localization of Adenosine Receptors in Rat Brain Using [3H] Cyclohexyladenosine. *J. Neurosci.* **1982,** *2,* 1230–1241.

Gould, E.; Tanapat, P.; McEwen, B. S.; Flügge, G.; Fuchs, E. Proliferation of Granule Cell Precursors in the Dentate Gyrus of Adult Monkeys is Diminished by Stress. *Proc. Natl. Acad. Sci.* **1998,** *95,* 3168–3171.

Gourley, S. L.; Wu, F. J.; Kiraly, D. D.; Ploski, J. E.; Kedves, A. T.; Duman, R. S.; Taylor, J. R. Regionally Specific Regulation of ERK MAP Kinase in a Model of Antidepressant-sensitive Chronic Depression. *Biol. Psychiatry.* **2008,** *63,* 353–359.

Grases G.; Pérez-Castelló J. A.; Sanchis P.; Casero A.; Perelló J.; Isern B.; Rigo E.; Grases F. Anxiety and Stress Among Science Students. Study of Calcium and Magnesium Alterations. *Magnes. Res.* **2006,** *192,* 102–106.

Greenberg, J. A.; Dunbar, C. C.; Schnoll, R.; Kokolis, R.; Kokolis, S.; Kassotis, J. Caffeinated Beverage Intake and the Risk of Heart Disease Mortality in the Elderly: A Prospective Analysis. *Am. J. Clin. Nutr.* **2007,** *85,* 392–398.

Grippo, A. J.; Sullivan, N. R.; Damjanoska, K. J.; Crane, J. W.; Carrasco, G. A.; Shi, J.; Van de Kar, L. D. Chronic Mild Stress Induces Behavioral and Physiological Changes, and may Alter Serotonin 1A Receptor Function, in Male and Cycling Female Rats. *Psychopharmacology* **2005,** *179,* 769–780.

Guaiana, G.; Barbui, C.; Churchill, R.; Cipriani, A.; McGuire, H. *Hydroxyzine for Generalised Anxiety Disorder.* Status and date: Edited (no change to conclusions), 2007.

Hamon, M.; Lanfumey, L.; Hajdahmane, S.; Jolas, T.; Kidd, E. J.; Bolanos, F.; Gozlan, H. In *Adaptation of Central 5-HT1A Receptors After Chronic Antidepressant Treatment in Rats,* 5th World Congress of Biological Psychiatry, Florence, Italy, 1991.

Hartley, T. R.; Lovallo, W. R.; Whitsett, T. L. Cardiovascular Effects of Caffeine in Men and Women. *Am. J. Cardiol.* **2004,** *93,* 1022–1026.

Hatano, Y.; Mizumoto, K.; Yoshiyama, T.; Yamamoto, M.; Iranami, H. Endothelium-dependent and-Independent Vasodilation of Isolated Rat Aorta Induced by Caffeine. *Am. J. Physiol. Heart Circ. Physiol.* **1995,** *269,* H1679–H1684.

Held, K.; Antonijevic, I. A.; Künzel, H.; Uhr, M.; Wetter, T. C.; Golly, I. C.; Steiger, A.; Murck, H. Oral Mg2+ Supplementation Reverses Age-related Neuroendocrine and Sleep EEG Changes in Humans. *Pharmacopsychiatry* **2002,** *354,* 135–143.

Heninger, G. R.; Charney, D. S. Monoamine Receptor Systems and Anxiety Disorders. *Psychiatr. Clin. North Am.* **1988.**

Hettema, J. M.; Kuhn, J. W.; Prescott, C. A.; Kendler, K. S. The Impact of Generalized Anxiety Disorder and Stressful Life Events on Risk for Major Depressive Episodes. *Psychol. Med.* **2006,** *36,* 789–795.

Higley, M. J.; Soler-Llavina, G. J.; Sabatini, B. L. Cholinergic Modulation of Multivesicular Release Regulates Striatal Synaptic Potency and Integration. *Nat. Neurosci.* **2009,** *12,* 1121.

Holtzman, S. G.; Mante, S. A. A. K. W. A.; Minneman, K. P. Role of Adenosine Receptors in Caffeine Tolerance. *J. Pharmacol. Exp. Therap.* **1991,** *256,* 62–68.

Holzer, J. C.; Gitelman, D. R.; Price, B. H. Efficacy of Buspirone in the Treatment of Dementia with Aggression. *Am. J. Psychiatry* **1995,** *152,* 812.

Huang, J. Y.; Arnold, D.; Qiu, C.; Miller, R. S.; Williams, M. A.; Enquobahrie, D. A. Association of Serum Vitamin D with Symptoms of Depression and Anxiety in Early Pregnancy. *J. Women. Health.* **2014,** *23,* 588–595.

Hurlemann, R.; Walter, H.; Rehme, A. K.; Kukolja, J.; Santoro, S. C.; Schmidt, C.; Kendrick, K. M. Human Amygdala Reactivity is Diminished by the β-noradrenergic Antagonist Propranolol. *Psychol. Med.* **2010,** *40,* 1839–1848.

Iezhitsa, I. N.; Spasov, A. A.; Kharitonova, M. V.; Kravchenko, M. S. Effect of Magnesium Chloride on Psychomotor Activity, Emotional Status, and Acute Behavioral Responses to Clonidine, *d*-amphetamine, Arecoline, Nicotine, Apomorphine, and L-5-hydroxytryptophan. *Nutr. Neurosci.* **2011,** *141,* 10–24.

Jann, M. W. Buspirone: An Update on a Unique Anxiolytic Agent. *Pharmacother. J. Human. Pharmacol. Drug. Ther.* **1988,** *8,* 100–116.

Janowsky, D.; Davis, J.; El-Yousef, M. K.; Sekerke, H. J. A Cholinergic-adrenergic Hypothesis of Mania and Depression. *Lancet* **1972,** *300,* 632–635.

Johnson, P. L.; Shekhar A. Panic-Prone State Induced in Rats with GABA Dysfunction in the Dorsomedial Hypothalamus is Mediated by NMDA Receptors. *J. Neurosci.* **2006,** *2626,* 7093–7104.

Juliano, L. M.; Ferre, S.; Griffiths, R. R. Caffeine: Pharmacol and Clinical Effects. In *Principles of Addiction Medicine,* 4th ed.; American Society of Addiction Medicine: Chevy Chase, MD, 2009.

Kaplan, G. B.; Greenblatt, D. J.; Ehrenberg, B. L.; Goddard, J. E.; Cotreau, M. M.; Harmatz, J. S.; Shader, R. I. Dose-dependent Pharmacokinetics and Psychomotor Effects of Caffeine in Humans. *J. Clin. Pharmacol.* **1997,** *37,* 693–703.

Kapur, S.; Zipursky, R.; Jones, C.; Shammi, C. S.; Remington, G.; Seeman, P. A Positron Emission Tomography Study of Quetiapine in Schizophrenia: A Preliminary Finding of an Antipsychotic Effect with Only Transiently High Dopamine D2 Receptor Occupancy. *Arch. Gen. Psychiatry* **2000,** *57,* 553–559.

Katz, R. J.; DeVeaugh-Geiss, J.; Landau, P. Clomipramine in Obsessive-compulsive Disorder. *Biol. Psychiatry* **1990,** *28,* 401–414.

Kaufmann, W. A.; Humpel, C.; Alheid, G. F.; Marksteiner, J. Compartmentation of Alpha 1 and Alpha 2 GABAA Receptor Subunits Within Rat Extended Amygdala: Implications for Benzodiazepine Action. *Brain. Res.* **2003,** *964,* 91–99.

Kelly, M. D.; Smith, A.; Banks, G.; Wingrove, P.; Whiting, P. W.; Atack, J.; Maubach, K. A. Role of the Histidine Residue at Position 105 in the Human A5 Containing

GABAA Receptor on the Affinity and Efficacy of Benzodiazepine Site Ligands. *Br. J. Pharmacol.* **2002,** *135*, 248–256.

Klatsky, A. L.; Hasan, A. S.; Armstrong, M. A.; Udaltsova, N.; Morton, C. Coffee, Caffeine, and Risk of Hospitalization for Arrhythmias. *Permanente. J.* **2011,** *15*, 19.

Klimek, V.; Zak-Knapik, J.; Mackowiak, M. Effects of Repeated Treatment with Fluoxetine and Citalopram, 5-HT Uptake Inhibitors, on 5-HT1A and 5-HT2 Receptors in the Rat Brain. *J. Psychiatr. Neurosci.* **1994,** *19*, 63.

Kocot, J.; Luchowska-Kocot, D.; Kiełczykowska, M.; Musik, I.; Kurzepa, J. Does Vitamin C Influence Neurodegenerative Diseases and Psychiatric Disorders? *Nutrients* **2017,** *9*, 659.

Kocsis, B.; Varga, V.; Dahan, L.; Sik, A. Serotonergic Neuron Diversity: Identification of Raphe Neurons with Discharges Time-Locked to the Hippocampal Theta Rhythm. *Proc. Natl. Acad. Sci. U.S.A.* **2006,** *103*, 1059–1064.

Koizumi, M.; Kondo, Y.; Isaka, A.; Ishigami, A.; Suzuki, E. Vitamin C Impacts Anxiety-Like Behavior and Stress-Induced Anorexia Relative to Social Environment in SMP30/GNL Knockout Mice. *Nutr. Res.* **2016,** *36*, 1379–1391.

Laarakker, M. C.; van Lith H. A.; Ohl, F. Behavioral Characterization of A/J and C57BL/6J Mice Using a Multidimensional Test, Association Between Blood Plasma and Brain Magnesium-Ion Concentration with Anxiety. *Physiol. Behav.* **2011,** *1022*, 205–219.

Lane, R.; Baldwin, D. Selective Serotonin Reuptake Inhibitor-induced Serotonin Syndrome. *J. Clin. Psychopharmacol.* **1997,** *17*, 208–221.

Lanfumey, L.; Hamon, M. 5-HT~ 1 Receptors. *Curr. Drug Target C.N.S. Neurol. Dis.* **2004,** *3*, 1–10.

Leathwood, P. D.; Pollet, P. Diet-induced Mood Changes in Normal Populations. *J. Psychiatr. Res.* **1982,** *17*, 147–154.

Leonard, B. E. Fundamentals of Psycopharmocology; Wiley and Sons: Chichester, 2003.

Lerner, V.; Miodownik, C.; Kaptsan, A.; Cohen, H.; Loewenthal, U.; Kotler, M. Vitamin B6 as Add-on Treatment in Chronic Schizophrenic and Schizoaffective Patients: A Double-Blind, Placebo-Controlled Study. *J. Clin. Psychiatry* **2000,** *63*, 54–58.

Li, G.; Mbuagbaw, L.; Samaan, Z.; Zhang, S.: Adachi, J. D.; Papaioannou, A.; Thabane, L. Efficacy of Vitamin D Supplementation in Depression in Adults: A Systematic Review Protocol. *Syst. Rev.* **2013,** *2*, 64.

Li, Q.; Muma, N. A.; Van de Kar, L. D. Chronic Fluoxetine Induces a Gradual Desensitization of 5-HT1A Receptors: Reductions in Hypothalamic and Midbrain Gi and G (o) Proteins and in Neuroendocrine Responses to a 5-HT1A Agonist. *J. Pharmacol. Exp. Therap.* **1996,** *279*, 1035–1042.

Locchi, F.; Dall'Olio, R.; Gandolfi, O.; Rimondini, R. Olanzapine Counteracts Stress-induced Anxiety-like Behavior in Rats. *Neurosci. Lett.* **2008,** *438*, 146–149.

Loke, W. H.; Hinrichs, J. V.; Ghoneim, M. M. Caffeine and Diazepam: Separate and Combined Effects on Mood, Memory, and Psychomotor Performance. *Psychopharmacology* **1985,** *87*, 344–350.

Lorenz, R. A.; Jackson, C. W.; Saitz, M. Adjunctive Use of Atypical Antipsychotics for Treatment-Resistant Generalized Anxiety Disorder. *Pharmacother. J. Human. Pharmacol. Drug. Therapy.* **2010,** *30*, 942–951.

Mahmud, A.; Feely, J. Acute Effect of Caffeine on Arterial Stiffness and Aortic Pressure Waveform. *Hypertension* **2001,** *38*, 227–231.

Maj, J.; Moryl, E. Effects of Fluoxetine Given Chronically on the Responsiveness of 5-HT Receptor Subpopulations to their Agonists. *Eur. Neuropsychopharmacol.* **1993,** *3*, 85–94.

Martens, E. J.; de Jonge, P.; Na, B.; Cohen, B. E.; Lett, H.; Whooley, M. A. Scared to Death? Generalized Anxiety Disorder and Cardiovascular Events in Patients with Stable Coronary Heart Disease: The Heart and Soul Study. *Arch. Gen. Psychiatry.* **2010,** *67*, 750–758.

Mattila-Evenden; Johan, Franck; Ulf Bergman, M. A Study of Benzodiazepine Users Claiming Drug-Induced Psychiatric Morbidity. *Nordic. J. Psychiatry.* **2001,** *55*, 271–278.

Mavissakalian, M. R. Imiprmaine vs. Sertraline in Panic Disorder: 24-week Treatment Completers. *Ann. Clin. Psychiatry.* **2003,** *15*, 171–180.

McCarty, M. F. High-dose Pyridoxine as an 'Anti-stress' Strategy. *Med. Hyp.* **2000,** *54*, 803–807.

McEwen, B. S. Stress and Hippocampal Plasticity. *Annu. Rev. Neurosci.* **1999,** *22*, 105–122.

McGehee, D. S.; Heath, M. J.; Gelber, S.; Devay, P.; Role, L. W. Nicotine Enhancement of Fast Excitatory Synaptic Transmission in CNS by Presynaptic Receptors. *Science* **1995,** *269*, 1692–1696.

Meltzer, H. Y.; W Massey, B.; Horiguchi, M. Serotonin Receptors as Targets for Drugs Useful to Treat Psychosis and Cognitive Impairment in Schizophrenia. *Curr. Pharm. Biotechnol.* **2012,** *13*, 1572–1586.

Mills, P. J.; Dimsdale, J. E. Cardiovascular Reactivity to Psychosocial Stressors: A Review of the Effects of Beta-blockade. *Psychosomatics.* **1991,** *32,* 209–220.

Mineka, S.; Watson, D.; Clark, L. A. Comorbidity of Anxiety and Unipolar Mood Disorders. *Annu. Rev. Psychol.* **1998,** *49,* 377–412.

Minkeviciene, R.; Banerjee, P.; Tanila, H. Cognition-enhancing and Anxiolytic Effects of Memantine. *Neuropharmacology* **2008,** *54,* 1079–1085.

Mocci, F.; Canalis, P.; Tomasi, P. A.; Casu, F.; Pettinato, S. The Effect of Noise on Serum and Urinary Magnesium and Catecholamines in Humans. *Occup. Med.* **2001,** *511,* 56–61.

Möhler, H. The Rise of a New GABA Pharmacology. *Neuropharmacology* **2011,** *60,* 1042–1049.

Morilak, D. A.; Frazer, A. Antidepressants and Brain Monoaminergic Systems: A Dimensional Approach to Understanding their Behavioral Effects in Depression and Anxiety Disorders. *Int. J. Neuropsychopharmacol.* **2004,** *7,* 193–218.

Morilak, D. A.; Barrera, G.; Echevarria, D. J.; Garcia, A. S.; Hernandez, A.; Ma, S.; Petre, C. O. Role of Brain Norepinephrine in the Behavioral Response to Stress. *Prog. Neuropsychopharmacol. Biol. Psychiatry* **2005,** *29,* 1214–1224.

Murck H.; Steiger A. Mg2+ Reduces ACTH Secretion and Enhances Spindle Power Without Changing Delta Power During Sleep in Men—Possible Therapeutic Implications. *Psychopharmacology* **1998,** *1373,* 247–252.

Murray, F.; Smith, D. W.; Hutson, P. H. Chronic Low Dose Corticosterone Exposure Decreased Hippocampal Cell Proliferation, Volume and Induced Anxiety And Depression like Behaviors in Mice. *Eur. J. Pharmacol.* **2008,** *583,* 115–127.

Myers, K. M.; Carlezon Jr, W. A.; Davis, M. Glutamate Receptors in Extinction and Extinction-Based Therapies for Psychiatric Illness. *Neuropsychopharmacology* **2011,** *36,* 274.

Nutt, D.; Allgulander, C.; Lecrubier, Y.; Peters, T.; Wittchen, H. U. Establishing Non-inferiority in Treatment Trials in Psychiatry—Guidelines From an Expert Consensus Meeting. *J. Psychopharmacol.* **2008,** *22,* 409–416.

O'Brien, C. P. Benzodiazepine Use, Abuse, and Dependence. *J. Clin. Psychiatry* **2005,** *66,* 28–33.

O'connor, R. M.; Finger, B. C.; Flor, P. J.; Cryan, J. F. Metabotropic Glutamate Receptor 7: At the Interface of Cognition and Emotion. *Eur. J. Pharmacol.* **2010,** *639,* 123–131.

Otto, M. W.; Basden, S. L.; Leyro, T. M.; McHugh, R. K.; Hofmann, S. G. Clinical Perspectives on the Combination of D-Cycloserine and Cognitive-Behavioral Therapy for the Treatment of Anxiety Disorders. *C.N.S. Spect.* **2007,** *12,* 51–61.

Palmer, K.; Berger, A. K.; Monastero, R.; Winblad, B.; Bäckman, L.; Fratiglioni, L. Predictors of Progression from Mild Cognitive Impairment to Alzheimer Disease. *Neurology* **2007,** *68,* 1596–1602.

Picciotto, M. R.; Caldarone, B. J.; King, S. L.; Zachariou, V. Nicotinic Receptors in the Brain: Links Between Molecular Biology and Behavior. *Neuropsychopharmacology* **2000,** *22,* 451–465.

Pitman, R. K.; Delahanty, D. L. Conceptually Driven Pharmacologic Approaches to Acute Trauma. *C.N.S. Spectr.* **2005,** *10,* 99–106.

Popa, D.; El Yacoubi, M.; Vaugeois, J. M.; Hamon, M.; Adrien, J. Homeostatic Regulation of Sleep in a Genetic Model of Depression in the Mouse: Effects of Muscarinic and 5-HT1A Receptor Activation. *Neuropsychopharmacology* **2006,** *31,* 1637–1646.

Popa, D.; Léna, C.; Alexandre, C.; Adrien, J. Lasting Syndrome of Depression Produced by Reduction in Serotonin Uptake During Postnatal Development: Evidence From Sleep, Stress, and Behavior. *J. Neurosci.* **2008,** *28,* 3546–3554.

PyndtJørgensen, B.; Winther, G.; Kihl, P.; Nielsen, D. S.; Wegener, G.; Hansen, A. K.; Sørensen, D. B. Dietary Magnesium Deficiency Affects Gut Microbiota and Anxiety-Like Behavior in C57BL/6N Mice. *Acta Neuropsychiatry* **2015,** *2705,* 307–311.

Raap, D. K.; Evans, S.; Garcia, F.; Li, Q.; Muma, N. A.; Wolf, W. A.; Van De Kar, L. D. Daily Injections of Fluoxetine Induce Dose-Dependent Desensitization of Hypothalamic 5-HT1A Receptors: Reductions in Neuroendocrine Responses to 8-OH-DPAT and in Levels of Gz And Gi Proteins. *J. Pharmacol. Exp. Ther.* **1999,** *288,* 98–106.

Raiteri, M.; Leardi, R.; Marchi, M. Heterogeneity of Presynaptic Muscarinic Receptors Regulating Neurotransmitter Release in the Rat Brain. *J. Pharmacol. Exp. Ther.* **1984,** *228,* 209–214.

Regier, D. A.; Rae, D. S.; Narrow, W. E.; Kaelber, C. T.; Schatzberg, A. F. Prevalence of Anxiety Disorders and their Comorbidity with Mood and Addictive Disorders. *Br. J. Psychiatry* **1998**.

Rhoads, D. E.; Huggler, A. L.; Rhoads, L. J. Acute and Adaptive Motor Responses to Caffeine in Adolescent and Adult Rats. *Pharmacol. Biochem. Behav.* **2011,** *99,* 81–86.

Ribeiro, J. A.; Sebastiao, A. M. Caffeine and Adenosine. *J. Alzheimer Dis.* **2010,** *20,* S3–S15.

Risch, S. C.; Cohen, R. M.; Janowsky, D. S.; Kalin, N. H.; Sitaram, N.; Gillin, J. C.; Murphy, D. L. Physostigmine Induction of Depressive Symptomatology in Normal Human Subjects. *Psychiatry Res.* **1981**, *4*, 89–94.

Rogers, P. J.; Heatherley, S. V.; Mullings, E. L.; Smith, J. E. Faster but not Smarter: Effects of Caffeine and Caffeine Withdrawal on Alertness and Performance. *Psychopharmacoly* **2013**, *226*, 229–240.

Rogers, P. J.; Hohoff, C.; Heatherley, S. V.; Mullings, E. L.; Maxfield, P. J.; Evershed, R. P.; Nutt, D. J. Association of the Anxiogenic and Alerting Effects of Caffeine with Adora2a and Adora1 Polymorphisms and Habitual Level of Caffeine Consumption. *Neuropsychopharmacology* **2010**, *35*, 1973.

Rosenbaum, J. F. Attitudes Toward Benzodiazepines over the Years. *J. Clin. Psychiatry* **2005**, *66*, 4–8.

Rossi, A.; Barraco, A.; Donda, P. Fluoxetine: A Review on Evidence Based Medicine. *Ann. Gen. Hosp. Psychiatry* **2004**, *3*, 2.

Rowlett, J. K.; Platt, D. M.; Lelas, S.; Atack, J. R.; Dawson, G. R. Different GABAA Receptor Subtypes Mediate the Anxiolytic, Abuse-Related, and Motor Effects of Benzodiazepine-Like Drugs in Primates. *Proc. Natl. Acad. Sci. U.S.A.* **2005**, *102*, 915–920.

Roy-Byrne, P. P.; Sullivan, M. D.; Cowley, D. S.; Ries, R. K. Adjunctive Treatment of Benzodiazepine Discontinuation Syndromes: A Review. *J. Psychiatry Res.* **1993**, *27*, 143–153.

Rudolph, T.; Knudsen, K. A Case of Fatal Caffeine Poisoning. *Acta Anaesthesiol. Scandinavica* **2010**, *54*, 521–523.

Rusted, J. Caffeine and Cognitive Performance: Effects on Mood or Mental Processing. *Caffeine Behav. Curr. View. Res. Trend.* **1999**, 221–229.

Salzman, C. Treatment of the Agitation of Late-life Psychosis and Alzheimer's Disease. *Eur. Psychiatry.* **2001**, *16*, 25–28.

Samuelsson, G.; McCamish-Svensson, C.; Hagberg, B.; Sundström, G.; Dehlin, O. Incidence and Risk Factors for Depression and Anxiety Disorders: Results from a 34-year Longitudinal Swedish Cohort Study. *Aging Mental Health* **2005**, *9*, 571–575.

Sartori, S. B.; Whittle, N.; Hetzenauer, A.; Singewald, N. Magnesium Deficiency Induces Anxiety and HPA Axis Dysregulation, Modulation by Therapeutic Drug Treatment. *Neuropharmacology* **2012**, *621*, 304–312.

Savitz, J.; Lucki, I.; Drevets, W. C. 5-HT1A Receptor Function in Major Depressive Disorder. *Prog. Neurobiol.* **2009**, *88*, 17–31.

Savoca, M. R.; MacKey, M. L.; Evans, C. D.; Wilson, M.; Ludwig, D. A.; Harshfield, G. A. Association of Ambulatory Blood Pressure and Dietary Caffeine in Adolescents. *Am. J. Hypertens.* **2005**, *18*, 116–120.

Sawyer, D. A.; Julia, H. L.; Turin, A. C. Caffeine and Human Behavior: Arousal, Anxiety, and Performance Effects. *J. Behav. Med.* **1982**, *5*, 415–439.

Schapira, K.; Roth, M.; Kerr, T. A.; Gurney, C. The Prognosis of Affective Disorders: The Differentiation of Anxiety States from Depressive Illnesses. *Br. J. Psychiatry* **1972**, *121*, 175–181.

Schelleman, H.; Brensinger, C. M.; Bilker, W. B.; Hennessy, S. Antidepressant-warfarin Interaction and Associated Gastrointestinal Bleeding Risk in A Case-Control Study. *PLoS One* **2011**, *6*, e21447.

Schmitt, R., Gazalle, F. K., Lima, M. S., Cunha, A., Souza, J., Kapczinski, F. The Efficacy of Antidepressants for Generalized Anxiety Disorder: A Systematic Review and Meta-analysis. *Rev. Bras. Psiquiatr.* **2005**, *27*, 18–24.

Schulz, M.; Schmoldt, A. Therapeutic and Toxic Blood Concentrations of More than 800 Drugs and Other Xenobiotics. *Pharmazie* **2003**, *58*, 447–474.

Sheline, Y. I. Hippocampal Atrophy in Major Depression: A Result of Depression-Induced Neurotoxicity? *Mol. Psychiatry* **1996**, *1*, 298–299.

Sieghart, W. Pharmacol of Benzodiazepine Receptors: An Update. *J. Psychiatr. Neurosci.* **1994**, *19*, 24.

Simon, F. E. R.; Simons, K. J. H$_1$ Antihistamines: Current Status and Future Directions. *World Allergy Org. J.* **2008**, *1*, 145–155.

Sinoff, G.; Werner, P. Anxiety Disorder and Accompanying Subjective Memory Loss in the Elderly as a Predictor of Future Cognitive Decline. *Int. J. Ger. Psychiatry* **2003**, *18*, 951–959.

Smythe, J. W.; Bhatnagar, S.; Murphy, D.; Timothy, C.; Costall, B. The Effects of Intrahippocampal Scopolamine Infusions on Anxiety in Rats as Measured by the Black–White Box Test. *Brain. Res. Bull.* **1998**, *45*, 89–93.

Sprouse, J.; Braselton, J.; Reynolds, L.; Clarke, T.; Rollema, H. Activation of Postsynaptic 5-HT(1A) Receptors by Fluoxetine Despite the Loss of Firing-Dependent Serotonergic Input: Electrophysiological and Neurochemical Studies. *Synapse* **2001**, *41*, 49–57.

Stanislav, S. W.; Fabre, T.; Crismon, M. L.; Childs, A. Buspirone's Efficacy in Organic-induced Aggression. *J. Clin. Psychopharmacol.* **1994**.

Stanton, A. L.; Revenson, T. A.; Tennen, H. Health Psychology: Psychological Adjustment to Chronic Disease. *Annu. Rev. Psychol.* **2007**, *58*, 565–592.

Starcevic, V.; Sammut, P.; Berle, D.; Hannan, A.; Milicevic, D.; Moses, K.; Eslick, G. D. Can Levels of a General Anxiety-prone Cognitive Style Distinguish Between Various Anxiety Disorders? *Comp. Psychiatry* **2012,** *53,* 427–433.

Stark, P.; Fuller, R. W.; Wong, D. T. The Pharmacologic Profile of Fluoxetine. *J. Clin. Psychiatry* **1985**.

Steffens, D. C.; McQuoid, D. R. Impact of Symptoms of Generalized Anxiety Disorder on the Course of Late-life Depression. *Am. J. Ger. Psychiatry* **2005,** *13,* 40–47.

Steimer, T. The Biology of Fear-and Anxiety-related Behaviors. *Dial. Clin. Neurosci.* **2002,** *4,* 231.

Stein, M. B.; Kerridge, C.; Dimsdale, J. E.; Hoyt, D. B. Pharmacotherapy to Prevent PTSD: Results From a Randomized Controlled Proof-of-concept Trial in Physically Injured Patients. *J. Traum. Stress* **2007,** *20,* 923–932.

Supinski, G. S.; Deal Jr, E. C.; Kelsen, S. G. The Effects of Caffeine and Theophylline on Diaphragm Contractility. *Am. Rev. Resp. Dis.* **1984,** *130,* 429–433.

Svenningsson, P.; Nomikos, G. G.; Fredholm, B. B. The Stimulatory Action and the Development of Tolerance to Caffeine is Associated with Alterations in Gene Expression in Specific Brain Regions. *J. Neurosci.* **1999,** *19,* 4011–4022.

Svenningsson, P.; Nomikos, G. G.; Ongini, E.; Fredholm, B. B. Antagonism of Adenosine A2A Receptors Underlies the Behavioral Activating Effect of Caffeine and is Associated with Reduced Expression of Messenger RNA for NGFI-A and NGFI-B in Caudate–putamen and Nucleus Accumbens. *Neuroscience* **1997,** *79,* 753–764.

Takase, B.; Akima, T.; Uehata, A.; Ohsuzu, F.; Kurita, A. Effect of Chronic Stress and Sleep Deprivation on Both Flow-Mediated Dilation in the Brachial Artery and the Intracellular Magnesium Level in Humans. *Clin. Cardiol.* **2004,** *274,* 223–227.

Thase, M. E. Antidepressant Treatment of the Depressed Patient with Insomnia. *J. Clin. Psychiatry* **1999,** *60,* 28–31.

Tucci, S.; Cheeta, S.; Seth, P.; File, S. E. Corticotropin Releasing Factor Antagonist, A-helical CRF 9–41, Reverses Nicotine-induced Conditioned, but not Unconditioned, Anxiety. *Psychopharmacology* **2003,** *167,* 251–256.

Tully, P. J.; Baker, R. A.; Knight, J. L. Anxiety and Depression as Risk Factors for Mortality After Coronary Artery Bypass Surgery. *J. Psychosom. Res.* **2008,** *64,* 285–290.

Tyrer, P. J.; Lader, M. H. Response to Propranolol and Diazepam in Somatic and Psychic Anxiety. *Br. Med. J.* **1974,** *2,* 14–16.

Ucok, A.; Gaebel, W. Side Effects of Atypical Antipsychotics: A Brief Overview. *World Psychiatry* **2008,** *7,* 58–62.

Uhlenhuth, E. H.; Starcevic, V.; Warner, T. D.; Matuzas, W.; McCarty, T.; Roberts, B.; Jenkusky, S. A General Anxiety-prone Cognitive Style in Anxiety Disorders. *J. Affect. Dis.* **2002,** *70,* 241–249.

Umemura, T.; Ueda, K.; Nishioka, K.; Hidaka, T.; Takemoto, H.; Nakamura, S.; Jitsuiki, D.; Soga, J.; Goto, C.; Chayama, K.; Yoshizumi, M. Effects of Acute Administration of Caffeine on Vascular Function. *Am. J. Cardiol.* **2006,** *98,* 1538–1541.

Van Hout, H. P.; Beekman, A. T.; De Beurs, E.; Comijs, H.; Van Marwijk, H.; De Haan, M.; Deeg, D. J. Anxiety and the Risk of Death in Older Men and Women. *Br. J. Psychiatry* **2004,** *185,* 399–404.

van Rijnsoever, C.; Täuber, M.; Choulli, M. K.; Keist, R.; Rudolph, U.; Mohler, H.; Fritschy, J. M.; Crestani, F. Requirement of α5-GABAA Receptors for the Development of Tolerance to the Sedative Action of Diazepam in Mice. *J. Neurosci.* **2004,** *24,* 785–6790.

van Stegeren, A. H.; Goekoop, R.; Everaerd, W.; Scheltens, P.; Barkhof, F.; Kuijer, J. P.; Rombouts, S. A. Noradrenaline Mediates Amygdala Activation in Men and Women During Encoding of Emotional Material. *Neuroimage* **2005,** *24,* 898–909.

Wess, J. Novel Insights into Muscarinic Acetylcholine Receptor Function Using Gene Targeting Technology. *Trend. Pharmacol. Sci.* **2003,** *24,* 414–420.

Wieland, H. A.; Lüddens, H.; Seeburg, P. H. A Single Histidine in GABAA Receptors is Essential for Benzodiazepine Agonist Binding. *J. Biol. Chem.* **1992,** *267,* 1426–1429.

Wonnacott, S. Presynaptic Nicotinic ACh Receptors. *Trend. Neurosci.* **1997,** *20,* 92–98.

World Health Organization. Depression and Other Common Mental Disorders: Global Health Estimates, 2017.

Wright, G. A.; Baker, D. D.; Palmer, M. J.; Stabler, D.; Mustard, J. A.; Power, E. F.; Borland, A. M.; Stevenson, P. C. Caffeine in Floral Nectar Enhances a Pollinator's Memory of Reward. *Science* **2013,** *339,* 1202–1204.

Zarrindast, M. R.; Valizadegan, F.; Rostami, P.; Rezayof, A. Histaminergic System of the Lateral Septum in the Modulation of Anxiety-like Behavior in Rats. *Eur. J. Pharmacol.* **2008,** *583,* 108–114.

Zhang, W.; Yamada, M.; Gomeza, J.; Basile, A. S.; Wess, J. Multiple Muscarinic Acetylcholine Receptor Subtypes Modulate Striatal Dopamine Release, as Studied with M1–M5 Muscarinic Receptor Knock-out Mice. *J. Neurosci.* **2002**, *22*, 6347–6352.

Zhao, Y.; Ma, R.; Shen, J.; Su, H.; Xing, D.; Du, L. A Mouse Model of Depression Induced by Repeated Corticosterone Injections. *Eur. J. Pharmacol.* **2008**, *581*, 113–120.

# CHAPTER 10

# Antimanic Drugs

AMAN UPAGANLAWAR*, ABDULLA SHERIKAR, and CHANDRASHEKHAR UPASANI

*SNJB's SSDJ College of Pharmacy, Maharashtra, India*

*Corresponding author. E-mail: amanrxy@gmail.com*

## ABSTRACT

Mania is a severe mental condition which is characterized by progression of inappropriate euphoria, hyperactive speech, locomotor behavior, insomnia, and better appetite. As the incident builds, the person experiences racing thoughts, tremendous agitation, delusions, hallucinations, and fear. Well-developed mania is a psychotic syndrome associated with hallucinations and depression. Antimanic drugs (mood stabilizers) stabilize mood by managing symptoms of mania such as irregular psychological state of excitement. Mood stabilizers are the drugs commonly employed to reversing the mood disorders such as intense and sustained mood shifts, typically bipolar disorder type I or type II or schizophrenia. Mood-stabilizing agents are useful for suppression of symptoms of bipolar disorders such as mania and depression, average individual character disturbances, and affective syndromes such as schizophrenia. There are several diverse kinds of antimanic drugs ranging from anticonvulsants to lithium. The most effective antimanic medications, which are recommended for the treatment of bipolar disorder includes simple salts, lithium chloride, or lithium carbonate. This chapter mainly deals with the pharmacological based mechanistic approach for the treatment of manic disorders.

## 10.1 INTRODUCTION

The definition of mood disorder involves alterations in the capability to control not only condition of mind, performance but also distress. The striking feature of depressive disorder involves lack of a manic or hypomanic experience and existence of frequent depression in families of bipolar individuals which is absent in bipolar disorder. As per the data presented by World Health Organization (WHO), the rank of unipolar major depression is the fourth among all diseases with respect to failure of compromised living years and by 2020 it would have proposed to be second in the world (Dennis et al., 2005). Change in mood condition (i.e., affective disorders) is manifested as mania relating joy or irritable mood, disturbed sleep, hyperactivity, uncontrollable thinking, and speech, as well as uncontrolled or violent behavior. On the contrarily, depression is characterized as unhappiness, failure of concentration and enjoyment, unimportance, blemish, physical and mental slowing, melancholia, and self-destructive ideation (Tripathi, 2013).

Generally, the bipolar disorder affects not only 1% of the inhabitant individuals classically among 20 and 30 years of age but also late childhood and early adolescence also experience premorbid symptoms. The rate of incidence of bipolar disorder is independent on sex, but the experiences of depression are more in women than men and men are more manic over a lifetime (Dennis et al., 2005). Bipolar disorder is recurrent involving several life span episodes (George and Thomas, 2003). Characteristically, the bipolar disorder involves erratic mood swing or mania (or hypomania) to sadness, elevated psychomotor doings with mania, unnecessary communal sociability, reduced will power for slumber, unprompted and destruction in decision and liberal, high-flying, and occasionally short-tempered mood. The aggravated condition of mania is characterized by occurrence of delusions and fearful thinking identical to schizophrenia. On the other hand, 50% cases of bipolar disorder experience a combination of psychomotor disturbances and establishment of dysphoria, nervousness, and bad temper (Dennis et al., 2005; Golan et al., 2012).

Bipolar I disorder is defined by existence of more than one manic episode, whereas bipolar II disorder involves hypomanic episodes and major depression. Bipolar I patients may experience both mania and hypomania along with major depressive disorders conversely, bipolar II patients experience only hypomania and key depressive attacks (Dennis et al., 2005). The bipolar II disorder involves the lack of the complete criterion for mania and the essential persistent depression is differentiated by attacks of miner stimulation and elevated power (hypomania). Less severe mood swings with numerous periods of highs and lows is not a mania or major depression but it is diagnosed as a cyclothymic disorder. The cyclothymic syndrome is highlighted with many short duration hypomanic periods and irregular gathering of depressive symptoms. It takes at least 2 years for the detection of severe mood fluctuations (Dennis et al., 2005).

Mania is a phase of steady liberal, prominent, or bad-tempered mood characterized by exaggerated self-esteem or high-flying thinking, marked cognitive changes such as distractibility or air travel of dream and thoughts, pressured speech, reduced will power for sleep, agitation or increased activity, and excessive potential harmful pleasurable activities. When similar signs are existed with minor alterations of functioning or require hospitalization, the condition is termed hypomania. The condition of mania is lasted for a week and hypomania for a period of 4 days. Mania can be very severe with psychotic experiences, extreme destructive behaviors and requires hospitalization, whereas hypomania may be brief, relatively mild, hard to detect or recall without need of hospitalization. Mixed episodes refer to seemingly simultaneous or rapidly alternating manic and depressive symptoms (George and Thomas, 2003; Laurence and Bennett, 2003; Sharma and Sharma, 2017).

The period required for appearance of chronic manic condition varies from hours to 8–12 months. The condition in which there are more than four attacks of either depression or mania per year is called as rapid cycling and the rate of existence of rapid cycling is 15% in women with the cases alterations thyroid function and expanded antidepressant treatment. On an average of 50% patients with rapid cycling are associated with continued hardness of work activities and psychosocial performance (Dennis et al., 2005). The fundamental mechanism in the alterations of physiological deep and periodic mood swings of bipolar disorder is remains unclear. The pathophysiology of bipolar disorder includes an alteration of volume of amygdala, elevated white matter responsiveness, activation and cellular modifications in membrane bound $Na^+/K^+$ ATPase activation, and disoriented signal transduction mechanisms such as phosphoinositol system and GTP-binding proteins. Neurophysiologic studies suggest that patients with bipolar disorder have altered circadian rhythmicity (Dennis et al., 2005). A drug having tendency to elevate dopamine or norepinephrine activity may cause precipitation of mania, whereas those having tendency to diminish activity of dopamine or norepinephrine may alleviate the condition of mania (Katzung et al., 2009). Antimanic (mood stabilizer) are useful in controlling the condition of mania and to break into cyclic affective disorders (Tripathi, 2013). This chapter focuses more on general perspective and brief information about drugs used in the management of mood disorders.

## 10.2 CLASSIFICATION OF ANTIMANIC DRUGS

As stated earlier, the condition of mania is highlighted by excessive desire, craziness, pathological ambitions, and excitements. There are various classes of drugs are available which are useful as antimanics and are given below (Table 10.1).

**TABLE 10.1**  Classification of Antimanic Drugs.

| Class | Examples |
|---|---|
| Lithium salts | Lithium |
| Antiepileptic drugs | Carbamazepine, gabapentin, zonisamide, topiramate, valproic acid, lamotrigine |
| Sedative BZDs | Lorazepam, clonazepam |
| Antipsychotic drugs | Olanzapine, quetiapine, risperidone, aripiprazole, ziprasidone |

**TABLE 10.1**  *(Continued)*

| Class | Examples |
|---|---|
| glutamate liberation inhibitors, AMPA stimulator | Riluzole |
| Novel protein kinase C inhibitors | Tamoxifen, ruboxistaurin, rottlerin, balanol, aprinocarsen |
| Inhibition of GSK | Zinc, indirubins, maleimides, hymenialdesine, paullones, thiadiazolidones, synthetic phosphorylated peptide, azole derivatives |
| Blockers of NMDA receptors | Ketamine, memantine, felbamate, zinc |
| Inhibitors of synthesis of glucocorticoid | Ketoconazole, aminoglutethimide, metyrapone |
| Bcl-2 enhancer | Pramipexole |

*Source*: Adapted from Bebchuk et al. (2000), Katzung et al. (2009), and Tripathi (2009).

## 10.3  ANTIMANIC DRUGS

### 10.3.1  *LITHIUM SALTS*

#### 10.3.1.1  *PROPERTIES*

Lithium ($Li^+$) characteristically is a lightest of the alkali metal compound (group Ia) and is a monovalent cation having similar properties as that of sodium ($Na^+$) and potassium ($K^+$) ions. An assay of lithium ion in biological fluid can be carried out voluntarily (Brunton et al., 2011; Wilson, 2004). The peculiar characteristic of lithium includes nonsedative, depressant or euphoriant making different from other psychotropic drugs (Jefferson et al., 1983) and transversely it has a comparatively tiny slope of circulation around biological membranes different from $Na^+$ and $K^+$ (Brunton et al., 2011). The various salts of lithium are shown in Table 10.2.

**TABLE 10.2**   Various Salts of Lithium with Their Uses.

| Lithium salt | Uses |
|---|---|
| Lithium urate | Treatment of gout |
| Lithium bromide | Sedative and anticonvulsant |
| Lithium chloride | Substitute for cardiac patient (hygroscopic and irritating to mucosa of gastrointestinal tract, therefore not used therapeutically) |
| Lithium carbonate (tetrahydrate) | Antimanic (therapeutically used) |
| Lithium citrate | Antimanic (therapeutically used) |

*Source*: Adapted from Cade (1949), Mitchell et al. (1999), and Wilson and Gisvold (2004).

### 10.3.1.2   PHARMACOKINETICS

The digestive absorption of lithium is ample and the maximum plasma concentration is observed about 6–8 h (Table 10.3).

**TABLE 10.3**   Pharmacokinetic Profile of Lithium.

| Pharmacokinetics | Effects |
|---|---|
| Absorption | • Slow but well or absolute within 6–8 h;<br>• Attain peak plasma levels within half an hour to 2 h |
| Distribution | • Initially scattered in the extracellular fluid followed by slow accumulation on different tissues that is, in total body water;<br>• Lack of plasma protein binding affinity<br>• Exhibit smaller concentration gradient across plasma membranes as compared to $Na^+$ and $K^+$ concentration;<br>• It shows variability in volume of distribution from initial concentration 0.5 L/kg to end concentration 0.7–0.9 L/kg;<br>• It has less blood brain barrier crossing tendency;<br>• It has 40–50% CSF and brain concentration respectively after steady state |

**TABLE 10.3**   *(Continued)*

| Pharmacokinetics | Effects |
|---|---|
| Metabolism | • No metabolism |
| Excretion | • Almost completely (95%) in urine;<br>• Clearance of lithium is almost 20% of creatinine as the reabsorption of filtered (80%) lithium is ranging between 15 and 30 mL/min from proximal convoluted tubules<br>• Plasma half-life is ranging in between 20 and 24 h;<br>• It is excreted by multiple route such as fecal (less than 1%), sweat gland route (4–5%) as well as secreted in breast milk;<br>• Secretion in saliva is twice those in plasma and follows similar rate of secretion from tears and plasma;<br>• lithium clearance is enhanced by $Na^+$ loading whereas its retention with $Na^+$ depletion<br>• Eliminated variably from the body as 40% of ingested lithium is quickly cleared during first 10 h followed by slow clearance of 60% over a period of 2 weeks (14 days). Due to the dual nature of elimination, it is essential to practice to determine the serum lithium level 12 h after the preceding dose |
| Dose regimen | • 0.6–1.4 mEq/L targeted plasma concentration;<br>• Steady state concentration; 0.5–0.8 mEq/L is considered optimum for maintenance therapy in bipolar disorder;<br>• 0.8–1.1 mEq/L is required for episodes of mania;<br>• Serum toxic levels is 1.5 mEq/L or above |

*Source*: Adapted from Katzung et al. (2009), Plenge et al. (1994), Riedl et al. (1997), Siegel (1998), Sharma and Sharma (2017), and Tripathi, (2013).

### 10.3.1.3  MECHANISM OF ACTION

Lithium interferes and blocks the breakdown of inositol-1-phosphate by interfering the activity of inositol monophosphatase (IP). This results in the reduced supply of free inositol for renewal of membrane phosphatidyl-inositides, the source of $IP_3$ and DAG as shown in Figure 10.1 (Katzung et al., 2009; Tripathi, 2013).

### 10.3.1.4  PHARMACOLOGICAL ACTION

Alternatively, lithium develops action potentials transversely replacing the membrane $Na^+$. Lithium slows down $Na^+$ substitution causing slow lithium-$Na^+$ exchange. At a concentration of 1 mmol/L, lithium fails to affect the $Na^+/Ca^{2+}$ exchange process or the $Na^+/K^+$ ATPase sodium pump (Katzung et al., 2009).

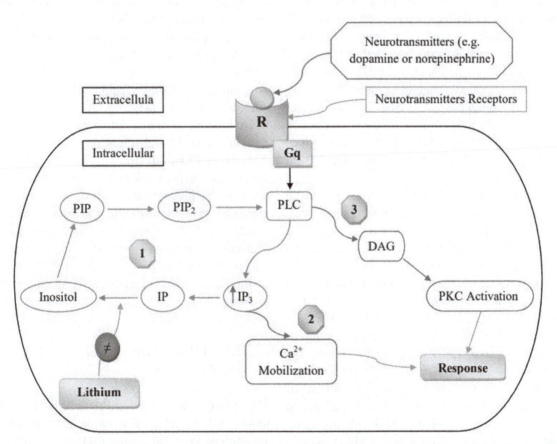

**FIGURE 10.1**  Mode of action of lithium. 1: How the IP3 and PLC are re-synthesized; 2 and 3: Mechanism by which PLC activation produces the response; $\neq$: Inhibition; IP: Inositol monophosphate; $IP_3$: Inositol triphosphate; PIP: Phosphatidylinositol monophosphate; PLC: Phospholipase C; DAG: Diacyl glycerol; PKC: Protein kinase C; Gq: Coupling Gq protein.

*Source*: Adapted from Tripathi (2013).

Contradictorily, lithium bears some properties of serotonin producing unpredictable effects on norepinephrine. The lithium produces antimanic effects by decreasing norepinephrine and dopamine turnover. Lithium blocks the progression of dopamine receptor super sensitivity similar to that of chronic antipsychotic therapy (Katzung et al., 2009). Lithium (concentrations of 1–10 mEq per liter) exhibits antagonistic effects on the liberation of norepinephrine (NE) and dopamine excluding serotonin from nerve terminals. Lithium momentarily enhances the discharge of serotonin from limbic system (Baldessarini et al., 1988). Lithium promotes the choline uptake into nerve terminals causing elevated synthesis of acetylcholine (Katzung et al., 2009).

Lithium also inhibits synthesis of $IP_1$ (inositol monophosphate) from $IP_2$ and further its conversion to inositol which causes reduction of phosphatidylinositol-4,5-bisphosphate ($PIP_2$), the membrane precursor of IP3 and DAG (Katzung et al., 2009).

Lithium has antidiuretic action on distal tubules thereby precipitating diabetes insipidus. It exerts insulin like action on glucose metabolism. Ingestion of lithium causes an increased level of leukocyte whereas reduces that of thyroxine by obstructing the iodination of tyrosine (Tripathi, 2009). As glycogen synthase kinase-3b (GSK-3b) play an important role in the neuronal and nuclear regulatory processes like restrictive expression of the regulatory protein β-catenin. Treatment with lithium and valproate causes prevention of glycogen synthase kinase-3b (GSK-3b) (Chen et al., 1999b; Manji et al., 1999b). By acting on peripheral targets, lithium also modifies actions of vasopressin and thyroid-stimulating hormone through adenylyl cyclase or phospholipase C second messenger system (Manji et al., 1999b; Urabe et al., 1991).

Lithium has ability to impede the action of both excitatory and inhibitory G proteins (Gs and Gi) by maintaining them in their less active αβδ trimeric state (Jope, 1999; Manji et al., 1999b). Treatment with lithium again and again reduces protein kinase function such as PKC in brain (Jope, 1999; Lenox and Manji, 1998). Myristoylated alanine-rich PKC-kinase substrate (MARCKS) protein acts as most important substrate for cerebral PKC which involved in synaptic and neuronal plasticity. The expression of MARCKS protein is reduced by both lithium and Valproic acid whereas carbamazepine or antipsychotic, antidepressant or sedative drugs (Watson, 1996; Watson, 1998).

Lithium and valproic acid both interact with nuclear regulatory factors such as increase in DNA binding of transcription factor activator protein-1 (AP-1) and alters the expression of AMI-1β or PEBP-2β (transcription factor) that affect gene expression (Chen, 1999a). Treatment with lithium and valproic acid causes increased expression of the regulatory protein B-cell lymphocyte protein-2 (bcl-2) providing protection against neuronal degeneration (Chen, 1999c; Manji et al., 1999c). The effect of lithium on various enzymes is summarized in Table 10.4.

### 10.3.1.5  MONITORING OF SERUM LEVEL AND DOSE OF DRUG

The safety margin of lithium (therapeutic index) is low down, so it requires continuous assessment of serum concentration of lithium (therapeutic drug monitoring). After repetitive administration of lithium, it is an essential and frequent practice which allows the measurement of the fluctuation in blood concentration from samples withdraws between 10 and 12 h

(Maj et al., 1986). Clinically, the analysis of concentration of lithium in serum helps in the dose adjustment for treating the acute mania and for prophylactic maintenance of mania (Katzung et al., 2009). The therapeutic dose, maintenance, and toxic dose for lithium is summarized in Table 10.5.

**TABLE 10.4** Effects of Lithium on Various Enzymes.

| Enzyme | Roles of enzyme | Effects on enzyme |
|---|---|---|
| Inositol monophosphatase | Inositol recycling rate limiting enzyme | Depletion of substrate for IP3 production |
| Inositol polyphosphate 1-phosphatase | Recycling of inositol | Depletion of substrate for IP3 production |
| Bisphosphate nucleotidase | Production of AMP | Lithium-induced nephrogenic diabetes insipidus |
| Fructose 1,6-bisphosphate | Synthesis of glucose | Inhibition |
| Phosphoglucomutase | Glycogenolysis | Inhibition |
| Glycogen synthase kinase-3 | Constitutively limit neurotrophic and neuroprotective processes | Inhibition |

*Source*: Adapted from Katzung et al. (2009).

**TABLE 10.5** Outline for Lithium Dose Range.

| Serum dose range | Use/precaution |
|---|---|
| 0.6–1.25 mEq/L | Effective and acceptably safe |
| 0.9–1.1 mEq/L | For attenuating acute mania or hypomania |
| 0.6–0.75 mEq/L | Adequate and safe concentration for long-term use for alleviating the recurrent manic-depressive illness |

**TABLE 10.5** *(Continued)*

| Serum dose range | Use/precaution |
|---|---|
| 0.5–0.8 mEq/L | Tolerated dose or maintenance dose in bipolar disorder |
| 0.4–0.9 mEq/L | Follow a clear dose-effect relationship |
| More than 1.5 mEq/L | Toxic symptoms |
| 900–1500 mg of lithium carbonate/day | For outdoor patients |
| 1200–2400 mg of lithium carbonate/day | For manic patients which are hospitalized |

*Source*: Reprinted with permission from Katzung et al. (2009), Maj et al. (1986), and Tripathi (2013).

### 10.3.1.6 THERAPEUTIC USES

Lithium salts are first and foremost used as a psychiatric medication. The therapeutic uses of lithium are listed in Figure 10.2.

### 10.3.1.7 CLINICAL TOXICITY

Intentional or unintentional acute and chronic use of lithium causes sudden increase in the toxic concentration of lithium which is correlated with the appearance of toxic effects. Lithium poisoning is classified as either acute, acute on chronic, or chronic. Acute poisoning of lithium is an accidental type in which the patients are not actually treated with lithium but occurs due to unintentional ingestion of lithium. Acute or chronic poisoning is an intentional type and occurs due to ingestion of high dose of lithium than prescribed dose. On the other hand, chronic poisoning due to lithium administration is observed fundamentally due to either decrease in renal function or elevation in the dose of lithium (Ellenhorn et al., 1997)

The clinical toxicity of lithium is summarized in Figures 10.3 and 10.4 (Baldessarini

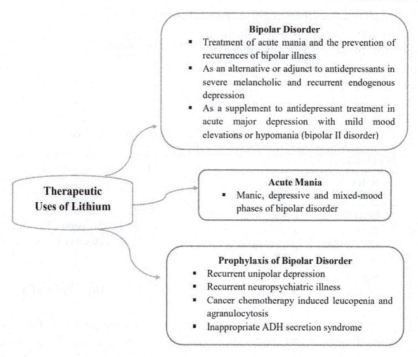

**FIGURE 10.2**  Therapeutic uses of lithium.

*Source*: Adapted from from Goodwin and Jamison (1990), Katzung et al. (2009), Licht (1998), Tohen and Zarate (1998), Tripathi (2013), and Tondo et al. (2001b).

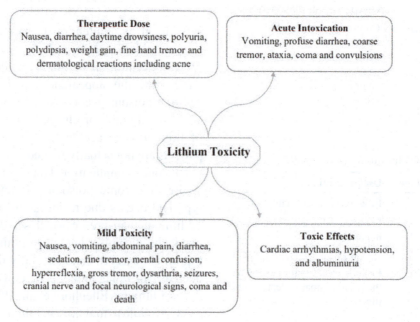

**FIGURE 10.3**  The toxicity of lithium.

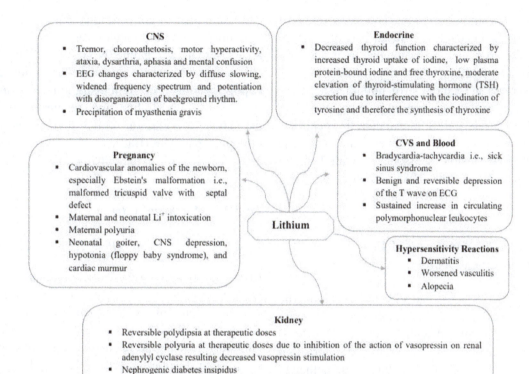

**CNS**
- Tremor, choreoathetosis, motor hyperactivity, ataxia, dysarthria, aphasia and mental confusion
- EEG changes characterized by diffuse slowing, widened frequency spectrum and potentiation with disorganization of background rhythm.
- Precipitation of myasthenia gravis

**Endocrine**
- Decreased thyroid function characterized by increased thyroid uptake of iodine, low plasma protein-bound iodine and free thyroxine, moderate elevation of thyroid-stimulating hormone (TSH) secretion due to interference with the iodination of tyrosine and therefore the synthesis of thyroxine

**Pregnancy**
- Cardiovascular anomalies of the newborn, especially Ebstein's malformation i.e., malformed tricuspid valve with septal defect
- Maternal and neonatal Li$^+$ intoxication
- Maternal polyuria
- Neonatal goiter, CNS depression, hypotonia (floppy baby syndrome), and cardiac murmur

**Lithium**

**CVS and Blood**
- Bradycardia-tachycardia i.e., sick sinus syndrome
- Benign and reversible depression of the T wave on ECG
- Sustained increase in circulating polymorphonuclear leukocytes

**Hypersensitivity Reactions**
- Dermatitis
- Worsened vasculitis
- Alopecia

**Kidney**
- Reversible polydipsia at therapeutic doses
- Reversible polyuria at therapeutic doses due to inhibition of the action of vasopressin on renal adenylyl cyclase resulting decreased vasopressin stimulation
- Nephrogenic diabetes insipidus
- Chronic interstitial nephritis and glomerulopathy with nephrotic syndrome
- Decreased glomerular filtration rate but no marked azotemia or renal failure
- Transient increase in the excretion of 17-hydroxycorticosteroids, Na$^+$, K$^+$ and water
- Edema due to Na$^+$ retention

**FIGURE 10.4** The clinical toxicity of lithium.

et al., 1988; Bauer and Whybrow, 1990; Baumgartner et al., 1994; Boton et al., 1987; Cohen et al., 1994; Iqbal et al., 2001; Lasser and Baldessarini, 1997; Neil et al., 1976; Pinelli et al., 2002; Siegel et al., 1998).

### 10.3.1.8  LITHIUM INTOXICATION

As such the precise fundamental remedy is not available for lithium intoxication but the practice of supportive remedy is followed taking care to avoid Na$^+$ and water depletion. In case of sever condition, that is, serum lithium level more than 4 mEq/L in acute upper dose or higher than 1.5 mEq/L in chronic upper dose, the means of

dialysis procedure is frequently employed to remove the drug from the body (Brunton et al., 2011).

### 10.3.1.9  DRUG INTERACTIONS

#### 10.3.1.9.1  Diuretics

Thiazide diuretic causes 25% reduction in renal removal of lithium from the body. Thiazide diuretics such as furosemide causes Na$^+$ loss thereby promoting proximal tubular reabsorption of both Na$^+$ and lithium and rising plasma levels of lithium. An ingestion of osmotic diuretics, acetazolamide, or aminophylline causes boost in kidney excretion of lithium. The reabsorption of lithium also occurs from distal part of nephron and the drug such as triamterene may increase excretion of lithium whereas spironolactone fails to increase the excretion of lithium. The occasional practice of condensed amount of lithium with amiloride and additional diuretic agents has been employed safely to reverse the syndrome of diabetes insipidus associated with Lithium therapy (Batlle et al., 1985; Boton et al., 1987; Katzung et al., 2009; Tripathi, 2013).

### 10.3.1.9.2   Nonsteroidal Anti-inflammatory Drugs

The NSAIDs have capacity to accelerate the renal proximal tubular resorption of lithium which is correlated with increased toxic level of lithium in plasma. This prominent interaction observed principally with indomethacin, ibuprofen, naproxen, and cyclooxygenase-2 (COX-2) inhibitors and possibly less so with sulindac, aspirin, and acetaminophen (Batlle et al., 1985; Boton et al., 1987; Katzung et al., 2009; Tripathi, 2013).

### 10.3.1.9.3   CNS Acting Drugs

Neuroleptics, including haloperidol, have been frequently used along with lithium without problem; sometimes the combination of haloperidol and lithium produces marked tremor and rigidity. The neuroleptic action appears to be potentiated by lithium (Batlle et al., 1985; Boton et al., 1987; Katzung et al., 2009; Tripathi, 2013).

### 10.3.1.9.4   Other Drugs

The accumulation of lithium concentration is observed with the drugs such as tetracyclines and angiotensin-converting-enzyme (ACE) inhibitors. Lithium antagonizes the pressure response to noradrenaline (NA). Lithium has an ability to enhance insulin/sulfonylurea-induced hypoglycaemia. An administration of drugs such as succinylcholine and pancuronium to the patients who are on lithium therapy causes prolonged paralysis (Batlle et al., 1985; Boton et al., 1987; Katzung et al., 2009; Tripathi, 2013).

### 10.3.2   VALPROIC ACID

#### 10.3.2.1   MECHANISM OF ACTION

Being as a well-known anticonvulsant drug, valproic acid has wide range of actions such as inhibition of voltage-sensitive sodium channels (VSSCs), potentiation of GABA inhibitory actions, and regulation of downstream signal transduction cascades. The underlying mechanism for potentiation of GABA inhibitory actions includes the increase in its release, decrease in its reuptake, or slowing its metabolic inactivation. The mood stabilizing effect of valproic acid is related to its binding to the regulatory site of VSSCs and causes a series of event such as alteration of phosphorylation of VSSCs and prevention of flow of ions through VSSCs (Katzung et al., 2009; Tripathi, 2013).

Valproic acid acts as novel histone deacetylase (HDAC) inhibitor and regulates gene expression. Histones are the small basic proteins and DNA complex. Histones present as histone acetylases (HATs) and HDACs forms. Acetylation of histones decreased their affinity toward DNA and is a major regulator of gene expression. A discriminating increase of DNA-methyltransferase 1 concentration in GABAergic neurons due to epigenetic hypermethylation of the respective promoters causes downregulation of reelin and GAD expression in cortical interneurons of schizophrenia and bipolar disorder patients. In addition to this valproic acid causes prevention of hypermethylation of reelin promoter which is methionine based mechanism and reelin mRNA downregulation as well as causes the correction of drawbacks of social interaction (Carlos et al., 2006). Valproic acid has antimanic and mood stabilizing effect as that of lithium (Manji et al., 1993). Valproic acid decreases an expression of myristoylated alanine-rich PKC-kinase substrate (MARCKS) protein, increases appearance of the regulatory protein β-cell lymphocyte protein-2 (bcl-2) (Chen et al., 1999c; Manji et al., 1999c), and inhibits glycogen synthase kinase-3b (GSK-3b) as that of lithium (Chen et al., 1999b; Manji et al., 1999b).

### 10.3.2.2   THERAPEUTIC USES

Valproic acid, an antiepileptic drug is a drug of choice for the control of acute manic condition (Katzung et al., 2009). It is used as an optional drug to antipsychotic or benzodiazepine medicaments. Those patients who are sensitive to lithium or patients with rapid cycling are successfully treated with the use of valproic acid. In addition to this those patients who became resistance to either drugs are treated with the use of combination of lithium and valproic acid. Valproic acid is well tolerated (Tripathi, 2013). The sodium salt derivative of valproic acid, divalproex is a FDA licensed drug for long-term prophylactic treatment of bipolar disorder patients and mania. The sodium salt of valproic acid is patented as an extended-release formulation which is effective for the acute manic phase of bipolar disorder due to its capacity to minimize the risk of gastrointestinal toxicity, sedation and possibly alopecia (Bowden et al., 2004).

### 10.3.3   INHIBITORS OF GLYCOGEN SYNTHASE KINASE-3 (GSK-3)

It is a serine/threonine kinase residue which is constitutively active in cells and is disabled by signals originating from numerous signaling pathways such as protein kinase C. GSK-3 signaling plays a significant role in bipolar disorder and the chemicals such as expression serotonin, dopamine, antidepressants, and psychostimulants act as GSK-3 signaling regulators. An augmented expression of GSK-3 acts as a pro-apoptotic, on the other hand, GSK-3 inhibitors prevents apoptosis. GSK-3 signaling cascade produces both antimanic and antidepressant effects in models of depression or mania. In patients with bipolar disorder, the genetic variations in GSK-3 signaling causes sleep deprivation which varies in age related response (Carlos et al., 2006).

### 10.3.4   CARBAMAZEPINE

#### 10.3.4.1   MECHANISM OF ACTION

The carbamazepine has different binding and inhibition capacity to the alpha subunit of VSSCs than that of valproic acid (Golan, 2012; Brunton et al., 2011; Katzung et al., 2009).

#### 10.3.4.2   THERAPEUTIC USES

Being as a second line mood stabilizer, it is recommended for the treatment of bipolar mania and maintenance therapy for its recurrences (Golan, 2012; Brunton et al., 2011; Katzung et al., 2009). Carbamazepine is also used in the prophylactic therapy of acute attack of mania. In refractory patients, carbamazepine is used as single or in amalgamation with lithium or occasionally with valproic acid (Katzung et al., 2009).

### 10.3.4.3   CLINICAL TOXICITY

The potential harmful effects of carbamazepine include bone marrow depression, sedation, and neural tube defects. Due to these toxicities, carbamazepine is used as a second line agent than that of valproic acid (Golan, 2012; Brunton et al., 2011; Katzung et al., 2009).

### 10.3.5   OXCARBAZEPINE

Oxcarbazepine shares the similar structural properties as that of carbamazepine and it is a prodrug. After administration, oxcarbazepine is immediately changed to its 10-hydroxy derivative, called as monohydroxy derivative (licarbazepine). Oxcarbazepine has similar mechanism of action property as that of carbamazepine, oxcarbazepine bind to the alpha subunit of VSSCs and causes the inhibition of VSSCs which is similar to carbamazepine. Being as less sedative and more tolerable, rate of incidence of bone marrow toxicity with oxcarbazepine is less as compared to carbamazepine, therefore oxcarbazepine is used as "off label" drug by many clinicians for the manic phase of bipolar disorder (Golan, 2012; Brunton et al., 2011; Katzung et al., 2009).

### 10.3.6   LAMOTRIGINE

### 10.3.6.1   MECHANISM OF ACTION

Lamotrigine have unique mechanism of action, after binding to VSSCs causes the reduction in the liberation of excitatory glutamate

neurotransmitter (Golan, 2012; Brunton et al., 2011; Katzung et al., 2009).

### 10.3.6.2   THERAPEUTIC USES

Lamotrigine is used as long-term prophylactic management in bipolar disorder without the risk of acute mania (Calabrese et al., 2002; Goldsmith et al., 2004). Long term treatment with olanzapine is useful for the management of bipolar I disorder. (Tohen and Zarate, 2003). Lamotrigine is approved as a mood stabilizer to prevent the recurrence of both mania and depression (Golan, 2012; Brunton et al., 2011; Katzung et al., 2009).

### 10.3.6.3   CLINICAL TOXICITY

As antidepressant drug causes instability in mood and increases suicide attempts in bipolar disorder, antidepressant drugs are replaced by lamotrigine as first line agents in bipolar depression treatment. Lamotrigine is well tolerated as anticonvulsant drug and it is rarely associated with Stevens-Johnson syndrome (toxic epidermal necrolysis) and it is minimized by slow up drug during initiation of therapy and avoiding drug interactions with valproic acid (Golan, 2012; Brunton et al., 2011; Katzung et al., 2009).

### 10.3.7   RILUZOLE

Riluzole is a drug used in the management of epilepsy and also in the treatment of amyotrophic lateral sclerosis (ALS)/Lou Gehrig's disease. Riluzole has similar mode of action as that of lamotrigine that is, binds to VSSCs and prevents glutamate release inhibiting motor neuron death due to glutamate induced excitotoxicity in

ALS. As riluzole inhibit glutamate release, it is used in the management of bipolar depression, unipolar depression resistance to treatment, and anxiety disorders. The disadvantage of riluzole is that it is expensive and causes abnormalities in liver function (Golan, 2012; Brunton et al., 2011; Katzung et al., 2009).

### 10.3.8 TOPIRAMATE

Topiramate is a novel anticonvulsant drug and used in the treatment of migraine and bipolar disorder. As it causes weight loss, so it is used as addition to mood stabilizers. Topiramate is diversely acting drug making different from other anticonvulsant drug, that is, causes potentiation of GABA activity and declining glutamate activity by interfering with sodium and calcium channel function. The topiramate has weak mood stabilizing property in manic phase and maintenance phase of bipolar disorder as that of valproic acid and carbamazepine. In addition, topiramate is a weak inhibitor of carbonic anhydrase (Golan, 2012; Brunton et al., 2011; Katzung et al., 2009).

### 10.3.9 ZONISAMIDE

Zonisamide is a sulfonamide derivative belonging to the class of anticonvulsant drug and is used in management of bipolar disorder. The zonisamide has similar mode of action

**TABLE 10.6** The Diagnostic Criteria of DSM-5 and ICD-10 for Mood Disorders.

| Disorders | | Diagnostic features |
|---|---|---|
| Manic episodes | Hypomania | • Low level of mania; |
| | | • Constant disturbances in mood and behavior related to cyclothymia; |
| | | • Absence of hallucinations or delusions; |
| | | • Constant mild rise in mood for brief period (days); |
| | | • Augmented power and action; |
| | | • Increased friendliness, chattiness, over knowledge, increased sexual power; |
| | | • Mild reduction in call for sleep; |
| | | • Irritability, self-importance and ill-mannered behavior and overjoyed friendliness |
| | Mania without psychotic symptoms | • Euphoria associated with augmented energy, over action; |
| | | • Reduction in call for sleep; |
| | | • Usual loss of social inhibitions and disappointment of attention; |
| | | • Noticeable distractibility; |
| | | • Exaggerated confidence and high-flying or over-optimistic ideas |
| | Mania with psychotic symptoms | • Exaggerated confidence and high-flying ideas progressing to delusion and bad temper; |
| | | • The dangerous states of dehydration often results from anger or violence, loss of attention for eating, drinking, and individual care |

**TABLE 10.6**    *(Continued)*

| Disorders | Diagnostic features |
|---|---|
| Bipolar I disorder   Manic episode | • Abnormal and persistent high-minded, liberal or short-tempered mood; |
| | • Abnormal and persistent augmented goal-directed activity or energy last for 7 days; |
| | • Present mostly on the day and virtually each day; |
| | • Inflated self-respect or lavishness; |
| | • Insomnia (e.g., feels rested after only 3 h of sleep); |
| | • Supplementary conversational and air travel of ideas; |
| | • Concentration to insignificant or neither here nor their external stimuli; |
| | • Augmented target oriented activity or psychomotor trouble; |
| | • More possibility for painful events such as sexual indiscretions or foolish business investments |
| | • Severe injury in communal and work-related functioning; |
| | • Need hospitalization |
| Hypomanic episode | • Abnormal and persistent high-minded, liberal, or short-tempered mood; |
| | • Abnormal and persistent augmented target oriented activity or energy last for four successive days; |
| | • Observed per day; |
| | • Inflated self-worth; |
| | • Wakefulness (e.g., feels rested after only 3 h of sleep); |
| | • More conversational and air travel of ideas; |
| | • Concentration to less important or meaningless external stimuli; |
| | • Augmented target oriented activity or psychomotor trouble; |
| | • More possibility for painful events such as sexual indiscretions or foolish business investments; |
| | • Unmistakable change in unusual functioning; |
| | • The interruption in mood and activities pointed by other; |
| | • No serious episodes; |
| | • No need for hospitalization |

**TABLE 10.6** *(Continued)*

| Disorders | | Diagnostic features |
|---|---|---|
| Bipolar II disorder | Major depressive episode | • Every day depressed mood observed by either subject himself, that is, heartbreak and discourage emotions or observed by other persons that is, weeping emotions; |
| | | • Every day loss of enjoyment emotions; |
| | | • Everyday decline or augment in body weight owing to diet with altered appetite; |
| | | • Almost restlessness or hypersomnia per day; |
| | | • Psychomotor anxiety and tiredness or powerless; |
| | | • Mind-set of insignificance or unnecessary or wrong shame; |
| | | • Moderate capacity to imagine or focus or decision; |
| | | • Repeated feelings of suicide; |
| | | • Destruction in communal and professional performance |
| | Hypomanic episode | • Unusual and steady prominent, open, or short-tempered mood; |
| | | • Abnormal and persistent enlarged movement or force lasted four consecutive days; |
| | | • Exaggerated self-esteem or showiness and decrease sleep and distractibility; |
| | | • Chattier than normal and voyage of ideas; |
| | | • Augmented target oriented activity or psychomotor trouble; |
| | | • Destruction in communal and professional performance; |
| | | • No severe destruction in public or work-related functioning; |
| | | • No need for hospitalization |
| | Major depressive episode | • Every day disheartened mood observed by either subject himself, that is, heartbreak and discourage emotions or observed by other persons that is, weeping emotions; |
| | | • Every day loss of enjoyment emotions; |
| | | • Everyday decrease or increase in body weight owing to diet with altered appetite; |
| | | • Almost restlessness or hypersomnia per day; |
| | | • Psychomotor anxiety and tiredness or powerless; |
| | | • Mind-set of insignificance or unnecessary or wrong shame; |
| | | • Moderate capacity to imagine or focus or decision; |
| | | • Repeated belief of suicide; |
| | | • Destruction in communal and professional performance |

**TABLE 10.6**   *(Continued)*

| Disorders | | Diagnostic features |
|---|---|---|
| precise for bipolar and related disorders | Manic or hypomanic episode with mixed characteristic | • Well-known dysphoria or depressed mood;<br>• Diminish curiosity or happiness in all activities;<br>• Psychomotor retardation each daytime;<br>• Fatigue or failure of energy and feelings of insignificance or unnecessary or wrong blame;<br>• Feelings of suicide or death |
| | Depressive episode with mixed features | • Prominent and liberal mood;<br>• Exaggerated self-respect or lavishness;<br>• More conversational than normal and flight of dreams;<br>• Augment in energy or target-oriented activity;<br>• Destruction in communal and professional performance;<br>• Feelings of restlessness |
| Cyclothymic disorder | | • Minimum 2 years' hypomanic symptoms and disheartened symptoms which fail to meet criteria for hypomanic and major depressive episodes;<br>• Physiological symptoms of a drug are due to its abuse or due to medical error or condition such as hyperthyroidism;<br>• Destruction in communal and professional performance |
| Substance or medication-induced bipolar and related disorder | | • A prominent and constant clinical interruption in frame of mind;<br>• Characterized by high, liberal, or short-tempered mood with or without disheartened mood or strikingly reduced willpower in all or almost all activities;<br>• Developed due to substance intoxication or withdrawal or after exposure to a medication<br>• The symptoms continue for a significant period of time, that is, about 1 month after the termination of acute removal or severe intoxication or due to a history of recurrent non-substance/medication-related episodes;<br>• Destruction in communal and professional performance |
| Bipolar and related disorder due to another medical condition | | • An important and constant clinical interruption in frame of mind;<br>• Symptoms developed due to pathophysiological outcome of another medical condition;<br>• Destruction in communal and professional performance |

*Source*: American Psychiatric Association (2013) and Kramer (1979).

as that of topiramate, that is, enhancement of GABA activity and reduction of glutamate function by interfering with both sodium and calcium channels. Being a sulfonamide derivative, zonisamide rarely causes rashes such as Stevens-Johnson syndrome, or toxic epidermal necrolysis (Golan, 2012; Brunton et al., 2011; Katzung et al., 2009).

## 10.3.10  ARIPIPRAZOLE

### 10.3.10.1  PHARMACOKINETICS

Aripiprazole achieved high and steady volume of distribution (404 L or 4.9 L/kg) following an intravenous ingestion which highlighted its wide extravascular distribution. It is 99% bound to plasma protein primarily to albumin. The biotransformation of aripiprazole is carried out by three major pathways such as dehydrogenation, hydroxylation, and N-dealkylation. The dehydrogenation and hydroxylation of aripiprazole is carried out by CYP3A4 and CYP2D6 enzymes and N-dealkylation carried out by CYP3A4 (Bristol-Myers, 2005).

### 10.3.10.2  MECHANISM OF ACTION

Being a psychotropic drug, aripiprazole exhibits more binding attraction to dopamine $D_2$ and $D_3$, serotonin 5-$HT_{1A}$ and 5-$HT_{2A}$ receptors. Aripiprazole exerts indistinct antimanic effects through a combination of partial agonist activity at $D_2$ and 5-$HT_{1A}$ receptors and antagonist activity at 5-$HT_{2A}$ receptors (Bristol-Myers, 2005).

### 10.3.10.3  CLINICAL TOXICITY

The potential harmful effects of aripiprazole include neuroleptic malignant syndrome, tardive dyskinesia, stroke, hyperglycemia, orthostatic hypotension, seizure, cognitive and motor impairment, and so on (Bristol-Myers, 2005).

## 10.3.11  PROTEIN KINASE C (PKC) INHIBITOR—TAMOXIFEN

PKC is an isozyme subspecies, heterogeneously scattered throughout the body (Casabona, 1997; Tanaka and Nishizuka, 1994). The 12 isoforms of PKC are discovered showing differences in construction, subcellular localization, tissue specificity, means of opening, and substrate specificity (Serova et al., 2006). According to their activation, these isoforms are further divided into three classes that is, traditional/usual, novel, and unusual. Conventional PKC isoforms such as α, βI, βII, γ are activated by calcium and diacylglycerol (DAG), novel PKC isoforms such as δ, ε, η, θ, and μ lacking C2 calcium-binding domain are activated by DAG whereas atypical PKC isoforms such as ζ, λ/ι are activated by lipid mediators such as phosphatidylinositol 3,4,5-triphosphate which lacks both C2 and DAG-binding C1 domains (Toker et al., 1998).

### 10.3.11.1  PHARMACOKINETICS

Tamoxifen is almost 100% bound to plasma protein. Tamoxifen undergoes biotransformation process through CYP2D6 and CYP 3A4/5 to 4-hydroxytamoxifen and N-desmethyltamoxifen, respectively, which are further metabolized into endoxifen. The half-life of tamoxifen is almost a week whereas N-desmethyltamoxifen is 14 days. Tamoxifen is removed from the body by both enterohepatic circulation (feces) and glomerular filtration (urine route) (Golan, 2012; Brunton et al., 2011; Katzung et al., 2009).

## 10.3.11.2   DISTRIBUTION AND ROLE

PKC is heterogeneously distributed in the brain and fundamentally control the presynaptic and postsynaptic neurotransmission release, excitation of neurons, ongoing modification in gene appearance and plasticity. PKC is found in the cytoplasmic and cell membrane compartments and it is activated when it is translocated from cytosol to membrane. The manic patients are characterized by increased percentage of platelet membrane-bound to cytosolic PKC behavior as well as enhanced serotonin-elicited platelet PKC translocation (Huang et al., 1986, Mackay and Twelves, 2007, Mellor and Parker, 1998; Takai et al., 1997).

The manic syndrome is characterized by exaggerated PKC stimulated release of dopamine. The treatment with tamoxifen in manic syndrome causes reduction in GAP-43 phosphorylation (GAP-43, vital neuronal PKC substrates), altered MAP kinases, oxidative stress, and mitochondrial permeability. The disadvantage of tamoxifen is that it causes loss of appetite (Haim et al., 2007).

## 10.3.12   OMEGA-3 FATTY ACIDS

Omega-3 fatty acids are long-chain polyunsaturated fatty acids having multiple double bonds. They are called omega-3 because the 3 carbon atoms contain first double bond according to nomenclature from the methyl end of the fatty acid. Omega-3 fatty acids contain two principle biological active compounds namely eicosapentaenoic acid (EPA) and docosahexaenoic acid (DHA). According to an epidemiological survey, the rate of development of major depression, prenatal depression and bipolar depression is low in those people who are consuming rich diet with omega-3-fatty acids (Wani et al., 2015).

## 10.3.12.1   MECHANISM OF ACTION

The thromboxane B2 and prostaglandin E2 have a vital function in aggravation of depression and bipolar disorders. The EPA and DHA have anti-inflammatory role in depression and bipolar disorders as these compounds fight with arachidonic acid amalgamation into membrane phospholipids dropping the both cellular and plasma concentration of arachidonic acid. Additionally, they cause inhibition of synthesis of proinflammatory eicosanoids such as prostaglandins, leukotrienes, and thromboxanes by competing with arachidonic acid. Adding more to above mentioned actions, EPA and DHA also blocks the discharge of proinflammatory cytokines such as IL-1$\beta$, IL-2, IL-6, interferon $\gamma$, and TNF-$\alpha$. The underlying mechanism in the neuroprotection causes encouragement of synaptic plasticity and neurotransmission through brain derived neurotrophic factor (Parker et al., 2006).

## 10.3.12.2   THERAPEUTIC USES

The disorders like schizophrenia and bipolar depression are successfully managed by the use of omega-3-oil. Omega-3 fatty acids are used as a supplement for the treatment of bipolar disorder related depression (Wani et al., 2015).

## 10.4   MOOD DISORDERS IN DSM-5 AND ICD-10

Diagnostic and Statistical Manual of Mental Disorders-5 (DSM-5) and International Classification of Diseases-10 (ICD-10) information is used for description of each disorder with the main clinical features, and also of important but less specific associated features. Diagnostic criteria are provided in most cases, indicating the number and balance of symptoms usually required before a confident diagnosis can be made.

## 10.5 FUTURE OPPORTUNITIES AND CHALLENGES

The exploration for the advancement in screening of novel mood stabilizers with added successful and rapid acting agents with few adverse effects produces enormous impact on the value on individual's health. The currently available mood stabilizers are paying attention on the basis of their mode of action. An extensive research is going on in the advancement of more effective, molecular target-specific, and biomarker selective drug therapy based on pathophysiology of mania. This explore the need and scope for the recent advancement in different pathophysiological targets such as different receptor types, molecular and cellular targets which helps in prominent treatment of manic disorders. An individual showing different pharmacological effects to various drugs needs to identify the different biological tests for efficient drug treatment. A biomarker based approaches which are target specific may be utilized for the successful treatment. Challenges are still there to identify both types of neuronal tissue targets in brain region and their activation or inhibition at various brain region.

## 10.6 CONCLUSION

In addition to the advancement in the management of acute mania as well as bipolar disorders, there are ample scopes in the research for new efficient and cost-effective therapeutic approaches based treatment. The existing treatments are focused on the acute mania, depressive conditions, and prevention of chronic recurrence and maintenance approaches. The drugs like lithium, valproic acid and carbamazepine are better and are used as prophylactic agents for the cure of acute mania. Dealing acute bipolar depressive condition alone with lithium is insufficient. Therefore, the combination therapy of mood stabilizer with other class of drugs such as anticonvulsant, antidepressant, and antipsychotics are more effective for mania treatment.

## KEYWORDS

- **bipolar disorder**
- **mania**
- **lithium salts**
- **mood stabilizers**
- **schizophrenia**

## REFERENCES

American Pscychiatric Association. *Diagnostic and Statistical Manual of Mental Disorders Fifth Edition DSM-5*™; Washington, DC London, England, 2013; Vol. 5, pp 123–154.

Baldessarini, R. J.; Vogt, M. Release of [³H] Dopamine and Analogous Monoamines from Rat Striatal Tissue. *Cell Mol. Neurobiol.* **1988,** *8,* 205–216.

Baldessarini, R. J.; Faedda, G. L.; Suppes, T. Treatment Response in Pediatric, Adult, and Geriatric Bipolar Disorder Patients. In *Mood Disorders Across the Life Span;* Shulman, K., Tohen, M., Kutcher, S. P., Eds.; Wiley-Liss: New York, 1996; pp 299–338.

Batlle, D. C.; Von Riotte, A. B.; Gaviria, M.; Grupp, M. Amelioration of Polyuria by Amiloride in Patients

Receiving Long-Term Lithium Therapy. *N. Engl. J. Med.* **1985,** *312*, 408–414.

Bauer, M. E.; Whybrow, P. C. Rapid Cycling Bipolar Affective Disorder II. Treatment of Refractory Rapid Cycling with High-Dose Levothyroxine: A Preliminary Study. *Arch. Gen. Psychiatry* **1990,** *47*, 435–440.

Baumgartner, A.; Bauer, M.; Hellweg, R. Treatment of Intractable Non-Rapid Cycling Bipolar Affective Disorder with High-Dose Thyroxine: An Open Clinical Trial. *Neuropsychopharmacology* **1994,** *10*, 183–189.

Bebchuk, J. M.; Arfken, C. L.; Dolan-Manji, S.; Murphy, J., Hasanat, K.; Manji, H. K. A Preliminary Investigation of a Protein Kinase C Inhibitor in the Treatment of Acute Mania. *Arch. Gen. Psychiatry* **2000,** *57*, 95–97.

Boton, R.; Gaviria, M.; Batlle, D. C. Prevalence, Pathogenesis and Treatment of Renal Dysfunction Associated with Chronic Lithium Therapy. *Am. J. Kidney Dis.* **1987,** *10*, 329–345.

Bowden, C. L.; Asnis, G. M.; Ginsberg, L. D.; Bentley B.; Leadbetter, R.; White, R. Safety and Tolerability of Lamotrigine for Bipolar Disorder. *Drug Saf.* **2004,** *27*, 173–184.

Bristol-Myers Squibb Company, Princeton, NJ 08543, USA. US Patent Nos 4,734,416 and 5,006,528, 2005.

Brunton, L. L.; Chabner, B. A.; Knollmann, B. C. *Goodman and Gillman: Pharmacological Basis of Therapeutics*, 12th ed.; McGraw-Hill: Medical Publishing Division New York, 2011.

Cade, J. F. J. Lithium Salts in the Treatment of Psychotic Excitement. *Med. J. Austral.* **1949,** *2*, 349–352.

Calabrese, J. R.; Shelton, M. D.; Rapport, D. J.; Kimmel, S. E.; Elhaj, O. Long-Term Treatment of Bipolar Disorder with Lamotrigine. *J. Clin. Psychiatry* **2002,** *63* (Suppl 10), 18–22.

Carlos, A.; Zarate, Jr.; Jaskaran, S.; Husseini, K. Cellular Plasticity Cascades: Targets for the Development of Novel Therapeutics for Bipolar Disorder. *Biol. Psychiatry* **2006,** *59*, 1006–1020.

Casabona, G. Intracellular Signal Modulation: A Pivotal Role for Protein Kinase C. *Prog. Neuropsychopharmacol. Biol. Psychiatry* **1997,** *21* (3), 407–25.

Chen, G.; Yuan, P. X.; Jiang, Y. M.; Huang, L. D.; Manji, H. K. Valproate Robustly Enhances Ap-1 Mediated Gene Expression. *Brain Res. Mol. Brain Res.* **1999a,** *64*, 52–58.

Chen, G.; Huang, L. D.; Jiang, Y. M.; Manji, H. K. The Mood-Stabilizing Agent Valproate Inhibits the Activity of Glycogen Synthase Kinase-3. *J. Neurochem.* **1999b,** *72*, 1327–1330.

Chen, G.; Zeng, W. Z.; Yuan, P. X.; Huang L. D., Jiang, Y. M.; Zhao, Z. H.; Manji, H. K. The Mood-Stabilizing Agents Lithium and Valproate Robustly Increase the Levels of the Neuroprotective Protein Bcl-2 in the CNS. *J. Neurochem.* **1999c,** *72*, 879–882.

Cohen, L. S.; Friedman, J. M.; Jefferson, J. W.; Johnson, E. M.; Weiner, M. L. A Reevaluation of Risk of in Utero Exposure to Lithium. *JAMA* **1994,** *271*, 146–150.

Dennis, L. K.; Eugene, B.; Anthony, S. F. *Harrison's Principles of Internal Medicine*, 16th ed.; Mcgraw-Hill Medical Publishing Division: USA, 2005.

Ellenhorn, M. J.; Schonwald, S.; Ordog, G.; Wasserberger, J. Lithium. In *Medical Toxicology: Diagnosis And Treatment Of Human Poisoning*, Ellenhorn, M. J., Schonwald, S., Ordog, G., Wasserberger, J., Eds.; Williams and Wilkins: Baltimore, 1997; pp 1579.

George, S.; Thomas, A. W. *Handbook of Psychology*; John Wiley & Sons, Inc., 2003; Vol. 8, pp 93–118.

Golan, D. E.; Tashijan, A. H.; Armstrong, E. J.; Armstrong, A. W. *Principles of Pharmacology. The Pathologic Basis of Drug Therapy*, 3rd ed.; Wolters Kluwer/Lippincott Wiliams and Wilkins, 2012.

Goldsmith, D. R.; Wagstaff, A. J.; Ibbotson, T.; Perry, C. M. Spotlight on Lamotrigine in Bipolar Disorder. *CNS Drugs.* **2004,** *18*, 63–67.

Goodwin, F. K.; Jamison, K. R. *Manic-Depressive Illness*; Oxford University Press: New York, 1990.

Haim, E.; Peixiong, Y.; Steven, T. S. Protein Kinase C Inhibition by Tamoxifen Antagonizes Manic-Like Behavior in Rats: Implications for the Development of Novel Therapeutics for Bipolar Disorder. *Neuropsychobiology* **2007,** *55*, 123–131.

Huang, K. P.; Nakabayashi, H.; Huang, F. L. Isozymic Forms of Rat Brain $Ca^{2+}$-Activated and Phospholipid Dependent Protein Kinase. *Proc. Natl. Acad. Sci.* **1986,** *83* (22), 8535–9.

Iqbal, M. M.; Gundlapalli, S. P.; Ryan, W. G.; Ryals, T.; Passman, T. E. Effects of Antimanic Mood-Stabilizing Drugs on Fetuses, Neonates, and Nursing Infants. *South. Med. J.* **2001,** *94*, 304.

Jefferson, J. W.; Greist, J. H; Ackerman, D. L. *Lithium Encyclopedia for Clinical Practice*; American Psychiatric Press: Washington, D.C., 1983.

Jope, R. S. A Bimodal Model of the Mechanism of Action of Lithium. *Mol. Psychiatry* **1999,** *4*, 21–25.

Katzung, B. G.; Masters S. B.; Trevor, A. T. *Basic and Clinical Pharmacology*, 11th ed.; Lange Medical Publications: California, 2009.

Kramer, M. The ICD-9 Classification of Mental Disorders: A Review of its Development And Contents. *Acta Psychiatrica Scanndinavica* **1979,** *59*, 241–262.

Lenox, R. H.; Manji, H. K. The American Psychiatric Press Textbook Of Psychopharmacology. In *American Psychiatric Press*; Schatzberg, A. F., Nemeroff, C. B., Eds.; Washington, D.C., 1998; pp 379–429.

Lasser, R. A.; Baldessarini, R. J. Thyroid Hormones in Depressive Disorders: A Reappraisal of Clinical Utility. *Harv. Rev. Psychiatry* **1997,** *4*, 291–305.

Laurence, D. R.; Bennett, P. N. *Clinical Pharmacology*, 9th ed.; Churchill Livingstone: Edinburgh, 2003.

Licht, R. W. Drug Treatment of Mania: A Critical Review. *Acta Psychiatr. Scand.* **1998,** *97*, 387–397.

Maj, M.; Starace, F.; Nolfe, G.; Kemali, D. Minimum Plasma Lithium Levels Required for Effective Prophylaxis in DSM III Bipolar Disorder: A Prospective Study. *Pharmacopsychiatry* **1986,** *19*, 420–423.

Manji, H. K.; Mcnamara, R.; Chen, G.; Lenox, R. H. Signaling Pathways in the Brain: Cellular Transduction of Mood Stabilization in the Treatment of Manic-Depressive Illness. *Aust. N.Z. J. Psychiatry* **1999b,** *33* (Suppl), S65–S83.

Manji, H. K.; Moore, G. J.; Chen, G. Lithium at 50: Have the Neuroprotective Effects of this Unique Cation Been Overlooked? *Biol. Psychiatry* **1999c,** *46*, 929–940.

Mackay, H. J.; Twelves, C. J. Targeting the Protein Kinase C Family: Are we There Yet? *Natl. Rev. Cancer* **2007,** *7* (7), 554–62.

Mellor, H.; Parker, P. J. The Extended Protein Kinase C Superfamily. *Biochem. J.* **1998,** *332* (Pt 2), 281–292.

Mitchell, P. B.; Hadzi-Pavlovic, D.; Manji, H. K. Fifty Years of Treatment for Bipolar Disorder: A Celebration of John Cade's Discovery. *Aust. N.Z. J. Psychiatry* **1999,** *33* (Suppl), S1–S122.

Neil, J. F.; Himmelhoch, J. M.; Licata, S. M. Emergence of Myasthenia Gravis During Treatment with Lithium Carbonate. *Arch. Gen. Psychiatry* **1976,** *33*, 1090–1092.

Parker, G.; Gibson, N. A.; Brotchie, H.; Heruc G, Rees, A. M.; Hadzi-Pavlovic, D. Omega-3 Fatty Acids and Mood Disorders. *Am. J. Psychiatry* **2006,** *163*, 969–978.

Pinelli, J. M.; Symington, A. J.; Cunningham, K. A.; Paes, B. A. Case Report And Review of the Perinatal Implications of Maternal Lithium Use. *Am. J. Obstet. Gynecol.* **2002,** *187*, 245.

Plenge, P.; Stensgaard, A.; Jensen, H. V.; Thomsen, C.; Mellerup, E. T.; Henriksen, O. 24-Hour Lithium Concentration in Human Brain Studied by 7Li Magnetic Resonance Spectroscopy. *Biol. Psychiatry* **1994,** *36*, 511–516.

Riedl, U.; Barocka, A.; Kolem, H.; Demling, J.; Kaschka, W. P.; Schelp, R.; Stemmler, M.; Ebert, D. Duration of Lithium Treatment And Brain Lithium Concentration in Patients with Unipolar and Schizoaffective Disorder: A Study With Magnetic Resonance Spectroscopy. *Biol. Psychiatry* **1997,** *41*, 844–850.

Serova, M.; Ghoul, A.; Benhadji, K. A.; et al. Preclinical and Clinical Development of Novel Agents that Target the Protein Kinase C Family. *Semin. Oncol.* **2006,** *33* (4), 466–478.

Sharma, H. L.; Sharma, K. K. *Principles of Pharmacology*, 3rd ed.; Paras Medical Publisher: Hyderabad, 2017.

Siegel, A. J.; Baldessarini, R. J.; Klepser, M. B.; Mcdonald, J. C. Primary and Drug-Induced Disorders of Water Homeostasis in Psychiatric Patients: Principles Of Diagnosis and Management. *Harv. Rev. Psychiatry* **1998,** *6*, 190–200.

Takai, Y.; Kishimoto, A.; Inoue, M.; Nishizuka, Y. Studies on A Cyclic Nucleotide-Independent Protein Kinase and its Proenzyme in Mammalian Tissues I: Purification and Characterization of An Active Enzyme from Bovine Cerebellum. *J. Biol. Chem.* **1977,** *252* (21), 7603–7609.

Tanaka, C.; Nishizuka, Y. The Protein Kinase C Family for Neuronal Signaling. *Ann. Rev. Neurosci.* **1994,** *17*, 551–567.

Tohen, M.; Zarate, C. A. Antipsychotic Agents and Bipolar Disorder. *J. Clin. Psychiatry* **1998,** *59* (Suppl 1), 38–48.

Tohen, M.; Ketter, T. A.; Zarate, C. A.; Suppes, T.; Frye, M.; Altshuler, L.; Zajecka, J.; Schuh, L. M.; Risser, R. C.; Brown, E.; Baker, R. Olanzapine versus Divalproex Sodium for the Treatment of Acute Mania and Maintenance of Remission: 47-Week Study. *Am. J. Psychiatry* **2003,** *160*, 1263–1271.

Toker, A. Signaling Through Protein Kinase C. Front. Biosci. **1998,** *3*, D1134–D1147.

Tondo, L.; Hennen, J.; Baldessarini, R. J. Reduced Suicide Risk with Long-Term Lithium Treatment in Major Affective Illness: A Meta-Analysis. *Acta. Psychiatr. Scand.* **2001b,** *104*, 163–172.

Tripathi, K. D. *Essentials of Medical Pharmacology*, 7th ed.; Jaypee Brothers, Medical Publishers: New Delhi, 2013.

Urabe, M.; Hershmann, J. M.; Pang, X. P.; Murakami, S.; Sugawara, M. Effect of Lithium on Function and Growth of Thyroid Cells in vitro. *Endocrinology* **1991,** *129*, 807–814.

Wani, A. L.; Bhat, S. A.; Ara, A. Omega-3 Fatty Acids and the Treatment of Depression: A Review of Scientific Evidence. *Integr. Med. Res.* **2015,** *4*, 132–141.

Watson, D. G.; Lenox, R. H. Chronic Lithium-Induced Down-Regulation of MARCKS in Immortalized Hippocampal Cells: Potentiation by Muscarinic Receptor Activation. *J. Neurochem.* **1996,** *67*, 767–777.

Watson, D. G.; Watterson, J. M.; Lenox, R. H. Sodium Valproate Down-Regulates the Myristoylated Alanine-Rich C Kinase Substrate (MARCKS) in Immortalized Hippocampal Cells: A Property of Protein Kinase C-Mediated Mood Stabilizers. *J. Pharmacol. Exp. Ther.* **1998,** *285*, 307–316.

*Wilson and Gisvold's Textbook of Organic Medicinal and Pharmaceutical Chemistry*, 11th ed.; Wilson and Gisvold, J. Lippincot Co.: Philadelphia, 2004.

# CHAPTER 11

# Sedative and Hypnotic Drugs

ARUP KUMAR MISRA* and PRAMOD KUMAR SHARMA

*All India Institute of Medical Sciences, Basni Industrial Area,
MIA 2nd Phase, Basni, Jodhpur 342005, Rajasthan, India*

*Corresponding author. E-mail: arup2003m@gmail.com*

## ABSTRACT

Anxiety and insomnia are the most commonly prevalent problems worldwide. These conditions affect the psychiatric, psychosocial, and medical wellbeing. Pharmacological agents by exerting calming and depressive effect on the central nervous system reduce anxiety and also induce the onset and maintenance of sleep. Benzodiazepines are the most commonly used sedative–hypnotics agents as they have favorable adverse effect profile. In spite of linear dose–response and wider safety margin of benzodiazepines, there is always a desire to search for novel therapies, which can be invented to be ideal sedative–hypnotics with acceptable adverse effects and tolerability profile. The basic goal of this chapter is to familiarize readers with current pharmacologic agents available for the management of sleep disorders with their pros and cons. Moreover, recent new therapies of sleep disorders exploring new molecules to different receptors and or targets to find a molecule that meet the requisite for a designer hypnotic drug with minimal adverse effects.

## 11.1 INTRODUCTION

"Sleep is the best meditation" is a well-said quote by Dalai Lama, the spiritual leader. In the modern world, worry has deprived human of rest, comfort, and sleep which resorted human to consume soothing substances derived from natural sources. Adolf von Baeyer, the chemist who in the late 19th century first synthesized barbituric acid. The sedative effect of the barbiturates was produced with addition of ethyl groups (Kauffman, 1980). At the start of the 20th century (1903), barbital was first introduced followed by phenobarbital (Luminal) in 1911 as it had both hypnotic and antiepileptic effects. Barbiturates due to its sedative effects flooded the market till 1960 (Norn et al., 2015).

Leo Sternbach, a chemist of Hoffmann-La Roche, was the first to identify chlordiazepoxide, the first benzodiazepine in 1960. It was marketed as Librium by Hoffmann-La Roche which was followed by Valium (diazepam) in 1963 (Wick, 2013). The discovery of benzodiazepines was enthusiastically greeted by the medical professionals as it has minimal toxicity and lesser dependence liability. It became mostly used sedatives and hypnotics in 1970s (Wick, 2013). Benzodiazepines are frequently prescribed drugs and have proportionately higher anxiolytic potency even though they have depressant action on brain. Ironically, an interaction between benzodiazepines and gamma-aminobutyric acid (GABA) function was detected many years later (Wick, 2013).

Sedative(anxiolytic)isknowntohavecalming effect and thus reduces anxiety. At therapeutic doses, there is minimal of depression of central nervous system (CNS). Hypnotics are the agents that facilitate the onset and maintenance of sleep and also produce drowsiness. CNS depression increases progressively as the dose is increased. Most sedative–hypnotics have the similar characteristics of graded dose-dependent depression. Drugs having a proportional effect relationship with dose are destined to have a linear slope. Graded increase in dose of the older sedative–hypnotics than the required hypnotic dose may land the patient to anesthesia, which can lead to coma and finally death due to the depressant action on cardiorespiratory center in the brain. Benzodiazepines have a flatter dose–response curve, which indicates that it requires greater increments in dose to have depressive action on CNS as compared with older sedatives–hypnotics. Newer approved hypnotics have similar dose–response characteristics when compared with benzodiazepines (Katzung et al., 2009).

Zolpidem, eszopiclone, and zaleplon are collectively known as "Z-drugs" though they have the same mechanism of action as benzodiazepines but they are structurally unrelated (Gregory, 2016). Ramelteon and tasimelteon are the melatonin receptor agonists which are approved by US FDA as newer hypnotics indicated for non-24-h sleep–wake disorders (Laudon, 2014). An orexin receptor antagonist, suvorexant was introduced in the market in August 2014 and is indicated to improve sleep duration. Buspirone is a slow-onset anxiolytic agent differing from the conventional sedative–hypnotics in their mechanism of action (Mendelson, 1990).

Some other classes of drugs, that is, antipsychotics, antidepressants, and antihistaminics are also used off-label for management of chronic anxiety disorders and insomnia (Misra and Sharma, 2017). The chapter gives an opportunity for the readers to review the current pharmacologic therapy for anxiety and insomnia. It also gives an insight how a novel molecule for insomnia with minimal adverse effects and desirable efficacy to be developed.

## 11.2 CHEMICAL STRUCTURE OF THE BENZODIAZEPINES

For decades, it is a known fact that the name benzodiazepines (Fig. 11.1) have been synonymous with anxiolytic in addition to other properties like muscle relaxant, anticonvulsant, and sleep-inducing activity. These activities are due to the seven-membered heterocyclic ring structure comprising the benzene ring (Kovacic et al., 2013). The biological activity

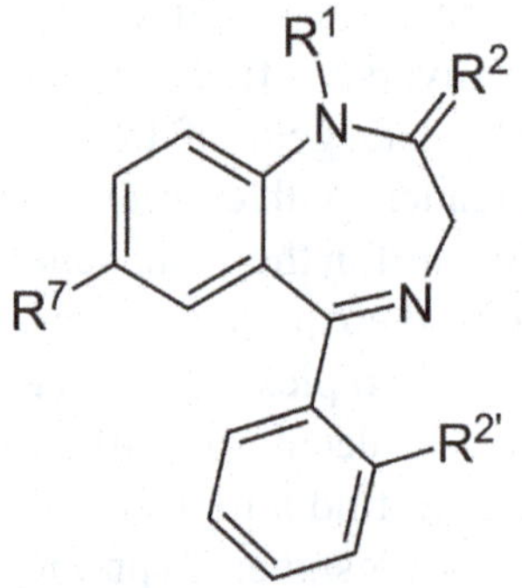

**FIGURE 11.1** The basic structure of benzodiazepines. Where, $R^1/R^2/R^{2'}/R^7$ means common locations of side chains that give different benzodiazepines.

of benzodiazepines is increased by addition or substitution of chemical groups in different positions like methyl group but its substitution with larger groups decreases its potency. It is found that substitution of halogen or nitro group in the seventh position of 1,4-benzodiazepines will produce the sedative–hypnotic activity (Gerecke, 1983).

Alkyl or aryl group added barbituric acid at fifth position increases its depressive action on CNS, thus confers sedative–hypnotic activity. At position C2, replacement of sulphur with oxygen confers increase in lipophilicity and thus accelerates its sedative–hypnotic action, increases its potency, and increases metabolic degradation (Brunton et al., 2018).

## 11.3 VERSATILITY OF GABA$_A$ RECEPTOR

Sleep–wake cycle is controlled by many important neurotransmitters including GABA, adenosine, galanin, and melatonin (Jonathan and Schwartz, 2008). On the other hand, neurotransmitters like norepinephrine, acetylcholine, dopamine, histamine, and orexin maintain wakefulness. Available drugs modulate the sleep–wake cycle by acting or modulating through these neurotransmitters (Zisapel, 2012).

Neurotransmitters bind to its specific receptor sites and induce the opening of the channel in postsynaptic receptors, thus making synaptic transmission possible. GABA$_A$–chloride channel receptor complex is shown in Figure 11.2, its specific interactions with GABA differentiate it from other super-family members. Its activation by GABA has a depressive effect in the CNS. The GABA$_A$ receptors are composed of multi-subunit proteins of pentameric proteins. Its isoform is generally made up of α1, β2, and δ2 subunits around a central anion-conducting channel

(Sigel and Steinmann, 2012). The structural subunits of the receptors are made up of short carboxy terminus, large extracellular amino terminus, and four transmembrane segments (M1–M4). The M2 segment of the receptor subunits constitutes the central conducting pore. The binding of GABA and benzodiazepines are at the specific sites of these subunits (Macdonald and Olsen, 1994).

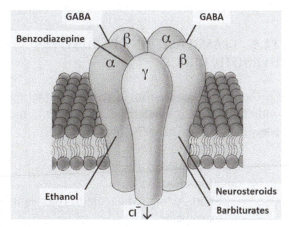

**FIGURE 11.2** Benzodiazepine–GABA$_A$–chloride ion channel receptor complex with specific site of drug interaction.
*Source*: Adapted from Katzung et al., 2009.

### 11.3.1 BENZODIAZEPINES VERSUS BARBITURATES

Barbiturates and benzodiazepines are the two classes of agents which act at the different concentration on the GABA$_A$ receptor with different potencies (Sigel and Steinmann, 2012). Benzodiazepines and barbiturates bind to sites different from GABA binding site (allosteric) on the GABA$_A$ receptor and enhance the GABA-stimulated Cl⁻channel entry inside the cell. Barbiturates require low micromolar concentrations to enhance GABA$_A$ receptor function, whereas benzodiazepines act with nanomolar affinity. At equi-effective concentrations, barbiturate has a greater enhancement of $GABA_A$ receptor and also at a higher concentration it can activate $GABA_A$ receptors directly. This mechanism of action of barbiturates causes greater depression of CNS. Benzodiazepines cannot activate the channel directly and also the requirement of presynaptic release of GABA at the receptor would make it as a safe drug (Kovacic et al., 2013).

## 11.4 CLASSIFICATION OF SEDATIVE–HYPNOTIC DRUGS

On the basis of mechanism of action (interacting with $GABA_A$–$Cl^-$ ion channel complex), the classifications are given below:

- Benzodiazepines: On the basis of duration of action.
  - Short-Acting: Midazolam, triazolam, oxazepam
  - Intermediate Acting: Estazolam, temazepam, nitrazepam, lorazepam, alprazolam
  - Long Acting: Diazepam, flurazepam, clonazepam, chlordiazepoxide
- Barbiturates: As sedative–hypnotic, it is obsolete and superseded by benzodiazepines. It is classified into the following groups according to duration of action:
  - Ultra-Short Acting: Methohexital, thiopental
  - Short-Acting: Amobarbital, pentobarbital
  - Long Acting: Mephobarbital, phenobarbital
- Non-Benzodiazepines Hypnotics (Z Drugs): Zolpidem, zaleplon, zopiclone, eszopiclone

- Melatonin Receptor Agonist: Melatonin, ramelteon, tasimelteon
- Orexin Receptor Antagonist: Suvorexant
- Off-label Agents: Diphenhydramine, doxylamine, trazodone, mirtazapine, amitriptyline, trimipramine, etomidate, doxepin
- Miscellaneous: Choral hydrate, meprobamate, carisoprodol, clomethiazole, propofol
- Agents Under Development: Pregabalin, ritanserin, agomelatine, volinanserin, eplivanserin, pruvanserin, lorediplon, esmirtazapine, lemborexant, piromelatine

## 11.5 BENZODIAZEPINES

Benzodiazepines are the group of drugs also addressed as minor tranquilizers. Benzodiazepines have high affinity and selectivity for specific site on $GABA_A$. Its molecular structure exhibits high hydrophobicity which potentiates its increase binding to plasma protein (Wingrove, 2002). This characteristic of increase binding to plasma protein reduces its free drug concentration and also decreases its movement through the blood–brain barrier and reduces toxicity. In cases of hepatotoxicity and liver failure, there is decrease production of albumin which will lead to dramatic increase in the potency and adverse effects of benzodiazepines (Goldberg, 2009).

Benzodiazepines act on $GABA_A$ receptors as modulators. In the presence of GABA, it activates the function of the channel. After modulation of the channel, it leads to increase frequency of channel opening which increases the net movement of chloride ion through the channel. $GABA_A$ receptors due to its high affinity for GABA and other agonists in the open state, thus benzodiazepines, increase the frequency of channel opening (Sigel and Steinmann, 2012).

In the absence of GABA, GABA$_A$ receptors are usually not activated by benzodiazepines but under certain situations, some mutant receptors may get activated due to its partial agonist activity. Benzodiazepines have the ability to increase the potency of GABA by three-fold as compared with other modulators like barbiturates or other general anesthetics (Golan et al., 2017).

## 11.5.1 PHARMACOKINETICS

The lipophilicity of benzodiazepines increases the rate of absorption of sedative–hypnotics. Due to its higher lipophilicity, the entry as well as redistribution in CNS becomes faster. All the benzodiazepines are given orally except midazolam (given IV or IM) For example, diazepam and triazolam have more rapid absorption than other benzodiazepines (Garzone and Kroboth, 1989). Benzodiazepines cross placental barrier and are secreted in breast milk which will cause depression of the fetal and neonate vital functions. Metabolic transformation in the liver forms water-soluble metabolites of sedative–hypnotics which is important for its clearance from the body. The active metabolites of sedative–hypnotics are formed by the active metabolizing enzyme systems which lead to the clearance of all benzodiazepines. Dealkylation and aliphatic hydroxylation are the primary metabolizing process for benzodiazepines which are followed by conjugation to form glucuronides which are water-soluble metabolites and thus are excreted through urine. Elimination of parent drugs in most of the cases does not get much effected by the renal functional status. Many benzodiazepines which undergone phase I metabolic process forms metabolites which are pharmacologically active and have longer half-lives compared with the parent molecule (Chouinard et al., 1999).

For example, desmethyldiazepam, diazepam, and clorazepate have an elimination half-life close to 40 h. Benzodiazepines having long half-lives usually show cumulative effects with repeated doses. Drugs with short half-lives and those which form inactive glucuronides like oxazepam, lorazepam, and estazolam do not show cumulative and residual effects (Sellers, 1978).

Elimination half-lives of many drugs may increase in liver failure patients as well as in geriatric age group which may lead to cumulative effects, thus increasing the depressive effects on CNS. The newer hypnotics and benzodiazepines normally have least effect on hepatic drug metabolizing enzyme activity (Chouinard et al., 1999).

## 11.5.2 PHARMACOLOGICAL PROPERTIES

Benzodiazepines is a centrally depressant drug with prominent actions of antianxiety, sedative, hypnotic, muscle relaxant, produces anterograde amnesia and also has anticonvulsant action. Coronary vasodilation on intravenous administration and neuromuscular blockade are the two peripheral actions with high doses with certain benzodiazepines (Soderpalm, 1987).

### 11.5.2.1 CNS EFFECTS

As benzodiazepines have less enhancing effects at GABA$_A$ receptors than other CNS depressant drugs like anesthetics compounds and barbiturates, but still they depress CNS activity. In the therapeutic class of benzodiazepines, there are some variations among the drugs regarding the selectivity to the binding sites of the receptors as well as their therapeutic uses (Wingrove et al., 2002). Increasing dose of benzodiazepine will lead to sedation due to depressive action on the CNS which may be followed by hypnosis which may further progress to stupor. These classes of drugs generally are not true general anesthesia as they do not abate the awareness in patients (O'Boyle, 1988).

### 11.5.2.1.1 Sedation

Benzodiazepines are the therapeutic choice for sedative regimens. Its high therapeutic index allows benzodiazepines to exert its antianxiety effects at even low doses as it is safe and have less chances of adverse effects. In order to produce minimal sedation, the required dose for benzodiazepines is below than that produce hypnosis. This might be not sure in the geriatric age group, liver failure patients, and in patients of obstructive sleep apnea as the metabolism of the drug is reduced and leads to accumulation of the drugs (Guilleminault, 1990). At an equivalent minimal dose for antianxiety effects, it disinhibits the effects of punishment-suppressed behavior like euphoria, perceiving judgment, and inability to self-control. Depending on the strength of the benzodiazepine dose, it also exhibits anterograde amnestic effects (Becker, 2012).

### 11.5.2.1.2 Hypnosis

Sleep is induced by higher doses of hypnotics. The stages of sleep are affected by the specific drug of the sedative–hypnotic class, their dose, dosage form, and the frequency of drug administration. Benzodiazepines modulate normal sleep patterns by inducing the following changes: it decreases the time duration to fall asleep, it increases non-rapid eye movement (NREM) stage of sleep (stage 2), and it also decreases duration of rapid eye movement and slow-wave sleep (stage 3 and 4, respectively). The most desirable effects of benzodiazepines are the effects on the rapid onset of sleep and prolongation of stage 2 (Kubicki et al., 1987). The positive impact of benzodiazepines is that they reduce the time to fall asleep, increases the total time period of sleep and reduce wakefulness. They are the therapeutic choice for management of short-term insomnia. At the lowest tolerated and intermittently dose, they are the most effective regimen. The drug should not be given beyond 2–4 weeks as the chance of risk of dependence increases (Ramakrishnan and Scheid, 2007).

### 11.5.2.1.3 Anesthesia

At high doses, some benzodiazepine drugs may depress the CNS function at the higher level that may lead to anesthetic effects. The anesthetic effect is the result of inhibition of GABAergic transmission in the higher level of CNS which is controlling memory, awareness, and sleep. The lipophilicity of the sedative–hypnotics will determine its rapidity and short duration of action which will determine its anesthetics properties (O'Boyle, 1988). Lorazepam and midazolam are the short-acting intravenously used benzodiazepines that are used for short procedure for its property of redistribution. Benzodiazepines used as adjuncts to general anesthetics in large doses may have adverse effects of depression of respiration post-anesthetically. Flumazenil, an antagonist, is usually used in the conditions to reverse the respiratory depression (Saari, 2011).

### 11.5.2.1.4  Anticonvulsant Effects

Only few drugs from sedative–hypnotics group have the ability to inhibit epileptic activity in the CNS. The only disadvantage with the anticonvulsant activity of the benzodiazepine is that they may cause psychomotor function impairment without depressing the CNS. Clonazepam, lorazepam, and diazepam are the few drugs from the benzodiazepines groups which have the ability in inhibiting the epileptic electrical activity efficiently (Ochoa and Kilgo, 2016). Although barbiturates are not first choice of drugs, but phenobarbital and metharbital are effectively used for management of petit-mal epilepsy. "Z" drugs are the safest and effective sedative–hypnotic but of their specific and selective binding on the GABA$_A$ receptor, they lack anticonvulsant property.

### 11.5.2.1.5  Muscle Relaxation

Diazepam is preferred among benzodiazepines for decreasing the spasticity of skeletal muscle. It acts by inhibiting the internuncial transmission and polysynaptic reflexes of central origin which helps in skeletal muscle spasticity. At high doses, it also depresses the transmission of the skeletal neuromuscular junction (Sharma and Sharma, 2017).

### 11.5.2.2  RESPIRATION SYSTEM

Hypnotic doses of benzodiazepines do not have effect on respiratory process. In patients with impaired hepatic or pulmonary function or in children, it may suppress alveolar ventilation and causes respiratory acidosis. This may cause hypoxia which will progress to alveolar hypoxia and CO$_2$ narcosis (Guilleminault,

1990). Benzodiazepines are usually given in higher doses during anesthesia or in combination with opioids which may further aggravate apnea. Respiratory assistance may be required for intoxicated patients. Few drugs from benzodiazepines are known to aggravate apnea in myocardial infarction patients during REM sleep (Guilleminault, 1990).

### 11.5.2.3  CARDIOVASCULAR SYSTEM

Benzodiazepines effect on cardiovascular effects is minimal. Diazepam at relatively low doses increases coronary flow by increasing its interstitial concentrations. Other benzodiazepines, clonazepam causes increase in coronary blood flow on a dose-dependent fashion (Pascoe et al., 1983). Diazepam and clonazepam effect on inotropy have a biphasic response, initially it causes decrease followed by increase which is dose-dependent effects. Benzodiazepines at preanesthetic doses cause fall in blood pressure and increase in heart rate (Zeegers, 1998).

### 11.5.3  THERAPEUTIC USES

Long-acting benzodiazepines are used anticonvulsants for their rapid entry into the brain and have efficacy for management of status epilepticus. Long-acting antianxiety agents are useful despite accumulation of drug and its propensity for risk of neuropsychological deficits. Hypnotic drugs with rapid action are taken before sleep to sustained or maintain sleep with no prolonged action by the next day (Cloos and Ferreira, 2009). Rapid acting agents such as midazolam usually have a short duration of action and disappearance from the body at a relatively rapid rate, hence it is prone to cause early morning insomnia and rebound insomnia on abrupt discontinuation of the drug (Table 11.1). Diazepam, long-acting benzodiazepines can be used for the management of alcohol

withdrawal symptoms and as hypnotics as it can be interchanged with other long-acting benzodiazepines (Brunton et al., 2018).

**TABLE 11.1** Relationship Among Route of Administration, Plasma Half-Life, and Therapeutic Uses of Benzodiazepines.

| Drug | Routes | Plasma half-life (h) | Therapeutic uses |
|---|---|---|---|
| Midazolam | IV, IM | 1.5–2 | Preanesthetic medication |
| Triazolam | Oral | 2–3 | Insomnia |
| Oxazepam | Oral | 6–10 | Anxiety |
| Estazolam | Oral | 10–24 | Insomnia |
| Temazepam | Oral | 5–17 | Insomnia |
| Lorazepam | Oral, IV, IM | 9–19 | Anxiety, preanesthetic medication |
| Alprazolam | Oral | 10–14 | Anxiety, agoraphobia |
| Diazepam | Oral, IV, IM, rectal | 30–56 | Anxiety, status epilepticus, preanesthetic medication, muscle relaxant |
| Flurazepam | Oral | 50–98 | Insomnia |
| Clonazepam | Oral | 18–28 | Seizure disorder, acute mania, movement disorders |
| Chlordiazepoxide | Oral, IM, IV | 7–13 | Anxiety, alcohol withdrawal, preanesthetic medication |

IM, intramuscular injection; IV, intravenous injection.
*Source*: Adapted from Brunton et al., 2018.

## 11.5.4 DRUG INTERACTIONS

Benzodiazepines in combination with other CNS depressant drugs cause additive effects which is the most common drug interactions (Abernethy et al., 1984). These additive interactions cause enhanced depression. The additive interactions are more common when used concomitantly with alcohol, anticonvulsants, antidepressants, opioid analgesics, and anticonvulsants. Although in anesthesia practice, these drugs are used as adjuvants which are useful additive interactions (Uzun et al., 2010).

## 11.5.5 UNTOWARD EFFECTS

Benzodiazepines at hypnotic doses may cause lassitude, light-headedness, motor incoordination, increased reaction time, confusion, impairment of motor functions, and anterograde amnesia. If combined with ethanol, it can affect the driving skills and other psychomotor skills (Terzano et al., 2003). Day-time sedation is the most common adverse effect seen during waking hours. With increasing age, these adverse effects of benzodiazepines are common. Headache, vertigo, nausea, epigastric distress, vomiting, and diarrhea are common side effects. Long-acting benzodiazepines used as anticonvulsants in epileptic patients are seen to increase the episodes of seizures (Terzano et al., 2003).

Benzodiazepines were found to cause some rare reactions like allergic reactions, liver toxicity, and hematological reactions in some cases. These rare conditions were associated mostly with the use of flurazepam, triazolam, and temazepam. Large doses of benzodiazepines may cause decrease temperature, muscle weakness, and mild respiratory depression in the neonate if the drug is taken before or during labor (Terzano et al., 2003).

## 11.5.6 TOLERANCE AND DEPENDENCE

On repeated exposure to benzodiazepines or its long-term use, responsiveness may come down due to the development of tolerance. Benzodiazepines given for a long duration for the treatment of insomnia will lead to tolerance to the dose which would necessitate to titrate the dose to higher level to maintain or promote sleep (Higgitt et al., 1988). Benzodiazepines have the features of partial cross-tolerance with other class of drugs, particularly with other sedative–hypnotics and ethanol. The definite mechanism for tolerance is not known but it is proposed that the increase activity of hepatic enzymes may accelerate the metabolism of drug which is known as metabolic tolerance or decrease response of the CNS (pharmacodynamic tolerance). Benzodiazepines, in particular, it was proposed that downregulation of benzodiazepine receptors is the mechanism for tolerance (Wong et al., 1986).

Sedative–hypnotic drugs are classified under the restrictive class of drugs in many countries and registered medical practitioner's prescription is required to procure the drug. Relief of anxiety,

inducing euphoria, and maintenance of sleep cycle are the benefits of these drugs which are abused by the vulnerable individuals (Brett and Murnion, 2015). Physiologic and psychological abuses are the common dependence of these classes of drugs. Initially, the psychological dependence to benzodiazepine abuse is mild and can be compared with neurotic behaviors like drinking coffee or smoking cigarette. It becomes worrisome when there is compulsion to consume sedative–hypnotic which may progress to physiologic dependence and tolerance. Psychological dependence appears prior to physical dependence (Higgitt et al., 1988).

Development of tolerance to drugs leads to physiologic dependence. Altered physiologic state might be the factor that leads to intense craving for the drug. In order to safeguard the patient, continuous drug administration may be required to avoid the development of an abstinence or withdrawal syndrome. Increased insomnia, anxiety, and convulsions may be a component of physical dependence due to benzodiazepines. Withdrawal symptoms severity may differ among drugs, dosage strength, and time period of treatment. Withdrawal symptoms are more with benzodiazepines as they are given for long term for management (Onyett, 1989). Short-acting sedative–hypnotics like triazolam if stopped abruptly may lead to daytime anxiety and withdrawal symptoms are more prone with drugs given in higher doses and if it is discontinued abruptly. Pharmacokinetics of long acting drugs differs as they have slow clearance and so they have fewer chances of withdrawal symptoms (Puntillo et al., 1997).

## 11.5.7 NOVEL BENZODIAZEPINE RECEPTOR AGONISTS—Z DRUGS

Novel benzodiazepine receptor agonists are also known as "Z compounds" or "Z drugs." This class of hypnotics includes zolpidem, zaleplon, zopiclone, and eszopiclone [S(+) enantiomer of zopiclone)]. They are structurally unrelated among themselves and also to benzodiazepines (Terzano et al., 2003). "Z compounds" act at allosteric site to benzodiazepines on the $GABA_A$ receptor to have the therapeutic benefit of hypnotics. Their specificity toward the α1 subunit of the benzodiazepines $GABA_A$ receptor decreases their efficacy as anticonvulsants or muscle relaxants as compared with benzodiazepines. Recently, Z compounds have been found as an alternate choice to benzodiazepines in the management of sleep disorders (Siriwardena et al., 2008). Nowadays, physicians and psychiatrists prefer these classes of drug for patient with insomnia. Its preference is due to its potential for less abuse and dependence compared with benzodiazepines. The drugs, zopiclone and zolpidem, if given for long duration, the patients do develop some tolerance and physical dependence. These data were collected from post-marketing Phase 4 clinical trial with Z drugs. Overdose of Z compounds have a similar toxicity and clinical profile to benzodiazepine. Flumazenil, a benzodiazepine antagonist, can be used to treat this condition (Atkin et al., 2018).

### 11.5.7.1 ZOLPIDEM

Zolpidem belongs to imidazopyridine family. Zolpidem has an agonist effect on α1subunit of benzodiazepine receptor site on $GABA_A$ receptors and thus have similar pharmacological actions. Zolpidem after absorption undergoes significant hepatic metabolism providing only 70% bioavailability (Swainston and Keating, 2005). The peak plasma concentration of the drug is reached within 1–3 h. Hepatic CYP3A4 enzymes rapidly metabolized zolpidem to inactive metabolites by oxidation and hydroxylation. In normal individuals, half-life of the drug is 2 h but depending on the age, sex, and clinical conditions, there may be some variation in the elimination half-life (Swainston and Keating, 2005). The elimination half-life may be significantly high in women, cirrhotic patients, renal failure patients, and elderly, thus requiring adjustment of dosage. Different dosage forms are available for the patient with insomnia like the sublingual and oral spray dosage forms. Zolpidem reduces onset to fall asleep and overall maintains the total time of sleep in patients of insomnia (Swainston and Keating, 2005). Zolpidem may exhibit mild rebound insomnia on the first night after stopping the drug but the benefit may last for approximately 1 week. Zolpidem is given in the therapeutic range of 5–10 mg for the short-term treatment of insomnia. Zolpidem exhibits less tolerance and physical dependence. Sedation on next morning, anterograde amnesia, and delayed reaction time may be observed if zolpidem is administered late-night (Atkin et al., 2018).

### 11.5.7.2 ZALEPLON

Zaleplon, pyrazolopyrimidine derivative binds on the allosteric site as zolpidem on α1 receptor benzodiazepine subunit of the $GABA_A$ receptors. Presystemic metabolism reduces its oral bioavailability to about 30%. In 1 h, the drug reaches its peak plasma level after rapid absorption (Rosen et al., 1999). It is metabolized to inactive metabolites partly by hepatic aldehyde oxidase and cytochrome P450 isoform CYP3A4. In hepatic impairment patients and in geriatric age group, adjustment of dose may be required. It has interaction with drugs whose metabolism involves cytochrome P450 and aldehyde oxidase pathway (Rosen et al., 1999). Enzymes inhibitors may markedly increase the peak plasma concentration of zaleplon as they inhibit CYP3A4 and aldehyde dehydrogenase. Oxidation and glucuronidation of zaleplon produced soluble metabolites which are excreted through urine and reducing the plasma half-life to about 1 h. The drug is available in 5, 10, and 20 mg doses. Chronic or transient insomnia patient treated with zaleplon may experience shorter periods of sleep onset latency (Atkin et al., 2018).

### 11.5.7.3   ESZOPICLONE

Eszopiclone, an S(+) enantiomer of zopiclone acts by enhancing the benzodiazepine binding site of $GABA_A$ receptor to promote its positive effects on sleep (Benjamin, 2006). It is usually used for long-term management of insomnia (nearly 12 months), thus differs its pharmacological actions from other Z compounds. It maintains sleep and also decreases the time to fall asleep. It has bioavailability of about 80%. Nearly 50–60% of its drug is bound to plasma proteins with a plasma half-life of nearly 6 h. Hepatic cytochromes CYP3A4 enzymes metabolized the drug to inactive metabolites, namely, N-oxide derivative and desmethyl eszopiclone, respectively (Benjamin, 2006). In elderly, the excretion of the drug is prolonged due to increase in elimination half-life. Eszopiclone administered with CYP3A4 inhibitors (e.g., ketoconazole, erythromycin, etc.) decrease its metabolism thus increasing its elimination half-life, whereas if coadministered with inducers of CYP3A4 (e.g., rifampin), the hepatic metabolism will increase leading to decrease in elimination half-life. The drug is available in the dosage of 1 mg, 2 mg, and 3 mg tablets. No tolerance or withdrawal syndrome is seen on discontinuation of the drug. Minor

symptoms of withdrawal like abnormal dreams, anxiety, nausea, and upset stomach are seen in less than 2% of the patients taking the drug for management of sleep disorder (Atkin et al., 2018).

### 11.5.8   LONG-TERM BENZODIAZEPINE TREATMENT

Benzodiazepine should not be stopped abruptly, rather it should be tapered slowly over a period of 2 weeks. The tapering of dose is important if the patients are taking short-acting hypnotics. It is always favorable to start treatment with long-acting hypnotic and it should be slowly tapered whenever there is requirement to stop the treatment. Long-acting hypnotics always have the advantages of delayed or slow onset of withdrawal symptoms (Janhsen et al., 2015).

### 11.5.9   FLUMAZENIL—AN ANTAGONIST

Flumazenil, an imidazobenzodiazepine binds strongly with the benzodiazepine binding site of the $GABA_A$ receptor. It competitively antagonizes the pharmacological effects of benzodiazepines and other relative ligands acting on the benzodiazepine binding site of the $GABA_A$ receptor (Whitwam and Amrein, 1995). Flumazenil is administered by a single bolus dose of 1 mg giving intravenously over a period of 1–3 min. Additional injection of the drug may be required if sedation reappears (Whitwam and Amrein, 1995). The given amount of dose is sufficient to reverse the action of benzodiazepines. It is rapidly absorbed orally and with bioavailability of only 25% due to extensive first-pass hepatic metabolism, due to its significant metabolism in the liver leading to produce inactive products lower the plasma half-life to about 1 h (Whitwam and Amrein, 1995). Flumazenil has only 30–60 min of pharmacological action so usually requires repetition of doses to have therapeutic effects in benzodiazepines overdosing. The IV dosage form of flumazenil is better as it does not cause headache and dizziness seen aptly with oral doses (Whitwam and Amrein, 1995).

Flumazenil is used for the treatment of benzodiazepine overdose and also indicated to reverse sedative action of benzodiazepines used during clinical procedures and general anesthesia. Flumazenil may exhibit seizures if administrated to patients who have developed benzodiazepine's tolerance or dependence and taking the drug for long duration (Hood et al., 2014).

## 11.6 MELATONIN CONGENERS

Melatonin is a circadian signalling molecule. Melatonin is the principal indoleamine produced in pineal gland whose main function is of imagination by forming a pigment. Melatonin is synthesis from serotonin by N-acetylation and O-methylation in the pineal gland of the hypothalamus and environmental light influenced its synthesis. It regulates biological rhythms especially in management of sleep disorders and in jet lag. Its analogues have been approved for the management of sleep disorders (Brzezinski, 1997).

### 11.6.1 RAMELTEON

Melatonin promotes and maintains sleep by inhibiting the wake-promoting signals in the suprachiasmatic nucleus. Melatonin levels follow a circadian cycle in the suprachiasmatic nucleus of the hypothalamus, its concentrations physiologically rise during sleep at night which is followed by plateau phase and then the concentration falls with the progression of the night (Tordjman et al., 2017). Melatonin has two G-protein coupled receptors, namely, MT1 and MT2, in the suprachiasmatic nucleus of the hypothalamus. Its mechanism of action is elicited by binding to MT1 receptors. It promotes or decreases the onset to sleep and looks after maintenance of circadian cycle of the body by binding with MT2 receptors (Tordjman et al., 2017).

Ramelteon is a melatonin receptor agonist (MT1 and MT2 receptors). It is used therapeutically to promote sleep onset and to treat sleep disorder. Somnolence, fatigue, etc., are some of the adverse effects of the drug (Masaomi, 2009). Ramelteon is specific for MT1 and MT2 only. Its side effects are less as it does not bind to any other classes of receptors, such as nicotinic ACh, neuropeptide, dopamine, and opiate receptors, or the benzodiazepine binding site on $GABA_A$ receptors (Masaomi, 2009).

It is absorbed rapidly from the gastrointestinal tract after oral administration. Its bioavailability is less than 2% as it undergoes significant hepatic first-pass metabolism (Masaomi, 2009). Hepatic cytochrome enzymes CYPs 1A2, 2C, and 3A4 largely metabolized the drug and make its plasma half-life of about 2 h. The dose is 8 mg and it should be taken about 30 min before bedtime. It is used for the management of transient and chronic insomnia. Absence of withdrawal symptoms or rebound insomnia is one of the advantages of using ramelteon on long duration. Cognitive dysfunction is usually absent the following day with ramelteon than other hypnotics (Laudon and Frydman-Marom, 2014).

## 11.6.2 TASIMELTEON

Tasimelteon is also melatonin receptors agonist approved recently for the treatment of non-24-h sleep–wake disorder. It is specially indicated in totally blind patients. Tasimelteon is used to improve the onset to fall asleep, thus improving the sleep timing but on discontinuation, the patient will continue with his/her previous sleep pattern within a month (Bonacci et al., 2015). It has affinity for MT2 receptor which gets reflected as it maintains the circadian rhythm of sleep. Inability to maintain mental alertness is an undesirable side effect which may affect the daily activities of the patients. Tasimelteon is given in the dose of 20 mg which is given before sleep. It reached its peak of action at 0.5–3 h after drug administration. The drug is metabolized in the liver by CYP1A2 and CYP3A4 (Vachharajani and Yeleswaram, 2003).

## 11.7 BARBITURATES

Barbiturates acts mainly on spinal cord, substantia nigra, cuneate nucleus cortex, thalamus reticular activating system, and cerebellum. It depresses neuronal activity in the brain by increasing GABAergic transmission by inhibiting the GABA$_A$ receptors and inhibits the activity of the reticular activating system. Sedation, anterograde amnesia, and loss of consciousness are some of the effects of barbiturates on CNS. It inhibits the GABAergic transmission, it suppresses the motor components of spinal cords, and relaxes the skeletal muscles (Saunders and Ho, 1990). Barbiturates differ from benzodiazepines as they have a variable binding on the GABA$_A$ receptors. In the mid-20th century, barbiturates was the most commonly used group of drugs but now it is less used with the growing popularity of benzodiazepines and newer sedative and hypnotics. It is used as inducing agents' in general anesthesia for the treatment of seizure and to control intracranial hypertension (Saunders and Ho, 1990).

Barbiturates acts by increasing the efficacy of GABA on the receptor and thus increases the total time of opening of the chloride channel by allowing greater influx of chloride ions. This act of barbiturate causes hyperpolarization and depresses the cell excitability. The barbiturates enhance GABA action more as compared with benzodiazepines. It also directly activates GABA$_A$ receptors and thus enhance GABA actions. Overdose of barbiturates may cause excessive potentiation of the receptor which may increase its toxicity and lead to sedation, hypnosis, respiratory depression, coma, and death (Saunders and Ho, 1990).

It may also cause excitatory or inhibitory neurotransmission by acting on other receptors. It decreases activation of glutamate-sensitive AMPA receptors causing depression of neuronal excitability, increasing the closed or inactivated state of voltage-dependent Na channels at anesthetic concentrations and also inhibits high-frequency neuronal ring (Haefely, 1977).

## 11.7.1 PHARMACOKINETICS

Barbiturates can be administered oral or parental routes. They undergo extensive metabolism in the liver and have reduced bioavailability. Due to the extensive first-pass metabolism, it forms metabolites that will ultimately get conjugated by glucuronidation before excreting through. CYP3A4, CYP3A5, and CYP3A are the enzymes that metabolize barbiturates (Pelkonen and Karki, 1973). Chronic barbiturate administration upregulates these enzymes which induces its own metabolism (metabolic tolerance). They also induce the metabolism of other sedative/ hypnotics and drugs like phenytoin, digoxin, oral contraceptives, etc. It penetrates the blood–brain barrier due to its lipid solubility (Pelkonen and Karki, 1973). It is rapidly redistributed from the brain to the splanchnic circulation followed by skeletal muscle and ultimately to the least perfused adipose tissue, thus terminating its action on the CNS. Barbiturates are taken up in adipose tissue due to its lipophilicity, thus increasing its high volume of distribution and ultimately longer elimination half-life. These properties are seen with the cumulative effects when drugs are given in multiple doses (Pelkonen and Karki, 1973). Hepatic failure and geriatric age group are more prone for greater CNS effects and toxicities as the drugs are not metabolized to inactive soluble metabolites. As the drugs are acidic in nature, the urine can be alkalized by administrating IV sodium bicarbonate which would increase their renal excretion. In different individuals, the elimination half-lives of phenobarbital are about 4–5 days, whereas the half-lives of secobarbital and pentobarbital may be upto 48 h (Kristensen, 1976).

### 11.7.2 THERAPEUTIC USES

Before the approval of benzodiazepines, barbiturates used to be the common choice among physicians for treatment of insomnia or anxiety. With the advent of benzodiazepines and the "Z" drugs, it has replaced barbiturates in most clinical applications. The newer classes of drugs are well tolerated, safe, and tolerance is less as they have less or none effects on drug-metabolizing enzymes. In the present scenario, barbiturates are used as inducting agents in general anesthesia to control seizures and for neuroprotection (Table 11.2) (Smith and Riskin, 1991).

Thiopental, methohexital, and pentobarbital are the lipid-soluble barbiturates. They have greater penetration in the CNS and have greater propensity for inducing general anesthesia. This property of lipid solubility allows it to rapidly enter brain and then redistribute to least perfused tissues like adipose tissue and skeletal muscles. It is short acting due to its property of redistribution (Smith and Riskin, 1991).

Phenobarbital, long-acting barbiturates is used epilepsy. Barbiturates perform its epileptic activity by rapidly reduce the depolarizing CNS neurons by enhancing its depressing activity on GABAergic transmission and also block the excitatory transmission by acting on AMPA glutamate receptor. Phenobarbital, long-acting barbiturates is used for the management of focal and tonic-clonic seizures with least side effects (Brodie and Kwan, 2012).

Electroencephalographic silence known as barbiturate coma is due to profound suppression of neuronal activity in the brain on administration of very high dose of barbiturates. It is useful condition as it protects the brain in ischemic conditions with the comorbid condition of decreased oxygen supply (e.g., profound anaemia, hypoxia, shock) or when the oxygen demand is more as in the cases of status epilepticus. It reduced consumption of oxygen in brain and decreases the blood flow in the brain when the demand rises in the pathological conditions (Smith and Riskin, 1991).

**TABLE 11.2**  Relationship Among Route of Administration, Plasma Half-Life and Therapeutic Uses of Barbiturates.

| Drug | Routes | Plasma half-life (h) | Therapeutic uses |
|------|--------|----------------------|------------------|
| Methohexital | IV | 3–5 | Anesthetic agent |
| Thiopental | IV | 8–10 | Preop sedation, anesthetic agent, seizures |

**TABLE 11.2**  *(Continued)*

| Drug | Routes | Plasma half-life (h) | Therapeutic uses |
|------|--------|---------------------|------------------|
| Amobarbital | IM, IV | 10–40 | Preop sedation, insomnia, seizure |
| Pentobarbital | Oral, IM, IV, rectal | 0–24 | Preop sedation, insomnia, seizure |
| Mephobarbital | Oral | 10–70 | Seizure, sedation |
| Phenobarbital | Oral, IV, IM | 80–120 | Seizure, sedation |

IM, intramuscular injection; IV, intravenous injection; Preop, preoperative.

*Source*: Adapted from Brunton et al., 2018.

## 11.7.3  ADVERSE EFFECTS

Barbiturates act on multiple sites with high efficiencies and enhancing high GABAnergic transmission by activating GABA$_A$ receptor that may enhance the adverse effects of these drugs which would decrease its therapeutic index. At high dose, it causes significant depression of CNS and respiratory centers. Pentobarbital, an inducing anesthetic agent has more depressive action on CNS as compared with phenobarbital, an antiepileptic barbiturate. Barbiturates have a fatal depressive action on CNS due to the synergistic effects with other CNS depressants when coadministered and hence should not be given together (Santos and Olmedo, 2017).

## 11.7.4  TOLERANCE AND DEPENDENCE

On chronic administration of barbiturate, cytochrome P450 enzymes will be activated which will increase its own metabolism. This phenomenon will produce self-metabolic tolerance and also cross-tolerance with related sedative–hypnotics like benzodiazepines, "Z" drugs, etc.

(Ito et al., 1996). Tolerance and physiologic dependence are seen if they are repeatedly and extensively misuse. Withdrawal syndrome characterized by tremors, anxiety, and insomnia is seen due to physiologic dependence to barbiturates. If untreated, these withdrawal symptoms can progress to serious clinical conditions like seizures and cardiac arrest (Santos and Olmedo, 2017).

## 11.7.5  DRUG INTERACTIONS

As discussed, barbiturates in combination or simultaneously given with CNS depressants (e.g., alcohol, antidepressants) have a greater depressive action on the CNS. Enzyme inhibitors like isoniazid, methylphenidate, and monoamine oxidase inhibitors when coadministered with barbiturates have profound CNS depressant effects as the former decrease the metabolism of barbiturates by inhibiting the cytochrome P450 pathway. It can ameliorate its own action by accelerating its own metabolism by inducing the hepatic enzymes (Mark, 1971). Hepatic CYPs enzymes induction by barbiturates can metabolize drugs and endogenous substances by increasing their hepatic metabolism. Barbiturates are known for their capability of inducing hepatic enzyme which may cause unwanted metabolism of physiological endogenous steroid hormones and also enhance the metabolism of oral contraceptives which may increase the chances of unwanted pregnancy (Jackson, 1969).

## 11.7.6  BARBITURATE POISONING

Barbiturate poisoning varies with its lethal dose. Severe poisoning with barbiturate happens when the dose is more than the 10 times of its therapeutic dose for hypnosis. If other depressant drugs or another sedative–hypnotics drugs are given concomitantly then the quantity of drug to be lethal is lowered. Barbiturate poisoning is most commonly seen in children due to accidental poisonings, suicidal bid, and drug abusers (Eugene et al., 1952). The patient is comatose with respiratory system affected early in severe barbiturate poisoning. Rapid and shallow or slow breathing pattern are characteristic of barbiturate poisoning. The lethal dose decreases the cardiac contractility leading to fall of blood pressure due to the depressive effect on the cardiovascular system. The medullary vasomotor centers and sympathetic ganglia are depressed by barbiturates leading to hypoxia (Eugene et al., 1952). Severe barbiturate poisoning may cause to kidney failure and pulmonary complications like atelectasis, edema, and bronchopneumonia. Management of barbiturate poisoning consists of supportive procedures like maintaining airway, breathing, and circulation. Forced diuresis, urine alkalinization, and maintaining of hydration will increase the renal excretion of barbiturate. In barbiturate poisoning, use of CNS stimulants are contraindicated (Santos and Olmedo, 2017).

## 11.8  MISCELLANEOUS SEDATIVE-HYPNOTIC DRUGS

With the changing lifestyle and sleep disorders, newer pharmacologic agents with good efficacy, safe, and unwanted side effects are the immediate requirement of the present scenarios. Many molecules are in the preliminary stage of drug development but the search for an ideal hypnotic molecule is still on. Sedative–hypnotic with diverse structures were used for a long time like chloral hydrate, meprobamate, and paraldehyde

and some are recently added, namely ramelteon (Masaomi, 2009). These miscellaneous agents should have some similarity with the sedative–hypnotic drugs and older hypnotics to be an ideal agent so that they can affect all the stages of sleep and causing less adverse effects on the respiratory and cardiovascular system. These drugs, given for a longer duration, may lead to physical dependence and tolerance which may be life threatening and deteriorate quality of life (Perry and Alexander, 1986).

### 11.8.1  CHLORAL HYDRATE

Chloral hydrate is used as replacement of benzodiazepines in treating patients with paradoxical reactions. It is metabolized in the liver by hepatic alcohol dehydrogenase to an active compound trichloroethanol. The metabolite, trichloroethanol, has a similar pharmacological to barbiturate. It acts by activating $GABA_A$ receptor. In the United States, it is mixed with alcoholic beverages to produce a cocktail that can render the person malleable or unconscious. "Mickey Finn" or "Mickey" is the name given to this cocktail (Gauillard et al., 2002).

### 11.8.2  MEPROBAMATE

Meprobamate, a bis-carbamate ester, is used as sedative–hypnotic. It has similar pharmacological action to benzodiazepines. It inhibits suppressed behaviors in animals. Meprobamate has unique properties of not producing anesthesia even though it causes depression of CNS (Rudolph and Dauss, 1974). Large doses of meprobamate cause severe depression of respiratory and cardiovascular system. It has the property of potentiating other analgesic drugs and has the mild analgesic therapeutic in musculoskeletal pain (Truter, 1998).

Meprobamate is well absorbed and undergoes metabolization in the liver by hydroxylation and glucuronidation. The elimination half-life of the drug is directly proportional to the amount of drug administered. There is increase in plasma half-life of meprobamate due to the cumulative effects of the drug given for a prolong duration. The adverse drug effects are drowsiness, ataxia, impair learning, motor incoordination, and prolong reaction time. Its abrupt discontinuation leads to withdrawal syndrome in the form of anxiety, insomnia, tremors, hallucinations, etc. (Ramchandani et al., 2006).

### 11.8.3  CARISOPRODOL

Carisoprodol is a skeletal muscle relaxant and also has abuse potential. Meprobamate is the active metabolite of carisoprodol which is commonly misused for its abuse potential. It is commonly known as "street drug" (Logan et al., 2000).

### 11.8.4  ETOMIDATE

Etomidate is commonly combined with an opioid, fentanyl, and used as an IV anesthetic. It can sometimes be used for sleep disorders. It has cardiac depressant activity but as such has no such effect on pulmonary and vascular activity (Janssen et al., 1975).

### 11.8.5  CLOMETHIAZOLE

Clomethiazole is a very safe drug with high therapeutic index and has least respiratory depression. It has sedative, muscle relaxant, and anticonvulsant properties (Ratz et al., 1999).

### 11.8.6  PROPOFOL

Propofol is a rapidly acting general anesthetic. It is used in surgical procedures for induction and maintenance of anesthesia. It has the property of maintaining sedation for a long duration. It is used to induce and maintain sedation during endoscopy in intensive care setting and also for transvaginal oocyte retrieval (Steinbacher, 2001).

### 11.9  OFF-LABEL HYPNOTIC DRUGS

These are some drugs which may be used as an alternative therapy for the treatment of insomnia but though approved for specific clinical conditions. These drugs are also not approved by regulatory authorities. Common off-label drugs for hypnotics are antihistaminics like diphenhydramine, doxylamine having sedative property. They are used mostly as over-the-counter (OTC) nonprescription sleep aids (Larry and Mark, 2015). These long acting antihistaminics due to its residual adverse effect of sleepiness are used for insomnia. Trazodone, mirtazapine, amitriptyline, and trimipramine are the commonly used antidepressant that has properties of sedation. These drugs have not shown or their long-term efficacy is unknown though sedation, motor incoordination, dependence and tolerance are some of the adverse effects seen with these drugs (Santos Moraes et al., 2011).

### 11.10  NEW AND EMERGING AGENTS IN THE PHARMACOTHERAPY OF INSOMNIA

### 11.10.1  OREXIN RECEPTOR ANTAGONIST

Orexin-A and orexin-B commonly known as hypocretin-1 and hypocretin-2, respectively, are neuropeptides that have affinity for OX-1 and OX-2 receptors and has pharmacological role on sleep pattern. They play a promoting role in sleep cycle especially wakefulness, metabolism effects, reward, stress, and autonomic function (Roecker et al., 2016). CNS neurons are quiescent during sleep but are active during wakefulness; thus antagonists at orexin receptors enhance the different stages of sleep. This class of drugs is a newly developed class of hypnotic used for the treatment of sleep disorder. Suvorexant, orexin 1 and 2 receptors antagonist was approved by FDA in late 2014 for the treatment of insomnia (Roecker et al., 2016). Its main pharmacological action is to decrease the time to fall asleep and maintain its total duration. The orexin receptor antagonist suvorexant is metabolized by the CYP3A4 enzymes and also it is a substrate of CYP3A4. Its half-life is prolonged by the presence of enzyme inhibitors like ketoconazole, clarithromycin, and verapamil. The drug strength is 10 mg and it is instructed to be taken before sleep. Daytime somnolence and worsening of depression or suicidal ideation are the most common adverse reaction seen with the drug. Its higher dose may disrupt the driving skills of the patients which may lead to road traffic accident. Some newer molecules are undergoing clinical trials (Vermeeren et al., 2015).

## 11.10.2   DOXEPIN

Doxepin, a tricyclic antidepressant, is approved by the FDA in 2010 with an additional property of antagonising histamine (H1/H2) receptors at low doses. It is approved for the management of maintenance of total duration of sleep. The approved initial doses are 6 mg (3 mg in the elderly) that should be taken within 30 min of bedtime for treatment of sleep disorders (Jigar

et al., 2013). Drowsiness and tiredness are some of the adverse effects seen with increasing dose. Abnormal thinking and behavior are the side effects observed with the drug which can worsen suicidal ideation and depression (Hajak et al., 2001).

## 11.10.3   PREGABALIN

Pregabalin, an anxiolytic agent that binds to $Ca^{2+}$ channel $\alpha2\delta$ subunits, have shown efficacy in clinical trials with properties of decreasing time to fall asleep and increases the time spent in slow-wave sleep. It has been found to be useful in generalized anxiety disorder and insomnia (Cho and Song, 2014).

## 11.10.4   RITANSERIN

Ritanserin and other 5HT2A/2C receptor antagonists have shown in clinical trial to promote slow-wave sleep in chronic primary insomnia or generalized anxiety disorder (Mayer, 2003).

## 11.10.5   AGOMELATINE

Agomelatine a melatonin receptor agonist has additional antagonist action on a serotonin receptor, $5HT_{2C}$. It is approved for use in depression and also used in sleep disorder associated with depression as it helps to maintain sleep duration (Levitan et al., 2015).

## 11.10.6   5-HT$_{2A}$ RECEPTOR INVERSE AGONIST

Volinanserin, eplivanserin, and pruvanserin are the new 5-HT$_{2A}$ serotonin receptor subtype agents.

As compared with existing antidepressants with sedative action, they have least affinity for dopamine, histamine, and adrenergic receptors. These drugs have shown efficacy in maintenance of sleep and also increases in slow wave pattern of sleep. Eplivanserin was discontinued during development stages for unknown reasons. The drug development of volinanserin and pruvanserin were stopped despite having positive efficacy data of phase III clinical trial (Teegarden et al., 2008).

### 11.10.7 LOREDIPLON

Lorediplon, a non-benzodiazepine belonging to pyrazolopyrimidine family is underdevelopment for the management for insomnia. Lorediplon acts by modulation of the $GABA_A$ receptor. Its long plasma half-life has the potential of hypnotic effect. It has the potential for maintenance of sleep and the sleep architecture. In the phase I pharmacodynamic study, it showed better efficacy in quality and maintenance of sleep as compared with zolpidem. It has been shown to be well tolerated, safe with no residual effects as compared with other sedative–hypnotic. It is undergoing Phase IIa of clinical trial (World Health Organization, 2011).

### 11.10.8 NEWER MELATONIN RECEPTOR AGONIST

Piromelatine known as Neu-P11 is a newer melatonin receptor agonist underdevelopmental stage. It acts on MT1, MT2, and also $5\text{-}HT_{1A}$/$5\text{-}HT_{1D}$ receptors. It has demonstrated improvement in the sleep cycle of primary insomnia patients and reducing sleep disturbances in Phase II clinical trial. It was found to have efficacy in maintaining deeper sleep, less arousal, and also preserving REM sleep pattern. In the clinical trial, piromelatine was well tolerated with no adverse effects on psychomotor performance the following day (She et al., 2014).

### 11.10.9 NEWER OREXIN RECEPTOR ANTAGONISTS

Filorexant (MK-6096) a promising agent acting antagonist at orexin receptor for the treatment of sleep disorders. It acts on OX-1 and OX-2 receptors as a dual antagonist. It improves the sleep pattern of the patient as it induces sleep and decreases locomotor activity. It was not developed beyond Phase II clinical trials which might be due to the cost in its drug development or due to its unwanted side effects (Kathryn et al., 2017).

Lemborexant is currently in Phase II trial and other molecules like MIN-202, SB-649868, and ACT-462206 are other drugs in the group in Phase I clinical trials. These drugs were proved to reduce latency to sleep, have sleep efficiency (SE), to decrease the phenomenon of wake after sleep onset (ASO), and increased the total sleep time (TST) or duration in relation to "Z" drugs and placebo. These drugs have better safety profile as compared with suvorexant (Misra and Sharma, 2017).

### 11.10.10 ESMIRTAZAPINE

Esmirtazapine is an antidepressant having antagonist action on $5\text{-}HT_{2A}$ and $H_1$ receptors. It showed good hypnotic effects in patients of insomnia. Esmirtazapine was found to improve the time to fall asleep, maintenance and total duration of sleep with course of 6 weeks. The drug is well tolerated, no rebound insomnia seen except for residual daytime effect as one of its side effects. A double-blind trial of 2 weeks was conducted in non-elderly adult patients with primary insomnia which showed improved sleep parameters as compared with placebo and the dose used was well tolerated and safe (Ivgy-May et al., 2015).

## 11.11  FUTURE OPPORTUNITIES AND CHALLENGES

In this modern world, sophistication and technology may be reflected in human's way of life. With increase in the standard of living, many problems become part and parcel of life. These potential problems are challenged by the widespread prevalence of insomnia. In the changing society and work pressure, people may suffer disturbances from abnormal circadian rhythm of sleep which will affect their mental, social, and personal life. Thus, the clinical practice of prescribing sedative–hypnotic became common with improvement in inducing sleep and its maintenance. Tolerance and rebound insomnia upon discontinuation are the common side effects seen when sedative–hypnotics are given for a long time. It is a challenge for the physicians for managing patient for long term. The best management for insomnia should be the combination of nonpharmacological and pharmacological. With the improvement of drug development and clinical research, vast opportunities await to explore new molecules with low side effects with improved efficacy which will bring down tolerance and rebound insomnia on discontinuation.

## 11.12  CONCLUSION

This chapter gives an insight about the development of sedative–hypnotics and its effects on the various psychosocial aspect of modern lifestyle. Insomnia is a chronic medical disorder affecting the medical and socioeconomic parameters. In the content of the problem, the course of treatment is itself a pain for the health professionals as it has significant adverse profile on long-term basis. Sedative–hypnotic deleterious side effect may change the course of treatment of the patients and will not improve the clinical conditions of the patients. With the discovery of new research molecules and advent of newer drugs, the treatment of insomnia will open up new avenues for the patients.

## KEYWORDS

- **benzodiazepines**
- **gamma-aminobutyric acid**
- **hypnotics**
- **sedatives**
- **insomnia**

## REFERENCES

Abernethy, D. R.; Greenblatt, D. J.; Ochs, H. R.; Shader, R. I. Benzodiazepine Drug–Drug Interactions Commonly Occurring in Clinical Practice. *Curr. Med. Res. Opin.* **1984,** *8,* 80–93.

Atkin, T.; Comai, S.; Gobbi, G. Drugs for Insomnia Beyond Benzodiazepines: Pharmacology, Clinical Applications, and Discovery. *Pharmacol. Rev.* **2018,** *70,* 197–245.

Becker, D. E. Pharmacodynamic Considerations for Moderate and Deep Sedation. *Anesth. Prog.* **2012,** *59,* 28–42.

Benjamin, D. Brielmaier. Eszopiclone (Lunesta): A New Nonbenzodiazepine Hypnotic Agent. *Proc (Bayl. Univ. Med. Cent.).* **2006,** *19,* 54–59.

Bonacci, J. M.; Venci, J. V.; Gandhi, M. A. Tasimelteon (Hetlioz™): A New Melatonin Receptor Agonist for the Treatment of Non-24-Hour Sleep-Wake Disorder. *J. Pharm. Pract.* **2015,** *28,* 473–478.

Brett, J.; Murnion, B. Management of Benzodiazepine Misuse and Dependence. *Aust. Prescr.* **2015,** *38,* 152–155.

Brodie, M. J.; Kwan, P. Current Position of Phenobarbital in Epilepsy and Its Future. *Epilepsia* **2012,** *53,* 40–46.

Brunton, L. L.; Hilal-Dandan, R.; Knollmann, B. C. *Hypnotics and Sedatives. Goodman and Gilman's: The Pharmacological Basis of Therapeutic,* 13th ed; McGraw Hill Education: New York, 2018, pp 457–480.

Brzezinski, A. Melatonin in Humans. *N. Engl. J. Med.* **1997,** *336,* 186–195.

Cho, Y. W.; Song, M. L. Effects of Pregabalin in Patients with Hypnotic-dependent Insomnia. *J. Clin. Sleep. Med.* **2014,** *10,* 545–550.

Chouinard, G.; Lefko-Singh, K.; Teboul, E. Metabolism of Anxiolytics and Hypnotics: Benzodiazepines, Buspirone, Zoplicone, and Zolpidem. *Cell. Mol. Neurobiol.* **1999,** *19,* 533–552.

Cloos, J. M.; Ferreira, V. Current Use of Benzodiazepines in Anxiety Disorders. *Curr. Opin. Psychiatry* **2009,** *22,* 90–95.

Connor, K. M.; Ceesay, P.; Hutzelmann, J.; Snavely, D.; Krystal, A. D.; Trivedi, M. H.; Thase, M.; Lines, C.; Herring, W. J.; Michelson, D. Phase II Proof-of-Concept Trial of the Orexin Receptor Antagonist Filorexant (MK-6096) in Patients with Major Depressive Disorder. *Int. J. Neuropsychopharmacol.* **2017,** *20,* 613–618.

Garzone, P. D.; Kroboth, P. D. Pharmacokinetics of the Newer Benzodiazepines. *Clin. Pharmacokinet.* **1989,** *16,* 337–364.

Gauillard, J.; Cheref, S.; Vacherontrystram, M. N.; Martin, J. C. Chloral Hydrate: A Hypnotic Best Forgotten? *Encephale* **2002,** *28,* 200–204.

Gerecke, M. Chemical Structure and Properties of Midazolam Compared with Other Benzodiazepines. *Br. J. Clin. Pharm.* **1983,** *16,* 11S–16S.

Golan, D. E.; Armstrong, E. J.; Armstrong, A. W. Pharmacology of GABAergic and Glutamatergic Neurotransmission. *Principles of Pharmacology: The Pathophysiologic Basis of Drug Therapy,* 4th ed; Wolters Kluwer: Philadelphia, 2017; pp 184–205.

Goldberg, R. Drugs Across the Spectrum. *Cengage Learning* **2009,** 195.

Gregory, M. A.; Thomas, M; Margaret, A. H. Pharmacotherapy Treatment Options for Insomnia: A Primer for Clinicians. *Int. J. Mol. Sci.* **2016,** *17,* 50.

Guilleminault, C. Benzodiazepines, Breathing, and Sleep. *Am. J. Med.* **1990,** *88,* 25S–28S.

Haefely, W. E. Synaptic Pharmacology of Barbiturates and Benzodiazepines. *Agents Actions* **1977,** *7,* 353–359.

Hajak, G.; Rodenbeck, A.; Voderholzer, U.; et al. Doxepin in the Treatment of Primary Insomnia: A Placebo-controlled, Doubleblind, Polysomnographic Study. *J. Clin. Psychiatry.* **2001,** *62,* 453–463.

Hargrove, E. A.; Bennett, A. E.; Ford, F. R. Acute and Chronic Barbiturate Intoxication—Recent Advances in Therapeutic Management. *Calif. Med.* **1952,** *77,* 383–386.

Higgitt, A.; Fonagy, P.; Lader, M. The Natural History of Tolerance to the Benzodiazepines. *Psychol. Med. Monogr. Suppl.* **1988,** *13,* 1–55.

Hood, S. D.; Norman, A.; Hince, D. A.; Melichar, J. K.; Hulse, G. K. Benzodiazepine Dependence and its Treatment with Low Dose Flumazenil. *Br. J. Clin. Pharmacol.* **2014,** *77,* 285–294.

Ito, T.; Suzuki, T.; Wellman, S. E.; Ho, I. K. Pharmacology of Barbiturate Tolerance/Dependence: GABAA Receptors and Molecular Aspects. *Life. Sci.* **1996,** *59,* 169–195.

Ivgy-May, N.; Roth, T.; Ruwe, F.; Walsh, J. Esmirtazapine in Non-elderly Adult Patients with Primary Insomnia: Efficacy and Safety From a 2-Week Randomized Outpatient Trial. *Sleep Med.* **2015,** *16,* 831–837.

Jackson, R. Barbiturates and their Drug Interactions. *Caementum* **1969,** *26,* 30–31.

Janhsen, K.; Roser, P.; Hoffmann, K. The Problems of Long-term Treatment with Benzodiazepines and Related Substances. *DtschArztebl. Int.* **2015,** *112,* 1–7.

Janssen, P. A.; Niemegeers, C. J.; Marsboom, R. P. Etomidate, a Potent Non-barbiturate Hypnotic. Intravenous Etomidate in Mice, Rats, Guinea-Pigs, Rabbits and Dogs. *Arch. IntPharmacodynTher.* **1975,** *214,* 92–132.

Jigar, K.; Ananda, K. K.; Jaykumar, J. S.; Marcelle, T.; Manish, M. Therapeutic Rationale for Low Dose Doxepin in Insomnia Patients. *Asian Pac. J. Trop. Dis.* **2013,** *3,* 331–336.

Jonathan, R. L.; Schwartz, T. R. Neurophysiology of Sleep and Wakefulness: Basic Science and Clinical Implications. *Curr. Neuropharmacol.* **2008,** *6,* 367–378.

Katzung, B. G.; Masters, S. B.; Trevor, A. J. Sedative-Hypnotics Drugs. *Basic and Clinical Pharmacology,* 11th ed; Lange Medical Publication: California, 2009, pp 369–383.

Kauffman, G. B. Adolf von Baeyer and the Naming of Barbituric Acid. *J. Chem. Educ.* **1980,** *57,* 222.

Kovacic, P.; Ott, N.; Cooksy, A. L. Benzodiazepines: Electron Affinity, Receptors and Cell Signaling–A Multifaceted Approach. *J. Recept. Signal. Transduct. Res.* **2013,** *33,* 338–343.

Kristensen, M. B. Drug Interactions and Clinical Pharmacokinetics. *Clin. Pharmacokinet.* **1976,** *1,* 351–372.

Kubicki, S.; Herrmann, W. M.; Holler, L.; Haag, C. On the Distribution of REM and NREM Sleep Under Two Benzodiazepines With Comparable Receptor Affinity But Different Kinetic Properties. *Pharmacopsychiatry* **1987,** *20,* 270–277.

Larry, C.; Mark, A. W. Over-the-Counter Agents for the Treatment of Occasional Disturbed Sleep or Transient Insomnia: A Systematic Review of Efficacy and Safety. *Prim. Care Companion CNS Disord.* **2015,** *17,* DOI: 10.4088/PCC.15r01798.

Laudon, M.; Frydman-Marom, A. Therapeutic Effects of Melatonin Receptor Agonists on Sleep and Comorbid Disorders. *Int. J. Mol. Sci.* **2014,** *15,* 15924–15950.

Levitan, M. N.; Papelbaum, M.; Nardi, A. E. Profile of Agomelatine and Its Potential in the Treatment of Generalized Anxiety Disorder. *Neuropsychiatr. Dis. Treat.* **2015,** *11,* 1149–1155.

Logan, B. K.; Case, G. A.; Gordon, A. M. Carisoprodol, Meprobamate, and Driving Impairment. *J. Forensic Sci.* **2000,** *45,* 619–623.

Macdonald, R. L.; Olsen, R. W. GABA$_A$ Receptor Channels. *Annu. Rev. Neurosci.* **1994,** *17,* 569–602.

Mark, L. C. Metabolism of the barbiturates. *Ann. R. Coll. Surg. Engl.* **1971,** *48,* 77–78.

Masaomi, M. Pharmacology of Ramelteon, a Selective MT$_1$/MT$_2$ Receptor Agonist: A Novel Therapeutic Drug for Sleep Disorders. *CNS Neurosci. Ther.* **2009,** 15, 32–51.

Mayer, G. Ritanserin Improves Sleep Quality In Narcolepsy. *Pharmacopsychiatry* **2003,** *36,* 150–155.

Mendelson, W. B.; Martin, J. V.; Rapoport, D. M. Effects of Buspirone on Sleep and Respiration. *Am. Rev. Respir. Dis.* **1990,** *141,* 1527–1530.

Misra, A. K.; Sharma, P. K. Pharmacotherapy of Insomnia and Current Updates. *J. Assoc. Physicians India.* **2017,** *65,* 43–47.

Norn, S.; Permin, H.; Kruse, E.; Kruse, P. R. On the History of Barbiturates. *Dan. Medicinhist. Arbog.* **2015,** *43,* 133–151.

O'Boyle, C. A. Benzodiazepine-induced Amnesia and Anaesthetic Practice: A Review. *Psychopharmacol. Ser.* **1988,** *6,* 146–165.

Ochoa, J. G.; Kilgo, W. A. The Role of Benzodiazepines in the Treatment of Epilepsy. *Curr. Treat. Options. Neurol.* **2016,** *18,* 18.

Onyett, S. R. The Benzodiazepine Withdrawal Syndrome and Its Management. *J. R. Coll. Gen. Pract.* **1989,** *39,* 160–163.

Pascoe, J. P.; Gallagher, M.; Kapp, B. S. Benzodiazepine Effects on Heart Rate Conditioning In the Rabbit. *Psychopharmacology (Berl).* **1983,** *79,* 256–261.

Pelkonen, O.; Karki, N. T. Effect of Physicochemical and Pharmacokinetic Properties of Barbiturates on the Induction of Drug Metabolism. *Chem. Biol. Interact.* **1973,** *7,* 93–99.

Perry, P. J.; Alexander, B. Sedative/Hypnotic Dependence: Patient Stabilization, Tolerance Testing, and Withdrawal. *Drug. Intell. Clin. Pharm.* **1986,** *20,* 532–537.

Puntillo, K.; Casella, V.; Reid, M. Opioid and Benzodiazepine Tolerance and Dependence: Application of Theory to Critical Care Practice. *Heart Lung* **1997,** *26,* 317–324.

Ramakrishnan, K.; Scheid, D. C. "Treatment Options for Insomnia." *Am. Family Physician* **2007,** *76,* 517–526.

Ramchandani, D.; Lopez-Munoz, F.; Alamo, C. Meprobamate-tranquilizer or Anxiolytic? A Historical Perspective. *Psychiatr. Q.* **2006,** *77,* 43–53.

Ratz, A. E.; Schlienger, R. G.; Linder. L.; Langewitz, W.; Haefeli, W. E. Pharmacokinetics and Pharmacodynamics of Clomethiazole After Oral And Rectal Administration In Healthy Subjects. *Clin. Ther.* **1999,** *21,* 829–840.

Roecker, A. J.; Cox, C. D.; Coleman, P. J. Orexin Receptor Antagonists: New Therapeutic Agents for the Treatment of Insomnia. *J. Med. Chem.* **2016,** *59,* 504–530.

Rosen, A. S.; Fournie, P.; Darwish, M.; Danjou, P.; Troy, S. M. Zaleplon Pharmacokinetics and Absolute Bioavailability. *Biopharm. Drug. Dispos.* **1999,** *20,* 171–175.

Rudolph, I.; Dauss, I. Premedication and General Anesthesia in Stomatologic Practice. *Stomatol. DDR* **1974,** *24,* 357–362.

Saari, T. I.; Uusi-Oukari, M.; Ahonen, J.; Olkkola, K. T. Enhancement of GABAergic Activity: Neuropharmacological Effects of Benzodiazepines and Therapeutic Use in Anesthesiology. *Pharmacol. Rev.* **2011,** *63,* 243–267.

Santos Moraes, W. A.; Burke, P. R.; Coutinho, P. L.; Guilleminault, C.; Bittencourt, A. G.; Tufik, S.; Poyares, D. Sedative Antidepressants and Insomnia. *Rev. Bras. Psiquiatr.* **2011,** *33,* 91–95.

Santos, C.; Olmedo, R. E. Sedative-Hypnotic Drug Withdrawal Syndrome: Recognition and Treatment. *Emerg. Med. Pract.* **2017,** *19,* 1–20.

Saunders, P. A.; Ho, I. K. Barbiturates and the GABAA Receptor Complex. *Prog. Drug. Res.* **1990,** *34,* 261–286.

Sellers, E. M. Clinical Pharmacology and Therapeutics of Benzodiazepines. *Can. Med. Assoc. J.* **1978,** *118,* 1533–1538.

Sharma, H. L.; Sharma, K. K. Anxiolytic and Hypnotics. *Principles of Pharmacology*, 2nd ed; Paras Medical Publisher: New Delhi, 2017, pp 442–450.

She, M.; Hu, X.; Su, Z.; Zhang, C.; Yang, S.; Ding, L.; et al. Piromelatine, A Novel Melatonin Receptor Agonist, Stabilizes Metabolic Profiles and Ameliorates Insulin Resistance in Chronic Sleep Restricted Rats. *Eur. J. Pharmacol.* **2014,** *727*, 60–65.

Sigel, E.; Steinmann, M. E. Structure, Function, and Modulation of $GABA_A$ Receptors. J. *Biol. Chem.* **2012,** *287*, 40224–40231.

Siriwardena, A. N.; Qureshi, M. Z.; Dyas, J. V. Hugh Middleton, Roderick Orner. Magic bullets for insomnia? Patients' Use and Experiences of Newer (Z Drugs) Versus Older (Benzodiazepine) Hypnotics for Sleep Problems in Primary Care. *Br. J. Gen. Pract.* **2008,** *58*, 417–422.

Smith, M. C.; Riskin, B. J. The Clinical Use of Barbiturates in Neurological Disorders. *Drugs.* **1991,** *42*, 365–378.

Soderpalm, B. Pharmacology of the Benzodiazepines; With Special Emphasis on Alprazolam. *Acta. Psychiatr. Scand. Suppl.* **1987,** *335*, 39–46.

Steinbacher, D. M. Propofol: a Sedative-hypnotic Anesthetic Agent for Use in Ambulatory Procedures. *Anesth. Prog.* **2001,** *48*, 66–71.

Swainston Harrison, T.; Keating, G. M. Zolpidem: A Review of Its Use in the Management of Insomnia. *CNS Drugs* **2005,** *19*, 65–89.

Teegarden, B. R.; Al, Shamma. H.; Xiong, Y. 5-HT(2A) Inverse-agonists for the Treatment of Insomnia. *Curr. Top. Med. Chem.* **2008,** *8*, 969–976.

Terzano, M. G.; Rossi, M.; Palomba, V.; Smerieri, A.; Parrino, L. New Drugs for Insomnia: Comparative Tolerability of Zopiclone, Zolpidem and Zaleplon. *Drug Saf.* **2003,** *26*, 261–282.

Tordjman, S.; Chokron, S.; Delorme, R.; Charrier, A.; Bellissant, E.; Jaafari, N.; Fougerou, C. Melatonin: Pharmacology, Functions and Therapeutic Benefits. *Curr. Neuropharmacol.* **2017,** *15*, 434–443.

Truter, I. An Investigation into Compound Analgesic Prescribing in South Africa, With Special Emphasis on Meprobamate-containing Analgesics. *Pharmacoepidemiol. Drug. Saf.* **1998,** *7*, 91–97.

Uzun, S.; Kozumplik, O.; Jakovljevic, M.; Sedic, B. Side Effects of Treatment with Benzodiazepines. *Psychiatr. Danub.* **2010,** *22*, 90–93.

Vachharajani, N. N.; Yeleswaram, K.; Boulton, D. W. Preclinical Pharmacokinetics and Metabolism of BMS-214778, A Novel Melatonin Receptor Agonist. *J. Pharm. Sci.* **2003,** *92*, 760–772.

Vermeeren, A.; Sun, H.; Vuurman, E. F, et al. On-the-Road Driving Performance the Morning after Bedtime Use of Suvorexant 20 and 40 mg: A Study in Non-elderly Healthy Volunteers. *Sleep* **2015,** *38*, 1803.

Whitwam, J. G.; Amrein, R. Pharmacology of Flumazenil. *Acta. Anaesthesiol. Scand. Suppl.* **1995,** *108*, 3–14.

Wick, J. Y. The History of Benzodiazepines. *Consult. Pharm.* **2013,** *28*, 538–548.

Wingrove, P. B.; Safo, P.; Wheat, L.; Thompson, S. A.; Wafford, K. A.; Whiting, P. J. Mechanism of Alpha-subunit Selectivity of Benzodiazepine Pharmacology at Gamma-aminobutyric Acid Type A Receptors. *Eur. J. Pharmacol.* **2002,** *437*, 31–39.

Wong, P. T.; Yoong, Y. L.; Gwee, M. C. Acute Tolerance to Diazepam Induced by Benzodiazepines. *Clin. Exp. Pharmacol. Physiol.* **1986,** *13*, 1–8.

World Health Organization. "International Nonproprietary Names for Pharmaceutical Substances (INN): Proposed INN: List 105" (PDF). *WHO Drug Information* **2011,** *25*, 179.

Zeegers, A.; van Wilgenburg, H.; Leeuwin, R. S. Cardiac Effects of Benzodiazepine Receptor Agonists and Antagonists in the Isolated Rat Heart: A Comparative Study. *Life Sci.* **1998,** *63*, 1439–1456.

Zisapel, N. Drugs for Insomnia. *Expert. Opin. Emerg. Drugs.* **2012,** *17*, 299–317.

# CHAPTER 12

# Antidepressant Drugs

VAIBHAV WALIA and MUNISH GARG*

*Maharshi Dayanand University, Rohtak, Haryana 124001, India*

*Corresponding author. E-mail: mgarg2006@gmail.com*

## ABSTRACT

Depression is characterized by the behavioral despair in the affected individuals which affects global population and increased the years of disability. Various theories suggested different mechanism leading to depression, but the exact cause of depression is still not known. Despite of the availability of the various drugs for the treatment but still there is not even a single drug which resolves the depressive symptoms completely. For example, the use of monoamine oxidase inhibitors (MAOIs) increase the risk of hypertensive crisis, tricyclic antidepressants (TCAs) increase cardiovascular toxicity and seizures while selective serotonin reuptake inhibitors (SSRIs) increase suicidality and sexual dysfunction, suggesting each category of antidepressants has its own advantages and limitations. Thus, special emphasis should be given while selecting antidepressant for the patients because the wrong choice may worsen the condition, increase the risk of relapse and drug resistance. In the present work, authors briefly describe depression, neurochemical implicated in the depression, and the several classes of antidepressants with their limitations.

## 12.1 INTRODUCTION

Depression is a psychiatric disorder which affects the people worldwide and is characterized by the hopelessness, helplessness, sleep disturbances, anhedonia, suicidal ideations, and attempts in the affected patients (WHO, 1992). It has been reported that the number of individuals suffering from the depression increases day by day and there is approximately 18.4% increase in the number of depression patients from 2005 to 2015 (WHO, 2017). The exact pathophysiology of depression is still not known. The most widely accepted theory is the monoaminergic theory, suggesting that the depression is accompanied by the reduction of monoamine levels in the synaptic cleft (Krishnan and Nestler, 2008). 5-HT regulates mood and thus its dysfunctioning evokes depressive symptoms (Southwick et al., 2005; Vashadze, 2007). Further the depressive individuals show the reduced levels of serotonin and its metabolite in the brain (Asberg, 1997; Meltzer, 1989; Placidi et al., 2001; Wester et al., 1990). Besides 5-HT, norepinephrine (NE) or noradrenaline (NA) also regulates mood (Drevets et al., 2002) and thus deficits in the functioning of NA/NE also provokes depressive symptoms (Briley and Moret, 2010). It has been reported that both the rate of metabolism of NE and density of NE transporters get reduced in the locus coeruleus region (Charney and Manji, 2004). Role of dopamine in the pathogenesis of depression is also not clear till today. Dopamine is known to regulate reward and motivation (Fibiger, 1995) and thus the impaired dopamine signaling is responsible for anhedonia and loss of motivation in the depression patients (Willner, 1995). Therefore, the potentiation of dopaminergic transmission in brain is responsible for the antidepressant effects (Arnt et al., 1984). Gamma-aminobutyric acid (GABA) is an inhibitory neurotransmitter (Sherif and Ahmed, 1995) and involved in the pathogenesis of depression (Sanacora et al., 1999). It has been reported that the patients suffering from depression exhibit reduced level of GABA in the brain (Bloom et al., 1971; Gold et al., 1980; Price et al., 2009; Sanacora et al., 2004, 2006). Furthermore, the treatment with the antidepressants produces a marked change in the GABAergic neurotransmission in brain (Sanacora and Saricicek, 2007). Stress and the stressful events make the individual susceptible to depression (Caspi et al., 2003; Vollmayr and Henn, 2003). It has been reported that the exposure to the stressors led to the alterations (Pace et al., 2006) responsible for the release of proinflammatory cytokines (Dowlati et al., 2010; Lee and Kim, 2006; Myint et al., 2005) further responsible for the activation of HPA axis (Gisslinger et al., 1993; Capuron et al., 2003; Anisman and Merali, 1999; Connor et al., 1998) and produces depressive symptoms (Nemeroff, 1996). Furthermore, the exposure to the stressor led to the release of nitric oxide (NO) in the brain (Madrigal et al., 2001; Walia and Gilhotra, 2016) which further negatively regulates the extracellular levels of DA and 5-HT (Kaehler et al., 1999; Wegener et al., 2000), negatively regulates the enzymes involved in 5-HT synthesis (Kuhn and Arthur, 1996), and increase the activity of 5-HT transporters (Miller and Hoffman, 1994). Furthermore, a recent study suggested that depression might result in the increased activities (MAO, COX, NOS, GSK, etc.) or decreased activities (TPH, GAD, etc.) of several enzymes (Walia, 2016). In this chapter, authors briefly describe the depression and the various categories of antidepressants with their advantages and limitations.

## 12.2 CLASSIFICATION OF ANTIDEPRESSANTS

Categories of the antidepressant drugs used for the treatments of depression (Anderson et al., 2008) include Figure 12.1, which represents the classes of antidepressants.

### 12.2.1 FIRST-GENERATION ANTIDEPRESSANTS

#### 12.2.1.1 MONOAMINE OXIDASEINHIBITORS (MAOIs)

MAO metabolizes monoamines and therefore the inhibition of the enzyme MAO by MAOIs results in the increased levels of brain monoamines. MAO-A inhibitors are generally used in the treatment of depression but generally reserved for the patients of atypical depression or refractory cases or in the patients where the other antidepressants do not work well (McGrath et al., 1993; Stewart et al., 1997; Thase et al., 2001). The use of MAOIs proves fatal when consumed with the milk products (Cohen, 1997). Adverse effects may include orthostatic hypotension, insomnia, weight gain, edema, muscle pains, myoclonus, paresthesias, hepatotoxicity, etc. (Evans et al., 1982; Rabkin et al., 1984; Gomez-Gil et al., 1996; Fava, 2000; Robinson, 2002).

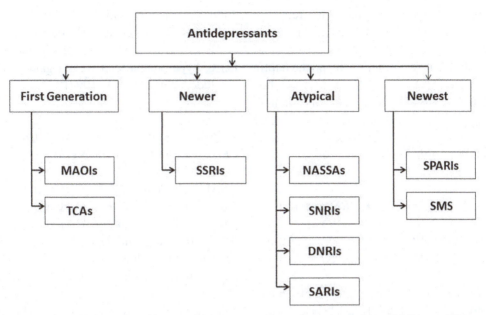

**FIGURE 12.1** Classification of antidepressant drugs. MAOIs: monoamine oxidase inhibitors; TCAs: tricyclic antidepressants; SSRIs: selective serotonin reuptake inhibitors; NASSAs: noradrenergic specific serotonergic agents; NDRIs: noradrenaline and dopamine reuptake inhibitors; SARIs: serotonin antagonist and reuptake inhibitors; SPARIs: serotonin partial agonist reuptake inhibitors; SMS: serotonin modulator and stimulators.

## 12.2.1.2 TRI-CYCLIC ANTIDEPRESSANTS (TCAs)

TCAs have three-ring chemical structure common to them (Santarsieri and Schwartz, 2015). TCAs inhibit the reuptake of NA, DA, and 5-HT in to the presynaptic neurons and exerts marked earlier response due to the dual inhibition of both 5-HT and NA reuptake (Thase et al., 2001; Thompson, 2002) and this is also further responsible for the unwanted side effects of TCAs (Pacher and Kecskemeti, 2004). The unwanted effects of these drugs is due to their action on various other receptors including adrenergic receptors $(\alpha_1)$, histamine $(H_1)$, and muscarinic (M) receptors (Glassman, 1984; Goodman et al., 2001; Pacher et al., 1999) and these unwanted effects result in the discontinuation of TCAs in approximately 27% patients (Montgomery et al., 1994). Despite of their effectiveness in the treatment of the depression, the adverse effects accompanied by

the TCAs imposed limitations regarding their use for the treatment of depression (Feighner, 1999; Holm and Markham, 1999). Adverse effects include dryness of mouth, blurred vision, dizziness, lethargy, sedation (Cohen, 1997), seizures, cardiac block or arrhythmias (Feighner, 1999), due to the blockade of cardiac sodium channels (Stahl, 2000).

## 12.2.2 NEWER ANTIDEPRESSANTS

### 12.2.2.1 SELECTIVE SEROTONIN REUPTAKE INHIBITORS (SSRIs)

SSRIs have the biggest advantage that they are 1500-fold more selective for 5-HT transporter and have minimal binding affinity for postsynaptic receptors (including $\alpha_1$, $\alpha_2$, $\beta$, $H_1$, M, and $D_2$) (Owens et al., 1997; Thomas et al., 1987). SSRIs inhibit the 5-HT transporter in the brain and inhibit the reuptake of 5-HT and thus increase the 5-HT levels (Papakostas, 2008). SSRIs have weaker or no direct pharmacological action at postsynaptic 5-HT receptors (Owens et al., 1997; Sanchez and Hyttel, 1999). Adverse events include aggression, nausea, vomiting, gastritis, sexual dysfunction, tremor, and cardiovascular toxicity (Briley et al., 1996; Detke et al., 2002).

### 12.2.3 ATYPICAL ANTIDEPRESSANTS

#### 12.2.3.1 NORADRENERGIC SPECIFIC SEROTONERGIC AGENTS (NASSAs)

NaSSAs, for example,. Mirtazapine (de Boer and Ruigt, 1995; De Montigny et al., 1995), enhances noradrenergic signaling and block central $\alpha_2$-adrenoceptors (Bengtsson et al., 2000; Barkin et al., 2000) and 5-HT receptors (including 5-HT$_2$ and 5-HT$_3$ subtype) (Bengtsson et al.,

2000) but has no significant effect on dopamine and muscarinic receptors (Barkin et al., 2000; de Boer, 1996). Adverse effects include somnolence, hyperphagia, and weight gain (Barkin et al., 2000; Stimmel et al., 1997).

#### 12.2.3.2 SELECTIVE NORADRENALINE REUPTAKE INHIBITORS (SNRIs)

SNRIs selectively inhibit the reuptake of NA only. For example, reboxetine a new antidepressants (Pinder, 1997, 2001) which activates 5-HT$_{1A}$ receptors and desensitizes $\alpha_2$-adrenergic heteroreceptors due to the inhibition of reuptake of noradrenaline (Szabo and Blier, 2001). However, there is no evidence regarding increase efficacy as compared with SSRIs (Brunello et al., 2002). Figure 12.2 represents the mechanism of venlafaxine, desvenlafaxine, levomilnacipran, and duloxetine.

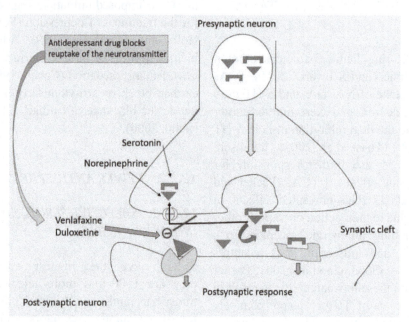

**FIGURE 12.2** Mechanism of action of SSRIs and SNRIs.
*Source*: Adapted from Harvey et al., 2011.

### 12.2.3.3  NORADRENALINE AND DOPAMINE REUPTAKE INHIBITORS (NDRIs)

NDRIs inhibit the reuptake of both NA and dopamine in the brain. For example, bupropion is an NDRI antidepressant mainly used in the cessation of smoking (Fava et al., 2005) and enhances the monoaminergic signaling with no appreciable serotonergic activity (Ascher et al., 1995; Stahl et al., 2004; Stahl, 1996; Richelson, 1996). Bupropion does not inhibit MAO and postsynaptic H, α, β, 5-HT, DA, or M receptors (Baldessarini, 2001; Ascher et al., 1995; Stahl et al., 2004; Dong and Blier, 2001) and therefore free from the sedation, cognitive impairment, or hypotensive effects (Stahl et al., 2004). Adverse effects include dryness of mouth, insomnia, headache, restlessness, anxiety, constipation, nasopharyngitis, and fatigue (Patel et al., 2016).

### 12.2.3.4  SEROTONIN ANTAGONIST AND REUPTAKE INHIBITORS (SARIs)

SARIs inhibit 5-HT transporters and show antagonism at $5\text{-HT}_{2A/2C}$ receptors (Pazzagli et al., 1999; Haria et al., 1994) and increase the levels of 5-HT in brain. For example, Trazodone a "second generation" antidepressant drug and used as an alternative to MAOIs and TCAs (Golden et al., 2009). Trazodone is used as a preferred treatment in MDD patients (Golden et al., 2009) and in other conditions including PTSD, bulimia nervosa, and adjustment disorders (Warner et al., 2001; Hudson et al., 1989; Rasavi et al., 1999).

### 12.2.4  NEWEST ANTIDEPRESSANTS

#### 12.2.4.1  SEROTONIN PARTIAL AGONIST REUPTAKE INHIBITORS (SPARIs)

SPARIs, for example, vilazodone inhibit the reuptake of 5-HT and shows partial $5\text{-HT}_{1A}$ receptor agonism (Schwartz and Stahl, 2011). Vilazodone appears to be safe, well tolerated, and have lower risk as compared with SSRIs or SNRIs (Robinson et al., 2011).

#### 12.2.4.2  SEROTONIN MODULATOR AND STIMULATORS (SMSs)

SMS, for example, vortioxetine block 5-HT transporters, shows agonism at $5\text{-HT}_{1A}$ receptors and antagonism at $5\text{-HT}_{1B/D}$, $5\text{-HT}_3$, and $5\text{-HT}_7$ receptors (Schatzberg and DeBattista, 2015) and has high efficacy and tolerability (Llorca et al., 2014) and low risk of side effects (Schatzberg and DeBattista, 2015) as compared with SSRIs.

## 12.3  FACTORS TO BE CONSIDERED BEFORE THE CHOICE OF AN ANTIDEPRESSANT

Various factors need to be considered while selecting an antidepressant. For example, it is better to start with the lower dose of the antidepressant and if possible switch to the antidepressant which confers the lowest risk (American Psychiatric Association, 2010). It is better to avoid SSRIs and SNRIs in case of the patients suffering from the sexual dysfunction (American Psychiatric Association, 2010). Of SSRIs class, fluoxetine and paroxetine impose the higher risk of the sexual dysfunction while (es)citalopram confers the lower risk (Kennedy et al., 2009). Bupropion is the preferred antidepressant followed by the mirtazapine for the patients suffering from sexual dysfunction (Korean Medication Guideline for Depressive Disorder, 2012). SSRIs can be used safely in the patients suffering from the cardiovascular diseases (American Psychiatric Association, 2010). SNRIs such as venlafaxine exacerbate cardiac arrhythmia and thus patients required monitoring for the cardiovascular functions (Suehs et al., 2008). TCAs imposed higher risk of seizures (Kennedy et al., 2009) while SSRIs such as citalopram/escitalopram confers the lowest risk of seizures (Korean Medication Guideline for Depressive Disorder, 2012). In diabetic TCAs can cause the worsening of glycemic control and thus SSRIs are preferred in these patients (American Psychiatric Association; 2010). Furthermore, (es) citalopram is the drug of primary choice while sertraline is the drug of second choice in case of diabetics (Korean Medication Guideline for Depressive Disorder, 2012).

## 12.4  PHASES OF ANTIDEPRESSANT TREATMENT

Clinical guidelines recommend that the antidepressant treatment should be continued for at least 6 months (Schulberg et al., 1998; Geddes and Butler, 2002). Treatment should aim to rapidly resolve the symptoms, prevention of relapse and recurrence. Antidepressant treatment includes three phases:

- Acute phase- The main aim of this phase is to resolve the symptoms, to restore the functioning, and to achieve remission (Lin et al., 1998, Rush et al., 2003).
- Continuation phase- The aim of this phase is to achieve the sustained remission. The failure of this phase is thus characterized by the adverse consequences, risk of relapse (Judd et al., 1998) and resistance to the antidepressant (Simon et al., 2000) sociality, and other conditions that led to morbidity.

- Maintenance phase- The maintenance phase of the antidepressant should last for of 3–5 years, however, there are no reported studies that clearly demonstrate the appropriate duration of long-term antidepressant drug treatment (Prien et al., 1984; Frank et al., 1990; Reynolds et al., 1999).

## 12.5  PROBLEMS ASSOCIATED WITH THE USE OF ANTIDEPRESSANTS

The main reason behind the problems regarding the use of the antidepressant is the lack of adequate long-term clinical efficacy and safety studies (Hughes and Cohen, 2009). Furthermore, the adequate use of the antidepressants also results in the high recurrence rate than the patients not treated with antidepressants (Bockting et al., 2008). Recent research showed the patients perspectives on antidepressants including sedation emotional blunting, and instability (Gibson et al., 2014) is not investigated and mentioned anywhere (Goldberg and Moncrieff, 2011). Furthermore, the use of the antidepressant drugs for the period greater than 3 years increased the risk of suicidality in the prescribed patients (Read et al., 2014; Walia, 2017). Another big problem is the poor compliance (Weich et al., 2007; Bockting et al., 2008) which further affect the rate of response and remission (Akerblad et al., 2006). The reasons for poor compliance may include low income, race/ethnicity, and age (Jeon-Slaughter, 2012) and lack of response (Sheehan et al., 2008). Another big problem is the lack of the response in the antidepressant treatment (ADT) tachyphylaxis (Byrne and Rothschild, 1998; Lieb, 1990; Nierenberg and Alpert, 2000) and is characterized by fatigue, cognitive deficits, sleep impairment, weight gain, and sexual dysfunction (Rothschild et al., 2009;

Rothschild, 2008). ADT tachyphylaxis is a form of relapse and is responsible for the development of tolerance to the antidepressant drugs (Lieb and Balter, 1984). ADT tachyphylaxis has been recognized for MAOIs (Cohen and Baldessarini, 1985; Lieb and Balter, 1984; Lieb, 1990; Mann, 1983) and SSRIs (Byrne and Rothschild, 1998; Solomon et al., 2005; Fava et al., 1995). Another problem is the discontinuation syndrome which is known to characterize by the reactions which arise rapidly, and reflect the signs of drug withdrawal (Baldessarini et al. 2010). These signs ate generally dependent upon the elimination half-life, treatment duration, and rate of metabolism (Blier and Tremblay, 2006; Bitter et al. 2011). The discontinuation syndrome can be misinterpret as the relapse of the underlying condition or in case of the switch to other drugs it is considered as the side effects of a newer drug (Haddad and Anderson, 2007). Discontinuation syndrome arises after the withdrawal of SSRIs treatment due to the rapid decrease in 5-HT level (Blier and Tremblay, 2006). The next main problem is the worsening of the depression by the use of the antidepressants such as SSRIs (Gunnell et al., 2005). It has been reported that the use of the SSRIs worsens the disease and is responsible for the suicidality in the prescribed patients (Fergusson al., 2005; Khan and Bernadt, 2011; Walia, 2017). Besides this, SSRIs which is considered as the safest antidepressant and is used widely in the treatment of depression but the use increases the sleep disturbances (Thase, 1999; Antai-Otong, 2004), seizures (Braitberg and Curry, 1995; Pisani et al., 2002), increases the risk of malformations (Diav-Citrin et al., 2008), sexual dysfunction in the patients (Woodrum and Brown, 1998) and risk of the breathing problems in fetus (Koren, 1996; Forsberg et al., 2014). It is demonstrated that the elevated levels of 5-HT reduces seizure threshold (Kondziella and Asztely, 2009) and pulmonary arterial smooth muscle cell proliferation resulting in the breathing problems (Helle et al., 2012; Sit et al., 2011). SSRIs therapy showed more incidence of sexual dysfunction in males as compared with the females, but sexual dysfunction in women is found to be more intense than men (Montejo-Gonzalez et al., 1997). Sexual dysfunction associated with the SSRIs therapy is characterized by the loss of libido, decreased partner desire, erectile dysfunction, delayed ejaculation, impaired ejaculation, anorgasmia, pain, loss of sensation, and decreased pleasure (Lorrain et al., 1997; Damis et al., 1999; Corona et al., 2009).

## 12.6 FUTURE OPPORTUNITIES AND CHALLENGES

According to International Classification of Diseases, depression is characterized by the three core symptoms which include low mood, low energy levels, and anhedonia. However, the other symptoms are also present and these include sleep disturbances, loss of appetite, low self-esteem, suicidal ideation and attempts. Accordingly, depression can be mild, moderate, and severe depending upon the symptoms (WHO, 1992). Researchers faced challenge in the understanding of depression, development of antidepressants, and the pharmaco-therapy of the depression. The biggest hurdle is limited knowledge of the disease, that is, the complete pathophysiology of the depression is not known still today. Furthermore, no clear biomarker of depression still exists and in the absence of a clear biomarker and the biological target, the appropriate pharmacotherapeutics of the depression is not possible (Tianmei and XIn, 2016). Furthermore, depression is also not a unitary disorder, it is a group of the disorder or the disorder accompanied by the other disorders of the varying nature. Therefore, the selection of the adequate drug for the treatment is crucial and again it is also important to optimize and individualize the therapy (Penn and Tracy, 2012). Furthermore, literature also does not provide the suitable information on it because of the increasing number of online published work available day by day. For example, numerous papers on the SSRIs suggested that the fluoxetine exert anxiogenic effect while some of the studies suggested the anxiolytic effect of fluoxetine. Such thing often led to the confusion and doubts regarding the concept. The antidepressant should be used only for the treatment of moderate and severe depression (Anderson et al., 2008). Furthermore, the biggest problem of the antidepressant therapy is the nonadherence to the therapy, various factors might be responsible for this and these includes incidence of sexual dysfunction, delayed onset of actions, substance abuse, fears of addiction, low follow-up, high cost of therapy, patients belief, etc. (Sansone and Saonsone, 2012). The future research should aim at the development of the appropriate biomarker for the depression. Furthermore, the research should be focused on the better understanding of the pathology because without the understanding of pathology depression cannot be resolved completely. The use of the antidepressant should be done accordingly with the patients since one drug does not fit for all. Also the patients on the antidepressant therapy should be monitored for the response and the incidence of the adverse drug reactions. It has been seen that the prescription of a psychopharmacological agent is often accompanied by the prescription of some multivitamins. Various minerals and vitamins has well-known therapeutic effect in the psychopharmacological conditions, therefore these should be further explored for their full potential in these conditions.

## 12.7   CONCLUSION

Antidepressants used for the treatment of the depression have its own limitations and merits. Therefore, care should be taken while selecting the antidepressants for the different categories of the patients. Furthermore, the increase in the age is a major risk of the depression, also the aging is characterized by the presence of the increased levels of the inflammatory cytokines, thus anti-inflammatory agents might be beneficial in such conditions. Also the vitamins are frequently prescribed with the antidepressants and these vitamins play a key role in the synthesis of the various neurotransmitter. Therefore, it might prove beneficial alone in the therapeutics of depression.

## KEYWORDS

- **depression**
- **suicidality**
- **monoamine oxidase inhibitors**
- **tricyclic antidepressants**

## REFERENCES

Akerblad, A. C.; Bengtsson, F.; von Knorring L.; Ekselius, L. Response, Remission and Relapse in Relation to Adherence in Primary Care Treatment of Depression: A 2-Year Outcome Study. *Int. Clin. Psychopharmacol.* **2006,** *21,* 117–124.

American Psychiatric Association. Treatment of Patients with Major Depression and Practical Guideline,3rd ed, Washington DC: American Psychiatric Association, 2010.

Anderson, I. M.; Ferrier, I. N.; Baldwin, R. C.; Cowen, P. J.; Howard, L.; Lewis, G.; Matthews, K.; McAllister-Williams, R. H.; Peveler, R. C.; Scott, J.; Tylee, A. Evidence-based Guidelines for Treating Depressive

Disorders with Antidepressants: A Revision of the 2000 British Association for Psychopharmacology Guidelines. *J. Psychopharmacol.* **2008**, *22*, 343–396.

Anisman, H.; Merali, Z. Anhedonic and Anxiogenic Effects of Cytokine Exposure. *Adv. Exp. Med. Biol.* **1999**, *461*, 199–233.

Antai-Otong, D. Antidepressant-induced Insomnia: Treatment Options. *Perspect. Psychiatr Care.* **2004**, *40*, 29–33.

Arnt, J.; Overo, K. F.; Hyttel, J.; Olsen, R. Changes in Rat Dopamine and Serotonin Function In Vivo After Prolonged Administration of the Specific 5-Ht Uptake Inhibitor Citalopram. *Psychopharmacology* **1984**, *84*, 457–465.

Asberg, M. Neurotransmitters and Suicidal Behavior. The Evidence from Cerebrospinal Fluid Studies. *Ann. N.Y. Acad. Sci.* **1997**, *836*, 158–181.

Ascher, J. A.; Cole, J. O.; Colin, J. N.; Feighner, J. P.; Ferris, R. M.; Fibiger, H. C.; Golden, R. N.; Martin, P.; Potter, W. Z.; Richelson, E. Bupropion: A Review of Its Mechanism of Antidepressant Activity. *J. Clin. Psychiatry* **1995**, *56*, 395–401.

Baldessarini, R.; Tondo, L.; Ghiani, C.; Lepri, B. Illness Risk Following Rapid Versus Gradual Discontinuation of Antidepressant. *Am. J. Psychiatry.* **2010**, *167*, 934–941.

Baldessarini, R. J. Drugs and the Treatment of Psychiatric Disorders: Depression and Anxiety Disorders. In: *Goodman and Gilman's The Pharmacological Basis of Therapeutics*; Hardman, J. G., Limbird, L. E., Eds; New York: McGraw-Hill, 2001, pp 447–483.

Barkin, R.L.; Schwer, W. A.; Barkin, S. J. Recognition and Management of Depression in Primary Care: A Focus on the Elderly: A Pharmacotherapeutic Overview of the Selection Process Among the Traditional and New Antidepressants. *Am. J. Ther.* **2000**, *7*, 205–226.

Bengtsson, H. J.; Kele, J.; Johansson, J.; Hjorth, S. Interaction of the Antidepressant Mirtazapine with Alpha2-Adrenoceptors Modulating the Release of 5-Ht in Different Rat Brain Regions In Vivo. *Naunyn. Schmiedebergs. Arch. Pharmacol.* **2000**, *362*, 406–412.

Bitter, I.; Filipovits, D.; Czobor, P. Adverse Reactions to Duloxetine in Depression. *Exp. Opin. Drug Saf.* **2011**, *10*, 839–850.

Blier P.; Tremblay P. Physiologic Mechanisms Underlying the Antidepressant Discontinuation Syndrome. *J. Clin. Psychiatry* **2006**, *67*, 8–13.

Bloom, F. E.; Iversen, L. L. Localizing 3H-GABA in Nerve Terminals of Rat Cerebral Cortex by Electron Microscopic Autoradiography. *Nature* **1971**, *229*, 628–630.

Bockting, C. L.; ten Doesschate, M. C.; Spijker, J.; Spinhoven, P.; Koeter, M. W.; Schene, A. H; DELTA study group. Continuation and Maintenance Use of Antidepressants in Recurrent Depression. *Psychother. Psychosom.* **2008**, *77*, 17–26.

Braitberg, G.; Curry, S. C. Seizure After Isolated Fluoxetine Overdose. *Ann. Emerg. Med.* **1995**, *26*, 234–237.

Briley, M.; Moret, C. Improvement of Social Adaptation in Depression with Serotonin and Norepinephrine Reuptake Inhibitors. *Neuropsychiatr. Dis. Treat.* **2010**, *6*, 647–655.

Briley, M.; Prost, J. F.; Moret, C. Preclinical Pharmacology of Milnacipran. *Int. Clin. Psychopharmacol.* **1996**, *11*, 9–14.

Brunello, N.; Mendlewicz, J.; Kasper, S.; Leonard, B.; Montgomery, S.; Nelson, J.; Paykel, E.; Versiani, M.; Racagni, G. The Role of Noradrenaline and Selective Noradrenaline Reuptake Inhibition in Depression. *Eur. NeuroPsychopharmacol.* **2002**, *12*, 461–475.

Byrne, S. E.; Rothschild, A. J. Loss of Antidepressant Efficacy During Maintenance Therapy: Possible Mechanisms and Treatments. *J. Clin. Psychiatry.* **1998**, *59*, 279–288.

Capuron, L.; Neurauter, G.; Musselman, D. L.; Lawson, D. H.; Nemeroff, C. B.; Fuchs, D.; Miller, A.H. Interferon-alpha-induced Changes in Tryptophan Metabolism, Relationship to Depression and Paroxetine Treatment. *Biol. Psychiatry* **2003**, *54*, 906–914.

Caspi, A.; Sugden, K.; Moffitt, T. E.; Taylor, A.; Craig, I. W.; Harrington, H.; McClay, J.; Mill, J.; Martin, J.; Braithwaite, A.; Poulton, R. Influence of Life Stress on Depression: Moderation by a Polymorphism in the 5-HT gene. *Science* **2003**, *301*, 386–389.

Charney, D. S.; Manji, H. K. Life Stress, Genes, and Depression: Multiple Pathways Lead to Increased Risk and New Opportunities for Intervention. *Sci. STKE.* (re5), **2004.**

Cohen, B.; Baldessarini, R. Tolerance to Therapeutic Effects of Antidepressants. *Am. J. Psychiatry.* **1985**, *142*, 489–490.

Cohen, L. J. Rational Drug Use in the Treatment of Depression. *Pharmacotherapy* **1997**, *17*, 45–61.

Connor, T. J.; Song, C.; Leonard, B. E.; Merali, Z.; Anisman, H. An Assessment of the Effects of Central Interleukin-1β, -2, -6, and Tumor Necrosis Factor—An Administration on Some Behavioural, Neurochemical, Endocrine and Immune Parameters in the Rat. *Neuroscience* **1998**, *84*, 923–933.

Corona, G.; Ricca, V.; Bandini, E.; Mannucci, E.; Lotti, F.; Boddi, V.; Rastrelli, G.; Sforza, A.; Faravelli, C.; Forti,

G.; Maggi, M. Selective Serotonin Reuptake Inhibitor-induced Sexual Dysfunction. *J. Sex. Med.* **2009**, *6*, 1259–1269.

Damis, M.; Patel, Y.; Simpson, G. M. Sildenafil in the Treatment of SSRI-induced Sexual Dysfunction: A Pilot Study. Prim. Care. Companion. *J. Clin. Psychiatry* **1999**, *1*, 184–187.

de Boer, T. The Pharmacologic Profile of Mirtazapine. *J. Clin. Psychiatry* **1996**, *57*, 19–25.

de Boer, T.; Ruigt, G. S. The Selective [Alpha]2-adrenoceptor Antagonist Mirtazapine (Org 3770) Enhances Noradrenergic and 5-Ht1a-mediated Serotonergic Neurotransmission. *CNS. Drugs.* **1995**, *4*, 29–38.

De Montigny, C.; Haddjeri, N.; Mongeau, R.; Blier, P. The Effects of Mirtazapine on the Interactions Between Central Noradrenergic And Serotonergic Systems. *CNS. Drugs* **1995**, *4*, 13–17.

Detke, M. J.; Lu, Y.; Goldstein, D. J.; Hayes, J. R.; Demitrack, M. A. Duloxetine, 60 mg Once Daily, for Major Depressive Disorder: A Randomized Double-blind Placebo-controlled Trial. *J. Clin. Psychiatry* **2002**, *63*, 308–315.

Diav-Citrin, O.; Shechtman, S.; Weinbaum, D.; Wajnberg, R.; Avgil, M.; Di Gianantonio, E.; Clementi, M.; Weber-Schoendorfer, C.; Schaefer, C.; Ornoy, A. Paroxetine and Fluoxetine in Pregnancy: A Prospective, Multicentre, Controlled, Observational Study. *Br. J. Clin. Pharmacol.* **2008**, *66*, 695–705.

Dong, J.; Blier, P. Modification of Norepinephrine and Serotonin, But Not Dopamine, Neuron Firing by Sustained Bupropion Treatment. *Psychopharmacology* **2001**, *155*, 52–57.

Dowlati, Y.; Herrmann, N.; Swardfager, W.; Liu, H.; Sham, L.; Reim, E. K.; Lanctôt, K. L. A meta-Analysis of Cytokines in Major Depression. *Biol. Psychiatry* **2010**, *67*, 446–457.

Drevets, W. C.; Bogers, W.; Raichle, M. E. Functional Anatomical Correlates of Antidepressant Drug Treatment Assessed Using Pet Measures of Regional Glucose Metabolism. *Eur. Neuro Psychopharmacol.* **2002**, *12*, 527–544.

Evans, D. L.; Davidson, J.; Raft, T. Early and Late Side-effects of Phenelzine. *J. Clin. Psychopharmacol.* **1982**, *2*, 208–210.

Executive Committee of Korean Medication Algorithm Project for Depressive Disorder 2012. Korean Medication Guideline for Depressive Disorder 2012. Seoul: ML communication, 2012.

Fava, M. Weight Gain and Antidepressants. *J. Clin. Psychiatry* **2000**, *61*, 37–41.

Fava, M.; Rappe, S. M.; Pava, J. A.; Nierenberg, A. A.; Alpert, J. E.; Rosenbaum, J. F. Relapse in Patients on Long-term Fluoxetine Treatment Respond to Increased Fluoxetine Dose. *J. Clin. Psychiatry* **1995**, *56*, 52–55.

Feighner, J. P. Mechanism of Action of Antidepressant Medications. *J. Clin. Psychiatry* **1999**, *60*, 4–11.

Fergusson, D.; Doucette, S.; Glass, K. C.; Shapiro, S.; Healy, D.; Hebert, P.; Brian Hutton, B. Association Between Suicide Attempts and Selective Serotonin Reuptake Inhibitors: Systematic Review of Randomised Controlled Trials. *Br. Med. J.* **2005**, *330*, 396.

Fibiger, H. C. Neurobiology of Depression: Focus on Dopamine. In *Depression and Mania: From Neurobiology to Treatment*, Gessa, G., Fratta, W., Pani, L., Serra, G., Eds.; Raven Press: New York, 1995, pp.1–17.

Forsberg, L.; Naver, L.; Gustafsson, L. L.; Wide, K. Neonatal Adaptation in Infants Prenatally Exposed to Antidepressants—Clinical Monitoring Using Neonatal Abstinence Score. *PLoS One* **2014**, *9*, e111327.

Frank, E.; Kupfer, D. J.; Perel, J. M.; Cornes, C.; Jarrett, D. B.; Mallinger, A. G.; Thase, M. E.; McEachran, A. B.; Grochocinski, V. J. Three-year Outcomes for Maintenance Therapies in Recurrent Depression. *Arch. Gen. Psychiatry* **1990**, *47*, 1093–1099.

Geddes, J. R.; Butler, R. *Depressive Disorders in Adults: Clinical Evidence*. London, UK: BMJ Publishing Group, 2002.

Gibson, K.; Cartwright, C.; Read, J. Patient-centered Perspectives on Antidepressant Use: A Narrative Review. *Int. J. Ment. Health.* **2014**, *43*, 81–99.

Gisslinger, H.; Svoboda, T.; Clodi, M.; Gilly, B.; Ludwig, H.; Havelec, L.; Luger, A. Interferon-alpha Stimulates the Hypothalamic-pituitary-adrenal Axis In Vivo and In Vitro. *Neuroendocrinology* **1993**, *57*, 489–495.

Glassman, A.H. Cardiovascular Effects of Tricyclic Antidepressants. *Ann. Rev. Med.* **1984**, *35*, 503–511.

Gold, B. I.; Bowers, M. B. Jr.; Roth, R. H.; Sweeney, D. W. GABA Levels in CSF of Patients with Psychiatric Disorders. *Am. J. Psychiatry* **1980**, *137*, 362–364.

Goldberg, L.; Moncrieff, J. The Psychoactive Effects of Antidepressants and Their Association with Suicidality. *Curr. Drug. Saf.* **2011**, *6*, 1–7.

Golden, R. N.; Dawkins, K.; Nicholas, L. Trazodone and Nefazodone. In: *The American Psychiatric Publishing Textbook of Psychopharmacology*, Schatzberg, A.F., Nemeroff, C. B., Eds, 4th ed; Arlington, VA: American Psychiatric Publishing, 2009.

Gomez-Gil, E.; Salmeron, J. M.; Mas, A. Phenelzine-induced Fulminant Hepatic Failure. *Ann. Intern. Med.* **1996,** *124*, 692–693.

Goodman, L. S.; Hardman, J. G.; Limbird, L. E.; Gilman, A. G. *Goodman and Gilman's the Pharmacological Basis of Therapeutics;* New York: McGraw-Hill, 2001, pp. 2148–10.

Gunnell, D.; Saperia, J.; Ashby, D. Selective Serotonin Reuptake Inhibitors (SSRIS) and Suicide in Adults: Meta-analysis of Drug Company Data From Placebo Controlled, Randomized Controlled Trials Submitted to the MHRA's Safety Review. *Br. Med. J.* **2005,** *330*, 385–388.

Harvey, R. A.; Clark, M. A.; Finkel, R.; Rey, J. A.; Whalen, K. *Lippincott's Illustrated Reviews: Pharmacology,* 5th ed; New York: Wolters Kluwer, 2011.

Haddad, P.; Anderson, I. Recognising and Managing Antidepressant Discontinuation Symptoms. *APT.* **2007,** *13*, 447–457.

Haria, M.; Fitton, A.; McTavish, D. Trazodone: A Review of Its Pharmacology, Therapeutic Use in Depression and Therapeutic Potential in Other Disorders. *Drug. Aging* **1994,** *4*, 331–355.

Holm, K. J.; Markham, A. Mirtazapine: A Review of Its Use in Major Depression. *Drugs* **1999,** *57*, 607–631.

Hudson, J. I.; Pope, H. G.; Keck, P. E.; McElroy, S. L. Treatment of Bulimia Nervosa with Trazodone: Short-term Response and Long-term Follow-up. *Clin. Neuropharmacol.* **1989,** *12*, S38–S46.

Hughes, S.; Cohen, D. A Systematic Review of Long-term Studies of Drug Treated and Non-drug Treated Depression. *J. Affect. Disord.* **2009,** *118* (1–3), 9–18.

Jeon-Slaughter, H. Economic Factors in of Patients' Nonadherence to Antidepressant Treatment. *Soc. Psychiatr. Psychiatr. Epidemiol.* **2012,** *47*, 1985–1998.

Judd, L. L.; Akiskal, H. S.; Maser, J. D.; Zeller, P. J.; Endicott, J.; Coryell, W.; Paulus, M. P.; Kunovac, J. L.; Leon, A. C.; Mueller, T. I.; Rice, J. A.; Keller, M. B. Major Depressive Disorder: A Prospective Study of Residual Subthreshold Depressive Symptoms as Predictor of Rapid Relapse. *J. Affect. Disord.* **1998,** *50*, 97–108.

Kaehler, S. T.; Singewald, N.; Sinner, C.; Philippu, A. Nitric Oxide Modulates the Release of Serotonin in the Rat Hypothalamus. *Brain Res.* **1999,** *835*, 346–349.

Kennedy, S. H.; Lam, R. W.; Parikh, S. V.; Patten, S. B.; Ravindran, A. V. Canadian Network for Mood and Anxiety Treatments (CANMAT) Clinical Guidelines for the Management of Major Depressive Disorder in Adults. Introduction. *J. Affect. Disord.* **2009,** *117* (1), S1–S2.

Khan, F.; Bernadt, M. Intense Suicidal Thoughts and Self-Harm Following Escitalopram Treatment. *Ind. J. Psychol. Med.* **2011,** *33*, 74–76.

Kieler, H.; Artama, M.; Engeland, A.; Ericsson, O.; Furu, K.; Gissler, M.; Nielsen, R. B.; Nørgaard, M.; Stephansson, O.; Valdimarsdottir, U.; Zoega, H.; Haglund, B. Selective Serotonin Reuptake Inhibitors During Pregnancy And Risk of Persistent Pulmonary Hypertension in the Newborn: Population Based Cohort Study From the Five Nordic Countries. *Br. Med. J.* **2012,** *344*, d8012.

Kondziella, D.; Asztely, F. Don't Be Afraid to Treat Depression in Patients with Epilepsy. *Acta. Neurol. Scand.* **2009,** *119*, 75–80.

Koren, G. First-trimester Exposure to Fluoxetine (Prozac). Does it Affect Pregnancy Outcome? *Can. Fam. Physician.* **1996,** *42*, 43–44.

Krishnan, V.; Nestler, E. The Molecular Neurobiology of Depression. *Nature* **2008,** *455*, 902–984.

Kuhn, D. M.; Arthur, R. E. J. Inactivation of Brain Tryptophan Hydroxylase by Nitric Oxide. *J. Neurochem.* **1996,** *67*, 1072–1077.

Lee, K. M.; Kim, Y. K. The Role of Il-12 and Tgf-Beta1 in the Pathophysiology of Major Depressive Disorder. *Int. Immunopharmacol.* **2006,** *6*, 1298–1304.

Lieb, J. Antidepressant Tachyphylaxis. *J. Clin. Psychiatry.* **1990,** *51*, 36.

Lieb, J.; Balter, A. Antidepressant Tachyphylaxis. *Med. Hypotheses.* **1984,** *15*, 279–291.

Lin, E. H.; Katon, W. J.; VonKorff, M.; Russo, J. E.; Simon, G. E.; Bush, T. M.; Rutter, C. M.; Walker, E. A.; Ludman, E. Relapse of Depression in Primary Care: Rate and Clinical Predictors. *Arch. Fam. Med.* **1998,** *7*, 443–449.

Llorca, P. M.; Lancon, C.; Brignone, M.; Rive, B.; Salah, S.; Ereshefsky, L.; Francois, C. Relative Efficacy and Tolerability of Vortioxetine Versus Selected Antidepressants by Indirect Comparisons of Similar Clinical Studies. *Curr. Med. Res. Opin.* **2014,** *30*, 2589–2606.

Lorrain, D. S.; Matuszewich, L.; Friedman, R. D.; Hull, E. M. Extracellular Serotonin in the Lateral Hypothalamic Area is Increased During the Post Ejaculatory Interval and Impairs Copulation in Male Rats. *J. Neurosci.* **1997,** *17*, 9361–9366.

Madrigal, J. L.; Moro, M. A.; Lizasoain, I.; Lorenzo, P.; Castrillo, A.; Bosca, L.; Leja, J. C. Inducible Nitric Oxide Synthase Expression in Brain Cortex After Acute Restraint Stress is Regulated by Nuclear Factor

Kappab-mediated Mechanisms. *J. Neurochem.* **2001,** *76,* 532–538.

Mann, J. J. Loss of Antidepressant Effect with Long-term Monoamine Oxidase Inhibitor Treatment Without Loss of Monoamine Oxidase Inhibition. *J. Clin. Psychopharmacol.* **1983,** *3,* 363–366.

McGrath, P. J.; Stewart, J. W.; Nunes, E. V.; Ocepek-Welikson, K.; Rabkin, J. G.; Quitkin, F. M.; Klein, D. F. A Double-blind Crossover Trial of Imipramine and Phenelzine for Outpatients with Treatment-refractory Depression. *Am. J. Psychiatry.* **1993,** *150,* 118–123.

Meltzer, H. Serotonergic Dysfunction in Depression. *Br. J. Psychiatry.* **1989,** *8,* 25–31.

Miller, K. J.; Hoffman, B. J. Adenosine 4 Receptors Regulate Serotonin Transport via Nitric Oxide and cGMP. *J. Biol. Chem.* **1994,** *269,* 27351–27356.

Montejo-Gonzalez, A. L.; Llorca, G.; Izquierdo, J. A.; Ledesma, A.; Bousoño, M.; Calcedo, A.; Carrasco, J. L.; Ciudad, J.; Daniel, E.; De la Gandara, J.; Derecho, J.; Franco, M.; Gomez, M. J.; Macias, J. A.; Martin, T.; Perez, V.; Sanchez, J. M.; Sanchez, S.; Vicens, E. SSRI-induced Sexual Dysfunction: Fluoxetine, Paroxetine, Sertraline, and Fluvoxamine in a Prospective, Multicenter, and Descriptive Clinical Study of 344 Patients. *J. Sex. Marital. Ther.* **1997,** *23,* 176–194.

Montgomery, S. A.; Henry, J.; McDonald, G.; Dinan, T.; Lader, M.; Hindmarch, I.; Clare, A.; Nutt, D. Selective Serotonin Reuptake Inhibitors: Meta-analysis of Discontinuation Rates. *Int. Clin. Psychopharmacol.* **1994,** *9,* 47–53.

Myint, A. M.; Leonard, B. E.; Steinbusch, H. W.; Kim, Y. K. Th1, Th2, and Th3 Cytokine Alterations in Major Depression. *J. Affect. Disord.* **2005,** *88,* 167–173.

Nemeroff, C. B. The Corticotropin-releasing Factor (Crf) Hypothesis of Depression: New Findings and New Directions. *Mol. Psychiatry.* **1996,** *1,* 336–342.

Nierenberg, A. A.; Alpert, J. E. Depressive Breakthrough. *Psychiatr. Clin. North. Am.* **2000,** *23,* 731–742.

Owens, M. J.; Morgan, W. N.; Plott, S. J.; Nemeroff, C. B. Neurotransmitter Receptor and Transporter Binding Profile of Antidepressants and Their Metabolites. *J. Pharmacol. Exp. Ther.* **1997,** *283,* 1305–1322.

Pace, T. W.; Mletzko, T. C.; Alagbe, O.; Musselman, D. L.; Nemeroff, C. B.; Miller, A. H.; Heim, C. M. Increased Stress Induced Inflammatory Response in Male Patients with Major Depression and Increased Early Life Stress. *Am. J. Psychiatr.* **2006,** *163,* 1630–1633.

Pacher, P.; Kecskemeti, V. Trends in the Development of New Antidepressants. Is There A Light at the End of the Tunnel?. *Curr. Med. Chem.* **2004,** *11,* 925–943.

Pacher, P.; Ungvari, Z.; Nanasi, P. P.; Furst, S.; Kecskemeti, V. Speculations on Difference Between Tricyclic and Selective Serotonin Reuptake Inhibitor Antidepressants on their Cardiac Effects. Is There Any? *Curr. Med. Chem.* **1999,** *6* (6), 469–480.

Papakostas, G. I. Tolerability of Modern Antidepressants. *J. Clin. Psychiatry* **2008,** *69,* 8–13.

Patel, K.; Allen, S.; Haque, M. N.; Angelescu, I.; Baumeister, D.; Tracy, D. K. Bupropion: A Systematic Review and Meta-analysis of Effectiveness as an Antidepressant. *Ther. Adv. Psychopharmacol.* **2016,** *6,* 99–144.

Pazzagli, M.; Giovannini, M. G.; Pepeu, G. Trazodone Increase Extracellular Serotonin Levels in the Frontal Cortex of Rats. *Eur. J. Pharmacol.* **1999,** *383,* 249–257.

Pinder, R. M. Designing a New Generation of Antidepressant Drugs. *Acta. Psychiatr. Scand.* **1997,** *96,* 7–13.

Pinder, R. M. On the Feasibility of Designing New Antidepressants. *Hum. Psychopharmacol.* **2001,** *16,* 53–59.

Pisani, F.; Oteri, G.; Costa, C.; Di Raimondo, G.; Di Perri, R. Effects of Psychotropic Drugs on Seizure Threshold. *Drug Saf.* **2002,** *25,* 91–110.

Placidi, G. P.; Oquendo, M. A.; Malone, K. M.; Huang, Y. Y.; Ellis, S. P.; Mann, J. J. Aggressivity, Suicide Attempts, and Depression: Relationship to Cerebrospinal Fluid Monoamine Metabolite Levels. *Biol. Psychiatry.* **2001,** *50,* 783–791.

Price, R. B.; Shungu, D. C.; Mao, X.; Nestadt, P.; Kelly, C.; Collins, K. A.; Murrough, J. W.; Charney, D. S.; Mathew, S. J. Amino Acid Neurotransmitters Assessed by Proton Magnetic Resonance Spectroscopy: Relationship to Treatment Resistance in Major Depressive Disorder. *Biol. Psychiatry.* **2009,** *65,* 792–800.

Prien, R. F.; Kupfer, D. J.; Mansky, P. A.; Small, J. G.; Tuason, V. B.; Voss, C. B.; Johnson, W. E. Drug Therapy in the Prevention of Recurrences in Unipolar and Bipolar Affective Disorders: A Report of the NIMH Collaborative Study Group Comparing Lithium Carbonate, Imipramine, and a Lithium Carbonate-Imipramine Combination. *Arch. Gen. Psychiatry.* **1984,** *41,* 1096–1104.

Rabkin, J.; Quitkin, F.; Harrison, W.; Tricamo, E.; McGrath, P. Adverse Reactions to Monoamine Oxidase Inhibitors. Part I. A Comparative Study. *J. Clin. Psychopharmacol.* **1984,** *4,* 270–278.

Rasavi, D.; Kormoss, N.; Collard, A.; Farvacques, C.; Delvaux, N. Comparative Study of the Efficacy and Safety of Trazodone Versus Clorazepate in the Treatment of Adjustment Disorders in Cancer Patients: A Pilot Study. *J. Int. Med. Res.* **1999**, *27*, 264–272.

Read, J.; Cartwright, C.; Gibson, K. Adverse Emotional and Interpersonal Effects Reported By 1829 New Zealanders While Taking Antidepressants. *Psychiatr. Res.* **2014**, *206*, 67–73.

Reynolds, C. F. 3rd, Frank, E.; Perel, J. M.; Imber, S. D.; Cornes, C.; Miller, M. D.; Mazumdar, S.; Houck, P. R.; Dew, M. A.; Stack, J. A.; Pollock, B. G.; Kupfer, D. J. Nortriptyline and Interpersonal Psychotherapy as Maintenance Therapies for Recurrent Major Depression: A Randomized Controlled Trial in Patients Older Than 59 Years. *JAMA.* **1999**, *281*, 39–45.

Richelson, E. Synaptic Effects of Antidepressants. *J. Clin. Psychopharmacol.* **1996**, *16*, 1S–9S.

Robinson, D. S. Monoamine Oxidase Inhibitors: A New Generation. *Psychopharmacol. Bull.* **2002**, *36*, 124–138.

Robinson, D. S.; Kajdasz, D. K.; Gallipoli, S.; Whalen, H.; Wamil, A.; Reed, C. R. A 1-year, Open-label Study Assessing the Safety and Tolerability of Vilazodone in Patients with Major Depressive Disorder. *J. Clin. Psychopharmacol.* **2011**, *31*, 643–646.

Rothman, R. B, Baumann, M. H.; Dersch, C. M.; Romero, D. V.; Rice, K. C.; Carroll, F. I.; Partilla, J. S. Amphetamine-type Central Nervous System Stimulants Release Norepinephrine More Potently Than They Release Dopamine and Serotonin. *Synapse.* **2001**, *39*, 32–41.

Rothschild, A. J. The Rothschild Scale for Antidepressant Tachyphylaxis: reliability and validity. *Comprehen. Psychiatry.* **2008**, *49*, 508–513.

Rothschild, A. J.; Dunlop, B. W.; Dunner, D. L.; Friedman, E. S.; Gelenberg, A.; Holland, P.; Kocsis, J. H.; Kornstein, S. G.; Shelton, R.; Trivedi, M. H.; Zajecka, J. M.; Goldstein, C.; Thase, M. E.; Pedersen, R.; Keller, M. B. Assessing Rates and Predictors of Tachyphylaxis During the Prevention of Recurrent Episodes of Depression with Venlafaxine ER For Two Years (Prevent) Study. *Psychopharmacol. Bull.* **2009**, *42*, 5–20.

Rush, A. J.; Crismon, M. L.; Kashner, T. M.; Toprac, M. G.; Carmody, T. J.; Trivedi, M. H.; Suppes, T.; Miller, A. L.; Biggs, M. M.; Shores-Wilson, K.; Witte, B. P.; Shon, S. P.; Rago, W. V.; Altshuler, K. Z.; TMAP Research Group. Texas Medication Algorithm Project, Phase 3 (TMAP-3): Rationale and Study Design. *J. Clin. Psychiatry.* **2003**, *64*, 357–369.

Sanacora, G.; Fenton, L. R.; Fasula, M. K.; Rothman, D. L.; Levin, Y.; Krystal, J. H.; Mason, G. F. Cortical Gamma-aminobutyric Acid Concentrations in Depressed Patients Receiving Cognitive Behavioral Therapy. *Biol. Psychiatry.* **2006**, *59*, 284–286.

Sanacora, G.; Gueorguieva, R.; Epperson, C. N.; Wu, Y. T.; Appel, M.; Rothman, D. L.; Krystal, J. H.; Mason, G. F. Subtype-specific Alterations of Gamma-aminobutyric Acid and Glutamate in Patients With Major Depression. *Arch. Gen. Psychiatry.* **2004**, *61*, 705–713.

Sanacora, G.; Mason, G. F.; Rothman, D. L.; Behar, K. L.; Hyder, F.; Petroff, O. A.; Berman, R. M.; Charney, D. S.; Krystal, J. H. Reduced Cortical Gamma-aminobutyric Acid Levels in Depressed Patients Determined by Proton Magnetic Resonance Spectroscopy. *Arch. Gen. Psychiatry.* **1999**, *56*, 1043–1047.

Sanacora, G.; Saricicek, A. GABAergic Contributions to the Pathophysiology of Depression and the Mechanism of Antidepressant Action. *CNS. Neurol. Disord. Drug. Targets.* **2007**, *6*, 127–140.

Sánchez, C.; Hyttel, J. Comparison of the Effects of Antidepressants and Their Metabolites on Reuptake of Biogenic Amines and on Receptor Binding. *Cell. Mol. Neurobiol.* **1999**, *19*, v467–489.

Santarsieri, D.; Schwartz, T. L. Antidepressant Efficacy and Side-effect Burden: A Quick Guide for Clinicians. *Drug. Context* **2015**, *4*, 212–290.

Schatzberg, A. F.; DeBattista, C. *Manual of Clinical Psychopharmacology*, 8th ed; Arlington, VA: American Psychiatric Publishing, 2015.

Schulberg, H. C.; Katon, W.; Simon, G. E.; Rush, A. J. Treating Major Depression in Primary Care Practice: An Update of the Agency for Health Care Policy and Research Practice Guidelines. *Arch. Gen. Psychiatry.* **1998**, *55*, 1121–1127.

Schwartz, T. L.; Stahl, S. M. Vilazodone: A Brief Pharmacologic and Clinical Review of the Novel SPARI (serotonin partial agonist and reuptake inhibitor). *Ther. Adv. Psychopharmacol.* **2011**, *1*, 81–87.

Sheehan, D. V.; Keene, M. S.; Eaddy, M.; Krulewicz, S.; Kraus, J. E.; Carpenter, D. J. Differences in Medication Adherence and Healthcare Resource Utilization Patterns: Older Versus Newer Antidepressant Agents in Patients with Depression and/or Anxiety Disorders. *CNS. Drugs.* **2008**, *22*, 963–973.

Sheldon, R. The Nature of the Discontinuation Syndrome Associated with Antidepressant Drugs. *J. Clin. Psychiatry.* **2006**, *67*, 3–7.

Sherif, F. M.; Ahmed, S. S. Basic Aspects of Gaba-transaminase in Neuropsychiatric Disorders. *Clin. Biochem.* **1995**, *28*, 145–154.

Simon, G. E.; Revicki, D.; Heiligenstein, J.; Grothaus, L.; VonKorff, M.; Katon, W. J.; Hylan, T. R. Recovery from Depression, Work Productivity, and Health Care Costs Among Primary Care Patients. *Gen. Hosp. Psychiatry.* **2000**, *22*, 153–162.

Sit, D.; Perel, J. M.; Wisniewski, S. R.; Helsel, J. C.; Luther, J. F.; Wisner, K. L. Mother-infant Antidepressant Levels, Maternal Depression and Perinatal Events. *J. Clin. Psychiatry.* **2011,** *72*, 994–1001.

Solomon, D. A.; Leon, A. C.; Mueller, T. I.; Coryell, W.; Teres, J. J.; Posternak, M. A.; Judd, L. L.; Endicott, J.; Keller, M. B. Tachyphylaxis in Unipolar Major Depressive Disorder. *J. Clin. Psychiatry.* **2005,** *66*, 283–290.

Southwick, S. M.; Vytilingham, S.; Charney, D. S. The Psychobiology of Depression and Resilience to Stress, Implications for Prevention and Treatment. *Annu. Rev. Clin. Psychol.* **2005,** *1*, 255–291.

Stahl, S. M. *Essential Psychopharmacology*, 1st ed; New York, NY: Cambridge University Press, 1996.

Stahl, S. M.; Pradko, J. F.; Haight, B. R. A Review of the Neuropharmacology of Bupropion, A Dual Norepinephrine and Dopamine Reuptake Inhibitor. *Prim. Care. Comp. J. Clin. Psychiatry.* **2004,** *6*, 159–166.

Stahl, S. M. The 7 Habits of Highly Effective Psychopharmacologists, Part 3: Sharpen the Saw with Selective Choices of Continuing Medical Education Programs. *J. Clin. Psychiatry.* **2000,** *61*, 401–402.

Stewart, J. W.; Tricamo, E.; McGrath, P. J.; Quitkinm F. M. Prophylactic Efficacy of Phenelzine and Imipramine in Chronic Atypical Depression: Likelihood of Recurrence on Discontinuation After 6 Months' Remission. *Am. J. Psychiatry.* **1997,** *154*, 31–36.

Stimmel, G. L.; Dopheide, J. A.; Stahl, S. M. Mirtazapine: An Antidepressant with Noradrenergic and Specific Serotonergic Effects. *Pharmacother. J. Hum. Pharmacol. Drug. Ther.* **1997,** *17*, 10–21.

Suehs, B.; Argo, T. R.; Bendele, S. D.; Crimson, M. L.; Trivedi, M. H.; Kurian, B. Texas Medication Algorithm Project Procedural Manual Major depressive disorder algorithms. Texas: Texas Department of State Health Services, 2008.

Szabo, S. T.; Blier, P. Effects of the Selective Norepinephrine Reuptake Inhibitor Reboxetine on Norepinephrine and Serotonin Transmission in the Rat Hippocampus. *Neuro Psychopharmacol.* **2001,** *25*, 845–857.

Thase, M. E. Antidepressant Treatment of the Depressed Patient with Insomnia. *J. Clin. Psychiatry* **1999,** *60*, 28–31.

Thomas, D. R.; Nelson, D. R.; Johnson, A. M. Biochemical Effects of the Antidepressant Paroxetine, A Specific 5-Hydroxytryptamine Uptake Inhibitor. *Psychopharmacology* **1987,** *93*, 193–200.

Thompson, C. Onset of Action of Antidepressants: Results of Different Analyses. *Hum. Psychopharmacol.* **2002,** *17*, S27–S32.

Tianmei, S. I.; Xin, Y. U. Current Problems in the Research and Development of more Effective Antidepressants. *Shanghai. Arch. Psychiatry.* 2016, *28* (3), 160–165.

Vashadze, ShV. Insomnia, Serotonin and Depression. *Georg. Med. News.* **2007,** *150*, 22–24.

Vollmayr, B.; Henn, F. A. Stress Models of Depression. *Clin. Neurosci. Res.* **2003,** *3*, 245–251.

Walia, V. Possible Role of Serotonin and Selective Serotonin Reuptake Inhibitors in Suicidal Ideations and Attempts. *J. Pharm. Sci. Pharmacol.* **2017,** *3*, 54–70.

Walia, V. Role of Enzymes in the Pathogenesis of Depression. *J. Crit. Rev.* **2016,** *3*, 1–6.

Walia, V.; Gilhotra, N. Nitriergic Influence in the Compromised Antidepressant Effect of Fluoxetine in Stressed Mice. *J. Appl. Pharm. Sci.* **2016,** *6*, 092–097.

Warner, M. D.; Dorn, M. R.; Peabody, C. A. Survey on the Usefulness of Trazodone in Patients with PTSD with Insomnia or Nightmares. *Pharmacopsychiatry* **2001,** *34*, 128–131.

Wegener, G.; Volke, V.; Rosenberg, R. Endogenous Nitric Oxide Decreases Hippocampal Levels of Serotonin and Dopamine In Vivo. *Br. J. Pharmacol.* **2000,** *130*, 575–580.

Weich, S.; Nazareth, I.; Morgan, L.; King, M. Treatment of Depression in Primary Care–Socio-economic Status, Clinical Need and Receipt of Treatment. *Br. J. Psychiatry.* **2007,** *191*, 164–169.

Wester, P.; Bergstrom, U.; Eriksson, A.; Gezelius, C.; Hardy, J.; Winblad, B. Ventricular Cerebrospinal Fluid Monoamine Transmitter and Metabolite Concentrations Reflect Human Brain Neurochemistry in Autopsy Cases. *J. Neurochem.* **1990,** *54*, 1148–1156.

Willner, P. Animal Models of Depression: Validity and Applications. *Adv. Biochem. Psychopharmacol.* **1995,** *49*, 19–41.

Woodrum, S. T.; Brown, C. S. Management of SSRI-induced Sexual Dysfunction. *Ann. Pharmacother.* **1998,** *32*, 1209–1215.

World Health Organization. Depression and Other Common Mental Disorders: Global Health Estimates, 2017.

World Health Organization. The ICD-10 Classification of Mental and Behavioural Disorders. Clinical Descriptions and Diagnostic Guidelines. Geneva, Switzerland: World Health Organization, 1992.

# CHAPTER 13

# Antipsychotic Drugs

HARLEEN KAUR[1], RAMNEEK KAUR[2], VARSHA RANI[3],
KANISHKA SHARMA[3], and PAWAN KUMAR MAURYA[4*]

[1]*Amity Institute of Biotechnology, Uttar Pradesh, India*

[2]*Jaypee Institute of Information Technology, Uttar Pradesh, India*

[3]*Amity Education Group, New York, USA*

[4]*Central University of Haryana, Mahendergarh, India*

[*]*Corresponding author. E-mail: pawanbiochem@gmail.com*

## ABSTRACT

Neuropsychiatric symptoms are mutual in older adults and can be associated with a decline in functional and cognitive status. Among the currently available atypical antipsychotics, olanzapine, haloperidol, risperidone, clozapine, and quetiapine have been the most widely studied in schizophrenia, bipolar disorder, and depression. Some controversy surrounds the use of different antipsychotic mediators in neuropsychiatric disorders with the submission that they may increase the occurrence of death or even stroke. Despite the potential for increased risk of harm from the use of these drugs, atypical antipsychotics are often effective in treating troublesome neuropsychiatric symptoms refractory to other treatments. Whenever possible, these atypical antipsychotic drug treatments should be integrated with nonpharmacological treatments to limit the dose and need of antipsychotic drugs and constant monitoring for potential harms should be maintained. The choice of atypical antipsychotic agent can be guided by the nature, severity of the target symptom, and the medication least likely to cause harm to the patient. This chapter highlights various antipsychotic drugs and their mechanism of action in neuropsychiatric diseases.

## 13.1 INTRODUCTION

Psychosis characteristically refers to a mental state involving a loss of contact with reality. According to the fifth edition of the article, "Diagnostic and Statistical Manual of Mental Disorders," psychosis is defined by the presence of delusions, hallucinations, disorganized thinking (speech), grossly disorganized or abnormal motor behavior (including catatonia), or negative symptoms (Seikkula et al., 2011). Psychosis may occur at any age, but in older population its etiology, manifestation, and treatment deserve special consideration.

Psychosis in the older population, may occur in the framework of early-onset schizophrenia that persists into later life, late-onset schizophrenia, delusion disorders, mood disorders with psychotic features, and various dementias including Alzheimer's and Parkinson's diseases. Additionally, it occurs as a result of drug use and withdrawal, both prescription and illicit, and in context of delirium, autoimmune disorders, stroke, brain tumors, metabolic disturbances, central nervous system infections, and various chronic neurological disorders. Further research indicates that a combination of genetic and environmental factors creates a situation where a person is vulnerable to, or at greater risk of developing psychosis.

A number of brain chemicals, including dopamine (DA) and serotonin, may play a role in psychosis development (Rizvi and Maurya, 2007). There is accumulating evidence that stress plays a role in the etiopathogenesis of mental disorders, particularly schizophrenia, and bipolar disorders (BDs), which commonly start at early ages with a first episode psychosis (Kapur, 2003). The conventional antipsychotics produce undesirable effects like hyperprolactinaemia, neuroleptic malignant syndrome (NMS) and extrapyramidal symptoms (EPS), which are associated with high doses. The difference between atypical antipsychotics and conventional antipsychotics can be characterized by its effectiveness, increased safety, and influence on behavior. Atypical antipsychotics possess a high rate of responders, lower risk of suicides, improved quality of life, favorable pharmacoeconomic profile, better functional capacity, and efficiency in patients with refractory disease. Many studies focus on the interaction of the receptor with the drugs as observed by their therapeutic actions.

The antipsychotic drugs bind mainly to DA, precisely dopamine receptor 2 (D2) as antipsychotic drugs can mediate through the potential site. The dopamine hypothesis is led by the association between neuroleptic drugs and D2, DA receptors for schizophrenia. Therefore, the development of drugs is focused to act at central DA receptors (Maurya et al., 2017). Recently, some atypical antipsychotic drugs, which are prototypic in nature like clozapine, focus on central receptors rather DA receptors. The reason behind the above fact is that as per the data indicated by in vitro binding clozapine was made to compare with most typical antipsychotic drugs and haloperidol and it was found that clozapine has low affinity for D2, DA receptors. Apart from D2 receptors the affinity of the site of action for clozapine also led to modification in DA and dopamine hypothesis for schizophrenia. The chapter focuses on the mechanism of action of various antipsychotic drugs like clozapine, raclopride, remoxipride, etc. Further, the comparison between first-generation antipsychotic and second-generation antipsychotic drugs has been described.

## 13.2   NEUROPSYCHIATRIC DISORDERS

### 13.2.1   *SCHIZOPHRENIA*

Schizophrenia is a mental disorder, severe chronic and complex disease that affects the thinking of a person, how he behaves and feels (Kota et al., 2015). A touch of reality has been forgotten by sufferers. However, this disease is not common as other disorders for the fact that this disease consists of varying of symptoms (Harvey et al., 2004). The symptoms are prevalent at the age of 16–30 (Mansur et al.,

2016). The symptoms are subcategorized into positive, negative, and cognitive.

### 13.2.1.1 POSITIVE SYMPTOM

It deals with psychotic behavior that is not widespread in vigorous person. Sufferers may establish lose touch with authenticity (Harvey et al., 2004). The symptoms are delusions, hallucinations, dysfunctional way of intellectual, and lack of speaking.

### 13.2.1.2 NEGATIVE SYMPTOM

It is associated with interruption of normal behavior plus emotion. The symptoms are reduced feeling of pleasure in daily routine, difficulty in commencing and sustaining activities, lack of speech, and lack of facial expressions (Erhart et al., 2006).

### 13.2.1.3 COGNITIVE SYMPTOMS

These symptoms may vary from person to person. Sufferers might experience subtle to severe changes in their intellect or any other aspects of thinking. Symptoms may consist of difficulty in concentration; problems with memory, lack of execution facility, that is, trouble in grasping information and making judgments (Simpson et al., 2010).

### 13.2.2 BIPOLAR DISORDER

Bipolar disorder (BD) is a mental disorder recognized by episodes of hypomania or maniac and depressions. The contribution of genetic factors to the susceptibility of BD ranges from 59% to 93%. BD is a complicated disorder. Symptoms of hypomanic or mania and depression define this disorder (Perlis et al., 2006). It is a psychotic disorder with the occurrence of 2–5%. Genetic factors, individual genes, contribute to this disorder with the range of the 59–93%. A T cell activation marker mainly interlukin-6 receptor (sIL-6R) are present in increased number during this episode. Alike, the immune-inflammatory profile is depicted in BD which is a major depression and is indicated by the escalation of soluble interleukin-2 receptor (sIL-2R), sIL-6R, IL-6 (Maurya et al., 2016). Therefore, maniac and BDs are accompanied due to increment of T cell activation and IL-6 transsignalling (Goldberg et al., 2009). Biological underpinnings of this disorder mainly focuses on nitrosative (NS) and oxidative stress (OD), neurotrophins, hypothalamic pituitary adrenal (HPA), circadian dysregulation, and alterations occurring in the course of BD (termed as neuroprogression) (Geddes and Miklowitz, 2013).

### 13.2.3 DEPRESSION

Disturbances in perception and pathological effects are considered to be a salient feature of depression. The characteristics of major depressions are a negative shift of emotional recognition and poor recognition of expressions. For example, judging happy faces as sad and sad faces as neutral (Van den Eynde et al., 2008).

### 13.2.3.1 CAUSES

The causes of depression include genetics, environmental, biological-changes in neurotransmitter level, psychological and social. Lack of control, anxiety, self-reported depression may not arise from single source (Hasler, 2010).

## 13.2.3.2   SYMPTOMS

Depressed moods, lack of interest in activities, insomnia or hypersomnia, fatigue, feeling of worthlessness, and lack of concentration are the various symptoms of depression as shown in Table 13.1 (Kreutzer et al., 2001). Many studies support neuroprogressive and dysregulated redox signals. Reactive nitrogen, and oxygen species (RNS, ROS) including superoxidase, peroxynitrite, nitric oxide, and peroxidases are generated via the interactions with fatty acids, proteins and DNA during a balanced physiological process, performing various roles to regulate cellular function (Mosley et al., 2006). Increased level of RNS/ROS can cause functional and structural changes leading to injury of cell. Due to a mechanism of intrinsic antioxidant, under optimum condition, these toxic effects are offset. NS and OD, can decrease the accessibility of antioxidant defense or/and elevate production of RNS and ROS. This result in the induction of dangerous autoimmune responses, destruction of cellular components and causes normal cell processes to fail.

Patients with bipolar and unipolar depression show impairment of redox signals (Vieta et al., 2013). According to studies, using animal and clinical models indicate that rise in the level of redox products like 8-iso-prostaglandin F2 and malondialdehyde. A lot of studies have demonstrated that in postmortem hippocampus oxidative damage to RNA in depression occurs, shortening of telomeres, oxidative damage to DNA due to a high level of 8-OHdG (8-hydroxy-2-deoxyguanosine) in serum.

Studies conducted in patients suffering from depression indicated to have increased concentration of NS and OD. This leads to decreased index of OD stress of serum,

lower concentration of n−3 fatty acids, fall in the antioxidant-functioning demonstrating decreased level of vitamin C and E concentration in plasma, reduced level of amino acids like tyrosine and tryptophan and antioxidant like coenzyme Q10, glutathione (GSH) and reduction in level of albumin. Antioxidant-enzyme modifications have also been reported. Low concentration of glutathione peroxidase (GPx) and superoxidase dismutase (SOD) have been reported in patients. An antioxidant enzyme paraoxonase (PON1) bound to high-density lipoprotein (HDL), was considered to be unipolar, not bipolar (Barim et al., 2009).

## 13.3   CLASSIFICATION OF ANTIPSYCHOTIC DRUGS

The use of antipsychotic drugs (Fig. 13.1) includes a difficult trade-off amid the benefit of lessening psychotic symptoms and the risk of a numerous adverse effects. Antipsychotic drugs are not curative and do not abolish chronic thought disorders, but they frequently decline the power of hallucinations and delusions and allow the person with schizophrenia to function in a caring situation (Harvey et al., 2011).

## 13.4   ANTIPSYCHOTICS AND MECHANISM OF ACTION

The conventional antipsychotics produce undesirable effects like hyperprolactinemia, NMS and EPS, these are associated with high doses (Haddad and Sharma, 2007). The difference between atypical antipsychotic and conventional antipsychotics can be characterized by its effectiveness, increased

**TABLE 13.1**  Outline of the Consequences of Drugs that Are Used as an Antipsychotic Drugs.

| Drug | Description | FDA approved | Bioavailability | Roots of administration | Effective against | References |
|---|---|---|---|---|---|---|
| Reserpine | Indole alkaloid, antihypertensive drug, antipsychotic drug | Yes | 50% | Oral | Schizophrenia, BD, depression, reduced blood pressure | Reserpine, 2018 |
| Risperidone | Antipsychotic drug | Yes | 70% | Injection into the muscle or oral | Schizophrenia, autism, manic and mixed episodes of BD | Risperidone, 2018 |
| Clozapine | Atypical antipsychotic | Yes | 60–70% | Oral | Schizophrenia, Parkinson's disease, depression, BD | Clozapine, 2018 |
| Olanzapine | Antipsychotic drug | Yes | 87% | Oral, intramuscular injection | Schizophrenia, BD | Olanzapine, 2018 |
| Haloperidol | Typical antipsychotic | Yes | 60–70% | Intravenously, oral | Schizophrenia, Tourette syndrome, mania in BD, nausea, hallucinations in alcohol withdrawal, delirium, acute psychosis | Haloperidol, 2018 |
| Quetiapine | Atypical antipsychotic | Yes | 100% | Oral | Schizophrenia, major depressive disorder, BD | Quetiapine, 2018 |

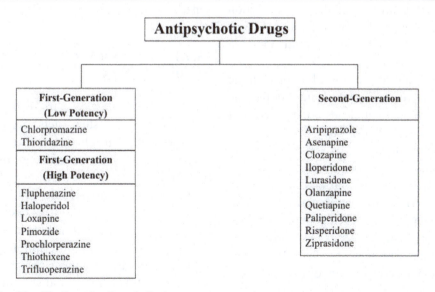

**FIGURE 13.1**  Classification of antipsychotic drugs.

safety, and influence on behavior (Kapur and Remington, 2001). Atypical antipsychotics possess a high rate of responders, lower risk of suicides, improved quality of life, favorable pharmacoeconomic profile, better functional capacity, and efficiency in patients with refractory disease. Many studies focus on the interaction of the receptor with the drugs as observed by their therapeutic actions. The antipsychotic drugs bind mainly to DA, precisely D2 as antipsychotic drugs can mediate through the potential site (Chiodo and Bunney, 1983). Association leads the dopamine hypothesis between neuroleptic drugs and D2 receptors for schizophrenia. Therefore, the development of drugs is focused to act at central DA receptors. Recently, some atypical antipsychotic drugs which are prototypic in nature like clozapine focus on central receptors rather DA receptors (Kuroki et al., 2008). The reason behind the above fact is that, as per the data indicated by in vitro binding clozapine was made to compare with most typical antipsychotic drugs and haloperidol and it was found that clozapine has low affinity for D2 receptors. Apart from D2 receptors the affinity of the site of action for clozapine also led to modification in DA and dopamine hypothesis for schizophrenia (Miyamoto et al., 2005).

The elemental target of all antipsychotic drugs is to attach with receptor names dopamine D2. By attaching to it, leads to induction of EPS (Shiloh et al., 2000). They may also bind to the receptor to increase serum prolactin. The clinically active range does not bring adverse effect if the unusual dosage is given. The disconnection of a distributed network of pyramidal neurons causes schizophrenia, which acts as a principal substrate for processing of information (Shiloh et al., 2000). The dysfunction in the connectivity of glutamatergic neurons is the proof of schizophrenia (Coyle, 2004), such as lowering in the density of prefrontal cortex of dendrites of pyramidal cells, the expression of mRNA as a synaptic density marker found in the postmortem in people suffering from this disease. Reduction in volume of brain and increases density of pyramidal neurons is seen in the sufferers. This can be stated by the fact that morphological disorders contain reduced number of fibers of neurons. Nonappearance of gliosis in the intellect of sufferers is supported by neurodevelopment etiology of structure findings but contradicted by the degenerative inflammatory process of schizophrenia (Wong and Van Tol, 2003). The trigger of all antipsychotic drugs is elementary monoaminergic receptors that are in the association of G proteins. These drugs have the ability to change the behavior of pyramidal cells. Such an outcome effect might be considered important. The atypical and typical drugs affect the glutamatergic system directly as a partial agonist at the NMDA (*N*-methyl-D-aspartate) receptor which is associated to recognition site of glycine and indirectly by glutamate transporters and blockade of glycine at the synaptic level (Wong and Van Tol, 2003).

## 13.4.1 NEUROPLASTIC EFFECT ON ANTIPSYCHOTICS

The term neuroplastic refers to the ability by which nervous system gets easily adapted by environmental changes. It may include synaptic plasticity and aids in the generation of new neuronal connection, as well as neurogenesis. Antipsychotic drugs appear to induce restructure of the network of neurons by inducing variation in neuroplastic. Thus, it subsidizes substantial description of contact between neurodevelopment variation in functions and antipsychotics and structure of brain of schizophrenia patients. It also helps in explaining the late onset of effect of antipsychotic drugs (Galletly, 2014).

## 13.5 ANTIPSYCHOTIC DRUGS

### 13.5.1 *SCHIZOPHRENIA*

The following are the drugs required to treat the disease. The drugs mainly are reserpine, risperidone, clozapine, haloperidol, quetiapine, olanzapine.

#### 13.5.1.1 *RESERPINE*

Reserpine (Fig. 13.2) synonymous names are (raudixin, serpalan), an indole alkaloid, antihypertensive, antipsychotic drug, which is essential for the relief of psychotic symptoms. *Rauvolfia serpentine* or *Indian snakeroot* is a species of a flower of family Apocynaceae. Reserpine is one of the indole components of this family only. However, today the use of the drug is limited because of the development of better drugs and due to numerous side effects of reserpine, it is now rarely used (Wilkins, 1961). These drugs have an ability to destruct catecholamines from peripheral sympathetic nerve endings. These drugs are associated to control the rate of heart, peripheral vascular resistance, and cardiac contraction. Furthermore, if a monoamine neurotransmitter is disrupted in the synapses act as a proof of depression in humans. Reserpine is associated to lowering the blood pressure by decreasing the nervous system which allows the reduction and widening of blood vessels (Halbreich et al., 2003). Thus, it helps the heart to beat slowly, and improving the flow of blood. Reserpine also contributes with treating a psychotic condition like schizophrenia.

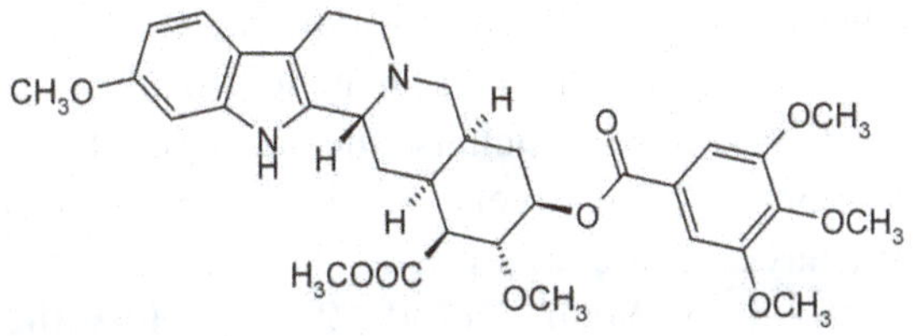

**FIGURE 13.2** Chemical structure of reserpine.

#### 13.5.1.1.1 *Therapeutic Effects*

The literature on the behavior of the mental disorder schizophrenia with drug named reserpine is now widely acknowledged. Reserpine is operative in chronic and acute cases of schizophrenia (Carlsson and Carlsson, 2006). The ratio of its conduction is narrated from 5% to 85%. However, the variation is dependent on factors like:

- nature of disease treated,
- medication and time period of treatment, and
- to what extent it constitutes 'improve' verdict by investigators.

Furthermore, it is difficult to say its consistency as to how one subtype of schizophrenia reacts better than any other subtypes. The reason behind this is due to variation of criteria patients are sited at specific types of schizophrenia (Schulze and Angermeyer, 2003). Hebephrenics is a type of schizophrenia which is categorized by disorganized behavior and speech (Ujike et al., 2002). It is also termed as disorganized schizophrenia and has the meager diagnosis if not treated well on time and its paranoids frequently have increased remission rate. If the drug acted to bring out the improvement one can expect the deprived reaction in hebephrenic. It is our imprint that the type of schizophrenia is a less important factor in predicting the result of the drug except that in generalized order of its response which is due to spontaneity. Although the recovery of all classes would not have been accomplished without the use of the drug (Maurya et al., 2016). More the degree of excitement and nervousness better will be the diagnosis. However, this is not different for the treatment of drug as it is also responsible for other procedures like lobotomy (it is a neurosurgical procedure, a form of psychosurgery). The known fact is not all stressed patients give a good reaction to the same. Many patients seem to maintain real improvement first after the completion of their therapy and the drug gets inhibited by the active administration. This holds true in the case of less stressed patients, therefore the period of evaluating treatment is generally 6 weeks to 2 months.

The literature does not tell us the difference of consistency with respect to the sex of patients. The drug in children is found to be meager as compared to adults but the general impression lays states that schizophrenic children do not react as favorably as an adult do. However, patients under hospitalization during all durations and in all conditions show a good reaction. It is rightly stated that the patients should not be denied for a therapeutic trial because it is said that the supplies of drug clinically are limited and therefore, it should be applied to the areas where they are most effective and long-term (Klein et al., 2006). Schizophrenia cannot be cured but it contains partial symptoms of remission (Masand et al., 2009). There are also patients living a glorious life without even being hospitalized. If careful examinations are performed in respect to psychological tests, chances are that the person might be schizophrenic. However, there would still be an unwillingness to tag them as disease-free.

### 13.5.1.2   RISPERIDONE

It is associated with the chemical class named benzisoxazole derivatives (Fig. 13.3) it has got a high affinity for D2 and 5-hydroxytryptamine (5-HT) receptors (de Leon et al., 2005). It primarily aims in treating positive and negative symptoms of schizophrenia. It also triggers irritability in children suffering from an autistic disorder, BD, inappropriate behavior in dementia, and manic episodes related to BD (Lam et al., 2010). Risperidone can either be taken orally or injected into muscles. The effect of injection lasts for about 2 weeks (Strickley, 2004). The common side effects include sleepiness, increased weight, and movement problems (Tempaku et al., 2016). Serious side effects include disorder in potentially permanent movement (tardive dyskinesia) and NMS, high blood sugar level. Risperidone is freeing soluble in methylene chloride, soluble in 0.1 N HCL and methanol and insoluble in water, it is off-white to beige powder. Its structural formula is $C_{23}H_{27}FN_4O_2$ and molecular weight is 410.49.

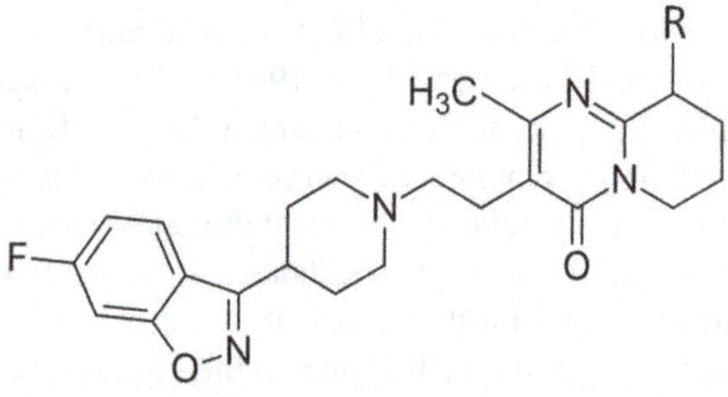

Risperidone (R = H)
Risperidone Active Metabolite (R = OH)

**FIGURE 13.3**   Chemical structure of risperidone and its active metabolite.

### 13.5.1.2.1   *Mode of Action*

Risperidone is an antipsychotic drug of second generation having an affinity for receptors like α1, α2, D2, H1, and 5-HT2A. Although the mechanism of action is not fully agreed, the recent theories focus mainly on receptors like 5-HT2A and D2 (Butini et al., 2008). From the perspective of pharmacodynamics, the antipsychotics share a common characteristic that is the ability to reduce neurotransmission of dopaminergic. According to the theory of dopamine in schizophrenia, overactivity of mesolimbic pathway explains the positive symptoms of schizophrenia. Cognitive and negative symptoms of schizophrenia are associated with dysfunction of this pathway. Antipsychotic drugs have an ability to inhibit dopamine receptors. A significant proportion of patient reacts poorly toward neuroleptics, especially sufferers with negative symptoms, which include social withdrawal and apathy. Some other neurotransmitters which include 5-hydroxytryptamine acts on receptor 5HT2,which aims to block ritanserin (Miyamoto et al., 2012).

Catecholamine receptor-like α1, α2, 5HT2, D2 adrenoceptors is blocked by risperidone. Risperidone contains antihistamine (H1) activity but no effect is seen on receptors like β adrenoceptors, peptidergic or cholinoceptors (Mishra et al., 2007). According to open studies, it has been suggested that therapeutic effect on schizophrenia with both negative and positive symptoms and it might be supportive in patients who lack response against conventional neuroleptics. Less frequent symptoms like extrapyramidal do not occur generally. Risperidone is effective against negative symptoms but it does cause fewer effects of extrapyramidal (Geddes et al., 2000). Risperidone can also be used as an alternative for clozapine (Geddes et al., 2000).

### 13.5.1.2.2 Therapeutic Effects

Supplementary treatment of schizophrenia with celecoxib has positive effects when treated with risperidone (Geddes et al., 2000). Also, treatment alongside immunomodulatory drug resulted to be helpful for dealing with the symptoms of schizophrenia indicating dysfunction of the immune system. It is associated to pathomechanism and is not just an epiphenomenon. The effect of celecoxib which is a nonimmunological therapeutic is mediated by the receptor of NMDA. Risperidone is considered to be well-established and a proven drug for the treatment of schizophrenia (Müller et al., 2002). The addition of celecoxib has a high impact in improvement of patients suffering from both negative and positive symptoms of schizophrenia (Miller et al., 2011).

The inflammation is mediated by activation of COX-2 and the brain, this COX-2 articulates. Cytokines, such as IL-6, IL-2, IL-10 induces activation of COX-2. Further, the inflammation is mediated by cytokine-activated COX-2 expression. In schizophrenic patients CSF level of sIL-2R and IL-2, soluble IL-6 receptors are a functional part of IL-10 and IL-6 system which are seen to be an increased amount.

Expression of adhesion molecules is regulated by blocking of COX-2. The reduction of the regulation of adhesion molecules is observed in schizophrenia, which leads to lack of communication and imbalance between CNS and peripheral immune system (Müller et al., 2002).

### 13.5.1.3 CLOZAPINE

Clozapine is an atypical antipsychotic which aims in triggering schizophrenia and its symptoms. It is a high-dose neuroleptic (Haddad and Sharma, 2007). The derivative of this drug is dibenzodiazepine. Chemical class of clozapine is dibenzoxazepine drug loxapine (Fig. 13.4). In schizoaffective disorder and schizophrenia, clozapine decreases the percentage of suicidal conducts. Clozapine is considered to be more active than typical antipsychotics. It is taken orally. Clozapine is related in lowering the leucocytes which result in death. So to overcome the problem sufferers should monitor their blood regularly (Moolman, 2013). It is considered to be the safest and effective medicine. Clozapine pertains superior therapeutic efficacy against Parkinson's disease although; it differs from loxapine with respect to pharmacological characteristics. Clozapine has several blocking activities like: adrenergic (α1), histaminergic (H1), serotonin (S2). It also has muscarinic acetylcholine receptor, which is an antagonist and is potent. The binding of this drug is weak with receptors like D2 and D1 If high amount of doses is given to patient's clozapine, then it produces little transient elevation observed in serum prolactin. The side effects produced by this drug are very dissimilar from typical neuroleptics. However, clozapine is considered to be superior to chlorpromazine because it produces fewer side effects. Clinically, clozapine has revealed its antipsychotic effects to be greater or same to haloperidol, levomepromazine, and chlorpromazine (Sim et al., 2004). This drug does not yield EPS like tardive dyskinesia or akathisia, parkinsonism, acute dystonia. Hypersalivation, tachycardia, orthostatic hypotension can be evoked as side effects of this drug.

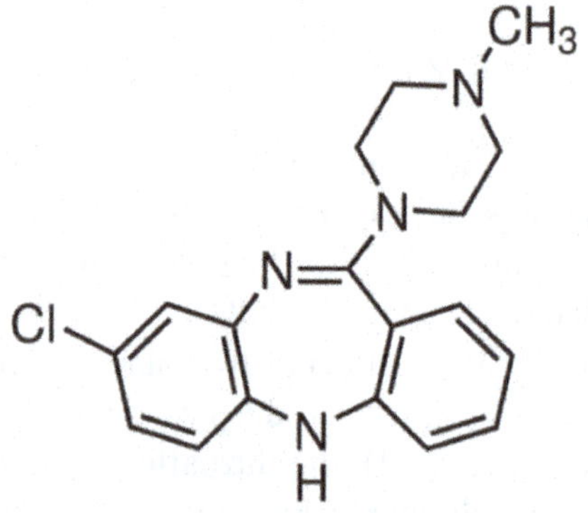

**FIGURE 13.4**  Chemical structure of clozapine.

### 13.5.1.3.1  *Mechanism of Action*

Clozapine produces serotonin (5HT2) and dopamine (D2) blockade receptor. These receptors have an ability to inhibit the increased cortisol secretion and growth hormone which is produced by MK-212 and apomorphine. They directly act on agonist like 5-HT2 and DA. Clozapine neither inhibits deceased level in plasma prolactin (PRL) concentration due to induction of prolactin nor does it increase PRL level, unlike D2 antagonist. Although clozapine drug increase DA release, evidenced by consistent PRL results (Sim et al., 2004). Clozapine decreases plasma homovanillic acid (HVA), basal plasma cortisol leaves and plasma tryptophan. According to rodent studies, it was suggested that clozapine is associated to an increase 5-HT release. However, later it was found that antagonism of $5\text{-HT}_2$ and D-2 receptors and increase of 5-HT and DA release are vital in minimizing schizophrenia symptoms both negative and positive without producing side effects like increase in plasma PRL or EPS. Dysregulation of D-2 and $5\text{-HT}_2$ mediated neurotransmission is involved in schizophrenia and clozapine partially restores the normal balance of dopaminergic and serotonergic neurotransmission. Clozapine treatment accelerates the level of 5-hydroxyindoleacetic acid (5-HIAA), a significant metabolite of 5-HT. Clozapine treatment marks level of tryptophan, which is the precursor of 5-HT (Manchia et al., 2017).

### 13.5.1.3.2  *Clozapine as Serotonin Antagonist*

Serotonergic effects of lysergic acid diethylamide are inhibited by clozapine and it also blocks the social effect of quipazine which is a 5-HT agonist (Glennon and Dukat, 2002). This is the proof that this drug is a serotonin antagonist (5-HT). It has also been discovered that clozapine inhibits hormone stimulating effect which is produced by MK-212 due to secretion of cortisol, as well as the hyperthermic effects. The treatment of this drug inhibits induced MK-212 increased secretion of plasma cortisol and therefore, its area under the curve in sufferers appeared to be less as compared to neuroleptic and nonmedicated patients. Thus, chlorpromazine inhibits induction of MK-212 cortisol like clozapine drug does but it is expected that clozapine is more effective and efficient in vivo than chlorpromazine as a 5-HT2 antagonist on the foundation of data for in vitro affinity (Abidi and Bhaskara, 2003). Otherwise, a response seen in a man due to induction of MK-212 cortisol is because of stimulation of receptor 5-HT of some other type which is not inhibited by chlorpromazine but is by clozapine.

### 13.5.1.3.3   *Neuroendocrine Effects*

Risperidone is associated to increase serum prolactin level unlike clozapine in humans which therefore differentiates it from a classical neuroleptic drug. The increased level of serum prolactin interferes with the menstrual cycle (Raggi, 2002).

### 13.5.1.3.4   *Therapeutic Effects*

Clozapine is efficient against schizophrenic patients who flop to react to typical neuroleptics (Marek, 2002). Clozapine minimizes psychiatric symptoms with high efficacy. It is medically proven that clozapine is superior to chlorpromazine. It differs from other conventional neuroleptics with respect to its pharmacological characteristics. Blockade of dopamine type-1 receptors are strong and blockade of dopamine type-2 receptors as compared to perphenazine which is a high potent neuroleptic. It acts as α2, α1 nonadrenaline receptor and serotonin receptor antagonist and has high affinity for dopamine type-4 receptors and anticholinergic effect.

No effects of clozapine on higher level of cognitive functions have been discovered in schizophrenic patients. According to a study, it was observed that some patients who received flupenthixol or haloperidol did not differ from patients receiving clozapine (Jones et al., 2006). However, one such issue is critical in clarifying this if cognitive impairment is an epiphenomenon or a primary manifestation. However, an improvement was seen due to clozapine but no change was observed in cognitive functions. Clinically, as clozapine patients were found to be more cooperative and they appeared to be much improved, however, many cognitive functions were observed to remain impaired.

The cognitive dysfunction in schizophrenia is because of interference occurred by epiphenomenon or by positive psychotic symptoms caused due to deficiency in motivation or cooperation. Bad effect of visual memory was seen in patients due to clozapine. The cholinergic and anticholinergic system plays vital role in memory (Trivedi, 2006). To summarize, clozapine did reduce symptoms of chronic schizophrenia but it lagged cognitive impairments, which appeared to be intrinsic feature for diseases. Vocational and social adjustments persisted to be on margin (McEvoy et al., 2006).

### 13.5.1.4   *OLANZAPINE*

It is also an atypical antipsychotic, which aims to trigger BD and early onset of schizophrenia.

It is also efficient in treating acute exacerbations. The structure of olanzapine (Fig. 13.5) is similar to quetiapine and clozapine but contains slightly different affinity at binding sites. Olanzapine has comparable structure relation with benzodiazepine anxiolytics, it has got minimum affinity for GABA$_A$ receptor and its effect is mediated via 5HT receptors and on dopamine. It is a derivative of thieno-benzodiazepine and is a dopamine antagonist. Its treatment results in increased glucose level, cholesterol level, and weight gain as compared to other antipsychotics of second generation except clozapine. This drug produces minimum sedation induction, anticholinergic symptoms, prolactin elevation, no traces of hematotoxicity, and extrapyramidal effects. Therefore, it is believed that olanzapine is the first drug for the treatment of schizophrenia (Leucht et al., 2009).

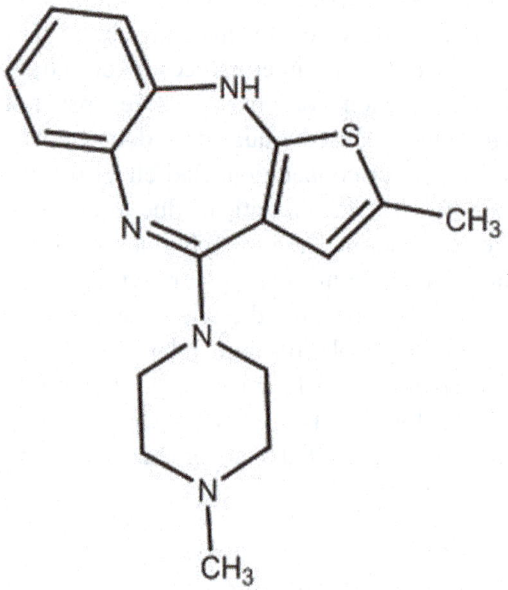

**FIGURE 13.5**　Chemical structure of olanzapine.

### 13.5.1.4.1　*Mechanism of Action*

Olanzapine possesses affinity at binding sites like serotonergic (5HT2, 3, 6), D1-D5, muscarinic (subtypes 1-5), adrenergic ($\alpha$1-2), and histaminergic (H1). This receptor is an alternative mediator in the histaminergic system, which aims for regulation of weight and behavior. Patients with less schizophrenia suffer from less side effects of olanzapine. But if the disorder is severe then it is to choose between atypical or typical drug. Olanzapine is efficient in dropping negative symptoms of this disorder but its clinical relevance is not clear yet (Geddes et al., 2000).

### 13.5.1.4.2　*Therapeutic Effects*

Between haloperidol and olanzapine, the latter showed better results on secondary measure including depression and overall improvement (Horacek et al., 2006). This drug is related less discontinuation of treatment because of lack of adverse events or deficiency of drug efficacy. However, it shows wider and greater spectrum of efficacy in patients and consists of favorable safety profile than haloperidol.

### 13.5.1.5　*HALOPERIDOL*

It is a typical antipsychotic (Fig. 13.6) and is used to cure tics in Tourette syndrome, mania in BD, vomiting, nausea, acute psychosis, hallucinations in alcohol withdrawal, and schizophrenia. It can either be consumed orally or intravenously, injecting into muscles. This drug takes 30–60 min to work (Rifkin et al., 1991). Tardive dyskinesia is a movement disorder which is caused by haloperidol and it might be permanent. This drug is considered to be the safest and effective. It exerts antiemetic and sedative activity. The pharmacological effect of haloperidol is parallel to piperazine-derivative phenothiazines. The drug acts at all the phases of central nervous system but majorly it acts at subcortical level-plus on multiple organ system. Haloperidol has feeble peripheral anticholinergic and strong antiadrenergic activity. The blocking action of ganglionic is slight. Haloperidol possesses little antiserotonin and antihistaminic activity (Gold, 1967).

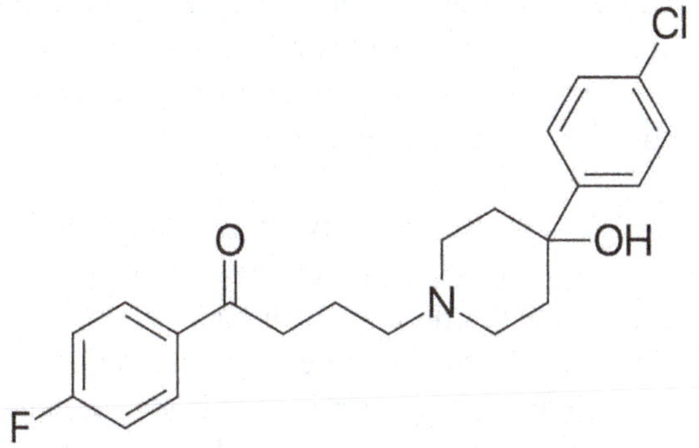

**FIGURE 13.6**   Chemical structure of haloperidol.

### 13.5.1.5.1   *Mechanism of Action*

The mechanism of production of therapeutic effect is not known yet, but this drug depresses CNS at mid brain, brainstem reticular formation and at subcortical level of brain. The ascending reticular activating system is blocked by this drug of brainstem via caudate nucleus, thus the impulse gets interrupted between cortex and diencephalon. Action of glutamic acid is antagonized by the drug inside the extrapyramidal system and blockage of catecholamine receptors also help in mechanism of haloperidol (Naidu et al., 2003). Reuptake of neurotransmitters of mid-brain gets hinder by haloperidol and appears to have feeble central anticholinergic and solid central antidopaminergic activity.

The drug inhibits spontaneous motor activity and produces catalepsy and conditions ignorant behavior in animals. Although antiemetic action of this drug has not been determined fully yet but the drug blocks dopamine receptor of chemoreceptor trigger zone (CTZ) and therefore, it affects directly (Naidu et al., 2003).

### 13.5.1.6   *QUETIAPINE*

It is an atypical antipsychotic (Fig. 13.7) and objectives to shoot BD, major depressive disorder and schizophrenia. It contributes as a sleep aid because of the presence of sedative effect. It can be taken orally only. Elder patients suffering from dementia are at risk of death because of quetiapine. If used in during late pregnancy, it might cause movement disorder in baby for a time period after the birth. Quetiapine blocks dopamine and serotonin receptor (Srisurapanont et al., 2004).

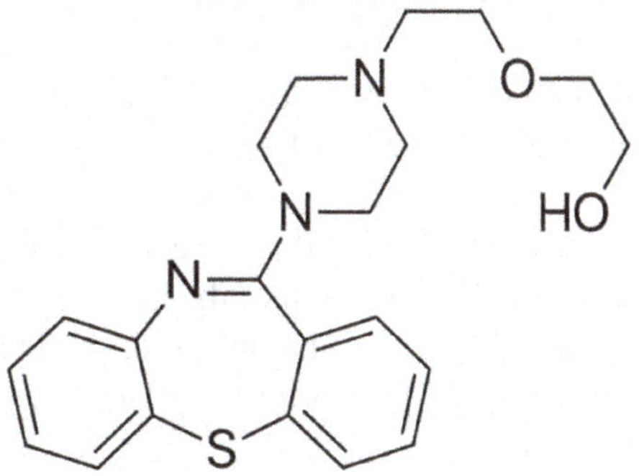

**FIGURE 13.7**   Chemical structure of quetiapine.

### 13.5.1.6.1   *Mechanism of Action*

It is a second-generation drug and has got affinity for receptors like 5-HT2A, D2. α1, H1, and 5-HT1A. Its mechanism has not been fully known. But antipsychotic effects are associated to lower dopaminergic neurotransmission in mesolimbic pathway (Seeman, 2002).

### 13.5.1.6.2    *Therapeutic Effects*

Quetiapine is effective for the treatment against positive symptoms of subchronic and chronic symptoms of schizophrenia. As measured by positive and negative syndrome scale (PANSS), this drug has enhanced negative symptoms of schizophrenia (Vardigan et al., 2010). Quetiapine was statistically related to minimize the level of akathisia. According to the study of population, this drug was well tolerated. The major worst events were either moderate or mild like insomnia or agitation (Van den Eynde et al., 2008). As compared to chlorpromazine, quetiapine was related with less clinically vital signs and low postural hypotension. This drug is not related to increase serum prolactin level.

### 13.5.2    *BIPOLAR DISORDERS*

Bipolar disorder (BD) is a mental disorder recognized by episodes of hypomania or maniac and depressions. It is a psychotic disorder. The following are the drugs required to treat the disease. The drugs mainly are: reserpine, risperidone, clozapine, haloperidol, quetiapine, and olanzapine (Maurya et al., 2016).

### 13.5.2.1    *RISPERIDONE*

Atypical antipsychotic drugs are used for curing BDs. Specifically, for bipolar mania, these drugs posses' good tolerability and high efficacy. Nowadays, atypical antipsychotics medications for bipolar mania are used as polytherapy or monotherapy. Usage of antipsychotic drugs is widespread in clinical settings (Benazzi, 2007).

A lot of BDs are centralized as antimaniac agents (Purcell et al., 2009). The first antipsychotic drug, that is, olanzapine received Food and Drug Administration (FDA) indication for bipolar mania, in 2000. Subsequently, in 2003 and 2004 FDA approvals were received for risperidone and for compounds like ziprasidone, aripiprazole, quetiapine for indication of bipolar mania.

Risperidone is considered to be second generation antipsychotic and is an atypical antipsychotic drug (McEvoy et al., 2006). Receptor binding profile of this drug includes antagonism potent to dopamine D2, α-adrenergic receptor and serotonin. Bipolar indications approved by FDA include:

- Combined therapy with valproate or lithium for treating mixed episodes or acute mania with disorder of bipolar I.
- Short-term treatment accompanied by monotherapy of mixed episodes or acute mania with disorder bipolar I.

### 13.5.2.1.1    *Pharmacokinetics and Pharmacodynamics*

Risperidone possess high affinity for receptor like D2, α2 adrenergic, α1, HI histaminergic receptor and 5-HT2 therefore, this drug is considered to be selective monoaminergic antagonist. This drug has low affinity for 5-HT1d, 5-HT1a, 5-HT1c receptors and feeble affinity for haloperidol sensitive site, D1 and no affinity for B2 adrenergic or cholinergic muscarinic receptor (Hirschfeld et al., 2003). This drug is absorbed easily, and food does not affect extent and rate of absorption. The oral bioavailability of this drug is 70% (Marder and Meibach, 1994).

Risperidone is bound to α1-glycoprotein acid and albumin. The plasma protein binding is 90% and 77% is the 9-hydroxyrisperidone bound. Metabolism of risperidone takes place in liver (Mauri et al., 2014). The hydroxylation through CYP2D6 enzyme of risperidone to 9-hydroxyrisperidone is the major metabolic pathway. Genetic polymorphism determines the rate of metabolism through the enzyme CYP2D6 (Kang et al., 2009).

### 13.5.3  DEPRESSION

#### 13.5.3.1  CLOZAPINE

It is an atypical antipsychotic drug used for treating not only schizophrenia but also for curing psychotic, mania, and nonpsychotic depression. This drug is effective against these three syndromes for long period of time. Although the mechanism of action of this drug is not same for all the syndromes (Kane et al., 1988). Clozapine is known to have high affinity for neurotransmitters like 5-HT2A, 5-HT6, 5-HT2C, α2, α1, muscarinic, adrenergic, H3, and H1. It has weak affinity for 5-HT1A, D1, D4, D2, 5-HT3. The efficacy of the drug in case of refractory depression in psychosis has proven the nonvital for D2 receptor. The reason behind is that clozapine produces low blockade than any other neuroleptics. For downregulating receptor 5-HT$_2$ clozapine possesses rapid and potent effect. Also, this drug is highly effective for treating schizophrenia and other major disorders (Rollema et al., 1997).

#### 13.5.3.2  RESERPINE

Reserpine, when complexed with diuretic, helps to lower arterial pressure. To increase the response of thiazide diuretic it is mixed with regimen. The uptake of this drug is 0.1 mg/day, the reason being its long half life. This drug is contraindicated in history of depression, renal failure, BD. During asthma, pregnancy, ulcerative colitis this drug is avoided. It may cause complications if used during hypothermia and pregnancy. As a precaution, less quantity of this drug is recommended. Thus, this drug is not measured as a drug of choice because of its numerous side effects. This drug is associated in causing harmful amendments in fetal brain (Dawson et al., 1958).

#### 13.5.3.3  RISPERIDONE

Risperidone is an antipsychotic drug of second generation having affinity for receptors like α1, α2, D2, H1, and 5-HT2A. Although the mechanism of action is not fully agreed, the recent theories focus mainly on receptors like 5-HT2A and D2 (Edwards, 1994). From the perspective of pharmacodynamics, the antipsychotics share a common characteristic that is the ability to reduce neurotransmission of dopaminergic. According to the theory of dopamine in depression, overactivity of mesolimbic pathway explains the positive symptoms of schizophrenia. Cognitive and negative symptoms of schizophrenia are associated to dysfunction of this pathway (Swerdlow and Koob, 1987).

Antipsychotic drugs have an ability to inhibit dopamine receptors. Significant proportion of patient reacts poorly toward neuroleptics, especially sufferers with negative symptoms which include social withdrawal and apathy. Some other neurotransmitters which include 5-hydroxytryptamine acts on receptor 5HT2 which aims to block ritanserin (de Leeuw and Westenberg, 2008).

### 13.5.3.4 HALOPERIDOL

This drug is used for delirium as a medication in an intensive care unit. In comparison with placebo this drug shows better result in reducing the duration of delirium. It is a widely used drug for neuroleptic patients. Due to its adverse effects (like NMS, dystonias, and extrapyramidal effects) its uptake is in low amount. This drug should be avoided to patients suffering with electrocardiographic. The class of this drug is butyrophenone of neuroleptics. Recently, this drug has gained favor due to acute treatment in delirium and psychosis (Rifkin et al., 1991).

### 13.5.3.5 QUETIAPINE

This drug binds strongly with 5-HT2A and weakly 5-HT2C or 5-HT1 receptors better than D2 receptor. This drug occupies 30% of D2 receptors. Dibenzothiazepine is a derivative of this drug. This drug is used for sedating patients as it has strong affinity for H1-histamine receptor. The active compound of this drug is norquetiapine, it possesses greater or similar potency toward receptors like its parent compound. The side effects of this drug are dyslipidemia, diabetogenesis, orthostatic hypotension, weight gain; the side effects are not adverse in comparison to olanzapine and clozapine. This is an antipsychotic drug which aids in treating psychosis (Srisurapanont et al., 2004).

### 13.5.3.6 OLANZAPINE

This drug is FDA approved and is used to treat BD. The major side effects of this drug are weight gain and sedation. The initial dose of this drug is 2.5–5 mg and can exceed to 20 mg. This drug is oxidized by CYP 1A2 and metabolized by glucuronidation. It's a monoaminergic antagonist. It can bind to multiple receptors like 5-HT6, 5-HT2/2C, D1-4, adrenergic receptors and H1 (Duggan et al., 2005).

## 13.6 FUTURE OPPORTUNITIES AND CHALLENGES

Extensive availability of antipsychotic drugs is observed. Psychotic diseases can be treated in the upcoming years, with the help of drugs, aiming in causing minimal side effects and drugs that cause less harm to the body of patient (Schiavone and Trabace, 2018). The new area of investigation is the translational research that encompasses the mutual application of innovative technologies that involve multiple disciplines of science including pathophysiology, genetics, physiology natural history of disease, and proof-of-concept studies of devices and drugs. Recent research breakthroughs, most importantly, completion of the human genome project, offer a pool of nonending opportunities for basic investigators to work and make further advancements in the areas of neuroscience. Other accomplishments like advances in biocomputing, information technology, high-throughput technologies for screening, identifying, and studying compounds of interest, and novel imaging capabilities also tend to provide immediate and direct payment for individual investigators and the institutions that support their work (Haefner and Maurer, 2006).

## 13.7 CONCLUSION

This chapter covers the use of common antipsychotic drugs (clozapine, reserpine, risperidone, olanzapine, haloperidol, quetiapine) in neuropsychiatric disorders.

Authors also discussed various mechanism of action of aforementioned drugs. Despite the controversy that surrounds the use of atypical antipsychotic drugs in neuropsychiatric disorders, these medications are frequently being prescribed for the treatment of neuropsychiatric diseases. The final choice of atypical antipsychotic drugs should be guided by the nature and severity of the target symptom being treated, and the medication least likely to cause harm to the patient. Whenever possible, these atypical antipsychotic drug treatments should be combined with non-pharmacological treatments to limit the need and dose of antipsychotic drugs.

## KEYWORDS

- **psychosis**
- **antipsychotics**
- **neuropsychiatric diseases**
- **schizophrenia**
- **reserpine**

## REFERENCES

Abidi, S.; Bhaskara, S. M. From Chlorpromazine to Clozapine—Antipsychotic Adverse Effects and the Clinician's Dilemma. *Can. J. Psychiatry* **2003**, *48*, 749–755.

Barim, A. O.; Aydin, S.; Colak, R.; Dag, E.; Deniz, O.; Sahin, İ. Ghrelin, Paraoxonase and Arylesterase Levels in Depressive Patients Before and After Citalopram Treatment. *Clin. Biochem.* **2009**, *42*, 1076–1081.

Benazzi, F. Bipolar Disorder—Focus on Bipolar II Disorder and Mixed Depression. *Lancet* **2007**, *369*, 935–945.

Butini, S.; Gemma, S.; Campiani, G.; Franceschini, S.; Trotta, F.; Borriello, M.; Ceres, N.; Ros, S.; Coccone, S. S.; Bernetti, M. Discovery of a New Class of Potential Multifunctional Atypical Antipsychotic Agents Targeting Dopamine D3 and Serotonin 5-HT1A and 5-HT2A Receptors: Design, Synthesis, and Effects on Behavior. *J. Med. Chem.* **2008**, *52*, 151–169.

Carlsson, A.; Carlsson, M. L. A Dopaminergic Deficit Hypothesis of Schizophrenia: The Path to Discovery. *Dialog. Clin. Neurosci.* **2006**, *8*, 137.

Chiodo, L. A.; Bunney, B. S. Typical and Atypical Neuroleptics: Differential Effects of Chronic Administration on the Activity of A9 and A10 Midbrain Dopaminergic Neurons. *J. Neurosci.* **1983**, *3*, 1607–1619.

Clozapine. https://en.wikipedia.org/wiki/Clozapine (accessed March 06, 2018).

Conley, R. R.; Buchanan, R. W. Evaluation of Treatment-Resistant Schizophrenia. *Schizophr. Bull.* **1997**, *23*, 663.

Coyle, J. T. The GABA-Glutamate Connection in Schizophrenia: Which is the Proximate Cause? *Biochem. Pharmacol.* **2004**, 68, 1507–1514.

Dawson, D.; Kernohan, G.; Knox, S. Reserpine in Schizophrenia. *Lancet* **1958**, *272*, 589.

de Leeuw, A. S.; Westenberg, H. G. Hypersensitivity of 5-HT2 Receptors in OCD Patients: An Increased Prolactin Response After a Challenge With Meta-Chlorophenylpiperazine and Pre-Treatment with Ritanserin and Placebo. *J. Psychiatr. Res.* **2008**, *42*, 894–901.

de Leon, J.; Susce, M. T.; Pan, R.-M.; Fairchild, M.; Koch, W. H.; Wedlund, P. J. The CYP2D6 Poor Metabolizer Phenotype may be Associated with Risperidone Adverse Drug Reactions and Discontinuation. *J. Clin. Psychiatry* **2005**, *66*, 15–27.

Duggan, L.; Fenton, M.; Rathbone, J.; Dardennes, R.; El-Dosoky, A.; Indran, S. Olanzapine for Schizophrenia. *Cochrane Lib.* **2005**, *18* (2), CD001359.

Edwards, J. G. Risperidone for Schizophrenia. *BMJ* **1994**, *308*, 1311.

Erhart, S. M.; Marder, S. R.; Carpenter, W. T. Treatment of Schizophrenia Negative Symptoms: Future Prospects. *Schizophr. Bull.* **2006**, *32*, 234–237.

Galletly, C. Role of Cognitive Enhancement in Schizophrenia. *Neuroscience and Neuroeconomics.* **2014**, *2014* (3), 75–85.

Geddes, J. R.; Miklowitz, D. J. Treatment of Bipolar Disorder. *Lancet* **2013**, *381*, 1672–1682.

Geddes, J.; Freemantle, N.; Harrison, P.; Bebbington, P. Atypical Antipsychotics in the Treatment of Schizophrenia: Systematic Overview and Meta-Regression Analysis. *BMJ* **2000**, *321*, 1371–1376.

Glennon, R. A.; Dukat, M. Serotonin Receptors and Drugs Affecting Serotonergic Neurotransmission. Philadelphia: Lippincott Williams & Wilkins, ©2008. *Foye's Principles of Medicinal Chemistry.* 2002, p 6.

Gold, M. I. Tranquilizers and Anesthesia. *Anesthesia Prog.* **1967,** *14*, 75.

Goldberg, J. F.; Perlis, R. H.; Bowden, C. L.; Thase, M. E.; Miklowitz, D. J.; Marangell, L. B.; Calabrese, J. R.; Nierenberg, A. A.; Sachs, G. S. Manic Symptoms During Depressive Episodes in 1,380 Patients with Bipolar Disorder: Findings from the STEP-BD. *Am. J. Psychiatry* **2009,** *166*, 173–181.

Haddad, P. M.; Sharma, S. G. Adverse Effects of Atypical Antipsychotics. *CNS Drugs.* **2007,** *21*, 911–936.

Haefner, H.; Maurer, K. Early Detection of Schizophrenia: Current Evidence and Future Perspectives. *World Psychiatry* **2006,** *5*, 130.

Halbreich, U.; Kinon, B.; Gilmore, J.; Kahn, L. Elevated Prolactin Levels in Patients with Schizophrenia: Mechanisms and Related Adverse Effects. *Psychoneuroendocrinology* **2003,** *28*, 53–67.

Haloperidol. https://en.wikipedia.org/wiki/Haloperidol (accessed Mar 06, 2018).

Harvey, P. D.; Green, M. F.; Keefe, R. S.; Velligan, D. I. Cognitive Functioning in Schizophrenia: A Consensus Statement on its Role in the Definition and Evaluation of Effective Treatments for the Illness. *J. Clin. Psychiatry.* **2004**.

Hasler, G. Pathophysiology of Depression: Do We Have Any Solid Evidence. *World Psychiatry.* **2010,** *9*, 155–161.

Harvey, R. A.; Clark, M. A.; Finkel, R.; Rey, J. A.; Whalen, K. Lippincott's Illustrated Reviews: Pharmacology, 5th ed., Wolters Kluwer: New York, 2011.

Hirschfeld, R.; Calabrese, J. R.; Weissman, M. M.; Reed, M.; Davies, M. A.; Frye, M. A.; Keck Jr, P. E.; Lewis, L.; McElroy, S. L.; McNulty, J. P. Screening for Bipolar Disorder in the Community. *J. Clin. Psychiatry.* **2003,** *64*, 53–59.

Horacek, J.; Bubenikova-Valesova, V.; Kopecek, M.; Palenicek, T.; Dockery, C.; Mohr, P.; Höschl, C. Mechanism of Action of Atypical Antipsychotic Drugs and the Neurobiology of Schizophrenia. *CNS Drugs* **2006,** *20*, 389–409.

Jones, P. B.; Barnes, T. R.; Davies, L.; Dunn, G.; Lloyd, H.; Hayhurst, K. P.; Murray, R. M.; Markwick, A.; Lewis, S. W. Randomized Controlled Trial of the Effect on Quality of Life of Second- vs First-Generation Antipsychotic Drugs in Schizophrenia: Cost Utility of the Latest Antipsychotic Drugs in Schizophrenia Study (CUtLASS 1). *Arch. Gen. Psychiatry* **2006,** *63*, 1079–1087.

Kane, J.; Singer, M.; Meltzer, M. Clozapine for the Treatment-Resistant. *Arch. Gen. Psychiatry* **1988,** *45*, 789–796.

Kang, R.-H.; Jung, S.-M.; Kim, K.-A.; Lee, D.-K.; Cho, H.-K.; Jung, B.-J.; Kim, Y.-K.; Kim, S.-H.; Han, C.; Lee, M.-S. Effects of CYP2D6 and CYP3A5 Genotypes on the Plasma Concentrations of Risperidone and 9-Hydroxyrisperidone in Korean Schizophrenic Patients. [*J. Clin. Psychopharmacol.* **2009,** *29*, 272–277.

Kapur, S. Psychosis as a State of Aberrant Salience: A Framework Linking Biology, Phenomenology, and Pharmacology in Schizophrenia. *Am. J. Psychiatry* **2003,** *160*, 13–23.

Kapur, S.; Remington, G. Dopamine D 2 Receptors and Their Role in Atypical Antipsychotic Action: Still Necessary and May Even be Sufficient. *Biol. Psychiatry* **2001,** *50*, 873–883.

Klein, D. J.; Cottingham, E. M.; Sorter, M.; Barton, B. A.; Morrison, J. A. A Randomized, Double-Blind, Placebo-Controlled Trial of Metformin Treatment of Weight Gain Associated With Initiation of Atypical Antipsychotic Therapy in Children and Adolescents. *Am. J. Psychiatry* **2006,** *163*, 2072–2079.

Kota, L. N.; Purushottam, M.; Moily, N. S.; Jain, S. Shortened Telomere in Unremitted Schizophrenia. *Psychiatry. Clin. Neurosci.* **2015,** *69*, 292–297.

Kreutzer, J. S.; Seel, R. T.; Gourley, E. The Prevalence and Symptom Rates of Depression After Traumatic Brain Injury: A Comprehensive Examination. *Brain Injury* **2001,** *15*, 563–576.

Kuroki, T.; Nagao, N.; Nakahara, T. Neuropharmacology of Second-Generation Antipsychotic Drugs: A validity of the Serotonin–Dopamine Hypothesis. *Prog. Brain Res.* **2008,** *172*, 199–212.

Lam, D. H.; Jones, S. H.; Hayward, P. *Cognitive Therapy for Bipolar Disorder: A Therapist's Guide to Concepts, Methods and Practice*; John Wiley & Sons; 2010; 101, 344.

Leucht, S.; Corves, C.; Arbter, D.; Engel, R. R.; Li, C.; Davis, J. M. Second-Generation Versus First-Generation Antipsychotic Drugs for Schizophrenia: A Meta-Analysis. *Lancet* **2009,** *373*, 31–41.

Manchia, M.; Carpiniello, B.; Valtorta, F.; Comai, S. Serotonin Dysfunction, Aggressive Behavior, and Mental Illness: Exploring the Link Using a Dimensional Approach. *ACS Chem. Neurosci.* **2017,** *8*, 961–972.

Mansur, R. B.; Cunha, G. R.; Asevedo, E.; Zugman, A.; Zeni-Graiff, M.; Rios, A. C.; Sethi, S.; Maurya, P. K.; Levandowski, M. L.; Gadelha, A. Socioeconomic Disadvantage Moderates the Association Between

Peripheral Biomarkers and Childhood Psychopathology. *PloS One* **2016,** *11*, e0160455.

Marder, S. R.; Meibach, R. C. Risperidone in the Treatment of Schizophrenia. *Am. J. Psychiatry* **1994,** *151*, 825.

Marek, G. Glutamate and Schizophrenia: Pathophysiology and Therapeutics. *Cent. Nerv. Syst. Agents Med. Chem.* **2002,** *2*, 29–44.

Masand, P. S.; Roca, M.; Turner, M. S.; Kane, J. M. Partial Adherence to Antipsychotic Medication Impacts the Course of Illness in Patients with Schizophrenia: A Review. *Prim. Care. Companion J. Clin. Psychiatry* **2009,** *11*, 147.

Mauri, M.; Paletta, S.; Maffini, M.; Colasanti, A.; Dragogna, F.; Di Pace, C.; Altamura, A. Clinical Pharmacology of Atypical Antipsychotics: An Update. *Excli. J.* **2014,** 13, 1163.

Maurya, P. K.; Noto, C.; Rizzo, L. B.; Rios, A. C.; Nunes, S. O.; Barbosa, D. S.; Sethi, S.; Zeni, M.; Mansur, R. B.; Maes, M. The Role of Oxidative and Nitrosative Stress in Accelerated Aging and Major Depressive Disorder. *Prog. Neuropsychopharmacol. Biol. Psychiatry* **2016,** *65*, 134–144.

Maurya, P. K.; Rizzo, L. B.; Xavier, G.; Tempaku, P. F.; Zeni-Graiff, M.; Santoro, M. L.; Mazzotti, D. R.; Zugman, A.; Pan, P.; Noto, C. Shorter Leukocyte Telomere Length in Patients at Ultra High Risk for Psychosis. *Eur. Neuropsychopharmacol.* **2017,** *27*, 538–542.

McEvoy, J. P.; Lieberman, J. A.; Stroup, T. S.; Davis, S. M.; Meltzer, H. Y.; Rosenheck, R. A.; Swartz, M. S.; Perkins, D. O.; Keefe, R. S.; Davis, C. E. Effectiveness of Clozapine Versus Olanzapine, Quetiapine, and Risperidone in Patients with Chronic Schizophrenia who did not Respond to Prior Atypical Antipsychotic Treatment. *Am. J. Psychiatry* **2006,** *163*, 600–610.

Miller, B. J.; Buckley, P.; Seabolt, W.; Mellor, A.; Kirkpatrick, B. Meta-Analysis of Cytokine Alterations in Schizophrenia: Clinical Status and Antipsychotic Effects. *Biol. Psychiatry* **2011,** *70*, 663–671.

Mishra, B.; Saddichha, S.; Kumar, R.; Akhtar, S. Risperidone-Induced Recurrent Giant Urticaria. *Br. J. Clin. Pharmaco.* **2007,** *64*, 558–559.

Miyamoto, S.; Duncan, G.; Marx, C.; Lieberman, J. Treatments for Schizophrenia: A Critical Review of Pharmacology and Mechanisms of Action of Antipsychotic Drugs. Nature Publishing Group: 2005. *Mol Psychiatry.* **2005,** *10* (1), 79–104.

Miyamoto, S.; Miyake, N.; Jarskog, L.; Fleischhacker, W.; Lieberman, J. Pharmacological Treatment of Schizophrenia: A Critical Review of the Pharmacology and Clinical Effects of Current and Future Therapeutic Agents. Nature Publishing Group: 2012. *Mol Psychiatry.* **2012, Dec**; *17* (12):1206–27.

Moolman, M.-S. Clozapine Usage in a Public Sector Psychiatric Hospital in the Nelson Mandela Metropole. 2013. Boloka Institutional Repository (https://repository.nwu.ac.za/handle/10394/15883)

Mosley, R. L.; Benner, E. J.; Kadiu, I.; Thomas, M.; Boska, M. D.; Hasan, K.; Laurie, C.; Gendelman, H. E. Neuroinflammation, Oxidative Stress, and the Pathogenesis of Parkinson's Disease. *J. Clin. Neurosci.* **2006,** *6*, 261–281.

Müller, N.; Riedel, M.; Scheppach, C.; Brandstätter, B.; Sokullu, S.; Krampe, K.; Ulmschneider, M.; Engel, R. R.; Möller, H.-J.; Schwarz, M. J. Beneficial Antipsychotic Effects of Celecoxib Add-On Therapy Compared to Risperidone Alone in Schizophrenia. *Am. J. Psychiatry* **2002,** *159*, 1029–1034.

Naidu, P. S.; Singh, A.; Kaur, P.; Sandhir, R.; Kulkarni, S. K. Possible Mechanism of Action in Melatonin Attenuation of Haloperidol-Induced Orofacial Dyskinesia. *Pharmacol. Biochem. Behav.* **2003,** *74*, 641–648.

Olanzapine. https://en.wikipedia.org/wiki/Olanzapine (accessed Mar 06, 2018).

Perlis, R. H.; Ostacher, M. J.; Patel, J. K.; Marangell, L. B.; Zhang, H.; Wisniewski, S. R.; Ketter, T. A.; Miklowitz, D. J.; Otto, M. W.; Gyulai, L. Predictors of Recurrence in Bipolar Disorder: Primary Outcomes from the Systematic Treatment Enhancement Program for Bipolar Disorder (STEP-BD). *Am. J. Psychiatry* **2006,** *163*, 217–224.

Purcell, S. M.; Wray, N. R.; Stone, J. L.; Visscher, P. M.; O'donovan, M. C.; Sullivan, P. F.; Sklar, P.; Ruderfer, D. M.; McQuillin, A.; Morris, D. W. Common Polygenic Variation Contributes to Risk of Schizophrenia and Bipolar Disorder. *Nature* **2009,** *460*, 748–752.

Quetiapine. https://en.wikipedia.org/wiki/Quetiapine (accessed Mar 06, 2018).

Raggi, M. Therapeutic Drug Monitoring: Chemical-Clinical Correlations of Atypical Antipsychotic Drugs. *Curr. Med. Chem.* **2002,** *9*, 1397–1409.

Reserpine. https://en.wikipedia.org/wiki/Reserpine (accessed Mar 06, 2018).

Resperidone. https://en.wikipedia.org/wiki/Risperidone (accessed Mar 06, 2018).

Rifkin, A.; Doddi, S.; Karajgi, B.; Borenstein, M.; Wachspress, M. Dosage of Haloperidol for Schizophrenia. *Arch. Gen. Psychiatry.* **1991,** *48*, 166–170.

Rizvi, S. I.; Maurya, P. K. Markers of Oxidative Stress in Erythrocytes During Aging in Humans. *Ann. N. Y. Acad. Sci.* **2007,** *1100,* 373–382.

Rollema, H.; Lu, Y.; Schmidt, A. W.; Zorn, S. H. Clozapine Increases Dopamine Release in Prefrontal Cortex by 5-HT1A Receptor Activation. *Eur. J. Pharm.* **1997,** *338,* R3–R5.

Schiavone, S.; Trabace, L. Small Molecules: Therapeutic Application in Neuropsychiatric and Neurodegenerative Disorders. *Molecules* **2018,** *23,* 411.

Schulze, B.; Angermeyer, M. C. Subjective Experiences of Stigma—A Focus Group Study of Schizophrenic Patients, Their Relatives and Mental Health Professionals. *Soc. Sci. Med.* **2003,** *56,* 299–312.

Seeman, P. Atypical Antipsychotics: Mechanism of Action. *Can. J. Psychiatry* **2002,** *47,* 29–40.

Seikkula, J.; Alakare, B.; Aaltonen, J. The Comprehensive Open-Dialogue Approach in Western Lapland: II. Long-term Stability of Acute Psychosis Outcomes in Advanced Community Care. *Psychosis* **2011,** *3,* 192–204.

Shiloh, R.; Stryjer, R.; Weizman, A.; Nutt, D. J. Atlas of Psychiatric Pharmacotherapy. CRC Press, Boca Raton, Florida, United States: 2000.

Sim, K.; Su, A.; Leong, J.; Yip, K.; Chong, M.-Y.; Fujii, S.; Yang, S.; Ungvari, G.; Si, T.; Chung, E. High Dose Antipsychotic use in Schizophrenia: Findings of the REAP (Research on East Asia psychotropic Prescriptions) Study. *Pharmacopsychiatry* **2004,** *37,* 175–179.

Simpson, E. H.; Kellendonk, C.; Kandel, E. A Possible Role for the Striatum in the Pathogenesis of the Cognitive Symptoms of Schizophrenia. *Neuron* **2010,** *65,* 585–596.

Srisurapanont, M.; Maneeton, B.; Maneeton, N.; Lankappa, S.; Gandhi, R. Quetiapine for Schizophrenia. The Cochrane Library, 2004.

Strickley, R. G. Solubilizing Excipients in Oral and Injectable Formulations. *Pharm. Res.* **2004,** *21,* 201–230.

Swerdlow, N. R.; Koob, G. F. Dopamine, Schizophrenia, Mania, and Depression: Toward a Unified Hypothesis of Cortico-Striatopallido-Thalamic Function. *Behavior. Brain Sci.* **1987,** *10,* 197–208.

Tempaku, P. F.; Mazzotti, D. R.; Hirotsu, C.; Andersen, M. L.; Xavier, G.; Maurya, P. K.; Rizzo, L. B.; Brietzke, E.; Belangero, S. I.; Bittencourt, L. The Effect of the Severity of Obstructive Sleep Apnea Syndrome on Telomere Length. *Oncotarget.* **2016,** *7,* 69216.

Trivedi, J. Cognitive Deficits in Psychiatric Disorders: Current Status. *Ind. J. Psychiatry* **2006,** *48,* 10.

Ujike, H.; Takaki, M.; Nakata, K.; Tanaka, Y.; Takeda, T.; Kodama, M.; Fujiwara, Y.; Sakai, A.; Kuroda, S. CNR1, Central Cannabinoid Receptor Gene, Associated with susceptibility to Hebephrenic Schizophrenia. *Mol. Psychiatry* **2002,** *7,* 515.

Van den Eynde, F.; Senturk, V.; Naudts, K.; Vogels, C.; Bernagie, K.; Thas, O.; Van Heeringen, C.; Audenaert, K. Efficacy of Quetiapine for Impulsivity and Affective Symptoms in Borderline Personality Disorder. *J. Clin. Psychopharmacol.* **2008,** *28,* 147–155.

Vardigan, J. D.; Huszar, S. L.; McNaughton, C. H.; Hutson, P. H.; Uslaner, J. M. MK-801 Produces a Deficit in Sucrose Preference that is Reversed by Clozapine, D-Serine, and the Metabotropic Glutamate 5 Receptor Positive Allosteric Modulator CDPPB: Relevance to Negative Symptoms Associated with Schizophrenia? *Pharmacol. Biochem. Behav.* **2010,** *95,* 223–229.

Vieta, E.; Popovic, D.; Rosa, A.; Solé, B.; Grande, I.; Frey, B.; Martinez-Aran, A.; Sanchez-Moreno, J.; Balanza-Martinez, V.; Tabares-Seisdedos, R. The Clinical Implications of Cognitive Impairment and Allostatic Load in Bipolar Disorder. *Eur. Psychiatry.* **2013,** *28,* 21–29.

Wilkins, R. W. Rauwolfia Alkaloids and Serotonin Antagonists in Hypertension. *Med. Clin. North Am.* **1961,** *45,* 361–373.

Wong, A. H. C.; Van Tol, H. H. Schizophrenia: From Phenomenology to Neurobiology. *Neurosci. Biobehav. Rev.* **2003,** *27,* 269–306.

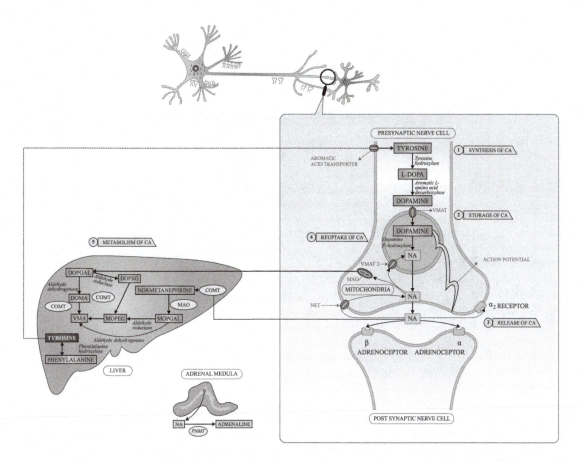

**FIGURE 3.1** The adrenergic neurotransmission mechanism: 1. Synthesis of CA: The synthesis of endogenous CA like NA, dopamine produced from tyrosine which crosses the blood–brain barrier via a carrier. 2. Storage of catecholamines: Dopamine, as well as noradrenaline synthesized, is translocated into synaptic vesicles via VMAT. 3. Release of CA: When subjected to action potential noradrenaline is liberated from the vesicle by exocytosis. In the cells of adrenal medulla, noradrenaline returns to the cytoplasm and gets converted into adrenaline with the aid of PNMT. Then Adrenaline is accumulated within the vesicle. 4. Reuptake of catecholamines: Reuptake of noradrenaline occurs by axonal uptake via NET and vesicular uptake via VMAT. 5. Metabolism of catecholamines are metabolized with the aid of the enzymes MAO and COMT. CA, catecholamines; DOPA, dihydroxyphenylalanine; NA, noradrenaline; VMAT, vesicular monoamine transporter; MAO, monoamine oxidase; NET, norepinephrine transporter; PNMT, phenylethanolamine N-methyl transferase; DOPGAL, 3,4-didydroxyphenylclycoaldehyde; DOMA, 3,4-dihydroxy mandelic acid; VMA, 3-methoxy 4-hydroxy mandelic acid; MOPEG, methoxy-4-hydroxyphenylethyl-glycol; DOPEG, 3,4-dihydroxyphenylethylene glycol; COMT, catechol-O-methyl transferase.

*Source*: Adapted from Tripathi (2014), Golan (2012), and Wevers et al. (1999).

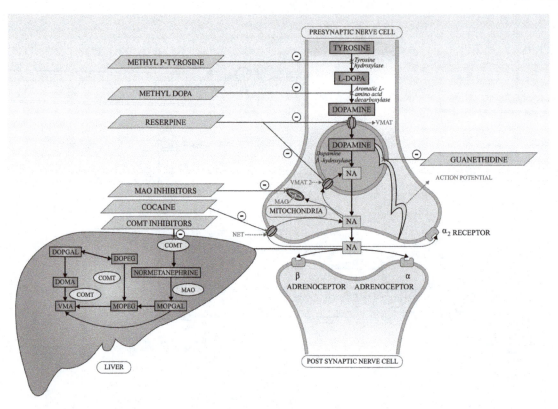

**FIGURE 3.2** Drugs modifying adrenergic neurotransmission. DOPA, dihydroxyphenylalanine; NA, noradrenaline; VMAT, vesicular monoamine transporter; MAO, monoamine oxidase; NET, norepinephrine transporter; DOPGAL, 3,4-didydroxyphenylclycoaldehyde; DOMA, 3,4-dihydroxy mandelic acid; VMA, 3-methoxy 4-hydroxy mandelic acid; MOPEG, methoxy-4-hydroxyphenylethyl-glycol; DOPEG, 3,4-dihydroxyphenylethylene glycol; COMT, catechol-O-methyl transferase.

*Source*: Adapted from Tripathi (2014) and Golan (2012).

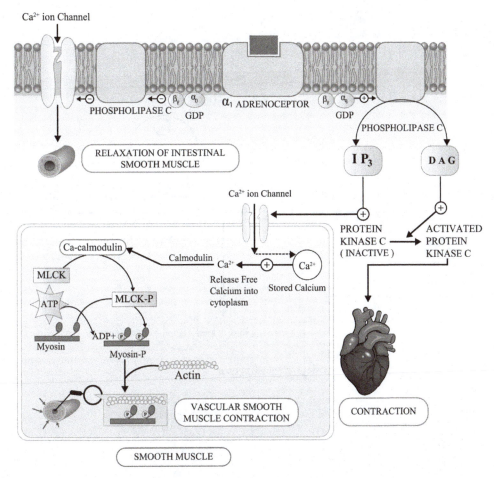

**FIGURE 3.3** Signal transduction and pharmacological action of the neurotransmitters, noradrenaline, adrenaline, and other exogenous sympathetic drugs on $\alpha_1$ adrenergic receptors. GDP, guanosine 5′-diphosphate; $IP_3$, inositol triphosphate; DAG, diacylglycerol; $Ca^{2+}$, calcium ions; ATP, adenosine triphosphate; ADP, adenosine diphosphate; MLCK, myosin light-chain kinase.

*Source*: Adapted from Katzung et al. (2014) and Rang et al. (2007). $\alpha_2$ adrenergic receptors (Fig. 3.4) are predominantly distributed in pancreatic $\beta$ cells, adrenergic nerve endings, platelets, and blood vessels and response of all $\alpha_{2A}$, $\alpha_{2B}$, and $\alpha_{2C}$ are mediated by $G_0$ or $G_i$ signaling mediators. $\alpha_2$ autoreceptors in the presynaptic neurons mediate reduced norepinephrine release by stimulating $G_0$ having reduced $\alpha$ action inhibiting the enzyme adenylyl cyclase thereby decreasing cAMP, hence reducing $Ca^{2+}$ influx and inhibiting exocytosis of the transmitter. Activation of the signaling mediator $G_i$ in the postsynaptic neuron results in inhibition of adenylyl cyclase thereby decreasing cAMP. Hence inactive protein kinase A cannot be activated which might be the probable process for the platelet aggregation and reduced insulin secretion (Bylund, 1992; Michelotti et al., 2000).

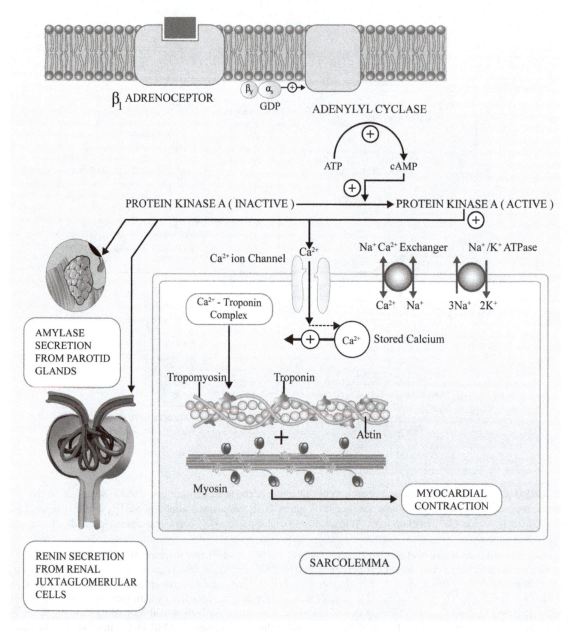

**FIGURE 3.5** Signal transduction and pharmacological action of the neurotransmitters, noradrenaline, adrenaline, and other exogenous sympathetic drugs on $\beta_1$ adrenergic receptors. cAMP, cyclic adenosine monophosphate; ATP, adenosine triphosphate; GDP, guanosine 5′-diphosphate; $Ca^{2+}$, calcium ions; $Na^+ Ca^{2+}$ exchanger, sodium–calcium exchanger or sodium–potassium pump; $Na^+ K^+$ ATPase, sodium–potassium adenosine triphosphatase.

*Source*: Adapted from Katzung et al. (2014) and Rang et al. (2007).

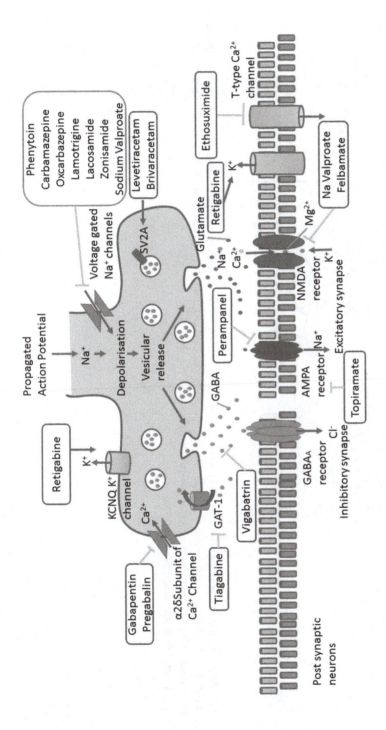

**FIGURE 14.6** Mechanism of action of antiepileptic drugs. SV2A, synaptic vesicle glycoprotein 2A; GABA, γ-aminobutyric acid; GAT-1, sodium- and chloride-dependent GABA transporter 1; AMPA, α-amino-3-hydroxy-5-methyl-4-isoxazole-propionic acid.

*Source*: Adapted from Shih et al., 2013.

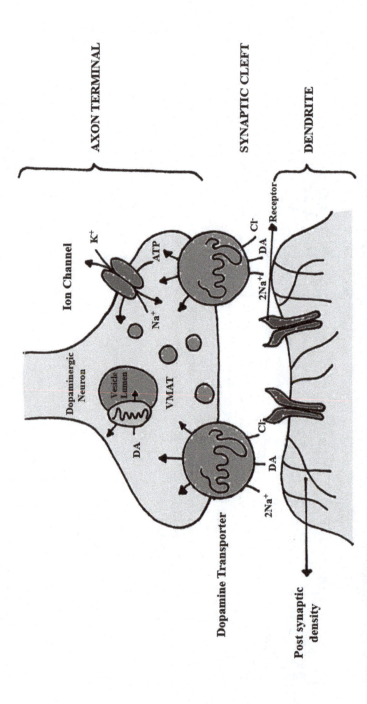

**FIGURE 20.1** Transporter proteins involved in uptake of dopamine (DA). The DA transporter uses the energy stored across the plasma membrane due to the occurrence of sodium gradient. Neurotransmitters diffuse through synaptic cleft and reach their respective receptors. Neurotransmitter once interior to the cell is taken to the vesicles through vesicular monoamine transporter (VMAT).

*Source:* Adapted from Amara and Sonders, 1998.

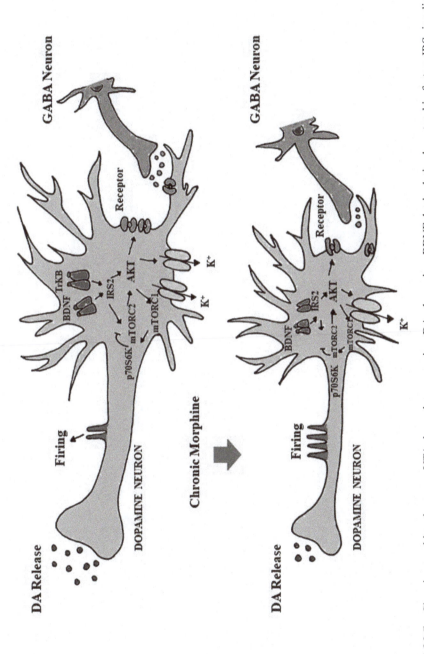

**FIGURE 20.5** Chronic morphine decreases VTA dopamine soma size. DA, dopamine; BDNF, brain-derived neutrophic factor; IRS, insulin receptor substance; mTORC, mammalian target of rapamycin complex; TrkB, tropomyosin receptor kinase B; AKT, protein kinase.

*Source:* Adapted from Nestler, 2013.

# CHAPTER 14

# Antiepileptic Drugs

AJAY GUPTA* and SURJIT SINGH

*All India Institute of Medical Sciences, Rajasthan, India*

*Corresponding author. E-mail: dr.ajaygupta24@gmail.com*

## ABSTRACT

Epilepsy is a neurological disorder with symptoms that can be adequately controlled in majority of patients with one or more antiseizure drugs. Based on location of origin of seizures, epilepsy can be classified as focal onset, generalized onset, and unknown onset. Phenobarbital was the first widely used drug for the control of epilepsy. The initial drugs used for treatment have narrow therapeutic index with wide range of side effects including teratogenicity. Later on, newer antiepileptic drugs (AEDs) with good efficacy, much better tolerability profile, and fewer adverse effects and no need for therapeutic drug monitoring were developed. Initially, the newer AEDs got marketing approval by the US Food and Drug Administration (FDA) as add-on therapy only, except topiramate and oxcarbazepine; nowadays, most of the drugs like levetiracetam and lamotrigine approved as add on in the past have been given approval to be used as monotherapy. Numerous clinical trials confirm that the efficacy of AEDs does not differ significantly for refractory focal seizures. In this chapter, the authors provide a synopsis of the different AEDs available for the management of epilepsy with the caution to be used carefully,

giving due consideration of potential adverse drug reactions.

## 14.1 INTRODUCTION

The oldest record of epilepsy in human history came from Babylonian era tablets dating back to 2000 BC. Many other ancient civilizations have description of epileptic seizures, including India, China, and Egypt. The word epilepsy has its origin from the Greek word "epilepsia" meaning "to take hold of" or "to seize" (Lowenstein, 2015). Nine out of 100 individuals can have lifetime likelihood of experiencing at least one epileptic seizure, and three out of 100 individuals have the lifetime likelihood of receiving a diagnosis of epilepsy. Nevertheless, active epilepsy is manifested in only about 0.8% of population (David, 2018).

Epilepsy enlists as the third most common brain disorder in the world. Cerebrovascular and Alzheimer's disease (AD) being the first two most common. Epilepsy is not a single disease entity but an array of varied seizures types and syndromes having different pathophysiology and mechanisms of origins of seizure activity, which is characteristically sudden in onset and have excessive, hypersynchronous discharge of cerebral neurons. This abnormal brain neuronal electrical activity may result in physical manifestations like abnormal body movements, atypical or odd behavior, and distorted perceptions and loss of consciousness. These events last for short duration but can recur if left untreated. Epilepsy is defined by International League Against Epilepsy (ILAE) and the International Bureau for Epilepsy (IBE) as a "brain disorder characterized by an enduring predisposition to generate epileptic seizures and by the neurobiological, cognitive, psychological, and social consequences of this condition" (Fisher et al., 2005). The definition of epilepsy requires the occurrence of at least one epileptic seizure.

One of the manifestations of neurologic or metabolic diseases can be epileptic seizures. Epileptic seizures have different etiologies like metabolic conditions such as hypoglycemia, genetic predisposition for abnormal brain activity, head trauma, ischemic and hemorrhagic stroke, brain tumors, alcohol withdrawal (delirium tremens) or drug withdrawal, degenerative disorders like AD. Epilepsy is marked by recurrent, unprovoked seizures (David, 2018). Therefore, repeated seizures with provocation (e.g., alcohol withdrawal or stroke) or identifiable cause do not constitute epilepsy. Clinically, epilepsy is diagnosed if there is occurrence of at least two unprovoked seizures (David, 2018). But most of the clinicians diagnose or label epilepsy when one unprovoked seizure occurs in the setting of a predisposing cause, such as a focal brain mass or injury, or a persistent genetic predisposition resulting in generalized interictal discharge. Seizures are the manifestation of abnormal hypersynchronous or hyperexcitable discharges of cortical neurons (David, 2018). The term epileptic seizure is used for differentiating this event from a nonepileptic seizure such as a psychogenic or pseudo seizure,

which involves abnormal clinical behavior of the subjects without any hypersynchronous neuronal discharge or brain activity.

Many drugs are approved for the treatment and control of seizures. The goal is to achieve seizure-free status in patients with epilepsy without having significant adverse effects. Studies have shown that only less than two-thirds of newly diagnosed epilepsy patients are free of seizures after 1 year of treatment (Ventola, 2014). AEDs with good efficacy, better tolerability profile, and fewer adverse effects especially teratogenicity and no need for therapeutic drug monitoring were developed. The objective of the chapter is to give an overview of pathophysiology of seizures along with the drugs which showed promise in the particular form of seizure. This chapter represents the mechanism of action, adverse drug reactions, and application of various AEDs.

## 14.2 PATHOPHYSIOLOGY OF SEIZURES

### 14.2.1 FOCAL SEIZURES

Focal onset seizures (formerly classified as partial seizures) mean that seizure initiated in just one focal region of the brain. It has propensity to spread to other regions of the brain and may progress to involve both cerebral hemispheres of the brain causing bilateral tonic–clonic seizure, also known as secondary generalized seizure. Focal seizures can often be unusual in nature. Usually goes unnoticed and many a times mistaken for daydreaming or being intoxicated (Wang et al., 2017). Focal onset seizures or focal seizures represent about 60% of epilepsy. Depending on person's awareness during a seizure, focal onset seizures can be further divided into focal aware and focal impaired awareness. In focal aware (formerly classified as simple partial seizures), the person is fully aware of his surroundings or what's happening in their vicinity but may not be able to respond to stimuli or talk to other person. Usually brief in duration, and are often misinterpreted or called as warning or "aura" (as thought that more significant seizure like tonic–clonic seizure activity may develop) but are actually part of the seizure. While in focal impaired awareness (formerly classified as a complex partial seizure), impaired awareness and the person seems to be confused or disorientated to their surroundings.

It occurs in three steps. First being initiation of seizure foci by an increase in electrical activity at the cellular level, then synchronous discharge of surrounding neurons which advance to the adjoining regions of the brain. Seizures are initiated as a paroxysmal depolarizing shift (PDS) which is marked by sudden onset or generation of an abnormally rapid train of action potentials within a group of neurons which lasts upto 200 ms (Lowenstein, 2015). Changes in the extracellular milieu or environment, attributable to various factors, can modify neuronal burst activity in brain. For example, increase in extracellular $K^+$ by the space-occupying lesion or other etiology could cause blunting of the effects of $K^+$ mediated after-hyperpolarization, lowering seizure threshold and increased firing of neurons (Lowenstein, 2015). Similarly, an upsurge in excitatory neurotransmitters or accentuation of NMDA and glutamate receptors by other exogenous molecules or abnormal channel conductance or altered membrane characteristics could also cause increase burst activity. It is because of surrounding inhibition that the local discharge of the neurons is often contained within the brain foci and do not results in synchronous

spread to adjacent brain regions or induce symptomatic seizures (Lowenstein, 2015).

However, repetitive firing of neurons which leads to increase in extracellular $K^+$ weakens $K^+$-mediated hyperpolarization resulting in diminishing the surrounding inhibition of neurons. Increase in extracellular $K^+$ results in synchronous rapid firing of neurons, allowing the seizure activity to spread. This increased extracellular $K^+$ results in opening of depolarization-sensitive N-methyl-D-aspartate (NMDA) channels and $Ca^{2+}$ accumulation in synaptic terminals, both of which accounts for increase likelihood of local synchronization of signals and seizure propagation (Wang et al., 2017). Degeneration of GABAergic neurons, or modulation at the GABA receptor level and decrease in $\gamma$-aminobutyric acid (GABA)-mediated inhibition because of exogenous factors are major contributing factors for synchronization and spread of a seizure focus and appears to be the most important mechanism. The patient may experience a sense of fear and confusion or an aura, a conscious "warning signals" of the spread of the seizure activity to the neighboring areas of brain. In addition, this may include an olfactory hallucination or disturbances in language or memory or, altered basic sensations. As the seizure spread to different brain regions, additional clinical manifestations appear depending on the brain regions involved (Extercatte et al., 2015). For example, Jacksonian march (named derived from English neurologist Hughlings Jackson) which is typically characterized, as the clinically manifesting as shaking of the hand and further advance to involve the arms and the legs as the other parts of the brain participate in seizure spread. These physical manifestations result from synchronous involvement across the motor homunculus (Extercatte et al., 2015).

### 14.2.2 SECONDARY GENERALIZED SEIZURES

Secondary generalized seizures result from focal seizures spread along brain parenchymal connections to involve both cerebral hemispheres. As the seizures involve both hemispheres, the loss of consciousness and therefore it is not classified as aware or unaware. But rather classified based on the motor activity as generalized motor and non-motor seizures similar to primary generalized seizures. Focal seizures may spread along subcortical U-fibers that connect adjacent gyri of the cortex; the corpus callosum mediates the spread between cerebral hemispheres and thalamocortical projections mediate diffuse synchronized spread throughout the brain. A patient usually loses consciousness as the seizure activity spread along both hemispheres (Blumenfeld et al., 2003).

Tonic–clonic seizures are one of the most common subtype in secondary generalized seizures. The abnormal channel activity and intracellular recording of neuronal bursts along with the drugs acting on various channels that can control the initiation and spread of seizures is shown in Figure 14.1. The symptoms can be explained by the modulation in the channel activity and neurotransmitter release. The tonic phase of the tonic–clonic seizure is associated with a sudden loss of GABAergic input, which exhibits as a long sustained, rapid firing of neurons, which manifests clinically as increased contraction of both the agonist and antagonist muscles. Eventually, excitatory transmission mediated by α-amino-3-hydroxy-5-methyl-4-isoxazolepropionic acid (AMPA)

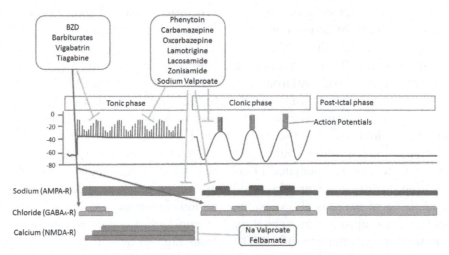

**FIGURE 14.1** Abnormal GABA, NMDA, and AMPA channel activity in the tonic–clonic seizure. In the tonic–clonic seizures, tonic phase is initiated by a sudden loss of GABA-mediated inhibition of surrounding neurons. This loss of inhibitory signals results in formation of rapid succession of action potentials within the seizure foci (increased activity of AMPA sodium channels and NMDA calcium channels), which manifests clinically as tonic contraction of the muscles. The GABA agonists and AMPA and NMDA receptor inhibitors will decrease the train of action potentials and hence controlling the seizures. As GABAergic innervation is restored, the GABA inhibitory component begins to alternate rhythmically with the NMDA and AMPA excitatory component, which manifests clinically as clonic movements. Enhanced GABA-mediated inhibition predominates in postictal phase, resulting in post-ictal depression or loss of consciousness.

*Source*: Adapted from Cornes et al., 2017.

and NMDA channels begins to oscillate with the inhibitory component mediated by GABAergic transmission, results in clonic or shaking (when involving the motor cortex) movements of the body (Meldrum, 1994; Barker-Haliski and White, 2015). During the postictal period, as the GABA-mediated inhibition prevails, the patient becomes flaccid as the muscle activity is lost and becomes unconscious (Cornes et al., 2017).

### 14.2.3 PRIMARY GENERALIZED SEIZURES

These have its origin from central part of the brain and then spread rapidly along corpus callosum to both hemispheres. These do not begin with an aura which differentiates these seizures from focal seizures with secondary generalization.

Generalized seizures involve both hemispheres resulting in loss of consciousness. It is further classified into generalized motor and non-motor seizures or absence seizures. Absence seizures occur as there is sudden lapse in awareness manifesting as staring spell or daydreaming. Motor generalized seizures can be further classified as tonic–clonic where the body stiffens during the tonic phase and then the limbs begin to jerk rhythmically in the clonic phase; myoclonic manifesting as sudden single jerks of a muscle or a group of muscles not lasting more than a second or two; tonic involves a brief stiffening of the body, back or torso or arms or legs, resulting in sudden fall if standing or sitting; atonic as brief seizures that cause a sudden loss of muscle tone resulting in persons fall or manifest as head nod if sitting and clonic causing jerking in various parts of the body (Liyanagedera and Williams, 2017).

One of the most common well understood mechanism of primary generalized seizures is of absence seizures or well known as petit mal seizures. Absence seizures are characterized by sudden interruptions in the level of consciousness without fall and are usually accompanied by a blank stare, or occasional motor symptoms, such as rapid blinking of eyes, lip smacking, and eyelid fluttering and chewing movements. Absence seizures are thought to result from abnormal synchronization between thalamocortical neurons and cortical cells (Fuentealba et al., 2005). The thalamocortical projections involved in the generation of spike wave discharges and generation of absence seizures is shown in Figure 14.2 (Liu et al., 1995; Steriade and Contreras, 1998; Fuentealba and Steriade, 2005). The electroencephalographic (EEG) readings in subjects experiencing absence seizures are similar to those observed during stage 3 sleep (Steriade and Contreras, 1995).

Absence seizures are associated with activation of T-type calcium channels (Fuentealba et al., 2004; Song et al., 2004). T-type calcium channels are active only when the cell is hyperpolarized. During the awake state, hyperpolarization of thalamic neurons is mediated by various mechanisms like an increase in inhibitory input from the reticular nucleus mediated by GABA neurotransmitter, increase in intracellular potassium, or a loss of excitatory neurotransmission mediated by glutamate and NMDA receptors (Ulrich and Huguenard, 1997; Sohal and Huguenard, 2003). Hyperpolarization results in activation of T-type calcium channels in thalamic relay as well as reticular neurons, resulting in synchronous depolarization of neurons in cerebral cortex via excitatory thalamocortical connections (Song et al., 2004). This activity of the T-type calcium channel in the relay neurons results in generation of 3-per-second spike-and-wave activity, characteristic of absence seizures. Thalamocortical rebound spikes result from inputs from cortical and reticular

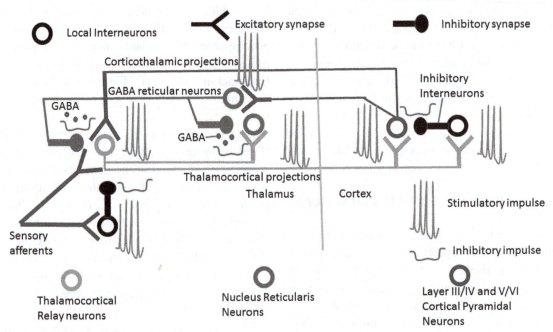

**FIGURE 14.2**   Simplified thalamocortical circuit and genesis and propagation of spike-wave discharges. Thalamic relay neurons (green color) receive sensory inputs and project onto the thalamic reticular nucleus neurons (inhibitory GABAergic transmission) and cortical pyramidal neurons in layers III/IV and V/VI in the cerebral cortex (blue color). Thalamocortical relay neurons synapse onto cortical inhibitory interneurons (black, cortex). In addition, sensory inputs project onto thalamic inhibitory interneurons (black, thalamus). Corticothalamic projections (Cortex layer VI) can reciprocally synapse onto thalamic relay neurons as well as onto the thalamic reticular nucleus neurons (pink color). Sensory inputs, thalamic inhibitory interneurons, and GABAergic reticular neurons can modulate the firing of thalamic relay neurons. As depicted in the figure, the GABAergic reticular neurons form a pacemaker sub circuit within the thalamus, having chemical and electrical synapses from cortex thalamic relay neurons. T-type calcium channels are highly expressed on reticular and relay neurons, which can exhibit rebound bursts during hyperpolarization.

*Source*: Adapted from Khosravani and Zamponi, 2006.

neurons. A large fraction of thalamocortical neurons cannot fire, resulting in disruption of information flow to cortex resulting in absence of phenotypic manifestation (Khosravani and Zamponi, 2006). The graphic representation of the above mechanism is depicted in Figure 14.3.

Activation of the T-type calcium channel in reticular and relay neurons play an important role in generation of clinical manifestation of absence phenotype, therefore, it is the primary target in the pharmacotherapy of absence seizures (Fig. 14.3). Ethosuximide and valproic acid which inhibit T-type calcium channels greatly reduce the high neuronal discharge or burst of thalamic relay neurons. The latter is essential for synchronous activation of cortical cells (Cornes et al., 2017). Benzodiazepine drug class, like clonazepam, potentiates $GABA_A$ chloride channels in the reticular thalamic nucleus, thus activating the inhibitory reticular neurons and decreasing the hyperpolarization of the thalamic relay neurons.

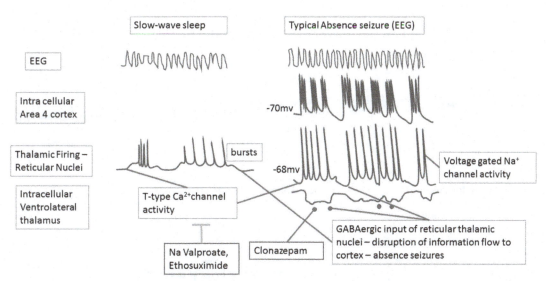

**FIGURE 14.3** The typical EEG and intracellular recordings of cortical and thalamic neurons during slow wave sleep and an epoch of spontaneous polyspike-wave seizure activity. During PDS, cortical neurons exhibit hypersynchronous bursting. This activity can be modulated by reticular neurons that exhibit T-type calcium channel mediated bursting activity. During spike-wave activity, the GABAergic influence of reticular neurons on thalamocortical cells marked by series of inhibitory postsynaptic potentials (red dots), results in tonic hyperpolarization of these neurons. As a result, a large number of thalamocortical neurons cells cannot fire, disrupting the flow of information to the cortex and clinically manifesting as absence phenotype during seizures. The drugs inhibiting T-type calcium channels are helpful in controlling the reticular thalamocortical neurons firing leading to disruption of absence seizures.

*Source*: Adapted from Khosravani and Zamponi, 2006.

## 14.3 CLASSIFICATION OF SEIZURES

The new classification of seizures by ILAE 2017 (Fisher et al., 2017) has been based on three main criteria. First, the part of the brain involved in the generation of seizure, what is the level of awareness during seizure, and other features of seizure like motor involvement or movement. The current classification of seizures is shown in Figure 14.4.

## 14.4 CLASSIFICATION OF ANTIEPILEPTIC DRUGS

In general, seizures can be managed by the help of one drug in about 75% of patients. For optimal control of seizures, patients may need more than one drug, and certain patients may never get total seizure control even with polytherapy (Harvey et al., 2011). A brief summary of AEDs is shown in Figure 14.5.

## 14.5 MECHANISM OF ACTION OF ANTIEPILEPTIC DRUGS

The action of drugs at synaptic cleft is shown in Figure 14.6. Sodium channel activation causes depolarization of neurons which leads to generation of seizures within the foci, drugs like phenytoin, carbamazepine, oxcarbazepine, lamotrigine, lacosamide, zonisamide, and sodium valproate inhibits these sodium

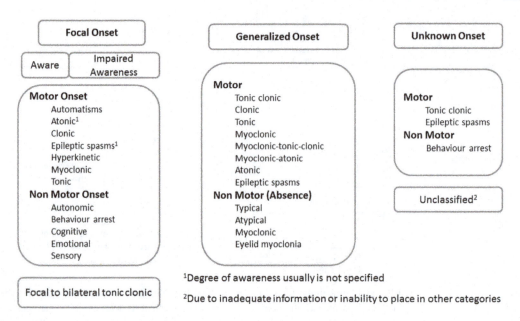

**FIGURE 14.4**   Classification of seizures by ILAE 2017.
*Source*: Adapted from Fisher et al., 2017.

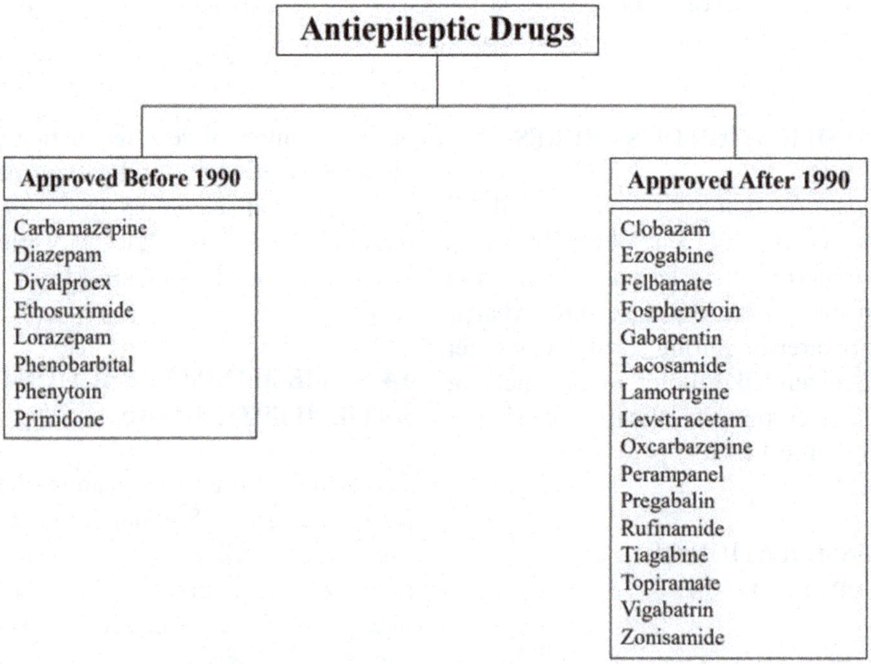

**Antiepileptic Drugs**

**Approved Before 1990**

Carbamazepine
Diazepam
Divalproex
Ethosuximide
Lorazepam
Phenobarbital
Phenytoin
Primidone

**Approved After 1990**

Clobazam
Ezogabine
Felbamate
Fosphenytoin
Gabapentin
Lacosamide
Lamotrigine
Levetiracetam
Oxcarbazepine
Perampanel
Pregabalin
Rufinamide
Tiagabine
Topiramate
Vigabatrin
Zonisamide

**FIGURE 14.5**   Classification of antiepileptic drugs.

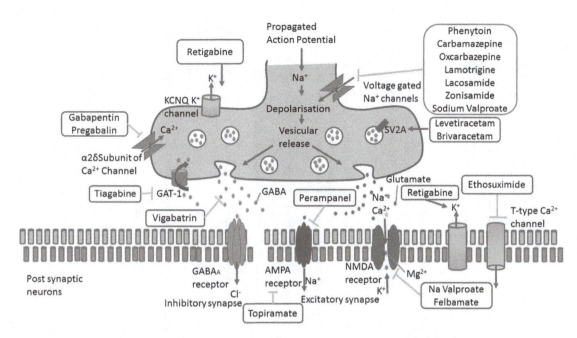

**FIGURE 14.6 (See color insert.)** Mechanism of action of antiepileptic drugs. SV2A, synaptic vesicle glycoprotein 2A; GABA, γ-aminobutyric acid; GAT-1, sodium- and chloride-dependent GABA transporter 1; AMPA, α-amino-3-hydroxy-5-methyl-4-isoxazole-propionic acid.
*Source*: Adapted from Shih et al., 2013.

channels, thereby decreasing the firing of neurons. Benzodiazepines like lorazepam, midazolam, clonazepam, and barbiturate like phenobarbital and thiopental act by increasing GABA chloride channel activity, thereby decreasing the generation of generalized tonic–clonic seizure as well as partial seizures. Tiagabine increase GABA concentration by inhibiting the GAT-1 transporters, thereby decreasing GABA reuptake while vigabatrin and valproate decrease the metabolism of GABA by inhibiting GABA aminotransferase. Gabapentin and pregabalin act by modulation of α2δ subunit of calcium channels, decreasing calcium entry through voltage-activated $Ca^{2+}$ channels, thereby decreasing glutamate release. Glutamate acts on NMDA, AMPA, and kainite receptors to generate excitatory postsynaptic potential (EPSP) in postsynaptic neuron.

Levetiracetam and brivaracetam may also affect release by binding to synaptic vesicles protein SV2A. Retigabine act by increasing the $K^+$ conductance along KCNQ/Kv7, thereby increasing the hyperpolarization and decreasing the generation of synchronous neural firing. Sodium valproate displays wide spectrum of activity by inhibiting sodium channels as well as T-type $Ca^{2+}$ channels and increasing GABA brain levels. Topiramate mechanism includes decreasing sodium channel and calcium channel activity, inhibiting glutamate receptors and increasing GABA activity in brain. Perampanel has its effects on both inhibitory (GABA–orange dots) and excitatory (Glutamate pink dots) nerve transmission (Shih et al., 2013).

The molecular mechanism of action of various AEDs that serve as complementary to the above figures is given in Table 14.1.

**TABLE 14.1** Proposed Mechanism of Action of Antiepileptic Drugs.

| Molecular targets | Drugs | Actions |
|---|---|---|
| Na$^+$ channel modulators: Act and enhance fast inactivation | Phenytoin, carbamazepine, lamotrigine, felbamate, oxcarbazepine, topiramate, valproate, eslicarbazepine, rufinamide | Block action potential propagation; neuronal membrane stabilization; decrease in focal firing, neurotransmitter release and spread of seizure |
| Act and enhance slow inactivation | Lacosamide | Increase spike frequency adaptation; decrease action potential bursts, focal firing, and spread of seizure; neuronal membrane stabilization |
| Ca$^{2+}$ channel blockers | Ethosuximide, valproate, lamotrigine | Decrease release of neurotransmitter (N- and P- types); decrease spike-wave discharges and slow-depolarization (T-type). |
| α2δ calcium channels ligands | Gabapentin, pregabalin | Modulate neurotransmitter release |
| GABA$_A$ receptor allosteric modulators | Benzodiazepines, phenobarbital, felbamate, primidone, topiramate, carbamazepine, oxcarbazepine, stiripentol, clobazam | Increase membrane hyperpolarization and seizure threshold; decrease focal firing benzodiazepines (BZDs)—attenuate spike-wave discharges; phenobarbital (PB), carbamazepine (CBZ), oxcarbazepine (OXC)—aggravate spike-wave discharges |
| GABA uptake inhibitors/ GABA transaminase inhibitors | Tiagabine, vigabatrin | Increase extrasynaptic GABA levels and membrane hyperpolarization; decrease focal firing; Increase spike-wave discharges |
| NMDA receptor antagonists | Felbamate | Decrease in excitatory neurotransmission and amino acid neurotoxicity; disruption of epileptogenesis |
| AMPA/kainate receptor antagonists | Phenobarbital, topiramate, perampanel | Fast excitatory neurotransmission decrease and decrease neuronal focal firing |
| Enhancers of HCN channel activity | Lamotrigine | Buffers large hyperpolarizing and depolarizing inputs; suppresses initiation of action potential by dendritic inputs |
| Positive allosteric modulator of KCNQ2-5 | Ezogabine | Suppresses bursts of action potentials; hyperpolarizes membrane potentials |
| SV2A protein | Levetiracetam, brivaracetam | Unknown; may decrease transmitter release |
| Inhibitors of brain carbonic anhydrase enzyme | Acetazolamide, topiramate, zonisamide | Increase HCN-mediated currents; decrease NMDA-mediated currents; increase GABA-mediated inhibition |

*Source:* Adapted from Leppik et al., 2006.

## 14.6 ANTIEPILEPTIC DRUGS

### 14.6.1 PHENYTOIN

#### 14.6.1.1 PHARMACOKINETICS

Phenytoin is available as two types of oral tablets—rapid release and extended-release tablets. Only extended-release formulations offer once daily dosing, due to differences in physical properties of these tablets such as dissolution, etc. The plasma levels of phenytoin may change when switching from one formulation to another (Porter et al., 2018). Out of different formulations of phenytoin, its equivalents should be taken into consideration to obtain comparable doses, but therapeutic drug monitoring is also necessary to ensure therapeutic safety. When route of administration is changed from oral to intramuscular (or vice versa), appropriate dose adjustments and blood level monitoring are recommended (Misty et al., 2018).

The pharmacokinetic properties of phenytoin are affected by:

- Protein binding in the plasma,
- Saturable elimination kinetics,
- Metabolism by hepatic cytochromes.

Phenytoin binds to plasma proteins, mainly albumin extensively. When it is administered along with valproic acid that displaces it from albumin and also inhibits its metabolism, free phenytoin levels are increased to toxic levels. Hence, free drug concentration should be measured in such scenarios rather than total phenytoin levels. The rate of elimination of phenytoin is nonlinear, that is, varies with change of its concentration. At low plasma concentrations, metabolism of phenytoin follows first-order kinetics; at higher concentrations, liver enzymes are saturated in its metabolizing capacity, and elimination follows zero-order kinetics which results in disproportionate increase plasma concentrations. CYP2C9 metabolizes most (95%) of phenytoin and CYP2C19 is the minor metabolizer. As the metabolism of phenytoin is saturable, other drugs metabolized by these CYP2C19 enzymes can inhibit its metabolism thereby increasing its plasma concentration (Misty et al., 2018). On the other hand, phenytoin inhibits the metabolism of drugs that are substrates for these CYPs. For example, warfarin administered in patients on phenytoin can lead to bleeding disorders due to inhibition of metabolism of warfarin by phenytoin (Misty et al., 2018).

Phenytoin also induces various CYPs like CYP3A4 which can lead to therapeutic failure of oral contraceptives, and hence unplanned pregnancy. Phenytoin has low water solubility which limits its intravenous use. Now a days a water-soluble drug fosphenytoin as an alternative is available in the market for intravenous use. Fosphenytoin is a prodrug (Misty et al., 2018) which is converted into phenytoin. It has a very short half-life of 8–15 min. High water solubility and short half-life are the two main properties for which it is being employed in the management of status epilepticus.

#### 14.6.1.2 MECHANISM OF ACTION

Three conformations of $Na^+$ channel exists closed, open, and inactivated. For action potential generation, $Na^+$ channels should be in the activated state. Phenytoin binds to $Na^+$ channels in the inactivated state and hence increases the duration of inactivation stage of $Na^+$ channels. Therefore, new action potential cannot be generated. So phenytoin increases the threshold of seizures and stabilizes the seizure focus by preventing the

PDS; latter being responsible for initiation of the focal seizure. It also prevents spread of neuronal discharge to other neurons, which accounts for its role in secondarily generalized seizures (Cornes et al., 2017).

Uses dependent $Na^+$ channels are those which are opened and closed at high frequency only. Phenytoin only binds to these high frequency firing channels and hence avoids adverse effects observed with $GABA_A$ facilitators which decreases spontaneous neuronal activity by indirectly decreasing the activity of non-use dependent $Na^+$ channels also (Cornes et al., 2017).

### 14.6.1.3 THERAPEUTIC USES

It is effective against focal and generalized tonic–clonic, focal-to-bilateral tonic–clonic, tonic–clonic of unknown onset (tonic–clonic), but not against generalized absence seizures (Porter et al., 2018). It is also used for the treatment of cardiac arrhythmias (Misty et al., 2018).

### 14.6.1.4 ADVERSE EFFECTS AND TOXICITY

The commonly observed dose related effects are related to CNS, which includes ataxia, diplopia, nystagmus, and vertigo (Misty et al., 2018). Other CNS adverse effects include slurred speech, decreased coordination, somnolence, and mental confusion drowsiness, behavioral alterations, hallucinations, disorientation, and rigidity. It can also cause nausea, epigastric pain, and vomiting, which can be minimized by taking the drug with meals (Pfizer, 2016; Misty et al., 2018). During chronic administration, following adverse effects can be seen:

### 14.6.1.4.1 Gingival Hyperplasia/Gum Hypertrophy

It is more common in younger patients. It is due to overgrowth of gingival collagen fibers. It can be minimized by good oral hygiene (Sharma and Sharma, 2017).

### 14.6.1.4.2 Hirsutism and Coarsening of Facial Features

It is troublesome in young girls (Pfizer, 2016).

### 14.6.1.4.3 Hypersensitivity Reactions

These include skin manifestations, toxic epidermal necrolysis (TEN), and Stevens–Johnson syndrome (SJS), fever, hepatitis (Pfizer, 2016).

### 14.6.1.4.4 Megaloblastic Anaemia

Phenytoin decreases folate absorption and increases its excretion (Porter et al., 2018).

### 14.6.1.4.5 Osteomalacia

Phenytoin interferes with metabolic activation of vitamin D and with calcium absorption/metabolism (Sharma and Sharma, 2017).

### 14.6.1.4.6 Hyperglycemia

It can inhibit insulin release and cause hyperglycemia (Misty et al., 2018).

### 14.6.1.4.7 Teratogenicity

During pregnancy, phenytoin can produce fetal hydantoin syndrome characterized by hypoplastic phalanges, cleft palate, hare lip, and microcephaly. Intake of phenytoin during pregnancy and presence of maternal epoxide (EPHX1) genotypes 113*H and 139*R are linked with increased risk of fetal hydantoin syndrome (Medscape, 1994–2018).

### 14.6.2 CARBAMAZEPINE

#### 14.6.2.1 PHARMACOKINETICS

It is metabolized to an active primary metabolite 10, 11-epoxide. The 10, 11-epoxide is converted to inactive chemicals by glucuronidation, which are excreted in the urine. In addition, carbamazepine is inactivated by hydroxylation and conjugation (Bertilsson, 1978). Biotransformation of carbamazepine is mainly through hepatic CYP3A4. Carbamazepine induces CYP3A, CYP2C, and UGT, therefore metabolism of those drugs which are substrate for these enzymes will increase (Bertilsson, 1978). Oral contraceptives, which are also metabolized by CYP3A4, are of particular importance in this regard. Hence, it is a drug with many clinically significant drug interactions. Therapeutic plasma concentration is in the range of 6–12 µg/mL, but considerable inter-individual variations can be seen. Frequency of CNS adverse effects increases at concentrations beyond 9 µg/mL (Misty et al., 2018).

#### 14.6.2.2 MECHANISMS OF ACTION

Carbamazepine is a $Na^+$ channel blocker that binds to $Na^+$ channels in the inactivated state and hence increases the duration of inactivation stage of $Na^+$ channels. Therefore, it delays the generation of action potential and hence increases the threshold of seizures (by preventing the PDS) as well as prevents rapid spread of neuronal discharge from the seizure focus (Cornes et al., 2017).

It is also related to tricyclic antidepressants chemically and is effective in bipolar disorder patients. Few patients suffering from bipolar disorder who do not show response to lithium carbonate also respond to this drug. Carbamazepine has antidiuretic effects that are attributed to increased plasma levels of antidiuretic hormone (ADH) (Misty et al., 2018).

#### 14.6.2.3 THERAPEUTIC USES

It is used for generalized tonic–clonic and focal seizures (both simple and complex), glossopharyngeal and trigeminal neuralgias, bipolar affective disorders, lightning/tabetic type pain associated with bodily wasting (Misty et al., 2018).

#### 14.6.2.4 ADVERSE EFFECTS AND TOXICITY

Respiratory depression, hyperirritability, convulsions, and stupor or coma can be seen with acute intoxication of carbamazepine. Chronic therapy, with carbamazepine can cause drowsiness, vertigo, diplopia, blurred vision, retention of water (cardiac disease and old age being greater risk factors), and ataxia. Over dosage increases the risk of seizure frequency. Other side effects are nausea, vomiting, elevation of pancreatic or hepatic enzymes, serious hematological reactions (agranulocytosis, aplastic anemia), and hypersensitivity reactions (splenomegaly, fatal skin reactions, eosinophilia, lymphadenopathy). Hepatic and renal function with complete blood count should be done whenever carbamazepine is used (Misty et al., 2018).

### 14.6.3 OXCARBAZEPINE

It is an analog of CBZ which is developed recently to get rid of limitations of CBZ use (drug interactions and auto induction properties).

#### 14.6.3.1 PHARMACOKINETICS

Oxcarbazepine (OXC) lacks adverse effects associated with CBZ due to epoxide metabolite because of former drug's structural modification. The active metabolite of OXC is S-licarbazepine (S+ enantiomer of eslicarbazepine). Glucuronide conjugation inactivates oxcarbazepine and it is eliminated by kidneys (May et al., 2003). Oxcarbazepine induces hepatic enzymes but to lesser extent as compared with carbamazepine. As compared with CBZ, oxcarbazepine does not reduce warfarin's anticoagulant effects, but it reduces blood levels of oral contraceptives by induction of CYP3A (May et al., 2003).

#### 14.6.3.2 MECHANISM OF ACTION

It binds to $Na^+$ channels in a similar way as that of carbamazepine. Oxcarbazepine is a $Na^+$ channel blocker that binds to $Na^+$ channels in the inactivated state and hence increases the duration of inactivation stage of $Na^+$ channels. Therefore, it delays the generation of action potential and hence increases the threshold of seizures (by preventing the PDS) as well as prevents rapid spread of neuronal discharge from the seizure focus (Cornes et al., 2017).

#### 14.6.3.3 THERAPEUTIC USES

It is used as single drug or in combination with other antiepileptics for focal seizures in adults; as monotherapy in focal seizures of children between age group 4–16; as adjuvant drug in children of age 2 years and older with epilepsy (Misty et al., 2018).

### 14.6.4 ESLICARBAZEPINE ACETATE

#### 14.6.4.1 PHARMACOKINETICS

Eslicarbazepine is a prodrug that is almost completely absorbed on oral administration. Both OXC and eslicarbazepine are transformed by liver to S-licarbazepine (active metabolite). As compared with CBZ and OXC, eslicarbazepine do not inhibit hepatic CYPs in a clinically relevant fashion. Eslicarbazepine is eliminated primarily by renal excretion (USFDA, 2013).

#### 14.6.4.2 MECHANISMS OF ACTION

Eslicarbazepine acts by similar mechanism of action as oxcarbazepine. It blocks $Na^+$ channels in the inactivated state and hence increases the duration of inactivation stage of $Na^+$ channels. Therefore, it delays the generation of action potential and hence increases the threshold of seizures (inhibits PDS) as well as prevents swift spread of neuronal discharge from the seizure focus (Cornes et al., 2017).

#### 14.6.4.3 THERAPEUTIC USE

Eslicarbazepine acetate is used in combination with other antiepileptics for focal seizures, primary and secondarily generalized tonic–clonic seizures (Epilepsy Society, 2014).

### 14.6.4.4 ADVERSE EFFECTS AND TOXICITY

The common effects are nausea, vomiting, dizziness, headache, vertigo, ataxia, somnolence, diplopia, fatigue, blurred vision, and tremors (Epilepsy Society, 2014).

## 14.6.5 EZOGABINE (RETIGABINE)

### 14.6.5.1 PHARMACOKINETICS

Ezogabine absorption is rapid and unaffected by food after oral administration. Ezogabine is metabolized by glucuronidation and acetylation (Tompson and Crean, 2013). Co-administration of carbamazepine or phenytoin reduces plasma concentrations of ezogabine; consequently, an increase in ezogabine dosage should be considered when adding phenytoin or carbamazepine (Tompson and Crean, 2013).

### 14.6.5.2 MECHANISMS OF ACTION

Ezogabine is a first $K^+$ channel opener, known as retigabine in the E.U. Ezogabine enhances transmembrane $K^+$ currents mediated by the potassium channel, voltage-gated (KQT) like sub-family of ion channels. Therefore, it stabilizes the resting membrane potential and reduces neuronal excitability (Orhan et al., 2012).

### 14.6.5.3 THERAPEUTIC USE

Ezogabine is useful in combination therapy for focal seizures in patients where other AEDs do not work adequately. The benefits over risk of retinal toxicity and visual acuity deficits must be taken into consideration before the start of this drug (Misty et al., 2018).

### 14.6.5.4 ADVERSE EFFECTS AND TOXICITY

The commonest effects associated with ezogabine include dizziness, somnolence, fatigue, confusion, and blurred vision (Misty et al., 2018). Vertigo, diplopia, memory impairment, gait disturbance, aphasia, dysarthria, and balance problems also may occur. Serious side effects include skin discoloration, QT prolongation, and neuropsychiatric symptoms, including suicidal thoughts and behavior, psychosis, and hallucinations. Due to the presence of Kv7.2–Kv7.5 in the bladder uroepithelium, ezogabine is also associated with urinary retention. Blue pigmentation of skin and lips occurs in as many as one-third of patients maintained on long-term ezogabine therapy (Misty et al., 2018). Chronic treatment with ezogabine may cause retinal abnormalities, independent of changes in skin coloration. So, baseline and periodic (every 6 months) visual monitoring (visual acuity and dilated fundus photography) is recommended by FDA in all patients who are on ezogabine (Misty et al., 2018).

## 14.6.6 FELBAMATE

### 14.6.6.1 MECHANISMS OF ACTION

It enhances GABA-induced response and inhibits NMDA-induced responses in animal experiments done on neurons of rats. Inhibition of excitatory neurotransmission as well as enhancement of inhibitory neurotransmission results in broad spectrum anti-seizure effects of felbamate (Rho et al., 1994).

### 14.6.6.2 THERAPEUTIC USES AND ADVERSE EFFECTS

Felbamate is indicated only as adjunctive therapy. It is useful for poorly controlled focal and secondarily generalized seizures and in patients of Lennox–Gastaut syndrome (Felbamate Study Group in Lennox–Gastaut Syndrome, 1993). The benefits over risk of drug-induced aplastic anemia or liver failure must be taken into consideration before the start of this drug (Sachdeo et al., 1992).

## 14.6.7 ETHOSUXIMIDE

### 14.6.7.1 PHARMACOKINETICS

Absorption of ethosuximide is complete after oral administration. The therapeutic plasma concentration is 40–100 µg/mL (Misty et al., 2018).

### 14.6.7.2 MECHANISMS OF ACTION

Ethosuximide in a voltage-dependent manner blocks low threshold T-type currents; without affecting the membrane potential across Na$^+$ channels (Porter et al., 2018).

### 14.6.7.3 THERAPEUTIC USES

It is used against absence seizures (Glauser et al., 2010) but not tonic–clonic seizures (Porter et al., 2018).

### 14.6.7.4 ADVERSE EFFECTS AND TOXICITY

The most commonly encountered adverse effects are dose-related and affect CNS (lethargy, drowsiness, dizziness, headache, and euphoria) and gastrointestinal system (nausea, vomiting, and anorexia) (Misty et al., 2018). Hypersensitivity reactions (Stevens–Johnson syndrome, systemic lupus erythematosus, other skin manifestations), hematologic abnormalities (eosinophilia, pancytopenia, aplastic anemia, leukopenia, thrombocytopenia) can also occur (Misty et al., 2018). Despite continuation of the drug, leukopenia may recover, but bone marrow depression resulted in several deaths (Misty et al., 2018).

## 14.6.8 VALPROIC ACID

It was discovered serendipitously as antiseizure drug when it was used as a lipophilic vehicle for evaluation of for antiseizure activity of other water-insoluble chemical entities by Pierre Eymard in France in 1962 (Löscher, 1999).

### 14.6.8.1 PHARMACOKINETICS

After oral administration, valproic acid displays completer absorption. The most of valproate (95%) is metabolized by liver (Gugler and von Unruh, 1980). In liver, its metabolism occurs mainly by β-oxidation and UGT enzymes. Valproate is a mainly metabolized by CYP2C9 and CYP2C19. Therefore, valproate has the propensity to inhibit the metabolism of drugs that are metabolized by CYP2C9. Valproate plasma concentrations associated with therapeutic effects are 30–100 µg/mL (Misty et al., 2018).

### 14.6.8.2 MECHANISMS OF ACTION

Similar to phenytoin and carbamazepine, valproic acid blocks Na$^+$ channels in the inactivated state and hence increases the duration of inactivation stage of Na+ channels. At slightly higher concentrations, valproic acid blocks low threshold T-type currents similar to ethosuximide. Thirdly, valproic acid acts at the level of GABA metabolism too (Cornes et al., 2017). In vitro, valproic acid enhances glutamic acid decarboxylase activity; the enzyme required for GABA synthesis and also inhibits the degradation of GABA. Both of the above factors increase GABA availability in the synaptic cleft and hence, GABA-mediated inhibition. Hence, valproic acid has broad spectrum anti-seizure properties (Porter et al., 2018).

### 14.6.8.3  THERAPEUTIC USES

Valproic acid is used as broad spectrum antiseizure drug for the treatment of tonic–clonic, absence, myoclonic, and focal seizure (Glauser et al., 2010).

### 14.6.8.4  ADVERSE EFFECTS AND TOXICITY

The most commonly encountered side effects are transient GI symptoms that include anorexia, nausea, and vomiting. CNS effects include sedation, ataxia, and tremor. Hepatic complications include elevation of hepatic transaminases in plasma, fulminant hepatitis (Dreifuss et al., 1989). Others include acute pancreatitis and hyperammonemia. It can also cause teratogenic effects such as neural tube defects (Sharma and Sharma, 2017).

## 14.6.9  LAMOTRIGINE

### 14.6.9.1  PHARMACOKINETICS

Its absorption after oral administration is complete. Glucuronidation is the main metabolic pathway. Lamotrigine and carbamazepine in combination can cause toxic levels of 10, 11-epoxide of carbamazepine in certain patients (Walker and Patsalos, 1995).

### 14.6.9.2  MECHANISMS OF ACTION

Lamotrigine inhibits glutamate release (excitatory amino acid) and also inhibits voltage-dependent sodium channel; both the actions result in neuronal membranes stabilization and hence control of seizures (Sharma and Sharma, 2017). Lamotrigine blocks Na+ channels in the inactivated state and hence increases the duration of inactivation stage of Na+ channels (Xie et al., 1995).

### 14.6.9.3  THERAPEUTIC USES

Lamotrigine is used as single drug and as add-on therapy for focal (Brodie et al., 1995; Steiner et al., 1999) and secondarily generalized tonic–clonic seizures in adults. It is also efficacious for Lennox–Gastaut syndrome in both children and adults (Motte et al., 1997).

### 14.6.9.4  ADVERSE EFFECTS AND TOXICITY

The commonly encountered effects are dizziness, nausea, vomiting, blurred vision, ataxia, and rash (Misty et al., 2018).

## 14.6.10  LEVETIRACETAM AND BRIVARACETAM

Both European Medicine Agency (EMA) and FDA approved brivaracetam recently on Jan 14, 2016 and on Feb 18, 2016, respectively (Kaur et al., 2016).

### 14.6.10.1   PHARMACOKINETICS

The drug is not bound to plasma proteins. Levetiracetam does not have any interactions with other AEDs, oral contraceptives, or anticoagulants, as it neither induces nor is a substrate for liver enzymes (Patsalos, 2004).

### 14.6.10.2   MECHANISM OF ACTION

Levetiracetam and brivaracetam bind with high affinity to synaptic vesicle glycoproteins, SV2A and also inhibit calcium channels located on the presynaptic membrane. Both of these molecular targets help to reduce neurotransmitter release and thereby act as a neuromodulator. It also inhibits neuronal voltage-gated $Na^+$ channels (Kenda et al., 2004; Zona et al., 2010).

### 14.6.10.3   THERAPEUTIC USES

Both levetiracetam and brivaracetam are used as adjunctive treatment of focal seizures in adults and children, for primary onset tonic–clonic seizures, and for myoclonic seizures of JME. Levetiracetam is found to be effective for refractory generalized myoclonic seizures as adjunctive therapy (Crespel et al., 2013).

### 14.6.10.4   ADVERSE EFFECTS AND TOXICITY

Both levetiracetam and brivaracetam are well tolerated. The commonly encountered effects associated with levetiracetam are somnolence, asthenia, ataxia, and dizziness. Behavioral and mood changes are serious, but less common (Helmstaedter, 2008). For brivaracetam, commonly encountered effects are similarly mild and include sedation, somnolence, dizziness, and GI upset. Brivaracetam dosage

needs adjustment in patients suffering from hepatic impairment (von Rosenstiel, 2007).

### 14.6.11   PERAMPANEL

### 14.6.11.1   PHARMACOKINETICS

Perampanel has a plasma half-life of about 105 h, which allows once daily administration. 95% of the drug is bound to plasma protein, mainly albumin, and is metabolized by hepatic glucuronidation and oxidation (Patsalos, 2015). Primary metabolism is mediated by hepatic CYP3A; thus, specific drug interactions and dose adjustments need to be considered. For example, perampanel may decrease the effectiveness of progesterone-containing hormone contraceptives, carbamazepine, clobazam, lamotrigine, and valproate, but it may increase the level of oxcarbazepine (Patsalos, 2015).

### 14.6.11.2   MECHANISMS OF ACTION

Perampanel is a one of the first selective, noncompetitive antagonist of the AMPA-type ionotropic glutamate receptor (Bialer and White, 2010; Stephen and Brodie, 2011). Unlike NMDA antagonists, which shorten the duration of repetitive discharges, AMPA receptor antagonists prevent repetitive neuronal firing. Perampanel decreases fast excitatory signaling critical to the seizure generation and spread (Tortorella et al., 1997). Perampanel has more inhibitory effect on seizure propagation than on seizure initiation.

### 14.6.11.3   THERAPEUTIC USE

Perampanel is used as an adjuvant drug for focal-onset seizures with or without secondarily generalized seizures in patients 12 years and older (Porter et al., 2018).

### 14.6.11.4 ADVERSE EFFECTS AND TOXICITY

Common effects are somnolence, anxiety, confusion, imbalance, double vision, dizziness, GI distress or nausea, and weight gain. Rare, but serious adverse behavioral reactions are seen including hostility, aggression, and suicidal thoughts (Misty et al., 2018).

## 14.6.12 RUFINAMIDE

### 14.6.12.1 PHARMACOKINETICS

Rufinamide is well absorbed orally, binds minimally to plasma proteins. Rufinamide is metabolized independent of CYPs and then excreted in the urine (Perucca et al., 2008).

### 14.6.12.2 MECHANISM OF ACTION

Rufinamide blocks $Na^+$ channels in the inactivated state and hence increases the duration of inactivation stage of $Na^+$ channels and hence inhibits sustained and repetitive firing of neurons (Misty et al., 2018).

### 14.6.12.3 THERAPEUTIC USE

It is effective in all seizure phenotypes of Lennox–Gastaut syndrome (Perucca et al., 2008).

### 14.6.12.4 ADVERSE EFFECTS AND TOXICITY

These are dizziness, somnolence, headache, fatigue, nausea, and dystonia (Perucca et al., 2008).

## 14.6.13 BARBITURATES

While most barbiturates have antiseizure properties, only some barbiturates, such as phenobarbital, exert their maximum antiseizure effects at doses below those that cause hypnosis. This therapeutic index determines a barbiturate's clinical utility as an antiseizure therapeutic drug. Two such useful antiseizure barbiturates—phenobarbital and primidone are discussed here.

### 14.6.13.1 PHENOBARBITAL

#### 14.6.13.1.1 Pharmacokinetics

Phenobarbital is completely absorbed after oral administration. Around 75% is principally metabolized by CYP2C9, with minor contribution by CYP2C19 and CYP2E1 in liver and remainder is excreted by kidneys in unchanged form (Porter et al., 2018). Phenobarbital is an enzyme inducer of CYP2C, CYP3A, and UGT. Enzyme induction by phenobarbital causes decreased plasma concentration of drugs such as oral contraceptives which are metabolized by CYP3A4. Therapeutic plasma concentration of phenobarbital is 10–35 µg/mL (Misty et al., 2018).

#### 14.6.13.1.2 Mechanism of Action

Phenobarbital binds to $GABA_A$ receptor and increase $Cl^-$ influx. The latter increase is due to increased duration of bursts of $GABA_A$ receptor-mediated currents. It does not alter the frequency of bursts (Twyman et al., 1989).

#### 14.6.13.1.3 Therapeutic Use

Phenobarbital is used for the treatment of focal, generalized tonic–clonic, focal-to-bilateral tonic–clonic seizures. Its reduced cost, high efficacy, and lower adverse effects make it a preferred agent for the treatment of various kinds of seizures especially in resource poor countries (Sharma and Sharma, 2017).

### 14.6.13.1.4    Adverse Effects and Toxicity

The initial effect is sedation which is present in almost all patients on initiation of therapy, but on chronic treatment, it disappears due to development of tolerance. Nystagmus and ataxia occur at excessive dosage (Misty et al., 2018). Phenobarbital can cause hyperactivity and irritability in children and, confusion and agitation in the elderly patients. Scarlatiniform or morbilliform rash, and other manifestations of drug allergy can occur. Hypoprothrombinemia with hemorrhage can occur in the newborn babies of mothers who receive phenobarbital during pregnancy; it can be prevented by administration of vitamin K prophylactically. Similar to phenytoin, megaloblastic anemia and osteomalacia can occur with long-term administration of phenobarbital therapy (Misty et al., 2018).

### 14.6.13.2    PRIMIDONE

Primidone is indicated in the United States for patients with focal or generalized epilepsy, but it has been replaced by carbamazepine and other newer AEDs that possess lower incidence of sedation in recent times. Primidone's antiseizure effects are due to its two active metabolites: phenobarbital and phenylethylmalonamide (PEMA) (Baumel et al., 1972).

### 14.6.14    BENZODIAZEPINES

Many of the benzodiazepines have broad antiseizure activities.

### 14.6.14.1    MECHANISM OF ACTION

Benzodiazepines have activity on $GABA_A$ receptors and increase the frequency of openings at GABA-activated $Cl^-$ channels (Twyman et al., 1989). It has no effect on duration of opening of these channels.

### 14.6.14.2    THERAPEUTIC USES

Clonazepam is used for the treatment of myoclonic seizures and absence of seizures in children. Intranasal spray of clonazepam is used as an orphan drug for recurrent acute repetitive seizures. Lorazepam and diazepam are efficacious in the treatment of status epilepticus. The latter is more frequently used because of less lipid solubility, more effectively confined to the vascular compartment, and has a longer effective half-life after a single dose (Porter et al., 2018). Clorazepate is effective in the treatment of focal seizures in combination with other antiepileptics. Clorazepate should be avoided in children less than 9 years of age (Porter et al., 2018). Clobazam is used in a variety of seizure phenotypes and is FDA approved for the treatment of Lennox–Gastaut syndrome in patients of age 2 years or more (Misty et al., 2018). Midazolam is used as an orphan drug for intermittent treatment of episodes of increased seizure activity in refractory epilepsy patients who are otherwise stable on other AEDs. More recently, midazolam was granted orphan drug designation in 2009 as a rescue agent for patients with intermittent bouts of increased seizure activity (i.e., acute repetitive seizure clusters), in 2012 for the treatment of nerve agent-induced seizures, and in 2016 for the treatment of status epilepticus and seizures induced by organophosphorus poisoning (Misty et al., 2018).

## 14.6.14.3 ADVERSE EFFECTS AND TOXICITY

The commonly encountered effects with chronic administration of clonazepam are lethargy and drowsiness. Behavior related effects are also seen, like aggression, hyperactivity, irritability, and difficulty in concentration (Sharma and Sharma, 2017). Both anorexia and hyperphagia have been reported. Increased secretion of saliva and bronchial secretions can occur which may be a deterrent as a treatment option for children. Abrupt discontinuation may result in exacerbation of seizures and status epilepticus (Porter et al., 2018).

## 14.6.15 GABAPENTIN AND PREGABALIN

### 14.6.15.1 PHARMACOKINETICS

Both these drugs do not bind to plasma proteins. They are not metabolized in our body and are excreted in the unchanged form by kidneys (Bockbrader et al., 2010). Appropriate dose adjustments are done in patients of renal insufficiency. These drugs do not cause drug interactions with other AEDs (Bockbrader et al., 2010).

### 14.6.15.2 MECHANISMS OF ACTION

Both these drugs have GABA molecule in their chemical structure. Their design as GABA agonists could not elicit GABA mimetic action on experiments conducted on the neurons in primary culture. Rather, it was found that these molecules bind with α2δ-1 subunit of the $Ca^{2+}$ channels (Brown and Gee, 1998). This interaction with the α2δ-1 protein may mediate the anticonvulsant effects of gabapentin and pregabalin by reducing neurotransmitter release.

## 14.6.15.3 THERAPEUTIC USES

Both are effective for focal onset seizures, with and without progression to bilateral tonic–clonic seizures, when used in addition to other AEDs. Gabapentin is also indicated in neuropathic pain due to postherpetic neuralgia in adults (Sharma and Sharma, 2017).

Pregabalin is also indicated for fibromyalgia and the neuropathic pain due to postherpetic neuralgia, spinal cord injury, or diabetic peripheral neuropathy (Misty et al., 2018).

### 14.6.15.4 ADVERSE EFFECTS AND TOXICITY

The common effects are dizziness, somnolence, ataxia, and fatigue (Porter et al., 2018).

## 14.6.16 STIRIPENTOL

### 14.6.16.1 PHARMACOKINETICS

Stiripentol is quickly absorbed; the drug is extensively bound to proteins in the plasma. Stiripentol's elimination kinetics is nonlinear. Plasma clearance decreases markedly at high doses and after repeated dosing, most likely due to saturation or inhibition of the CYPs responsible for stiripentol metabolism (Chiron, 2007). Stiripentol has diverse pharmacokinetic and pharmacodynamic interactions with concomitantly administered drugs. It is a potent inhibitor of CYPs 3A4, 1A2, and 2C19. Thus, adjunctively administered AEDs may require dose adjustments due to the potent inhibition of CYPs involved in their hepatic metabolism (Chiron, 2007).

### 14.6.16.2 MECHANISMS OF ACTION

Stiripentol inhibits uptake of GABA and inhibition of GABA transaminase in the synapse. This leads to enhancement of GABAergic inhibitory neurotransmission required for the control of seizures like barbiturates (Quilichini et al., 2006; Fisher, 2011).

### 14.6.16.3 THERAPEUTIC USES

Stiripentol is used clinically as an add-on therapy for Dravet syndrome (refractory generalized tonic–clonic seizures in patients with severe myoclonic epilepsy in infancy) in patients where valproate and clobazam are unable to control symptoms satisfactorily (Plosker, 2012; Aneja and Sharma, 2013). It is still not approved by USFDA for Dravet syndrome because of many pharmacokinetic and pharmacodynamic drug interactions (Misty et al., 2018). Stiripentol also decreases the severity and frequency of tonic–clonic seizures as well as status epilepticus with a variety of epilepsy syndromes in infants and children (Goossens et al., 1999; Perez et al., 1999; Inoue et al., 2009).

### 14.6.16.4 ADVERSE EFFECTS AND TOXICITY

The commonly encountered effects are drowsiness, ataxia, anorexia, weight loss, insomnia, hypotonia, and dystonia (Chiron, 2007).

### 14.6.17 TIAGABINE

### 14.6.17.1 PHARMACOKINETICS

Oral absorption of tiagabine is fast, it is largely bound to plasma proteins, and metabolized in liver, mainly by CYP3A (Perucca and Bialer, 1996).

### 14.6.17.2 MECHANISM OF ACTION

Tiagabine inhibits uptake of GABA into neurons by inhibiting GABA transporter, GAT-1. This leads to enhancement of GABAergic inhibitory neurotransmission required for the control of seizures (Sharma and Sharma, 2017).

### 14.6.17.3 THERAPEUTIC USES

Tiagabine is efficacious as adjuvant drug for focal seizures with or without secondary generalization in patients who do not respond to other first line drugs (Porter et al., 2018).

### 14.6.17.4 ADVERSE EFFECTS AND TOXICITY

The most commonly encountered adverse effects are dizziness, somnolence, and tremor. Tiagabine is contraindicated in patients suffering from generalized absence seizures, as it causes exacerbations of spike and wave discharges on EEG (Misty et al., 2018).

### 14.6.18 VIGABATRIN

### 14.6.18.1 PHARMACOKINETICS

An oral dose is well absorbed; the presence of food prolongs absorption. Vigabatrin is excreted unmetabolized by the kidney, and the administered dose must be decreased in patients with renal insufficiency. Vigabatrin induces CYP2C9 (Walker and Patsalos, 1995).

### 14.6.18.2 MECHANISMS OF ACTION

It is a structural analog of GABA that inhibits GABA transaminase irreversibly, the enzyme that degrades majority of GABA. Thus, it causes increased concentration of GABA in the brain which potentiates GABAergic inhibitory neurotransmission required for the control of seizures (Porter et al., 2018).

### 14.6.18.3 THERAPEUTIC USE

Vigabatrin is used as an adjuvant drug for refractory focal seizures with impaired awareness in adults. It is also an orphan drug for patients of infantile spasms (Appleton et al., 1999).

### 14.6.18.4 TOXICITY AND ADVERSE EFFECTS

Vigabatrin can cause permanent visual loss (Eke et al., 1997) and therefore has been kept as reserve drug, where other first-line treatment fails. The most common side effects include weight gain, concentric visual field constriction, fatigue, somnolence, dizziness, hyperactivity, and seizures. Vigabatrin is secreted in the milk of nursing mothers and has been classified as pregnancy category C (Misty et al., 2018).

### 14.6.19 TOPIRAMATE

#### 14.6.19.1 PHARMACOKINETICS

Important properties include rapid oral absorption, minimal protein binding, and excretion in the urine in the unchanged form. Higher dose of oral contraceptives should be given when administered along with topiramate (Garnett, 2000).

#### 14.6.19.2 MECHANISMS OF ACTION

Topiramate has broad spectrum antiseizure activity. It inhibits voltage-dependent sodium and calcium channels. It also inhibits the excitatory neurotransmission of glutamate pathway and increases inhibitory neurotransmission of GABA (Cornes et al., 2017).

### 14.6.19.3 THERAPEUTIC USES

It is used for focal-onset seizures, primary generalized tonic–clonic seizures, and Lennox–Gastaut syndrome. It is also used in the prophylaxis of migraine (Garnett, 2000).

### 14.6.19.4 ADVERSE EFFECTS AND TOXICITY

The most commonly encountered effects are fatigue, nervousness, somnolence, weight loss, and cognitive impairment. It inhibits carbonic anhydrase enzyme which may predispose the patients on topiramate to development of renal calculus (Dodgson, 2000).

### 14.6.20 ZONISAMIDE

#### 14.6.20.1 PHARMACOKINETICS

It is completely absorbed after oral administration. Most of the drug is excreted unchanged in urine. Remaining is metabolized by CYP3A4 which can lead to drug interactions with enzyme inducers and inhibitors (Misty et al., 2018).

#### 14.6.20.2 MECHANISM OF ACTION

Zonisamide is a $Na^+$ channel blocker that binds to $Na^+$ channels in the inactivated state and hence increases the duration of inactivation stage of $Na^+$ channels. Therefore, it delays the generation of action potential. In addition, zonisamide inhibits T-type $Ca^{2+}$ currents and reduces the influx of calcium (Cornes et al., 2017).

### 14.6.20.3   THERAPEUTIC USE

Zonisamide is used as adjuvant drug for focal seizures in adults 12 years or older (Misty et al., 2018).

### 14.6.20.4   ADVERSE EFFECTS AND TOXICITY

The commonly observed effects are dizziness, somnolence, cognitive impairment, ataxia, anorexia, nervousness, and fatigue. Potentially serious skin rashes are rare but may occur. Very rarely patients may develop kidney stones during treatment, which is due to inhibition of carbonic anhydrase by zonisamide (De Simone et al., 2005). As a carbonic anhydrase inhibitor, zonisamide

may also cause metabolic acidosis. Thus, zonisamide administration must be considered in younger patients predisposed to conditions like severe respiratory disorders, renal disease, diarrhea, surgery, ketogenic diet, which may pose a greater risk for metabolic acidosis (De Simone et al., 2005). Measurement of serum bicarbonate is recommended. Female patients of childbearing age are at twice the risk for spontaneous abortions and congenital abnormalities (Misty et al., 2018).

### 14.6.21   LACOSAMIDE

#### 14.6.21.1   PHARMACOKINETICS

Food consumption does not affect the absorption. 95% excretion is in the urine, about half of which is the unchanged parent compound. Patients with renal or hepatic impairment may experience a significant increase in lacosamide concentration

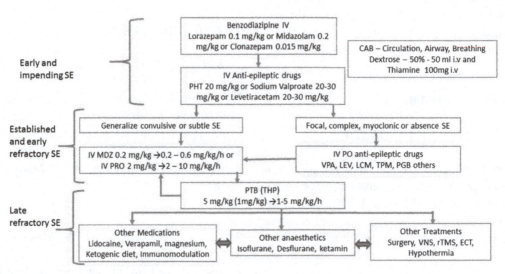

**FIGURE 14.7**  Flow diagram for the management of status epilepticus. MDZ, midazolam; PHT, phenytoin or fosphenytoin; PRO, propofol; VPA, valproic acid; LEV, levetiracetam; LCM, lacosamide; TPM, topiramate; PTB, phenobarbital; THP, thiopental; PGB, pregabalin; VNS, vagal nerve stimulation; ECV, electroconvulsive therapy; rTMS, repetitive transcranial magnetic stimulation.

*Source*: Adapted from Lowenstein, 2015.

who is taking inhibitors of CYP3A4 or CYP2C9 (Doty et al., 2007).

### 14.6.21.2 MECHANISM OF ACTION

Lacosamide blocks voltage gated $Na^+$ channels in the inactivation stage and prolongs the slow inactivation of $Na^+$ channels, rather than fast inactivation; latter being the action of all other antiepileptic drugs that block $Na^+$ channels (a unique characteristic) (Rogawski et al., 2015). Lacosamide modulates collapsin response mediator protein-2 (CRMP-2), which causes neuronal differentiation and axonal outgrowth and can form abnormal CNS connections. CRMP-2 decreases activity of NMDA receptor subtype NR2B, which is involved in epileptogenesis (Doty et al., 2007).

### 14.6.21.3 THERAPEUTIC USES

Lacosamide is approved in patients 4 years and older for focal-onset seizures both as single drug and as add on drug (Ben-Menachem et al., 2007).

### 14.6.21.4 ADVERSE EFFECTS AND TOXICITY

Minor adverse effects include headache, dizziness, double vision, nausea, vomiting, fatigue, tremor, loss of balance, and somnolence (Doty et al., 2007). Lacosamide may cause suicidal ideations and suicide, like most available AEDs. As a consequence, the FDA has mandated a black-box labeling for this agent (Misty et al., 2018). Patients who are taking concomitant drugs that prolong the PR interval should have a baseline electrocardiogram before starting lacosamide and be closely monitored for the risk of bradycardia or AV block (Krause et al., 2011).

## 14.7 STATUS EPILEPTICUS

It is defined by Working Group of Status Epilepticus of the Epilepsy Foundation of America as "more than 30 min of continuous seizure or two or more sequential seizures without full recovery of consciousness." This states that if the patients have continuous two episodes of seizures without regaining consciousness, then it is a case of status epilepticus. In clinical practice, even a single seizure episode lasting greater than 5 min should be considered as status epilepticus (Betjemann and Lowenstein, 2015). The flow diagram for the management of status epilepticus is shown in Figure 14.7.

It is essential to follow circulation, airway, and breathing algorithm in case of unconscious patient with epilepsy as it is an emergency. In addition to vital signs, neurological assessment should be conducted initially and correction of blood glucose should be done, if hypoglycemia is present. In order to avoid depletion of thiamine stores in body and preventing an acute Wernicke encephalopathy, patients should be given intravenous thiamine 500 mg before or along with blood glucose correction (Sechi and Serra, 2007). Intravenous lorazepam is the preferred initial drug due to its rapidity in action onset, and low respiratory depression, although RAMPART (Silbergleit et al., 2012) study gave preference to intramuscular midazolam. In a randomized double-blind study, there is proven efficacy of intravenous lorazepam as compared with other combinations of barbiturates and benzodiazepines (Treiman et al., 1998). Lorazepam abolished an attack of status epilepticus in almost two-third of total subjects, with comparable efficacy as observed with other combinations of barbiturates and benzodiazepines but is superior to phenytoin alone. A recommendation is to treat status epilepticus with intravenous lorazepam, which can be repeated up to three times and followed by long acting drugs like intravenous phenytoin or fosphenytoin for the adequate maintenance or control of seizures (Betjemann and Lowenstein, 2015). Despite being costly, fosphenytoin is preferred over phenytoin. Both the drugs have similar risk of arrhythmia and hypotension (Browne et al., 1996). Initial weight-based dose of phenytoin and fosphenytoin (20 mg/kg) is administered and (approximate dose is 1000 mg intravenously) is usually sufficient (Osorio and Reed, 1989). If seizures are not adequately controlled with the initial dose of fosphenytoin, a second dose of 5–10 mg/kg can be given. Phenobarbital can be used if there is inadequate control with the use of fosphenytoin. Parenteral valproic acid and levetiracetam are emerging second-line antiepileptic alternatives for the management of status epilepticus (Misra et al., 2006; Alvarez et al., 2011; Misra et al., 2012). It is recommended to use intravenous anesthetics, if seizures continue. Some researchers recommend an early initiation of intravenous anesthetics, that is, within 30–60 min of seizure onset rather than giving further trials of second line AEDs.

## 14.8 APPLICATIONS OF ANTIEPILEPTIC DRUGS

The main application of AEDs is prevention and treatment of epilepsy. But because of their pleiotropic effects, the AEDs have varied application in the disorders having involvement of neurological pathways like neuralgias, neuropathic pain, and personality disorders.

### 14.8.1 EPILEPSY

Various drugs approved for different forms of epilepsy are described in detail in the text under the specific drugs. Drugs preferred for particular form of epilepsy is detailed in Table 14.2.

**TABLE 14.2** Antiepileptic Drugs Preferred in Different Epilepsies.

| Tonic–clonic and focal seizures | Absence seizure | Myoclonic seizures | Back up and adjunctive drugs |
|---|---|---|---|
| Carbamazepine, lamotrigine, phenytoin, valproic acid | Valproic acid, clonazepam, ethosuximide | Clonazepam, valproic acid, lamotrigine | Brivaracetam, clobazam, ezogabine, felbamate, gabapentin, levetiracetam, lacosamide, perampanel, pregabalin, rufinamide, tiagabine, topiramate, vigabatrin, zonisamide, phenobarbital. |

*Source*: Adapted from Lowenstein, 2015.

### 14.8.2 BIPOLAR DISORDER

Some of the antiepileptics like valproate, carbamazepine, lamotrigine, and topiramate are used in bipolar disorder as maintenance therapy. Carbamazepine is effective in patients having no response to lithium therapy and is rapid cycling bipolar disorder. Valproate is effective in treating and preventing mania while lamotrigine is preferred in depressed phase of bipolar disorder. Topiramate has off-label indication (Stephen, 2018).

### 14.8.3 TRIGEMINAL NEURALGIA

Carbamazepine is USFDA approved for the management of trigeminal neuralgia. 100 mg produced significant relief within 2 h. In addition, gabapentin, lamotrigine, oxcarbazepine has shown effectiveness in the management of trigeminal neuralgia. If there is no partial response with carbamazepine, reconsider the diagnosis of trigeminal neuralgia. Phenytoin is less effective but can be used. Topiramate have shown efficacy in an open label uncontrolled study and can be used as second line agent (Singh, 2016).

### 14.8.4 NEUROPATHIC PAIN

Gabapentin and pregabalin have shown evidence in improving diabetic neuropathy and postherpetic neuralgia. Pregabalin is effective in central neuropathic pain and fibromyalgia. Phenytoin and valproic have shown no evidence in decreasing pain. Similarly, carbamazepine, oxcarbazepine, topiramate, lamotrigine, and lacosamide have either no or low quality evidence to be effective in reducing pain (Wiffen et al., 2013).

### 14.8.5 PAINFUL DIABETIC NEUROPATHY

Pregabalin and gabapentin are excellent in painful diabetic neuropathy (PDN), lamotrigine has shown efficacy in PDN patients. Duloxetine (selective norepinephrine reuptake inhibitor) is one of the most preferred antidepressant approved for PDN (Quan, 2018).

### 14.8.6 BORDERLINE PERSONALITY DISORDER

Carbamazepine and oxcarbazepine have shown significant improvement in impulsivity while valproate, topiramate, and lamotrigine have shown to be effective in treating the symptoms of aggression. In addition, topiramate is effective in treating interpersonal problems and anxiety (Olabi and Hall, 2010).

### 14.8.7 ANXIETY AND SOCIAL PHOBIAS

Selective serotonin reuptake inhibitors (SSRIs) are drug of choice for anxiety and phobias. Although it has been observed that pregabalin, valproic acid, and gabapentin have shown promise in controlling the symptoms of phobias and anxiety (Adrian, 2014).

### 14.8.8 ADDICTION/CRAVING

Topiramate, gabapentin, and lamotrigine are used in the treatment of withdrawal of opioids and dependence on alcohol (Johnson et al., 2004).

### 14.8.9 MIGRAINE

Divalproex sodium, topiramate have been shown to decrease migraine attacks (Jasvinder, 2018). Valproic acid decrease migraine attacks by 50%. Topiramate is indicated for migraine prophylaxis. Gabapentin, levetiracetam, carbamazepine, and zonisamide may be used later as second line agents (Shahien and Beiruti, 2012; Bagnato and Good, 2016).

### 14.8.10 OBESITY

Zonisamide and topiramate cause significant reduction in weight (Antel and Hebebrand, 2012).

### 14.9 NEWER DRUG TARGETS AND DRUGS IN PIPELINE WITH THEIR CURRENT CLINICAL TRIAL STATUS

The development of AEDs lays emphasis on drug targets associated with pathophysiology of epilepsy during last few decades, but the efficacy of most of the available AEDs is questionable or doubtful in patients where these drugs do not show their response (Kaur et al., 2016). So, AED discovery process needs to focus on developing antiepileptics particularly for management of refractory seizures (Kaur et al., 2016). The newer AED research should focus on novel compounds with more therapeutic effectiveness and less toxicity. Currently, researchers in the field of AED development focused on three main approaches for the development of novel compounds. First being the adaptation of existing drugs to make them more safe, second is development of molecules against newer targets, and last being screening of drug molecules without relation to its mechanism; both in vitro and in vivo animal models like primates (Pollard and French, 2006; Simonato et al., 2013).

Table 14.3 summarizes the molecules being developed for different epilepsies in various stages of clinical trials (Kaur et al., 2016).

**TABLE 14.3**  Newer Antiepileptic Drugs (Recently Approved and in Different Phases of Clinical Trials).

| Drug/chemical molecule | Approved/ phase of clinical development | Therapeutic target/receptor | Indications |
|---|---|---|---|
| Eslicarbazepine acetate | Approved | Inhibition of sodium channel | Epilepsy with focal seizures; refractory focal seizures |
| Brivaracetam (ucb 34714) | Approved | Modulation of synaptic (SV2A) vessel protein | Epilepsy |
| Pregabalin | Approved | Modulation of $\alpha2\delta$ subunit of calcium channel | Primary generalized tonic-clonic seizures; focal onset seizures; |
| Cannabidiol (GWP42003-P) | Phase 2 Phase 3 | – | Dravet syndrome; myoclonic epilepsy; Lennox–Gastaut syndrome |
| Docosahexaenoic acid | Phase 3 | – | Refractory seizures |
| Carisbamate (RWJ-333369) | Phase 3 | Not elucidated | Focal epilepsy |
| Thalidomide | Phase 2 | – | Refractory epilepsy |
| Buspirone | Phase 2 | $5\text{-HT}_{1A}$ receptor partial agonist | Localized epilepsy |
| YKP3089 | Phase 2 | Sodium channel modulation; increase GABA release | Resistant focal onset seizures |
| PRX-00023 | Phase 2 | Serotonin (5-HT) receptor | Temporal lobe epilepsy; focal epilepsy |
| Bumetanide | Phase 1 | Na-K-Cl cotransporter (NKCC1) nhibition | Neonatal seizures |
| Muscimol | Phase 1 | $\text{GABA}_A$ receptor | Epilepsy |

*Source*: Adapted from Kaur et al. 2016.

## 14.10   FUTURE OPPORTUNITIES AND CHALLENGES

Despite the availability of newer AEDs, many hurdles or challenges remain in the treatment of epilepsy. These include lack of effective agents for the management of drug-resistant epilepsy, persistence of adverse effects, and drug interactions with available AEDs, inability to treat comorbidities along with epilepsy, inability to improve existing animal models, the lack of disease modifying agents or agents that can modify the process of epileptogenesis (Ventola, 2014). Lifestyle modifications of humans have led to increased risk of post-stroke epilepsies. Considering the above-mentioned challenges, conventional drug discovery and development process requires serious introspection. There is an urgent need to improve existing animal models so that newer AEDs with lesser side effects and drug interactions could be developed. About $6 \times 10^{15}$ molecules are neuroactive; these compounds should be screened for the development of antiseizure compounds with novel chemical structure and novel therapeutic targets with lesser potential of side effects and drug interactions (Weaver, 2013). Advancements in the field of biotechnology and genetics can help in development of newer in vitro models of epilepsies. Moreover, new molecular targets with better understanding of pathophysiology of seizure should be taken up for drug development. As more and more information of genetics comes to forefront, the application of a more rational, mechanism-based pharmacology will become increasingly possible (Lason et al., 2013).

## 14.11   CONCLUSION

This chapter represents an outline of the diverse drugs used for epilepsy. In last few decades,

enhancement of knowledge regarding neuronal signaling pathways, their pathophysiology in the CNS resulted in more comprehensive and holistic understanding of the current AEDs. It also led to better design of novel agents and hence their discovery. Newer drug targets such as CRMP-2, SVA-2, KCNQ channels, and NMDA receptors have added enormously to the existing arsenal of drugs against epilepsy. Increased molecular profiling helped us to achieve better grip of the mechanisms of certain seizure types. Despite newer molecular targets, there exists a gap between the scientific knowledge and actual ongoing insult to the brain in various seizure types. Hence, current drug therapy is mostly empirical rather than guided by known molecular mechanisms.

## KEYWORDS

- **epilepsy**
- **seizures**
- **antiepileptic drugs**
- **phenytoin**
- **status epilepticus**

## REFERENCES

Adrian, P. Phobic Disorders Medication. *Medscape* 2014. (accessed June 15, 2018). https://emedicine.medscape.com/article/288016-medication#9.

Alvarez, V.; Januel, J. M.; Burnand, B.; Rossetti, A. O. Second-line Status Epilepticus Treatment: Comparison of Phenytoin, Valproate, and Levetiracetam. *Epilepsia* **2011**, *52* (7), 1292–1296.

Aneja, S.; Sharma, S. Newer Anti-epileptic Drugs. *Indian Pediatr.* **2013**, *50* (11), 1033–1040.

Antel, J.; Hebebrand, J. Weight-reducing Side Effects of the Antiepileptic Agents Topiramate and Zonisamide. *Handb. Exp. Pharmacol.* **2012**, *209*, 433–466.

Appleton, R. E.; Peters, A. C. B.; Mumford, J. P.; Shaw, D. E. Randomised, Placebo-Controlled Study of Vigabatrin as First-Line Treatment of Infantile Spasms. *Epilepsia* **1999,** *40* (11), 1627–1633.

APTIOM® (eslicarbazepine acetate) Tablets, for Oral Use. USFDA. 2013. (accessed June 8, 2018). https://www.accessdata.fda.gov/drugsatfda_docs/label/2013/022416s000lbl.pdf.

Bagnato, F.; Good, J. The Use of Antiepileptics in Migraine Prophylaxis. *Headache* **2016,** *56* (3), 603–615.

Barker-Haliski, M.; White, H. S. Glutamatergic Mechanisms Associated with Seizures and Epilepsy. *Cold Spring Harb. Perspect. Med.* **2015,** *5* (8), a022863.

Baumel, I. P.; Gallagher, B. B.; Mattson, R. H. Phenylethylmalonamide (PEMA): An Important Metabolite of Primidone. *Arch. Neurol.* **1972,** *27* (1), 34–41.

Ben-Menachem, E.; Biton, V.; Jatuzis, D.; Abou-Khalil, B.; Doty, P.; Rudd, G. D. Efficacy and Safety of Oral Lacosamide as Adjunctive Therapy in Adults with Partial-onset Seizures. *Epilepsia* **2007,** *48* (7), 1308–1317.

Bertilsson, L. Clinical Pharmacokinetics of Carbamazepine. *Clin. Pharmacokinet.* **1978,** *3* (2), 128–143.

Betjemann, J. P.; Lowenstein, D. H. Status Epilepticus in Adults. *Lancet Neurol.* **2015,** *14* (6), 615–624.

Bialer, M.; White, H. S. Key Factors in the Discovery and Development of New Antiepileptic Drugs. *Nat. Rev. Drug. Discov.* **2010,** *9* (1), 68–82.

Blumenfeld, H.; Westerveld, M.; Ostroff, R. B.; Vanderhill, S. D.; Freeman, J.; Necochea, A.; Uranga, P.; Tanhehco, T.; Smith, A.; Seibyl, J. P.; Stokking, R. Selective Frontal, Parietal, and Temporal Networks in Generalized Seizures. *Neuroimage* **2003,** *19* (4), 1556–1566.

Bockbrader, H. N.; Wesche, D.; Miller, R.; Chapel, S.; Janiczek, N.; Burger, P. A Comparison of the Pharmacokinetics and Pharmacodynamics of Pregabalin and Gabapentin. *Clin. Pharmacokinet.* **2010,** *49* (10), 661–669.

Brodie, M. J.; Richens, A.; Yuen, A. W. UK Lamotrigine/Carbamazepine Monotherapy Trial Group. Double-blind Comparison of Lamotrigine and Carbamazepine in Newly Diagnosed Epilepsy. *Lancet.* **1995,** *345* (8948), 476–479.

Brown, J. P.; Gee, N. S. Cloning and Deletion Mutagenesis of the Alpha2 Delta Calcium Channel Subunit from Porcine Cerebral Cortex. Expression of a Soluble Form of the Protein That Retains [3h]Gabapentin Binding Activity. *J. Biol. Chem.* **1998,** *273* (39), 25458–25465.

Browne, T. R.; Kugler, A. R.; Eldon, M. A. Pharmacology and Pharmacokinetics of Fosphenytoin. *Neurology* **1996,** *46* (6 Suppl 1), S3–S7.

Chiron, C. Stiripentol. *Neurotherapeutics* **2007,** *4* (1), 123–125.

Cornes, S. B.; Griff n, Jr., E. A.; Lowenstein, D. H. Pharmacology of Abnormal Electrical Neurotransmission in the Central Nervous System. In: *Principles of Pharmacology: The Pathophysiologic Basis of Drug Therapy*; Golan, D. E., Armstrong, E. J., Armstrong, A.W., 4th ed, Wolters Kluwer: Philadelphia, 2017, pp. 249–264.

Crespel, A.; Gelisse, P.; Reed, R. C.; Ferlazzo, E.; Jerney, J.; Schmitz, B.; Genton, P. Management of Juvenile Myoclonic Epilepsy. *Epilepsy Behav.* **2013,** *28* (Suppl 1), S81–S86.

David, Y. K. Epilepsy and Seizures. *Medscape* 2018. (accessed June, 9 2018). https://emedicine.medscape.com/article/1184846-overview#a2.

De Simone, G.; Di Fiore, A.; Menchise, V.; Pedone, C.; Antel, J.; Casini, A.; Scozzafava, A.; Wurl, M.; Supuran, C. T. Carbonic Anhydrase Inhibitors. Zonisamide is an Effective Inhibitor of the Cytosolic Isozyme II and Mitochondrial Isozyme V: Solution and X-Ray Crystallographic Studies. *Bioorg. Med. Chem. Lett.* **2005,** *15* (9), 2315–2320.

DILANTIN-125-phenytoin Suspension. *Pfizer* 2016. (accessed June 8, 2018). http://labeling.pfizer.com/showlabeling.aspx?id=560.

Dodgson, S. J.; Shank, R. P.; Maryanoff, B. E. Topiramate as an Inhibitor of Carbonic Anhydrase Isoenzymes. *Epilepsia* **2000,** *41* (s1), 35–39.

Doty, P.; Rudd, G. D.; Stoehr, T.; Thomas, D. Lacosamide. *Neurotherapeutics* **2007,** *4* (1), 145–148.

Dreifuss, F. E.; Langer, D. H.; Moline, K. A.; Maxwell, J. E. Valproic Acid Hepatic Fatalities. II. US Experience Since 1984. *Neurology* **1989,** *39* (2 Pt 1), 201–207.

Eke, T.; Talbot, J. F.; Lawden, M. C. Severe Persistent Visual Field Constriction Associated with Vigabatrin. *BMJ* **1997,** *314* (7075), 180.

Eslicarbazepine Acetate. *Epilepsy Society* 2014. (accessed June 20, 2018). https://www.epilepsysociety.org.uk/eslicarbazepine-acetate#.WypHulOFOT8.

Extercatte, J.; de Haan, G. J.; Gaitatzis, A. Teaching Video NeuroImages: Frontal opercular seizures with jacksonian march. *Neurology* **2015,** *84* (11), e83–e84.

Felbamate Study Group in Lennox-Gastaut Syndrome. Efficacy of Felbamate in Childhood Epileptic Encephalopathy (Lennox-Gastaut syndrome). *N. Engl. J. Med.* **1993,** *328* (1), 29–33.

Fisher, J. L. The Effects of Stiripentol on GABA(A) receptors. *Epilepsia* **2011**, *52* (Suppl 2), 76–78.

Fisher, R. S.; Cross, J. H.; D'souza, C.; French, J. A.; Haut, S. R.; Higurashi, N.; Hirsch, E.; Jansen, F. E.; Lagae, L.; Moshé, S. L.; Peltola, J. Instruction Manual for the ILAE 2017 Operational Classification of Seizure Types. *Epilepsia* **2017**, *58* (4), 531–542.

Fisher, R. S.; Van Emde Boas, W.; Blume, W.; Elger, C.; Genton, P.; Lee, P.; Engel, J. Epileptic Seizures and Epilepsy: Definitions Proposed by the International League Against Epilepsy (ILAE) and the International Bureau for Epilepsy (IBE). *Epilepsia* **2005**, *46* (4), 470–472.

Fuentealba, P.; Steriade M. The Reticular Nucleus Revisited: Intrinsic and Network Properties of a Thalamic Pacemaker. *Prog. Neurobiol.* **2005**, *75* (2), 125–141.

Fuentealba, P.; Timofeev, I.; Bazhenov, M.; Sejnowski, T. J.; Steriade, M. Membrane Bistability in Thalamic Reticular Neurons During Spindle Oscillations. *J. Neurophysiol.* **2005**, *93* (1), 294–304.

Fuentealba, P.; Timofeev, I.; Steriade, M. Prolonged Hyperpolarizing Potentials Precede Spindle Oscillations in the Thalamic Reticular Nucleus. *Proc. Natl. Acad. Sci. U S A.* **2004**, *101* (26), 9816–9821.

Garnett, W. R. Clinical Pharmacology of Topiramate: A Review. *Epilepsia* **2000**, *41* (s1), 61–65.

Glauser, T. A.; Cnaan, A.; Shinnar, S.; Hirtz, D. G.; Dlugos, D.; Masur, D.; Clark, P. O.; Capparelli, E. V.; Adamson, P. C. Ethosuximide, Valproic Acid, and Lamotrigine in Childhood Absence Epilepsy. *N. Engl. J. Med.* **2010**, *362* (9), 790–799.

Goossens, A.; Beck, M. H.; Haneke, E.; McFadden, J. P.; Nolting, S.; Durupt, G.; Ries, G. Adverse Cutaneous Reactions to Cosmetic Allergens. *Contact Dermatitis.* **1999**, *40* (2), 112–113.

Gugler, R.; von Unruh, G. E. Clinical Pharmacokinetics of Valproic Acid. *Clin. Pharmacokinet.* **1980**, *5* (1), 67–83.

Harvey, R. A.; Clark, M. A.; Finkel, R.; Rey, J. A.; Whalen, K. *Lippincott's Illustrated Reviews: Pharmacology*, 5th ed, New York: Wolters Kluwer, 2011.

Helmstaedter, C.; Fritz, N. E.; Kockelmann, E.; Kosanetzky, N.; Elger, C. E. Positive and Negative Psychotropic Effects of Levetiracetam. *Epilepsy and Behav.* **2008**, *13* (3), 535–541.

Inoue, Y.; Ohtsuka, Y.; Oguni, H.; Tohyama, J.; Baba, H.; Fukushima, K.; Ohtani, H.; Takahashi, Y.; Ikeda, S. Stiripentol Open Study in Japanese Patients with Dravet syndrome. *Epilepsia* **2009**, *50* (11), 2362–2368.

Jasvinder, C. Migraine Headache Treatment & Management. *Medscape* 2018. (accessed June 21, 2018). https://emedicine.medscape.com/article/114255 6treatment?pa=ADYX2VZzid TH0K1ZthGO6kmtshiiF AnExmzvwgyH%2FWUZLKEq%2F9he07g0%2Bw92 3t57CGw%2BRfN0z%2FS0aIlf6xwngjXN99AvxzL5C uuwz1VEyh4%3D.

Johnson, B. A.; Swift, R. M.; Ait-Daoud, N.; DiClemente, C. C.; Javors, M. A.; Malcolm, R. J. Development of Novel Pharmacotherapies for the Treatment of Alcohol Dependence: Focus on Antiepileptics. *Alcohol. Clin. Exp. Res.* **2004**, *28* (2), 295–301.

Kaur, H.; Kumar, B.; Medhi, B. Antiepileptic Drugs in Development Pipeline: A Recent Update. *E. Neurological. Sci.* **2016**, *4*, 42–51.

Kenda, B. M.; Matagne, A. C.; Talaga, P. E.; Pasau, P. M.; Differding, E.; Lallemand, B. I.; Frycia, A. M.; Moureau, F. G.; Klitgaard, H. V.; Gillard, M. R.; Fuks, B. Discovery of 4-substituted Pyrrolidone Butanamides as New Agents with Significant Antiepileptic Activity. *J. Med. Chem.* **2004**, *47* (3), 530–549.

Khosravani, H.; Zamponi, G. W. Voltage-gated Calcium Channels and Idiopathic Generalized Epilepsies. *Physiol. Rev.* **2006**, *86* (3), 941–966.

Krause, L. U.; Brodowski, K. O.; Kellinghaus, C. Atrioventricular Block Following Lacosamide Intoxication. *Epilepsy Behav.* **2011**, *20* (4), 725–727.

Lason, W.; Chebicka, M.; Rejdak, K. Research Advances in Basic Mechanisms of Seizures and Antiepileptic Drug Action. *Pharmacol. Rep.* **2013**, *65* (4), 787–801.

Leppik, I. E.; Kelly, K. M.; Patrylo, P. R.; DeLorenzo, R. J.; Mathern, G. W.; White, H. S. Basic Research in Epilepsy and Aging. *Epilepsy Res.* **2006**, *68*, 21–37.

Liu, X. B.; Honda, C. N.; Jones, E. G. Distribution of Four Types of Synapse on Physiologically Identified Relay Neurons in the Ventral Posterior Thalamic Nucleus of the Cat. *J. Comp. Neurol.* **1995**, *352* (1), 69–91.

Löscher, W. The discovery of Valproate. In: *Valproate. Milestones in Drug Therapy*, Löscher W. Ed., Birkhäuser, Basel, 1999, pp. 1–3.

Lowenstein, D. H. Seizures and Epilepsy. In: *Harrison's Manual of Medicine*, Kasper, D. L.; Fauci, A. S.; Hauser, S. L.; Longo, D. L.; Jameson, J. L.; Loscalzo, J. Eds, 19th ed; Mc Graw Hill Education: New York, USA, 2015, pp 2542–2558.

May, T. W.; Korn-Merker, E.; Rambeck, B. Clinical Pharmacokinetics of Oxcarbazepine. *Clin. Pharmacokinet.* **2003**, *42* (12), pp 1023–1042.

Meldrum, B. S. The Role of Glutamate in Epilepsy and Other CNS disorders. *Neurology* **1994,** *44* (11 Suppl 8), S14–S23.

Misra, U. K.; Kalita, J.; Maurya, P. K. Levetiracetam Versus Lorazepam in Status Epilepticus: A Randomized, Open Labeled Pilot Study. *J. Neurol.* **2012,** *259* (4), 645–648.

Misra, U. K.; Kalita, J.; Patel, R. Sodium Valproate Vs Phenytoin in Status Epilepticus: A Pilot Study. *Neurology* **2006,** *67* (2), 340–342.

Misty, D. M.; Metcalf, C. S.; Wilcox, K. S. Pharmacotherapy of the Epilepsies. In: *Goodman and Gilman's: The Pharmacological Basis of Therapeutic*, Brunton, L. L.; Hilal-Dandan, R.; Knollmann, B. C., Eds, 13th ed; McGraw Hill Education: New York, 2018, pp. 314–338.

Motte, J.; Trevathan, E.; Arvidsson, J. F.; Barrera, M. N.; Mullens, E. L.; Manasco, P. Lamotrigine for Generalized Seizures Associated with the Lennox-Gastaut Syndrome. Lamictal Lennox-Gastaut Study Group. *N. Engl. J. Med.* **1997,** *337* (25), 1807–1812.

Olabi, B.; Hall, J. Borderline Personality Disorder: Current Drug Treatments and Future Prospects. *Ther. Adv. Chronic. Dis.* **2010,** *1* (2), 59–66.

Orhan, G.; Wuttke, T. V.; Nies, A. T.; Schwab, M.; Lerche, H. Retigabine/Ezogabine, a KCNQ/KV7 Channel Opener: Pharmacological and Clinical Data. *Exp. Opin. Pharmacother.* **2012,** *13* (12), 1807–1816.

Osorio, I.; Reed, R. C. Treatment of Refractory Generalized Tonic-clonic Status Epilepticus with Pentobarbital Anesthesia After High-dose Phenytoin. *Epilepsia* **1989,** *30* (4), 464–471.

Patsalos, P. N. Clinical Pharmacokinetics of Levetiracetam. *Clin. Pharmacokinet.* **2004,** *43* (11), 707–724.

Patsalos, P. N. The Clinical Pharmacology Profile of the New Antiepileptic Drug Perampanel: A Novel Noncompetitive AMPA Receptor Antagonist. *Epilepsia* **2015,** *56* (1), 12–27.

Perez, J.; Chiron, C.; Musial, C.; Rey, E.; Blehaut, H.; d'Athis, P.; Vincent, J.; Dulac, O. Stiripentol: Efficacy and Tolerability in Children with Epilepsy. *Epilepsia* **1999,** *40* (11), 1618–1626.

Perucca, E.; Bialer, M. The Clinical Pharmacokinetics of the Newer Antiepileptic Drugs. *Clin. Pharmacokinet.* **1996,** *31* (1), 29–46.

Perucca, E.; Cloyd, J.; Critchley, D.; Fuseau, E. Rufinamide: Clinical Pharmacokinetics and Concentration–Response Relationships in Patients with Epilepsy. *Epilepsia* **2008,** *49* (7), 1123–1141.

Phenytoin (Rx). *Medscape.* 1994–2018. (accessed June 8, 2018). https://reference.medscape.com/drug/dilantin-phenytek-phenytoin-343019#10.

Plosker, G. L. Stiripentol: In Severe Myoclonic Epilepsy of Infancy (Dravet Syndrome). *CNS Drugs* **2012,** *26* (11), 993–1001.

Pollard, J. R.; French, J. Antiepileptic Drugs in Development. *Lancet Neurol.* **2006,** *5* (12), 1064–1067.

Porter, R. J.; Rogawski, M. A. Antiseizure Drugs. In: *Basic & Clinical Pharmacology*, Katzung, B. G. Ed, 14th ed; Mc-Graw Hill Education: New York, USA, 2018, pp. 409–439.

Quan, D. Diabetes Neuropathy Medication. *Medscape.* 2018 (accessed June 15, 2018). https://emedicine.medscape.com/article/1170337-medication#4.

Quilichini, P. P.; Chiron, C.; Ben-Ari, Y.; Gozlan, H. Stiripentol, A Putative Antiepileptic Drug, Enhances the Duration of Opening of Gaba-A Receptor Channels. *Epilepsia* **2006,** *47* (4), 704–716.

Rogawski, M. A.; Tofighy, A.; White, H. S.; Matagne, A.; Wolff, C. Current Understanding of the Mechanism of Action of the Antiepileptic Drug Lacosamide. *Epilepsy Res.* **2015,** *110*, 189–205.

Rho, J. M.; Donevan, S. D.; Rogawski, M. A. Mechanism of Action of the Anticonvulsant Felbamate: Opposing Effects on N-Methyl-D-Aspartate and Gamma-aminobutyric acidA Receptors. *Ann. Neurol.* **1994,** *35* (2), 229–234.

Sachdeo, R.; Kramer, L. D.; Rosenberg, A.; Sachdeo, S. Felbamate Monotherapy: Controlled Trial in Patients with Partial Onset Seizures. *Ann. Neurol.* **1992,** *32* (3), 386–392.

Sechi, G.; Serra, A. Wernicke's Encephalopathy: New Clinical Settings and Recent Advances in Diagnosis and Management. *Lancet. Neurol.* **2007,** *6* (5), 442–455.

Shahien, R.; Beiruti, K. Preventive Agents for Migraine: Focus on the Antiepileptic Drugs. *J. Cent. Nerv. Syst. Dis.* **2012,** *4*, 37–49.

Sharma, H. L.; Sharma, K. K. Antiepileptic Drugs. *Principles of Pharmacology*, 2nd ed; Paras Medical Publisher: New Delhi, 2017, pp. 523–538.

Shih, J. J.; Tatum, W. O.; Rudzinski, L. A. New Drug Classes for the Treatment of Partial Onset Epilepsy: Focus on Perampanel. *Ther. Clin. Risk. Manag.* **2013,** *9*, 285–293.

Silbergleit, R.; Durkalski, V.; Lowenstein, D.; Conwit, R.; Pancioli, A.; Palesch, Y.; Barsan, W. Intramuscular Versus Intravenous Therapy for Prehospital Status Epilepticus. *N. Engl. J. Med.* **2012,** *366* (7), 591–600.

Simonato, M.; French, J. A.; Galanopoulou, A. S.; O'Brien, T. J. Issues for New Antiepilepsy Drug Development. *Curr. Opin. Neurol.* **2013**, *26* (2), 195–200.

Singh, M. K. Trigeminal Neuralgia Medication. *Medscape* 2016. (accessed June 15, 2018). https://emedicine.medscape.com/article/1145144-medication#2.

Sohal, V. S.; Huguenard, J. R. Inhibitory Interconnections Control Burst Pattern and Emergent Network Synchrony in Reticular Thalamus. *J. Neurosci.* **2003**, *23* (26), 8978–8988.

Song, I.; Kim, D.; Choi, S.; Sun, M.; Kim, Y.; Shin, H. S. Role of the α1G T-type Calcium Channel in Spontaneous Absence Seizures in Mutant Mice. *J. Neurosci.* **2004**, *24* (22), 5249–5257.

Steiner, T. J.; Dellaportas, C. I.; Findley, L. J.; Gross, M.; Gibberd, F. B.; Perkin, G. D.; Park, D. M.; Abbott, R. Lamotrigine Monotherapy in Newly Diagnosed Untreated Epilepsy: A Double-blind Comparison with Phenytoin. *Epilepsia* **1999**, *40* (5), 601–607.

Stephen, L. J.; Brodie, M. J. Pharmacotherapy of Epilepsy: Newly Approved and Developmental Agents. *CNS Drugs* **2011**, *25* (2), 89–107.

Stephen, S. Bipolar Affective Disorder Medication. *Medscape.* **2018**, Retrieved 15 June, 2018, from https://emedicine.medscape.com/article/286342-medication#4.

Steriade, M.; Contreras, D. Relations Between Cortical and Thalamic Cellular Events During Transition From Sleep Patterns to Paroxysmal Activity. *J. Neurosci.* **1995**, *15* (1 Pt 2), 623–642.

Steriade, M.; Contreras, D. Spike-wave Complexes and Fast Components of Cortically Generated Seizures. I. Role of Neocortex and Thalamus. *J. Neurophysiol.* **1998**, *80* (3), 1439–1455.

Tompson, D. J.; S. Crean, C. S. Clinical Pharmacokinetics of Retigabine/Ezogabine. *Curr. Clin. Pharmacol.* **2013**, *8* (4), 319–331.

Tortorella, A.; Halonen, T.; Sahibzada, N.; Gale, K. A Crucial Role of the a-amino-3-hydroxy-5-methylisoxazole-4-propionic Acid Subtype of Glutamate Receptors in Piriform and Perirhinal Cortex for the Initiation and Propagation of Limbic Motor Seizures. *J. Pharmacol. Exp. Ther.* **1997**, *280* (3), 1401–1405.

Treiman, D. M.; Meyers, P. D.; Walton, N. Y.; Collins, J. F.; Colling, C.; Rowan, A. J.; Handforth, A.; Faught, E.; Calabrese, V. P.; Uthman, B. M.; Ramsay, R. E. A Comparison of Four Treatments for Generalized Convulsive Status Epilepticus. *N. Engl. J. Med.* **1998**, *339* (12), 792–798.

Twyman, R. E.; Rogers, C. J.; Macdonald, R. L. Differential Regulation of Γ-Aminobutyric Acid Receptor Channels by Diazepam and Phenobarbital. *Ann. Neurol.* **1989**, *25* (3), 213–220.

Ulrich, D.; Huguenard, J. R. GABA(A)-receptor-mediated Rebound Burst Firing and Burst Shunting in Thalamus. *J. Neurophysiol.* **1997**, *78* (3), 1748–1751.

Ventola, C. L. Epilepsy Management: Newer Agents, Unmet Needs, and Future Treatment Strategies. *Pharm. Ther.* **2014**, *39* (11), 776.

von Rosenstiel, P. Brivaracetam (ucb 34714). *Neurotherapeutics* **2007**, *4* (1), 84–87.

Walker, M. C.; Patsalos, P. N. Clinical Pharmacokinetics of New Antiepileptic Drugs. *Pharm. Ther.* **1995**, *67* (3), 351–384.

Wang, Y.; Trevelyan, A. J.; Valentin, A.; Alarcon, G.; Taylor, P. N.; Kaiser, M. Mechanisms Underlying Different Onset Patterns of Focal Seizures. *PLoS Computational Biol.* **2017**, *13* (5), e1005475.

Weaver, D. F. Design of Innovative Therapeutics For Pharmacoresistant Epilepsy: Challenges And Needs. *Epilepsia* **2013**, *54* (suppl 2), 56–59.

Wiffen, P. J.; Derry, S.; Moore, R. A.; Aldington, D.; Cole, P.; Rice, A. S.; Lunn, M. P.; Hamunen, K.; Kalso, E. A. Antiepileptic Drugs for Neuropathic Pain and Fibromyalgia. Status and Date. New published in 2013 (6).

Xie, X.; Lancaster, B.; Peakman, T.; Garthwaite, J. Interaction of the Antiepileptic Drug Lamotrigine with Recombinant Rat Brain Type IIA Na+ Channels and with Native Na+ Channels in Rat Hippocampal Neurones. *Pflügers Archiv.* **1995**, *430* (3), 437–446.

Zona, C.; Pieri, M.; Carunchio, I.; Curcio, L.; Klitgaard, H.; Margineanu, D. G. Brivaracetam (ucb 34714) Inhibits Na+ Current in Rat Cortical Neurons in Culture. *Epilepsy Res.* **2010**, *88* (1), pp 46–54.

Liyanagedera, S., Williams, R. P. The New Classification of Seizures: An Overview for the General Physician. *J. R. Coll. Physicians Edinb.* **2017**, *47* (4), 336–338.

# CHAPTER 15

# Local Anesthetics

ELENA GONZÁLEZ BURGOS*, LUIS GARCÍA-GARCÍA,
M. PILAR GÓMEZ-SERRANILLOS, and FRANCISCA GÓMEZ OLIVER

*University Complutense of Madrid, Madrid, Spain*

*Corresponding author. E-mail: elenagon@ucm.es*

## ABSTRACT

Local anesthetics (LAs) are endowed with the capacity to induce a transitory loss of sensation by blocking neuronal voltage-gated sodium channels (VGSCs) impeding the nerve to reach an action potential and the propagation of the signal toward the specific part of the body. This chapter reviews the up-to-date physicochemical characteristics of LAs, key contributors to their pharmacologic properties. The chapter comments on the structural and functional characteristics of primary pharmacologic target that mediates the action of Las, and it briefly points at other mechanisms identified as potential contributors to the LAs-mediated effects. The authors' overview the LAs pharmacokinetics together with the benefits and disadvantages associated to the use of coadjuvant drugs which are aimed to prolong the LAs effect and to either prevent or reduce eventual undesirable effects. It comments on the main characteristics of the different categories of anesthesia as well as the main LAs commonly used in clinical practice. Certainly, the adverse reactions, toxic local and systemic effects of LAs, and the appropriated therapeutic interventions to be implemented in each situation are fully addressed. Finally, some conclusions and suggestions on potential fields of research that might result in improve of the efficacy and safety of the LAs.

## 15.1 INTRODUCTION

Since cocaine was first introduced in 1884, local anesthetics (LAs) have been widely employed in many clinical contexts to induce local anesthesia and as analgesics to manage pain. These properties are due to the LAs capacity to produce a temporary and reversible lack of sensation in a part of the body through their reversible binding to neuronal voltage-gated sodium channels (VGSCs) and the consequent peripheral and neural conduction of sensory inputs inhibition (Catterall and Swanson, 2015). LAs can be used alone as a single form of anesthesia/analgesia for local procedures, in combination with general anesthesia allowing for reducing the dose of general anesthetic and its risk as well as to provide postoperative analgesia after general anesthesia (Berde et al., 2005; Valverde et al., 2005; White, 2005). The main disadvantages of local anesthesia/analgesia are the need of high expertise to perform certain types of anesthesia (e.g., epidural), short duration of effects, even of long-acting LAs and their potential systemic toxicity causing among others, depression in at the level of the central nervous system (CNS), and myocardial electrical impulses blockade.

The major development of LAs has focused on amide type of LAs because the ester type is more unstable when in solution, they have a shorter shelf-life and they are predisposed to degradation when exposed to high temperatures. In addition, as it will be further address, LAs of the ester type tend to cause more allergic reactions than the amide type (Becker and Reed, 2006).

Research on experimental and clinical effects of LAs performed during the last decades has provided novel roles of LAs as antimicrobial and anti-inflammatory agents (Hollmann and Durieux, 2000; Johnson et al., 2008; Rimback et al., 1988; Taniguchi et al., 1996). Nevertheless, these aspects are outside of the scope of this chapter and therefore will not be addressed. The aim of this chapter is an updated overview of the LAs focused on their physicochemical characteristics, pharmacokinetics properties together with the benefits and disadvantages associated to the use of coadjuvant drugs, and structural and functional characteristics of the primary pharmacologic target that mediates their action. Moreover, this chapter includes the adverse reactions, toxic local and systemic effects of Las, and the appropriated therapeutic interventions to be implemented in each situation are fully addressed. Furthermore, the main features of the distinct types of anesthesia and the main LAs commonly used in clinical practice are also commented in the present chapter.

## 15.2 PHYSICOCHEMICAL CHARACTERISTICS OF LAs

The pharmacologic activity of LAs relates intimately with their physicochemical characteristics, being a function of their liposoluble nature, their percent of ionization at physiologic pH, and their capacity to pass through the tissues and affinity to bind to tissue proteins as well as their vasoactive properties.

### 15.2.1 COMMON PHYSICOCHEMICAL CHARACTERISTICS OF LAs

LAs are amphiphilic molecules (i.e., they have polar and nonpolar regions) that share a basic chemical structure consisting of three parts:

1.  An aromatic lipophilic (nonpolar) group, commonly being an unsaturated benzene ring;
2.  An intermediate hydrocarbon chain and
3.  A hydrophilic (polar) tertiary amine.

The intermediate link can be either an ester (-CO-) or amide (-HNC-) and based upon that LAs are classified into two main groups: the amine esters group and the amine amides group (Fig. 15.1). Due to the ability of the tertiary amine group to accept a proton yielding a cationic quaternary amine, LAs behave as weak bases when in solution. Because most clinically used LAs have $pK_a$ values ranging from 7 to 10 (Table 15.1), when at physiologic pH (approximately 7.4), the molecules of the LAs can be found in both uncharged (B) and charged ($BH^+$) forms. In addition, the three-dimensional molecular structure of LAs can exist as enantiomers, optical isomers, or mirror images. These chemical properties markedly contribute to the pharmacologic features of LAs such as potency, duration of action, and safety profile (Table 15.2) (Butterworth and Strichartz, 1990; Chernoff and Strichartz, 1990; Strichartz et al., 1990).

**TABLE 15.1**  Relation Between the pK$_a$ Value and Onset of Action of Most Frequently Used Local Anesthetics.

| Drug | pK$_a$ | Onset (min) | Amino ester | pK$_a$ | Onset (min) |
|------|--------|-------------|-------------|--------|-------------|
| Bupivacaine | 8.1 | 5–8 | Benzocaine | 3.5* | |
| Etidocaine | 7.7 | 2–4 | Chloroprocaine | 9.0 | 10–15 |
| Levobupivacaine | 8.1 | 2–4 | Cocaine | 8.7 | |
| Lidocaine | 7.8 | 2–4 | Procaine | 8.9 | 14–18 |
| Mevipacaine | 7.6 | 2–4 | Tetracaine | 8.2 | 10–15 |
| Prilocaine | 7.8 | 2–4 | *pK$_a$ value < to the mean value of physiologic pH (7.4). (To be used only on surfaces) | | |
| Ropivacaine | 8.1 | 2–4 | | | |

Lidocaine and Procaine: Prototypes of the LAs within their respective groups.

**TABLE 15.2**  Physicochemical and Pharmacologic Attributes of Local Anesthetics and Their Consequences on the Potency, Onset and Duration of Action.

| Attribute–consequence | Brief overview |
|-----------------------|----------------|
| Liposolubility–potency | A more liposoluble molecule will cross the membranes easier resulting in higher potency. Chemically, the higher number of carbons in the molecule increases the lipid solubility. |
| pK$_a$–onset | A higher percentage of the molecules of a given LA will be in the nonionized (BH, tertiary amine) form as the value of the pK$_a$ of the LA is lower. Only the nonionized molecules are able to passively diffuse through biological membranes (nerve sheath), therefore diminishing the onset of action. |
| Binding to tissue–onset | The higher the ability of the drug to diffuse through the tissue the faster the LA will reach the site of action. |
| Binding to VGSCs–potency Binding to VGSCs–duration | Once in the axoplasm (with a more acidic pH), a given LA will be mainly in its ionized form (BH$^+$, quaternary amine), able to block the pore of the VGSCs. The strongest the binding, the longer the duration of the action. |

The LAs properties can be improved by coadjuvant drugs (see text).

## 15.2.2 PHYSICOCHEMICAL CHARACTERISTICS AND PHARMACOLOGIC PROFILE OF LAs

### 15.2.2.1 ESTERS VERSUS AMIDES

The chemical composition (ester vs. amide, see Fig. 15.1) of LAs determines the class of biotransformation and metabolites formed as well as the prospect to cause more or less hypersensitivity reactions.

### 15.2.2.2 AMPHIPHILIC NATURE, LIPID/WATER PARTITION COEFFICIENT, AND LIPOSOLUBILITY

Even though LAs have a polar region, the amphiphilic nature of LAs endowed them with a lipid/water partition coefficient higher than 1 (Butterworth and Strichartz, 1990; Chernoff and Strichartz, 1990; Strichartz et al., 1990). Therefore, LAs are able to diffuse into the hydrophobic region of the membrane

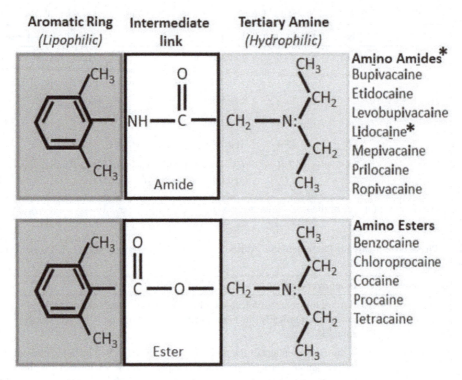

**FIGURE 15.1**   Chemical structure based upon which LAs are classified as amino amides or amino esters.

by nonpolar interactions, particularly van der Waals forces (Lipkind and Fozzard, 2005) and to further induce dipoles (anesthetic–hydrocarbonated tails) causing an increase of the membrane fluidity (Efimova et al., 2016). In fact, the lipid/water partition coefficient can be considered as an index of the LAs potency because they act once they are inside the cell (see the mechanism of action). Thus, the higher their liposolubility, the better their ability to diffuse throughout the biological membranes.

### 15.2.2.3   WEAK BASES, pKa, AND pH

LAs are weak bases and therefore they can exist as charged and uncharged molecules depending on the pH of the medium and the drug's pKa value. The percentage of ionization can be calculated by the equation developed by Henderson and Hasselbalch:

$$pH = pKa(*) + \log\left([B]/[BH+]\right)$$

Where, (*)pKa: pH of the medium at which the forms of the drug molecules are ionized and nonionized in a 50%.

Both, the uncharged (BH) and charged (BH$^+$) forms of the drug are key to determine its pharmacological properties. Thus, whereas the BH form is the one that diffuses through nerve sheaths and cell membranes, once in the axoplasm, BH$^+$ form of the drug is the one that binds to the inner pore blocking the VGSCs (Fig. 15.2). Hence, the onset of action and potency is markedly influenced by both, the pH of the medium and the pKa of the drug (Frazier et al., 1970; Narahashi et al., 1970; Strobel and

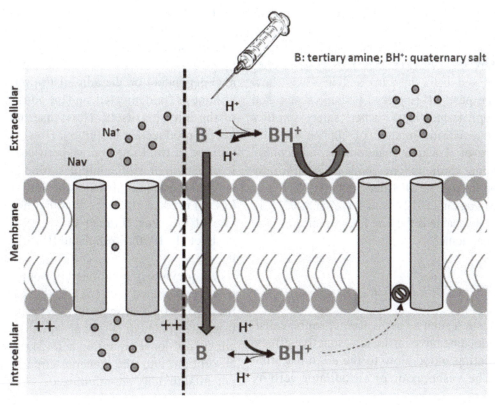

**FIGURE 15.2** Schematic representation depicting the mechanism of action of local anesthetics and the relative role of their nonionized and ionized forms.

Bianchi, 1970a, b). In general, and considering that the pKa of LAs varies from 7 to 9, the lower the pKa the greater percentage of nonionized, liposoluble molecules of the drug able to enter the cell. There are nonetheless two exceptions. On one hand, chloroprocaine has a rapid onset of action despite its high pKa value. On the other hand, benzocaine, with a very low pKa, does not exist in ionized form acting by alternative mechanisms (Frazier et al., 1970; Narahashi et al., 1970; Strobel and Bianchi, 1970a, b).

Clinically, it is essential to consider that commercially prepared solutions of LAs that do not contain any vasoconstrictor agent use to have a pH that varies from 5.5 to 7. Nevertheless, it is common that, once the solution is in the location of administration, the pH increases to the physiological value by the buffering effects of physiologic fluids. By contrast, commercially solutions of LAs that do contain a vasoconstrictor agent (generally epinephrine) are prepared with a more acidic pH. The aim is to delay epinephrine oxidation, but it also results in its longer action duration. Another well-known aspect is that the efficacy of LAs is reduced when the drug is applied in inflamed tissues, mainly due to the acidosis (a reduction on the pH of approximately 0.5–1.0) at and around the inflamed area, therefore impairing the access of the drug to its pharmacologic target (Punnia-Moorthy, 1987).

### 15.2.2.4  THREE-DIMENSIONAL STRUCTURE

Also noteworthy is the contribution of the 3D structure of LA (Fig. 15.3). Several studies carried out both, in vitro and in vivo, show that the pure left-isomers L(-)-bupivacaine and L(-)-ropivacaine have better safety profiles than the racemic mixture of bupivacaine (1:1 mixture of R and S enantiomers), particularly regarding cardiotoxicity, without significantly compromising the anesthetic potency (Tsuchiya and Mizogami, 2012; Tsuchiya et al., 2011). In fact, ropivacaine is the first LA marketed as pure L(-) enantiomer.

### 15.2.2.5  OTHER FACTORS

Various LA such as bupivacaine, ropivacaine, and lidocaine have intrinsic vasoactive effects modulating blood flow to the administration site. The vasopressor or vasodilator activity of LA usually depends on dose. Generally, administration of low LA doses results in vasoconstriction while high doses lead to vasodilation. Nonetheless, these effects are also dependent on the administration site, the volume of fluid injected, and the consequences of the injection itself. Thus, insertion of the needle produces vasodilation. However, if the volume of fluid injected is significantly high, it can temporarily compress blood vessels. Nevertheless, the intrinsic vasoactivity of LA has not marked effect modulating the abortion rate of the drug (Frazier et al., 1970; Narahashi et al., 1970; Strobel and Bianchi, 1970a, b).

As previously mentioned, vasopressors are usually combined with LA anesthetics to delay the absorption. Consequently, two main advantages are obtained: (1) the anesthetic/analgesic effect is of longer duration and (2) the risk for absorption into the general circulation and potential toxicity are minimized.

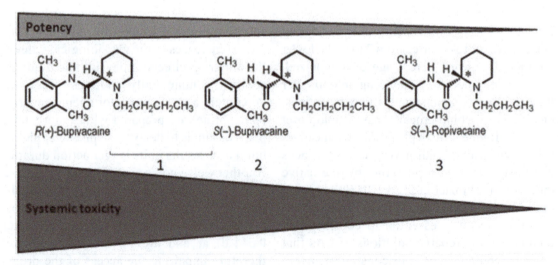

*chiral center:  a carbon atom at the center to which four other compounds are attached.

**FIGURE 15.3**  The 3D chemical structure of local anesthetics and associated systemic toxicity.

## 15.3 PHARMACOKINETICS OF LAs

### 15.3.1 ABSORPTION

The clinical use of LAs has the main purpose to induce local or regional anesthesia and accordingly, they are predictable to act nearby the administration site. Therefore, the classical pharmacokinetic processes of systemic absorption, distribution, metabolism, and excretion have an almost negligible impact to the action of the drug. When systemic absorption of the LA occurs, the main consequences are a shorter duration of the local effect and an augmented risk of cardiac and central toxicity. The systemic absorption of LAs depends on various aspects.

First, the physicochemical characteristics of the drug such as dose, pKa, liposolubility, percentage of ionized molecules, and the protein binding together with the characteristics of the drug-containing solution such as pH, presence or absence of adjuvant vasoconstrictor, alkalinizing, or tissue diffusion enhancing agents. Nonetheless, the blood flow that irrigates the site of injection has a key impact on the degree of absorption being the greatest amount of uptake to the least: IV> tracheal> intercostal> caudal> paracervical> epidural> brachial> sciatic> subcutaneous. For example, when 400 mg of lidocaine alone is administered in the intercostal space, the peak blood concentration obtained is of 7 μg/mL that can be toxic, whereas the same

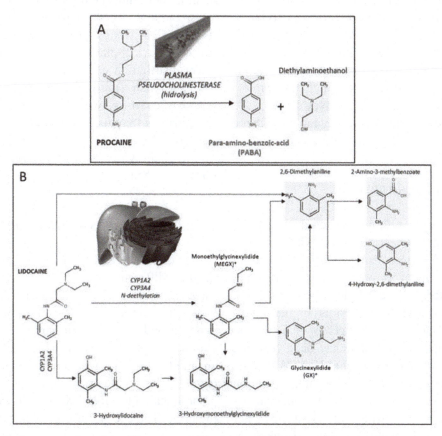

**FIGURE 15.4** Representation of the different metabolic pathways in function of the type of LA.

dose in the brachial plexus leads to a plasma concentration maximum of 3 µg/mL which is not generally toxic. As before mentioned, the use of vasopressor agents (i.e., epinephrine) is generally employed to reduce vascular absorption, but its effect does not work the same for all anesthetics in all the sites of injection. Thus, addition of 5 µg/mL of epinephrine in the epidural space reduces systemic absorption of lidocaine but not bupivacaine, whereas the reduction is obtained for both LAs when used to block peripheral nerves (Gasser and Erlanger, 1929; Franz and Perry, 1974, Raymond et al., 1989).

Since the initial studies, the sort of nerve fibers (diameter, length affected and presence or absence of myelin) has been known to be relevant determining the effect of LAs (Gasser and Erlanger, 1929; Franz and Perry, 1974, Raymond et al., 1989). Thus, the degree of sensitive to blockade by LAs from low to high is as follows: small myelinated fibers (Aα motor) < Aα type Ia; Aα type Ib; Aβ type II; Aγ < Aδ sensory fibers < small, nonmyelinated C fibers, and partially myelinated B fibers.

### 15.3.2   DISTRIBUTION

LAs are frequently administered closed or into the compartment where the site of action is located. A proper distribution of the drug in this compartment is necessary to obtain the clinical effect. For example, injection within the subarachnoid space results in the distribution of the LA in the cerebrospinal fluid (CSF) compartment that it has been reported to be potentially affected by at least 25 factors. Among them, we can mention administration location, needle angle, CSF volume and density, and patient position when considering whether the anesthetic solution is hypobaric, isobaric, or

hyperbaric. Thus, along with the subarachnoid space, once the patient adopts a stand-up position, hyperbaric solutions will tend to descend, the isobaric ones to stay more or less static, whereas hypobaric solutions will tend to ascend (Greene, 1985).

When LAs are in plasma they bind to plasma proteins, mainly α1-acid glycoprotein. However, they also bind to albumin and are taken up by red blood cells. The percentage of LA binding to plasma components seems to correlate with the liposolubility of the drug being higher in highly lipophilic LAs such as etidocaine and bupivacaine (Greene, 1985).

Systemic distribution of LAs is better predicted by a pharmacokinetic two-compartment model. Thus, an α-phase characterized by a fast and high distribution into highly irrigated tissues (CNS, lungs, heart, liver, kidneys) is followed by a β-phase reflecting a slower and lower distribution of the drug into tissues with low perfusion (muscle and gut). The rapid uptake of the drug by the lungs (phase α) seems to contribute to reduce the initial blood concentrations. The uptake of the drug by muscle (phase β) provides a large reservoir for LAs (Greene, 1985).

### 15.3.3   METABOLISM AND EXCRETION

Amino esters and amino amide LAs are metabolized by different pathways (Fig. 15.4). Amino esters LAs are rapidly and extensively hydrolyzed in the plasma by a cholinesterase, also denominated pseudocholinesterase or butyrylcholinesterase (BuChE). The metabolism yields an alkylamine and a para-amino-benzoic acid (PABA). Those are inactive, more hydrophilic polar metabolites that are further excreted in the urine. The exception is cocaine that besides ester hydrolysis also undergoes liver N-methylation (Higuchi et al., 2013; Maimo and Redick, 2004).

Patients with a deficient BuChE activity, whatever cause, are more vulnerable to toxicity because of LA ester accumulation. Patients with the genetic polymorphism known as atypical cholinesterase are one example. Because most genetic variants of plasma BuChE can be inhibited by the amide type LA dibucaine, the dibucaine number assay is employed to determine BuChE phenotypical functionality. Thus, in control patients, dibucaine inhibits enzyme activity at 80% (this value corresponds to dibucaine number of 80). Approximately 4% of the population are heterozygous for atypical plasma BuChE and are characterized by a dibucaine number that ranges from 30 to 65. The presence of homozygous atypical plasma BuChE (approximately 0.04% of the population) relates to a dibucaine number of 20 (Elamin, 2003; Parnas et al., 2011). Apart from a deficient metabolism of genetic origin, plasma BuChE levels are also modulated by physiopharmacologic factors. Thus, alcoholic or obese patients have higher levels of BuChE. By contrast, a significant reduction occurs during pregnancy, in patients with liver pathologies and can also be reduced by pharmacological interactions with other drugs such as neostigmine, pyridostigmine, edrophonium, and mono-amine-oxidase inhibitors (Higuchi et al., 2013; Maimo and Redick, 2004).

The amide type LAs are primarily metabolized by hepatocytes microsomal cytochrome P-450 enzymes by N-dealkylation and hydroxylation yielding polar metabolites more easily excreted in the urine. The main isoforms involved are CYP2E1A and CY3A4 and the rate of metabolism varies among the different amide type LAs being from rapid to low as follows: prilocaine> lidocaine> mepivacaine> ropivacaine> bupivacaine. Prilocaine and lidocaine have been shown to occasionally induce methemoglobinemia due to accumulation of metabolites derived from toluidine and xylidine (Higuchi et al., 2013; Maimo and Redick, 2004). Nevertheless, it should always be kept in mind that patients with severe liver and/or kidney dysfunctions might be at more risk of toxicity.

Regarding metabolism, another clinical aspect to consider is that the hydrolysis of some type ester LA such as procaine leads to the production of the metabolite PABA which has the potential for allergenicity and consequently, esters have been generally related with a higher risk of allergic reactions. However, independently of the ester or amide type, most times are the additives and not the drug responsible for the allergic reaction. Thus, methylparaben, used as a preservative in preparations of amide-type LA, is also metabolized to PABA (Higuchi et al., 2013; Maimo and Redick, 2004).

## 15.4 PHARMACODYNAMICS OF LAs

The main molecular mechanism of action of LAs is the reversible blockade of the inner pore of the VGSCs. The blockade impedes $Na^+$ influx through the channel, a required step to enable nerve cell membranes to reach an action potential as well as the consequent signal transmission along the nerve. In doing so, LAs produce a temporary insensitivity (anesthesia and analgesia) in the area innervated by the blocked nerve. Hence, to understand how LAs act, we will briefly outline some structural and functional aspects of the VGSC, their main pharmacologic target (Catterall and Swanson, 2015).

### 15.4.1 VOLTAGE-GATED SODIUM CHANNELS AND MEMBRANE POTENTIAL

When at rest, excitable membranes of nerve axons (as well as neurons or cardiac muscle cells) maintain a resting transmembrane potential around −90 mV to −60 mV. This potential is achieved thanks to the electrochemical gradient maintained by the sodium–potassium pump. At the resting potential, VGSCs are at resting nonconducting state. When an excitatory signal drives the transmembrane potential to more positive values and the threshold potential is reached, VGSCs open becoming active and ensuring the conductance of $Na^+$ to the axoplasm in favor of the electrochemical gradient. The inward $Na^+$ current ultimately results in the depolarization of the membrane (+40 mV). At this potential, VGSCs are blocked impeding any further inward current of $Na^+$, whereas the potassium channels open allowing the outward $K^+$ current. The outflow of $K^+$ eventually repolarizes the membrane reaching a potential of approximately −95 mV. At this point, VGSCs return to their resting state being able to respond to a new excitatory signal. The conduction of neural impulses are blocked by LAs, therefore preventing the conductivity of $Na^+$ through the VGSCs without altering the resting membrane potential (Catterall and Swanson, 2015).

### 15.4.2 STRUCTURE OF VOLTAGE-GATED SODIUM CHANNELS

The VGSCs are macroproteins that, embedded in the axon cell membrane, are formed of multiple domains assembled in a pseudotetrameric manner. Thus, the VGSC is composed of one main α-subunit that can be related to other accessory β-subunits. Even though β-subunits are not required for the main function of the VGSCs, when present, modulate voltage-dependent activation/inactivation kinetics of the α-subunit. Thus, β- subunits are fundamental constituents of the structure of VGSCs located at the Ranvier's nodes, at the initial segments of axons as well as at cardiac levels, where they can be found in the intercalated disks. In those VGSCs, the β-subunits contribute to regulate the action potential propagation by key communication events at molecular and cellular levels also acting as cell adhesion molecules (CAMs) (Patino and Isom, 2010; Payandeh et al., 2011).

The α-subunit has a homotetrameric structure having four clearly differentiated domains (DI–DIV). Each one of the four domains is in turn conformed by six transmembrane homologous, but not identical, α-helices segments (S1–S6) (Fig. 15.5A). In each domain, segments S1–S4 are known as the "voltage sensor domains (VSDs)." Nonetheless, the S4 segments seem to be most crucial due to their primary structure consisting of a sequence of three amino acid residues, such as arginines and lysines, that positively charged, are repeated with an in between separation by two nonpolar amino acid residues. The membrane depolarization drives the S4 segments outward opening the pore (Patino and Isom, 2010; Payandeh et al., 2011).

The primary structure of segments S5 and S6 constitute the central ion-conducting pore module (PM) that contributes to the specificity of $Na^+$ permeation. At the intracellular side of the PM, there is the so-called "activation gate" or "m-gate" that alternates between opened and closed conformations. The S6 segments of the α-subunit DI, DIII, and DIV domains contain the binding site for LAs. The linker loop between the S4 and S5 domains also known as "fast inactivation gate" or "h-gate" can alter its conformation. The fast inactivation gate closes, by moving into the inner part of the pore, when the membrane is depolarized and therefore blocks any further $Na^+$ conductance (Clairfeuille et al., 2017; Duclohier, 2009; Goldin, 2003; Ulbricht, 2005).

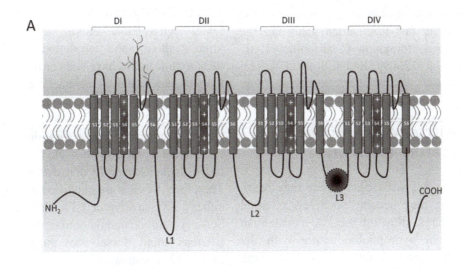

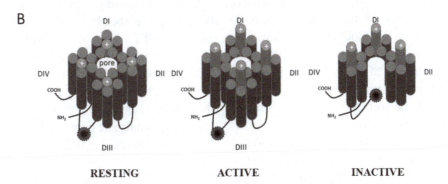

RESTING          ACTIVE          INACTIVE

**FIGURE 15.5** Structure and function of the voltage-gated sodium channels (VGSCs). (A) Representative schematic structure of the VGSCs showing the four domains with their six transmembrane segments. As seen in the image, the transmembrane segments number 4, rich in positively charged aminoacids are the voltage sensor of the channel. (B) Representation of the tridimensional structure of the VGSCs and their respective three distinct functional states.

Repolarization of the membrane is required for its conformational change to open.

Thus, VGSCs alternate their conformation in a voltage-sensitive manner yielding three different structural conformations that correlate with three distinct functional states (Fig. 15.5B) (Bagneris et al., 2015; Catterall and Swanson, 2015). At membrane potentials below −70 mV, the channel pore is blocked by the activation gate (resting state) so that Na$^+$ ions cannot permeate but this conformation is susceptible to change to the active open state if the membrane potential reaches the threshold potential (−40 mV). The Na$^+$ influx further depolarizes the membrane and the "fast inactivation gate" closes blocking the inner pore of the channel (Clairfeuille et al., 2017; Duclohier, 2009; Goldin, 2003; Ulbricht, 2005) resulting in the inactive, nonconducting state refractory to any further excitatory signal. The subsequent inactive VGSC-mediated declining in Na$^+$ influx, together with the outward K$^+$ current through voltage-gated K$^+$ channels, ultimately results in the transmembrane repolarization, the opening of the fast inactivation gate, and the resting state conformation (Clairfeuille et al., 2017; Duclohier, 2009; Goldin, 2003; Ulbricht, 2005).

### 15.4.3 VOLTAGE-GATED SODIUM CHANNELS BLOCKADE BY LAs

Because LAs are weak bases, once they reach the axoplasm they become positively charged at both endings allowing the LA to establish interactions of hydrophobic nature with the amino acid residues of the inner pore blocking the VGSC (Becker and Reed, 2006). To achieve the therapeutic goal of impeding propagation of sensory inputs the blockade of VGSCs by the drug, it is required that the VGSC blockade occurs along a critical length of the nerve. In the case of myelinated nerves, the required length seems to be around 2–3 Ranvier's nodes (Becker and Reed, 2006).

It is important to note that LAs have a different affinity for the distinct VGSC conformations and consequently their functional state resulting in as the so called "use-dependent block" (Sheets et al., 2010). Most LAs have less affinity to bind VGCSs in their resting state (resting membrane potential nonfiring axons) than for the active or inactive conformations. Consequently, most LAs will have a more rapid and strong effect in firing axons (positive membrane potentials). Furthermore, the LA can be "locked" inside the inner pore by the fast inactivation gate in the inactive conformation resulting in a VGSC blockade 10–1000 times longer and therefore longer duration of the effect than that the one obtained by blocking the VGSC in its active conformation (Sheets et al., 2010).

VGSCs can be also blocked by many biologic toxins such as batrachotoxin, veratridine, aconitidine, saxitoxin, and tetrodotoxin (TTX) and in fact, organic toxins have been proven to be useful tools to understand the channel mechanisms and structure (Morales-Lazaro et al., 2015). Among the different toxins, TTX has been particularly useful allowing the characterization and further classification of the VGSCs into two main classes based on their sensitivity or resistance to the toxin. Therefore, VGSCs are classified as belonging to either TTX-sensitive or TTX-resistant channels. Sensory neurons mainly contain TTX-S VGSCs, whereas small dorsal root ganglions contain more TTX-resistant VGSCs. This distinction has potential clinical implications in the development of local more specific LAs. Furthermore, TTX blocks the TTX-sensitive VSGCs acting on the channel extracellular surface yielding effects alike to those induced by LAs with blockade of the conduction of the impulse without affecting the resting membrane potential. Furthermore, a therapeutic analgesic potential has been proposed for TTX (Nieto et al., 2012).

### 15.4.4 ACTIONS OF LAs OTHER THAN VOLTAGE-GATED SODIUM CHANNELS BLOCKADE

Even though VGSCs are the primary target of LAs, the affinity is actually low and furthermore rather unspecific. Thus, when administered at therapeutically efficient doses, LAs are able to act on a variety of other targets. Among others those include:

1.  Voltage-gated $K^+$ channels, voltage-gated $Ca^{2+}$ channels;
2.  $Ca^{2+}$-ATPase pump in the sarcoplasmic reticulum;

3. Ionotropic glutamate receptors type NMDA (N-methyl-d-aspartate) and type AMPA (α-amino-3-hydroxy-5-methyl-4-isoxazolepropionic acid) as well as metabotropic targets such as GTP-binding protein coupling receptors (Sanchez et al., 2010; Yanagidate and Strichartz, 2006, 2007). Albeit these actions might contribute to the overall therapeutic as well as to the toxic effects mediated by LAs, those are still unexplored fields open to further experimental and clinical inquiry.

## 15.5 COADJUVANT DRUGS

LAs are frequently administered along with other drugs (Brummet and Williams, 2011). The aims of using adjuvant medication are quite varied including the achievement of a prolonged duration of the anesthetic effect, to reduce systemic absorption ensuing a lesser likelihood of adverse effects, to increase analgesic effect and, to potentiate the extent of the area treated (regional anesthesia).

### 15.5.1 VASOCONSTRICTOR AGENTS

As mentioned in the section regarding pharmacokinetics, most LAs have vasodilator properties augmenting blood flow to the location of drug administration and consequently the absorption at systemic level. By adding epinephrine, an α1 receptor agonist, vasoconstriction of peripheral arterioles is achieved and consequently, by delaying absorption, three main goals are achieved (Bailard et al., 2014). First, the eventual systemic toxicity of LAs is reduced. Second, lower doses of LA can be administered, also contributing to a reduced likelihood of local toxicity. And third, the duration of anesthesia at the site of injection is increased (Bailard et al., 2014).

Concentrations of epinephrine as low as 1:1,000,000 have proven effective inducing vasoconstriction but there is long latency period and the vasoconstriction is of short duration. This is the motive why many of the commercial preparations contain epinephrine at concentration ranging from 1:100,000 to 1:400,000. At higher concentrations, epinephrine improves hemostasis reducing bleeding being a benefit in certain surgical interventions (e.g., dentistry) (Bailard et al., 2014).

With medium-duration LA (i.e., lidocaine and mepivacaine), a standard epinephrine concentration of 5 μg/mL is used. By contrast, epinephrine addition has no noteworthy effect on anesthetic effect duration of long-acting LAs (i.e., ropivacaine and bupivacaine) (Bailard et al., 2014).

Although epinephrine remains the vasoconstrictor agent most widely used together with LAs, it can induce toxic effects, mainly of cardiovascular nature. In fact, it might be a potential trigger for cardiac arrhythmias in patients suffering from heart disease or when epinephrine is concomitantly used under halothane anesthesia. Besides, the vasoconstrictor effect can induce hypertension in patients with some condition, even triggering hypertensive crises. Lastly, it has been attributed some role for epinephrine in the LA-induced neurotoxicity and myotoxicity (Neal, 2003). Therefore, its use, even at low doses, should be evaluated in those patients at risk such as those suffering cardiopathologies. Furthermore, it is particularly necessary during dentistry, minor surgical interventions in which epinephrine is used at high concentrations (i.e., 10–20 μg/mL) to prevent bleeding.

On a more bright side, the significant cardiovascular effects induced by epinephrine are also considered as a valuable indicator of undesired intravascular injection of the LA. Actually, for some experts, the eventual early detection of LA injection into the blood outweighs the potential risk of toxic effects. Anyhow, there are other available alternatives for patients with cardiopathologies such as levonordefrin (a vasopressor similar to norepinephrine) or felypressin (a vasopressin analog) (Malamed, 2015). Nevertheless, the possible harmful actions on blood pressure of these vasoconstrictor agents should not be ignored.

We also have to mention that, other anesthetics, including the obsolete cocaine and the widely used lidocaine and ropivacaine are vasoconstrictors. Thus, when administering these LAs, epinephrine addition should be avoided or at least monitored.

### 15.5.2   REDUCING BURNING PAIN AND IMPROVING PATIENT COMFORT—SODIUM BICARBONATE

The pH of LA solutions is generally set around values of 4–5 to prolong the shelf-life of the preparations. Because the acidity of the solution often leads patients to experience burning at the moment of injection, addition of bicarbonate is common. Thus, this procedure alleviates the burning and improves patient comfort, principally when patient is awake during the procedure (Hanna et al., 2009). In addition, sodium bicarbonate alkalinizes the solution, increasing the number of uncharged base molecules and their liposolubility providing a more rapid onset of the effect (Best et al., 2015). For example, a formulation of lidocaine with epinephrine and sodium bicarbonate (pH = 7.2) has an onset of 2 min, whereas the onset is delayed to about 5 min

when the same formulation without bicarbonate (pH = 4.6) is administered. Furthermore, the presence of bicarbonate augments the duration of anesthesia. In this sense, a recent study by Warren et al. 2017, reveals that blocking of the mandibular nerve with the combination of lidocaine, epinephrine, and bicarbonate reduces in a 50% the dose of lidocaine needed to achieve the same duration of pulpal anesthesia in comparison with that induced by lidocaine and epinephrine without bicarbonate (Hanna et al., 2009).

Adding bicarbonate to local anesthesia preparation also has some drawbacks if the pH value obtained is too high. In these cases, the LA precipitates due to its poor solubility in aqueous solution (Peterfreund et al., 1989). Furthermore, its stability is similarly reduced leading to shorter shelf-life, mainly if the formulation is prepared and stored for several years. Finally, alkalinization of epinephrine-free anesthetic solutions offers no benefit, since even in some cases a decrease in anesthetic/analgesic magnitude and duration has been described in experimental peripheral nerve block (Sinnott et al., 2000).

### 15.5.3   INCREASING TISSUE PENETRATION AND SPREADING OF LAs— HYALURONIDASE

Hyaluronidase is an enzyme considered as a "spreading factor" as it breaks down the intercellular cement of connective tissues by degrading the hyaluronic acid, the main component of the extracellular matrix. Nowadays, and to increase the capacity of the LA to diffuse through the tissue, recombinant human hyaluronidase is used as local adjuvant. The first study using hyaluronidase together with LAs was reported in 1949 by W.S. Atkinson (Atkinson, 1949). When hyaluronidase is supplemented to an

LA-containing solution, its action on connective tissue results in an enhanced and fast spread of the anesthetic leading to a more efficient, fast, and extensive anesthesia that facilitates patient throughput. Other potential advantages seem to include a reduction of the inflammation caused by the local infiltration as well as an increase in the LA solubility (Roberts et al., 1993) and any improved anesthesia and akinesia. The later, mostly due to that hyaluronidase preparations are buffered with phosphates and their addition to the LA solution leads to the alkalinization of solution (e.g., pH values from 5.3 to 6.3 when hyaluronidase is added to bupivacaine solution). Nonetheless, some studies have described lack of significant benefits from the routine use of hyaluronidase in anesthetic techniques, such as peribulbar anesthesia, using standard anesthetic agents and epinephrine. Hence, excellent results can be consistently achieved without using hyaluronidase. Furthermore, hyaluronidase action on connective tissue also results in increased absorption that in turn might contribute to reduce effect duration, to induce adverse unpredicted long-term effects as consequence of the interstitial tissue barriers destruction and last, but not least, to eventually increase the incidence of systemic adverse reactions. In this regard, there are some concerns due to the small but inherent risk of hyaluronidase to result in a potentially serious allergic reaction due to its antigenic nature (Bowman et al., 1997). Altogether, there is not a clear consensus regarding the balance benefit-inconvenient (including the economic burden) associated to the addition of hyaluronidase to LA-containing solutions.

### 15.5.4   *OTHERS ADDITIVES*

Others drugs such as α2-agonists (i.e., clonidine and dexmedetomidine) are employed in combination with LAs in peripheral nerve block (Brummett and Williams, 2011; Hu et al., 2017). Those drugs have been stated to prolong analgesia, although both two have a modest effect. In addition, another benefit is that both drugs reduced local anesthetic-mediated cardiotoxicity.

The opioid agonist buprenorphine, the analgesic tramadol, and the corticosteroid dexamethasone seem to offer a potential benefit as adjuvants to induce local anesthesia. Yet, the real beneficial outcome remains unclear.

## 15.6   TYPES OF LOCAL ANESTHESIA

LAs when used, as the name indicates, are employed to get an anesthetic effect in a specific part of the body. LAs can be administered by different ways to achieve a local or regional anesthesia (Rawal, 2001). Thus, the way of administration ultimately depends on the clinical goal.

### 15.6.1   *LOCAL ANESTHESIA*

This type of anesthesia refers more particularly to the anesthesia of small, defined areas of the body.

- Topical administration by using creams, gels, or sprays. For example, administration on mucous membranes (such as oral, nasal, or rectal mucosa) or onto the skin surface. Unlike infiltration, which requires injection which can be painful, topical anesthetics are applied painlessly without needles (Jenkins et al., 2014). Besides, the topical administration avoids the tissue distortion that occurs with infiltrated anesthetics even reducing the amount of LA required for the anesthetic effect.

- Infiltration, that is, administration by injection of the drug can be achieved by different procedures and it is the best option for pain control and minor dental and surgical procedures (Press, 2015).
  - Subcutaneous and perineural infiltration are done by injecting the LA solution into the skin, nearby peripheral nerve endings. These procedures are useful for intravenous (IV) line placement, superficial biopsy, suturing, etc
  - Submucosal infiltration is commonly used for dental procedures, laceration repairs, etc.
  - Wound infiltration is performed when postoperative pain control at the incision site is needed.
  - Intraarticular injections of LAs are intended to control postsurgical or arthritic joint pain (Rawal, 2001).

## 15.6.2 REGIONAL ANESTHESIA

This term is used when the objective is to target a larger area (arm, leg, abdomen…). It has the advantage over general anesthesia that the surgical procedure is done without being unconscious. In this kind of anesthesia, the anesthetic is injected close to a nerve bundle to anesthetize the area. Regional anesthesia can be classified into two main types in function of the location anesthetic injection: surroundings of a cluster of peripheral nerves (i.e., peripheral nerve block) and closed to the spinal cord (i.e., spinal and epidural anesthesia procedures) (Rawal, 2001).

Peripheral nerve block, the LA-containing solution is injected around a specific nerve or close to a major nerve trunk (nerve bundle).

The most commons are the femoral nerve and the brachial plexus blocks. In the first one, the injection is administered in the leg, whereas in the second one the anesthetic is injected in the arm and shoulder. The goals are to allow for surgical, dental, diagnostic, and pain control management areas such as the knee, shoulder, arm, leg, or even face (Rawal, 2001).

Sciatic and popliteal nerve block is usually used for surgery on areas innervated by the sciatic nerve.

Other important type of regional anesthesia is the paravertebral block (types unilaterally or bilaterally). This type of blockade affects the dorsal and ventral rami of the spinal nerve and the ganglion chain of the sympathetic autonomous nervous system. It is commonly used at thoracic level although any vertebral level is effective. Thus, blockade at neck level is done for thyroid gland or carotid artery surgery, whereas the blockade at the levels of either chest or abdomen is suited for diverse kinds of surgery on the abdominal, breast, and thoracic areas. When paravertebral block occurs at hip level, then the areas affected are hip, knee, and thigh (Rawal, 2001).

Besides, whereas some types of procedures can benefit from unilateral paravertebral blocks such as thoracotomy, breast surgery, and open nephrectomy among others (Richardson et al., 2011), the bilateral paravertebral block has confirmed to be beneficial for midline abdominal surgery (Richardson et al., 2011).

The general complications of regional nerve blocks are the following (Davies et al., 2014): bleeding and hematoma, infection, failure due to incomplete anesthesia, pain or paresthesia, nerve injury, systemic LA toxicity, postoperative injury, and ischemia possibly caused by epinephrine.

### 15.6.2.1 SPINAL (SUBARACHNOID) AND EPIDURAL ANESTHESIA

The LA injection is achieved in the back, near spinal cord also affecting the nerves that connect to the spinal cord. Consequently, both sensitivity and pain in a complete area of the body (belly, hips, or legs) is achieved. Hence, the LA is injected into the subarachnoid space in the spinal anesthesia. Instead, in the epidural anesthesia, the drug is administered inside the epidural space of the spinal cord. Since the drug does not directly reach the CSF in the epidural anesthesia, this kind of anesthesia is characterized by a delayed onset and the need of larger doses as compared with spinal anesthesia. These types of anesthesia are chosen frequently to alleviate pain during labor and childbirth, prostate surgery, and other types of surgeries such as laparotomy, hysterectomy, and hip replacement, among other indications (Rawal, 2001).

### 15.6.2.2 INTRAVENOUS REGIONAL ANESTHESIA (IVRA) OR BIER BLOCK

This type of anesthesia was first introduced by August Bier in 1908. The anesthetic is administered intravenously to get a regional effect (Stancil, 2014). To avoid any eventual pass of the drug into the systemic circulation, the technique usually involves two extra procedures:

1. Exsanguination of the region, extracting the blood from the region and
2. Tourniquets to avoid blood from either flowing in or out of the exsanguinated region.

For this reason, this special type of regional anesthesia is most suited for surgeries on the extremities such as legs, feet, arms, and hands. The LA is injected into the limb while the tourniquets retain it within the desired region. It is mainly beneficial to perform short duration surgeries such as wrist or hand ganglionectomy and carpal tunnel release. The IVRA of the upper limb is considered a reliable, cost-effective, safe, and simple technique (Rawal, 2001). Nowadays, no other technique provides such a good control on essential features of blockade like onset and duration of the block and the recovery from its effects. It has been estimated a success rate of 94–100% (Rawal, 2001).

## 15.7 LAs TOXICITY

In general, the use of LAs is quite safe; however, toxicity might occur when inappropriately administered and, in some circumstances, unintended effects due to idiosyncratic reactions might appear even when the drug has been properly administered.

Systemic toxicity by LAs is mostly attributed to high plasma concentrations. High plasma concentrations of a LA can be due to: (1) administration of an excessive dose (e.g., by dosing error or when multiple injections are performed); (2) a rapid absorption rate of the drug when injected in an area rich in blood vessels; (3) an accidental intravenous injection as well as (4) a badly performed anesthetic procedure (e.g., in Bier block anesthesia where the drug is intravenously injected into the patient extremities). The lowest IV toxic dose of LAs in humans is displayed in the Table 15.3 (Schwartz and Kaufman, 2015).

**TABLE 15.3**  Minimum Toxic Doses of Some of the Most Commonly Used Local Anesthetics.

| Agent | Minimum toxic dose (mg/kg body weight) |
|---|---|
| Procaine | 19.2 |
| Tetracaine | 2.5 |
| Chloroprocaine | 22.8 |
| Lidocaine | 6.4 |
| Mepivacaine | 9.8 |
| Bupivacaine | 1.6 |
| Etidocaine | 3.4 |

Manifestations of LAs toxicity typically occur in the first 5 min after the injection; however, onset is more variable (from 30 s to 1 h after the injection) (Neal et al., 2018). The toxicity caused by LAs can be classified in:

- Local toxicity, when affecting the site of the drug administration and nearby areas. Local toxicity includes myotoxicity and neurotoxicity involving cytotoxicity and transient neurological symptoms.
- Systemic toxicity, when affecting the cardiovascular and CNS.
- Immune-related toxicity, when either allergic or hematologic reactions appear as consequence of the individual exaggerated and abnormal immune-related response to the proper drug administration (Rawal, 2001).

### 15.7.1  LOCAL TOXICITY

As above mentioned, local toxicity induced by LAs manifests in two main forms: neurotoxicity or myotoxicity.

### 15.7.1.1  NEUROTOXICITY

LAs can induce directly neurotoxic effects. The neurotoxic potential differs among the different LAs. For example, lidocaine is more neurotoxic than bupivacaine. Neurotoxicity-induced by LAs can be noticeable as persistent neurological injury and transient neurological symptom complex (Johnson, 2000).

- Persistent neurological injury is caused by intrathecal injection of a high dose of lidocaine (Drasner, 1993). Nevertheless, continuous administration of a lower dose by using microcatheters also seems to contribute to induce persistent neurological injury. Furthermore, the cause of this syndrome seems to be a defective dilution of the LA in the CSF. The lack of an appropriated dilution of the drug on the CSF might impede the correct flow of the LA, therefore regionally accumulating in the cauda equina up to toxic concentrations (Reina et al., 1999). Interestingly, the mechanism by which LAs induce neurotoxicity seems to be independent from the LAs action blocking the VSGCs. Instead, the blockade of $Ca^{2+}$ channels, the increased glutamate-induced excitotoxicity, and oxidative stress are the mechanisms that seemingly underpin neurotoxicity (Butterworth and Strichartz, 1990).
- Transient neurological symptoms (TNS) are symptoms that include pain and dysesthesia and they appear after recovery from spinal anesthesia, mainly affecting the buttocks or legs and it is frequently associated with pain in the lower back (Pollock et at, 1999). The pain and the transient radicular irritation, typically occurs in the first 12 h after

administration. Usually, those symptoms do not last more than 7 days. However, approximately 30% of patients suffer severe debilitation and are readmitted into the hospital for further analgesic therapy. In most reported cases of TNS, the LA used was lidocaine (Freedman et al., 1998). It seems that that the TNS induced by lidocaine might be a slight transitory and direct form of neurotoxicity that can be exacerbated when epinephrine is also used, but the mechanisms have not been elucidated.

## 15.7.1.2 MYOTOXICITY

Myotoxicity, first described more than 55 years ago by Brun (Brun, 1959), after peri- and retrobulbar block during ophthalmologic procedures relates to a great occurrence of muscle dysfunction. Even though all clinically used LAs may cause skeletal muscle injury, even leading to muscle necrosis, tetracaine, and procaine are less disposed to myotoxicity. By contrast, bupivacaine seems to be the LA with higher toxicity. Nevertheless, the myonecrosis of skeletal muscle is an unusual side effect that in most cases is reversible (Dippenaar, 2007). The severity of the damage is dose-dependent. Furthermore, it worsens when the LA is administered by continuous infusion or when serial injections are done. After injection, there is hypercontraction of the myofibrils that over the next 1–2 days is followed by lytic degeneration of the sarcoplasmic reticulum (SR), myocyte edema, and necrosis. In most cases, the muscle fibers regenerate within 3–4 weeks. Nonetheless, the concomitant administration of epinephrine and corticosteroids potentiates the severity of myotoxicity, even leading to permanent muscle damage.

The LA injection causes toxicity resulting in trauma because of: (1) the injected volume; (2) the pressure of injection, and (3) chemically by varying intracellular concentrations of calcium (Nair, 2017). Regarding the later, increased free cytosolic $Ca^{2+}$ concentration seems to be the primary cause. The mechanism of myotoxicity caused by LAs remains uncertain but it seems that multiple mechanisms are involved. Thus, LAs can bind and activate the ryanodine receptors (RyR) located in the sarcoplasmic reticulum membrane releasing more calcium from the SR to the cytosol and therefore leading to muscle contraction. Nevertheless, activation of the RyRs by LAs is both concentration- and pH-dependent. In fact, the charged form of the LA is the one able to activate the RyR. Inhibition of the ATPase responsible of the calcium reuptake into the SR (SERCA) by LAs also seems to contribute the disturbance of the calcium cellular homeostasis (Dippenaar, 2007). Two of the LAs with myotoxic properties are bupivacaine and ropivacaine, being the former more myotoxic than the latter. The probable reason for this difference in myotoxicity seems to be the higher liposolubility of bupivacaine compared with ropivacaine.

A third mechanism is the effect of LAs uncoupling oxidative phosphorylation which causes a decrease of intracellular ATP that further contributes to $Ca^{2+}$ dysregulation. Overall, and independently of the mechanism, the high intracellular levels of free $Ca^{2+}$ derives in the activation of several proteases culminating in cell necrosis.

As previously mentioned, myotoxicity is an infrequent adverse effect when LAs are locally applied, such as it may be affectation extraocular muscles in ophthalmic surgery. Furthermore, permanent damage is highly uncommon. However, cases of diplopia (complication of regional anesthesia) have been described when bupivacaine has been used during cataract surgery (van Rooyen, 2010).

## 15.7.2  SYSTEMIC TOXICITY

Systemic toxicity induced by LAs manifests as the blood concentration of the drug increases and the main organ systems susceptible to LA toxicity are the CNS and the heart, both with high oxygen demands. Consequently, systemic toxicity manifests as a progressive spectrum from neurological to cardiologic symptoms paralleling the increase of LA concentrations in blood (Goyal and Shukla, 2012). Usually, systemic toxicity induced by LAs first affects the CNS and in a later stage the cardiovascular system. Thus, the onset of CNS signs usually requires lower concentrations of LAs than those needed to elicit cardiovascular depression. Nevertheless, cardiovascular toxicity (which may lead to cardiac arrest) sometimes occurs without previous toxic CNS signs. This is important because in comparison to CNS toxicity, cardiovascular toxicity can yield more severe consequences that are in turn more difficult to treat. That is the case for some anesthetics which show a low CNS/heart ratio toxicity such as bupivacaine. Besides, in patients treated with drugs that depress the CNS, such as benzodiazepines, the CNS toxicity signs can be masked.

The site of injection is key contributing to systemic toxicity, since the blood levels of LA vary in function of the type of regional block. The approximate order (from most to least) of the associated circulating concentration is depicted in Figure 15.6 (Goyal and Shukla, 2012; van Rooyen, 2010).

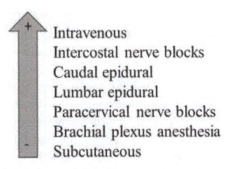

+

Intravenous
Intercostal nerve blocks
Caudal epidural
Lumbar epidural
Paracervical nerve blocks
Brachial plexus anesthesia
Subcutaneous

-

**FIGURE 15.6** Diagram depicting how the type of administration contribute in lesser or higher extent to systemic toxicity induced by local anesthetics.

### 15.7.2.1  CNS TOXICITY

CNS toxicity has two phases or "two-stage process." Thus, there is an initial excitatory phase followed by a depressive one (Christie et al., 2015; Dippenaar, 2007). The first symptoms are of sensory nature including numbness of face muscles or of the tongue, alterations in the sense of taste (metallic taste) and in the auditory sense, typically tinnitus. Those are followed by excitatory neurological symptoms such as agitation, confusion, and muscle twitching that can ultimately culminate in seizures with generalized convulsions. Nonetheless, the aforementioned excitatory signs are not always present. The symptoms corresponding to the depressive phase are characterized by loss of consciousness, coma, and cardiorespiratory depression.

This two-stage process (initial excitation followed by depression) links to the VGSC channels blockade time course (van Rooyen, 2010). Thus, the initial excitatory phase is mostly due to the prevalence of VGSC blockade located at the inhibitory GABAergic neurons. Blocking the inhibitory GABAergic neurotransmission leaves the excitatory neurons without restraint or opposition resulting in a mainly excitatory state, which climaxes in tonic–clonic convulsions. In the second phase, when higher concentrations of LA are present, the VGSCs blockade affects both inhibitory and excitatory neurotransmission resulting in a general depression of the CNS. It is in this stage, when coma and cardiorespiratory depression occur.

As before mentioned, it is remarkable that the excitatory symptoms may be masked in patients pretreated with anticonvulsants or other depressant drugs such as benzodiazepines or barbiturates. However, because these drugs do not have main effects altering the signs of cardiovascular toxicity, the consequences of their use might be misleading, therefore preventing the detection of the early warning signs of LA toxicity. The symptoms related to LAs-induced CNS toxicity in the different phases are shown in Figure 15.7.

CNS effects depend on various clinical factors including:

- CNS active state in patients receiving depressant drugs. Under these conditions, the excitatory stage could be veiled and signs of seizure might not manifest. As stated before, those drugs do not affect the cardiac toxicity, abolishing the early warnings of potential toxicity.
- Hypercarbia. High levels of $PaCO_2$ reduce the seizure threshold to LA. The alteration in the pH by $PaCO_2$ increase leads to the trapping of the charged

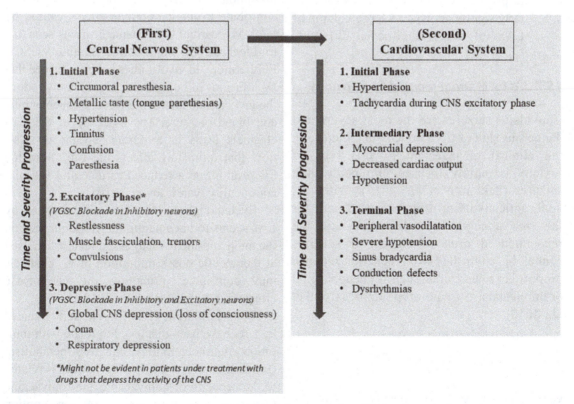

**FIGURE 15.7** Types and time-dependent pattern of toxicity signs induced by local anesthetics in the central nervous system (CNS) as well as on cardiovascular function.

form of drug into the CNS as well as, in an augmented blood flow to the brain, delivering more anesthetic to the CNS. Besides, the acidosis caused by hypercarbia decreases the binding of LAs to proteins making more drug available to the CNS.

- It has been stated that coadministration with epinephrine may decrease the convulsion threshold.
- Patient factors such as renal of hepatic failure, respiratory acidosis, or the presence of cardiac conditions.
- Considering that LAs bind to α1 glycoprotein (AAG), the toxicity may increase when blood concentrations of this plasma protein are reduced as it happens during in pregnancy. This factor being particularly significant when a large volume of LA is infiltrated (van Rooyen, 2010).

### 15.7.2.1.1  *Management of CNS Toxicity*

Convulsive seizures are the most severe and dangerous sign of CNS toxicity and therefore, the clinical management of this type of toxicity is mainly focused into controlling seizures (Sekimoto et al., 2017). With this goal, anticonvulsant drugs are used and, for the first attempt, benzodiazepines are the recommended drug of choice. If seizures cannot be controlled with benzodiazepines, propofol or a barbiturate such as phenobarbital or thiopental are used to stop seizures (Neal et al., 2010).

### 15.7.2.2  *CARDIOVASCULAR TOXICITY*

Cardiovascular signs of toxicity induced by LAs are related to more serious consequences (Bourne et al., 2010; Sekimoto et al., 2017). In

general, the cardiotoxic potency, lipid solubility, and nerve blocking potency of the LA correlate in a positive manner. The ranking of toxicity being the following: prilocaine < lidocaine (lignocaine) < mepivacaine < ropivacaine < levobupivacaine < racemic bupivacaine < R-bupivacaine < etidocaine < tetracaine. The dissociation time constant for bupivacaine from its binding to the VGSCs is at least over 10-fold than that of lignocaine (van Rooyen, 2010), resulting in a greater cardiac depression by bupivacaine. In this regard, bupivacaine is known to affect the cardiovascular system more than the nervous system. Apart from the blockade of VGSC, other mechanisms have been involved (Bourne et al., 2010; Dippenaar, 2007).

In this context and similar to neurotoxicity, inhibition of $Ca^{2+}$ ion channels seems to contribute to the cardiotoxic effects caused by LAs. Myocardial $Ca^{2+}$ channels are as sensitive to blockade by bupivacaine as the VGSCs. Furthermore, in vivo studies have confirmed that IV infusions of bupivacaine significantly reduce the myocardial contractile force, frequency rate, and blood pressure. The blockade of the $Ca^{2+}$ channels leads to a diminished inward $Ca^{2+}$ flow that ultimately affects the $Ca^{2+}$ from the SR resulting in a reduced cardiac and vascular contractility (van Rooyen, 2010).

Even though the inhibition of $K^+$ channels by LAs seems to have negligible consequences on the main anesthetic/analgesic effects when used at therapeutic doses, inhibition of $K^+$ channels may contribute promoting the cardiotoxic effects of LAs (van Rooyen, 2010).

Both cardiotoxicity and myotoxicity induced by LAs have been displayed to relate with inhibition of mitochondrial oxidative metabolism (see Section 15.7.1.2). The mechanism is related to the capacity of bupivacaine and ropivacaine to uncouple the processes of $O_2$ consumption and phosphorylation of ADP in the complex I

(NADH ubiquinones reductase). As a result, ATP concentrations decrease in myocytes, ultimately contributing to the myocardial depression.

The cardiac lipid metabolism is decreased by bupivacaine which also inhibits the enzyme carnitine-acylcarnitine transferase (CACT). The CACT regulates the transport of acylcarnitines throughout the mitochondrial membranes. This is essential for the transport chain of fatty acid in mitochondrial respiration phase I, an essential step for aerobic metabolism. Finally, inhibitory actions of bupivacaine on the brainstem center controlling cardiovascular function are another mechanism responsible for its cardiotoxic effects (Bourne et al., 2010; Dippenaar, 2007).

### 15.7.2.2.1 Manifestation of Cardiovascular Toxicity

The types and pattern of the toxic effects of LAs on cardiovascular function are represented in Figure 15.7.

- Reduced myocardial contractility, mainly due to the blockade of $Ca^{2+}$ channels and to the interference with myocardial metabolism.
- LAs act in two ways: (1) LAs reduce cardiac automaticity by blocking the $Ca^{2+}$-channel blockade during depolarization in the phase 4, resulting in a slower activity of the pacemaker cells in diastole and (2) the cardiac impulse conduction also slows down due to the blockade of the VGSCs. Altogether, a slow impulse conduction prompts to unidirectional block and re-entry events, which may cause ventricular tachycardia and fibrillation. Besides, intracranial (and cervical intra-arterial) administration of bupivacaine can activate the sympathetic

nervous system response to efferent CNS inputs resulting in ventricular arrhythmias.

- LAs have direct and indirect actions on the blood vessels. Direct actions are rather variable, with vasoconstriction or vasodilatation, depending on the drug, dose, and model. Catecholamine release and vasoconstriction during the initial phase or in response to mild intoxication may cause hypertension and tachycardia. Further sympathetic blockade (neuraxial block), depression of the medullary vasomotor center, hypoxia, acidosis, or direct vasodilation would lead to hypotension and cardiovascular collapse (van Rooyen, 2010).

### 15.7.2.2.2 Management of Cardiac Toxicity

#### 15.7.2.2.2.1 Initial Management

The Advance Cardiovascular Life Support (ACLS) 2015 guidelines (Singletary et al., 2015, van Rooyen, 2010) indicate that the course of action before any other intervention includes the airway management and the suppression of seizure activity. The goal is to effectively reduce the exacerbating consequences of hypercarbia, hypoxia, and acidosis on LAs-induced toxicity.

#### 15.7.2.2.2.2 Specific Therapies for Cardiotoxicity

In this category are included sympathomimetic agents which raise coronary perfusion and cardiac contractility. The dose of epinephrine indicated for this situation is 2–3 times higher than the usual doses. However, in cases such as is patients with signs of bupivacaine toxicity, vasopressin might be an alternative to epinephrine when considering the potential of epinephrine to aggravate arrhythmias. In any case, it seems that the arrhythmias are best managed with amiodarone. It is key to highlight that lidocaine, being a LA, is not an optional antiarrythmic treatment of intoxication by Las (van Rooyen, 2010).

The administration of insulin is another treatment that improves myocardial repolarization. Insulin by increasing inward flow of $K^+$ counteracts the bupivacaine-induced inhibition of K+ channels. Besides, the administration of insulin and glucose may ultimately improve the cellular metabolism in the face of the lack of alternative substrates due to the effects of bupivacaine inhibiting lipid substrate utilization (van Rooyen, 2010).

In the past, cardiopulmonary bypass was the only effective method for refractory cardiac arrest treatment from overdose with LAs. It was not until 1998, when the IV lipid emulsion therapy was proposed as an alternative with undoubtedly benefits on the therapeutic management of LA toxicity (Bourne, 2010; Sekimoto, 2017; Udelsmann, 2012). However, the mechanisms of lipid emulsion on cardiac toxicity caused by LAs are not well understood (Christie et al. 2015; Udelsmann, 2012; van Rooyeen, 2010). Thus, it has been indicated that lipid emulsions act as a "lipid sink," where the lipophilic drugs are sequestered, making LAs unavailable to the tissues. The lipid emulsion impedes LAs for getting in contact with the tissue and even recovers the drug from the tissue. Other hypotheses propose that the effect of lipid emulsions might act as a source of free fatty acids, re-enabling a normal oxygen–energy coupling in the cardiomyocytes and therefore improving of the cell metabolism previously impaired by LAs (bupivacaine inhibits CACT). Finally, lipid emulsion has been attributed with direct positive inotropic effects.

Among the distinct lipid emulsions, the most usually employed are the ones deriving from soybean oil and coconut oil with long- or medium-chain triglycerides as well as those composed by oils derived from olive and fish. Although lipid emulsion therapy is usually well-tolerated and widely used in parenteral nutrition, some complication has been stated such as augmented infection risk, thrombolytic production, allergic reactions (even reaching to anaphylaxis), and pulmonary emboli (Bourne, 2010). Accordingly, some associations such as the Association of Anaesthetists of Great Britain and Ireland (AAGBI) have published guidelines for the safest and fastest management in those cases. Finally, and even though propofol may be useful for counteracting LA-induced hypotension and seizure activity, thus it cannot be as a substitute for lipid emulsion therapy.

### 15.7.2.3 HYPERSENSITIVITY REACTIONS TO LAs

The hypersensitivity reactions comprise adverse effects associated to idiosyncratic bizarre immune reactions. These reactions include allergic and hematologic ones (Sekimoto et al., 2017). These types of adverse effects do not depend on action mechanism and doses of the LAs and only relate to an individual immune reaction of the patient. Though not very common, hypersensitivity reactions may occur with use of LAs and caused by the LA and other components added to the preparation (van Rooyen, 2010).

Allergic reactions are infrequent for amino ester LAs and almost exceptional in the ones that belong to amino amide group. In general, many of the reactions to LAs administrations are a consequence of the patient (anxiety, panic attack, or vasovagal responses) but they can be

**Amide-type local anesthetics**

Articaine

Bupivacaine

Etidocaine

Levobupivacaine

Lidocaine

Mepivacaine

Prilocaine

Ropivacaine

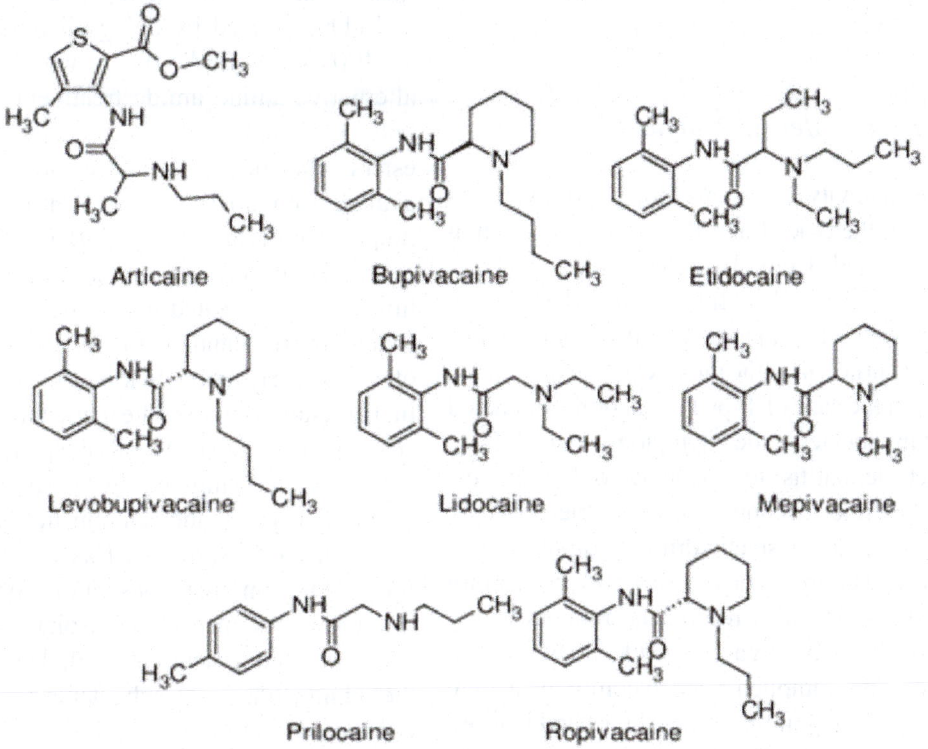

**Ester-type local anesthetics**

Benzocaine

Chloroprocaine

Tetracaine

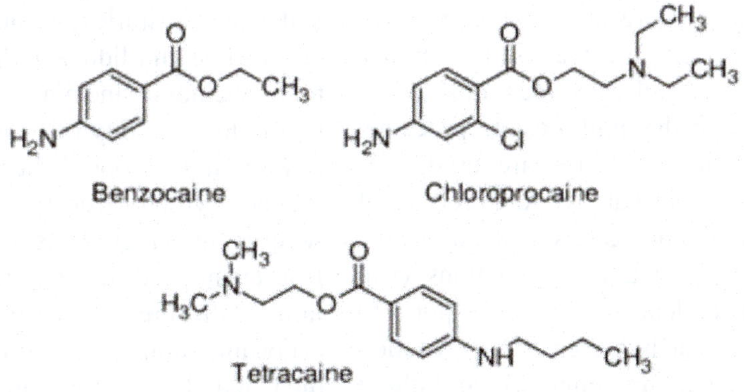

**FIGURE 15.8** Representation of the chemical structure of the main types of local anesthetics.

also the result of an unintended intravascular injection. Indeed, less than 1% of all reactions to LAs are considered as true allergic reactions (van Rooyen, 2010).

### 15.7.2.3.1  Allergic Reactions

Hypersensitivity reactions triggered by LAs belong to the types I and IV reactions according to Gells and Coombs classification. Type I reactions, mediated by IgE, are manifested by a wide variety of symptoms ranging from local or systemic urticaria to the most severe anaphylactic shock. In the case of Type IV reactions, mediated by lymphocytes, the symptoms vary from contact dermatitis to anaphylactoid reactions. These allergic reactions can be restricted to the area where the anesthetic drug was applied with signs of rash and urticaria (hives), or be more generalized. When the reaction becomes systemic affecting various organs is called anaphylaxic reaction and, although very uncommon, it may happen resulting in rhinorrhea, hives and in more severe symptoms affecting the cardiovascular system (hypotension, tachycardia, and angioedema) as well as the respiratory system (dysnea, bronchospasm, with wheezing) (van Rooyen, 2010).

Allergic reactions are more common when LAs of amino ester type are used. Previous reports show that allergic reactions occur up to a 30% when the amino esters procaine, tetracaine, and chloroprocaine are used. The accepted hypothesis is that PABA, a produce of the metabolism of amino esters, is the molecule associated with acute allergic reactions (van Rooyen, 2010). Patients known to respond to amino esters with a hypersensitivity reaction could be treated with an amino amide without risk. In general, hypersensitivity reactions to amino amides are exceptional and, when occurring, they are not elicit by the amino

amide itself that is not metabolized to PABA but by methylparaben or by metabisulphites (preservative agents added to the preparation) and characterized by being chemically similar to PABA. Accordingly, when choosing an alternative amino amide treatment for patients known to have allergic reactions to amino esters, the best is to use preservative-free available commercial amino amide anesthetic preparations (van Rooyen, 2010).

In the occurrence of an allergic reaction, this has to be treated according to the severity following the standard treatments used for any other allergic reaction (Sekimoto et al., 2017). The first measure comprises the immediate cease of the surgical procedure and treats the patient to reduce the signs and symptoms of the hypersensitivity reaction. This includes administration of oral or IV anti-histamine drugs such as diphenhydramine for mild skin reactions and corticosteroids (prednisone or methylprednisolone). In case of severe allergic reaction (anaphylactic reaction), the administration of subcutaneous epinephrine is required (van Rooyen, 2010).

### 15.7.2.4  HEMATOLOGICAL REACTIONS— METHEMOGLOBINEMIA

Prilocaine and in a lesser extend other LAs, mainly benzocaine and lidocaine have been associated to methemoglobinemia. This condition refers to a reduced ability of hemoglobin to transport oxygen in the blood. When mild, this condition can be asymptomatic but, when the degree of severity increases the symptoms can progress from cyanosis, discoloration of the skin, dyspnea, fatigue, dizziness to even syncope. Importantly, prilocaine-induced methemoglobinemia is mediated by orthotoluidine. Orthotoluidine, hepatic metabolite of prilocaine, oxidizes hemoglobin converting it into methemoglobin.

The oxidative potency of orthotoluidine results in a methemoglobinemia that does not respond appropriately to supplemental oxygen at high flow. Although the doses of prilocaine required to reduce the oxygen saturation levels are typically higher than the standard doses used in clinical regional anesthesia, some countries have prohibited the use of prilocaine in regional anesthesia procedures. Furthermore, to prevent potential complications, preparations containing prilocaine (e.g., EMLA cream) should not be used in particularly susceptible patients (e.g., suffering from either congenital or idiopathic methemoglobinaemia) neither in children younger than 1 year old that are under drug treatments such as sulphonamides or phenytoin characterized by their well-known ability to induce methemoglobinemia (Berkman, 2012).

Although methemoglobinemia is usually reversible and short-lasting, the reversal can be accelerated by a slow IV infusion of a methylene blue solution (Sekimoto et al., 2017). Methylene blue is as a substrate for the enzyme erythrocyte methemoglobin reductase. Methylene blue, by being reduced to leukomethylene blue allows methemoglobin to revert into oxyhemoglobin, therefore restoring the normal ability of hemoglobin for carrying oxygen.

## 15.8 CLASSIFICATION OF LAs

### 15.8.1 AMINO-AMIDE TYPE LAs

#### 15.8.1.1 ARTICAINE

Articaine (Fig. 15.8), one of the most recently approved LAs, is noted for its effectiveness, rapid action, and tolerability. Articaine is available commercially as a single injection of 4% articaine solution associated with the vasoconstrictor epinephrine (1:100,000 and

1:200,000) and it is indicated for anesthesia in minor dentistry procedures by infiltration and nerve block in adults and children over 4 years of age. The pulpal anesthesia duration of action ranged from 45 to 75 min and the duration of action of soft tissue is around 120–140 min (Malamed et al., 2001; Yapp et al., 2011). In a meta-analysis study, it was observed that articaine 4% was as potent local anesthetic as lidocaine 2% by infiltration route of administration to provide pulpal anesthesia (Brandt et al., 2011).

Adverse reactions related to articaine, but which are rare or very rare, include cardiovascular disorders (hypotension, decreased heart rate), nervous system disorders (nausea, vomiting, nervousness), respiratory disorders (tachypnea), and allergic reactions. Articaine binds highly to plasma proteins (95%). It is metabolized by plasma cholinesterases and eliminated in the urine (AEMPS).

From the structural point of view, this amino amide type local anesthetic differs from those of its category in that it has a thiophene group instead of a benzene ring, which gives it greater lipid solubility and potency. Moreover, articaine has an ester group which gives it a lower plasma half-life (20 min) since this compound is rapidly hydrolyzed by the action of nonspecific plasma esterases to its inactive metabolite articainic acid. This rapid metabolism is associated with a reduced risk of systemic toxicity and overdose compared with other amino amide local anesthetics (Yapp et al., 2011). However, several epidemiological studies suggest that articaine has a higher occurrence of causing paresthesia. Paresthesia is an abnormal sensation, usually pricking, tickling, numbness, burning or itching which occurs in hands, feet, toes, fingers, or another part of the body. Comparing articaine 4% and lidocaine 2% associated both local anesthetic drugs with epinephrine 1:100,000, the ratio of suffering from paresthesias after their administration of in a dental procedure is 1/49 for articaine and 1/63 for lidocaine and the duration of paresthesia varies from a few hours to around 20 days (Haas, 2006; Hopman et al., 2017; Malamed et al., 2001).

### 15.8.1.2   BUPIVACAINE

Bupivacaine (Fig. 15.8) is a longer-acting amide-type local anesthetic marketed as racemic mixture [R(+)-bupivacaine and S(-)-bupivacaine] indicated for local infiltration and for peripheral, sympathetic, and epidural nerve block. This drug is on the list of essential local anesthetics (World Health Organization, WHO) as 0.25% and 0.5% bupivacaine hydrochloride in vial for injection and as 0.5% bupivacaine hydrochloride in 4 mL ampoule to be mixed with 7.5% glucose solution for spinal anesthesia. Bupivacaine has narrow therapeutic index and it presents a high plasma protein binding (95%). Concomitant administration of bupivacaine with nonsteroidal anti-inflammatory drugs and heparin may increase the tendency to hemorrhage (AEMPS). Its major clinical limitation is related to its cardiotoxicity (palpitations, alterations of tension, arrhythmias, and even cardiac arrest) and its neurotoxicity (tremors, tinnitus, paresthesias, and convulsions) problems. To reduce this toxicity in recent years, different controlled release systems have been developed including nanoparticles and liposomes microparticles. The United States Food and Drug Administration (FDA) approved EXPAREL® which encapsulates bupivacaine into an aqueous suspension of multivesicular liposome composed of phospholipids, triglycerides, and cholesterol (DepoFoam technology) for single-dose injection to manage postsurgical analgesia in different surgical procedures including total knee arthroplasty, mammoplasty, transversus abdominis plane block, inguinal hernia repair, hemorrhoidectomy, and bunionectomy (Beiranvand et al., 2016; Malik et al., 2017; World Health Organization, 2017).

### 15.8.1.3   ETIDOCAINE

Etidocaine (Fig. 15.8) is a long-acting and rapid onset local anesthetic with a higher in vitro potency and longer duration of action than lidocaine. Etidocaine is indicated to mitigate and to reduce postoperative pain after dental procedures. Etidocaine 1.5% with epinephrine resulted to be a similar potent local anesthetic than lidocaine 2% with epinephrine when administered by inferior alveolar nerve block. However, its use in clinic is limited since it is associated with an increased risk of intraoperative and postoperative bleeding (etidocaine has vasodilator properties) and it does not produce a differential blockade of motor fibers and sensory fibers. Moreover, etidocaine is used associated with epinephrine for other diagnostic and surgical procedures (i.e., obstetrical and ophthalmic) by percutaneous infiltration, caudal anesthesia, lumbar peridural anesthesia, and peripheral nerve block. Etidocaine, particularly when it is administered by an IV injection, could produce cardiac depressant action (hypotension, bradycardia, and even heart failure). In addition, etidocaine could have neurotoxicity effects that include depression, anxiety, and neuropathy (Deupree, 2007; Rood, 1989).

### 15.8.1.4   LEVOBUPIVACAINE

Levobupivacaine (Fig. 15.8) is the S(-)-enantiomer of bupivacaine. Comparing to bupivacaine, levobupivacaine has similar potency, greater duration of action in neural tissue (10–15%) and less cardiac toxicity and central nervous toxicity (30% fewer toxic). This long-acting local anesthetic is used for major and minor surgical anesthesia (i.e., spinal anesthesia, epidural anesthesia) and, for analgesia (i.e., labor analgesia and postsurgical pain). Levobupivacaine is highly bound to plasma proteins (97%). It is metabolized in the liver through CYP3A4 and CYP1A2 isoforms CYP450 enzymes to the inactive metabolites 3-hydroxy levobupivacaine and desbutyl levobupivacaine. Levobupivacaine is mainly excreted as its metabolites in urine (mostly) and feces (Burlacu and Buggy, 2008; Foster and Markham, 2000; Leone et al., 2008; Sanford and Keating, 2010). Taking into account that this LA is metabolized via cytochrome P450 isoforms, its metabolism may be attenuated by the concomitant administration of CYP3A4 inhibitors such as ketoconazole and CYP142 inhibitors such as methylxanthines, increasing the plasma concentrations of levobupivacaine (AEMPS).

## 15.8.1.5 LIDOCAINE

Lidocaine (Fig. 15.8), synthesized for the first time in 1943, is a short-acting rapid onset amide-type local anesthetics widely employed in clinical which has analgesic, anti-inflammatory, and anti-hyperalgesic properties. This drug is included in the World Health Organization (WHO) list of essential medicines of LAs together with bupivacaine (World Health Organization, 2017). Lidocaine is frequently administered in association with the localLA prilocaine for topical anesthesia in superficial surgical interventions and punctures. Moreover, lidocaine patch 5% is first-line treatment for the relief of pain associated

with postherpetic neuralgia (Davies and Galer, 2004; Daykin, 2017; Yousefshahi et al., 2017). In addition to its use as a local anesthetic, lidocaine is also employed intravenously as a second-line therapy antiarrhythmic (class Ib antiarrhythmic) for the treatment of ventricular arrhythmias (Collinsworth et al., 1974). For this medical purpose, lidocaine has been also included in the list of essential antiarrhythmic medicines of the WHO as 20 mg (hydrochloride)/mL in 5 mL ampoule for injection (World Health Organization, 2017). It is available commercially in various formulations including cream, oropharyngeal tablets, solution for injection, gel, and skin patch. The half-life of lidocaine is 1–3 h and its effect starts at 1–5 min when administered for superficial anesthesia of mucous membranes, dental anesthesia, spinal anesthesia, infiltration anesthesia, and anesthesia for peripheral and sympathetic nerve blocks. The induction time of LA could increase up to 4–5 min when lidocaine is administered using other techniques than those previously mentioned. Lidocaine is metabolized in the liver by the cytochrome P450 isoforms, CYP1A2 and CYP3A4, via oxidative N-deethylation to yield monoethyl glycine xylidide (MEGX) as major metabolite and via hydroxylation to yield 3-OH-lidocaine as minor metabolite. This drug is mainly excreted in form of metabolites by the kidneys (Wang et al., 2000). The most characteristic adverse event of lidocaine is transient neurologic symptoms (TNS). This painful and/or dysesthesia condition in lower limbs, lower back, and buttock occurs in the early postoperative period after spinal anesthesia and it may last from a few minutes to days. Comparing with other LAs such as mepivacaine, prilocaine, procaine, and bupivacaine, the relative risk for transient neurologic symptoms following spinal anesthesia may increase 2–10-fold when lidocaine is used (Zaric et al., 2003). Moreover, several clinical cases have reported that lidocaine can produce methemoglobinemia as adverse reaction. In vitro studies using whole blood have evidenced that the metabolite 4-hydroxyxylidine is the one major responsible to cause methemoglobinemia (Barash et el., 2015; Hartman et al., 2017).

### 15.8.1.6   MEPIVACAINE

Mepivacaine (Fig. 15.8) is an intermediate-acting and fast onset local amide anesthetic belonging to pipecoloxylidide family (PPX) like bupivacaine and ropivacaine. Mepivacaine has an anesthetic capacity comparable to that of lidocaine but with a lower vasodilator power, which gives it a longer duration of action. Mepivacaine is marketed as an injectable solution without or with vasoconstrictor (epinephrine). It is indicated for LA in dentistry by infiltration and truncal block as well as for therapeutic and diagnostic purposes by central and peripheral nerve block, infiltration, epidural anesthesia, and regional IV anesthesia. Regarding its pharmacokinetic properties, its percentage of plasma protein binding for mepivacaine is around 75%, it is metabolized by liver microsomal enzymes to p-hydroxymepivacaine and m-hydroxymepivacaine, and it is mainly excreted as glucuronide conjugates by bile (Brockmann, 2014). Sensitivity to mepivacaine can be increased in prolonged treatments with anticonvulsants, antiarrhythmics, and psychotropic drugs (AEMPS).

### 15.8.1.7   PRILOCAINE

This intermediate-acting and fast onset local anesthetics with secondary amine structure is restricted used via intrathecal route in short-term surgeries at hospitals. Moreover, it is also frequently employed for dental procedures as injectable and gel forms. Furthermore,

prilocaine–lidocaine cream combined (eutectic mixture of local anesthetics, EMLA®) is used for topical anesthesia in skin, genital mucosa, and ulcers in the lower extremities. The eutexia is a physical phenomenon whereby the mixture of two active components in certain concentrations has a lower melting point than any of them isolated or mixed in any other proportion. EMLA® is an oil-in-water emulsion containing a combination of 2.5% lidocaine and 2.5% prilocaine, being its melting point around 18°C. This mixture allows an effective penetration of the local anesthetics in the skin of 5 mm maximal depth after 120 min application with fewer side effects.

Of all amino amide type LAs, prilocaine (Fig. 15.8) presents the highest renal clearance which results in low plasma concentration and consequently, in lower toxicity. On the other hand, lidocaine is metabolized by hepatic amidases to σ-toluidine and *N*-propylalanine. The metabolite σ-toluidine has been involved in the serious adverse reaction methemoglobinemia; this pathological condition is produced by high levels of hemoglobin containing oxidized form of iron ($Fe^{3+}$) which reduces the total blood oxygen capacity and manifests as cyanosis, fatigue, anxiety, tachycardia and even coma, and death. Methylene blue is used primarily for the treatment of methemoglobinemia. This specific antidote increases the capacity of NADPH reductases to metabolize methemoglobin to leucomethylene blue (Higuchi et al., 2013; Manassero et al., 2017).

### 15.8.1.8   ROPIVACAINE

Ropivacaine (Fig.15.8) began to be commercialized before levobupivacaine and just like this LA is a left-isomer of bupivacaine with a reduced risk potential of toxicity in cardiovascular and CNS than its racemic bupivacaine (AEMPS). Structurally, ropivacaine differs from levobupivacaine in that it has a propyl group instead of a butyl group on the piperidine nitrogen atom of the molecule, which gives it a lower lipophilicity and in consequence, a lower brain and heart toxicity even than levobupivacaine. Moreover, previous studies associate this lower lipophilicity with a minor capacity to penetrate motor fibers and therefore a lesser power motor block potency and of shorter duration than bupivacaine and levobupivacaine as evidenced in minimum local analgesia concentration (MLAC) and median effective dose (ED50) parameters (Leone et al., 2008; Li et al., 2014; Sisk, 1992).

Ropivacaine is marketed as an injectable solution for perineural and epidural administration for the treatment of acute pain and for anesthesia in surgery (i.e., peripheral nerve and ocular block; epidural administration for abdominal surgery, cesarean section and lower limb surgery, and intrathecal administration for orthopedics and urological surgeries). Ropivacaine has pharmacokinetic properties very similar to those of levobupivacaine: high binding to plasma proteins, hepatic drug metabolism with CYP1A2 and CYP3A4 isoenzymes involvement giving rise to the metabolites 3'-hydroxy-ropivacaine and 2',6'-pipecoloxylidide, and primary renal excretion in form of these metabolites (Leone et al., 2008; Li et al., 2014; Sisk, 1992).

The toxic systemic effects of ropivacaine may be increased when it is administered concomitantly with other similar structural drugs such as lidocaine and mexiletine. Moreover, the coadministration of ropivacaine and inhibitors of CYP1A2 and CYP3A4 may raise adverse reactions (AEMPS).

### 15.8.2  ESTER-TYPE LAs

#### 15.8.2.1  BENZOCAINE

Benzocaine (Fig. 15.8) is available as dental solution and oral gel for the relief of oral mucosal pain and dental pain in adults and children. Moreover, benzocaine is marketed in association with other active compounds, usually chlorhexidine and naloxone as tablets for local and temporary symptomatic relief of mild infectious and inflammatory processes of the mouth and throat. Furthermore, benzocaine is associated with ephedrine as rectal ointments for symptomatic relief of pain and itching associated with hemorrhoids in adults.

Benzocaine is metabolized by plasma and hepatic cholinesterases to yield the renal metabolite p-aminobenzoic acid. The main potentially fatal complication of the use of benzocaine is methemoglobinemia. Most of the reported methemoglobinemia cases are related to its prolonged used or overdoes as well as the use of benzocaine oral spray for orogastric intubation. Previous in vitro studies have demonstrated that the metabolite benzocaine hydroxylamine (BenzNOH) is responsible to form methemoglobinemia. Moreover, the blood levels of methemoglobin (MetHb) are greater with benzocaine than with lidocaine (Cooper, 1997).

Benzocaine interacts with hyaluronidase (increasing systemic reactions of benzocaine), sulfonamides (antagonizing antibacterial effects), and cholinesterase inhibitors (inhibiting benzocaine metabolism). In addition, cross-sensitization reactions between benzocaine and other drugs, including LAs, paraminobenzoic acid, and mepricaine, among others, may also occur (AEMPS).

## 15.8.2.2   CHLOROPROCAINE

Chloroprocaine (Fig. 15.8) is an ultrashort-acting and rapid onset amino ester LA. Chloroprocaine as injectable solution is used for spinal anesthesia in adults and for epidural anesthesia in infants and children. Regarding pharmacokinetic properties of chloroprocaine, it does not bind to plasma proteins and that it undergoes enzymatic degradation by plasmatic pseudocholinesterases resulting in the renal metabolites beta-dimethylaminoethanol and 2-chloro-4-aminobenzoic acid (Veneziano and Tobis, 2017).

## 15.8.2.3   TETRACAINE (AMETHOCAINE)

This ester-type LA is included in the WHO list of essential medicines of LAs for ophthalmological preparations as tetracaine hydrochloride 0.5% w/v eye drops solution (World Health Organization, 2017). In addition, tetracaine (Fig. 15.8), which is available in different topical formulations (i.e., ointment, gel, and solution), is employed not only as a LA in ophthalmology but also in bronchoscopy, esophagoscopy, otorhinolaryngology, dentistry, and gastroscopy. In pediatric populations, tetracaine (4% w/w) is as effective as the combination of lidocaine–prilocaine (5% w/w) in reducing pain during cutaneous procedures after its application for 30–60 min (Taddio et al., 2002). In adults, it has been also demonstrated that lidocaine–tetracaine (7% w/w) as cream is an efficient and safe drug combination to relieve pain in dermatological procedures such as collagen injections, laser treatments, and cryotherapy (Alster, 2007).

Tetracaine binds to plasma proteins around 75%. It is metabolized by the action of plasma cholinesterases (due to the presence of an ester group in its structure) giving rise to the active metabolites para-aminobenzoic acid and diethylaminoethanol which are eliminated through urine (O'Brien et al., 2005).

The main drug interactions of tetracaine include cholinesterase inhibitors which inhibit tetracaine metabolism and sulfonamides whose antibacterial effects can be reduced or antagonized (AEMPS).

## 15.9   FUTURE OPPORTUNITIES AND CHALLENGES

Nowadays, a wide variety of LAs are available offering rapid anesthetic/analgesic effects with appropriated lasting effects. Nevertheless, the research keeps ongoing trying to enhance the anesthetic effect and pain management while reducing the occurrence of undesirable side effects and toxicity. In this context, new delivery devices and administration techniques are being studied (Saxena et al., 2013). There are being designed both organic and inorganic nanoparticles formulations with different LAs with the aim to increase their action duration and reduce their adverse reactions such as cardiotoxicity and neurotoxicity (Andreu and Arruebo, 2018). As example, in vitro and in vivo studies have demonstrated that lipid-based systems increase bioavailability of the LAs ropivacaine, prilocaine, and lidocaine (Cereda et al., 2004; Cereda et al., 2006). However, the clinical use of all these carrier systems with LAs is very limited and focuses mainly on the field of dentistry. The results available to date that evaluate the effectiveness of LAs are very varied, so more clinical trials are required to confirm the in vitro and in vivo results (Franz-Montan et al., 2012; Tofoli et al., 2011). Likewise, the development and study of distinct isomers of LAs are relevant in determining the pharmacological profile of these drugs, helping to better understand their mechanisms of action, efficacy, and safety properties (Mitra and Chopra, 2011). Furthermore, apart from the classical and main mechanism of action of LAs blocking the VGSCs, the potential contribution of other mechanisms involved in mediating either the therapeutic and/or the toxic effects of LAs, remains to be elucidated. Finally, the antimicrobial and anti-inflammatory properties which with LAs seem to be endowed and their potential contribution to the overall outcome of interventions involving the use of LAs are new grounds for future research. Regarding their anti-inflammatory properties, LAs have demonstrated to reduce inflammation that accompanies various diseases such as ulcerative proctitis, burn injuries, and arthritis by acting on various steps of the inflammatory cascade (leucocyte adhesion and migration, phagocytosis, and inflammatory mediators). There have been even demonstrated that the potential anti-inflammatory effect of LAs is higher than that of nonsteroidal anti-inflammatory drugs (NSAIDs) (Cassuto et al., 2016). On the other hand, several studies have showed the antimicrobial properties of LAs (antibacterial, antiviral, and antifungal effects). The antibacterial action against a wide range of bacteria including *P. aeruginosa, E. coli, H. influenza, and M. tuberculosis* may be related to the interaction of LAs with bacterial wall by alternating structural proteins on membrane and membrane fluidity. Moreover, there have been demonstrated that LAs have antiviral effects against herpes simplex virus by inhibiting ATPase, release lysozymes, and free radicals as well as antifungal effects against *Candida albicans* by blocking calcium channels (Cassuto et al., 2016). Given the increasing antimicrobial resistance of the usual antibiotics, it would be of great interest to continue investigating the role of LAs as agents against infections of viruses, bacteria, and fungi.

## 15.10 CONCLUSION

The major development of LAs has been formed on amide type because of their greater physical and chemical stability and their longer duration of action. LAs act as primary target on VGSCs blockage. However, LAs are also able to act on other targets which include voltage-gated $K^+$ channels, voltage-gated $Ca^{2+}$ channels, and $Ca^{2+}$-ATPase pump in the sarcoplasmic reticulum, among others. This could explain other therapeutically effects attributed to LAs and their adverse reactions. New roles of LAs have been developed in the last recent years and they include antimicrobial and anti-inflammatory properties. Regarding adverse reactions, LAs can cause systemic toxicity attributed to their vasodilators properties by increasing blood flow on the site of action. LAs are usually associated with adjuvants such as epinephrine to reduce adverse reactions, to reduce local anesthetics doses, and to prolong anesthetic effect duration. Currently, the main lines of research with LAs are focused on finding new adjuvants and new biological targets.

## KEYWORDS

- local anesthesia
- action potential
- voltage-gated sodium channels
- amino esters local anesthetics
- coadjuvant drugs
- amino amides local anesthetics

## REFERENCES

AEMPS. Agencia Española del Medicamento y Productos Sanitarios. CIMA: Centro de Información Online de Medicamentos. https://www.aemps.gob.es/cima/publico/home.html.

Alster, T. S. The Lidocaine/Tetracaine Peel: A Novel Topical Anesthetic for Dermatologic Procedures in Adult Patients. *Dermatol. Surg.* **2007,** *33* (9), 1073–1081.

Andreu, V.; Arruebo, M. Current Progress and Challenges of Nanoparticle-based Therapeutics in Pain Management. *J. Control. Rel.* **2018,** *269,* 189–213.

Atkinson, W. S. Use of Hyaluronidase with Local Anesthesia in Ophthalmology; Preliminary Report. *Arch. Ophthal.* **1949,** *42* (5), 628–633.

Bagneris, C.; Naylor, C. E.; McCusker, E. C.; Wallace, B. A. Structural Model of the Open-closed-inactivated Cycle of Prokaryotic Voltage-gated Sodium Channels. *J. Gen. Physiol.* **2015,** *145,* 5–16.

Bailard, N. S.; Ortiz, J.; Flores, R. A. Additives to Local Anesthetics for Peripheral Nerve Blocks: Evidence, Limitations, and Recommendations. *Am. J. Health Syst. Pharm.* **2014,** *71* (5), 373–385.

Barash, M.; Reich, K. A.; Rademaker, D. Lidocaine-induced Methemoglobinemia: A Clinical Reminder. *J. Am. Osteopath. Assoc.* **2015,** *115* (2), 94–98.

Becker, D. E.; Reed, K. L. Essentials of Local Anesthetic Pharmacology. *Anesth. Prog.* **2006,** *53,* 98–108; 109–110.

Beiranvand, S.; Eatemadi, A.; Karimi, A. New Updates Pertaining to Drug Delivery of Local Anesthetics in Particular Bupivacaine Using Lipid Nanoparticles. *Nanoscale Res. Lett.* **2016,** *11* (1), 307.

Berde, C. B.; Jaksic, T.; Lynn, A. M.; Maxwell, L. G.; Soriano, S. G.; Tibboel, D. Anesthesia and Analgesia During and After Surgery in Neonates. *Clin. Ther.* **2005,** *27,* 900–921.

Berkman, S.; MacGregor, J.; Alster, T. Adverse Effects of Topical Anesthetics for Dermatologic Procedures. *Expert Opin. Drug Saf.* **2012,** *11* (3), 415–423.

Best, C. A.; Best, A. A.; Best, T. J.; Hamilton, D. A. Buffered Lidocaine and Bupivacaine Mixture—The Ideal Local Anesthetic Solution? *Plast. Surg. (Oakv).* **2015,** *23* (2), 87–90.

Bourne, E.; Wright, C.; Royse, C. A Review of Local Anesthetic Cardiotoxicity and Treatment with Lipid Emulsion. *Local Reg. Anesth.* **2010,** *3,* 11–19.

Bowman, R. J.; Newman, D. K.; Richardson, E. C.; Callear, A. B.; Flanagan, D. W. Is Hyaluronidase Helpful for Peribulbar Anaesthesia? *Eye (Lond).* **1997,** *11* (Pt 3), 385–388.

Brandt, R. G.; Anderson, P. F.; McDonald, N. J.; Sohn, W.; Peters, M. C. The Pulpal Anaesthetic Efficacy of Articaine Versus Lidocaine in Dentistry: A Meta-analysis. *J. Am. Dent. Assoc.* **2011,** *142,* 493–504.

Brockmann, W. G. Mepivacaine: A Closer Look at Its Properties and Current Utility. *Gen. Dent.* **2014,** *62* (6), 70–75.

Brummett, C. M.; Williams, B. A. Additives to Local Anesthetics for Peripheral Nerve Blockade. *Int. Anesthesiol. Clin.* **2011,** *49* (4), 104–116.

Brun, A. Effect of Procaine, Carbocain and Xylocaine on Cutaneous Muscle in Rabbits and Mice. *Acta Anaesthesiol Scand.* **1959,** *3* (2), 59–73.

Burlacu, C. L.; Buggy, D. J. Update on Local Anesthetics: Focus on Levobupivacaine. *Ther. Cin, Risk Mang.* **2008,** *4* (2), 381–392.

Butterworth, J. F. Strichartz, G. R. Molecular Mechanisms of Local Anesthesia: A Review. *Anesthesiology* **1990,** *72* (4), 711–734.

Cassuto, J.; Sinclair, R.; Bonderovic, M. Anti-inflammatory Properties of Local Anesthetics and Their Present and Potential Clinical Implications. *Acta Anaesthesiol. Scand.* **2006,** *50,* 265–282.

Catterall, W. A.; Swanson, T. M. Structural Basis for Pharmacology of Voltage-gated Sodium and Calcium Channels. *Mol. Pharmacol.* **2015,** *88,* 141–150.

Cereda, C.M.S.; Brunetto G. B.; de Araújo, D. R.; de Paula, E. Liposomal Formulations of Prilocaine, Lidocaine and Mepivacaine Prolong Analgesic Duration. *Can. J. Anaesth.* **2006,** *53,* 1092–1097.

Cereda, C. M. S.; de Araújo, D. R.; Brunetto G. B.; de Paula, E. Liposomal Prilocaine: Preparation, Characterization, and In Vivo Evaluation. *J. Pharm. Pharm. Sci.* **2004,** *7,* 235–240.

Chernoff, D. M.; Strichartz, G. R. Kinetics of Local Anesthetic Inhibition of Neuronal Sodium Currents. pH and Hydrophobicity Dependence. *Biophys. J.* **1990,** *58,* 69–81.

Christie, L. E.; Picard, J.; Weinberg, G. L. Local Anaesthetic Systemic Toxicity. *BJA Education* **2015,** *15* (3), 136–142.

Clairfeuille, T.; Xu, H.; Koth, C. M.; Payandeh, J. Voltage-gated Sodium Channels Viewed Through a Structural Biology Lens. *Curr. Opin. Struct. Biol.* **2017,** *45,* 74–84.

Collinsworth, K. A.; Kalman, S. M.; Harrison, D. C. The Clinical Pharmacology of Lidocaine as an Antiarrhythmic Drug. *Circulation* **1974,** *50* (6), 1217–1230.

Cooper, H. A. Methemoglobinemia caused by benzocaine topical spray. *South Med. J.* **1997,** *90* (9), 946–948.

Davies, P. S.; Galer, B. S. Review of Lidocaine Patch 5% Studies in the Treatment of Postherpetic Neuralgia. *Drugs* **2004,** *64* (9), 937–947.

Davies, T.; Karanovic, S.; Shergill, B. Essential Regional Nerve Blocks for the Dermatologist: Part 1. *Clin. Exp. Dermatol.* **2014,** *39* (8), 861–867.

Daykin, H. The efficacy and safety of intravenous lidocaine for analgesia in the older adult: a literature review. *Br. J. Pain.* **2017,** *11*(1), 23-31.

Deupree, J. Etidocaine. *xPharm: The Comprehensive Pharmacology Reference.* 2007, 1–7.

Dipenaar, J. M. Local Anaesthetic Toxicity. *South Afr. J. Anaesth. Analg.* **2007,** *13* (3), 23–28.

Drasner, K. Models for Local Anesthetic Toxicity from Continuous Spinal Anesthesia. *Reg. Anesth.* **1993,** *18* (6), 434–438.

Duclohier, H. Structure-function Studies on the Voltage-gated Sodium Channel. *Biochim. Biophys. Acta.* **2009,** *1788*, 2374–2379.

Efimova, S. S.; Zakharova, A. A.; Schagina, L. V.; Ostroumova, O. S. Local Anesthetics Affect Gramicidin A Channels via Membrane Electrostatic Potentials. *J. Membr. Biol.* **2016,** *249*, 781–787.

Elamin, B. Dibucaine Inhibition of Serum Cholinesterase. *J. Biochem. Mol. Biol.* **2003,** *36*, 149–153.

Foster, R. H.; Markham, A. Levobupivacaine: A Review of Its Pharmacology and Use as a Local Anaesthetic. *Drugs* **2000,** *59* (3), 551–579.

Franz-Montan, M., de Paula, E., Groppo, F. C., Ranali, J., Volpato, M. C. Efficacy of Liposome-encapsulated 0.5% Ropivacaine in Maxillary Dental Anaesthesia. *Br. J. Oral Maxillofac. Surg.* **2012,** *50*, 454–458.

Frazier, D. T., Narahashi, T.; Yamada, M. The Site of Action and Active Form of Local Anesthetics. II. Experiments with Quaternary Compounds. *J. Pharmacol. Exp. Ther.* **1970,** *171*, 45–51.

Freedman, J. M.; Li, D. K.; Drasner, K.; Jaskela, M. C.; Larsen, B.; Wi, S. Transient Neurologic Symptoms After Spinal Anesthesia: An Epidemiological Study of 1863 Patients. *Anesthesiology* **1998,** *89*, 633 –641.

Goldin, A. L. Mechanisms of Sodium Channel Inactivation. *Curr. Opin. Neurobiol.* **2003,** *13*, 284–290.

Goyal, R.; Shukla, R. N. Local Anesthetic Systemic Toxicity (LAST)—Should We Not Be Concerned? *MJAFI* **2012,** *68*, 371–375.

Haas, D. A. Articaine and Paresthesia: Epidemiological Studies. *J. Am. Coll. Dent.* **2006,** *73* (3), 5–10.

Hanna, M. N.; Elhassan, A.; Veloso, P. M.; Lesley, M.; Lissauer, J.; Richman, J. M.; Wu C. L.. Efficacy of Bicarbonate in Decreasing Pain on Intradermal Injection of Local Anesthetics: A Meta-analysis. *Reg. Anesth. Pain Med.* 2009, *34* (2), 122–125.

Hartman, N.; Zhou, H.; Mao, J.; Mans, D.; Boyne, M.; Patel, V.; Colatsky, T. Characterization of the Methemoglobin Forming Metabolites of Benzocaine and Lidocaine. *Xenobiotica* **2017,** *47* (5), 431–438.

Higuchi, R.; Fukami, T.; Nakajima, M.; Yokoi, T. Prilocaine- and Lidocaine-induced Methemoglobinemia is Caused by Human Carboxylesterase-, CYP2E1-, and CYP3A4-mediated Metabolic Activation. *Drug Metab. Dispos.* **2013,** *41*, 1220–1230.

Hollmann, M. W.; Durieux, M. E. Local Anesthetics and the Inflammatory Response: A New Therapeutic Indication? *Anesthesiology* **2000,** *93*, 858–875.

Hopman, A. J. G.; Baart, J. A.; Brand, H. S. Articaine and Neurotoxicity—A Review. *Br. Dent. J.* **2017,** *223* (7), 501–506.

Hu, X.; Li, J.; Zhou, R.; Wang, Q.; Xia, F.; Halaszynski, T.; Xu, X. Dexmedetomidine Added to Local Anesthetic Mixture of Lidocaine and Ropivacaine Enhances Onset and Prolongs Duration of a Popliteal Approach to Sciatic Nerve Blockade. *Clin. Ther.* **2017,** *39* (1), 89–97.

Jenkins, M. G.; Murphy, D. J.; Little, C.; McDonald, J.; McCarron, P. A. A Non-inferiority Randomized Controlled Trial Comparing the Clinical Effectiveness of Anesthesia Obtained by Application of a Novel Topical Anesthetic Putty with the Infiltration of Lidocaine for the Treatment of Lacerations in the Emergency Department. *Ann. Emerg. Med.* **2014,** *63* (6), 704–710.

Johnson, M. E. Potential Neurotoxicity of Spinal Anesthesia with Lidocaine. *Mayo Clin. Proc.* **2000,** *75* (9), 921–932.

Johnson, S. M.; Saint John, B. E.; Dine, A. P. Local Anesthetics as Antimicrobial Agents: A Review. *Surg. Infect (Larchmt).* **2008,** *9*, 205–213.

Leone, S.; Di Cianni, S.; Casati, A.; Fanelli, G. Pharmacology, Toxicology, and Clinical Use of New Long Acting Local Anesthetics, Ropivacaine and Levobupivacaine. *Acta Biomed.* **2008,** *79* (2), 92–105.

Li, M.; Wan, L.; Mei, W.; Tian, Y. Update on the Clinical Utility and Practical Use of Ropivacaine in Chinese Patients. *Drug Des. Devel. Ther.* **2014,** *8*, 1269–1276.

Lipkind, G. M.; Fozzard, H. A. Molecular Modeling of Local Anesthetic Drug Binding by Voltage-gated Sodium Channels. *Mol. Pharmacol.* **2005,** *68*, 1611–1622.

Maimo, G., Redick, E. Recognizing and Treating Methemoglobinemia: a Rare But Dangerous Complication of Topical Anesthetic or Nitrate Overdose. *Dimens. Crit. Care Nurs.* **2004,** *23*, 116–118.

Malamed, S. *Drug Oversode Reactions in Medical Emergencies in the Dental Office*; Mosby, 7th ed, 2515. https://doi.org/10.1016/B978-0-323-17122-9.09987-X.

Malamed, S. F.; Gagnon, S.; Leblanc, D. Articaine Hydrochloride: A Study of the Safety of A New Amide Local Anaesthetic. *J. Am. Dent. Assoc.* **2001**, *132*, 177–185.

Malik, O.; Kaye, A. D.; Kaye, A.; Belani, K.; Urman, R. D. Emerging Roles of Liposomal Bupivacaine in Anesthesia Practice. *J. Anaesthesiol. Clin. Pharmacol.* **2017**, *33* (2), 151–156.

Manassero, A.; Fanelli, A. Prilocaine Hydrochloride 2% Hyperbaric Solution for Intrathecal Injection: A Clinical Review. *Local Reg. Anesth.* **2017**, *10*, 15–24.

Mitra, S.; Chopra, P. Chirality and anaesthetic drugs: A review and an update. *Indian J. Anaesth.* **2011**, *55*(6), 556-562.

Morales-Lazaro, S. L.; Hernandez-Garcia, E.; Serrano-Flores, B.; Rosenbaum, T. Organic Toxins as Tools to Understand Ion Channel Mechanisms and Structure. *Curr. Top. Med. Chem.* **2015**, *15*, 581–603.

Nair, A. S. Local Anaesthetic Myotoxicity Due to Fascial Plane Blocks: A Brief Review. *Glob. J. Anesth.* **2017**, *4* (1), 001–003.

Narahashi, T.; Frazier, T.; Yamada, M. The Site Of Action And Active Form Of Local Anesthetics. I. Theory and pH Experiments with Tertiary Compounds. *J. Pharmacol. Exp. Ther.* **1970**, *171*, 32–44.

Neal, J. M. Effects of Epinephrine in Local Anesthetics on the Central and Peripheral Nervous Systems: Neurotoxicity and Neural Blood Flow. *Reg. Anesth. Pain Med.* **2003**, *28* (2), 124–134.

Neal, J. M.; Bernards, C. M.; Butterworth, J. F.; Di Gregorio, G.; Drasner, K.; Hejtmanek, M. R.; Mulroy, M. F., Rosenquist, R. W.; Weinberg, G. L. ASRA Practice Advisory On Local Anesthetic Systemic Toxicity. *Reg. Anesth. Pain Med.* **2010**, *35* (2), 152–161.

Neal, J. M.; Woodward, C. M.; Harrison, T. K. The American Society of Regional Anesthesia and Pain Medicine Checklist for Managing Local Anesthetic Systemic Toxicity: 2017 Version. *Reg. Anesth. Pain Med.* **2018**, *43* (2), 150–153.

Nieto, F. R.; Cobos, E. J.; Tejada, M. A.; Sanchez-Fernandez, C.; Gonzalez-Cano, R.; Cendan, C. M. Tetrodotoxin (TTX) as a Therapeutic Agent For Pain. *Mar Drugs* **2012**, *10*, 281–305.

O'Brien, L.; Taddio, A.; Lyszkiewicz, D. A.; Koren, G. A Critical Review of the Topical Local Anesthetic Amethocaine (Ametop) For Pediatric Pain. *Paediatr. Drugs* **2005**, *7* (1), 41–54.

Parnas, M. L.; Procter, M.; Schwarz, M. A.; Mao, R.; Grenache, D. G. Concordance of Butyrylcholinesterase Phenotype With Genotype: Implications For Biochemical Reporting. *Am. J. Clin. Pathol.* **2011**, *135*, 271–276.

Patino, G. A.; Isom, L. L. Electrophysiology and Beyond: Multiple Roles Of Na+ Channel Beta Subunits In Development And Disease. *Neurosci. Lett.* **2010**, *486*, 53–59.

Payandeh, J.; Scheuer, T.; Zheng, N.; Catterall, W. A. The Crystal Structure of a Voltage-Gated Sodium Channel. *Nature* **2011**, *475*, 353–358.

Peterfreund, R. A.; Datta, S.; Ostheimer, G. W. pH Adjustment Of Local Anesthetic Solutions With Sodium Bicarbonate: Laboratory Evaluation Of Alkalinization And Precipitation. *Reg. Anesth.* **1989**, *14* (6), 265–270.

Pollock, J. E.; Liu, S. S.; Neal, J. M.; Stephenson, C. A. Dilution of Spinal Lidocaine Does Not Alter The Incidence Of Transient Neurologic Symptoms. *Anesthesiology* **1999**, *90*, 445–450.

Press, C. D. Infiltrative Administration of Local Anesthetic Agents. In Medscape Clinical Procedures, https://emedicine.medscape.com/article/149178-overview.

Punnia-Moorthy, A. A Study of the Effectiveness of Dental 2% Lignocaine Local Anaesthetic Solution at Different pH Values. *Br. Dent. J.* **1987**, *163*, 314.

Rawal, N. Analgesia for Day-case Surgery. *Br. J. Anaesth.* **2001**, *87* (1), 73–87.

Reina, M. A.; López, A.; de Andrés, J. A. Hypothesis Concerning the Anatomical Basis of Cauda Equina Syndrome and Transient Nerve Root Irritation After Spinal Anesthesia. *Rev. Esp. Anestesiol. Reanim.* **1999**, *46* (3), 99–105.

Richardson, J.; Lönnqvist, P. A.; Naja Z. Bilateral Thoracic Paravertebral Block: Potential and Practice. *Br. J. Anaesth.* **2011**, *106* (2), 164–171.

Rimback, G.; Cassuto, J.; Wallin, G.; Westlander, G. Inhibition of Peritonitis by Amide Local Anesthetics. *Anesthesiology* **1988**, *69*, 881–886.

Roberts, J. E.; MacLeod, B. A.; Hollands, R. H. Improved Peribulbar Anaesthesia with Alkalinization and Hyaluronidase. *Can. J. Anaesth.* **1993**, *40* (9), 835–838.

Rood, J. P. Etidocaine in Dentistry. *Anesth. Prog.* **1989**, *36* (4–5), 190–191.

Sanchez, G. A.; Takara, D.; Alonso, G. L. Local Anesthetics Inhibit Ca-ATPase in Masticatory Muscles. *J. Dent. Res.* **2010**, *89*, 372–377.

Sanford, M.; Keating, G. M. Levobupivacaine: A Review of Its Use in Regional Anaesthesia and Pain Management. *Drugs.* **2010**, *70* (6), 761–791.

Saxena, P.; Gupta, S. K.; Newaskar, V.; Chandra, A. Advances in Dental Local Anesthesia Techniques And Devices: An Update. *Natl. J. Maxillofac. Surg.* **2013,** *4* (1), 19–24.

Schwartz, D. R.; Kaufman, B. Local Anesthetics. In: *Goldfrank's Toxicologic Emergencies*; Hoffman, R. S., Howland, M. A., Lewin, N. A., Nelson, L. S., Goldfrank, L. S., Eds; 10e, New York: McGraw-Hill Education, 2015.

Sekimoto, K.; Tobe, M.; Saito, S. Local Anesthetic Toxicity: Acute and Chronic Management. *Acute Med. Surg.* **2017,** *4* (2), 152–160.

Sheets, M. F.; Fozzard, H. A.; Lipkind, G. M.; Hanck, D. A. Sodium Channel Molecular Conformations and Antiarrhythmic Drug Affinity. *Trends Cardiovasc. Med.* **2010,** *20,* 16–21.

Singletary, E. M.; Charlton, N. P.; Epstein, J. L.; Ferguson, J. D.; Jensen, J. L.; MacPherson, A. I.; Pellegrino, J. L.; Smith, W. W.; Swain, J. M.; Lojero-Wheatley, L. F.; Zideman, D. A. Part 15: First Aid: 2015 American Heart Association and American Red Cross Guidelines Update for First Aid. *Circulation* **2015,** *132* (18 Suppl 2), S574–S589.

Sinnott, C. J.; Garfield, J. M.; Thalhammer, J. G.; Strichartz, G. R. Addition of Sodium Bicarbonate to Lidocaine Decreases the Duration of Peripheral Nerve Block in the Rat. *Anesthesiology* **2000,** *93* (4), 1045–1052.

Sisk, A. L. Long-acting Local Anesthetics in Dentistry. *Anesth. Prog.* **1992,** *39* (3), 53–60.

Stancil, S. A. A Bier Block Implementation Protocol. *US Army Med. Dep. J.* **2014,** 53–56.

Strichartz, G. R.; Sanchez, V.; Arthur, G. R.; Chafetz, R.; Martin, D. Fundamental Properties of Local Anesthetics. II. Measured Octanol: Buffer Partition Coefficients and pKa Values of Clinically Used Drugs. *Anesth. Analg.* **1990,** *71,* 158–170.

Strobel, G. E.; Bianchi, C. P. The effects of pH Gradients on the Action of Procaine and Lidocaine in Intact and Desheathed Sciatic Nerves. *J. Pharmacol. Exp. Ther.* **1970a,** *172,* 1–17.

Strobel, G. E.; Bianchi, C. P. The Effects of pH Gradients on the Uptake and Distribution of C14-Procaine and Lidocaine in Intact and Desheathed Sciatic Nerve Trunks. *J. Pharmacol. Exp. Ther.* **1970b,** *172,* 18–32.

Taddio, A.; Gurguis, M. G.; Koren, G. Lidocaine-prilocaine Cream Versus Tetracaine Gel for Procedural Pain in Children. *Ann. Pharmacother.* **2002,** *36* (4), 687–692.

Taniguchi, T.; Shibata, K.; Yamamoto, K.; Kobayashi, T.; Saito, K.; Nakanuma, Y. Lidocaine Attenuates the Hypotensive and Inflammatory Responses to Endotoxemia in Rabbits. *Crit. Care Med.* **1996,** *24,* 642–646.

Tofoli, G. R.; Cereda, C. M. S.; Groppo, F. C.; Volpato, M. C.; Franz-Montan, M.; Ranali, J.; de Araújo, D. R.; de Paula, E. Efficacy of Liposome-encapsulated Mepivacaine for Infiltrative Anesthesia in Volunteers. *J. Liposome Res.* **2011,** *21,* 88–94.

Tsuchiya, H.; Mizogami, M. R(+)-, Rac-, and S(-)-bupivacaine Stereostructure-specifically Interact with Membrane Lipids at Cardiotoxically Relevant Concentrations. *Anesth. Analg.* **2012,** *114,* 310–312.

Tsuchiya, H.; Ueno, T.; Mizogami, M. Stereostructure-based Differences in the Interactions of Cardiotoxic Local Anesthetics with Cholesterol-containing Biomimetic Membranes. *Bioorg. Med. Chem.* **2011,** *19,* 3410–3415.

Udelsmann, A.; Dreyer, E.; Melo, MdS.; Bonfim, M. R.; Borsoi, L. F. A.; Oliveira, T. G. Lipids in Local Anesthetic Toxicity. *Arq. Bras. Cir. Dig.* **2012,** *25* (3), 169–172

Ulbricht, W. Sodium Channel Inactivation: Molecular Determinants and Modulation. *Physiol. Rev.* **2005,** *85,* 1271–1301.

Valverde, A.; Gunkelt, C.; Doherty, T. J.; Giguere, S.; Pollak, A. S. Effect of a Constant Rate Infusion of Lidocaine on the Quality of Recovery From Sevoflurane or Isoflurane General Anaesthesia in Horses. *Equine Vet. J.* **2005,** *37,* 559–564.

Van Rooyen, H. Local Anaesthetic Agent Toxicity. *South Afr. J. Anaesth. Analg.* **2010,** *16* (1), 83–88.

Veneziano, G.; Tobias J. D. Chloroprocaine for Epidural Anesthesia in Infants and Children. *Paediatr. Anaesth.* **2017,** *27* (6), 581–590.

Wang, J. S.; Backman, J. T.; Taavitsainen, P.; Neuvonen, P. J.; Kivistö, K. T. Involvement of CYP1A2 and CYP3A4 in Lidocaine N-Deethylation and 3-Hydroxylation in Humans. *Drug Metab. Dispos.* **2000,** *28* (8), 959–965.

Warren, V. T.; Fisher, A. G.; Rivera, E. M.; Saha, P. T.; Turner, B.; Reside, G.; Phillips, C.; White, R. P. Jr. Buffered 1% Lidocaine With Epinephrine Is as Effective as Non-buffered 2% Lidocaine With Epinephrine for Mandibular Nerve Block. *J. Oral Maxillofac. Surg.* **2017,** *75* (7), 1363–1366.

White, S. M. GMC Guidance Upheld. *Anaesthesia* **2005,** *60,* 1248.

WHO Model List of Essential Medicines (20th List). World Health Organization. June 2017.

Yapp, K. E.; Hopcraft, M. S.; Parashos, P. Articaine: A Review of the Literature. *Br. Dent. J.* **2011,** *210* (7), 323–329.

Yanagidate, F.; Strichartz, G. R. Bupivacaine Inhibits Activation of Neuronal Spinal Extracellular Receptor-activated Kinase Through Selective Effects on Ionotropic Receptors. *Anesthesiology* **2006,** *104,* 805–814.

Yanagidate, F.; Strichartz, G. R. Local anesthetics. *Handb. Exp. Pharmacol.* **2007,** 95–127.

Yousefshahi, F.; Predescu, O.; Francisco Asenjo, J. The Efficacy of Systemic Lidocaine in the Management of Chronic Pain: A Literature Review. *Anesth. Pain Med.* **2017,** *7*(3), e44732.

Zaric, D.; Christiansen, C.; Pace, N. L.; Punjasawadwong, Y. Transient Neurologic Symptoms (TNS) Following Spinal Anaesthesia with Lidocaine Versus Other Local Anaesthetics. *Cochrane Database Syst. Rev.* **2003,** *2,* CD003006.

# CHAPTER 16

# General Anesthetics

AMAN UPAGANLAWAR*, ABDULLA SHERIKAR, and CHANDRASHEKHAR UPASANI

*SNJB's SSDJ College of Pharmacy, Maharashtra, India*

*Corresponding author. E-mail: amanrxy@gmail.com*

## ABSTRACT

In the 19th century, the usefulness of general anesthetics such as diethyl ether and chloroform (as inhalational vapors) as a therapeutic tool came into picture. In addition to general anesthetics, there are other drugs which have the capacity to obstruct the conduction mechanism in sensory as well as motor neurons. The examples of such drugs include skeletal muscle relaxants, anticonvulsants, and CNS depressants. In order to produce balanced anesthesia, that is, controlled muscle relaxation, sedation, and pain management with minimal adverse effects, the combination therapy has been employed. The anesthetic agents have wide range of structural diversity which permits the physician to standardize and optimize the practice of patient specific anesthetic therapy. In case of major surgical procedure, the combination therapy of general anesthetics with other classes of adjunctive medications followed in order to reduce the possible clinical toxicity of anesthetics. While doing surgical procedure, the general anesthetics should minimize the possible adverse reaction and maintain the acceptable level of physiological condition. It is expected that general anesthetic agents should reduce postoperative complications related to the surgical procedure. This chapter is organized in order to give full pharmacology of general anesthetic agents.

## 16.1 INTRODUCTION

Anesthesia is nothing but a state of failure of feelings with or with no failure of consciousness. Although differences in the chemical structure and properties, the general anesthetics are able to produce the balanced state of anesthesia. Because of the reduction in central nervous system (CNS) property to an acceptable level, nowadays general anesthetics are routinely used for wide range of surgical and medical procedures (Brunton et al., 2011; Rang et al., 2011). Earlier in 1864, William Morton at Massachusetts General Hospital used the diethyl ether as anesthetic agents for experimental surgical procedure (Georg and Craig, 2010).

Elementarily general anesthetic causes differed rate of analgesia, loss of memory, loss of all feelings, reflexes and perception, relaxation of muscle with partial loss of respiratory and cardiovascular functions. A model anesthetic drug should provoke the unconsciousness smoothly and rapidly and should provide recovery also after its discontinuation (Brunton et al., 2011; Rang et al., 2011). No single agent is able to achieve all these desirable effect without risk of toxicity. Therefore, in anesthetic practice a number of different categories of drug are utilized concomitantly to produce a "balanced anesthesia" with respect to their both beneficial and minimized toxicity producing effects. While performing the surgery, it is general practice to use the pre-anesthetic medication followed by administration of intravenous thiopental or propofol for rapid and soft initiation of anesthesia. In addition to this therapy, combinations of inhaled and intravenous anesthetics agents are administering to maintain the balanced state of anesthesia. There are other drugs such as neuromuscular blockers which facilitate intubation and suppress muscle tone. In case of minor surgery, the low doses of general anesthetics along with local anesthetics are given. This technique is called as "monitored anesthesia" where distinct analgesia is achieved with no loss of respiratory functions as well as allowing the verbal communication of the patient (Katzung et al., 2009; Sharma and Sharma, 2017).

The ideally general anesthetic agents must produce controlled state of anesthesia and their selection and route of ingestion will be depending on pharmacokinetic profile. The common properties of general anesthetic agents include loss of all sensations, sleep, that is, unconsciousness and amnesia, abolition of somatic and autonomic reflexes, low therapeutic index, production of similar anesthetic effects and dissimilar side effects, that is, secondary effects on other body organ (Brunton et al., 2011; Rang et al., 2011). In the present book chapter, the authors tried to focus on the interplay of anesthesia and body as well as the pharmacotherapy of general anesthetic medications.

## 16.2 UNIVERSAL IDEOLOGY OF SURGICAL ANESTHESIA

There are number of examples of drugs are available with anesthetic property but very few are successfully clear the test of the requirement of surgical anesthesia. Therapeutically anesthetics agent should be harmless to the body, without altering the normal body physiology and with minimized toxicity producing ability (Alice, 1949). Fundamental, surgical anesthetic agents should produce following effects (Brunton et al., 2011):

- Reduction in direct and indirect adverse effects.
- Maintain controlled body physiological balance during surgical procedures (such as blood function, ischemia and reperfusion of tissue, and body response to a cold environment).
- Minimize possible postoperative effects.

## 16.3 STAGES OF GENERAL ANESTHESIA

The fundamental function of general anesthetic is depending on their ability to cause the failure of sensation and development of analgesia ability. The suppression of neurons in the specific area of the brain such as cortical region and inhibitory/excitatory stellate cells causes the development of anesthetic state. Guedel in 1883 observed and describe the four stages of general anesthetics (Fig. 16.1).

## 16.4 PHARMACOLOGICAL EFFECTS OF GENERAL ANESTHETICS

### 16.4.1 HEMODYNAMIC EFFECTS

Being as a undeviating vasodilators, depressant of myocardium, dull controller of baroreceptor, and depressors of central sympathetic tempo, an intravenous and inhalation ingestion of general anesthetics produces varying degree of hypertensive effects. The sympathetic outflow controls the intravascular volume; the traumatic and normal individuals are characterized by decreased volume of intravascular volume therefore the general anesthetics such as ketamine and etomidate are cautiously used in these patients (Sellgren et al., 1990).

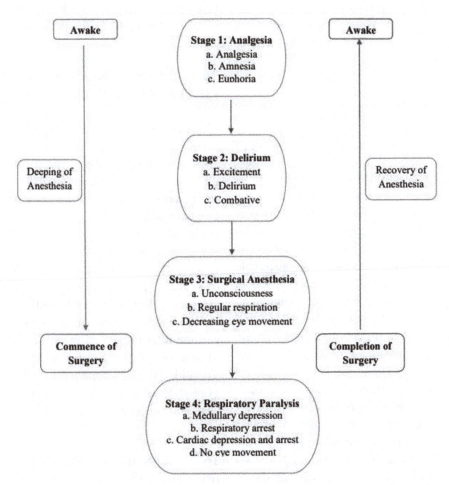

**FIGURE 16.1**  Different stages of the general anesthesia. Where, Stage 1: Characterized by inhibition of spinothalamic tract, absence of memory loss, presence of consciousness and touch reflex and exaggerated hearing senses. Stage 2: Characterized by amnesic state as well as presence of violent argumentative behavior and irregular respiratory and blood pressure rate. Stage 3: Relaxation of skeletal muscle and regular respiration occurs in this state. Additionally, it is divided in four planes. Plane 1—no roving movement of eyeballs, skeletal muscle tone and respiration are normal. Plane 2—most surgical processes are performed in this plane. Plane 3—respiration is mainly abdominal, marked muscle relaxation, pupils are dilated. Plane 4—plane of complete muscle relaxation. Respiration is only abdominal, complete loss of light, corneal and laryngeal reflexes. Stage 4: Characterized by chronic suppression of the respiratory and vasomotor center in the medulla. This stage is fatal and death can guarantee.

*Source*: Sharma and Sharma, 2017.

### 16.4.2   RESPIRATORY EFFECTS

The general anesthetics causes' reduction in ventilator force and the reflexes which require controlled ventilation during surgery. In addition to this, the general anesthetics are also known to cause failure of gag reflex, dullness of cough stimuli, decline in lower esophageal sphincter tone, and growth of passive and active regurgitation (Brunton et al., 2011; Tripathi, 2013).

### 16.4.3   HYPOTHERMIA

General anesthetic frequently develops hypothermia (body temperature <36°C) during surgery due to various causes such as low down ambient temperature, showing body parts to the outer environment, intravenous administration of cold fluids, decreased metabolic rate, and altered temperature regulation (Sessler, 2000). The condition such as hypothermia may precipitate the condition such as change in cardiac function, wound infection, and damaged coagulation (Franks et al., 1998; Kurz et al., 1996). During anesthesia, in order to put core body temperature, the various measures are used such as lukewarm intravenous fluids, circuit of heat exchangers anesthesia, forced-warm-air covers, and water-filled garments with microprocessor feedback. At the time of induction of anesthesia neuromuscular blockers are used usually to make possible the smooth airway endotracheal intubation, to reduce the risk of cough or gagging during laryngoscopy, and to decrease the risk of aspiration before endotracheal tubing (Brunton et al., 2011; Rang et al., 2011).

### 16.4.4   NAUSEA AND VOMITING

The different chemical mediators such as serotonin, histamine, dopamine, and acetylcholine the activity of chemoreceptor trigger zone of vomiting. The postoperative use 5 HT-3 receptor antagonist such as ondansetron, droperidol, metoclopramide, and dexamethasone causes the inhibition of nausea and vomiting due to stimulation of chemoreceptor trigger zone and brainstem by general anesthetics (Brunton et al., 2011; Tripathi, 2013; Sharma, 2017).

### 16.4.5   OTHER EMERGENCE AND POSTOPERATIVE PHENOMENA

The general anesthetic causes precipitation of myocardial ischemic condition in patients with coronary artery diseases. The use these agents are also associated with 5–30% development of increase in heart rate, restlessness, crying, moaning and thrashing, and other neurological symptoms. The use of meperidine (12.5 mg) is useful in controlling the post anesthesia shivering due to general anesthetic. The reduction of pulmonary function and hypoxemia are characteristic effects of general anesthetics (Brunton et al., 2011). The pain due to the use of general anesthetic is prevented by the intravenous use of ketorolac (30–60 mg; Brunton et al., 2011; Tripathi, 2013, Sharma, 2017). The summary of pharmacological effects produced by general anesthetics is shown in Table 16.1.

### 16.5   CLASSIFICATION OF GENERAL ANESTHETICS

The general anesthetic agents are mainly categorized into three subcategories such as inhalation anesthesia, intravenous anesthesia, and adjuvant anesthesia. The general classification of the general anesthetics is depending on their route of ingestion and anesthesia inducing capacity (Table 16.2; Brunton et al., 2011; Rang et al., 2011; Sharma and Sharma, 2017).

**TABLE 16.1** Outline of Various Pharmacological Actions of General Anesthetics.

| Pharmacological actions of general anesthetics | | | | |
|---|---|---|---|---|
| Hemodynamic effects:<br>• Decreases systemic arterial blood pressure (hypotension) | Respiratory effects:<br>• Lower the ventilator force and the reflexes<br>• Failure of gag reflex<br>• Blunt cough stimulus<br>• Decrease lower esophageal sphincter tone<br>• Passive and active regurgitation | Decreased body temperature | Nausea and vomiting | Other emergence and postoperative phenomena:<br>• Hypertension and increased heart rate<br>• Myocardial ischemia<br>• Appearance of enthusiasm<br>• Shivering after anesthesia<br>• Airway obstruction<br>• Decreased pulmonary activity<br>• Hypoxemia |

**TABLE 16.2** Classification of General Anesthetics.

| Class | Examples |
|---|---|
| A. Inhalational anesthetics | |
| a. Gas | Nitrous oxide |
| b. Halogenated liquids | Halothane, methoxyflurane, enflurane, sevoflurane, desflurane |
| B. Intravenous anesthetics | |
| a. Fast inducers | |
| I. Barbiturates | Thiopental, methohexital |
| II. Non-barbiturates | Propofol, etomidate |
| b. Slow inducers | Diazepam, lorazepam, midazolam |
| c. Dissociative anesthetics | Ketamine |
| C. Adjuvant anesthetics | Benzodiazepine, α2 adrenergic agonists, neuromuscular blockers, analgesics |

## 16.6 MECHANISMS OF GENERAL ANESTHETICS

### 16.6.1 STATE OF ANESTHESIA

These agents produced a state of common end point behavioral changes, such as general anesthesia are defined as a reversible depression of working of CNS producing the breakdown of answer to the complete types of external stimuli. The state is explained by memory loss, unconsciousness, silence in response to noxious stimuli, and reduction in autonomic responses to noxious stimulation (Brunton et al., 2011; Rang et al., 2011; Tripathi, 2013).

The development of anesthetic state varies among the individual anesthetic agents. As amnesia is an important characteristic symptom of anesthesia, the agents such as barbiturates are highly effective in developing the amnesic condition and failure of consciousness but are not effective as analgesics.

### 16.6.2 MEASUREMENT OF ANESTHETIC POTENCY

The ability of general anesthetic to block the movement due to surgical stimulation determines their potency in developing the state of general anesthesia (Stoelting et al., 1970).

#### 16.6.2.1 INHALATIONAL ANESTHETICS

Anesthetic potency of these agents are expressed as the minimum alveolar concentration (MAC) which blocks the movement due to surgical stimulation in 50% of subjects which is measured by either infrared or mass spectroscopy. This method provides a direct correlation of measurement of free anesthetic concentration in the CNS and therefore allows the achievement of clinical goal (Brunton et al., 2011). In addition to this the other parameters such as ability to respond to verbal commands ($MAC_{awake}$) and the ability to form memories are also used in the determination of anesthetic potency of these agents (Stoelting et al., 1970). The ideal properties of various general anesthetics are given in Table 16.3.

**TABLE 16.3**    The Ideal Properties of Various General Anesthetics.

| Anesthetic agent | MAC (Vol %) | $MAC_{awake}$ (Vol %) | $EC_{50}$ for suppression of memory (Vol %) | Vapor pressure (mmHg at 20°C) | Partition coefficient at 37°C | | | | | Special characteristics |
|---|---|---|---|---|---|---|---|---|---|---|
| | | | | | Blood: Gas | Brain: Blood | Fat: Blood | Oil: Gas | Muscle: Blood | |
| Halothane | 0.75 | 0.41 | – | 243 | 2.3 | 2.9 | 51 | 224 | 3.4 | Intermediate onset and recovery |
| Isoflurane | 1.2 | 0.4 | 0.24 | 250 | 1.4 | 2.6 | 45 | 99 | 2.9 | Most potent; Comparatively slow onset and recovery; Pungent odor; Not suitable |
| Enflurane | 1.6 | 0.4 | – | 175 | 1.8 | 1.4 | 36 | 98 | 1.7 | Medium speed of onset and recovery |
| Sevoflurane | 2 | 0.6 | – | 160 | 0.65 | 1.7 | 48 | 65 | 3.1 | Frequently used; Fast onset and recovery; Non-pungent; Appropriate for generation of anesthesia |
| Desflurane | 6 | 2.4 | – | 664 | 0.45 | 1.3 | 27 | 19 | 2.0 | Extremely quick onset and recovery; Pungent odor; irritates airways; Inappropriate for development of anesthesia; |

**TABLE 16.3** *(Continued)*

| Anesthetic agent | MAC (Vol %) | MAC$_{awake}$ (Vol %) | EC$_{50}$ for suppression of memory (Vol %) | Vapor pressure (mmHg at 20°C) | Partition coefficient at 37°C | | | | | Special characteristics |
| | | | | | Blood: Gas | Brain: Blood | Fat: Blood | Oil: Gas | Muscle: Blood | |
|---|---|---|---|---|---|---|---|---|---|---|
| Nitrous oxide | 105 | 60.0 | 52.5 | Gas | 0.47 | 1.1 | 2.3 | 1.4 | 1.2 | Usually insufficient if used alone; Frequently give with more potent inhalational anesthetic to achieve the "second gas effect"; Rapid onset and recovery |
| Xenon | 71 | 32.6 | – | Gas | 0.12 | – | – | – | – | – |
| Methoxyflurane | 0.2 | – | – | – | 12 | – | – | 970 | – | – |
| Diethyl ether | 3.2 | – | – | – | – | – | – | – | 1.3 | – |
| Chloroform | 0.5 | – | – | – | – | – | – | – | – | – |

Where, MAC values are expressed as Vol %. MAC greater than 100% produces hyperbaric conditions would be required. MAC$_{awake}$ is the concentration at which suitable responses to instructions are lost. EC$_{50}$ is the concentration that produces memory suppression in 50% of patients.

### 16.6.2.2 INTRAVENOUS ANESTHETIC AGENTS

The continuous measurement of blood or plasma concentration as well as free concentration of anesthetic agents at the site of action is difficult. Therefore, the potency of intravenous anesthetic agent is measured as the free plasma concentration (at equilibrium) that produces failure of response to surgical incision (or other end points) in 50% of subjects (Dwyer et al., 1992). The summary of measurement of anesthetic potency is given in Table 16.4.

**TABLE 16.4** Summary of Measurement of Anesthetic Potency Criteria of General Anesthetics.

| Anesthetic potency of general anesthetics | |
|---|---|
| **Inhalational anesthetics** | **Intravenous anesthetics** |
| • Immobilization; <br> • Capability to respond to vocal instructions (MAC$_{awake}$); <br> • Appearance memory capacity; <br> • Expressed as MAC (minimum alveolar concentration) | • Complicated to measure; <br> • Expressed as free plasma concentration (at equilibrium) |

### 16.6.3   MECHANISMS OF ANESTHESIA

#### 16.6.3.1   ANATOMIC SITES

Anatomically general anesthetic causes an alteration in the ongoing activities of nervous system at myriad levels involving the peripheral sensory neurons, spinal cord, brainstem, and the cerebral cortex. In response to surgical procedure, an inhalational anesthetic exerts actions of spinal cord and therefore causes the state of immobilization. The GABA analogs such as pentobarbital and propofol produce sedative actions through $GABA_A$ receptors in the tuberomammillary nucleus whereas dexmedetomidine a α2 adrenergic-receptor agonist exerts sedative effects via locus ceruleus. An inhalational anesthetic causes loss of consciousness by depressing the excitability of thalamic neurons and inhibition of thalamocortical communication. Both intravenous and inhalational anesthetic produce amnesic effect through depression of hippocampal neurotransmission (Koblin et al., 1994).

#### 16.6.3.2   CELLULAR MECHANISMS

The general anesthetics have noteworthy effects on synaptic transmission rather than action-potential generation or propagation (Nicoll and Midson, 1982). The inhalational anesthetic causes an inhibition of responses due to excitatory synapses and augments that of inhibitory synapses (Larrabee and Posternak, 1952). Isoflurane an inhalational types of anesthetic agent blocks the liberation of neurotransmitter therefore reduces influx of presynaptic $Ca^{2+}$ at excitatory synapses. The intravenous anesthetics are associated with a narrow range of physiological effects whereas ketamine inhibits predominantly excitatory glutaminergic neurotransmission (Wu, 2004b).

#### 16.6.3.3   MOLECULAR MECHANISM

Both halogenated inhalational and intravenous anesthetic agent acts mainly through inhibitory $GABA_A$ chloride channel. Clinically general anesthetic causes stimulation of actions of GABA by enhancing the sensitivity of $GABA_A$ receptor therefore facilitates the inhibitory neurotransmission and causes depression of nervous system activity (Krasowski and Harrison, 1999). The $\beta_2$ and $\beta_3$ subunit of $GABA_A$ receptors are responsible for the development of sedative effects and reducing the response to noxious stimuli respectively. The propofol and etomidate acts mainly on the $\beta_3$ subunit of the $GABA_A$ receptor and therefore causes inhibition of reply to the noxious stimuli (Krasowski and Harrison, 1999; Mihic et al., 1997). An inhalational anesthetic such as propofol and barbiturates causes inhibition of neurotransmission at spinal cord and brainstem level by facilitating the glycine-gated chloride channels (glycine receptors) transmission. The sub anesthetic concentration of inhalational agent causes the inhibition of few types of neuronal nicotinic acetylcholine receptors (Hales and Lambert, 1988). The selective calcium glutamate NMDA receptor agents such as ketamine, nitrous oxide, cyclopropane and xenon modulate the long term synaptic potentiation and glutamate mediated neurotoxicity by inhibiting NMDA receptors. Ketamine shows its activity by inhibiting phencyclidine site on the NMDA-receptor protein (Mennerick et al., 1998; Jevtovic-Todorovic, 1998). The $K^+$ channels plays asignificant role in the creation of resting membrane potential in neurons which is stimulated by halogenated inhalational anesthetics. The inhalational anesthetic causes presynaptic inhibition in hippocampus and generates amnesic effects (Gray et al., 1998; Patel et al., 1999). The summary of mechanism of anesthesia produced by general anesthetics is shown in Table 16.5.

## 16.7 PARENTERAL ANESTHETICS

### 16.7.1 PHARMACOKINETICS

These agents are small, lipid loving, and aromatic as well as heterocyclic compounds. Being as lipophilic agents, the rate of penetration into highly perfused tissues like brain and spinal cord is more after a single intravenous bolus administration and produces anesthesia within a single circulation time. This is followed by rapid fall in CNS concentration which results in efflux of drug into blood from CNS. The rate of diffusion across the poorly perfused tissues like muscle and viscera is slow except adipose tissues. After redistribution, the fall in blood level and half-life of anesthetic agents is dependent on their rate of metabolism, amount of the drug present in peripheral tissue and lipophilicity of the drug (Shafer and Stanski, 1992). Pharmacological properties of various parenteral anesthetic agents are shown in Table 16.6.

### 16.7.2 SPECIFIC PARENTERAL AGENTS

Parenteral agents are used for numerous reasons. Pharmacological properties of various agents used in parenteral anesthetic are shown in Tables 16.7–16.10.

**TABLE 16.5** Outline of the Mechanism of General Anesthesia.

| Mechanism of general anesthetics | | |
|---|---|---|
| **Anatomical sites** | **Molecular mechanism** | **Cellular mechanism** |
| • Peripheral sensory neurons, the spinal cord, the brainstem, and the cerebral cortex. | • GABA$_A$ chloride channels, for example, halogenated inhalational anesthetics and intravenous agents. | • Hyperpolarization of neurons, for example, inhalational anesthetics. |
| • Tuberomammillary nucleus (GABA$_A$ receptors) for example, pentobarbital and propofol. | • β$_3$ subunit of the GABA$_A$ receptor, for example, propofol and etomidate. | • Synaptic transmission and action-potential generation or propagation, for example, inhalational and intravenous anesthetics. |
| • Locus ceruleus, for example, dexmedetomidine. | • Glycine-gated chloride channels, for example, propofol and barbiturates. | |
| • Thalamic neurons, for example, inhalational anesthetics. | • Neuronal nicotinic acetylcholine receptors, for example, inhalational anesthetics. | • Inhibition of excitatory synapses and enhancement of inhibitory synapses, for example, inhalational and intravenous anesthetics. |
| • Hippocampus, for example, inhalational and intravenous anesthetics. | • NMDA receptor, for example, ketamine, nitrous oxide, cyclopropane and xenon. | • Inhibition of neurotransmitter release, for example, inhalational anesthetics. |
| | • K$^+$ channels, for example, Xenon, nitrous oxide, and cyclopropane (halogenated). | • Inhibition of excitatory neurotransmission, for example, ketamine. |
| | • Protein complex, for example, inhalational anesthetics. | |

**TABLE 16.6** Pharmacological Properties of Parenteral Anesthetics.

| Drug | Formulation | Dose for IV induction (mg/kg) | Minimal hypnotic level (mg/mL) | Induction duration (min) | $t_{1/2}$ (h) | CL (mL/min/kg) | Protein binding (%) | Vss (L/kg) | Notes |
|---|---|---|---|---|---|---|---|---|---|
| Thiopental | 25 mg/mL of Thiopental sodium in aqueous solution add 1.5 mg/mL of sodium bicarbonate, maintain pH = 10–11 | 3–5 | 15.6 | 5–8 | 12.1 | 3.4 | 85 | 2.3 | Recovery period: A day |
| Methohexital | 10 mg/mL in aqueous solution + 1.5 mg/mL $Na_2CO_3$; pH = 10–11 | 1–2 | 10 | 4–7 | 3.9 | 10.9 | 85 | 2.2 | In lower cardiac output patients, the rate of relative perfusion and delivery of anesthetic to the brain is greater; therefore patients with septic shock or with cardiomyopathy generally require lower doses of anesthetic. As elder patients have small initial volume of distribution, they requires smaller dose of anesthetic agents. |
| Propofol | 10 mg/mL in 10% soybean oil, 2.25% glycerol, 1.2% egg PL, 0.005% EDTA or 0.025% Na-MBS; pH = 4.5–7 | 1.5–2.5 | 1.1 | 4–8 | 1.8 | 30 | 98 | 2.3 | |
| Etomidate | 2 mg/mL in 35% PG; pH = 6.9 | 0.2–0.4 | 0.3 | 4–8 | 2.9 | 17.9 | 76 | 2.5 | |
| Ketamine | 10, 50, or 100 mg/mL in aqueous solution; pH = 3.5–5.5 | 0.5–1.5 | 1 | 10–15 | 3.0 | 19.1 | 27 | 3.1 | |

*Source:* Adapted from Homer and Stanski (1985), Shafer and Stanski (1992), and Wada et al. (1997).

**TABLE 16.7** Pharmacological Properties of Barbiturates.

| Drug | Formulation | Reconstitution (%) | Stability | Potency | Pharmacokinetics | ADR | Pain | Notes |
|---|---|---|---|---|---|---|---|---|
| Sodium thiopental (frequently used) | Sodium salts with 6% sodium carbonate with pH 10 to 11 (alkaline) | Water or isotonic saline to produce 2.5% thiopental | – | – | Slow clearance therefore longer unconsciousness. More psychomotor harm which is maintained up to 8 h. Elimination: By liver and kidney. A minor amount of thiopental metabolized by desulfuration to the longer-acting hypnotic pentobarbital | Reduced blood pressure due to dilation of veins. Dull baroreceptor reflex. Respiratory depression. Reduced cerebral metabolic rate. Porphyria. | Little or painless | Reduces venous irritation due to lidocaine. Thiopental produces garlic taste before anesthesia. Methohexital ingestion is associated with enthusiasm which is characterized by muscle tremor, hypertonus, and hiccups. Methohexital frequently preferred for outpatient procedures but replaced by propofol. Thiopental provides protective mechanism against cerebral metabolism. |
| Thiamylal | | Water or isotonic saline to produce 2% thiamylal | 1 week with refrigeration | Equipotent as that of thiopental | Slow clearance therefore longer unconsciousness. Elimination: By liver and kidney. | | Little or painless | The patients with coronary artery diseases are well tolerated the thiopental ingestion as thiopental do not disturb the ratio of myocardial oxygen supply to demand. |
| Methohexital | | Water or isotonic saline to produce 1% methohexital | 6 week with refrigeration | Three fold as that of thiopental but similar in onset and duration of action as with thiopental | Rapid clearance. Less psychomotor impairment lasting up to 8 h. Elimination: By liver and kidney. | | Mild Pain | The concurrent administration of barbiturates with more acidic drugs should not be practiced or cautiously used due to precipitation of barbiturate toxicity. |

*Source:* Adapted from Andrews and Mark (1982), Beskow et al. (1995), Grounds et al. (1987), Nussmeier et al. (1986), Reiz et al. (1981), Schwilden and Stoeckel (1990), and Stullken et al. (1977).

**TABLE 16.8** Pharmacological Properties of Propofol.

| Chemistry and formulation | Pharmacokinetics | Dosage and clinical uses | ADR |
|---|---|---|---|
| • The energetic component of propofol is 2, 6-diisopropylphenol.<br>• Frequently used as parenteral (i.v.) anesthetic.<br>• Insoluble in aqueous solutions.<br>• Formulated as a 1% (10 mg/ml) emulsion in 10% soybean oil, 2.25% glycerol, and 1.2% purified egg phosphatide.<br>• Propofol formulation requires the addition of disodium EDTA (0.05 mg/ml) or sodium metabisulfite (0.25 mg/mL) to inhibit bacterial growth. | • Pharmacokinetically similar to barbiturates and<br>• Similar in onset and duration of action as that of thiopental.<br>• Rate of elimination is higher producing short duration of action, lesser hangover effect and rapid recovery.<br>• Hepatic metabolism and kidney elimination.<br>• More protein binding.<br>• Penetrates into placental membranes. | • Elder patients require reduced dose.<br>• In the presence of other sedatives requires less amount of drug.<br>• Young children requires increased dose.<br>• In healthy adult: 1.5 to 2.5 mg/kg.<br>• Short elimination half-life, therefore preferred for maintenance and induction of anesthesia.<br>• Sedative doses: 20–50% of those required for general anesthesia.<br>• Suitable for cerebral ischemic patients.<br>• Significant anti-emetic action.<br>• Pain on injection at larger arm and antecubital veins so lidocaine is used. | • Reduces the cerebral blood flow.<br>• Depressed intracranial as well as intraocular pressures.<br>• CVS: Dose dependent hypotension due to both vasodilation and mild depression of myocardial contractility.<br>• Blunt baroreceptor reflex or is directly vagotonic associated with smaller increases in heart rate and fall in blood pressure.<br>• Respiratory & Other: Associated with more respiratory depression, less bronchospasm, occurrence of anaphylactic reactions and release of histamine. |

*Source*: Adapted from Blouin et al. (1991), Langley and Heel (1988), Ravussin and Tribolet (1993), and Simons et al. (1988).

**TABLE 16.9** Pharmacological Properties of Etomidate.

| Chemistry and formulation | Pharmacokinetics | Dosage and clinical uses | ADR |
|---|---|---|---|
| • It is an active D-isomer of substituted imidazole derivative. | • Characterized by rapid onset and short duration of action. | • Preferred for hypotensive patients for induction of anesthesia. | • Reduces the cerebral blood flow. |
| • Inadequately soluble in water. | • Hepatic metabolism. | • Maintenance dose is 10 mg/kg per minute and sedative dose is 5 mg/kg per minute. | • Depressed intracranial as well as intraocular pressures. |
| • Formulated as a 2 mg/ml solution in 35% propylene glycol. | • 78% of drug is eliminated by kidney and 22% by excretion into biliary salt. | • Rectal dose is 6.5 mg/kg with an onset of about 5 min. | • CVS: Minute increase in heart rate. |
| • Preferred for induction of anesthesia along with neuromuscular blockers or other drugs. | • Repeated ingestion increases duration of action. | • Used as a protectant against cerebral ischemia. | • Respiratory: Characterized by hiccups, nausea and vomiting but no significant histamine release. |
| | • High plasma protein binding but less than barbiturates and propofol | • Best suited for maintaining CVS functions in patients with coronary artery disease, cardiomyopathy, cerebral vascular disease or hypovolemia. | • Not recommended for longer use as it inhibits an adrenal biosynthetic enzyme required for the production of cortisol and some other steroids. |
| | | | • High incidence of pain on injection, lidocaine is helpful. |
| | | | • Associated with myoclonic seizures, benzodiazepines or opiates are helpful. |

*Source:* Adapted from Criado et al. (1980), Gooding et al. (1976), Heykants et al. (1975), Modica et al. (1992), and Wagner et al. (1984).

**TABLE 16.10**  Pharmacological Properties of Ketamine.

| Chemistry and formulation | Pharmacokinetics | Dosage and clinical uses | ADR |
|---|---|---|---|
| • Arylcyclohexylamine, a phencyclidine derivative.<br>• Supplied as a racemic mixture.<br>• S-isomer is more potent.<br>• More lipophilic than thiopental.<br>• Water soluble.<br>• Available as 10, 50, and 100 mg/mL solutions in NaCl and benzethonium chloride. | • Metabolized by liver to CNS inactive norketamine and excreted by both kidney and bile.<br>• Vd is larger and fast elimination rate so it is best option for thiopental.<br>• Lesser protein binding. | • Causes indirect sympathetic stimulation and bronchodilatory activity.<br>• Associated with fairly dissimilar hypnotic state.<br>• Produces dissociative anesthesia which is associated with deep analgesia, loss of response to commands, amnesia, open eyes, involuntarily movements of limbs, and spontaneous breathing.<br>• Administered by parenteral, oral, and rectal routes.<br>• The induction dose is 0.5 to 1.5 mg/kg I.V, 4 to 6 mg/kg IM, and 8 to 10 mg/kg PR.<br>• I.V administration: Similar onset of action to that of the other parenteral anesthetics but longer duration of anesthesia.<br>• It does not bring out pain on injection or excitation as that of methohexital. | • CNS: Elevates cerebral blood flow and intracranial pressure with fewer effects on cerebral metabolism.<br>• Frequent delirium characterized by delirium, vivid dreams, and illusions.<br>• CVS: hypertension, tachycardia and increased cardiac output due to inhibition of both central and peripheral catecholamine reuptakes.<br>• Increases myocardial oxygen consumption.<br>• Respiratory: Small and transient decreases in minute ventilation but less respiratory depression. |

*Source:* Adapted from Chang and Glazko (1974), Mayberg et al. (1995), Modica et al. (1992), Whitacre (1994), and White and Ellis (1982).

## 16.8 SPECIFIC PARENTERAL ANESTHETIC AGENTS

### 16.8.1 HALOTHANE

#### 16.8.1.1 CHEMISTRY AND FORMULATIONS

In 1956, halogen was brought into clinical practice and is a 2-bromo-2-chloro-1,1,1-trifluoroethane. Being as a volatile liquid, nonflammable, nonexplosive, and halogenated compound, halothane must be packed in sealed amber colored bottle at room condition (Brunton et al., 2011; Wilson and Gisvold, 2004).

#### 16.8.1.2 PHARMACOKINETICS

Halothane has slow induction rate and relatively far above the ground blood: gas and fat: blood partition coefficient. Due to its greater solubility, halothane accumulates in body and fat tissue delaying the speed of recovery from state of anesthesia. The principle organ for the elimination of halothane is lungs which clears the nearly 60–80% ingested halothane from the body as unchanged after 1 day. Liver cytochrome enzyme system causes the removal of Br and Cl⁻ ions from halothane to trifluoroacetic acid as a major product which can be detected in the urine of patient (Gruenke et al., 1988). An immune reaction such as fulminate halothane-induced hepatic disorders occurs due to trifluoroacetation of liver proteins due to trifluoroacetylchloride, an intermediate oxidative metabolite of halothane (Kenna et al., 1988) as shown in Figure 16.2.

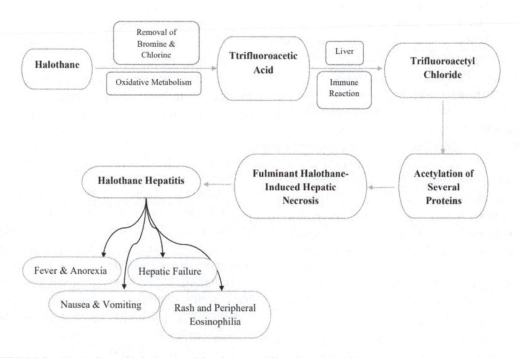

**FIGURE 16.2**   Metabolism of halothane and development of halothane hepatitis.

### 16.8.1.3 PHARMACOLOGICAL EFFECTS

Arterial hypotension due to halothane is associated dose dependent condensed cardiac output and shrinking of intracellular calcium transients due to depolarization (Lynch, 1997). Being as a direct vasodilator, halothane has ability to accelerate the elevation in blood flow of cerebra, responsible for percolation of skin and poorly ventilated regions of the lung as well as causes elevation of alveolar: arterial oxygen gradient. Discharges from sinoatrial node are depressed due to halothane which results in development of benign sinus bradycardia and atrioventricular rhythms (Sumikawa et al., 1983). Acting as stimulator of endogenous adrenal synthesis, halothane elevates plasma epinephrine levels precipitating premature ventricular contractions and sustained ventricular tachycardia through synergistic effects on $\alpha 1$ and $\alpha 2$ adrenergic receptors (Hayashi et al., 1988). Halothane causes deep and fast spontaneous respiration and also decreases alveolar ventilation. Halothane dose dependently causes suppression of central chemoceptor effect which is associated with blockade of $CO_2$ ventilatory response and elevation of $CO_2$ (Knill and Gelb, 1978). The use of halothane in patients with increased intracranial pressure is contraindicated as it dilates the cerebral vasculature and increases cerebral blood flow. Halothane also causes failure of cerebral blood flow autoregulatory mechanism and therefore it develops condition in which marked arterial hypotension is associated with reduction in blood flow of cerebra. It also causes decrease in metabolic oxygen consumption by the cerebra and well defined toleration of moderate decrease in blood flow cerebra (Brunton et al., 2011; Rang et al., 2011).

In the interest of relaxation, halothane potentiates the duration and magnitude of action of nondepolarizing skeletal muscle relaxants such as curariform. In genetic susceptible patients, it may also associated with condition such as malignant hyperthermia, a state of exaggerated contraction of muscle, fast progression of hyperthermic condition, and a huge increase in metabolic rate. Halothane also exhibits a helpful property of relaxation of uterus which is advantageous for exploitation of the fetus (version) in the prenatal period and for delivery of postnatal retained placenta. On the other side during parturition, halothane causes inhibition of uterine contractions which results in labor prolongation and increased blood loss. For this reason, halothane is not practiced as pain killer or to induce anesthesia during labor and vaginal delivery (Brunton et al., 2011; Rang et al., 2011). In anesthetized patients, halothane causes reversible decrease in both blood flow of kidney and rate of filtration of glomerulus as well as produces little level of concentrated urine (Brunton et al., 2011; Tripathi, 2013). Halothane induced decrease in pressure is linked with reduced splanchnic and hepatic blood flow. The rate of incidence of halothane induced fulminant hepatic necrosis is less (Brunton et al., 2011; Tripathi, 2013).

### 16.8.1.4 DOSAGE AND CLINICAL USE

Halothane is potent anesthetic agents used at end-tidal concentrations of 0.7–1% for maintenance of anesthesia. Because children have increased capacity to tolerate the dose of halothane and the serious side effects associated with halothane is less, so it is extensively used in these children. In developed countries, halothane is preferred agent for anesthesia because of its low cost (Brunton et al., 2011; Wilson and Gisvold, 2004).

## 16.8.1.5 ADVERSE DRUG REACTION

Halothane generally causes hypotension, benign sinus bradycardia and atrioventricular rhythms, premature ventricular contractions and sustained ventricular tachycardia, malignant hyperthermia, and fulminant hepatic necrosis (Brunton et al., 2011; Tripathi, 2013).

## 16.8.2 ISOFLURANE

### 16.8.2.1 CHEMISTRY AND PHYSICAL PROPERTIES

Isoflurane is 1-chloro-2, 2, 2-trifluoroethyl difluoromethyl ether. Characteristically, isoflurane is a potent and non-inflammable anesthetic agent and brought in to clinical practice in 1981 by the US (Brunton et al., 2011; Wilson and Gisvold, 2004).

### 16.8.2.2 PHARMACOKINETICS

Being lower blood:gas partition coefficient, isoflurane has superiority in speedy development and reversal of anesthesia over halothane or enflurane. Almost 100% of breathed isoflurane is unaffectedly removed from the body through the lungs and near about 0.2% of absorbed isoflurane undergoes oxidative metabolism by CYP2E1 (Kharasch et al., 1993).

### 16.8.2.3 PHARMACOLOGICAL EFFECTS

The vasodilator effect of isoflurane is observed in vascular beds such as skin and muscle. It also causes a concentration-dependent arterial hypotension due to reduced vascular resistance and compensatory tachycardia. Characteristically, the elevated blood flow of cerebra due to vasodilation and reduced myocardial oxygen utilization bearing capacity, the isoflurane has gain popularity as a safe and well tolerable drug in patients with ischemic heart illness. As isoflurane causes increased blood flow to the well perfused area of myocardium rather than poor perfused area ("coronary steal"), it may precipitate the condition such as myocardial ischemia (Buffington et al., 1988).

The noncompliance administration schedule of isoflurane develops a condition in which the tidal volume is decreased in association with significant decrease in ventilation of alveoli and increased burden of arterial $CO_2$ occurs. In state of hypercapnia and hypoxia, the ventilatory response is reduced by administration of isoflurane. The symptoms such as coughing and laryngospasm are observed while developing the anesthetic state with isoflurane because of its bronchodilator and irritant properties (Hirshman et al., 1977). Because of its dilatory effects on cerebral vasculature, ingestion of isoflurane causes elevated cerebral blood flow, likelihood of elevated intracranial pressure, and decreased cerebral metabolic $O_2$ consumption. Moreover, isoflurane also causes less vasodilatory effect which make it best use for various neurological procedures (Drummond et al., 1983). On the part of its central effects, isoflurane not only relaxes skeletal muscle relaxation but also accelerates effects of both depolarizing and nondepolarizing muscle relaxants. In addition to this isoflurane shares the property of relaxation of uterine smooth muscle relaxation because of which it is not advisable to practice as an analgesic or to develop anesthesia during labor pain and vaginal delivery (Brunton et al., 2011; Barar, 2004; Tripathi, 2013). The concentrated urine is produced after ingestion of isoflurane because it is associated with the reduction in flow of blood to the kidney as well as rate of glomerular filtration and reverses this condition after its withdrawal (Brunton et al., 2011; Rang et al., 2011). Because administration of isoflurane causes a state of systemic hypotension, its dose dependently reduces splanchnic and hepatic blood flows (Brunton et al., 2011; Rang et al., 2011; Golan, 2012).

### 16.8.2.4   CLINICAL USE

Isoflurane is a generally used worldwide drug of choice for maintenance of anesthetic state with other agents because of its strong odor (Brunton et al., 2011; Rang et al., 2011; Golan, 2012).

### 16.8.3   ENFLURANE

### 16.8.3.1   CHEMICAL AND PHYSICAL PROPERTIES

Enflurane is 2-chloro-1,1,2-trifluoroethyl difluoromethyl ether. At room temperature, enflurane is a plain, neutral liquid and sugary smell, brought in to clinical practice in 1973 by the United States. The air or oxygen containing mixture of enflurane is nonflammable and nonexplosive in nature (Brunton et al., 2011; Wilson and Gisvold, 2004).

### 16.8.3.2   PHARMACOKINETICS

Being relative far above the ground blood:gas partition coefficient, enflurane produces induction and recovery from anesthesia at a slower rare and undergo oxidative metabolism in the liver by CYP2E1 as a fluoride byproduct (Kharasch et. al, 1994; Mazze, 1982).

### 16.8.3.3   PHARMACOLOGICAL EFFECTS

Enflurane produces dose dependent arterial hypotension which is a result of depressed contraction of myocardium and peripheral vasodilation in association with less effect on heart rate (Brunton et al., 2011; Tripathi, 2013). Being in similar with halothane on respiratory system, enflurane produces rapid and shallow breathing and it has useful bronchodilator effect (Hirshman et al., 1977). In spite of its vasodilatory effects on cerebral, enflurane causes increased intracranial pressure, reduces cerebral metabolic $O_2$ consumption and characteristically generates high voltage and frequency electroencephalographic (EEG) electrical seizure (Brunton et al., 2011; Tripathi, 2013). Enflurane exhibit skeletal and uterine smooth muscle relaxation property and accelerates the actions of nondepolarizing muscle relaxants (Brunton et al., 2011; Rang et al., 2011; Golan, 2012).

Enflurane reversibly is responsible for reduction in kidney blood flow, glomerular filtration rate, and urine production. After metabolism of enflurane, the significant plasma levels of fluoride ions are elevated (20–40 mmol), therefore the prolonged administration of enflurane causes transient urinary-concentrating defects (Mazze et al., 1982). Enflurane reduces splanchnic and hepatic blood flow and neither is a hepatotoxic nor alter the liver function (Brunton et al., 2011; Rang et al., 2011).

### 16.8.3.4   CLINICAL USE

At a dose of 1.5–3.0%, enflurane is typically used for maintenance of anesthesia. It induced a state of surgical anesthesia within 10 min when inhaled at a dose of 4% with oxygen (Brunton et al., 2011; Golan, 2012).

## 16.8.4  DESFLURANE

### 16.8.4.1  CHEMISTRY AND PHYSICAL PROPERTIES

Desflurane is difluoromethyl 1-fluoro-2, 2, 2-trifluoromethyl ether. At room temperature, desflurane is a pungent, non-inflammatory, non-corrosive, and highly volatile liquid (vapor pressure=681 mmHg) brought in to clinical practice in 1992 by the USA. It is at room temperature. Generally heated vaporizers are used for delivery of a precise concentration of desflurane as a pure vapor in combination *i*th other gases ($O_2$, air, or $N_2O$; Brunton et al., 2011; Wilson and Gisvold, 2004).

### 16.8.4.2  PHARMACOKINETICS

Despites its very short blood:gas partition coefficient (0.42) and insolubility in fat or other peripheral tissues, produces rapid anesthesia due to 80% rise in the alveolar (and blood) concentration within 5 min (Smiley et al., 1991). With the use of desflurane, the time to awake can be reduced to half time as that of halothane or sevoflurane. A little amount of absorbed desflurane undergoes oxidative metabolism by hepatic CYPs (Koblin et al., 1988).

### 16.8.4.3  PHARMACOLOGICAL EFFECTS

Dose dependently, desflurane produces milder negative inotropic effect and decreases systemic vascular resistance which is linked to hypotensive condition (Eger et al., 1994). While developing the state of anesthesia, desflurane usually causes significant tachycardia due to stimulation of the sympathetic outflow (Ebert and Muzi., 1993). On the part of respiratory system, desflurane dose dependently is responsible with elevated rate of respiration as well as a suppression of tidal volume. Above I MAC concentration, it significantly suppresses the ventilation along with elevated arterial $CO_2$ tension ($PaCO_2$). Because of its bronchodilator and airway irritant property, desflurane is not advisable as an inducer of anesthesia and its administration is associated with phenomenon such as coughing, breath-holding and laryngospasm, and exaggerated secretion from respiratory system (Lockhart et al., 1991). Desflurane reduces cerebral vascular resistance and consumption of cerebral metabolic $O_2$. Desflurane shares the skeletal muscle relaxant property of nondepolarizing and depolarizing neuromuscular blocking and accelerates their effects (Brunton et al., 2011; Golan, 2012).

### 16.8.4.4  CLINICAL USE

Being as a rapid onset and recovery from anesthesia, desflurane is a preferably used for outpatient surgery (Brunton et al., 2011; Golan, 2012).

## 16.8.5  SEVOFLURANE

### 16.8.5.1  CHEMISTRY AND PHYSICAL PROPERTIES

Sevoflurane is fluoromethyl 2,2,2-trifluoro-1-[trifluoromethyl] ethyl ether. Being as a plain, neutral volatile liquid, nonflammable, and nonexplosive, sevoflurane must be must be packed in sealed amber colored bottle with mixture of air or oxygen. Sevoflurane when reacted with desiccated $CO_2$ absorbent, an exothermic reaction occurs which causes airway burns or spontaneous ignition, explosion and in addition to this it also reacts with fire as well as

CO which is hazardous to the patient (Fatheree and Leighton, 2004; Wu, 2004a).

### 16.8.5.2   PHARMACOKINETICS

Being lower blood and other tissue solubility, sevoflurane produces anesthesia at a rapid rate. Sevoflurane (3% of absorbed) undergoes hepatic metabolism by CYP2E1 to hexafluoroisopropanol as a major metabolite as well as produces inorganic fluoride. The compound A, that is, pentafluoroisopropenyl fluoromethyl ether is produced due to an interaction of sevoflurane with soda lime as a decomposition products (Hanaki et al., 1987; Kharasch and Thummel, 1995).

### 16.8.5.3   PHARMACOLOGICAL EFFECTS

Because of its dose dependent hypotensive and bradycardia producing effects, sevoflurane is preferred as an inducer of anesthesia in myocardial ischemic disease (Brunton et al., 2011). The spontaneously breathing patients are associated with the risk of suppressed tidal volume and elevated rate of respiration with dose dependent use of sevoflurane. Among the other inhalational anesthetics, clinically sevoflurane is considered as non-irritatant to airway as well as a potent dilator of bronchi (Doi, 1987). With respect to isoflurane and desflurane, sevoflurane produces similar effects on nervous system (Brunton et al., 2011). In addition to this, sevoflurane causes relaxation of skeletal muscle and accelerates actions of nondepolarizing and depolarizing neuromuscular blocking agents (Brunton et al., 2011). Neither sevoflurane produces renal toxicity nor toxicity of hepatic tissue (Brunton

et al., 2011, Eger et al., 1997; Golan, 2012; Mazze et al., 2000).

### 16.8.5.4   CLINICAL USE

Because of its rapid recovery and non-irritating profile at a dose of 2–4%, sevoflurane is a preferably used for outpatient surgery and in children (Brunton et al., 2011; Golan, 2012).

### 16.8.6   NITROUS OXIDE

### 16.8.6.1   CHEMICAL AND PHYSICAL PROPERTIES

Commonly called as laughing gas, Nitrous oxide (dinitrogen monoxide; $N_2O$) is a neutral, fragrance-free, nonflammable, and nonexplosive gas at room condition. It is supplied under calibrated flow meters' steel cylinders provided on all anesthesia machines (Brunton et al., 2011).

### 16.8.6.2   PHARMACOKINETICS

Fundamentally, the use of nitric oxide produces rapid anesthetic state and risk of emergency because it has very low solubility in blood and other body tissues and provides speedy equilibration between delivered and anesthetic concentration of alveoli. The concomitant administration of other halogenated anesthetics with nitrous oxide causes "second gas effect" which involving its rapid uptake from alveolar gas which accelerated the rate of induction of anesthesia. The stoppage of $N_2O$ ingestion is responsible for diffusion of nitrous oxide gas from blood to the alveoli diluting oxygen carrying capacity of the lung. This can produce an effect called diffusional hypoxia which can be avoided by 100% $O_2$ administration rather than air. Almost 99.9% of ingested nitrous oxide is removed from the body through lungs and it does not undergo metabolism by enzymatic system in humans. The interaction between vitamin B12 and nitrous oxide in intestinal flora causes inactivation of methionine synthesis producing vitamin B12 deficient megaloblastic anemia and peripheral neuropathy, therefore, nitrous oxide is not used as a sedative in critical conditions (Brunton et al., 2011; Sharma and Sharma, 2017).

### 16.8.6.3 PHARMACOLOGICAL EFFECTS

Being as an activator of sympathetic nervous system, administration of nitrous oxide with halogenated inhalational anesthetics causes tachycardia, hypertension and increases cardiac outflow. On the other hand, when co-administered with an opioid, it generally produces hypotensive effects and decreases cardiac output. It also raises venous tone in both the peripheral and pulmonary vasculature (Schulte-Sasse, 1982).

In spontaneous breathing patients, nitrous oxide causes increase in rate of respiration and decreases tidal volume without affecting minute ventilation and $PaCO_2$. In hypoxic condition, the higher concentration of nitrous oxide markedly depresses the ventilation (Brunton et al., 2011). It has vasodilator effect on cerebral blood flow that is, increases both blood flow of cerebra and intracranial pressure. The concomitant administration of nitrous oxide along with other intravenous anesthetic agent results in decrease in blood flow of cerebra on the other hand when co-administered with halogenated inhalational anesthetics; it reduces the cerebral blood flow that is, reduced vasodilatory effect (Brunton et al., 2011). Nitrous oxide neither relaxes skeletal muscle nor enhances the effects of neuromuscular blocking drugs. It does not trigger malignant hyperthermia (Brunton et al., 2011; Sharma and Sharma, 2017).

### 16.8.6.4 CLINICAL USE

As a weak anesthetic agent, Nitrous oxide shows consistent anesthetic effects during surgery under hyperbaric situation. It produces marked analgesia at 20% concentrations and sedation at 30% and 80% concentrations. It also produces analgesia and sedation in outpatient dentistry at 50% concentration. Nitrous oxide is often used along with other agents like inhalational anesthetics or intravenous anesthetics. In the body cavity having hollow space, nitrous oxide exchange with $N_2$ as it produces different blood: a gas partition coefficient which is consider as one of the disadvantage of nitrous oxide. The formed nitrous oxide enters into body cavity at a faster rate than escape of nitrogen thereby raising the volume and/or pressure inside the body cavity. The increased pressure and/or pressure causes development of pneumothorax which is fundamentally associated with hindered middle ear, embolus ofair, a hindered bowel loop, an air bubble in inner ocular cavity, a pulmonary bulla, and intracranial air (Brunton et al., 2011; Sharma and Sharma, 2017).

### 16.8.7 XENON

Xenon is an inert gas and in 1951, it was brought into practice as a first anesthetic agent. Because of difficulty in production and extraordinary nature, it must prepare from air which increases its cost. Being insoluble in blood and other tissues, xenon produces anesthetic state at a faster rate and rapid emergency from anesthesia. When administer with 30% oxygen, xenon gas produces acceptable level of anesthesia. It has no significant effects on cardiac outflow, rhythm of heart, and systemic vascular resistance and does not affect lung function. Xenon is devoid of liver or kidney dysfunctioning and it is not biotransformed within human body (Lynch, 2000).

## 16.9   INHALATIONAL ANESTHETICS

Variety of volatile liquids and gases can produce anesthesia. All Inhalational anesthetics possess difference in their physical characteristics and show low margin of safety which directs their pharmacokinetic summary. Inhalational anesthetic having therapeutic indices in the range of 2–4, this makes the agents dangerous for clinical uses. The choice of anesthetic agents depends on patient's pathphysiology and the unwanted effects of the drug, because each anesthetic agent varies in their side effect profile. Inhalational anesthetics have advantages that they generate a fast initiation of anesthesia and a fast revival of anesthesia once it is stop (Brunton et al., 2011; Rang et al., 2011, Golan, 2012).

### *16.9.1   PHARMACOKINETICS*

This anesthetic agent mostly acts as gas rather than as liquid. The action of this agent is completely based on its distribution pattern between tissues or between blood and gas. The balance achieved within tissue when the partial pressure of inhalation anesthetic is equal in the two tissues with dissimilar concentration. It means that the partial pressure of the inhalation anesthetic in all tissues will be identical to the partial pressure of the inhalation anesthetic in inspired gas. The partition coefficient of the inhalation anesthetic is calculated as the ratio of anesthetic concentration in two tissues when the

partial pressures of anesthetic are equal in the two tissues. Based on the partition coefficient profile, these agents are highly soluble in tissues like fat than others tissues like blood (Brunton et al., 2011). The partition coefficient profile for inhalational anesthetics is shown in Table 16.4.

### *16.9.1.1   EQUILIBRIUM*

Equilibrium attained when the partial pressure in the inspired gas is same as that of end tidal that is, alveolar gas. In another words, the equilibrium point is the point in which there is no net uptake of anesthetic from the alveoli into the blood. If the agents are less soluble in blood and other tissue, equilibrium is reached immediately, conversely, if the agent is highly soluble in tissue equilibrium takes hours to reach. This occurs because fat represents a huge anesthetic reservoir that will be filled slowly because of the modest blood flow to fat (Brunton et al., 2011; Katzung et al., 2009).

### *16.9.1.2   SPEED OF INDUCTION OF ANESTHESIA*

MAC of anesthetic agents suggests the speed of its induction. The induction of anesthesia in brain (highly perfused organ) depends when partial pressure is equal or greater than MAC. As a result, the anesthesia level is achieved in a minute once alveolar partial pressure reaches MAC. The rate of rise of alveolar pressure for those anesthetics which are rapidly soluble in blood and other tissues will be slower which limits its speed of induction and is overcome by higher delivering of higher inspired partial pressures of the anesthetic agents (Brunton et al., 2011; Rang et al., 2011; Golan, 2012).

## 16.9.1.3 ELIMINATION OF ANESTHESIA

Elimination of anesthesia by inhalational anesthetics is a reverse process. For agents with little blood and tissue solubility, recovery from anesthesia should be parallel to anesthetic induction irrespective of duration of its administration whereas for agents with high blood and tissue solubility, recovery from anesthesia depend on the duration of anesthetic administration because the accumulated amounts of anesthetic in the fat reservoir will prevent blood (and therefore alveolar) partial pressures from falling rapidly. Patients will be arousal when alveolar partial pressure reaches $MAC_{awake}$, a partial pressure somewhat lower than MAC (Brunton et al., 2011; Rang et al., 2011).

## 16.10 ANESTHETIC ADJUNCTS

### 16.10.1 BENZODIAZEPINES

These are the agents used as anxiolytic, amnesia, and sedation before the initiation of anesthesia as well as used for sedation in the procedures which is independent of general anesthetics. Some of the examples of benzodiazepine include diazepam, midazolam, and lorazepam. The solubility of Midazolam in water is good, administered parentally as well as orally. Oral route of midazolam is preferred for producing sedation in young children. This drug has fast induction and small duration of action. Midazolam at the dose of 0.01–0.07 mg/kg, intravenously used as sedative drug and the peak effect observed within 2 min and remains for about 30 min (Reves et al., 1985). The hepatic clearance rate of midazolam is 6–11 ml/min/kg and it is about twenty times greater than diazepam and seven times greater lorazepam. Benzodiazepines are effective anticonvulsants and are sometimes recommended in the management of status epilepticus (Modica and Tempelhoff, 1990). Benzodiazepine also produces decrease in blood pressure and respiratory functions, infrequently produces apnea (Reves et al., 1985).

### 16.10.2 α2 ADRENERGIC AGONISTS

Dexmedetomidine, an imidazole derivative stimulates α2A adrenergic receptor (Kamibayashi and Maze, 2000) and produces sedation, analgesia but does not dependably provide general anesthesia (Aho et al., 1992). Dexmedetomidine has distribution as well as terminal half-life is 6 min and 2 h, highly protein bound drug, metabolize primarily by hepatic glucuronide and methyl conjugates, and excreted via kidney (Khan et al., 1999). The drug shows reduced blood pressure and bradycardia which might be due to the activation of α2A adrenergic receptor. At higher concentration, it also stimulates α2B-subtype receptor which results in hypertension followed by reduced heart rate and cardiac output (Lakhlani et al., 1997). Dexmedetomidine has good property to induce sedation and analgesic effect along with minimum respiratory problems (Belleville et al., 1992). The drug is available in the form of hydrochloride salt which has to be diluted with normal saline solution before administration. The suggested dose of the drug is 1 mg/kg, the infusion is not recommended for more than 24 h as it produces rebound hypertension (Khan et al., 1999).

### 16.10.3 ANALGESICS

Analgesics are the agents which are usually recommended along with general anesthetic agent to reduce the dose of anesthetic, reduced hemodynamic alterations produced by painful stimuli and sufficient analgesia during minor surgical procedures. NSAIDs, inhibitors of Cyclo-oxygenase-2 enzymes and opioids are routinely used during surgical process as an analgesic (Brunton et al., 2011). Because of fast and deep analgesic activity opioids are used as drug of choice as analgesic during surgery. Various opioids are listed (Pasternak, 1993) are sufentanil, remifentanil, fentanil, alfentanil, morphine, and meperidine (Clotz, 1991). Remifentanil produce slight analgesic effects as it is ultrashort acting (~10 min) drug (Glass et al., 1993). Fentanil, alfentanil, and sufentanil are intermediate acting analgesics having 30, 20, and 15 min duration of action, respectively (Shafer and Stanski., 1991). All opioids undergoes metabolism in liver and excretion through kidney except the drug remifentanil which is hydrolyzed by esterase present in tissue and plasma (Tegeder et al., 1991). Metabolites of morphine also showed significant analgesic and hypnotic effects when administered for prolong period (Christrup, 1997). Opioid are administered by intrathecal and epidural route in the management of pain. These classes of drugs often produce minor reduction in respiration and reduction in blood pressure (Bowdle, 1998). Adverse drug reaction (ADR) produced by opioids includes nausea, vomiting, and pruritus. The adverse effects of Meperidine (reduction in shivering) are considered as advantages for the induction of anesthesia (Brunton et al., 2011).

### 16.10.4   *NEUROMUSCULAR BLOCKERS*

Neuromuscular blockers are useful during induction of anesthesia so as to facilitate muscle relaxation. Depolarizing neuromuscular blocker such as succinylcholine and nondepolarizing muscle relaxants such as pancuronium are regularly used to facilitate laryngoscopy and endotracheal intubation during surgery, as they relax muscles of jaws, neck, and airway. The muscarinic receptor antagonists such as glycopyrrolate

or atropine are administered along with cholinesterase blockers such as neostigmine or edrophonium to counteract the muscarinic activation due to esterase inhibition if muscle paralysis is not necessary. Succinylcholine causes bradycardia, hyperkalemia, and severe myalgia and malignant hyperthermia (Brunton et al., 2011, Sharma and Sharma, 2017).

## 16.11   CONCLUSION

For various surgical as well as diagnostic processes general anesthetic agents are widely used. The class of agents having anesthetic potential is skeletal muscle relaxants, anticonvulsants, and CNS depressants. As different agents possess different pharmacokinetic property it is difficult to administer and monitor the anesthesia. The onset of action and duration of action of these agents depends upon their pharmacokinetic profile. Anesthetic adjuvants are the supportive therapy which is required as pre-anesthetic medication, they help to develop smooth anesthesia to the patient.

## KEYWORDS

- **barbiturates**
- **consciousness**
- **general anesthesia**
- **sensation**
- **surgical procedure**

## REFERENCES

Aho, M.; Erkola, O.; Kallio, A.; Scheinin, H.; Korttila K. Dexmedetomidine Infusion for Maintenance of Anesthesia in Patients Undergoing Abdominal Hysterectomy. *Anesth. Analg.* **1992,** *75,* 940–946.

Alice, M. Hunt., Anesthesia:Principles and Practice; G.P. Putnam Sons: New York, 1949.

Andrews, P. R.; Mark, L. C. Structural Specificity of Barbiturates and Related Drugs. *Anesthesiology* **1982,** *57,* 314–320.

Barar, F. S. K. Essentials of Pharmacotherapeutics; S. Chand and Co.: New Delhi, 2004.

Belleville, J. P.; Ward, D. S.; Bloor, B. C.; Maze, M. Effects of Intravenous Dexmedetomidine in Humans. I. Sedation, Ventilation, and Metabolic Rate. *Anesthesiology* **1992,** *77,* 1125–1133.

Beskow, A.; Werner, O.; Westrin, P. Faster Recovery After Anesthesia in Infants After Intravenous Induction with Methohexital Instead of Thiopental. *Anesthesiology* **1995,** *83,* 976–979.

Blouin, R. T.; Conard, P. F.; Gross, J. B. Time Course of Ventilatory Depression Following Induction Doses of Propofol and Thiopental. *Anesthesiology* **1991,** *75,* 940–944.

Bowdle, T. A. Adverse Effects of Opioid Agonists and Agonist-Antagonists in Anesthesia. *Drug Saf.* **1998,** *19,* 173–189.

Buffington, C. W.; Davis, K. B.; Gillespie, S.; Pettinger, M. The Prevalence of Steal-Prone Coronary Anatomy in Patients with Coronary Artery Disease: An Analysis of the Coronary Artery Surgery Studio Registry. *Anesthesiology* **1988,** *69,* 721–727.

Chang, T.; Glazko, A. J. Biotransformation and Disposition of Ketamine. *Int. Anesthesiol. Clin.* **1974,** *12,* 157–177.

Christrup, L. L. Morphine Metabolites. *Acta. Anaesthesiol. Scand.* **1997,** *41,* 116–122.

Clotz, M. A.; Nahata, M. C. Clinical Uses of Fentanyl, Sufentanil, and Alfentanil. *Clin. Pharm.* **1991,** *10,* 581–593.

Criado, A.; Maseda, J.; Navarro, E.; Escarpa, A.; Avello, F. Induction of Anaesthesia with Etomidate: Haemodynamic Study of 36 Patients. *Br. J. Anaesth.* **1980,** *52,* 803–806.

Doi, M.; Ikeda, K. Respiratory Effects of Sevoflurane. *Anesth. Analg.* **1987,** *66,* 241–244.

Drummond, J. C.; Todd, M. M.; Toutant, S. M.; Shapiro, H. M. Brain Surface Protrusion During Enflurane, Halothane, and Isoflurane Anesthesia in Cats. *Anesthesiology* **1983,** *59,* 288–293.

Dwyer, R.; Bennett, H. L.; Eger, E. I.; Peterson, N. Isoflurane Anesthesia Prevents Unconscious Learning. *Anesth. Analg.* **1992,** *75,* 107–112.

Ebert, T. J.; Muzi, M. Sympathetic Hyperactivity During Desflurane Anesthesia in Healthy Volunteers. A Comparison with Isoflurane. *Anesthesiology* **1993,** *79,* 444–453.

Eger, E. I.; Koblin, D. D.; Bowland, T.; et al. Nephrotoxicity of Sevoflurane Versus Desflurane Anesthesia in Volunteers. *Anesth. Analg.* **1997,** *84,* 160–168.

Eger, E.I. II. New Inhaled Anesthetics. *Anesthesiology* **1994,** *80,* 906–922.

Fatheree, R. S.; Leighton, B. L. Acute Respiratory Distress Syndrome after an Exothermic Baralymeò-Sevoflurane Reaction. *Anesthesiology* **2004,** *101,* 531–533.

Franks, N. P.; Dickinson, R.; De Sousa, S. L.; Hall, A. C.; Lieb, W. R. How does Xenon Produce Anaesthesia? *Nature* **1998,** *396,* 324.

George, M. N.; Craig, W. S. *Pharmacology,* 3rd ed.; Elsevier Publication, 2010.

Glass, P. S.; Hardman, D.; Kamiyama, Y.; et al. Preliminary Pharmacokinetics and Pharmacodynamics of an Ultra-Short-Acting Opioid: Remifentanil (GI87084B). *Anesth. Analg.* **1993,** *77,* 1031–1040.

Gooding, J. M.; Corssen, G. Etomidate: An Ultrashort-Acting Nonbarbiturate Agent for Anesthesia Induction. *Anesth. Analg.* **1976,** *55,* 286–289.

Golan, D. E.; Tashijan, A. H.; Armstrong, E. J.; Armstrong, A. W. *Principles of Pharmacology. The Pathologic Basis of Drug Therapy,* 3rd ed.; Wolters Kluwer/Lippincott Wiliams and Wilkins: New Delhi, 2012.

Brunton, L. L.; Chabner, B. A.; Knollmann, B. C. *Goodman and Gillman: Pharmacological Basis of Therapeutics,* 12th ed.; Mcgraw-Hill, Medical Publishing Division: New York, 2011.

Gray, A. T.; Winegar, B. D.; Leonoudakis, D. J.; et al. TOK1 is a Volatile Anesthetic Stimulated K+ Channel. *Anesthesiology* **1998,** *88,* 1076–1084.

Grounds, R. M.; Maxwell, D. L.; Taylor, M. B.; Aber, V.; Royston, D. Acute Ventilatory Changes During I.V. Induction of Anaesthesia with Thiopentone or Propofol in Man. Studies Using Inductance Plethysmography. *Br. J. Anaesth.* **1987,** *59,* 1098–1102.

Gruenke, L. D.; Konopka, K.; Koop, D. R.; Waskell, L. A. Characterization of Halothane Oxidation by Hepatic Microsomes and Purified Cytochromes P-450 Using a Gas Chromatographic Mass Spectrometric Assay. *J. Pharmacol. Exp. Ther.* **1988,** *246,* 454–459.

Guedel.; Arthur Ernest. *Inhalation Anesthesia: A Fundamental Guide*; The Macmillan Company: New York, 1883..

Hales, T. G.; Lambert, J. J. Modulation of the $GABA_A$ Receptor By Propofol. *Br. J. Pharmacol.* **1988,** *93,* 84.

Hanaki, C.; Fujii, K.; Morio, M.; Tashima, T. Decomposition of Sevoflurane by Soda Lime. *Hiroshima J. Med. Sci.* **1987**, *36*, 61–67.

Hayashi, Y.; Sumikawa, K.; Tashiro, C.; Yamatodani, A.; Yoshiya, I. Arrhythmogenic Threshold of Epinephrine During Sevoflurane, Enflurane, and Isoflurane Anesthesia in Dogs. *Anesthesiology* **1988**, *69*, 145–147.

Heykants, J. J.; Meuldermans, W. E.; Michiels, L. J.; Lewi, P. J.; Janssen, P. A. Distribution, Metabolism and Excretion of Etomidate, A Short-Acting Hypnotic Drug, In The Rat. Comparative Study Of (R)-(+)-(-)-Etomidate. *Arch. Int. Pharmacodyn. Ther.* **1975**, *216*, 113–129.

Hirshman, C. A.; Mccullough, R. E.; Cohen, P. J.; Weil, J. V. Depression of Hypoxic Ventilatory Response by Halothane, Enflurane and Isoflurane in Dogs. *Br. J. Anaesth.* **1977**, *49*, 957–963.

Homer, T. D.; Stanski, D. R. The Effect of Increasing Age on Thiopental Disposition and Anesthetic Requirement. *Anesthesiology* **1985**, *62*, 714–724.

Jevtovic-Todorovic, V.; Todorovic, S. M.; Mennerick, S., et al. Nitrous Oxide (Laughing Gas) is an NMDA Antagonist, Neuroprotectant and Neurotoxin. *Nat. Med.* **1998**, *4*, 460–463.

Kamibayashi, T.; Maze, M. Clinical Uses of A2-Adrenergic Agonists. *Anesthesiology* **2000**, *93*, 1345–1349.

Katzung, B. G.; Masters, S. B.; Trevor A. T. *Basic and Clinical Pharmacology*, 11th ed.; Lange Medical Publications: California, 2009.

Kenna, J. G.; Satoh, H.; Christ, D. D.; Pohl, L. R. Metabolic Basis Of Drug Hypersensitivity: Antibodies in Sera from Patients with Halothane Hepatitis Recognize Liver Neoantigens that Contain the Trifluoroacetyl Group Derived From Halothane. *J. Pharmacol. Exp. Ther.* **1988**, *2435*, 1103–1109.

Khan, Z. P.; Munday, I. T.; Jones, R. M., et al. Effects of Dexmedetomidine on Isoflurane Requirements in Healthy Volunteers. 1: Pharmacodynamic and Pharmacokinetic Interactions. *Br. J. Anaesth.* **1999**, *83*, 372–380.

Kharasch, E. D.; Thummel, K. E. Identification of Cytochrome P450 2E1 as the Predominant Enzyme Catalyzing Human Liver Microsomal Defluorination of Sevoflurane, Isoflurane, and Methoxyflurane. *Anesthesiology* **1993**, *79*, 795–807.

Kharasch, E. D.; Armstrong, A. S.; Gunn, K.; et al. Clinical Sevoflurane Metabolism and Disposition. II. The Role of Cytochrome P450 2E1 in Fluoride and Hexafluoroisopropanol Formation. *Anesthesiology* **1995**, *82*, 1379–1388.

Kharasch, E. D.; Thummel, K. E.; Mautz, D.; Bosse, S. Clinical Enflurane Metabolism by Cytochrome P450 2E1. *Clin. Pharmacol. Ther.* **1994**, *55*, 434–440.

Knill, R. L.; Gelb, A. W. Ventilatory Responses to Hypoxia and Hypercapnia During Halothane Sedation and Anesthesia in Man. *Anesthesiology* **1978**, *49*, 244–251.

Koblin, D. D.; Chortkoff, B. S.; Laster, M. J.; et al. Polyhalogenated and Perfluorinated Compounds that Disobey the Meyer-Overton Hypothesis. *Anesth. Analg.* **1994**, *79*, 1043–1048.

Koblin, D. D.; Eger, E. I.; Johnson, B. H.; Konopka, K.; Waskell, L. I-653 Resists Degradation in Rats. *Anesth. Analg.* **1988**, *67*, 534–538.

Krasowski, M. D.; Harrison, N. L. General Anesthetic Actions on Ligand-Gated Ion Channels. *Cell. Mol. Life Sci.* **1999**, *55*, 1278–1303.

Kurz, A.; Sessler, D. I.; Lenhardt, R. Perioperative Normothermia to Reduce the Incidence of Surgical-Wound Infection and Shorten Hospitalization. Study of Wound Infection and Temperature Group. *N. Engl. J. Med.* **1996**, *334*, 1209–1215.

Lakhlani, P. P.; Macmillan, L. B.; Guo, T. Z.; et al. Substitution of a Mutant A2a-Adrenergic Receptor Via "Hit And Run" Gene Targeting Reveals the Role of this Subtype in Sedative, Analgesic, and Anesthetic-Sparing Responses in vivo. *Proc. Natl. Acad. Sci. USA* **1997**, *94*, 9950–9955.

Langley, M. S.; Heel, R. C. Propofol. A Review of its Pharmacodynamic and Pharmacokinetic Properties and use as an Intravenous Anaesthetic. *Drugs.* **1988**, *35*, 334–372.

Larrabee, M. G.; Posternak, J. M. Selective Action of Anesthetics on Synapses and Axons in Mammalian Sympathetic Ganglia. *J. Neurophysiol.* **1952**, *15*, 91–114.

Laurence, D. R.; Bennett, P. N. *Clinical Pharmacology*, 9th Ed.; Churchill Livingstone: Edinburgh, 2003.

Lockhart, S. H.; Rampil, I. J.; Yasuda, N.; Eger, E. I. II & Weiskopf, R. B. Depression of Ventilation by Desflurane in Humans. *Anesthesiology* **1991**, *74*, 484–488.

Lynch, C. Myocardial Excitation-Contraction Coupling. In: *Anesthesia: Biologic Foundations*; Yaksh, T. L., Lynch, C. III, Zapol, W. M., Eds.; Lippincott-Raven: Philadelphia, 1997; pp 1047–1079.

Lynch, C.; Baum, J.; Tenbrinck, R. Xenon Anesthesia. *Anesthesiology* **2000**, *92*, 865–868.

Mayberg, T. S.; Lam, A. M.; Matta, B. F.; Domino, K. B.; Winn, H. R. Ketamine does not Increase Cerebral Blood Flow Velocity or Intracranial Pressure During Isoflurane/Nitrous Oxide Anesthesia in Patients

Undergoing Craniotomy. *Anesth. Analg.* **1995**, *81*, 84–89.

Mazze, R. I.; Callan, C. M.; Galvez, S. T.; Delgado-Herrera, L.; Mayer, D. B. The Effects of Sevoflurane on Serum Creatinine and Blood Urea Nitrogen Concentrations: A Retrospective, Twenty-Two-Center, Comparative Evaluation of Renal Function in Adult Surgical Patients. *Anesth. Analg.* **2000**, *90*, 683–688.

Mazze, R. I.; Woodruff, R. E.; Heerdt, M. E. Isoniazid-Induced Enflurane Defluorination in Humans. *Anesthesiology* **1982**, *57*, 5–8.

Mennerick, S.; Jevtovic-Todorovic, V.; Todorovic, S. M.; et al. Effect of Nitrous Oxide on Excitatory and Inhibitory Synaptic Transmission in Hippocampal Cultures. *J. Neurosci.* **1998**, *18*, 9716–9726.

Mihic, S. J.; Ye, Q.; Wick, M. J.; et al. Sites of Alcohol and Volatile Anaesthetic Action on GABA$_A$ and Glycine Receptors. *Nature* **1997**, *389*, 385–389.

Modica, P. A.; Tempelhoff, R. Intracranial Pressure During Induction of Anaesthesia and Tracheal Intubation with Etomidate-Induced EEG Burst Suppression. *Can. J. Anaesth.* **1992**, *39*, 236–241.

Modica, P. A.; Tempelhoff, R.; White, P. F. Pro- and Anticonvulsant Effects of Anesthetics. *Anesth. Analg.* **1990**, *70*, 433–444.

Nicoll, R. A.; Madison, D. V. General Anesthetics Hyperpolarize Neurons in the Vertebrate Central Nervous System. *Science* **1982**, *217*, 1055–1057.

Nussmeier, N. A.; Arlund, C.; Slogoff, S. Neuropsychiatric Complications after Cardiopulmonary Bypass: Cerebral Protection by a Barbiturate. *Anesthesiology* **1986**, *64*, 165–170.

Pasternak, G. W. Pharmacological Mechanisms of Opioid Analgesics. *Clin. Neuropharmacol.* **1993**, *16*, 1–18.

Patel, A. J.; Honore, E.; Lesage, F.; et al. Inhalational Anesthetics Activate Two-Pore-Domain Background K+ Channels. *Nat. Neurosci.* **1999**, *2*, 422–426.

Rang, H. P.; Dale, M. M.; Ritter, J. M.; Flower, R. J.; Henderson, G. Rang & Dale's Pharmacology, 7th Ed.; Churchill Livingstone: Edinbergh, 2011.

Ravussin, P.; De Tribolet, N. Total Intravenous Anesthesia with Propofol for Burst Suppression in Cerebral Aneurysm Surgery: Preliminary Report Of 42 Patients. *Neurosurgery* **1993**, *32*, 236–240.

Reiz, S.; Balfors, E.; Friedman, A.; Haggmark, S.; Peter, T. Effects of Thiopentone on Cardiac Performance, Coronary Hemodynamics and Myocardial Oxygen Consumption in Chronic Ischemic Heart Disease. *Acta Anaesthesiol. Scand.* **1981**, *25*, 103–110.

Reves, J. G.; Fragen, R. J.; Vinik, H. R.; Greenblatt, D. J. Midazolam: Pharmacology and Uses. *Anesthesiology* **1985**, *62*, 310–324.

Schulte-Sasse, U.; Hess, W.; Tarnow, J. Pulmonary Vascular Responses to Nitrous Oxide in Patients with Normal and High Pulmonary Vascular Resistance. *Anesthesiology* **1982**, *57*, 9–13.

Schwilden, H.; Stoeckel, H. Effective Therapeutic Infusions Produced by Closed-Loop Feedback Control of Methohexital Administration During Total Intravenous Anesthesia with Fentanyl. *Anesthesiology* **1990**, *73*, 225–229.

Sellgren, J.; Ponten, J.; Wallin, B. G. Percutaneous Recording of Muscle Nerve Sympathetic Activity During Propofol, Nitrous Oxide, and Isoflurane Anesthesia in Humans. *Anesthesiology* **1990**, *73*, 20–27.

Sessler, D. I. Perioperative Heat Balance. *Anesthesiology* **2000**, *92*, 578–596.

Shafer, S. L.; Stanski, D. R. Improving the Clinical Utility of Anesthetic Drug Pharmacokinetics. *Anesthesiology* **1992**, *76*, 327–330.

Shafer, S. L.; Varvel, J. R. Pharmacokinetics, Pharmacodynamics, and Rational Opioid Selection. *Anesthesiology* **1991**, *74*, 53–63.

Sharma, H. L.; Sharma, K. K. *Principles of Pharmacology*, 3rd ed.; Paras Medical Publisher: Hyderabad, 2017.

Simons, P. J.; Cockshott, I. D.; Douglas, E. J.; et al. Disposition in Male Volunteers of a Subanaesthetic Intravenous Dose of an Oil in Water Emulsion of 14C-Propofol. *Xenobiotica* **1988**, *18*, 429–440.

Smiley, R. M.; Ornstein, E.; Matteo, R. S.; Pantuck, E. J.; Pantuck, C. B. Desflurane and Isoflurane in Surgical Patients: Comparison of Emergence Time. *Anesthesiology* **1991**, *74*, 425–428.

Stoelting, R. K.; Longnecker, D. E.; Eger, E. I. II. Minimum Alveolar Concentration in Man on Awakening from Methoxyflurane, Halothane, Ether and Fluroxene Anesthesia: Macawake. *Anesthesiology* **1970**, *33*, 5–9.

Stullken, E. H.; Jr. Milde, J. H.; Michenfelder, J. D.; Tinker, J. H. The Nonlinear Responses of Cerebral Metabolism to Low Concentrations of Halothane, Enflurane, Isoflurane, and Thiopental. *Anesthesiology* **1977**, *46*, 28–34.

Sumikawa, K.; Ishizaka, N.; Suzaki, M. Arrhythmogenic Plasma Levels of Epinephrine During Halothane, Enflurane, and Pentobarbital Anesthesia in the Dog. *Anesthesiology* **1983**, *58*, 322–325.

Tegeder, I.; Lotsch, J.; Geisslinger, G. Pharmacokinetics of Opioids in Liver Disease. *Clin. Pharmacokinet.* **1999**, *37*, 17–40.

Tripathi, K. D. *Essentials of Medical Pharmacology*, 7th ed.; Jaypee Brothers, Medical Publishers: New Delhi, 2013.

Wada, D. R.; Bjorkman, S.; Ebling, W. F.; et al. Computer Simulation of the Effects of Alterations in Blood Flows and Body Composition on Thiopental Pharmacokinetics in Humans. *Anesthesiology* **1997,** *87*, 884–899.

Wagner, R. L.; White, P. F.; Kan, P. B.; Rosenthal, M. H.; Feldman, D. Inhibition of Adrenal Steroid Genesis by the Anesthetic Etomidate. *N. Engl. J. Med.* **1984,** *310*, 1415–1421.

Whitacre, M. M.; Ellis, P. P. Outpatient Sedation for Ocular Examination. *Surv. Ophthalmol.* **1984,** *28*, 643–652.

White, P. F.; Way, W. L.; Trevor, A. J. Ketamine and its Pharmacology and Therapeutic Uses. *Anesthesiology* **1982,** *56*, 119–136.

Wilson; Gisvold. *Wilson And Gisvold's Textbook of Organic Medicinal and Pharmaceutical Chemistry*, 11th Ed.; J. Lippincot Co.: Philadelphia 2004.

Wu, J.; Previte, J. P.; Adler, E.; Myers, T.; Ball, J.; et al. Spontaneous Ignition, Explosion and Fire with Sevoflurane and Barium Hydroxide Lime. *Anesthesiology* **2004a,** *101*, 534–537.

Wu, X.-S.; Sun, J.-Y.; Evers, A. S.; Crowder, M.; Wu, L.-G. Isoflurane Inhibits Transmitter Release and the Presynaptic Action Potential. *Anesthesiology* **2004b,** *100*, 663–670.

# CHAPTER 17

# Alcohol Pharmacology and Pharmacotherapy of Alcoholism

MANOJ GOVINDARAJULU, SINDHU RAMESH, ELLERY JONES,
VISHNU SUPPIRAMANIAM, TIMOTHY MOORE, and
MURALIKRISHNAN DHANASEKARAN*

*Auburn University, Auburn, USA*

*Corresponding author. E-mail: dhanamu@auburn.edu*

## ABSTRACT

Ethanol, the most extensively used alcohol, is a resourceful, multipurpose molecule, with different effects on various biological systems. Due to its relative easiness to synthesize, availability, and its rewarding effects, it has been consumed by humans for ages for recreational purposes. Excessive alcohol consumption with poor lifestyle has drastically increased globally in developing countries and western world. Chronic alcohol consumption for years is liable to contract diseases, affect several organs, ultimately leading to organ damage and eventually death, prompting for in-depth awareness and information about its pharmacology and toxicology. Alcoholism is described as obsessive utilization despite explicit, harmful social and physiological results. This chapter emphasizes on the effects of alcohol on various organs, novel molecular signaling mechanisms involved, and currently available approved management for alcohol use disorder. In addition, potential future targets for drug discovery and development and re-purposing of approved treatment by the Food and Drug Administration are summarized.

## 17.1 INTRODUCTION

Alcohol (ethanol) has played an important role in human culture since documented history. The tradition of consuming alcohol as a means of socialization are predominantly because of its pleasurable effects as it reduces anxiety and induces a feeling of well-being. Alcohol use is widespread globally, mostly in the developed countries wherein the United States more than 8 million people are alcohol dependent (Kosten and O'Connor, 2003). Modernization has created a drastic increase in consumption of alcohol worldwide causing substantial influence on social and health issues. An estimated 2.5 million deaths every single year have been identified, making alcohol a direct causal agent in several diseases and damages; constituent reason in numerous other diseases (Rehm et al., 2009). Around 4% of the deaths and 2.4% of disability-adjusted life years

(DALY) are because of conditions recognized by alcoholism (Parry et al., 2011). Since alcohol is most widely abused drug, a detailed picture of how alcohol influences various organ systems in relation to development of various disorders so as to improve preventive and therapeutic interventions.

Alcoholism or alcohol use disorder (AUD) is a chronic condition characterized by obsessive, intense alcohol-seeking behavior with lack of ability to restrain from consumption and an undesirable emotional state deprived of alcohol intake (American Psychiatric Association, 2013). These disorders are complex with hereditary and environment-related factors contributing to its pathophysiology. Besides, conventional FDA-approved drugs for treating AUD, significant progress in research has been performed over the previous years to understand novel mechanisms involved in AUD which have opened new avenues for newer pharmacological treatments (Akbar et al., 2018). This involves various other neurotransmitters and chemical messenger systems in the brain, hormones, and cytokines. Finally, study on pathophysiology of mechanisms incorporated in molecular genetics will consequently enhance the possibilities for detecting targets for interventions. Hence, drugs developed with high affinity to the specific target of interest provide effective therapeutic management for those with AUD. Therefore, this chapter provides synopsis of the pathophysiological alterations in acute and chronic alcohol consumption; future therapies and challenges faced in treating alcoholism.

## 17.2   PHARMACOKINETICS OF ETHANOL

Differences in the pharmacokinetics of alcohol have important consequences in clinical and legal medicine.

### 17.2.1   ABSORPTION

Rapid absorption occurs from the gastrointestinal tract attaining highest blood alcohol concentration (BAC) within 30 min of alcohol intake in the fasting state. Absorption is rapid from small intestine than gastric region and hence delay in the process of gastric emptying due to presence of food particles in the stomach slows ethanol absorption (Wagner et al., 1977).

### 17.2.2   DISTRIBUTION

Alcohol readily disperses into all tissues and body fluids from the blood because it is not bound to plasma proteins. The relative content of water in the tissue determines the amount of alcohol present in the tissue; the distribution of alcohol in the tissue quickly equilibrates with the concentration of alcohol in the plasma. The volume of distribution is proportional to total body water (0.5–0.7 L/kg). Women due to their relatively low total body water content and differences in first pass metabolism mentioned in subsequent section have higher peak concentration compared to men (Cole-Harding and Wilson, 1987; Frezza et al., 1990). In organs with increased blood supply such as brain, alcohol passes through biological membranes and causes a rapid rise in concentration of alcohol across these tissues.

### 17.2.3   METABOLISM

Major portion of alcohol (90%) is oxidized in the liver with remaining expelled through pulmonary route and in urine. There are two major pathways that metabolize alcohol to acetaldehyde, and a final metabolic process further oxidizes acetaldehyde to acetate.

### 17.2.3.1 ALCOHOL DEHYDROGENASE PATHWAY

Alcohol dehydrogenase (ADH) belongs to a family of cytosolic enzymes with zinc, found mostly in the hepatic area, gastrointestinal tract, brain, kidneys, testes, and uterus. It oxidizes exogenously consumed alcohol, endogenous alcohol formed by gut microflora, and also in substrates involved in bile acid/steroid metabolism (Cederbaum, 2012). ADH enzyme in presence of nicotinamide adenine dinucleotide (NAD) as a cofactor brings about the transformation of alcohol to acetaldehyde. Genetic variations in these enzymes can modify the degree of alcohol metabolism and hence predispose to individual's susceptibility to alcohol abuse disorders as described below in Table 17.2. For instance, genetic polymorphism in one ADH allele- *ADH1B*2,* more common in East Asians leads to hasty transformation of alcohol to acetaldehyde results in protection against alcohol dependence (Suddendorf, 1989). ADH is also present in the stomach and contributes partly to metabolism of alcohol. Women have low levels of gastric enzyme which cause dissimilarities in BAC in men and women (Lieber, 2000; Schuckit, 2006b).

### 17.2.3.2 MICROSOMAL ETHANOL-OXIDIZING SYSTEM

This primarily consists of cytochrome P4502E1, 1A2, and 3A4 and termed as mixed function oxidase system. Microsomal ethanol-oxidizing system (MEOS) increase in chronic consumption especially CYP2E1 and with higher ethanol concentrations leading to increased rate of alcohol metabolism and elimination of other drugs metabolized by this cytochrome P450 (Lieber, 2004).

### 17.2.3.3 ACETALDEHYDE METABOLISM

The above two systems produce acetaldehyde which is further oxidized by the mitochondrial enzyme NAD-dependent aldehyde dehydrogenase (ALDH) to form acetate, advances to be metabolized to $CO_2$ and water in the liver (Edenberg, 2007). Disulfiram, a medication to manage alcohol dependence causes inhibition of this enzyme. The mechanisms that underlie hepatobiliary disease resulting from excessive alcohol use reflect an intricate blend of the metabolic factors such as: stimulation of CYP2E1, increased activation of toxin (acetaldehyde), generation of $H_2O_2$ and oxygen radicals, and perhaps heightened release of endotoxin because of alcohol's effect on gram-negative organisms in the gastrointestinal tract. The effects of excessive use of alcohol on different organs in the body are briefed in the following section. Damage to the body tissues from malabsorption and deficiency of vitamins (A, D, thiamine) most probably reflects the poor nutritional status of alcoholics due to immuno-compromised and a variety of other generalized effects (Rossi et al., 2015).

### 17.2.3.4 OTHER ENZYMES ASSOCIATED WITH ALCOHOL METABOLISM

Catalase, an enzyme which catalyzes the removal of hydrogen peroxide plays an insignificant role in oxidation of alcohol (Kuo et al., 2008).

### 17.2.3.5 GENETIC VARIATION IN ALCOHOL METABOLISM

Genetic variations in the ADH, CYP2E1 enzymes alter the degree of alcohol metabolism and leads to alcohol effects. ADH is an important enzyme and exists in three forms *ADH1A,*

*ADH1B*, and *ADH1C* on chromosome 4q22. They are in charge for majority of alcohol metabolizing capacity at BACs of 22 mM (i.e., ~0.10 g/dL). Polymorphisms in these isoforms alter the degree of alcohol metabolism (Quertemont, 2004). There are no polymorphisms associated with *ADH1A* gene; however, polymorphism of *ADH1B* called as *ADH1B*2* (arginine 47 residue is substituted by histidine) has higher Vmax than *ADH1B* by 40-fold resulting in rapid metabolism of alcohol. This leads to higher blood acetaldehyde levels and consequently decreased possibility of heavy consumption and alcohol-associated problems in East Asians (Edenberg et al., 2006). 30% of African is known to have a second polymorphism related to *ADH1B*3* (arginine 269 supplanted by cysteine), which has a 30-fold higher Vmax and related with lower hazards of excessive alcohol consumption. *ADH1C* gene exhibits two polymorphisms in high linkage disequilibrium, the γ1 and γ2 forms. Both have higher Vmax and are connected with

a slightly rapid metabolism of ethanol. The ADH cluster on chromosome 4 also contains class II ADHs which contribute approximately 30% of the ethanol metabolizing ability noted above. At higher BACs, catalase and CYP2E1 are also associated with metabolism of ethanol to acetaldehyde (Kuo et al., 2008). Lastly, 5–10% of Japanese, Chinese, and Koreans carry homozygous mutations in the *ALDH2* gene (12q24), called as *ALDH2*2* (substitution of glycine 487 with lysine) leading to a nonfunctional ALDH thereby causing severe adverse reactions after consumption of less than one drink. Hence, the hazards of severe repetitive heavy drinking are close to zero. Heterozygous polymorphism (*ALDH2*2,2*1*) occurs in about 30–40% of Asian people who tend to develop flushing of the face and an increased sensitivity to ethanol, but without other unfavorable bearings (Wall and Ehlers, 1995). The role of genetic variations in alcohol metabolism is illustrated in Figure 17.1.

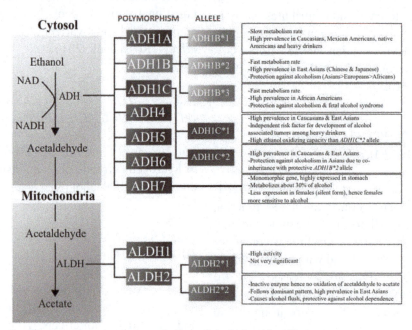

**FIGURE 17.1** Genetic variations related to metabolism of alcohol and its effects.

### 17.2.4 ELIMINATION

Alcohol undergoes zero-order elimination wherein a constant amount from body is eliminated at a steady rate per unit time and independent of alcohol concentration. Decreased alcohol concentration causes ADH saturation because the Km of most ADH isozymes for ethanol is diminished (around 1 mM); as a result, the evacuation process proceeds at maximal velocity and is independent of alcohol concentration. Nevertheless, a linear trend is not detected at minimal alcohol concentrations because there is no ADH saturation with ethanol. Alcohol elimination can be characterized by Michaelis–Menten energy, with the rate of change in alcohol concentration subject to concentration of alcohol and the kinetic constants, Km and Vmax (Matsumoto and Fukui, 2002; Salaspuro and Lieber, 1978).

## 17.3 PHARMACODYNAMICS OF ALCOHOL

### 17.3.1 CENTRAL NERVOUS SYSTEM

Alcohol initially has a brief stimulatory activity followed principally by a depressant action. Sedation, anxiety relief, and behavioral disinhibition are noted at mild to moderate amounts of ethanol. Attention deficits, information processing skills, motor skills which are obligatory for activities like driving motor vehicles are impaired at higher doses. In the United States, DUI is one of the primary causes of death with 30–40% of all accidents involving a minimum of one person with BAC above the legal levels of intoxication (Oslin et al., 1998). Higher concentrations cause slurring of speech, ataxia, impaired judgment, feeling spiritual to uncontrolled mood swings with violent component and disinhibition behavior termed as intoxication. Severe intoxication is characterized by impairment of central nervous system (CNS), general anesthesia followed by respiratory depression. Anterograde amnesia known as "alcoholic blackouts" is seen with heavy consumption of alcohol resulting in incapability to remember entirely or portion of the experiences during alcohol consumption. Disturbances in sleep architecture, with frequent awakenings and restless sleep, are noted at minor doses; vivid and disturbing dreams are noted at high doses due to suppression of rapid eye movement state. Heavy drinking, in elderly group, is associated with sleep apnea because of the effects of alcohol on respiration and muscle relaxing property (Sakurai et al., 2007). Consequently, hangover or next morning syndrome are characterized by nausea, headache, thirst, and cognitive impairment are related to heavy ethanol consumption, which results in inability to work and attend school. The mechanisms underlying it mimic mild alcohol withdrawal, dehydration, and mild acidosis (Stephens et al., 2008). The CNS effects noted with increased BAC levels are described in Table 17.1.

**TABLE 17.1** BAC Levels and Impairments in Non-tolerant Individuals.

| BAC level (mg/dL) | Dose specific effects |
| --- | --- |
| <50 | Mild euphoria |
| 50–100 | Feeling of high (subjectively), slower reaction times, sedative effect |
| 100–200 | Slurring of speech, severely impaired motor coordination and balance (ataxia) |
| 200–300 | Needs assistance in walking, stupor |
| 300–400 | Loss of consciousness/coma |
| >400 | Respiratory depression and death |

Chronic intake is accompanied by tolerance, dependence, and craving. Tolerance specifies that high amount of alcohol is essential to manifest the CNS effects. For example, a BAC of 300–400 as indicated above might result in coma in a nontolerant individual, whereas a chronic alcoholic may look like sober or marginally inebriated. Chronic heavy alcohol consumption in youth and adults leads to permanent cognitive deficit, often known as "alcoholic dementia." However, the symptoms and signs of cognitive impairment and brain atrophy often recuperate, the consequent weeks to months following abstinence (Bartsch et al., 2006). Chronic drinking is associated with thiamine depletion causing Wernicke-Korsakoff syndromes characterized by ataxia and ophthalmoparesis of Wernicke's, and the severe anterograde and retrograde amnesias of Korsakoff's. Additional severe neurological syndromes include cerebellar degeneration with cerebellar vermis atrophy (1% of alcoholics), and a peripheral neuropathy (10% of alcoholics) (Alexander-Kaufman et al., 2007; Peters et al., 2006). Heavy doses of alcohol are associated with several temporary but disturbing "alcohol-induced" psychiatric syndromes (Schuckit, 2006a). Severe alcohol-related depressive symptoms with suicidal thoughts, panic attacks, and generalized anxiety, are noted in alcohol-dependent individuals during the withdrawal syndrome. Auditory hallucinations and paranoid delusions which are temporary, resemble schizophrenia, are noted beginning during episodes of heavy alcohol consumption; all these psychotic symptoms prospectively progress within numerous days or a month of abstinence. Chronic alcohol abuse causes reduction of both white and gray matter leading to impaired cognition and judgment (Kril and Halliday, 1999). Older individuals are susceptible to chronic alcohol abuse, affecting predominantly the frontal lobes as compared to younger ones (Pfefferbaum et al., 1998). Besides causing malnutrition and vitamin deficiencies,

ethanol by itself is neurotoxic and results in brain damage, playing an important role in the development and progression of Wernicke's encephalopathy and Korsakoff's psychosis. In addition, positron-emission tomography has displayed reduced brain metabolism in chronic alcoholics with age of the person and the number of years of alcohol intake directly correlating with magnitude of decreased metabolic rate (Volkow et al., 1994). The imbalance in the brain, between excitatory and inhibitory neurotransmission, is amended by ethanol. These are brought about by effects on ligand-gated, voltage-gated ion channels, and various receptors including G-protein coupled receptors (GPCRs) systems which are illustrated in the following sections.

### 17.3.1.1   ALCOHOL ON ION CHANNELS

The ligand-gated gamma-aminobutyric acid A ($GABA_A$) receptors present in the CNS mediates inhibitory neurotransmission and alcohol augments its effects. GPCR regulates the increased GABA released by ethanol. Stimulation of this ligand-gated $Cl^-$ channel system contributes to hypersomnia, muscle relaxation, and the acute anticonvulsant properties associated with all GABA related drugs (Krystal et al., 2006). Acute alcohol intake indicates the release of GABA; the pattern of gene expression of $GABA_A$ subunits can be drastically altered with chronic alcohol consumption (Dick et al., 2006a). Several $GABA_A$ receptor gene polymorphisms correlate with a predisposition toward heavy chronic alcoholism. Alcohol augments the action of large conductance, $Ca^{2+}$ activated $K^+$ channels in neuro-hypophyseal terminals, causing reduced release of oxytocin and vasopressin (Dopico et al., 1999). Additionally, ethanol inhibits N- and

P/Q-type $Ca^{2+}$ channels that may be antagonized by channel phosphorylation by protein kinase A (Solem et al., 1997). BK channels (maxi calcium-activated potassium, slo1) too are a target for alcohol action (Davies et al., 2003). Conversely, G protein-gated inwardly rectifying $K^+$ channels (GIRK or Kir channels) can be activated by the $\beta\gamma$ subunits of the Gi/Go family, by $PIP_2$ by alcohol (Aryal et al., 2009).

## 17.3.1.2 MODULATION OF NEUROTRANSMITTERS AND ITS RECEPTORS

Alcohol heavily sensitizes nicotinic acetylcholine receptor (nAChR). Increase in acetylcholine (ACh) in the ventral tegmental area, with subsequent increase in neurotransmitter dopamine (DA) in the nucleus accumbens, is noted following acute consumption (Joslyn et al., 2008). In addition, alcohol regulates nAChR expression, functioning as an allosteric regulator by increasing or decreasing expression, stabilization, desensitization, or internalization (Dopico and Lovinger, 2009). Ethanol dependence and reward are a result of the stimulatory activity of the mesolimbic DA signaling (Smith et al., 1999). Furthermore, Ethanol inhibits the *N*-methyl-d-aspartate subtype of glutamate receptors (NMDA) in a concentration-dependent manner. Experiments utilizing a fast solution exchange technique to measure NMDA-evoked currents showed ethanol to have a very rapid (less than 100 ms) onset of action (Peoples and Stewart, 2000; Wirkner, 2000). α-amino-3-hydroxy-5-methyl-4-isoxazolepropionic acid (AMPA) receptors are essentially unaffected by alcohol (Carta et al., 2003).

Alcohol reinforcement is characterized by involvement of various neurotransmitters of brain reward systems (Koob et al., 1994). In the ventral tegmental area, alcohol modulates GABAergic transmission leading to disinhibition of dopamine release. Consequently, there is an increase in the DA release in nucleus accumbens (area involved in reward circuit) thereby increasing reward process. In contrast, alcohol withdrawal causes decline in DA release and function in chronic alcoholics, which has an impact on withdrawal symptoms, psychiatric manifestations, and also relapse (Koob and Kreek, 2007; Volkow et al., 2007). Stimulation of μ opioid receptors in the nucleus accumbens and ventral tegmental area (VTA) are associated with the release of DA due to release of β endorphins from acute alcohol ingestion (Job et al., 2007). Thus, effects of alcohol on reward systems relate to alterations in opioid systems (Pastor and Aragon, 2006). Similarly, acute ethanol intake increases synaptic serotonin (5-HT) and continued use causes up-regulation of 5-HT receptors. Higher intake of alcohol is associated with lower synaptic 5-HT possibly because of rapid uptake of 5-HT transporter thereby causing less intensity of reaction to alcohol (Barr et al., 2004). There is dense representation of $CB_1$ cannabinoid receptor which is a GPCR in the VTA, nucleus accumbens and prefrontal cortex. CB1 receptors are activated due to acute alcohol intake results in altered release of DA, GABA, and glutamate thereby affecting reward circuits in the brain. The influence of alcohol on dopaminergic systems is blocked by rimonabant, an antagonist of CB1 receptors.

## 17.3.1.3 ALCOHOL EFFECTS ON SECOND MESSENGER SIGNALING

Dopaminergic D1 receptors coupled to $G_s$ and $G_{olf}$ are activated by alcohol which causes downstream signaling events through activating adenylyl cyclase (AC) and protein kinase A. Conversely, inhibition of several AC isoforms due to activation of D2 receptors coupled to $G_{i/o}$ can occur. Protein kinase C, especially γ isoform is important and lack of this isoform causes reduced effects of alcohol measured behaviorally (Harris et al., 1995). Tyrosine kinases, mitogen-activated protein kinase (MAPK), and neurotrophic factor receptors are intracellular signal transduction cascades which are also affected by ethanol. Translocation of protein kinase C and A (PKC and PKA) between subcellular partitions also are sensitive to alcohol. Ethanol increases the activities of several isoforms of adenylyl cyclase, with AC7 existing as the most sensitive (Tabakoff and Hoffman, 1998). This causes amplified release of cyclic adenosine monophosphate (AMP) and thus enhanced activity of PKA. Activation of $G_s$, promotion of the interaction between $G_s$ and adenylyl cyclase—these activities are facilitated by ethanol's action.

## 17.3.2  CARDIOVASCULAR SYSTEM

Ethanol intake (> than three standard drinks/ day) increase the likelihood of cardiovascular disease and related strokes. Increased occurrence of arrhythmias, cardiomyopathy, hypertension, and congestive heart failure are prominent with daily consumption of alcohol in large quantities (Goel et al., 2018).

### 17.3.2.1  CORONARY HEART DISEASE

Epidemiological data implies that moderate ethanol consumption (20–30 g ethanol per day) confers a cardioprotective effect and prevents cardiovascular diseases like coronary heart disease (CHD), stroke and peripheral arterial

disease (Goldberg et al., 2001; Mukamal et al., 2006). Around 10–40% have decreased risk of CHD as compared to non-alcoholics, in those who consume 1–3 drinks per day. This association between mortality and dose of the ethanol has a J-shaped dose-mortality curve. Ethanol decreases CHD risk through its effect on blood lipids. Ethanol increases the levels of high-density lipoprotein (HDL) which further decreases plasma cholesterol levels and the accumulation of cholesterol in the walls of the arteries, thereby decreasing the risk of ischemia. Additionally, red wine has flavonoids that protect low-density lipoprotein (LDL) from oxidative stress thereby contributing to an antiatherogenic role. Eventually, ethanol decreases clot formation by increasing the levels of tissue plasminogen activator (clot-dissolving enzyme) and by decreasing fibrinogen concentrations (Rimm et al., 1999). Despite its benefits, there has yet to be randomized clinical trials conducted to test the efficacy of daily alcohol use in decreasing rates of cardiovascular diseases and mortality. Patients with the possibility of progressing CHD should be advised to modify lifestyle changes or medical treatments.

### 17.3.2.2  HYPERTENSION

The diastolic and systolic blood pressure is increased with heavy ethanol intake, irrespective of the age, gender, smoking status, or the consumption of oral contraceptives. Increase in systolic and diastolic blood pressure by 1.5–2.3 mmHg is noted in those who consume 30 g of alcohol per day (Briasoulis et al., 2012).

### 17.3.2.3  CARDIAC ARRHYTHMIAS

The outcome of chronic alcohol intake results in supraventricular tachycardia, atrial fibrillation, and atrial flutter (George and Figueredo, 2010). The frequent cause of sudden cardiac arrest are ventricular tachycardia studied in patients who are alcohol-dependent (Kupari and Koskinen, 1998). These arrhythmias are resilient to conventional treatments and people with recurrent or refractory arrhythmias are required to be cautioned about alcohol intake.

### 17.3.2.4 STROKE

Both hemorrhagic and ischemic stroke are more commonly seen in persons who drink more than 40–60 g ethanol on daily basis (Hansagi et al., 1995) and is pursed by persistent binge drinking. This can occur either due to cardiac arrhythmias and accompanied dislodging of thrombi, high blood pressure, and head trauma from chronic alcohol ingestion.

### 17.3.2.5 CARDIOMYOPATHY

Left ventricular dysfunction and decreased cardiac contractility, subsequently leading to cardiomyopathy is noted in chronic alcoholics. Alcohol-dependent patients who discontinue drinking are seen to have improved prognosis associated with alcoholic cardiomyopathy. Men have decreased risk alcoholic cardiomyopathy when compared to women (Urbano-Márquez et al., 1995). Abstinence remains the principal treatment as most of alcoholic cardiomyopathy patients who prolong drinking habits die within 3–5 years.

### 17.3.3 SKELETAL MUSCLE

Chronic alcoholics develop decreased skeletal muscle strength with irreversible damage, characterized by increase in serum creatine kinase. Muscle biopsies from excessive alcohol consumption patients illustrate the following pathologies, diminished glycogen stores, and decreased pyruvate kinase activity with type II fiber atrophy (Vernet et al., 1995; Wassif et al., 1993).

### 17.3.4 GASTROINTESTINAL SYSTEM

### 17.3.4.1 ORAL CAVITY

Alcohol, resembling an immunocompromised state, is correlated with the accumulation of pathological microorganisms within the oropharyngeal mucosa proceeding to chronic infections, predisposition to carcinogens and cell proliferation in the mucosa along with genetic changes thus resulting in the increased risk of developing dysplasia, leucoplakia, and carcinoma (Baan et al., 2007; Goldstein et al., 2010).

### 17.3.4.2 ESOPHAGUS

Alcohol consumption causes esophageal reflux, traumatic rupture of the esophagus, Mallory–Weiss tears, Barrett's esophagus, and ultimately cancer of the esophagus. Smokers who are alcoholics have a 10-fold elevated risk of acquiring esophageal carcinoma (Bosetti et al., 2000; Yang et al., 2017).

### 17.3.4.3 STOMACH

Heavy alcohol consumption (greater than 40% alcohol) affects acid production in the stomach, disrupts the barrier of gastric mucosa, and also induces inflammation, resulting in acute and chronic gastritis. Alcohol enhances the release of gastrin and histamine hence increasing the gastric secretions. In addition, the chances of developing gastric cancer are correlated with the concentration of alcohol ingestion (Ma et al., 2017).

### 17.3.4.4 INTESTINES

Malabsorption in the small intestine results in chronic diarrhea seen in many alcoholics (Addolorato et al., 1997). Chronic diarrhea, in turn, leads to development of anal fissures and pruritus ani. Pathophysiological alterations noted in the small intestine are flattened villi in the mucosa of the intestine and decreased digestive enzymes (Papa et al., 1998).

### 17.3.4.5 PANCREAS

Acute and chronic pancreatitis occurs due to heavy ethanol consumption. Increased rate of alcohol metabolism generates acetaldehyde, highly reactive oxygen species, and fatty acid ethyl esters which destroy pancreatic acinar cells predisposing to develop hemorrhagic pancreatitis leading to shock, renal failure, respiratory failure, and death (Vonlaufen et al., 2007).

### 17.3.4.6 LIVER

Fatty liver progressing to hepatitis and cirrhosis are most common dose-related lethal consequences of alcohol intake (Gao and Bataller, 2011). In chronic alcoholics, the excess NADH generated due to the oxidation of ADH and ALDH affects the tricarboxylic acid cycle and the oxidation of fat thereby promoting fatty liver. Over a period, cellular necrosis and chronic inflammation lead to irreversible changes such as fibrosis and scarring, which are the hallmark of alcoholic cirrhosis. This is due to

alteration of stellate cells into collagen-forming, myofibroblast-like cells, subsequently leading to collagen deposition around portal triad (Lieber, 2000). Molecular and cellular mechanisms for alcoholic cirrhosis include alterations in phospholipid peroxidation, release of harmful free radicals and oxidative stress (Parrish et al., 1991). Diminished oxidase activity and oxygen consumption in hepatic mitochondria can be seen due to decrease in levels of phosphatidylcholine because of ethanol consumption. Cytokines increase rates of fibrinogenesis and fibrosis in the hepatocytes. Additionally, highly reactive compound such as acetaldehyde and some free radicals causes depletion of glutathione an important antioxidant, vitamins, trace metals, adduct formation, reduced transport, and secretion of proteins due to inhibition of tubulin polymerization.

### 17.3.5 VITAMINS AND MINERALS

Nutritional deficiencies in alcoholics can occur due to diminished consumptions and absorption, or reduced utilization of nutrients. Vitamin B complex deficiencies manifest predominantly as peripheral neuropathy, with severe cases progressing to Wernicke's encephalopathy and Korsakoff's psychosis (Harper, 1998). Osteoporosis attributed to reduced osteoblastic activity can occur. Decrease in serum parathormone (PTH) and $Ca^{2+}$ levels results in rebound elevation of PTH without affecting $Ca^{2+}$ levels in initial phase of acute alcohol consumption. Alcohol-induced osteopenia improves with abstinence (Alvisa-Negrin et al., 2009). Hydroxylation of vitamin D takes place in the liver that can be affected in chronic alcoholics with liver damage leading to defective absorption of $Ca^{2+}$ from the intestine and the kidneys.

## 17.3.6 ENDOCRINE SYSTEM

Research indicates a decreased risk hyperglycemia (type 2 diabetes) attributed to improvement in insulin sensitivity on ingestion of modest amounts of alcohol (Baliunas et al., 2009; Hendriks, 2007). Other possible causative factors include elevated levels of acetaldehyde and acetate (Sarkola et al., 2002); anti-inflammatory effects of alcohol (Imhof et al., 2001) and improvement in HDL (Rimm et al. 1999). Conversely, increased intake leads to increase risk of diabetes mellitus due to its dose-dependent effects. This is due to increase in body weight, blood triglycerides, and hypertension (Wannamethee et al., 2003a,b). Impotence is known to appear in men with both acute and chronic alcohol ingestion. Hypothalamic dysfunction and toxic effect of alcohol on Leydig cells result in Testicular atrophy and decreased fertility. Gynecomastia which occurs due to increased metabolism of testosterone and cellular response to estrogen is seen in alcoholic liver disease. Menstrual cycle abnormalities, decreased libido, and low fertility rates are noted in alcoholic women. For those who abstain from alcohol, the prognosis is favorable in regard to hepatic or gonadal failure (O'Farrell et al., 1997).

## 17.3.7 IMMUNOLOGICAL AND HEMATOLOGICAL EFFECTS

Alcohol affects erythrocytes, leucocytes, and thrombocytes (Ballard, 1997). Chronic alcohol use causes microcytic anemia (due to iron deficiency or chronic blood loss), macrocytic anemia (vitamin deficiencies), or normochromic anemia (effect of chronic illness on hematopoiesis). Reversible thrombocytopenia due to trapping of platelets in the spleen and bone marrow has been proposed (Peltz, 1991). Leukopenia, decrease in T-cell number and function, reduced immunoglobin production, and dysfunctional lymphocytes are noted. Chronic alcoholics have poor immunity to certain infections (e.g., *Klebsiella pneumoniae,* listeriosis, and tuberculosis) which is attributed to diminished leukocyte migration into the inflammation sites (Trevejo-Nunez et al., 2015). Altered cytokine regulation especially interleukin2 (IL-2) is also seen in alcoholics. Acquired immunodeficiency syndrome (AIDS) contracted from human immunodeficiency virus-1 (HIV) infection is frequent in alcoholics and in vitro studies with human lymphocytes suggests suppression of CD4 T lymphocytes concanavalin A–stimulated IL-2 production induced by alcohol which enhances the viral replication. Additionally, there are increased rates of high-risk sexual behavior in prolonged alcohol consumption (Stueve and O'Donnell, 2005).

## 17.3.8 NEOPLASMS

Increased risk of cancers involving upper and lower gastrointestinal tract and female breast have been reported in individuals with chronic high intake of alcohol (Fedirko et al., 2011; Islami et al., 2010; Turati et al., 2010). This can be due to specific variants of ADH, ALDH, and CYP2E1 enzymes that metabolize alcohol; increased estrogen levels; dysfunctional folic acid metabolism and DNA repair (Boffetta and Hashibe, 2006). Interestingly, this lessens the possibility of acquiring certain cancers, for instance, hypernephroma (Bellocco et al., 2012; Song et al., 2012), Hodgkin's lymphoma (Tramacere et al., 2012a), and non-Hodgkin's lymphoma (Tramacere et al. 2012b) are noted with alcohol intake. Moderate alcohol ingestion leading to increased insulin sensitivity has been attributed to alcohol's protection against renal cancer (Davies et al., 2002; Facchini et al. 1994; Joosten et al., 2008). However, the mechanisms involved in Hodgkin's lymphoma and non-Hodgkin's lymphoma are currently unidentified (Tramacere et al., 2012a, b). With the biological mechanism remaining undefined, these witnessed, protective effects should be understood cautiously.

### 17.3.9   TERATOGENIC EFFECTS

During pregnancy, alcohol intake ensues in distinguishing physical and mental defects in the fetus, known as fetal alcohol syndrome (FAS). CNS, craniofacial abnormalities, dysfunction, and pre/postnatal stunting of growth are characteristic triad noted in the neonates (Jones and Smith, 1973). As the child grows, hearing and speech disorders are apparent (Church and Kaltenbach, 1997). Craniofacial abnormalities include micro-cephaly, shortened palpebral fissures, a flat midface, a long and smooth philtrum, cleft lip and palate, and epicanthal folds. Other characteristic features noticed in these patients are diminished volumes in the cerebrum, cerebellum, basal ganglia, and corpus callosum resulting in atrophy (Mattson et al., 1992). CNS dysfunction includes hyperactivity, attention deficits, learning disabilities, and mental retardation. Adolescents who abuse alcohol have been attributed to fetal alcohol exposure (Baer et al., 1998). Spontaneous abortion can occur during first trimester on consumption of large quantities of alcohol. Lastly, fetal damage and mortality are augmented by concurrent substance abuse, such as consumption of ethanol and cocaine.

### 17.3.10   OTHER EFFECTS

Ethanol augments cutaneous and gastric circulation thereby causing a feeling of warmth. Increased sweating and heat loss occur, leading to decrease in internal body temperature. Intake of large quantities leads to depressed central temperature regulating system and hypothermia. Alcohol is one of the leading risk factors for hypothermia, as reported by many studies (Albiin and Eriksson, 1984). Diuresis occurs due to suppression of vasopressin (antidiuretic hormone) and volume loading. Tolerance occurs for the diuretic effects and manifested by decreased urine output in response to trial dose with ethanol (Collins et al., 1992). Upon alcohol withdrawal, alcoholics experience increased vasopressin release, which leads to greater water retention and dilutional hyponatremia.

### 17.4   ACUTE ETHANOL INTOXICATION

BAC is a quantitative measure of how much alcohol is present in the bloodstream and expressed as the weight of ethanol (g) in 100 mL of blood. Each of the beverages alcoholic content is quantified relative to its proof which is twice the alcoholic content. For example, a glass of 24 proof wine would be 12% alcohol (Stockwell and Honig, 1990). The content of alcohol in wine ranges from 10% to 15%, in beer from 4% to 6% (volume/volume), and 40% or more in distilled spirits. Determination of alcohol content in the body is usually done by measuring in blood is primarily indicated for medicolegal purposes. More common indication to perform alcohol content is to inspect for driving under the influence (DUI) as there has been an underlying correlation between excess alcohol consumption and road traffic accidents. The content of alcohol at which an individual is considered lawfully impaired varies among different countries. In the United States, 0.08% BAC is set as the legitimate constrain for DUI (Fell et al., 2013). A BAC of 0.04% can lead to DUI conviction for commercial drivers and for those less than 21 years, even trace quantities of alcohol lead to DUI arrest indicating there is a zero-tolerance limit (Kelley-Baker et al., 2013). Other commonly measured way of estimating alcohol in the body is by breath analysis. Since the proportion of alcohol in blood and end-expiratory alveolar air is somewhat same, BAC can be measured by estimating the alcohol levels in end-expiratory alveolar expired air with a relentless proportion of 1:2100. For instance, a breathalyzer estimation of 0.10 mg/L of alcohol would designate a 0.021 g/dL of alcohol in blood (Cederbaum, 2012).

Ethanol intoxication occurs in the subsequent 2 to 3 drinks, through the peak time of BAC ~30–60 min following intake on an empty stomach. The dose-related specific effects in association with BAC levels have been concise in Table 17.1. Concomitant administration of alcohol, along with other CNS depressants, or other medications known to cause sleepiness can exacerbate the severity of the symptoms and can appear at a lower BAC. BAC of greater than 400 mg/dL is considered fatal, conversely, alcohol-tolerant individuals endure the equivalent blood alcohol levels (Perper et al., 1986). The characterization of intoxication differs across various countries. In most of the states of the United States, legal intoxication is 80 mg/dL. The BAC is mainly reliant upon the body weight, gender, and degree of absorption from the GI tract. In men, BAC of 67–92 mg/dL and about 30–53 mg/dL is attained after mixed meal or on fasting state preceding intake of three standard drinks (42 g of ethanol). In women who consume same amount of alcohol as men, BAC attained is relatively more than

men due to lesser weight, less body water per unit of weight, and lastly due to less gastric ADH activity (Frezza et al., 1990). The characteristic breath odor noted in alcoholics is due to impurities in alcoholic beverages; however, suspected alcoholism may be perplexing as there are other causes of similar breath odor. Hence, to confirm alcohol intoxication, BAC determination has to be performed (Schuckit, 2006b). Lastly, violent crimes like homicide and suicide (28–86%), assault (24–37%), robbery (7–72%), and sexual offenses (13–60%) are strongly associated with binge drinking (Brewer and Swahn, 2005).

Stabilizing patient's hemodynamic condition and vitals is the primary treatment involved in acute alcohol intoxication; furthermore, accelerating the removal of alcohol, symptomatic treatment of various manifestations and management depends upon the severity of CNS and respiratory depression. Comatose patients experiencing respiratory depression will likely require intubation to protect and increase ventilation through the airway. A gastric lavage is administered, but precautions to avert pulmonary aspiration have to be undertaken. Alcohol is easily miscible in water and its ability to diffuse through biological membranes, can be readily eliminated from blood by hemodialysis (Schuckit, 2006b). Clinical care includes observing the patient for 4–6 h, for the duration of which the patient metabolizes the consumed ethanol. BAC decreases by 15 mg/dL every hour (Cederbaum, 2012). Some patients are susceptible to extremely violent behavior during this period and may require sedatives or antipsychotics. When administering sedatives cautiously, to those who have consumed excessive amounts of alcohol; administering multiple CNS depressants result in synergistic effects.

## 17.5 TOLERANCE, DEPENDENCE, AND CHRONIC ETHANOL USE

Tolerance is the diminished social, behavioral, or physiological response to the corresponding dose of ethanol because of repeated use and characterized by downregulation of receptors. Acute tolerance is noted soon after alcohol intake (Tabakoff et al., 1986). Chronic tolerance occurs in long term and characterized by alterations in alcohol-metabolizing enzymes. Physical dependence is established by how the body experiences physiological adaptation of withdrawal syndrome on alcohol termination. Psychological dependence is characterized by craving and drug-seeking behavior (Baconi et al., 2015). The molecular mechanisms involved in tolerance and dependence include changes in synaptic and intracellular signaling due to changes in gene expression. Chronic actions of alcohol comprise alterations in glutamate and GABA receptor neurotransmitter signaling and downstream signaling pathways. For instance, hyperactivation of NMDA-receptors in chronic alcoholics attribute to the CNS hyper excitability and neurotoxicity during ethanol withdrawal (Becker and Redmond, 2002).

Addiction means compulsive utilization of drugs for non-medical purposes. Hence, addiction is a kind of behavior whereas tolerance and dependence are physiological changes. The neurobiological basis of the change from controlled, volitional alcohol use to compulsive, and uncontrolled addiction remains obscure and is presently being investigated. Alterations in the dopaminergic reward system, prefrontal cortex (important for judgment and emotion) are highly sensitive to impairment from alcohol and compromised in the alcoholics (Boileau et al., 2003; Pfefferbaum et al., 1998). Thus, damage to executive function in cortical regions may be responsible for lack of judgment and control

that is stated as obsessive alcohol consumption. The decrease in the brain volume and impairment of function in chronic alcoholics is at least partially reversible by abstinence but worsen with continued drinking. Early diagnosis and treatment of alcoholism are crucial to preventing brain damage that can ultimately lead to serious addiction.

### 17.5.1 GENETIC POLYMORPHISMS ASSOCIATED WITH ALCOHOL USE DISORDERS

Social, cultural, biological, and interpersonal factors are the important risk factors contributing to the risk of developing AUD (Schuckit, 2009), of which 60% develop AUD because of genetic or hereditary factors. Polymorphisms in the enzymes involved in ethanol metabolism are validated by the role of genes as noted in Asian populations offer protection from alcoholism. Interestingly, genetic variations of ADH that exhibit increased activity and ALDH variants with diminished activity provide protection against excessive drinking. The above-mentioned effect is because of accumulation of acetaldehyde which causes disagreeable effects in these individuals (Li, 2000). Conversely, polymorphisms in $GABA_A$ receptors (Dick et al., 2006a, b), an *ADH4* variant (Edenberg et al., 2006), as well as a muscarinic receptor gene, *CHRM2* (Jones et al., 2006) are correlated with greater risk of drinking and impulsivity. In addition, genetic alterations for decreased response have been recognized for two $GABA_A$ subunits (Dick et al., 2006a, b), polymorphism in the 5-HT transporter' promoter region involves lesser 5-HT in synaptic space (Barr et al., 2004), a polymorphism of the $K^+$ channel-related *KCNMA1*, and a variant nicotinic ACh receptor that is accompanied to have a greater risk for intense smoking (Joslyn et al., 2008).

Polymorphisms in several 5-HT receptors are involved in antisocial alcoholism (Ducci et al., 2009). Polymorphisms in dopaminergic and opioid systems and in genes that correlate with a predisposition to bipolar disorder, schizophrenia, and several anxiety conditions likewise influence a diversity of drug abuse disorders. All these genes which are involved in the developing risk of alcoholism are condensed in Table 17.2.

**TABLE 17.2** Outline of the Phenotype and Genes Involved in Alcohol Use Disorder.

| Phenotype | Genes involved |
|---|---|
| Unpleasant effects (high acetaldehyde) and facial flushing | *ALDH2, ADH1B, ADH1C* |
| Increased impulsivity | *GABRA2, ADH2, CHRM2* |
| Low levels of response to alcohol | *GABRA1, GABRA6, 5HTT* promoter, *KCNMA1* and *CHRN* |
| Antisocial activates | Several genes of 5-HT receptors |

## 17.6 PHARMACOTHERAPY OF ALCOHOLISM

### 17.6.1 TREATMENT OF ALCOHOL WITHDRAWAL SYNDROME

Alcohol dependence individual discontinuing alcohol abruptly experience anxiety, insomnia, and reduced seizure threshold. Mild forms of withdrawal syndrome are distinguished by tachycardia, tremors, anxiety, and insomnia after 6–8 h after discontinuation (Kattimani and Bharadwaj, 2013), following which these effects decrease over time. Severe reactions are characterized by withdrawal seizures or hallucinations during initial period of withdrawal. A severe form of withdrawal termed as delirium tremens occurs in the next few days and manifests as delirium, autonomic hyperactivity, and eventually leading to cardiovascular collapse. Treatment of alcohol withdrawal mainly focuses on preventing seizures, delirium, and cardiovascular manifestations (Dixit et al., 2016). Seizures can be treated with benzodiazepines with lorazepam most frequently chosen medication because of its prolonged duration of action, hence offering benefit from recurrent seizures. Additionally, it is not as much metabolized through hepatic route compared to other benzodiazepines. Interestingly, impaired hepatic function in chronic alcoholics validates its use. Restorating potassium, magnesium, and phosphate equilibrium should occur rapidly. In all cases, thiamine therapy is administered (Fisher, 2009). Further pharmacological therapy is not required for individuals experiencing mild alcohol withdrawal (Mirijello et al., 2015).

Strategies to treat serious cases include replacing alcohol with a long-acting sedative-hypnotic drug and steadily decreasing its dose. Benzodiazepines are quite safe and can be administered as a symptom-triggered or fixed-schedule regimen with or without loading, making them a preferred treatment. Symptom-triggered therapy utilizes a validated protocol for drug administration called "The Clinical Institute Withdrawal of Alcohol Scale, revised (CIWA-Ar)." This protocol involves only taking medication once the CIWA-Ar score is greater than 8 points (Sullivan et al., 1989). Symptom-triggered therapy exhibits a better efficacy profile than fixed-schedule dosing for patients undergoing detoxification (Cassidy et al., 2012). Chlordiazepoxide and diazepam, which are long-acting benzodiazepines, are superior because they require less frequent dosing. These long-acting drugs exhibit a tapering effect due to the slow elimination of their pharmacologically active metabolites; this characteristic is disadvantageous to those with impaired liver function because the drugs and their active metabolites are susceptible to accumulation (Sachdeva et al., 2015). In comparison, lorazepam and oxazepam which are rapidly acting agents, do not accumulate due to their rapid conversion to inactive water-soluble metabolites. These short-acting drugs are an appropriate therapy for alcoholic patients with liver disorders (Greenblatt, 1981). Oral benzodiazepines are administered to mild and moderate cases, whereas, those with severe withdrawal reactions may require parenteral administration. Following critical treatment of alcohol withdrawal syndrome, treatment with benzodiazepines must be tapered over a period of several weeks. Thorough detoxification resulting in return of nervous system to its normal functions may require several months of alcohol abstinence rather than a couple of days.

### 17.6.2 MANAGEMENT OF ALCOHOL DEPENDENCE

Complete abstinence is by far the most beneficial therapy of alcohol dependence. The principal management of alcohol dependence following detoxification is psychosocial or behavioral therapy, improvements in personal health, and social functioning, either in intensive inpatient or in outpatient rehabilitation programs. Currently, AA (alcoholics anonymous) method has been utilized and involves psychosocial techniques (motivational interviewing, cognitive behavioral therapy, family therapy, self-help groups, and contingency management) for changing behavior (Kelly, 2017). Untreated psychiatric problems, such as depression, maniacs, panic disorders, can increase the probability of relapse in detoxified alcoholics (Overall et al., 1973). Hence,

counseling and medications to treat psychiatric problems can potentially decrease the relapse for alcoholic patients. Currently only three drugs are FDA approved for adjunctive management of alcohol dependence that is naltrexone, disulfiram, besides acamprosate (Mann, 2004). The objectives of using these medicines are to provide support to the alcoholic patients in maintaining abstinence and are discussed as follows.

### 17.6.2.1 NALTREXONE

Naltrexone is an opioid-receptor antagonist which lessens craving for alcohol and urge to drink in addition to decreasing the volume and number of alcohol drinking days (Rösner et al., 2010). A strong connection concerning alcohol consumption and opioids were noted in experimental animals, wherein opioids increased alcohol intake whereas opioid antagonists inhibited self-administration of alcohol. Naltrexone blocks dopaminergic influence on brain endorphins (critical to reward) which are released after alcohol consumption. Cognitive behavioral therapy, or some other psychosocial therapy, is recommended in association with naltrexone (Anton et al., 1999). An extended release formulation as intramuscular injection given monthly was FDA approved in 2006. Nausea is the most frequent side effect observed; however, at high doses, liver damage can occur. Severe hepatotoxicity is seen when naltrexone and disulfiram are taken together, hence this combination should be avoided. Finally, the overall benefits of naltrexone seem modest in nature (Donoghue et al., 2015; Garbutt, 2010).

### 17.6.2.2 ACAMPROSATE

Acamprosate is an analog of GABA (Plosker, 2015), which mainly acts by stabilizing the neurotransmitters equilibrium in the brain, which is altered in chronic alcoholics (Williams, 2005). The primary mechanism is through a weak NMDA-receptor antagonist activity as its bind to polyamine site and as a $GABA_A$ activator (Naassila et al., 1998). Additionally, it decreases the elevated glutamate levels as well as inhibits metabotropic glutamate receptor 5 (mGlur5). Acamprosate at a dose of 1.3–2 g/day reduces rate of drinking as well as relapse in those abstinent. Most common side effect is diarrhea, yet abuse liability is not noted. Acamprosate with psychosocial interventions have shown potential improvements in alcohol consumption outcomes (Plosker, 2015). Increase efficacy of acamprosate is demonstrated when given besides disulfiram with no adverse drug interactions (Besson et al., 1998).

### 17.6.2.3 DISULFIRAM

In 1951, FDA approved oldest drug for AUD was disulfiram. It primarily acts by irreversibly inhibiting ALDH enzyme through competing for binding sites with NAD coenzyme. In patients taking disulfiram, ingestion of even minute quantities of alcohol produces nausea, vomiting, hot flushing, throbbing headache, sweating, and thirst. Severe cases lead to respiratory depression, arrhythmias, cardiovascular collapse, seizures, coma, and death. This is attributed to acetaldehyde being accumulated is immediately oxidized by ALDH. Disulfiram as a therapeutic agent is unsafe, which prompts for using it under strict medical supervision (Skinner et al., 2014). Prior to administration of disulfiram, 12 h of abstinence is mandatory. In the early phase, loading dose of 500 mg is administered for initial few (1 to 2) weeks, monitored by a maintenance dose (125–500 mg) daily subject subjected to adverse

effects. Based on the time-consuming rate of return of ALDH, sensitization to alcohol may be there for 14 days after the final ingestion of disulfiram (Johansson, 1992). Disulfiram can interfere with the metabolism of phenytoin, chlordiazepoxide, barbiturates, warfarin, and other drugs by inhibiting various hepatic microsomal cytochromes (Enghusen Poulsen et al., 1992).

Inhibition of dopamine-β-hydroxylase (by chelation of copper) which converts dopamine to norepinephrine in noradrenergic neurons is because of a metabolite of disulfiram- diethyldithiocarbamate (Gaval-Cruz and Weinshenker, 2009) leading to psychosis at rare instances (Mohapatra and Rath, 2017). Addiction to cocaine can also be treated with disulfiram (Carroll et al., 1998). Disulfiram's clinical efficacy in diminishing cravings and avoiding drinking is mostly due to secondary CNS actions, which arise from regulation of catecholamine neurotransmission (Goldstein and Nakajima, 1967).

### 17.6.2.4 FDA-APPROVED DRUGS WITH POTENTIAL REPURPOSING FOR ALCOHOL USE DISORDERS

Topiramate, an anticonvulsant drug has been demonstrated in improving drinking behavior by decreasing heavy drinking episodes as well as maintaining abstinence. Topiramate enhances $GABA_A$-mediated inhibitory action as well as antagonize glutamate receptors. Other mechanisms include inhibition of carbonic anhydrase and inhibition of sodium and calcium channels (Dodgson et al., 2000; White et al., 2000; Zhang et al., 2000). The reinforcing effects seen in alcohol abuse are due to enhanced dopamine release from the nucleus accumbens. This is inhibited through glutamate antagonism mediated by topiramate (Johnson and Ait-Daoud, 2010). The increase in GABA neuronal activity in nucleus accumbens is inhibited by topiramate. Through these mechanisms, topiramate may decrease dopamine activity in the reward pathway with alcohol ingestion and reduce withdrawal symptoms (Paparrigopoulos et al., 2011). Furthermore, decreased craving for alcohol and markedly decrease in serum gamma-glutamyl transpeptidase (GGT) levels (an important biomarker of chronic alcohol abuse) were observed with topiramate (Johnson et al., 2003a). Adverse events include burning sensation of skin (paresthesia), alteration of taste sensation, anorexia, inability to concentrate and focus. Dose-related adverse effects restrict the use of topiramate, thereby, requiring weekly dosage titration to effective doses of 100–300 mg/day. Pharmacotherapy with topiramate in those with alcohol abuse should be accompanied by counseling or psychotherapy for best outcomes.

Treatment of alcoholism is moving to a new paradigm of tailoring to individual's genetic profile. For example, polymorphisms (*5'HTTLPR* and *rs1042173*) in the serotonin 5-HTT gene has been linked to early onset alcoholism (Virkkunen and Linnoila, 2002). Ondansetron is a serotonin receptor antagonist with selectivity against 5-HT3 subtype which is utilized for managing chemotherapy-induced nausea and vomiting. It has shown efficacy in improving drinking concerns in early-onset alcoholics responding poorly to behavioral treatment (Kranzler et al., 2003). This action is brought through modulating cortico-mesolimbic dopamine system modulation. Decrease in alcohol consumption and reduction in subjective effects like desire to drink is some of the effects noted on ondansetron administration. Since there are genetic polymorphisms involved, genetic screening beforehand to match a treatment with a genotype is a significant influence to tailoring treatments for alcoholism. Numerous other drugs currently under study and investigation for the management of alcoholism and which have depicted positive results have been summarized in Table 17.3 including its mechanism of action.

**TABLE 17.3** Drugs with Potential Benefit in Treatment of Alcohol Use Disorder/Alcoholism.

| Drug | Mechanism of actions | Effects | References |
|------|---------------------|---------|------------|
| **Anticonvulsants** | | | |
| Topiramate | Block voltage-gated sodium and calcium channels; Enhances GABA receptor; Blocks AMPA/kainate receptors | Induces abstinence, reduces relapse | Porter et al., 2012; Baltieri et al., 2008; Falk et al., 2010 |
| Gabapentin | Interaction with high-affinity binding sites in brain membranes (auxiliary subunit of voltage-sensitive calcium channels) | Induces abstinence, reduces relapse, dependence and craving | Taylor, 1997; Stock et al., 2013; Mason et al., 2014 |
| Pregabalin | Voltage-gated calcium channels (α2δ subunit) | Reduces relapse, craving, and alcohol withdrawal syndrome | Martinotti et al., 2010; Förg et al., 2012; Addolorato and Leggio, 2010 |
| **Antipsychotics** | | | |
| Aripiprazole | Partial dopamine agonist, and antagonist of 5-HT2A and 5-HT7 receptors | Reduced alcohol craving in MDD patients, inhibits cue-induced heavy drinking, decreased volume and number of drinks in low but not in high impulsivity patients | Han et al., 2013; Myrick et al., 2010; Anton et al., 2017 |

**TABLE 17.3** *(Continued)*

| Drug | Mechanism of actions | Effects | References |
|------|----------------------|---------|------------|
| Quetiapine | Antagonist of serotonin-dopamine and adrenergic receptors, partial agonist on 5-HT1A receptors | Reduced alcohol intake in open-label/retrospective study in AD patients, reduced akathisia and depression, improved sleep, but was not effective multisite RCT placebo control trial in heavy alcohol drinking patients | Monnelly et al., 2004; Martinotti et al., 2008; Kurlawala and Vatsalya, 2016; Litten et al., 2012 |
| Other class of drugs | | | |
| Ondansetron | Selective antagonist of 5-HT3 receptors | Reduced alcohol intake in early-onset alcoholics, reduced depression, anxiety, and hostility | Kranzler et al., 2003; Johnson et al., 2003b |
| Nalmefene | Opioid receptor antagonist | Reduced alcohol dependence, heavy drinking days, Borderline personality disorders and AUD | Mason et al., 1999; Di Nicola et al.,2017; Martín-Blanco et al., 2017 |
| Baclofen | GABA-B agonist | Combined or ineffective in reducing alcohol intake, craving, abstinence also conflicting results with alcohol dependence | Imbert et al., 2015; Beraha et al., 2016; Reynaud et al., 2017 |

## 17.6.2.5 *NOVEL ALCOHOL USE DISORDER MEDICATIONS AND THEIR SIGNALING PATHWAYS*

AUD is a complex heterogeneous disease differentiated by multiple neurotransmitters, hormones and neuromodulatory systems. Hence the currently FDA approved medications are unlikely to benefit and show efficacy in every individual. Hence, intense research on other signaling mechanisms and development of novel therapeutic targets are the main objectives. The emphasis is on developing drugs with high selectivity to reduce alcohol-mediated effects on the organ systems, and less incidence of adverse effects. Some of the medications described exist in various stages of clinical trials and show abundant prospective in the future for the management of AUD and listed in Table 17.4. Some of the promising candidate compounds are listed below.

## 17.7 FUTURE OPPORTUNITIES AND CHALLENGES

Due to the wide variability in the response to currently available treatments, search for new information and approaches related to mechanisms of molecular genetics implicated in neurocircuitry of the CNS would open wide avenue for identifying new targets for interventions. An integrated network based system has been proposed by National Institutes of Health Library of Integrated Network-based Cellular Signatures (LINCS) program (Keenan et al., 2018). Alterations in expression of genes and additional molecular, cellular process are labeled to understand the biological mechanisms in a better way. Finally these diverse information can be integrated with computational tools to have a complete knowledge of normal and the disease processes which would help in development of novel biomarkers and therapeutics (Oulas et al., 2019).

**TABLE 17.4** Novel Drugs Under Investigation for Treatment of Alcohol Use Disorder.

| Drug | Mechanisms of actions | Effects | References |
|---|---|---|---|
| ABT-436 | Vasopressin receptor (type 1B) antagonist | Increases abstinence of alcohol, prevents relapse | Katz et al., 2016; Ryan et al., 2017 |
| Ibudilast | Phosphodiesterase 4 and 10 inhibitor; Inhibits macrophage migration inhibitory factor (MIF) | Decrease mood altering effects of alcohol | Ray et al., 2017; Bell et al., 2015 |
| Orexin | Agonist for orexin receptor | Prospective biomarker of alcohol relapse | Ziółkowski et al., 2016; von der Goltz et al., 2011 |
| Oxytocin | GABA receptor agonist | Reverses tolerance and reduces alcohol withdrawal syndrome | Pedersen et al., 2013 |
| Pexacerfont | CRF receptor-1 antagonist | Reversal of dependence and craving in animal models. No significant effects on alcohol craving in human studies | Roberto et al., 2010; Spierling and Zorilla, 2017 |
| Prazosin | Alpha 1 receptor antagonist | Decrease in alcohol consumption | Simpson et al., 2015; Petrakis et al., 2016 |
| Varenicline | Nicotine Acetylcholine receptor α4β2 partial agonist | Decreased alcohol craving | Verplaetse et al., 2016; Litten et al., 2013 |

Various novel modalities for evaluating function of brain in vivo are being developed in association with Brain Research through Advancing Innovative Neurotechnologies (BRAIN) initiative in the United States and other nations worldwide (Braininitiative.nih. gov and Humanbrainproject.eu, accessed 2018). For instance, techniques like CLARITY which involves removing specific components of the tissue and replacing with exogenous components to increase the accessibility and functionality (Chung and Deisseroth, 2013) and optogenetics are being developed (Deisseroth, 2012). These newer techniques either alone or in combination would help in studying the functionality and dysfunction in the nervous system thereby providing newer targets to explore neurocircuitry (Day et al., 2015; Nussinov and Tsai, 2015). A recent initiative by the National Institutes of Health termed "Illuminating the Druggable Genome (IDG) program aims at improving the knowledge of the properties, biological functions of various proteins that are

not well known within usually drug-targeted families of the protein (commonfund.nih.gov/ IDG, assessed 2018). The above mentioned innovative programs equip researchers with the knowledge and resources to facilitate critical discoveries in drug development, supporting efforts to combat intricate diseases and improve health overall. Genome-wide association studies have enabled the identification of multiple genetic markers, and current studies are attempting to recognize precise single nucleotide polymorphisms (SNPs) in humans. These SNPs will be instrumental in predicting a patient's predisposition for development of addiction and in determining their response to medications.

Excessive alcohol consumption phenotypes have been correlated with multiple polymorphisms in corticotropin-releasing factor (CRF) system in humans, with specific associations observed in relation to stress. SNPs of *CRF1* gene is associated with alcohol dependent individuals as well as binge drinking in young people (Treutlein et al., 2006). One of these SNPs potentially influences transcription of the *CRF1* receptor gene. In adolescents homozygous for a SNP of the *Crhr1* gene, a history of stress was associated with early inception of ingestion and amplified risk of future alcohol consumption (Blomeyer et al., 2008; Schmid et al., 2010). Furthermore, SNPs in the *Crhr1* gene has strong correlation with excessive alcohol ingestion in previously addicted people (Treutlein et al., 2006).

SNPs in the *NPY2R* gene is associated with alcohol dependence and addiction, withdrawal symptoms to alcohol as well as concomitant alcohol and cocaine dependence (Wetherill et al., 2008). Similarly, neuropeptide Y (*NPY*) gene with specific polymorphisms is correlated to alcohol dependence (Bhaskar et al., 2013). Hence, some genetic subgroups would potentially be more responsive to a drug that inhibits *CRF* or stimulates *NPY* neurotransmission. Supporting this line of reasoning, a clinical study performed with acamprosate for AUD treatment showed standard serum glutamate levels to be greater in responders than non-responders. The levels of serum glutamate came back to normal in responders subsequently with acamprosate treatment while no significant change existed in non-responders. This suggests that responsiveness to acamprosate is derived from genetic predisposition. Ultimately, SNP polymorphisms are a means of expecting which alcoholics are probably to respond to specific medications (Nam et al., 2015).

## 17.8 CONCLUSION

This chapter has summarized the physiological effects of alcohol and pharmacological therapies for AUD, including current and newly emerging drugs. Exploring the neurophysiological

and pharmacological mechanisms involved in alcohol abuse has been more beneficial than the conventional syndromic approach because it has resulted in the discovery of new therapeutic targets. Despite these advancements in understanding the pharmacology of alcohol, further research must be conducted to investigate the role of genes, environmental elements, and complex interactions between alcohol and other substances of abuse. With AUDs persistently being a major socioeconomic and health concern internationally, there is a critical prerequisite to conduct more applied, translational research for improved therapeutic strategies.

## KEYWORDS

- alcohol dehydrogenase
- acute and chronic alcoholics
- alcohol use disorder
- alcoholism
- drug abuse
- ethanol

## REFERENCES

Addolorato, G.; Leggio, L. Pregabalin Similar to Lorazepam for Alcohol Withdrawal Symptoms. *Evid. Based. Med.* **2010,** *15* (3), 73.

Addolorato, G.; Montalto, M.; Capristo, E.; Certo, M.; Fedeli, G.; Gentiloni, N.; Stefanini, G. F.; Gasbarrini, G. Influence of Alcohol on Gastrointestinal Motility: Lactulose Breath Hydrogen Testing in Orocecal Transit Time in Chronic Alcoholics, Social Drinkers and Teetotaler Subjects. *Hepatogastroenterology.* **1997,** *44* (16), 1076–1081.

Akbar, M.; Egli, M.; Cho, Y.-E.; Song, B.-J.; Noronha, A. Medications for Alcohol Use Disorders: An Overview. *Pharmacol. Ther.* **2018,** *185*, 64–85.

Albiin, N.; Eriksson, A. Fatal Accidental Hypothermia and Alcohol. *Alcohol Alcohol* **1984,** *19* (1), 13–22.

Alexander-Kaufman, K.; Harper, C.; Wilce, P.; Matsumoto, I. Cerebellar Vermis Proteome of Chronic Alcoholic Individuals. *Alcohol. Clin. Exp. Res.* **2007,** *31* (8), 1286–1296.

Alvisa-Negrin, J.; Gonzalez-Reimers, E.; Santolaria-Fernandez, F.; Garcia-Valdecasas-Campelo, E.; Valls, M. R. A.; Pelazas-Gonzalez, R.; Duran-Castellon, M. C.; de los Angeles Gomez-Rodriguez, M. Osteopenia in Alcoholics: Effect of Alcohol Abstinence. *Alcohol Alcohol* **2009,** *44* (5), 468–475.

American Psychiatric Association. *Diagnostic and Statistical Manual of Mental Disorders*; American Psychiatric Association, 2013.

Anton, R. F.; Moak, D. H.; Waid, L. R.; Latham, P. K.; Malcolm, R. J.; Dias, J. K. Naltrexone and Cognitive Behavioral Therapy for the Treatment of Outpatient Alcoholics: Results of a Placebo-Controlled Trial. *Am. J. Psychiatry* **1999,** *156* (11), 1758–1764.

Anton, R. F.; Schacht, J. P.; Voronin, K. E.; Randall, P. K. Aripiprazole Suppression of Drinking in a Clinical Laboratory Paradigm: Influence of Impulsivity and Self-Control. *Alcohol. Clin. Exp. Res.* **2017,** *41* (7), 1370–1380.

Aryal, P.; Dvir, H.; Choe, S.; Slesinger, P. A. A Discrete Alcohol Pocket Involved in GIRK Channel Activation. *Nat. Neurosci.* **2009,** *12* (8), 988–995.

Baan, R.; Straif, K.; Grosse, Y.; Secretan, B.; El Ghissassi, F.; Bouvard, V.; Altieri, A.; Cogliano, V.; WHO International Agency for Research on Cancer Monograph Working Group. Carcinogenicity of Alcoholic Beverages. *Lancet. Oncol.* **2007,** *8* (4), 292–293.

Baconi, D. L.; Ciobanu, A.-M.; Vlăsceanu, A. M.; Negrei, C. Current Concepts on Drug Abuse and Dependence. *J. Mind Med. Sci.* **2015,** *2* (1), 18–33.

Baer, J. S.; Barr, H. M.; Bookstein, F. L.; Sampson, P. D.; Streissguth, A. P. Prenatal Alcohol Exposure and Family History of Alcoholism in the Etiology of Adolescent Alcohol Problems. *J. Stud. Alcohol* **1998,** *59* (5), 533–543.

Baliunas, D. O.; Taylor, B. J.; Irving, H.; Roerecke, M.; Patra, J.; Mohapatra, S.; Rehm, J. Alcohol as a Risk Factor for Type 2 Diabetes: A Systematic Review and Meta-Analysis. *Diabetes Care* **2009,** *32* (11), 2123–2132.

Ballard, H. S. The Hematological Complications of Alcoholism. *Alcohol Health Res. World* **1997,** *21* (1), 42–52.

Baltieri, D. A.; Daró, F. R.; Ribeiro, P. L.; de Andrade, A. G. Comparing Topiramate with Naltrexone in the Treatment of Alcohol Dependence. *Addiction* **2008,** *103* (12), 2035–2044.

Barr, C. S.; Newman, T. K.; Lindell, S.; Shannon, C.; Champoux, M.; Lesch, K. P.; Suomi, S. J.; Goldman, D.; Higley, J. D. Interaction Between Serotonin Transporter Gene Variation and RearingCondition in Alcohol Preference and Consumption in Female Primates. *Arch. Gen. Psychiatry* **2004,** *61* (11), 1146.

Bartsch, A. J.; Homola, G.; Biller, A.; Smith, S. M.; Weijers, H.-G.; Wiesbeck, G. A.; Jenkinson, M.; Stefano, N. D.; Solymosi, L.; Bendszus, M. Manifestations of Early Brain Recovery Associated with Abstinence from Alcoholism. *Brain* **2006,** *130* (1), 36–47.

Becker, H. C.; Redmond, N. *Role of Glutamate in Alcohol Withdrawal Kindling*, 2002; pp 375–387.

Bell, R. L.; Lopez, M. F.; Cui, C.; Egli, M.; Johnson, K. W.; Franklin, K. M.; Becker, H. C. Ibudilast Reduces Alcohol Drinking in Multiple Animal Models of Alcohol Dependence. *Addict. Biol.* **2015,** *20* (1), 38–42.

Bellocco, R.; Pasquali, E.; Rota, M.; Bagnardi, V.; Tramacere, I.; Scotti, L.; Pelucchi, C.; Boffetta, P.; Corrao, G.; La Vecchia, C. Alcohol Drinking and Risk of Renal Cell Carcinoma: Results of a Meta-Analysis. *Ann. Oncol.* **2012,** *23* (9), 2235–2244.

Beraha, E. M.; Salemink, E.; Goudriaan, A. E.; Bakker, A.; de Jong, D.; Smits, N.; Zwart, J. W.; Geest, D. van; Bodewits, P.; Schiphof, T.; et al. Efficacy and Safety of High-Dose Baclofen for the Treatment of Alcohol Dependence: A Multicentre, Randomised, Double-Blind Controlled Trial. *Eur. Neuropsychopharmacol.* **2016,** *26* (12), 1950–1959.

Besson, J.; Aeby, F.; Kasas, A.; Lehert, P.; Potgieter, A. Combined Efficacy of Acamprosate and Disulfiram in the Treatment of Alcoholism: A Controlled Study. *Alcohol. Clin. Exp. Res.* **1998,** *22* (3), 573–579.

Bhaskar, L. V. K. S.; Thangaraj, K.; Kumar, K. P.; Pardhasaradhi, G.; Singh, L.; Rao, V. R. Association between Neuropeptide Y Gene Polymorphisms and Alcohol Dependence: A Case-Control Study in Two Independent Populations. *Eur. Addict. Res.* **2013,** *19* (6), 307–313.

Blomeyer, D.; Treutlein, J.; Esser, G.; Schmidt, M. H.; Schumann, G.; Laucht, M. Interaction between CRHR1 Gene and Stressful Life Events Predicts Adolescent Heavy Alcohol Use. *Biol. Psychiatry* **2008,** *63* (2), 146–151.

Boffetta, P.; Hashibe, M. Alcohol and Cancer. *Lancet Oncol.* **2006,** *7* (2), 149–156.

Boileau, I.; Assaad, J.-M.; Pihl, R. O.; Benkelfat, C.; Leyton, M.; Diksic, M.; Tremblay, R. E.; Dagher, A.

Alcohol Promotes Dopamine Release in the Human Nucleus Accumbens. *Synapse* **2003,** *49* (4), 226–231.

Bosetti, C., La Vecchia, C., Negri, E., Franceschi, S. Wine and other Types of Alcoholic Beverages and the Risk of Esophageal Cancer. *Eur. J. Clin. Nutr.* **2000,** *54* (12), 918–920.

Brewer, R. D.; Swahn, M. H. Binge Drinking and Violence. *JAMA* **2005,** *294* (5), 616.

Briasoulis, A.; Agarwal, V.; Messerli, F. H. Alcohol Consumption and the Risk of Hypertension in Men and Women: A Systematic Review and Meta-Analysis. *J. Clin. Hypertens.* **2012,** *14* (11), 792–798.

Carroll, K. M.; Nich, C.; Ball, S. A.; Mccance, E.; Rounsavile, B. J. Treatment of Cocaine and Alcohol Dependence with Psychotherapy and Disulfiram. *Addiction* **1998,** *93* (5), 713–727.

Carta, M.; Ariwodola, O. J.; Weiner, J. L.; Valenzuela, C. F. Alcohol Potently Inhibits the Kainate Receptor-Dependent Excitatory Drive of Hippocampal Interneurons. *Proc. Natl. Acad. Sci. USA.* **2003,** *100* (11), 6813–6818.

Cassidy, E. M.; O'Sullivan, I.; Bradshaw, P.; Islam, T.; Onovo, C. Symptom-Triggered Benzodiazepine Therapy for Alcohol Withdrawal Syndrome in the Emergency Department: A Comparison with the Standard Fixed Dose Benzodiazepine Regimen. *Emerg. Med. J.* **2012,** *29* (10), 802–804.

Cederbaum, A. I. Alcohol Metabolism. *Clin. Liver Dis.* **2012,** *16* (4), 667–685.

Chung, K.; Deisseroth, K. Clarity for Mapping the Nervous System. *Nat. Methods* **2013,** *10* (6), 508–513.

Church, M. W.; Kaltenbach, J. A. Hearing, Speech, Language, and Vestibular Disorders in the Fetal Alcohol Syndrome: A Literature Review. *Alcohol. Clin. Exp. Res.* **1997,** *21* (3), 495–512.

Cole-Harding, S.; Wilson, J. R. Ethanol Metabolism in Men and Women. *J. Stud. Alcohol* **1987,** *48* (4), 380–387.

Collins, G. B.; Brosnihan, K. B.; Zuti, R. A.; Messina, M.; Gupta, M. K. Neuroendocrine, Fluid Balance, and Thirst Responses to Alcohol in Alcoholics. *Alcohol. Clin. Exp. Res.* **1992,** *16* (2), 228–233.

Davies, M. J.; Baer, D. J.; Judd, J. T.; Brown, E. D.; Campbell, W. S.; Taylor, P. R. Effects of Moderate Alcohol Intake on Fasting Insulin and Glucose Concentrations and Insulin Sensitivity in Postmenopausal Women: A Randomized Controlled Trial. *JAMA* **2002,** *287* (19), 2559–2562.

Davies, A. G.; Pierce-Shimomura, J. T.; Kim, H.; VanHoven, M. K.; Thiele, T. R.; Bonci, A.; Bargmann, C. I.; McIntire, S. L. A Central Role of the BK Potassium Channel in Behavioral Responses to Ethanol in C. Elegans. *Cell* **2003,** *115* (6), 655–666.

Day, J. J.; Kennedy, A. J.; Sweatt, J. D. DNA Methylation and its Implications and Accessibility for Neuropsychiatric Therapeutics. *Annu. Rev. Pharmacol. Toxicol.* **2015,** *55* (1), 591–611.

Deisseroth, K. Optogenetics and Psychiatry: Applications, Challenges, and Opportunities. *Biol. Psychiatry* **2012,** *71* (12), 1030–1032.

Di Nicola, M.; De Filippis, S.; Martinotti, G.; De Risio, L.; Pettorruso, M.; De Persis, S.; Maremmani, A. G. I.; Maremmani, I.; di Giannantonio, M.; Janiri, L. Nalmefene in Alcohol Use Disorder Subjects with Psychiatric Comorbidity: A Naturalistic Study. *Adv. Ther.* **2017,** *34* (7), 1636–1649.

Dick, D. M.; Plunkett, J.; Wetherill, L. F.; Xuei, X.; Goate, A.; Hesselbrock, V.; Schuckit, M.; Crowe, R.; Edenberg, H. J.; Foroud, T. Association Between GABRA1 and Drinking Behaviors in the Collaborative Study on the Genetics of Alcoholism Sample. *Alcohol. Clin. Exp. Res.* **2006a,** *30* (7), 1101–1110.

Dick, D. M.; Bierut, L.; Hinrichs, A.; Fox, L.; Bucholz, K. K.; Kramer, J.; Kuperman, S.; Hesselbrock, V.; Schuckit, M.; Almasy, L. et al. The Role of GABRA2 in Risk for Conduct Disorder and Alcohol and Drug Dependence across Developmental Stages. *Behav. Genet.* **2006b,** *36* (4), 577–590.

Dixit, D.; Endicott, J.; Burry, L.; Ramos, L.; Yeung, S. Y. A.; Devabhakthuni, S.; Chan, C.; Tobia, A.; Bulloch, M. N. Management of Acute Alcohol Withdrawal Syndrome in Critically Ill Patients. *Pharmacother. J. Hum. Pharmacol. Drug Ther.* **2016,** *36* (7), 797–822.

Dodgson, S. J.; Shank, R. P.; Maryanoff, B. E. Topiramate as an Inhibitor of Carbonic Anhydrase Isoenzymes. *Epilepsia* **2000,** *41* (Suppl 1), S35–9.

Donoghue, K.; Elzerbi, C.; Saunders, R.; Whittington, C.; Pilling, S.; Drummond, C. The Efficacy of Acamprosate and Naltrexone in the Treatment of Alcohol Dependence, Europe versus the Rest of the World: A Meta-Analysis. *Addiction* **2015,** *110* (6), 920–930.

Dopico, A. M.; Chu, B.; Lemos, J. R.; Treistman, S. N. Alcohol Modulation of Calcium-Activated Potassium Channels. *Neurochem. Int.* **1999,** *35* (2), 103–106.

Dopico, A. M.; Lovinger, D. M. Acute Alcohol Action and Desensitization of Ligand-Gated Ion Channels. *Pharmacol. Rev.* **2009,** *61* (1), 98–114.

Ducci, F.; Enoch, M.-A.; Yuan, Q.; Shen, P.-H.; White, K. V; Hodgkinson, C.; Albaugh, B.; Virkkunen, M.; Goldman, D. HTR3B Is Associated with Alcoholism with Antisocial Behavior and Alpha EEG Power—an

Intermediate Phenotype for Alcoholism and Co-Morbid Behaviors. *Alcohol* **2009,** *43* (1), 73–84.

Edenberg, H. J. The Genetics of Alcohol Metabolism: Role of Alcohol Dehydrogenase and Aldehyde Dehydrogenase Variants. *Alcohol Res. Health* **2007,** *30* (1), 5–13.

Edenberg, H. J.; Xuei, X.; Chen, H.-J.; Tian, H.; Wetherill, L. F.; Dick, D. M.; Almasy, L.; Bierut, L.; Bucholz, K. K.; Goate, A.; et al. Association of Alcohol Dehydrogenase Genes with Alcohol Dependence: A Comprehensive Analysis. *Hum. Mol. Genet.* **2006,** *15* (9), 1539–1549.

Enghusen Poulsen, H.; Loft, S.; Andersen, J. R.; Andersen, M. Disulfiram Therapy—Adverse Drug Reactions and Interactions. *Acta Psychiatr. Scand. Suppl.* **1992,** *369,* 59-65-6.

Facchini, F.; Chen, Y. D.; Reaven, G. M. Light-to-Moderate Alcohol Intake Is Associated with Enhanced Insulin Sensitivity. *Diabetes Care* **1994,** *17* (2), 115–119.

Falk, D.; Wang, X. Q.; Liu, L.; Fertig, J.; Mattson, M.; Ryan, M.; Johnson, B.; Stout, R.; Litten, R. Z. Percentage of Subjects with No Heavy Drinking Days: Evaluation as an Efficacy Endpoint for Alcohol Clinical Trials. *Alcohol. Clin. Exp. Res.* **2010,** *34* (12), 2022–2034.

Fedirko, V.; Tramacere, I.; Bagnardi, V.; Rota, M.; Scotti, L.; Islami, F.; Negri, E.; Straif, K.; Romieu, I.; La Vecchia, C.; et al. Alcohol Drinking and Colorectal Cancer Risk: An Overall and Dose-Response Meta-Analysis of Published Studies. *Ann. Oncol.* **2011,** *22* (9), 1958–1972.

Fell, J.; Auld-Owens, A.; Snowden, C. Evaluation of Impaired Driving Assessments and Special Management Reviews in Reducing Impaired Driving Fatal Crashes in the United States. *Ann. Adv. Automot. Med. Assoc. Adv. Automot. Med. Annu. Sci. Conf.* **2013,** *57,* 33–44.

Fisher, C. M. Prompt Responses to the Administration of Ethanol in the Treatment of the Alcohol Withdrawal Syndrome (AWS). *Neurologist* **2009,** *15* (5), 242–244.

Förg, A.; Hein, J.; Volkmar, K.; Winter, M.; Richter, C.; Heinz, A.; Müller, C. A. Efficacy and Safety of Pregabalin in the Treatment of Alcohol Withdrawal Syndrome: A Randomized Placebo-Controlled Trial. *Alcohol Alcohol.* **2012,** *47* (2), 149–155.

Frezza, M.; di Padova, C.; Pozzato, G.; Terpin, M.; Baraona, E.; Lieber, C. S. High Blood Alcohol Levels in Women. *N. Engl. J. Med.* **1990,** *322* (2), 95–99.

Gao, B.; Bataller, R. Alcoholic Liver Disease: Pathogenesis and New Therapeutic Targets. *Gastroenterology* **2011,** *141* (5), 1572–1585.

Garbutt, J. C. Efficacy and Tolerability of Naltrexone in the Management of Alcohol Dependence. *Curr. Pharm. Des.* **2010,** *16* (19), 2091–2097.

Gaval-Cruz, M.; Weinshenker, D. Mechanisms of Disulfiram-Induced Cocaine Abstinence: Antabuse and Cocaine Relapse. *Mol. Interv.* **2009,** *9* (4), 175–187.

George, A.; Figueredo, V. M. Alcohol and Arrhythmias: A Comprehensive Review. *J. Cardiovasc. Med.* **2010,** *11* (4), 221–228.

Goel, S.; Sharma, A.; Garg, A. Effect of Alcohol Consumption on Cardiovascular Health. *Curr. Cardiol. Rep.* **2018,** *20* (4), 19.

Goldberg, I. J.; Mosca, L.; Piano, M. R.; Fisher, E. A.; Nutrition Committee, Council on Epidemiology and Prevention, and Council on Cardiovascular Nursing of the American Heart Association. AHA Science Advisory: Wine and Your Heart: A Science Advisory for Healthcare Professionals from the Nutrition Committee, Council on Epidemiology and Prevention, and Council on Cardiovascular Nursing of the American Heart Association. *Circulation* **2001,** *103* (3), 472–475.

Goldstein, B. Y.; Chang, S.-C.; Hashibe, M.; La Vecchia, C.; Zhang, Z.-F. Alcohol Consumption and Cancers of the Oral Cavity and Pharynx from 1988 to 2009: An Update. *Eur. J. Cancer Prev.* **2010,** *19* (6), 431–465.

Goldstein, M.; Nakajima, K. The Effect of Disulfiram on Catecholamine Levels in the Brain. *J. Pharmacol. Exp. Ther.* **1967,** *157* (1), 96–102.

Greenblatt, D. J. Clinical Pharmacokinetics of Oxazepam and Lorazepam. *Clin. Pharmacokinet.* **1981,** *6* (2), 89–105.

Han, D. H.; Kim, S. M.; Choi, J. E.; Min, K. J.; Renshaw, P. F. Adjunctive Aripiprazole Therapy with Escitalopram in Patients with Co-Morbid Major Depressive Disorder and Alcohol Dependence: Clinical and Neuroimaging Evidence. *J. Psychopharmacol.* **2013,** *27* (3), 282–291.

Hansagi, H.; Romelsjö, A.; Gerhardsson de Verdier, M.; Andréasson, S.; Leifman, A. Alcohol Consumption and Stroke Mortality. 20-Year Follow-up of 15,077 Men and Women. *Stroke* **1995,** *26* (10), 1768–1773.

Harper, C. The Neuropathology of Alcohol-Specific Brain Damage, or Does Alcohol Damage the Brain? *J. Neuropathol. Exp. Neurol.* **1998,** *57* (2), 101–110.

Harris, R. A.; McQuilkin, S. J.; Paylor, R.; Abeliovich, A.; Tonegawa, S.; Wehner, J. M. Mutant Mice Lacking the Gamma Isoform of Protein Kinase C Show Decreased Behavioral Actions of Ethanol and Altered Function of Gamma-Aminobutyrate Type A Receptors. *Proc. Natl. Acad. Sci. USA.* **1995,** *92* (9), 3658–3662.

Hendriks, H. F. J. Moderate Alcohol Consumption and Insulin Sensitivity: Observations and Possible Mechanisms. *Ann. Epidemiol.* **2007**, *17* (5), S40–S42.

Imbert, B.; Alvarez, J.-C.; Simon, N. Anticraving Effect of Baclofen in Alcohol-Dependent Patients. *Alcohol. Clin. Exp. Res.* **2015**, *39* (9), 1602–1608.

Imhof, A.; Froehlich, M.; Brenner, H.; Boeing, H.; Pepys, M. B.; Koenig, W. Effect of Alcohol Consumption on Systemic Markers of Inflammation. *Lancet* **2001**, *357* (9258), 763–767.

Islami, F.; Tramacere, I.; Rota, M.; Bagnardi, V.; Fedirko, V.; Scotti, L.; Garavello, W.; Jenab, M.; Corrao, G.; Straif, K.; et al. Alcohol Drinking and Laryngeal Cancer: Overall and Dose–risk Relation–A Systematic Review and Meta-Analysis. *Oral Oncol.* **2010**, *46* (11), 802–810.

Job, M. O.; Tang, A.; Hall, F. S.; Sora, I.; Uhl, G. R.; Bergeson, S. E.; Gonzales, R. A. Mu (Mu) Opioid Receptor Regulation of Ethanol-Induced Dopamine Response in the Ventral Striatum: Evidence of Genotype Specific Sexual Dimorphic Epistasis. *Biol. Psychiatry* **2007**, *62* (6), 627–634.

Johansson, B. A Review of the Pharmacokinetics and Pharmacodynamics of Disulfiram and its Metabolites. *Acta Psychiatr. Scand. Suppl.* **1992**, *369*, 15–26.

Johnson, B. A.; Ait-Daoud, N.; Bowden, C. L.; DiClemente, C. C.; Roache, J. D.; Lawson, K.; Javors, M. A.; Ma, J. Z. Oral Topiramate for Treatment of Alcohol Dependence: A Randomised Controlled Trial. *Lancet* **2003a**, *361* (9370), 1677–1685.

Johnson, B. A.; Ait-Daoud, N.; Ma, J. Z.; Wang, Y. Ondansetron Reduces Mood Disturbance Among Biologically Predisposed, Alcohol-Dependent Individuals. *Alcohol. Clin. Exp. Res.* **2003b**, *27* (11), 1773–1779.

Johnson, B. A.; Ait-Daoud, N. Topiramate in the New Generation of Drugs: Efficacy in the Treatment of Alcoholic Patients. *Curr. Pharm. Des.* **2010**, *16* (19), 2103–2112.

Jones, K. L.; Smith, D. W. Recognition of the Fetal Alcohol Syndrome in Early Infancy. *Lancet.* **1973**, *302* (7836), 999–1001.

Jones, K. A.; Porjesz, B.; Almasy, L.; Bierut, L.; Dick, D.; Goate, A.; Hinrichs, A.; Rice, J. P.; Wang, J. C.; Bauer, L. O.; et al. A Cholinergic Receptor Gene (CHRM2) Affects Event-Related Oscillations. *Behav. Genet.* **2006**, *36* (5), 627–639.

Joosten, M. M.; Beulens, J. W. J.; Kersten, S.; Hendriks, H. F. J. Moderate Alcohol Consumption Increases Insulin Sensitivity and ADIPOQ Expression in Postmenopausal Women: A Randomised, Crossover Trial. *Diabetologia* **2008**, *51* (8), 1375–1381.

Joslyn, G.; Brush, G.; Robertson, M.; Smith, T. L.; Kalmijn, J.; Schuckit, M.; White, R. L. Chromosome 15q25.1 Genetic Markers Associated with Level of Response to Alcohol in Humans. *Proc. Natl. Acad. Sci. USA.* **2008**, *105* (51), 20368–20373.

Kattimani, S.; Bharadwaj, B. Clinical Management of Alcohol Withdrawal: A Systematic Review. *Ind. Psychiatry J.* **2013**, *22* (2), 100–108.

Katz, D. A.; Locke, C.; Liu, W.; Zhang, J.; Achari, R.; Wesnes, K. A.; Tracy, K. A. Single-Dose Interaction Study of the Arginine Vasopressin Type 1B Receptor Antagonist ABT-436 and Alcohol in Moderate Alcohol Drinkers. *Alcohol. Clin. Exp. Res.* **2016**, *40* (4), 838–845.

Keenan, A. B.; Jenkins, S. L.; Jagodnik, K. M.; Koplev, S.; He, E.; Torre, D.; Wang, Z.; Dohlman, A. B.; Silverstein, M. C.; Lachmann, A.; et al. The Library of Integrated Network-Based Cellular Signatures NIH Program: System-Level Cataloging of Human Cells Response to Perturbations. *Cell Syst.* **2018**, *6* (1), 13–24.

Kelley-Baker, T.; Lacey, J. H.; Voas, R. B.; Romano, E.; Yao, J.; Berning, A. Drinking and Driving in the United States: Comparing Results from the 2007 and 1996 National Roadside Surveys. *Traffic Inj. Prev.* **2013**, *14* (2), 117–126.

Kelly, J. F. Is Alcoholics Anonymous Religious, Spiritual, Neither? Findings from 25 Years of Mechanisms of Behavior Change Research. *Addiction* **2017**, *112* (6), 929–936.

Koob, G. F.; Rassnick, S.; Heinrichs, S.; Weiss, F. Alcohol, the Reward System and Dependence. *EXS* **1994**, *71*, 103–114.

Koob, G.; Kreek, M. J. Stress, Dysregulation of Drug Reward Pathways, and the Transition to Drug Dependence. *Am. J. Psychiatry* **2007**, *164* (8), 1149–1159.

Kosten, T. R.; O'Connor, P. G. Management of Drug and Alcohol Withdrawal. *N. Engl. J. Med.* **2003**, *348* (18), 1786–1795.

Kranzler, H. R.; Pierucci-Lagha, A.; Feinn, R.; Hernandez-Avila, C. Effects of Ondansetron in Early- Versus Late-Onset Alcoholics: A Prospective, Open-Label Study. *Alcohol. Clin. Exp. Res.* **2003**, *27* (7), 1150–1155.

Kril, J. J.; Halliday, G. M. Brain Shrinkage in Alcoholics: A Decade on and What Have We Learned? *Prog. Neurobiol.* **1999**, *58* (4), 381–387.

Krystal, J. H.; Staley, J.; Mason, G.; Petrakis, I. L.; Kaufman, J.; Harris, R. A.; Gelernter, J.; Lappalainen, J.

γ-Aminobutyric Acid Type A Receptors and Alcoholism. *Arch. Gen. Psychiatry* **2006,** *63* (9), 957.

Kuo, P.-H.; Kalsi, G.; Prescott, C. A.; Hodgkinson, C. A.; Goldman, D.; van den Oord, E. J.; Alexander, J.; Jiang, C.; Sullivan, P. F.; Patterson, D. G.; et al. Association of ADH and ALDH Genes with Alcohol Dependence in the Irish Affected Sib Pair Study of Alcohol Dependence (IASPSAD) Sample. *Alcohol. Clin. Exp. Res.* **2008,** *32* (5), 785–795.

Kupari, M.; Koskinen, P. Alcohol, Cardiac Arrhythmias and Sudden Death. *Novartis Found. Symp.* **1998,** *216,* 68-79-85.

Kurlawala, Z.; Vatsalya, V. Heavy Alcohol Drinking Associated Akathisia and Management with Quetiapine XR in Alcohol Dependent Patients. *J. Addict.* **2016,** *2016,* 6028971.

Li, T. K. Pharmacogenetics of Responses to Alcohol and Genes That Influence Alcohol Drinking. *J. Stud. Alcohol* **2000,** *61* (1), 5–12.

Lieber, C. S. Alcohol and the Liver: Metabolism of Alcohol and Its Role in Hepatic and Extrahepatic Diseases. *Mt. Sinai J. Med.* **2000,** *67* (1), 84–94.

Lieber, C. S. The Discovery of the Microsomal Ethanol Oxidizing System and Its Physiologic and Pathologic Role. *Drug Metab. Rev.* **2004,** *36* (3–4), 511–529.

Litten, R. Z.; Ryan, M. L.; Fertig, J. B.; Falk, D. E.; Johnson, B.; Dunn, K. E.; Green, A. I.; Pettinati, H. M.; Ciraulo, D. A.; Sarid-Segal, O.; et al. A Double-Blind, Placebo-Controlled Trial Assessing the Efficacy of Varenicline Tartrate for Alcohol Dependence. *J. Addict. Med.* **2013,** *7* (4), 277–286.

Litten, R. Z.; Fertig, J. B.; Falk, D. E.; Ryan, M. L.; Mattson, M. E.; Collins, J. F.; Murtaugh, C.; Ciraulo, D.; Green, A. I.; Johnson, B.; et al. A Double-Blind, Placebo-Controlled Trial to Assess the Efficacy of Quetiapine Fumarate XR in Very Heavy-Drinking Alcohol-Dependent Patients. *Alcohol. Clin. Exp. Res.* **2012,** *36* (3), 406–416.

Ma, K.; Baloch, Z.; He, T.-T.; Xia, X. Alcohol Consumption and Gastric Cancer Risk: A Meta-Analysis. *Med. Sci. Monit.* **2017,** *23,* 238–246.

Mann, K. Pharmacotherapy of Alcohol Dependence: A Review of the Clinical Data. *CNS Drugs* **2004,** *18* (8), 485–504.

Martín-Blanco, A.; Patrizi, B.; Soler, J.; Gasol, X.; Elices, M.; Gasol, M.; Carmona, C.; Pascual, J. C. Use of Nalmefene in Patients with Comorbid Borderline Personality Disorder and Alcohol Use Disorder. *Int. Clin. Psychopharmacol.* **2017,** *32* (4), 231–234.

Martinotti, G.; Andreoli, S.; Di Nicola, M.; Di Giannantonio, M.; Sarchiapone, M.; Janiri, L. Quetiapine Decreases Alcohol Consumption, Craving, and Psychiatric Symptoms in Dually Diagnosed Alcoholics. *Hum. Psychopharmacol. Clin. Exp.* **2008,** *23* (5), 417–424.

Martinotti, G.; di Nicola, M.; Frustaci, A.; Romanelli, R.; Tedeschi, D.; Guglielmo, R.; Guerriero, L.; Bruschi, A.; De Filippis, R.; Pozzi, G.; et al. Pregabalin, Tiapride and Lorazepam in Alcohol Withdrawal Syndrome: A Multi-Centre, Randomized, Single-Blind Comparison Trial. *Addiction* **2010,** *105* (2), 288–299.

Mason, B. J.; Salvato, F. R.; Williams, L. D.; Ritvo, E. C.; Cutler, R. B. A Double-Blind, Placebo-Controlled Study of Oral Nalmefene for Alcohol Dependence. *Arch. Gen. Psychiatry* **1999,** *56* (8), 719–724.

Mason, B. J.; Quello, S.; Goodell, V.; Shadan, F.; Kyle, M.; Begovic, A. Gabapentin Treatment for Alcohol Dependence. *JAMA Intern. Med.* **2014,** *174* (1), 70.

Matsumoto, H.; Fukui, Y. Pharmacokinetics of Ethanol: A Review of the Methodology. *Addict. Biol.* **2002,** *7,* 5–14.

Mattson, S. N.; Riley, E. P.; Jernigan, T. L.; Ehlers, C. L.; Delis, D. C.; Jones, K. L.; Stern, C.; Johnson, K. A.; Hesselink, J. R.; Bellugi, U. Fetal Alcohol Syndrome: A Case Report of Neuropsychological, MRI and EEG Assessment of Two Children. *Alcohol. Clin. Exp. Res.* **1992,** *16* (5), 1001–1003.

Mirijello, A.; D'Angelo, C.; Ferrulli, A.; Vassallo, G.; Antonelli, M.; Caputo, F.; Leggio, L.; Gasbarrini, A.; Addolorato, G. Identification and Management of Alcohol Withdrawal Syndrome. *Drugs* **2015,** *75* (4), 353–365.

Mohapatra, S.; Rath, N. R. Disulfiram Induced Psychosis. *Clin. Psychopharmacol. Neurosci.* **2017,** *15* (1), 68–69.

Monnelly, E. P.; Ciraulo, D. A.; Knapp, C.; LoCastro, J.; Sepulveda, I. Quetiapine for Treatment of Alcohol Dependence. *J. Clin. Psychopharmacol.* **2004,** *24* (5), 532–535.

Mukamal, K. J.; Chung, H.; Jenny, N. S.; Kuller, L. H.; Longstreth, W. T.; Mittleman, M. A.; Burke, G. L.; Cushman, M.; Psaty, B. M.; Siscovick, D. S. Alcohol Consumption and Risk of Coronary Heart Disease in Older Adults: The Cardiovascular Health Study. *J. Am. Geriatr. Soc.* **2006,** *54* (1), 30–37.

Myrick, H.; Li, X.; Randall, P. K.; Henderson, S.; Voronin, K.; Anton, R. F. The Effect of Aripiprazole on Cue-Induced Brain Activation and Drinking Parameters in Alcoholics. *J. Clin. Psychopharmacol.* **2010,** *30* (4), 365–372.

Naassila, M.; Hammoumi, S.; Legrand, E.; Durbin, P.; Daoust, M. Mechanism of Action of Acamprosate. Part I. Characterization of Spermidine-Sensitive Acamprosate Binding Site in Rat Brain. *Alcohol. Clin. Exp. Res.* **1998,** *22* (4), 802–809.

Nam, H. W.; Karpyak, V. M.; Hinton, D. J.; Geske, J. R.; Ho, A. M. C.; Prieto, M. L.; Biernacka, J. M.; Frye, M. A.; Weinshilboum, R. M.; Choi, D.-S. Elevated Baseline Serum Glutamate as a Pharmacometabolomic Biomarker for Acamprosate Treatment Outcome in Alcohol-Dependent Subjects. *Transl. Psychiatry* **2015,** *5* (8), e621–e621.

Nussinov, R.; Tsai, C.-J. The Design of Covalent Allosteric Drugs. *Annu. Rev. Pharmacol. Toxicol.* **2015,** *55* (1), 249–267.

O'Farrell, T. J.; Choquette, K. A.; Cutter, H. S.; Birchler, G. R. Sexual Satisfaction and Dysfunction in Marriages of Male Alcoholics: Comparison with Nonalcoholic Maritally Conflicted and Nonconflicted Couples. *J. Stud. Alcohol* **1997,** *58* (1), 91–99.

Oslin, D.; Atkinson, R. M.; Smith, D. M.; Hendrie, H. Alcohol Related Dementia: Proposed Clinical Criteria. *Int. J. Geriatr. Psychiatry* **1998,** *13* (4), 203–212.

Oulas, A.; Minadakis, G.; Zachariou, M.; Sokratous, K.; Bourdakou, M. M.; Spyrou, G. M. Systems Bioinformatics : Increasing Precision of Computational Diagnostics and Therapeutics through Network-Based Approaches. *Brief Bioinform.* **2019,** *20* (3), 806–824.

Overall, J. E.; Brown, D.; Williams, J. D.; Neill, L. T. Drug Treatment of Anxiety and Depression in Detoxified Alcoholic Patients. *Arch. Gen. Psychiatry* **1973,** *29* (2), 218.

Papa, A.; Tursi, A.; Cammarota, G.; Certo, M.; Cuoco, L.; Montalto, M.; Cianci, R.; Papa, V.; Fedeli, P.; Fedeli, G.; et al. Effect of Moderate and Heavy Alcohol Consumption on Intestinal Transit Time. *Panminerva Med.* **1998,** *40* (3), 183–185.

Paparrigopoulos, T.; Tzavellas, E.; Karaiskos, D.; Kourlaba, G.; Liappas, I. Treatment of Alcohol Dependence with Low-Dose Topiramate: An Open-Label Controlled Study. *BMC Psychiatry* **2011,** *11* (1), 41.

Parrish, K. M.; Higuchi, S.; Dufour, M. C. Alcohol Consumption and the Risk of Developing Liver Cirrhosis: Implications for Future Research. *J. Subst. Abuse* **1991,** *3* (3), 325–335.

Parry, C. D.; Patra, J.; Rehm, J. Alcohol Consumption and Non-Communicable Diseases: Epidemiology and Policy Implications. *Addiction* **2011,** *106* (10), 1718–1724.

Pastor, R.; Aragon, C. M. G. The Role of Opioid Receptor Subtypes in the Development of Behavioral Sensitization

to Ethanol. *Neuropsychopharmacology* **2006,** *31* (7), 1489–1499.

Pedersen, C. A.; Smedley, K. L.; Leserman, J.; Jarskog, L. F.; Rau, S. W.; Kampov-Polevoi, A.; Casey, R. L.; Fender, T.; Garbutt, J. C. Intranasal Oxytocin Blocks Alcohol Withdrawal in Human Subjects. *Alcohol. Clin. Exp. Res.* **2013,** *37* (3), 484–489.

Peltz, S. Severe Thrombocytopenia Secondary to Alcohol Use. *Postgrad. Med.* **1991,** *89* (6), 75–76, 85.

Peoples, R. W.; Stewart, R. R. Alcohols Inhibit N-Methyl-D-Aspartate Receptors via a Site Exposed to the Extracellular Environment. *Neuropharmacology* **2000,** *39* (10), 1681–1691.

Perper, J. A.; Twerski, A.; Wienand, J. W. Tolerance at High Blood Alcohol Concentrations: A Study of 110 Cases and Review of the Literature. *J. Forensic Sci.* **1986,** *31* (1), 212–221.

Peters, T. J.; Kotowicz, J.; Nyka, W.; Kozubski, W.; Kuznetsov, V.; Vanderbist, F.; De Niet, S.; Marcereuil, D.; Coffiner, M. Treatment of Alcoholic Polyneuropathy with Vitamin B Complex: A Randomised Controlled Trial. *Alcohol Alcohol.* **2006,** *41* (6), 636–642.

Petrakis, I. L.; Desai, N.; Gueorguieva, R.; Arias, A.; O'Brien, E.; Jane, J. S.; Sevarino, K.; Southwick, S.; Ralevski, E. Prazosin for Veterans with Posttraumatic Stress Disorder and Comorbid Alcohol Dependence: A Clinical Trial. *Alcohol. Clin. Exp. Res.* **2016,** *40* (1), 178–186.

Pfefferbaum, A.; Sullivan, E. V; Rosenbloom, M. J.; Mathalon, D. H.; Lim, K. O. A Controlled Study of Cortical Gray Matter and Ventricular Changes in Alcoholic Men over a 5-Year Interval. *Arch. Gen. Psychiatry* **1998,** *55* (10), 905–912.

Plosker, G. L. Acamprosate: A Review of its use in Alcohol Dependence. *Drugs* **2015,** *75* (11), 1255–1268.

Porter, R. J.; Dhir, A.; Macdonald, R. L.; Rogawski, M. A. Mechanisms of Action of Antiseizure Drugs. In *Handbook of clinical neurology*, 2012; Vol. 108, pp 663–681.

Quertemont, E. Genetic Polymorphism in Ethanol Metabolism: Acetaldehyde Contribution to Alcohol Abuse and Alcoholism. *Mol. Psychiatry* **2004,** *9* (6), 570–581.

Ray, L. A.; Bujarski, S.; Shoptaw, S.; Roche, D. J.; Heinzerling, K.; Miotto, K. Development of the Neuroimmune Modulator Ibudilast for the Treatment of Alcoholism: A Randomized, Placebo-Controlled, Human Laboratory Trial. *Neuropsychopharmacology* **2017,** *42* (9), 1776–1788.

Rehm, J.; Mathers, C.; Popova, S.; Thavorncharoensap, M.; Teerawattananon, Y.; Patra, J. Global Burden of Disease and Injury and Economic Cost Attributable to Alcohol Use and Alcohol-Use Disorders. *Lancet* **2009**, *373* (9682), 2223–2233.

Reynaud, M.; Aubin, H.-J.; Trinquet, F.; Zakine, B.; Dano, C.; Dematteis, M.; Trojak, B.; Paille, F.; Detilleux, M. A Randomized, Placebo-Controlled Study of High-Dose Baclofen in Alcohol-Dependent Patients—The ALPADIR Study. *Alcohol Alcohol* **2017**, *52* (4), 439–446.

Rimm, E. B.; Williams, P.; Fosher, K.; Criqui, M.; Stampfer, M. J. Moderate Alcohol Intake and Lower Risk of Coronary Heart Disease: Meta-Analysis of Effects on Lipids and Haemostatic Factors. *BMJ* **1999**, *319* (7224), 1523–1528.

Roberto, M.; Cruz, M. T.; Gilpin, N. W.; Sabino, V.; Schweitzer, P.; Bajo, M.; Cottone, P.; Madamba, S. G.; Stouffer, D. G.; Zorrilla, E. P.; et al. Corticotropin Releasing Factor–Induced Amygdala Gamma-Aminobutyric Acid Release Plays a Key Role in Alcohol Dependence. *Biol. Psychiatry* **2010**, *67* (9), 831–839.

Rösner, S.; Hackl-Herrwerth, A.; Leucht, S.; Vecchi, S.; Srisurapanont, M.; Soyka, M. Opioid Antagonists for Alcohol Dependence. *Cochrane Database Syst. Rev.* **2010**, *8* (12), CD001867.

Rossi, R. E.; Conte, D.; Massironi, S. Diagnosis and Treatment of Nutritional Deficiencies in Alcoholic Liver Disease: Overview of Available Evidence and Open Issues. *Dig. Liver Dis.* **2015**, *47* (10), 819–825.

Ryan, M. L.; Falk, D. E.; Fertig, J. B.; Rendenbach-Mueller, B.; Katz, D. A.; Tracy, K. A.; Strain, E. C.; Dunn, K. E.; Kampman, K.; Mahoney, E.; et al. A Phase 2, Double-Blind, Placebo-Controlled Randomized Trial Assessing the Efficacy of ABT-436, a Novel V1b Receptor Antagonist, for Alcohol Dependence. *Neuropsychopharmacology* **2017**, *42* (5), 1012–1023.

Sachdeva, A.; Choudhary, M.; Chandra, M. Alcohol Withdrawal Syndrome: Benzodiazepines and Beyond. *J. Clin. Diagn. Res.* **2015**, *9* (9), VE01–VE07.

Sakurai, S.; Cui, R.; Tanigawa, T.; Yamagishi, K.; Iso, H. Alcohol Consumption Before Sleep Is Associated With Severity of Sleep-Disordered Breathing Among Professional Japanese Truck Drivers. *Alcohol. Clin. Exp. Res.* **2007**, *31* (12), 2053–2058.

Salaspuro, M. P.; Lieber, C. S. Non-Uniformity of Blood Ethanol Elimination: Its Exaggeration after Chronic Consumption. *Ann. Clin. Res.* **1978**, *10* (5), 294–297.

Sarkola, T.; Iles, M. R.; Kohlenberg-Mueller, K.; Eriksson, C. J. P. Ethanol, Acetaldehyde, Acetate, and Lactate

Levels after Alcohol Intake in White Men and Women: Effect of 4-Methylpyrazole. *Alcohol. Clin. Exp. Res.* **2002**, *26* (2), 239–245.

Schmid, B.; Blomeyer, D.; Treutlein, J.; Zimmermann, U. S.; Buchmann, A. F.; Schmidt, M. H.; Esser, G.; Rietschel, M.; Banaschewski, T.; Schumann, G.; et al. Interacting Effects of CRHR1 Gene and Stressful Life Events on Drinking Initiation and Progression among 19-Year-Olds. *Int. J. Neuropsychopharmacol.* **2010**, *13* (6), 703–714.

Schuckit, M. A. Alcohol-Use Disorders. *Lancet* **2009**, *373* (9662), 492–501.

Schuckit, M. A. Comorbidity between Substance Use Disorders and Psychiatric Conditions. *Addiction* **2006a**, *101*, 76–88.

Schuckit, M. A. *Drug and Alcohol Abuse : A Clinical Guide to Diagnosis and Treatment*; Springer, 2006b.

Simpson, T. L.; Malte, C. A.; Dietel, B.; Tell, D.; Pocock, I.; Lyons, R.; Varon, D.; Raskind, M.; Saxon, A. J. A Pilot Trial of Prazosin, an Alpha-1 Adrenergic Antagonist, for Comorbid Alcohol Dependence and Posttraumatic Stress Disorder. *Alcohol. Clin. Exp. Res.* **2015**, *39* (5), 808–817.

Skinner, M. D.; Lahmek, P.; Pham, H.; Aubin, H.-J. Disulfiram Efficacy in the Treatment of Alcohol Dependence: A Meta-Analysis. *PLoS One* **2014**, *9* (2), e87366.

Smith, B. R.; Horan, J. T.; Gaskin, S.; Amit, Z. Exposure to Nicotine Enhances Acquisition of Ethanol Drinking by Laboratory Rats in a Limited Access Paradigm. *Psychopharmacology (Berl)* **1999**, *142* (4), 408–412.

Solem, M.; McMahon, T.; Messing, R. O. Protein Kinase A Regulates Regulates Inhibition of N- and P/Q-Type Calcium Channels by Ethanol in PC12 Cells. *J. Pharmacol. Exp. Ther.* **1997**, *282* (3), 1487–1495.

Song, D. Y.; Song, S.; Song, Y.; Lee, J. E. Alcohol Intake and Renal Cell Cancer Risk: A Meta-Analysis. *Br. J. Cancer* **2012**, *106* (11), 1881–1890.

Spierling, S. R.; Zorrilla, E. P. Don't Stress about CRF: Assessing the Translational Failures of CRF1 antagonists. *Psychopharmacology (Berl)* **2017**, *234* (9–10), 1467–1481.

Stephens, R.; Ling, J.; Heffernan, T. M.; Heather, N.; Jones, K. A Review of the Literature on the Cognitive Effects of Alcohol Hangover. *Alcohol Alcohol* **2008**, *43* (2), 163–170.

Stock, C. J.; Carpenter, L.; Ying, J.; Greene, T. Gabapentin Versus Chlordiazepoxide for Outpatient Alcohol Detoxification Treatment. *Ann. Pharmacother.* **2013**, *47* (7–8), 961–969.

Stockwell, T.; Honig, F. Labelling Alcoholic Drinks: Percentage Proof, Original Gravity, Percentage Alcohol or Standard Drinks? *Drug Alcohol Rev.* **1990,** *9* (1), 81–89.

Stueve, A.; O'Donnell, L. N. Early Alcohol Initiation and Subsequent Sexual and Alcohol Risk Behaviors Among Urban Youths. *Am. J. Public Health* **2005,** *95* (5), 887–893.

Suddendorf, R. F. Research on Alcohol Metabolism among Asians and Its Implications for Understanding Causes of Alcoholism. *Public Health Rep.* **1989,** *104* (6), 615–620.

Sullivan, J. T.; Sykora, K.; Schneiderman, J.; Naranjo, C. A.; Sellers, E. M. Assessment of Alcohol Withdrawal: The Revised Clinical Institute Withdrawal Assessment for Alcohol Scale (CIWA-Ar). *Br. J. Addict.* **1989,** *84* (11), 1353–1357.

Tabakoff, B.; Hoffman, P. L. Adenylyl Cyclases and Alcohol. *Adv. Second Messenger Phosphoprotein Res.* **1998,** *32*, 173–193.

Tabakoff, B.; Cornell, N.; Hoffman, P. L. Alcohol Tolerance. *Ann. Emerg. Med.* **1986,** *15* (9), 1005–1012.

Taylor, C. P. Mechanisms of Action of Gabapentin. *Rev. Neurol. (Paris)* **1997,** *153 Suppl*, S39–45.

Tramacere, I.; Pelucchi, C.; Bonifazi, M.; Bagnardi, V.; Rota, M.; Bellocco, R.; Scotti, L.; Islami, F.; Corrao, G.; Boffetta, P.; et al. A Meta-Analysis on Alcohol Drinking and the Risk of Hodgkin Lymphoma. *Eur. J. Cancer Prev.* **2012a,** *21* (3), 268–273.

Tramacere, I.; Pelucchi, C.; Bonifazi, M.; Bagnardi, V.; Rota, M.; Bellocco, R.; Scotti, L.; Islami, F.; Corrao, G.; Boffetta, P.; et al. Alcohol Drinking and Non-Hodgkin Lymphoma Risk: A Systematic Review and a Meta-Analysis. *Ann. Oncol.* **2012b,** *23* (11), 2791–2798.

Treutlein, J.; Kissling, C.; Frank, J.; Wiemann, S.; Dong, L.; Depner, M.; Saam, C.; Lascorz, J.; Soyka, M.; Preuss, U. W.; et al. Genetic Association of the Human Corticotropin Releasing Hormone Receptor 1 (CRHR1) with Binge Drinking and Alcohol Intake Patterns in Two Independent Samples. *Mol. Psychiatry* **2006,** *11* (6), 594–602.

Trevejo-Nunez, G.; Kolls, J. K.; de Wit, M. Alcohol Use As a Risk Factor in Infections and Healing: A Clinician's Perspective. *Alcohol Res.* **2015,** *37* (2), 177–184.

Turati, F.; Garavello, W.; Tramacere, I.; Bagnardi, V.; Rota, M.; Scotti, L.; Islami, F.; Corrao, G.; Boffetta, P.; La Vecchia, C.; et al. A Meta-Analysis of Alcohol Drinking and Oral and Pharyngeal Cancers. Part 2: Results by Subsites. *Oral Oncol.* **2010,** *46* (10), 720–726.

Urbano-Márquez, A.; Estruch, R.; Fernández-Solá, J.; Nicolás, J. M.; Paré, J. C.; Rubin, E. The Greater Risk of Alcoholic Cardiomyopathy and Myopathy in Women Compared with Men. *JAMA* **1995,** *274* (2), 149–154.

Vernet, M.; Cadefau, J. A.; Balagué, A.; Grau, J. M.; Urbano-Márquez, A. U.; Cussó, R. Effect of Chronic Alcoholism on Human Muscle Glycogen and Glucose Metabolism. *Alcohol. Clin. Exp. Res.* **1995,** *19* (5), 1295–1299.

Verplaetse, T. L.; Pittman, B. P.; Shi, J. M.; Tetrault, J. M.; Coppola, S.; McKee, S. A. Effect of Varenicline Combined with High-Dose Alcohol on Craving, Subjective Intoxication, Perceptual Motor Response, and Executive Cognitive Function in Adults with Alcohol Use Disorders: Preliminary Findings. *Alcohol. Clin. Exp. Res.* **2016,** *40* (7), 1567–1576.

Virkkunen, M.; Linnoila, M. Serotonin in Early-Onset Alcoholism. In *Recent Developments in Alcoholism*; Springer: Boston, MA, **2002;** pp 173–189.

Volkow, N. D.; Wang, G. J.; Hitzemann, R.; Fowler, J. S.; Overall, J. E.; Burr, G.; Wolf, A. P. Recovery of Brain Glucose Metabolism in Detoxified Alcoholics. *Am. J. Psychiatry* **1994,** *151* (2), 178–183.

Volkow, N. D.; Fowler, J. S.; Wang, G.-J.; Swanson, J. M.; Telang, F. Dopamine in Drug Abuse and Addiction. *Arch. Neurol.* **2007,** *64* (11), 1575.

von der Goltz, C.; Koopmann, A.; Dinter, C.; Richter, A.; Grosshans, M.; Fink, T.; Wiedemann, K.; Kiefer, F. Involvement of Orexin in the Regulation of Stress, Depression and Reward in Alcohol Dependence. *Horm. Behav.* **2011,** *60* (5), 644–650.

Vonlaufen, A.; Wilson, J. S.; Pirola, R. C.; Apte, M. V. Role of Alcohol Metabolism in Chronic Pancreatitis. *Alcohol Res. Health* **2007,** *30* (1), 48–54.

Wagner, J. G.; Kay, D. R.; Sedman, A. J.; Sakmar, E.; Wilkinson, P. K. *J. Pharmacokin. Biopharm.* [Plenum Pub. Corp.], **1977,** *5* (3), 207–24.

Wall, T. L.; Ehlers, C. L. Alcohol Health and Research World. *Genet. Alcohol.* **1995,** *19* (3), 184–189.

Wannamethee, S. G.; Shaper, A. G. Alcohol, Body Weight, and Weight Gain in Middle-Aged Men. *Am. J. Clin. Nutr.* **2003a,** *77* (5), 1312–1317.

Wannamethee, S. G.; Camargo, C. A.; Manson, J. E.; Willett, W. C.; Rimm, E. B. Alcohol Drinking Patterns and Risk of Type 2 Diabetes Mellitus Among Younger Women. *Arch. Intern. Med.* **2003b,** *163* (11), 1329.

Wassif, W. S.; Preedy, V. R.; Summers, B.; Duane, P.; Leigh, N.; Peters, T. J. The Relationship between Muscle Fibre Atrophy Factor, Plasma Carnosinase Activities and Muscle RNA and Protein Composition in Chronic Alcoholic Myopathy. *Alcohol Alcohol* **1993,** *28* (3), 325–331.

Wetherill, L.; Schuckit, M. A.; Hesselbrock, V.; Xuei, X.; Liang, T.; Dick, D. M.; Kramer, J.; Nurnberger Jr., J. I.; Tischfield, J. A.; Porjesz, B.; et al. Neuropeptide Y Receptor Genes Are Associated With Alcohol Dependence, Alcohol Withdrawal Phenotypes, and Cocaine Dependence. *Alcohol. Clin. Exp. Res.* **2008,** *32* (12), 2031–2040.

White, H. S.; Brown, S. D.; Woodhead, J. H.; Skeen, G. A.; Wolf, H. H. Topiramate Modulates GABA-Evoked Currents in Murine Cortical Neurons by a Nonbenzodiazepine Mechanism. *Epilepsia* **2000,** *41* (Suppl 1), S17-20.

Williams, S. H. Medications for Treating Alcohol Dependence. *Am. Fam. Physician* **2005,** *72* (9), 1775–1780.

Wirkner, K.; Eberts, C.; Poelchen, W.; Allgaier, C.; Illes, P. Mechanism of Inhibition by Ethanol of NMDA and AMPA Receptor Channel Functions in Cultured Rat Cortical Neurons. *Naunyn. Schmiedebergs. Arch. Pharmacol.* **2000,** *362* (6), 568–576.

Yang, X.; Chen, X.; Zhuang, M.; Yuan, Z.; Nie, S.; Lu, M.; Jin, L.; Ye, W. Smoking and Alcohol Drinking in Relation to the Risk of Esophageal Squamous Cell Carcinoma: A Population-Based Case-Control Study in China. *Sci. Rep.* **2017,** *7* (1), 17249.

Zhang, X.; Velumian, A. A.; Jones, O. T.; Carlen, P. L. Modulation of High-Voltage-Activated Calcium Channels in Dentate Granule Cells by Topiramate. *Epilepsia* **2000,** *41* (Suppl 1), S52–60.

Ziółkowski, M.; Czarnecki, D.; Budzyński, J.; Rosińska, Z.; Żekanowska, E.; Góralczyk, B. Orexin in Patients with Alcohol Dependence Treated for Relapse Prevention: A Pilot Study. *Alcohol Alcohol* **2016,** *51* (4), 416–421.

# CHAPTER 18

# Cognition Enhancers

RAMNEEK KAUR[1], RASHI RAJPUT[1], SACHIN KUMAR[1],
HARLEEN KAUR[2], RACHANA[1], and MANISHA SINGH[1*]

[1]Jaypee Institute of Information Technology, Uttar Pradesh, India

[2]Amity Institute of Biotechnology, Uttar Pradesh, India

*Corresponding author. E-mail: manishasingh1295@gmail.com

## ABSTRACT

Cognition is a physiological process of knowing which includes perception, judgment, reasoning, and awareness. Cognitive functions can be categorized into attention, intelligence, creativity, and memory. Being subjective in nature, it is usually affected by factors including hypertension, stress, aging, and various pathological conditions like Alzheimer's disease, Parkinson's disease, human immunodeficiency virus, cancer, and schizophrenia. Cognitive impairment is the major concern in the mentioned neural diseases and normal aged life. Cognitive enhancers (CE) are used to facilitate the attention acquisition and abilities, storage and retrieval of information, and lessen the cognitive deficits that are associated with age and neurodegenerative disorders (NDs). With much research in this field, various signaling molecules and neurotransmitters are identified which can be utilized as a therapeutic target. New and conventional molecules are tried against them. The research has validated a number of targets like PDE4 (phosphodiesterase-4), calcium ion channel blockers, nicotinic receptors, and 5HT6 (5-hydroxytryptamine 6 receptor) which have immense therapeutic importance as smart drugs. This chapter covers all the perspective of cognitive enhancement ranging from various memory pathways, methods to improve cognition, various strategies, and compounds used as CE/smart drugs to side effects for the same.

## 18.1 INTRODUCTION

Cognitive enhancement refers to the improvements in emotional, motivational and cognitive. The drugs which enhance the cognition are usually called "cognition enhancers," "memory enhancing drugs," "nootropic drugs," and "smart drugs" (Giurgea, 1972). CE are the drugs that affect the cognitive functions positively like vigilance, learning, attention, executive functions, and memory. Psychostimulants are the category of drugs which are extensively used as a cognitive enhancer; yet, nootropic drugs have particular neuroprotective effect without producing stimulation or sedation. Cognitive enhancement is a prevalent subject that attracts attention both from the scientists and general public (Eickenhorst et al., 2012, Heinz et al., 2012). The stimulants which are prescribed like amphetamines and methylphenidate (MPH) are regularly used as a smart drug in college campuses, wherein 5–35% of students have reported of consuming them for cognitive enhancement (Wilens et al., 2008). Though, the use of MPH is just not limited to students, but also surgeons because of its use like wakefulness and high cognitive performance (Franke et al., 2013). The placebo-controlled trial substantiates the advantages of some neuroenhancers, but claims are not tested formally. Additionally, the concerns include toxicity, adverse consequences, and addiction as these drugs are consumed for a long term and without any medical follow-ups.

The utilization of drugs and other techniques to improve the cognition is quite old. Caffeine is utilized as a stimulant for around thousand years back and is consumed in high doses. Nicotine also helps to promote cognitive ability (Rezvani and Levin, 2001). A study at Duke University discovered that nicotine patches considerably improved the age-related cognitive disability (Newhouse et al., 2012). Amphetamines were used by the armed forces in the Korean War and the World War II (Stoil, 1990) and are still used by the military forces of the United States. Students have long utilized amphetamines for the study aid (Schrage, 1985), with MPH being the existing cognitive enhancer in the campuses of US colleges (Babcock and Byrne, 2000; Farah, 2002; Zielbauer, 2000).

Pharmaceutical CE are the major concern for public health. Recent reports have focused on the ethical debate instead of efficacy and effectiveness whereas, the scientists claim that research on efficacy and safety should be the rate-determining step. There are some meta-analysis and systemic reviews on various drugs (Repantis et al., 2010a; Repantis et al.,

2010b; Heishman et al., 2010), yet no synthetic and comprehensive review is available that concerns over a few enhancers. Thus, cognitive enhancement can take diverse and many forms. Different approaches of cognitive enhancement have various consequences in the near future. Further, it raises a variety of ethical issues. Like, these technologies interact with the good life, role of medicines and concept of authenticity in our lives. Anticipated and present means of cognitive enhancement confronts for regulation and public policy. This chapter discusses the mechanism of cognition enhancement and various drugs which are used for it and the subsequent signaling involved in it.

## 18.2   COGNITION ENHANCEMENT

Cognitive enhancement is presently linked with a vast range of prevailing, developing, and futuristic biomedical technology that expect to develop the cognitive status of human beings and animals (Bostrom and Sandberg, 2009). Mostly, the efforts for enhancing the cognition are banal, and some are used and practiced since many years. The major example is training and education, in which the goal is to convey a particular information or skill and develop the general abilities like thinking, memory, and concentration. Additionally, martial arts, yoga, creativity courses, and meditation are other practices of mental training. Training, education, and the utilization of devices which help to process the external signal and are considered as conventional ways of increasing cognition (Serruya and Kahana, 2008); these are culturally well-recognized. The unconventional ways of cognition which are gene therapy, nootropic drugs (drugs which increase the functioning of brain), and neural implants are under experimental stage at this time (Singh and Narang, 2014). The examples of the drugs which are used as CE are provided below (Table 18.1). Chemically, cognition can be enhanced by certain compounds. Caffeine is widely used to increase the alertness. Herbal extracts are supposed to enhance the memory and are prevalent, with the sale of million dollars of the plant *Ginkgo biloba* annually (Van Beek, 2002). Huperzine-A which is derived from *Huperzia serrata,* Chinese moss is an example of natural cognitive enhancer (McDougall Jr et al., 2005). Acetyl-L-carnitine is another example of brain booster that helps in maintaining the brain cells.

## 18.3 SIGNALING PATHWAYS INVOLVED IN COGNITION ENHANCEMENT

Consuming CE is a common approach for treating the consequences of conditions that involve cognitive impairments like Alzheimer's disease (AD), Parkinson's disease (PD), Down's syndrome, and traumatic brain injury. The main objective of CE intake, in daily life is, to develop a treatment which does not necessarily influence a cure, but provide the support and helps in improving the life quality in patients undergoing from neural ailments (Williams and Kemper, 2010). It involves pharmacological and nutraceutical interventions for promoting the cognitive ability and memory, targeting both mental ability and cognitive disorders (Daffner, 2010). Such drugs will be quite significant even, if the effective therapeutics for these diseases are found. Since, halting the pathogenesis of disease at a later stage of NDs will be of minimal importance without a way of improving the perceptive and intellectual functioning and repairing the accumulated

**TABLE 18.1**  Outline of the Consequences of the Drugs Which Are Used as CE.

| Cognitive enhancer | Neuromodulatory mechanisms | Improved functions | Brain parts most affected | Clinical use |
|---|---|---|---|---|
| Nicotine | Acts as an antagonist of nicotinic cholinergic receptor | Episodic memory, attention, working memory | Medial temporal lobe, default mode network, frontoparietal attentional system | – |
| Modafinil | Not known, but likely to affect noradrenaline, orexin, and dopamine systems | Episodic memory working memory, attention | Frontal lobe attentional systems | Eugeroic |
| Memantine | Low-affinity, non-competitive, N-methyl-D-aspartate (NMDA) receptor blocker | Attention, episodic memory | Parietal and frontal lobe | Alzheimer's disease |
| Caffeine | Antagonist of adenosine receptor | Incidental learning and working memory vigilance | Frontal lobe attentional systems | – |
| Amphetamine, methylphenidate | Noradrenaline and Dopamine reuptake inhibitors | Working memory, response inhibition, vigilance, attention | Striatum, default mode networks, frontoparietal attentional systems | Eugeroic, attention deficit hyperactivity disorder (ADHD) |

injuries. The approaches of finding cognitive enhancer are:

1. The identification of molecular signaling pathways for memory and learning, thereafter, examination of compounds that helps in activating the specific component of the pathway (Menges et al., 2015). The targets include the receptors of cell surface, protein kinases, neural signaling mechanism, synaptic transmission component, or, enzymes in the signaling pathways. The approach involves the stimulation of enzymes which are involved in the synthesis of transmitters, increasing the contact between post- and pre-synaptic proteins or stimulating synaptic maturation.

2. The activation of normal repair mechanism of neurons, so as to repair the synaptic pathways which were lost during the disease progression. This is again governed by understanding the signaling pathways and its interaction. The example can be a drug that stimulates synaptic development.

### 18.3.1 INVOLVEMENT OF SIGNALING PATHWAYS IN LEARNING

An important characteristic of cognition enhancers is the capability to expedite the process of learning and memory enhancement/retention. It is also a known fact that calcium cations and ion channels are associated in the signal transduction pathway of learning (Menard et al., 2015). This section highlights the signaling pathways involved in cognition enhancement by two approaches:

- Recognition of molecular signaling pathways of memory and

- Activation of normal repair mechanism of neuron.

For these two approaches, prospective therapeutic agents which act on the pathways, namely, small molecule drugs, and their advantages and shortcomings are discussed.

### 18.3.1.1 PROTEIN KINASE C

Protein Kinase C (PKC) is an essential component of memory and learning. An enhancement in PKC was identified in the crude membrane fraction of CA1 hippocampus, prepared by eye blink conditioning (EBC) process on rabbits (Bank et al., 1988), and was linked with the escalation in [3H] phorbol-12,13-dibutyrate binding (a ligand of PKC) (Olds et al., 1989). It exhibited an increase in phosphorylation of PKC substrates and consequently, associative learning in *Hermissenda* (Neary et al., 1981; Nelson et al., 1990). It was noticed that PKC regulates A-type potassium channels that are linked with classical conditioning in rabbits (Alkon and Rasmussen, 1988; Etcheberrigaray et al., 1992) and *Hermissenda* (Alkon et al., 1982). Also, Bryostatin 1 causes the activation of PKC that boosts memory and learning yet, increases dendritic spines, double-synapse presynaptic boutons, perforated postsynaptic densities (PSD) related with spines. An inhibitor of PKC is Ro 31-8220 and it helps in prevention of effects (Hongpaisan and Alkon, 2007). The activation of PKC induces the formation of other proteins when it's been provided before training (Alkon et al., 2005). It was detected in a study that the synthesis of proteins reduced the training events number needed for attainment of memory in *Hermissenda* and an extended withholding to approximately a week from 7 min, which indicates a decreased threshold for long-lasting consolidation (Alkon et al., 1982).

The synthesis of protein has been documented as an important event in the long-lasting acquisition of memory. The significant protein synthesis occurs locally in the neuronal dendrites, thereby, conferring specificity to artificial neural network (associative memory) (Jiang and Schuman, 2002; Govindarajan et al., 2011). Phorbol esters, bryostatin, and other activators of PKC like pnicogens and ingenol, bind to C1A and C1B domain of the PKC (DeChristopher et al., 2012; Nelson and Alkon, 2009). 1,2-diacylglycerol is the natural ligand for C1 domain (Blumberg et al., 2008).

### 18.3.1.2   OTHER C1 DOMAIN PROTEINS

Like PKC, there are many other proteins that have a C1 domain. Many proteins including mammalian uncoordinated (Munc)-13 family proteins (Wojcik and Brose, 2007) and rat sarcoma guanyl releasing protein 1 (Lorenzo et al., 2000) binds to 1,2-diacylglycerol and have homologous domains. These listed proteins are present in neurons and contribute in the growth of dendrites and release of synaptic vesicles. Therefore, they cannot be disregarded as a prospective target for various cognition enhancers and bryostatin; then, there are numerous isoforms of Munc13. Munc13 isoform is present in Purkinje and cerebellum granule cells, where it helps in controlling the motor learning. Munc13-1 is the profuse isoform which is present throughout the brain and is a prospective target of cognition enhancers as it is a synaptic vesicle C which colocalizes with the presynaptic marker synaptophysin (Augustin et al., 2001) and therefore, it is well-located to alter the synaptic efficiency. It was observed that phorbol dibutyrate binds to Munc13 and decreases the energy barrier for

fusion of synaptic vesicle (Basu et al., 2007). Phorbol esters and diacylglycerol (DAG) increases the release of neurotransmitters in the hippocampal neurons; Munc13 (but not PKC) is the protein which is accountable of causing this effect (Rhee et al., 2002).

The proteins that activate the C-2 domain like, DCP-LA methyl ester and DCP-LA (8-[2-(2-pentyl-cyclopropylmethyl)-cyclopropyl]-octanoic acid) cause advantageous behavioral and morphological effects thereby, specifying the potential value as cognition enhancers (Hongpaisan et al., 2011; Hongpaisan et al., 2013). In 5XFAD and Tg2576 transgenic mice model for AD, DCP-LA averts cognitive deficiency and synaptic loss (Hongpaisan et al., 2011).

### 18.3.1.3   PKC SUBSTRATES

A lot of PKC substrates are implicated in normal and cognitive learning like, myristoylated alanine-rich C-kinase substrate (MARCKS) which is an acidic membrane bound protein and can be phosphorylated by PKC which prevents the association with plasma membrane. Further, it increases F-actin activation and the destabilization of dendritic spines along with low capability to crosslink F-actin (Calabrese and Halpain, 2005). The spine destabilization which is induced by PKC can be averted by generating a mutant of non-phosphorylatable MARCKS. Approximately, 80% of the F-actin in dendritic spines turns in every minute, thus they are extremely dynamic in nature (Koleske, 2013). Therefore, PKC along with the signaling molecules which promotes stabilization might be specifically involved in restructuring of dendrimer during the development or adulthood when the dendritic spine is stabilized largely.

Growth associated protein-43 (GAP-43) is another PKC substrate which plays an integral role in cognitive functioning and is linked with the neuronal growth. GAP-43 interacts with actin (just like MARCKS). The phosphorylation of GAP-43 is done by PKC and thereafter it stabilizes the filaments (He et al., 1997). The activation of PKC upsurges autophosphorylated calcium/calmodulin-dependent protein kinase II (CaMKII) level and promotes the association with N-methyl-D-aspartate (NMDA) receptors, probably by phosphorylation of CaMK protein (like neuromodulin) (Yan et al., 2011). The interaction between PKC, calcium/calmodulin-dependent protein kinase II, and cytoskeletal proteins (like actin filaments) changes the dendritic spines geometry and thus, alters the connectivity and function of neuronal network which is responsible for cognition.

## 18.3.2 INVOLVEMENT OF SIGNALING PATHWAYS IN NEURONAL REPAIR

Another method of cognitive enhancement is inhibition of cellular and synaptic losses which occurs because of disease and injury. CE also reduce the synaptic losses which in turn prove to be advantageous for treating head injuries and ND; yet having no influence on healthy individuals. This category is therefore more specific to disease and has less possibility of abuse as compared to the general CE (Nelson et al., 2015).

### 18.3.2.1 NEUROTROPHINS

Neurotrophins (NTs) are the family of proteins that help in functioning, survival, and development of neurons. Levi-Montalcini and Cohen discovered the nerve growth factor (NGF) (the first NT) in 1950s. After 30 years, brain-derived neurotrophic factor (BDNF) was

discovered which was followed by NT-3, NT-4, then central neurotrophic factor (CNF), and members of glial cell line neurotrophic factor family. The widely studied NT pertaining to restoration of synapse is BDNF. The BDNF messenger ribonucleic acid (mRNA) are reduced during AD (Murray et al., 1994; Phillips et al., 1991; Connor et al., 1997) and process of aging (Calabrese et al., 2013). Similarly, another receptor of BDNF was tropomyosin receptor kinase B (TrkB). Phosphorylated TrkB is downregulated in aged rats (Calabrese et al., 2013). PKC phosphorylates co-activator associated arginine methyltransferase 1 (CARM1), which is a protein that methylates HuD (RNA binding protein) that enhances the expression and stability of NT-3, NGF, and BDNF. The expression of these proteins has displayed an increase in development of hippocampal neurons in the culture (Lim and Alkon, 2012). Therefore, BDNF can be increased by modulating the receptor of NT and NT mRNA pharmacologically.

Even though the synthesis of BDNF is controlled by PKC, the actions of TrkB receptors and BDNF receptors seem to depend on ERK/Ras, phospholipase Cγ (PLCγ), and phosphatidylinositol-3 kinase (PI3K)/Akt pathways (Xia et al., 2010; Leal et al., 2014). Akt (also called protein kinase B) is a kinase which is concerned in a range of inhibiting pathways related to apoptosis. Environmental enrichment increases BDNF, number of dendritic spines, and Akt levels thereby, enhancing neurogenesis and cognition (Ramirez-Rodriguez et al., 2014).

As BDNF is associated in activity-dependent maturation and short-range functioning of synapse, whose damage is one of the initial changes in AD, and BDNF will be quite efficient as a cognition enhancer. Therefore, the interest is to up-regulate the BDNF levels in the brain. But, BDNF is unable to cross the blood-brain barrier (BBB) (Pardridge et al., 1998). Another activator of TrkB, 7, 8-dihydroxyflavone (Fig. 18.1) is able to cross the BBB and affects the emotional learning in amygdala (Andero et al., 2011; Zeng et al., 2012). Therefore, the emphasis has been on the discovery of small molecules which are the agonists of Trk receptors and small molecules that synthesize NTs. 4-methylcatechol (4-MC) is one such molecule which is reported to synthesize NGF (Kaechi et al., 1993; Kaechi et al., 1995), to increase the mRNA of BDNF, and immunoreactivity by 50% in the brain when administered peripherally to the 10-day old rat (Fukumitsu et al., 1999). Further, when 4-MC was administered intracerebroventricularly, it enhanced memory and spatial learning in rats and caused an antidepressant effect (Sun and Alkon, 2008). The process by which BDNF is induced is still not known, but, on the whole, it can be stimulated by various transcription factors (TFs) like cAMP response element binding protein (CREB), calcium-response factor and neuronal Per Arnt Sim domain protein 4 (NPAS4) (Calabrese et al., 2013). Evidences suggest that 4-MC stimulates MAP kinases and NT receptors of Trk family directly (Sometani et al., 2002).

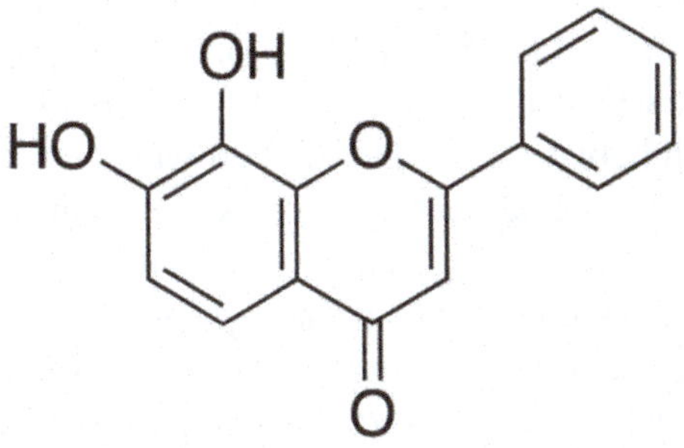

**FIGURE 18.1**   Chemical structure of 7,8-dihydroxyflavone.

### 18.3.2.2   ACTIVATORS OF NT RECEPTORS

Though, 4-MC given in Figure 18.2 is able to cross the BBB of young mice (Fukumitsu et al.,

1999; Fukuhara et al., 2012), another activator of TrkB, 7,8-dihydroxyflavone crosses the BBB and has an effect on emotional learning in the brain (specifically amygdala portion) (Andero et al., 2011; Zeng et al., 2012). Even though the specificity and affinity of flavone is quite low, it can be used as a preliminary point for development of drug. An alternative method can be activation of the pathways downstream of TrkB which would lead to the increase in specificity. There are three pathways (listed below) are activated by TrkB that play an important role in memory and growth of synapse (Yamada and Nabeshima, 2003). These are as follows:

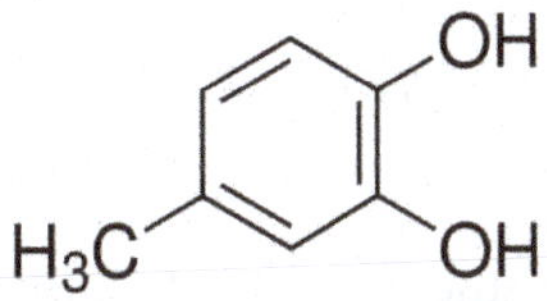

**FIGURE 18.2**   Chemical structure of 4-methylcatechol.

#### 18.3.2.2.1   MAPK Signaling

This signaling is a mitogen-activated protein kinase (MAPK) signaling (Garrington and Johnson, 1999). MARKS are associated in directing cellular responses to a diverse array of stimuli, for example mitogens, heat shock, osmotic stress, etc.

#### 18.3.2.2.2   NMDA Signaling

Activated TrkB receptors result in phosphorylation of N-methyl D-aspartate receptor subtype 2B (NR2B) and N-methyl D-aspartate receptor subtype 1 (NR1) subunits of NMDA receptor which cause the increase in release of glutamate (McGee and Abdel-Rahman, 2016).

### 18.3.2.2.3 Akt/PKB Signaling Pathway

In protein kinase B (PKB) signaling pathway, TrkB acts over sequences of phosphorylation steps involving Akt, mammalian target of rapamycin (which controls translation of mRNA), and PI3K. It involves phospholipase Cγ whose function is to cleave the phospholipids at phosphate site, thus, forming 1, 2-diacylglycerol and inositol 1, 4, 5-triphosphate (IP3). IP3 is an essential secondary messenger for the IP3 receptor on endoplasmic reticulum (ER) that is a main constituent of ryanodine receptor (RyR) and calcium-induced calcium release (Song et al., 2005).

### 18.3.2.3 INHIBITORS OF APOPTOSIS

If a neuron is stressed or injured, it might undergo apoptosis. It may be either extrinsic (which can be started by activating the receptors of cell surface) or intrinsic (involving the ER and the mitochondria). Cell death can be triggered by either of the losses of factors responsible for cell survival. Further, damage of DNA, which may thereby, cause the pro-apoptotic proteins from the mitochondria to stimulate caspase proteases and eventually, caspase activated DNase. Apoptosis can also be induced in caspase independent manner by triggering apoptosis-inducing factor (AIF) which is protein present in the intermembrane of the mitochondria. Attenuation of cell death can be triggered by stimulation of PKCγ in the hippocampus. Therefore, it is suggested that the activators of PKCγ inhibits apoptosis (Sun et al., 2009), thereby, increasing the usefulness as CE that can act on the patients specifically suffering from stroke, brain injury, and acute radiation sickness.

There are many evidences which suggest that CREB hinders apoptosis (Walton and Dragunow, 2000) and the CREB phosphorylation increase after ischemic stress and injury (Kitagawa, 2007), thereby, proposing that CREB can be a survival factor.

### 18.3.2.4 PKC ACTIVATORS AS NEUROPROTECTIVE AGENTS

Various apoptosis inhibitors are known till now, but none has been approved for clinical use because inhibiting the cell death is quite questionable. But, the idea of utilizing CE as apoptosis inhibitors is not dead. As there are few evidences that support this theory like— PKCγ protein is known to inhibit apoptosis (Basu and Sivaprasad, 2007; Ding et al., 2002; Gillespie et al., 2005; Okhrimenko et al., 2005) and it raises the view of using the activators of PKCγ like DCP-LA as inhibitors of central nervous system (CNS) apoptosis. Similarly, isozymes like PKCα, β, βII, and γ also participate in this process and are overexpressed in cochlea hair cells during the regeneration phase after the damage caused by glutamate analog AMPA (Lerner-Natoli et al., 1997). Among the isozymes, PKCδ and PKCε are the important ones which determine the struggle of life and death between apoptosis and growth.

The interaction between PKC isozymes and apoptosis pathways is quite convoluted. Though, PKCδ is considered as an apoptotic isozyme whereas, PKCδ restrains apoptosis which is induced by irradiation (Bluwstein et al., 2013) or oxidized low-density lipoprotein (LDL) (Larroque-Cardoso et al., 2013). Also, PKCε is an antagonist of PKCδ, the suppression of PKCε increases the cell death by obstructing the ability of PKC to inhibit BCL2 (B-cell lymphoma 2, an anti-apoptotic factor). Furthermore, caspases can cleave all types of PKC ε, δ, ζ, and θ (Basu and Sivaprasad, 2007; Mizuno et al., 1997; Datta et al., 1997; Smith et al., 2000). Therefore, further investigation is required to know how neuroprotective activity of PKC is dependent on apoptosis.

## 18.4 METHODS OF COGNITION ENHANCEMENT

It has become increasingly recognized that various cellular and molecular processes in neurons can be modified to increase the cognitive functions and treat the mental illness. With the accelerating identification of molecular mechanisms in the area of memory and learning allows us to suggest various manipulations that might enhance the acquisition of information or/and retention (Jorgenson et al., 2015). These manipulations help to apprehend the relation between cognitive behavior and neuronal communication. Whereas, if cognitive enhancement is achieved, it can be anticipated that these same manipulations will be suitable for treating the disorders where cognitive functioning is weakened, though the pathogenic alterations are not apprehended. So, the understanding on the cognitive enhancement mechanism requires both translational and basic research perspectives.

The study on animal models has been proposed by various molecular variations that can result in increased cognition (Levin and Buccafusco, 2006). Mainly, the genetic alterations are related more or less to the process called synaptic plasticity, specifically long-term depression (LTD) and long-term potentiation (LTP) (Lüscher and Malenka, 2012). So, by improving or facilitating synaptic mechanism processes cognition enhancement can be achieved, which is not a minor task. Synaptic plasticity is also like learning and an activity

(experience) mediated process (Takeuchi et al., 2014). Therefore, just by increasing the strength of synapse (surpassing the physiological stimulus), it cannot be expected that there will be increase in cognition, rather these manipulations will turn out to be more dangerous for the cognitive function. On the contrary, it can be proposed that the ideal forms of cognition enhancers do not alter the synaptic plasticity, yet, functioning on the modulatory features of these mechanisms (Lee and Silva, 2009a). This section covers a few examples that focus on the mechanisms that are the targets for cognition enhancement.

### 18.4.1 TARGETING THE INDUCTION OF NEURONAL PLASTICITY (D-SERINE, GLYCINE)

NMDA receptors are the significant originators of plasticity underlying memory and learning (Morris et al., 1986; Tsien et al., 1996). They are the target of genetic and pharmacological manipulations to examine the cognitive performance and the possibility of cognitive enhancement (Collingridge et al., 2013; Lee and Silva, 2009b). This section concentrates on the two molecules: D-serine and glycine, jointly with glutamate acts as co-agonists and activates NMDA receptor causing the induction of plasticity.

#### 18.4.1.1 MECHANISM OF ACTION

The role of glycine and D-serine as a co-agonist was determined more than 25 years ago (Johnson and Ascher, 1987; Kleckner and Dingledine, 1988). These amino acids bind to GluN1 component of the NMDA receptor and glutamate (the classic agonist) binds to GluN2 subunit. Simultaneous binding of these subunits, that is, GluN1 and GluN2 causes the full activation of NMDA receptor (Laube et al., 1997). There is a release of glutamate from the synaptic bouton during the synaptic transmission and it is anticipated that the obtainability of glutamate extracellular spaces regulates the functioning of the receptor and thereby, affecting the synaptic plasticity. It is applicable for both D-serine (Yang et al., 2003) and glycine (Martina et al., 2004). With aging, the concentrations of D-serine and serine racemase (an enzyme that forms D-serine from L-serine) decrease in hippocampus region (Turpin et al., 2011). This shortcoming can be amended by the addition of D-serine, which helps in restoring the plasticity in hippocampus from mouse model suffering from SAMP8 (senescence accelerated mouse-prone 8) and aged rats (Yang et al., 2005). Further, an addition of D-serine up-regulates the neurogenesis in vitro and in vivo (Sultan et al., 2013).

The fact that D-serine and glycine seem to have same effect on NMDA receptor functioning that has made the recognition of endogenous ligands (for GluN1 site) intricate. The consequences of D-serine and glycine on synaptic plasticity are converted into cognitive function. Therefore, by inhibiting the formation of D-serine by knocking out serine racemase inhibits the memory that is restored by treatment with D-serine (Balu et al., 2013).

### 18.4.2 MODIFICATION OF SYNAPSE WITH GHRELIN

Ghrelin is a recognized hunger hormone which is produced by endocrine cells in mucous membrane of stomach responding to fasting (Inui et al., 2004). The hormone increases the production of signals and energy directly to the hypothalamic regulatory nuclei that controls homeostasis of energy (Inui, 2001). Yet, the ghrelin receptor (growth-hormone secretagogue type 1a receptor

(GHS-R1a)) is expressed profusely in ventral tegmental area, substantia nigra, and hippocampus (Guan et al., 1997), regions of brain undertaking intensive plasticity which is related to memory, reward, and learning.

### 18.4.2.1 MECHANISM OF ACTION

Ghrelin is a ligand for the GHS-R1a, which is a G-protein-coupled receptor (GPCR). After binding of ghrelin to GHS-R1a, a signal is sent through $G_{q/11}$ for the stimulation of phospholipase C (PLC) directing the calcium mobilization from intracellular stores, activating PKC and production of diacylglycerol (Camina, 2006). The activation of GHS-R1a is linked to activation of various pathways like phosphatidylinositol-3 kinase, PKA, and MAPK pathways (Camina, 2006), yet the mechanism is still not understood and might comprises heterodimerization with various G-protein-coupled receptors (Schellekens et al., 2013). In spite of the uncertainties in signaling pathways, ghrelin has a strong effect on neuronal and synaptic function in hippocampus. Therefore, circulating ghrelin enters the hippocampus, where the effect is synaptogenic in cortical cultures of neurons (Diano et al., 2006) and in vivo. The effect can be facilitated by actin cytoskeleton remodeling inside the dendritic spine (Berrout and Isokawa, 2012; Cuellar and Isokawa, 2011). The growth stimulating action of ghrelin is facilitated by Akt pathway and is connected with potentiation of synaptic transmission (Chen et al., 2011).

### 18.5 STRATEGIES OF COGNITION ENHANCEMENT

Insufficiencies in intellectual function are concomitant with lots of neurological disorders and mental diseases. Autism, mental retardation, attention deficit, depression, schizophrenia, AD, PD, Huntington's disease, and along with other NDs; all of these disorders do have cognitive impairments along with the existing pathologies. In addition, memory impairment and cognitive decline are associated with changes related to age in the brain and may specify the beginning of dementia (Bishop et al., 2010). Cognitive enhancement is regarded as a tactic for the treatment of these diseases or allows delaying the effect of aging.

In today's scenario, enhancing the cognition with the help of pharmacological agents is not a new practice and the psychostimulants which are self-administered like, nicotine and caffeine are already an established agent in enhancing cognitive performance. MPHs along with other medications are efficient in treatment of attention disorders. Nevertheless, not all memory and learning problems can be treated by stimulants as there are a number of possible reasons for cognitive disorders. More precise and effective therapeutic strategies are required. Irrespective of the different causes of cognitive disorders, an effective approach for the treatment of various forms of cognitive insufficiency will be to target key methods which have a positive effect on enhancing the cognition (Lee and Silva, 2009b).

## 18.5.1 GENES

The understanding about the genetic makeup of an individual helps in identifying the etiology of cognitive disorders, just like most non-infectious diseases. Genetic makeup has an influence on cognitive ability, but a myriad of interconnected factors like psychological state and environment affect the capability and comprehension to dedicate the experiences to memory. Additionally, the state of physiological conditions or health

influences the processes that facilitate learning, besides this hormone present in the periphery and brain affects the process of learning, and the neuroendocrine, humoral systems and endocrine "talk" to the brain. Stress impairs cognition and leads to various mental disorders. Exercise might stimulate the process of cognition (Lautenschlager et al., 2008) and both of those affects the plasticity by influencing neuronal replacement by neurogenesis (Van Praag, 2008).

## 18.5.2 PROTEIN TRANSLATION

The synthesis of protein is essential for memory and synaptic plasticity (Cajigas et al., 2010). The synthesized mRNA is moved to distal synapses by dendritic-targeting elements. The protein translation is induced in response to synaptic activity and the synapse translation profile, which is being remodeled function as a synaptic memo consolidation snapshot. Inhibition of protein translation weakens cognition. Many cognitive enhancement strategies are found to control mRNA translocation and the synaptic protein translation. With respect to this, a transgenic technology is developed, known as BacTRAP, which permits the isolation of translated cell type specific mRNAs (Doyle et al., 2008; Heiman et al., 2008; Dougherty et al., 2010). This technology is used to outline the activated synapse profile and helps to compare the profile between mice modeling for the cognitive disorder and normal mice (Silverman et al., 2010).

## 18.5.3 PROTEIN DEGRADATION

The remodeling of synapse includes alterations in synaptic proteome. Additionally, the synthesis of protein, selective degradation is an utmost feature in cognition. The ubiquitin proteasome system locates to synapse (Bingol et al., 2010) and the activation during the activity of synapse is integral to learning and plasticity (Cajigas et al., 2010). The activation of NMDA receptor facilitates the proteasome subunit redistribution. Furthermore, protein phosphatases which are activated by calcium ions (for example Calpain) are involved in synaptic remodeling inherently. Calpain cleaves NFκB (Schölzke et al., 2003), the protein that regulates actin and striatal-enriched protein tyrosine phosphatase (STEP) (Xu et al., 2009), p35, NR2B subunit of NMDA receptor (Guttmann et al., 2002), and an array of pre- and post-synaptic proteins (Liu et al., 2006).

## 18.6   DOWNSIDE OF COGNITION ENHANCERS

The cognition enhancers can have various side effects via body system other than the brain. Memantine (NMDA receptor antagonist) and acetylcholinesterase inhibitors (AChEIs) are used as a standard treatment therapy for various NDs like AD and PD can cause nausea or gastrointestinal upset, thereby, leading to discontinue the treatment altogether. These consequences have the ability to counteract the positive effect of the drug on the overall performance. Donepezil and rivastigmine are used to treat dementia with Parkinson's disease dementia (PDD) and Lewy bodies (DLB). It is seen that rivastigmine in healthy patients improve learning and helps in making association between digits and symbols, but at the same time, it can impair visual and verbal episodic memory. Bromocriptine is a dopamine agonist that can increase spatial working memory (WM) yet, also impair probabilistic reversal learning. The outcome echoes the

outcomes in patients with PD, the dopamine receptor agonist helps to improve the WM and task-set switching tasks yet, degrades the reversal learning (RL) (Swainson et al., 2000; Cools et al., 2001). It is hypothesized that these divergent effects are because of the replenishment of dopamine in dorsal striatal areas required for the anterior and overdosing of ventral striatal areas involved in the end. Therefore, the doses of dopaminergic medication play an integral role to improve the functioning of motor neuron and some features have the ability to worsen others.

The conclusion is well suited to the recent reports that PD patients on the dopaminergic agonist medication had acquired imprudent behavior such as hypersexuality, gambling, and compulsive shopping (Weintraub et al., 2010; Weintraub et al., 2006). It is seen that these behaviors in PD is linked to the presence of involuntary movements, dyskinesias which is because of extreme stimulation of dopamine (Voon et al., 2009). Reduction in the dopaminergic drug dosage can lead to reduction in impulsivity. Thus, it shows that the dopamine agonist has multiple consequences, such as both harmful and beneficial on behavior and cognition.

## 18.7   FUTURE OPPORTUNITIES AND CHALLENGES

In each neurological illness, the cognitive impairment may be the result of dysfunctioning in neurotransmitter system and brain area (Xu et al., 2012). Thus, various pharmacological strategies are required for the restoration of cognitive functioning in these disorders. Much research is focused on the treatments to restore the impairment, and it has also achieved success for treating the cognitive deficits in various neurological and neuropsychiatric disorders. Yet, further research is required to recognize the neuro-biological substrates of neuropsychiatric disorders for the development of new drugs/treatments with improved efficacy for restoration of cognitive functioning.

In the near future, scientists and researchers should carefully consider the information they provide about the benefits and risks of using CE and how they present the information. The scientists should be careful of not oversimplifying the laboratory findings to real life scenarios. The research in this field needs to consider the ethical issues of raising the awareness about the illegal behaviors and harmful effects by stressing on the desirability or effectiveness of smart drugs for enhancing the cognition (Ragan et al., 2013).

## 18.8 CONCLUSION

This chapter represents perception of cognitive enhancement and compounds used as CE. In a world where life span and work span are increasing altogether, the cognitive enhancement strategies can be quite beneficial to enhance the work productivity, delay the age-related cognitive impairments, and improve the quality of life. Effective and safe CE will profit both the society and the individual. But, the research on the use of cognition enhancers on regular basis is still lacking while some are positive than others. It is still challenging to know the long-term effects of CE which are new, like nootropics and attention deficit drugs. These CE have been the most successful in treating the brain diseases like AD and PD. Especially in NDs, the risk of using relatively new and unknown enhancers may be overshadowed by their ability to improve health.

## KEYWORDS

- **cognition**
- **cognitive enhancers**
- **cognitive impairment**
- **intelligence**
- **neuronal plasticity**

## REFERENCES

Alkon, D. L.; Epstein, H.; Kuzirian, A.; Bennett, M. C.; Nelson, T. J. Protein Synthesis Required for Long-term Memory is Induced by PKC Activation on Days Before Associative Learning. *Proc. Natl. Acad. Sci. U. S. A.* **2005**, *102*, 16432–16437.

Alkon, D. L.; Lederhendler, I.; Shoukimas, J. J. Primary Changes of Membrane Currents During Retention of Associative Learning. *Science* **1982**, *215*, 693–695.

Alkon, D. L.; Rasmussen, H. A Spatial-temporal Model of Cell Activation. *Science* **1988**, *239*, 998–1005.

Etcheberrigaray, R.; Matzel, L. D.; Lederhendler, I. I.; Alkon, D. L. Classical Conditioning and Protein Kinase C Activation Regulate the Same Single Potassium Channel in Hermissenda Crassicornis Photoreceptors. *Proc. Natl. Acad. Sci. USA* **1992**, *89*, 7184–7188.

Andero, R.; Heldt, S. A.; Ye, K.; Liu, X.; Armario, A.; Ressler, K. J. Effect of 7, 8-dihydroxyflavone, a Small-molecule TrkB Agonist, on Emotional Learning. *Am. J. Psychiatry* **2011**, *168*, 163–172.

Augustin, I.; Korte, S.; Rickmann, M.; Kretzschmar, H. A.; Südhof, T. C.; Herms, J. W.; Brose, N. The Cerebellum-specific Munc13 Isoform Munc13-3 Regulates Cerebellar Synaptic Transmission and Motor Learning in Mice. *J. Neurosci.* **2001**, *21*, 10–17.

Babcock, Q.; Byrne, T. Student Perceptions of Methylphenidate Abuse at a Public Liberal Arts College. *J. Am. Coll. Health* **2000**, *49*, 143–145.

Balu, D. T.; Li, Y.; Puhl, M. D.; Benneyworth, M. A.; Basu, A. C.; Takagi, S.; Bolshakov, V. Y.; Coyle, J. T. Multiple Risk Pathways for Schizophrenia Converge in Serine Racemase Knockout Mice, a Mouse Model of NMDA Receptor Hypofunction. *Proc. Natl. Acad. Sci. USA* **2013**, *110*, E2400–E2409.

Bank, B.; DeWeer, A.; Kuzirian, A. M.; Rasmussen, H.; Alkon, D. L. Classical Conditioning Induces Long-term Translocation of Protein Kinase C in Rabbit Hippocampal CA1 Cells. *Proc. Natl. Acad. Sci. USA* **1988**, *85*, 1988–1992.

Basu, J.; Betz, A.; Brose, N.; Rosenmund, C. Munc13-1 C1 Domain Activation Lowers the Energy Barrier for Synaptic Vesicle Fusion. *J. Neurosci.* **2007**, *27*, 1200–1210.

Basu, A.; Sivaprasad, U. Protein Kinase Cε Makes the Life and Death Decision. *Cell. Signal* **2007**, *19*, 1633–1642.

Berrout, L.; Isokawa, M. Ghrelin Promotes Reorganization of Dendritic Spines in Cultured Rat Hippocampal Slices. *Neurosci. Lett.* **2012**, *516*, 280–284.

Bingol, B.; Wang, C.-F.; Arnott, D.; Cheng, D.; Peng, J.; Sheng, M. Autophosphorylated CaMKIIα Acts as a Scaffold to Recruit Proteasomes to Dendritic Spines. *Cell* **2010**, *140*, 567–578.

Bishop, N. A.; Lu, T.; Yankner, B. A. Neural Mechanisms of Ageing and Cognitive Decline. *Nature* **2010**, *464*, 529–535.

Bluwstein, A.; Kumar, N.; Leger, K.; Traenkle, J.; Van Oostrum, J.; Rehrauer, H.; Baudis, M.; Hottiger, M. PKC Signaling Prevents Irradiation-induced Apoptosis of Primary Human Fibroblasts. *Cell Death Dis.* **2013**, *4*, e498.

Bostrom, N.; Sandberg, A. Cognitive Enhancement: Methods, Ethics, Regulatory Challenges. *Sci. Eng. Ethics* **2009**, *15*, 311–341.

Cajigas, I. J.; Will, T.; Schuman, E. M. Protein Homeostasis and Synaptic Plasticity. *EMBO J.* **2010**, *29*, 2746–2752.

Calabrese, F.; Guidotti, G.; Racagni, G.; Riva, M. A. Reduced Neuroplasticity in Aged Rats: A Role for the Neurotrophin Brain-derived Neurotrophic Factor. *Neurobiol. Aging* **2013**, *34*, 2768–2776.

Camina, J. Cell Biology of the Ghrelin Receptor. *J. Neuroendocrinol.* **2006**, *18*, 65–76.

Calabrese, B.; Halpain, S. Essential Role for the PKC Target MARCKS in Maintaining Dendritic Spine Morphology. *Neuron.* **2005**, *48*, 77–90.

Chen, L.; Xing, T.; Wang, M.; Miao, Y.; Tang, M.; Chen, J.; Li, G.; Ruan, D. Y. Local Infusion of Ghrelin Enhanced Hippocampal Synaptic Plasticity and Spatial Memory through Activation of Phosphoinositide 3-kinase in the Dentate Gyrus of Adult Rats. *Eur. J. Neurosci.* **2011**, *33*, 266–275.

Collingridge, G. L.; Volianskis, A.; Bannister, N.; France, G.; Hanna, L.; Mercier, M.; Tidball, P.; Fang, G.; Irvine, M. W.; Costa, B. M. The NMDA Receptor as a Target for Cognitive Enhancement. *Neuropharmacology* **2013**, *64*, 13–26.

Connor, B.; Young, D.; Yan, Q.; Faull, R.; Synek, B.; Dragunow, M. Brain-derived Neurotrophic Factor is Reduced in Alzheimer's Disease. *Mol. Brain Res.* **1997**, *49*, 71–81.

Cools, R.; Barker, R. A.; Sahakian, B. J.; Robbins, T. W. Enhanced or Impaired Cognitive Function in Parkinson's Disease as a Function of Dopaminergic Medication and Task Demands. *Cereb Cortex* **2001**, *11*, 1136–1143.

Cuellar, J. N.; Isokawa, M. Ghrelin-induced Activation of cAMP Signal Transduction and its Negative Regulation by Endocannabinoids in the Hippocampus. *Neuropharmacology* **2011**, *60*, 842–851.

Daffner, K. R. Promoting Successful Cognitive Aging: A Comprehensive Review. *J. Alzheimers Dis.* **2010**, *19*, 1101–1122.

Datta, R.; Kojima, H.; Yoshida, K.; Kufe, D. Caspase-3-mediated Cleavage of Protein Kinase C θ in Induction of Apoptosis. *J. Biol. Chem.* **1997**, *272*, 20317–20320.

DeChristopher, B. A.; Fan, A. C.; Felsher, D. W.; Wender, P. A. "Picolog," a Synthetically-available Bryostatin Analog, Inhibits Growth of MYC-induced Lymphoma in vivo. *Oncotarget* **2012**, *3*, 58.

Blumberg, P.; Kedei, N.; Lewin, N.; Yang, D.; Czifra, G.; Pu, Y.; Peach, M.; Marquez, V. Wealth of Opportunity-the C1 Domain as a Target for Drug Development. *Curr. Drug Targets* **2008**, *9*, 641–652.

Diano, S.; Farr, S. A.; Benoit, S. C.; McNay, E. C.; da Silva, I.; Horvath, B.; Gaskin, F. S.; Nonaka, N.; Jaeger, L. B.; Banks, W. A. Ghrelin Controls Hippocampal Spine Synapse Density and Memory Performance. *Nat. Neurosci.* **2006**, *9*, 381–388.

Ding, L.; Wang, H.; Lang, W.; Xiao, L. Protein Kinase C-ε Promotes Survival of Lung Cancer Cells by Suppressing Apoptosis Through Dysregulation of the Mitochondrial Caspase Pathway. *J. Biol. Chem.* **2002**, *277*, 35305–35313.

Dougherty, J. D.; Schmidt, E. F.; Nakajima, M.; Heintz, N. Analytical Approaches to RNA Profiling Data for the Identification of Genes Enriched in Specific Cells. *Nucleic Acids Res.* **2010**, *38*, 4218–4230.

Doyle, J. P.; Dougherty, J. D.; Heiman, M.; Schmidt, E. F.; Stevens, T. R.; Ma, G.; Bupp, S.; Shrestha, P.; Shah, R. D.; Doughty, M. L. Application of a Translational Profiling Approach for the Comparative Analysis of CNS Cell Types. *Cell.* **2008**, *135*, 749–762.

Eickenhorst, P.; Vitzthum, K.; Klapp, B. F.; Groneberg, D.; Mache, S. Neuroenhancement Among German University Students: Motives, Expectations, and

Relationship with Psychoactive Lifestyle Drugs. *J. Psychoacti. Drugs* **2012**, *44*, 418–427.

Farah, M. J. Emerging Ethical Issues in Neuroscience. *Nature Neurosci.* **2002**, *5*, 1123.

Zielbauer, P. New Campus High: Illicit Prescription Drugs. *New York Times* **2000**, B8.

Franke, A. G.; Bagusat, C.; Dietz, P.; Hoffmann, I.; Simon, P.; Ulrich, R.; Lieb, K. Use of Illicit and Prescription Drugs for Cognitive or Mood Enhancement Among Surgeons. *BMC Med.* **2013**, *11*, 102.

Fukuhara, K.; Ishikawa, K.; Yasuda, S.; Kishishita, Y.; Kim, H.-K.; Kakeda, T.; Yamamoto, M.; Norii, T.; Ishikawa, T. Intracerebroventricular 4-methylcatechol (4-MC) Ameliorates Chronic Pain Associated with Depression-like Behavior via Induction of Brain-derived Neurotrophic Factor (BDNF). *Cell Mol. Neurobiol.* **2012**, *32*, 971–977.

Fukumitsu, H.; Sometani, A.; Ohmiya, M.; Nitta, A.; Nomoto, H.; Furukawa, Y.; Furukawa, S. Induction of a Physiologically Active Brain-derived Neurotrophic Factor in the Infant Rat Brain by Peripheral Administration of 4-methylcatechol. *Neurosci. Lett.* **1999**, *274*, 115–118.

Garrington, T. P.; Johnson, G. L. Organization and Regulation of Mitogen-activated Protein Kinase Signaling Pathways. *Curr. Opin. Cell Biol.* **1999**, *11*, 211–218.

Giurgea, C. Pharmacology of Integrative Activity of the Brain. Attempt at Nootropic Concept in Psychopharmacology. *Actual. Pharmacol.* **1972**, *25*, 115.

Govindarajan, A.; Israely, I.; Huang, S.-Y.; Tonegawa, S. The Dendritic Branch is the Preferred Integrative Unit for Protein Synthesis-dependent LTP. *Neuron.* **2011**, *69*, 132–146.

Guan, X.-M.; Yu, H.; Palyha, O. C.; McKee, K. K.; Feighner, S. D.; Sirinathsinghji, D. J.; Smith, R. G.; Van der Ploeg, L. H.; Howard, A. D. Distribution of mRNA Encoding the Growth Hormone Secretagogue Receptor in Brain and Peripheral Tissues. *Mol. Brain Res.* **1997**, *48*, 23–29.

Guttmann, R. P.; Sokol, S.; Baker, D. L.; Simpkins, K. L.; Dong, Y.; Lynch, D. R. Proteolysis of theN-Methyl-d-Aspartate Receptor by Calpain in Situ. *[J. Pharmacol. Exp. Ther.* **2002**, *302*, 1023–1030.

He, Q.; Dent, E. W.; Meiri, K. F. Modulation of Actin Filament Behavior by GAP-43 (neuromodulin) is Dependent on the Phosphorylation Status of Serine 41, the Protein Kinase C Site. *J. Neurosci.* **1997**, *17*, 3515–3524.

Heiman, M.; Schaefer, A.; Gong, S.; Peterson, J. D.; Day, M.; Ramsey, K. E.; Suárez-Fariñas, M.; Schwarz, C.; Stephan, D. A.; Surmeier, D. J. A Translational Profiling Approach for the Molecular Characterization of CNS Cell Types. *Cell* **2008**, *135*, 738–748.

Heinz, A.; Kipke, R.; Heimann, H.; Wiesing, U.; Cognitive Neuroenhancement: False Assumptions in the Ethical Debate. *J. Med. Ethics* **2012**, *38*, 372–375.

Heishman, S. J.; Kleykamp, B. A.; Singleton, E. G. Meta-analysis of the Acute Effects of Nicotine and Smoking on Human Performance. *Psychopharmacology* **2010**, *210*, 453–469.

Hongpaisan, J.; Alkon, D. L. A Structural Basis for Enhancement of Long-term Associative Memory in Single Dendritic Spines Regulated by PKC. *Proc. Natl. Acad. Sci. USA.* **2007**, *104*, 19571–19576.

Hongpaisan, J.; Sun, M.-K.; Alkon, D. L. PKC ε Activation Prevents Synaptic Loss, Aβ Elevation, and Cognitive Deficits in Alzheimer's Disease Transgenic Mice. *J. Neurosci.* **2011**, *31*, 630–643.

Hongpaisan, J.; Xu, C.; Sen, A.; Nelson, T. J.; Alkon, D. L. PKC Activation During Training Restores Mushroom Spine Synapses and Memory in the Aged Rat. *Neurobiol. Dis.* **2013**, *55*, 44–62.

Inui, A.; Asakawa, A.; Bowers, C. Y.; Mantovani, G.; Laviano, A.; Meguid, M. M.; Fujimiya, M. Ghrelin, Appetite, and Gastric Motility: the Emerging Role of the Stomach as an Endocrine Organ. *The FASEB J.* **2004**, *18*, 439–456.

Inui, A. Ghrelin: An Orexigenic and Somatotrophic Signal from the Stomach. *Nat. Rev. Neurosci.* **2001**, *2*, 551–560.

Jiang, C.; Schuman, E. M. Regulation and Function of Local Protein Synthesis in Neuronal Dendrites. *Trends Biochem. Sci.* **2002**, *27*, 506–513.

Johnson, J.; Ascher, P. Glycine Potentiates the NMDA Response in Cultured Mouse Brain Neurons. *Nature.* **1987**, *325*, 529–531.

Jorgenson, L. A.; Newsome, W. T.; Anderson, D. J.; Bargmann, C. I.; Brown, E. N.; Deisseroth, K.; Donoghue, J. P.; Hudson, K. L.; Ling, G. S. F.; MacLeish, P. R.; Marder, E.; Normann, R. A.; Sanes, J. R.; Schnitzer, M. J.; Sejnowski, T. J.; Tank, D. W.; Tsien, R. Y.; Ugurbil, K.; Wingfield, J. C. The BRAIN Initiative: Developing Technology to Catalyse Neuroscience Discovery. *Philos. Trans. R. Soc. Lond. B. Biol. Sci.* **2015**, *370*, 20140164.

Kaechi, K.; Furukawa, Y.; Ikegami, R.; Nakamura, N.; Omae, F.; Hashimoto, Y.; Hayashi, K.; Furukawa, S. Pharmacological Induction of Physiologically Active Nerve Growth Factor in Rat Peripheral Nervous System. *J. Pharmacol. Exp. Ther.* **1993**, *264*, 321–326.

Kaechi, K.; Ikegami, R.; Nakamura, N.; Nakajima, M.; Furukawa, Y.; Furukawa, S. 4-Methylcatechol, An Inducer of Nerve Growth Factor Synthesis, Enhances Peripheral Nerve Regeneration Across Nerve Gaps. *J. Pharmacol. Exp. Ther.* **1995**, *272*, 1300–1304.

Kleckner, N. W.; Dingledine, R. Requirement for Glycine in Activation of NMDA-receptors Expressed in Xenopus Oocytes. *Science.* **1988**, *241*, 835.

Kitagawa, K. CREB and cAMP Response Element-mediated Gene Expression in the Ischemic Brain. *FEBS J.* **2007**, *274*, 3210–3217.

Koleske, A. J. Molecular Mechanisms of Dendrite Stability. *Nat. Rev. Neurosci.* **2013**, *14*, 536.

Larroque-Cardoso, P.; Swiader, A.; Ingueneau, C.; Negre-Salvayre, A.; Elbaz, M.; Reyland, M.; Salvayre, R.; Vindis, C. Role of Protein Kinase C δ in ER Stress and Apoptosis Induced by Oxidized LDL in Human Vascular Smooth Muscle Cells. *Cell Death Dis.* **2013**, *4*, e520.

Laube, B.; Hirai, H.; Sturgess, M.; Betz, H.; Kuhse, J. Molecular Determinants of Agonist Discrimination by NMDA Receptor Subunits: Analysis of the Glutamate Binding Site on the NR2B Subunit. *Neuron.* **1997**, *18*, 493–503.

Lautenschlager, N. T.; Cox, K. L.; Flicker, L.; Foster, J. K.; van Bockxmeer, F. M.; Xiao, J.; Greenop, K. R.; Almeida, O. P. Effect of Physical Activity on Cognitive Function in Older Adults at Risk for Alzheimer Disease: A Randomized Trial. *JAMA.* **2008**, *300*, 1027–1037.

Leal, G.; Comprido, D.; Duarte, C. B. BDNF-induced Local Protein Synthesis and Synaptic Plasticity. *Neuropharmacology* **2014**, *76*, 639–656.

Levin, E. D.; Buccafusco, J. J. *Animal Models of Cognitive Impairment*; CRC Press, **2006**.

Lim, C. S.; Alkon, D. L. Protein Kinase C Stimulates HuD-mediated mRNA Stability and Protein Expression of Neurotrophic Factors and Enhances Dendritic Maturation of Hippocampal Neurons in Culture. *Hippocampus.* **2012**, *22*, 2303–2319.

Lee, Y.-S.; Silva, A. J. The Molecular and Cellular Biology of Enhanced Cognition. *Nat. Rev. Neurosci.* **2009a**, *10*, 126–140.

Lerner-Natoli, M.; Ladrech, S.; Renard, N.; Puel, J.-L.; Eybalin, M.; Pujol, R. Protein Kinase C May be Involved in Synaptic Repair of Auditory Neuron Dendrites After AMPA Injury in the Cochlea. *Brain Res.* **1997**, *749*, 109–119.

Liu, M. C.; Akle, V.; Zheng, W.; Dave, J. R.; Tortella, F. C.; Hayes, R. L.; Wang, K. K. Comparing Calpain-and Caspase-3-mediated Degradation Patterns in Traumatic Brain Injury by Differential Proteome Analysis. *Biochem. J.* **2006**, *394*, 715–725.

Lorenzo, P. S.; Beheshti, M.; Pettit, G. R.; Stone, J. C.; Blumberg, P. M. The Guanine Nucleotide Exchange Factor RasGRP is a High-affinity Target for Diacylglycerol and Phorbol Esters. *Mol. Pharmacol.* **2000**, *57*, 840–846.

Lüscher, C.; Malenka, R. C. NMDA Receptor-Dependent Long-Term Potentiation and Long-Term Depression (LTP/LTD). *Cold Spring Harb. Perspect. Biol.* **2012**, *4*, a005710.

Martina, M.; Gorfinkel, Y.; Halman, S.; Lowe, J. A.; Periyalwar, P.; Schmidt, C. J.; Bergeron, R. Glycine Transporter Type 1 Blockade Changes NMDA Receptor-mediated Responses and LTP in Hippocampal CA1 Pyramidal Cells by Altering Extracellular Glycine Levels. *J. Physiol.* **2004**, *557*, 489–500.

McDougall Jr, G. J.; Austin-Wells, V.; Zimmerman, T. Utility of Nutraceutical Products Marketed for Cognitive and Memory Enhancement. *J. Holist. Nurs.* **2005**, *23*, 415–433.

McGee, M. A.; Abdel-Rahman, A. A. N-methyl-D-aspartate Receptor Signaling and Function in Cardiovascular Tissues. *J. Cardiovasc. Pharmacol.* **2016**, *68*, 97.

Menard, C.; Gaudreau, P.; Quirion, R. Signaling Pathways Relevant to Cognition-enhancing Drug Targets. *Handb. Exp. Pharmacol.* **2015**, *228*, 59–98.

Menges, S. A.; Riepe, J. R.; Philips, G. T. Latent Memory Facilitates Relearning Through Molecular Signaling Mechanisms that are Distinct from Original Learning. *Neurobiol. Learn Mem.* **2015**, *123*, 35–42.

Mizuno, K.; Noda, K.; Araki, T.; Imaoka, T.; Kobayashi, Y.; Akita, Y.; Shimonaka, M.; Kishi, S.; Ohno, S. The Proteolytic Cleavage of Protein Kinase C Isotypes, Which Generates Kinase and Regulatory Fragments, Correlates with Fas-Mediated and 12-O-Tetradecanoyl-Phorbol-13-Acetate-Induced Apoptosis. *The FEBS J.* **1997**, *250*, 7–18.

Morris, R.; Anderson, E.; Lynch, G. A.; Baudry, M. Selective Impairment of Learning and Blockade of Long-term Potentiation by an N-methyl-D-aspartate Receptor Antagonist, AP5. *Nature.* **1986**, *319*, 774–776.

Murray, K.; Gall, C.; Jones, E.; Isackson, P. Differential Regulation of Brain-derived Neurotrophic Factor and Type II Calcium/Calmodulin-dependent Protein Kinase Messenger RNA Expression in Alzheimer's Disease. *Neuroscience* **1994**, *60*, 37–48.

Neary, J. T.; Crow, T.; Alkon, D. L. Change in a Specific Phosphoprotein Band Following Associative Learning in Hermissenda. *Nature* **1981**, *293*, 658–660.

Nelson, T.; Alkon, D. Neuroprotective Versus Tumorigenic Protein Kinase C Activators. *Trends Biochem. Sci.* **2009**, *34*, 136–145.

Nelson, T. J.; Collin, C.; Alkon, D. L. Isolation of a G Protein that is Modified by Learning and Reduces Potassium Currents in Hermissenda. *Science.* **1990**, *247*, 1479–1484.

Nelson, T. J.; Sun, M.-K.; Alkon, D. L., Signaling Pathways Involved in Cognitive Enhancement. In *Cognitive Enhancement*, Elsevier; 2015, pp 11–42.

Newhouse, P.; Kellar, K.; Aisen, P.; White, H.; Wesnes, K.; Coderre, E.; Pfaff, A.; Wilkins, H.; Howard, D.; Levin, E. Nicotine Treatment of Mild Cognitive Impairment a 6-month dDouble-blind Pilot Clinical Trial. *Neurology* **2012**, *78*, 91–101.

Okhrimenko, H.; Lu, W.; Xiang, C.; Hamburger, N.; Kazimirsky, G.; Brodie, C. Protein Kinase C-ε Regulates the Apoptosis and Survival of Glioma Cells. *Cancer Res.* **2005**, *65*, 7301–7309.

Olds, J. L.; Anderson, M. L.; McPhie, D. L.; Staten, L. D.; Alkon, D. L. Imaging of Memory-specific Changes in the Distribution of Protein Kinase C in the Hippocampus. *Science.* **1989**, *245*, 866–870.

Pardridge, W. M.; Wu, D.; Sakane, T. Combined Use of Carboxyl-directed Protein Pegylation and Vector-mediated Blood-brain Barrier Drug Delivery System Optimizes Brain Uptake of Brain-derived Neurotrophic Factor Following Intravenous Administration. *Pharm. Res.* **1998**, *15*, 576–582.

Phillips, H. S.; Hains, J. M.; Armanini, M.; Laramee, G. R.; Johnson, S. A.; Winslow, J. W. BDNF mRNA is Decreased in the Hippocampus of Individuals with Alzheimer's Disease. *Neuron* **1991**, *7*, 695–702.

Ragan, C. I.; Bard, I.; Singh, I. What Should We Do About Student Use of Cognitive Enhancers? An Analysis of Current Evidence. *Neuropharmacology* **2013**, *64*, 588–595.

Ramirez-Rodriguez, G.; Ocaña-Fernández, M.; Vega-Rivera, N.; Torres-Pérez, O.; Gómez-Sánchez, A.; Estrada-Camarena, E.; Ortiz-López, L. Environmental Enrichment Induces Neuroplastic Changes in Middle Age Female BalbC Mice and Increases the Hippocampal Levels of BDNF, p-Akt and p-MAPK1/2. *Neuroscience.* **2014**, *260*, 158–170.

Repantis, D.; Laisney, O.; Heuser, I. Acetylcholinesterase Inhibitors and Memantine for Neuroenhancement in Healthy Individuals: A Systematic Review. *Pharmacol. Res.* **2010b**, *61*, 473–481.

Repantis, D.; Schlattmann, P.; Laisney, O.; Heuser, I. Modafinil and Methylphenidate for Neuroenhancement in Healthy Individuals: A Systematic Review. *Pharmacol. Res.* **2010a**, *62*, 187–206.

Rezvani, A. H.; Levin, E. D. Cognitive Effects of Nicotine. *Biol. Psychiatry.* **2001**, *49*, 258–267.

Schrage, M. Soon Drugs May Make Us Smarter, but If We Use These "Brain Steroids," Will Nobel Laureates Need Urine Tests. *Washington Post* **1985**, *3*, C1.

Rhee, J.-S.; Betz, A.; Pyott, S.; Reim, K.; Varoqueaux, F.; Augustin, I.; Hesse, D.; Südhof, T. C.; Takahashi, M.; Rosenmund, C. β Phorbol Ester-and Diacylglycerol-induced Augmentation of Transmitter Release is Mediated by Munc13s and Not by PKCs. *Cell* **2002**, *108*, 121–133.

Schellekens, H.; Dinan, T. G.; Cryan, J. F. Taking Two to Tango: A Role for Ghrelin Receptor Heterodimerization in Stress and Reward. *Front Neurosci.* **2013**, *7*.

Schölzke, M. N.; Potrovita, I.; Subramaniam, S.; Prinz, S.; Schwaninger, M. Glutamate Activates NF-κB through Calpain in Neurons. *Eur. J. Neurosci.* **2003**, *18*, 3305–3310.

Serruya, M. D.; Kahana, M. J. Techniques and Devices to Restore Cognition. *Behav. Brain Res.* **2008**, *192*, 149–165.

Silverman, J. L.; Yang, M.; Lord, C.; Crawley, J. N. Behavioural Phenotyping Assays for Mouse Models of Autism. *Nat. Rev. Neurosci.* **2010**, *11*, 490–502.

Singh, M.; Narang, M. Cognitive Enhancement Techniques. *J. Int. Inform. Tech. & Knowledge Mgmt.* **2014**, *7*, 98–107.

Smith, L.; Chen, L.; Reyland, M. E.; DeVries, T. A.; Talanian, R. V.; Omura, S.; Smith, J. B. Activation of Atypical Protein Kinase C ζ by Caspase Processing and Degradation by the Ubiquitin-proteasome System. *J. Biol. Chem.* **2000**, *275*, 40620–40627.

Sometani, A.; Nomoto, H.; Nitta, A.; Furukawa, Y.; Furukawa, S. 4-Methylcatechol Stimulates Phosphorylation of Trk Family Neurotrophin Receptors and MAP Kinases in Cultured Rat Cortical Neurons. *J. Neurosci. Res.* **2002**, *70*, 335–339.

Song, G.; Ouyang, G.; Bao, S. The Activation of Akt/PKB Signaling Pathway and Cell Survival. *J. Cell. Mol. Med.* **2005**, *9*, 59–71.

Stoil, M. Amphetamine Epidemics: Nothing New. *Addiction and Recovery* **1990**, *10*, 9.

Swainson, R.; Rogers, R.; Sahakian, B.; Summers, B.; Polkey, C.; Robbins, T. Probabilistic Learning and Reversal Deficits in Patients with Parkinson's Disease or Frontal or Temporal Lobe Lesions: Possible Adverse Effects of Dopaminergic Medication. *Neuropsychologia* **2000**, *38*, 596–612.

Sultan, S.; Gebara, E. G.; Moullec, K.; Toni, N. D-serine Increases Adult Hippocampal Neurogenesis. *Front. Neurosci.* **2013**, *7*.

Sun, M.-K.; Alkon, D. L. Effects of 4-methylcatechol on Spatial Memory and Depression. *Neuroreport* **2008,** *19,* 355–359.

Sun, M.-K.; Hongpaisan, J.; Alkon, D. L. Postischemic PKC Activation Rescues Retrograde and Anterograde Long-term Memory. *Proc. Natl. Acad. Sci. US A.* **2009,** *106,* 14676–14680.

Takeuchi, T.; Duszkiewicz, A. J.; Morris, R. G. M. The Synaptic Plasticity and Memory Hypothesis: Encoding, Storage and Persistence. *Philos. Trans. R. Soc. Lond. B. Biol. Sci.* **2014,** *369,* 20130288.

Tsien, J. Z.; Huerta, P. T.; Tonegawa, S. The Essential Role of Hippocampal CA1 NMDA Receptor–dependent Synaptic Plasticity in Spatial Memory. *Cell* **1996,** *87,* 1327–1338.

Turpin, F.; Potier, B.; Dulong, J.; Sinet, P.-M.; Alliot, J.; Oliet, S.; Dutar, P.; Epelbaum, J.; Mothet, J.-P.; Billard, J.-M. Reduced Serine Racemase Expression Contributes to Age-related Deficits in Hippocampal Cognitive Function. *Neurobiol. Aging.* **2011,** *32,* 1495–1504.

Van Beek, T. A. Chemical Analysis of Ginkgo Biloba Leaves and Extracts. *[J. Chromatogr. A* **2002,** *967,* 21–55.

Van Praag, H. Neurogenesis and Exercise: Past and Future Directions. *Neuromolecular Med.* **2008,** *10,* 128–140.

Voon, V.; Fernagut, P.-O.; Wickens, J.; Baunez, C.; Rodriguez, M.; Pavon, N.; Juncos, J. L.; Obeso, J. A.; Bezard, E. Chronic Dopaminergic Stimulation in Parkinson's Disease: From Dyskinesias to Impulse Control Disorders. *Lancet Neurol.* **2009,** *8,* 1140–1149.

Walton, M. R.; Dragunow, M. Is CREB a Key to Neuronal Survival? *Trends Neurosci.* **2000,** *23,* 48–53.

Weintraub, D.; Koester, J.; Potenza, M. N.; Siderowf, A. D.; Stacy, M.; Voon, V.; Whetteckey, J.; Wunderlich, G. R.; Lang, A. E. Impulse Control Disorders in Parkinson Disease: A Cross-sectional Study of 3090 Patients. *Arch. Neurol.* **2010,** *67,* 589–595.

Weintraub, D.; Siderowf, A. D.; Potenza, M. N.; Goveas, J.; Morales, K. H.; Duda, J. E.; Moberg, P. J.; Stern, M. B. Association of Dopamine Agonist Use with Impulse Control Disorders in Parkinson Disease. *Arch. Neurol.* **2006,** *63,* 969–973.

Wilens, T. E.; Adler, L. A.; Adams, J.; Sgambati, S.; Rotrosen, J.; Sawtelle, R.; Utzinger, L.; Fusillo, S. Misuse and Diversion of Stimulants Prescribed for ADHD: A Systematic Review of the Literature. *J. Am. Acad. Child Adolesc. Psychiatry.* **2008,** *47,* 21–31.

Williams, K.; Kemper, S. Exploring Interventions to Reduce Cognitive Decline in Aging. *J. Psychosoc. Nurs. Ment. Health Serv.* **2010,** *48,* 42–51.

Wojcik, S. M.; Brose, N. Regulation of Membrane Fusion in Synaptic Excitation-secretion Coupling: Speed and Accuracy Matter. *Neuron* **2007,** *55,* 11–24.

Yan, J.-Z.; Xu, Z.; Ren, S.-Q.; Hu, B.; Yao, W.; Wang, S.-H.; Liu, S.-Y.; Lu, W. Protein Kinase C Promotes N-methyl-D-aspartate (NMDA) Receptor Trafficking by Indirectly Triggering Calcium/Calmodulin-dependent Protein Kinase II (CaMKII) Autophosphorylation. *J. Biol. Chem.* **2011,** *286,* 25187–25200.

Xia, Y.; Wang, C. Z.; Liu, J.; Anastasio, N. C.; Johnson, K. M. Brain-derived Neurotrophic Factor Prevents Phencyclidine-induced Apoptosis in Developing Brain by Parallel Activation of Both the ERK and PI-3K/Akt Pathways. *Neuropharmacology* **2010,** *58,* 330–336.

Yamada, K.; Nabeshima, T. Brain-derived Neurotrophic Factor/TrkB Signaling in Memory Processes. *[J. Pharm. Sci.* **2003,** *91,* 267–270.

Xu, J.; Kurup, P.; Zhang, Y.; Goebel-Goody, S. M.; Wu, P. H.; Hawasli, A. H.; Baum, M. L.; Bibb, J. A.; Lombroso, P. J. Extrasynaptic NMDA Receptors Couple Preferentially to Excitotoxicity Via Calpain-mediated Cleavage of STEP. *J. Neurosci.* **2009,** *29,* 9330–9343.

Xu, Y.; Yan, J.; Zhou, P.; Li, J.; Gao, H.; Xia, Y.; Wang, Q. Neurotransmitter Receptors and Cognitive Dysfunction in Alzheimer's Disease and Parkinson's Disease. *Prog. Neurobiology* **2012,** *97,* 1–13.

Yang, Y.; Ge, W.; Chen, Y.; Zhang, Z.; Shen, W.; Wu, C.; Poo, M.; Duan, S. Contribution of Astrocytes to Hippocampal Long-term Potentiation through Release of D-serine. *Proc. Natl. Acad. Sci. U. S. A.* **2003,** *100,* 15194–15199.

Yang, S.; Qiao, H.; Wen, L.; Zhou, W.; Zhang, Y. D-serine Enhances Impaired Long-term Potentiation in CA1 Subfield of Hippocampal Slices from Aged Senescence-accelerated Mouse Prone/8. *Neurosci. Lett.* **2005,** *379,* 7–12.

Zeng, Y.; Liu, Y.; Wu, M.; Liu, J.; Hu, Q. Activation of TrkB by 7, 8-dihydroxyflavone Prevents Fear Memory Defects and Facilitates Amygdalar Synaptic Plasticity in Aging. *J. Alzheimers Dis.* **2012,** *31,* 765–778.

# CHAPTER 19

# Opioids Analgesics and Antagonists

RACHANA*, TANYA GUPTA, SAUMYA YADAV, and MANISHA SINGH

*Jaypee Institute of Information Technology, Uttar Pradesh, India*

*Corresponding author. E-mail: rachana.dr@gmail.com*

## ABSTRACT

Opioids are majorly utilized for the treatment of moderate, acute as well as chronic pains, along with many other pathological conditions. The term opioid is derived from the plant opium, from where many narcotic drugs, like morphine are obtained since old days. These opiates can either be natural or synthetic and they act upon opioid receptors majorly present in central nervous system which are also found in peripheral nervous system (PNS) but, in less number. They are also present at few other locations in the body, such as: gastrointestinal tract, heart and so on. There are three categories for these receptors: δ (delta), κ (kappa), and μ (mu) opioid receptors which, correspond to agonists, antagonists, and partial agonists. μ receptors are considered to be of utmost importance as, majority of the opioids bind to μ opiate receptors as agonists and, elicit their analgesic functions. Along with discovering new analgesic molecules, scientists are also working upon many combinations of opioids to obtain better pharmacological effects, avoiding the side effects. Scientists have achieved many desirable goals while working in this direction but, still a lot research has to be done to get the desired efficiency with no toxicity. Present chapter describes the journey of opioids development, types of receptors, mechanism of action, uses, and pharmacokinetics of important opioids with side effects.

## 19.1 INTRODUCTION

The term opioid is derived from the term "opium" which is used for the plant extract of *Papaver somniferum*. Opium is known to cause euphoria, pain relief, and sedation in humans. The term opiates include the natural products derived from thebaine or opium and the natural alkaloids found in poppies only whereas, the term opioids includes all the natural and synthetic medicines derived from opium (Cruz et al., 2015).

Opioid analgesics have four naturally occurring alkaloids such as morphine (10% in opium), codeine (0.5% in opium), thebaine (0.2% in opium), and benzoisoquinoline (derivative and papaverine; 1%) (Blakemore and White, 2002). Following Sertürner's isolation, a range of semi-synthetic clinically useful opioids (diamorphine, dihydrocodeine, buprenorphine, nalbuphine, naloxone, and oxycodone) can be yielded from the chemical modifications of these four basic opiate alkaloids (Charlton, 2005).

During the 17th century, when the use of tobacco was banned in China, opium could grow its popularity. Mid-19th century saw the use of opiates in surgery to manage the pain (Niewijk, 2017). In 1806, morphine, the active part of opium, was extracted and isolated in America and, then it was started using as an anesthetics during the surgeries in 1850s and therefore, lessened the use of chloroform as an anesthetic (Brownstein, 1993). 1898 was the year when heroin, the alkaloid part of opium, was synthesized from morphine (Brownstein, 1993; Niewijk, 2017). It was found to be a highly potent analgesic and anti-tussive agent but, on the other hand it was highly addictive as well, and rose to commercialization by end of 19th century. Narcotics act was passed in 1914 for this reason to control the use of opium and its products (Cruz et al., 2015). In 1994, heroin was banned in theUnited States, but scientists started to develop similar drugs for pain relief as its alternatives (Lecia Bushak, 2016). The year 1939 saw the introduction of the first fully synthesized opium, meperidine (Duarte, 2005). 1980s was called as Opiophobia decade, because there were not recommended to the patients as they were addictive in nature (Lecia Bushak, 2016).

Opioids are classified in various following ways: their manufacturing processes, their presence inside or outside the body, and according to how they act on the various opioid receptors. They are in use as potent analgesics; however, their side effects like addiction and respiratory depression, limit their clinical use to a very great extent (Morgan and Christie, 2011). Complex interactions of the opioids with the three receptors, they bind namely: μ receptor, δ receptor, and κ receptor, result into different pharmacological effects of opioids (Pathan and Williams, 2012). These receptors are distributed throughout the body and so have complex impact on all of them.

Opioids can work as, partial agonists, agonists or as antagonists as well, depending upon their binding to respective receptors. As mentioned earlier, the effect of opioids on μ receptor is considered of utmost importance one as, μ opiate (MOP) receptors are the target for a number of opioids and exert analgesic function. When opioids bind to G-protein coupled receptors, they can result into a condition of cellular hyperpolarization (Al-Hasani and Bruchas, 2011). Some examples of opioid agonists are cocaine, morphine, tramadol, and so on.

Partial agonists, also target the receptors of agonist but, the potential of receptor activation remains low if, compared with the agonists, and the conformational change caused to the receptor also remains less significant. Raised doses of partial agonists can lead to side effects rather than the benefits (Helm et al., 2008). Most commonly used partial agonist is buprenorphine which acts on MOP receptor.

Antagonists are the chemicals which hinder or minimize the outcome of agonists as they can interact with other molecule(s) involved in activation process of the receptors or molecules ahead, thus they have the capability to suppress the effects of the opioids. Some general examples for opioid antagonists are: naloxone and naltrexone. They block the neurotransmission as neuroreceptor does not get activated even if the agonists are present (Chahl, 1996) and that is why opioid addiction can be treated using antagonists and as their antidotes as well. The current chapter gives an overall view of different types of opioid analgesics and antagonists along with their mechanism of action with respect to their interaction with different opioid receptors, their side effects, and future opportunities and challenges to be faced.

## 19.2 OPIOID RECEPTORS

Opioids bind to their respective receptors to show the particular effect. Opioid receptors are divided into three different categories which were originally named as μ, δ, and κ receptors. All three receptors are G-protein receptors (McDonald and Lambert, 2005). The above mentioned names were further modified by the IUPHAR (International Union of Pharmacology) in 1996 to OP1 (for δ receptor), OP2 (for κ receptor), and OP3 (for μ receptor). These names were further modified to DOP, KOP, and MOP in 2000 (Pathan and Williams, 2012). However, presently IUPHAR recommends using the Greek nomenclature namely δ, κ, and μ and the DOP, KOP, and MOP classification of 2000 (Pathan and Williams, 2012). The concept behind this nomenclature of the receptors was based on interactions between morphine and nalorphine. The "M" or μ receptors were based on morphine and the "N" or κ receptors were based on nalorphine. The δ receptors were based upon the enkephalins (Pasternak, 2013). Both, the endogenously produced opioids and exogenous opioids, can activate these receptors (Waldhoer et al., 2004).

### 19.2.1 δ RECEPTOR

The first proof for the presence of δ receptors was given in 1977 (Traynor and Elliott, 19993). They have the greatest affinity and bind primarily to enkephalins (Evans et al., 1992). They are mostly present in brain and have dysphoric and psychomimetic effects (Trescot et al., 2008). In various experiments with knockout mice, the researchers had shown the presence of δ receptor endogenously and their role in reducing chronic pain. The data showed that δ opioid agonists can be a very effective alternative of μ analgesics which, are being presently used for chronic pain treatment (Pradhan et al., 2011).

δ receptors were further divided into $\delta_1$ and $\delta_2$ subtypes, based on their various pharmacological studies (Corbett et al., 1999). Rijn and Whistler (2009) studied their effects the relation between alcohol consumption where, these two types receptors were found to have opposite effects thus, indicating the distinct molecular targets of these two subtypes. In vitro and in vivo experiments performed by them indicated that δ1 and δ2 subtype are μ–δ heterodimer and δ homomer, respectively (Rijn and Whistler, 2009).

### 19.2.2 κ RECEPTOR

These receptors were first characterized in brain using [³H]-ethylketocyclazocine (EKC) by Attali et al. (1982). They are present mostly in spinal cord, limbic areas, and brain stem. They are useful for spinal analgesia, dyspnea, and sedation (Trescot et al., 2008) and other distinctive functions like drinking modulation, water balance, intake of food, gut motility, controlling temperature, and many endocrine functions; however, these receptors are not primarily reinforced for analgesia. These receptors have low levels of expression in rats and are in abundance in guinea pigs, human, and other higher primates (Attali et al., 1982). An endogenous opioid peptide precursor such as prodynorphin, which posttranslationally produces many endogenous opioid peptides, shows very high affinity toward these receptors (Riters et al., 2017). Various synthetic pharmacological agents such as benzomorphans (ex bremazocine) and arylacetamides (ex U-50,488) also interact with these receptors. High degree of selectivity is shown by arylacetamides for these receptors (Meng et al., 1993).

The two major subtypes of κ receptors are KOR-1 and KOR-2, of which KOR-1 is specified based on its preference to bind to arylacetamide-like agonists (U69, 593; U50, 488; CI-977, 441) and the KOR antagonist norbinaltorphimine (nor-BNI) while the characterization of KOR-2 is done on the basis that this receptor is insensitive to U69, 593 (putative $κ_1$-selective ligand) and also shows lower affinity as much as 100-fold for nor-BNI than KOR-1. KOR-2 prototypical agonists are Bremazocine and GR89, 696 (Brissett et al., 2012).

### 19.2.3  μ RECEPTOR

MOR-1 or MOP-1 were the terms given to the first clone of this receptor in 1993 by Chen et al. Out of all four exons of the mRNA for this receptor, it consists of, N-terminus and, the first transmembrane domain is reported to be encoded by first exon and, an additional three transmembrane domains is encoded by second and third exon, each thus, producing a seven-transmembrane structure of G-protein coupled receptor. The 12 amino acids found at the terminal of the intracellular C-terminus of the receptor are encoded by the fourth exon (Pasternak, 2013). These amino acids are present mainly in the brain-stem, medial thalamus and are responsible for respiratory depression, supraspinal analgesia, sedation, euphoria, decreased gastrointestinal motility, and physical dependency (Trescot et al., 2008). These receptors are found abundantly in central nervous system (CNS) and bind to drugs like morphine and morphine-like other drugs and also to several opioid peptides, found endogenously (Meng et al., 1993).

MOR-1A and MOR-1B are the two splice variants of MOR-1, of which the very first description of MOR-1A was given from

human neuroblastoma cells and MOR-1B was given from rat μ opioid receptor (rMOR1B). MOR-1A contains only three exons where, the reading frame of third exon is extended and encodes for four amino acids in place of 12 encoded by fourth exon. In MOR-1B, exon 5 replaces the fourth exon. Further cloning of MOR-1A and MOR-1B showed that the 1, 2, and 3 exons of all these variants are intact, which indicates that the amino termi-nuses of all these variants are same and, the transmembrane domains of all are identical? The only difference is present at the intracellular C-terminus tip of the receptor that is, different numbers of amino acids at C terminal (Pasternak, 2014).

### 19.3  OPIOIDS AND THEIR RECEPTOR SELECTIVITY

The opioids have their own significant pharmacologic differences and these differences are because of their varying and complex interactions with the three different opioid receptors they bind. In 2010, in an experiment, where genetic approach on mice was used, it was seen that those mice in which the μ receptors were absent, did not react to morphine and therefore, show no results of analgesia, respiratory depression, physical dependence and immunosuppression and so on (Bilkei-Gorzo et al., 2010)

Numerous studies revealed agonist like: morphine binds mainly to μ opioid receptors, leading to analgesic effects and naloxene and other opioid antagonists bind to all the three receptors, having most affinity for μ opioid receptors. Other than that, it can lead to inhibition or reversion of morphine—like agonist effects as well. The third group comprises agonist–antagonist drugs. The specificity of these drugs relies on condition of the patient and is capable of acting as agonist by displaying activity at the κ receptor or exhibiting antagonist activity at the μ opioid receptors (Inturrisi, 2002).

Various groups of drugs having selectivity for discrete receptors are given in Table 19.1.

**TABLE 19.1** Opioids Having Preferential Selectiveness for Discrete Opioid Receptors.

| Drug | Receptor types | | | |
|---|---|---|---|---|
| | **MOP** | **KOP** | **DOP** | **NOP** |
| **Agonists** | | | | |
| Morphine | +++ | + | + | – |
| Pethidine | +++ | + | + | – |
| Diamorphine | +++ | + | + | – |
| Fentanyl | +++ | + | – | – |
| **Partial agonists** | | | | |
| Buprenorphine | ++ | + | – | – |
| Pentazocine | – | ++ | – | – |
| **Antagonists** | | | | |
| Naloxone | +++ | ++ | ++ | – |
| Naltrexone | +++ | ++ | ++ | – |

+, low affinity; ++, moderate affinity; +++, high affinity; –, no affinity; MOP, Mu opioid peptide receptor; KOP, kappa opioid peptide receptor; DOP, delta opioid peptide receptor; NOP, nociceptin opioid peptide receptor.

## 19.4 CLASSIFICATION OF OPIOIDS

Opioids are categorized on various basis such as, their manufacturing processes, the kind of receptor they work upon and, also on the basis of their presence in the body.

### 19.4.1 BASED ON THE SYNTHESIS PROCESSES

#### 19.4.1.1 NATURALLY OCCURRING

The opioids which are formed and found directly in poppy plant are classified as naturally occurring opioids and the examples are morphine, thebaine, codeine, and papaverine (Pathan and Williams, 2012).

#### 19.4.1.2 SEMI-SYNTHETIC

The opioids derived from poppy plant but had to be processed with other materials for their formation which is termed as semi-synthetic opioids and the examples are diamorphine (heroin), dihydromorphine, and buprenorphine (Pathan and Williams, 2012).

#### 19.4.1.3 SYNTHETIC

Synthetic opioids are generally derived from the manmade processes and possess opiate like effects and the examples pethidine, fentanyl, and methadone (Pathan and Williams, 2012).

### 19.4.2 BASED ON THE ACTION ON RECEPTORS

The pharmacological effects of the opioids are the result of their interaction with their respective receptors present in CNS and peripheral tissues (Trescot et al., 2008; Akhondzadeh et al., 2001). The two characteristic qualities of these compounds are affinity and efficacy toward the receptor. Affinity determines the extent of interaction strength of the compound to its receptor, while efficacy is measure of the response or result observed after this binding at the site if receptor (Helm et al., 2008). In accordance with above two criteria, opioids are grouped as agonists, partial agonists and antagonists, as depicted in Table 19.2 (Pathan and Williams, 2012).

**TABLE 19.2** Categorization of Opioids Premised on Their Work on Receptors.

| Criteria | Agonists | Partial agonists/ antagonists | Antagonists |
|---|---|---|---|
| Affinity | High | Moderate | Low |
| Efficacy | Prominent | Partial | No |
| Functional response | Maximal | Restricted | No |
| Others | Act as analgesics | Does not depend on the amount of drug administered | Also prevents agonist from binding |
| Examples | Heroin and morphine | Nalorphine, pentazocine, and nalbuphine | Naloxone and naltrexone |

### 19.4.3   BASED ON THE LOCATION IN THE BODY

A broad classification of opioids includes exogenous and endogenous opioids (Cruz and Soto et al., 2015). Exogenous opioids are formed and found outside the body, for example morphine and heroin while, endogenous are formed and found within the body and generally termed as neurotransmitters, for example endorphins and enkephalins (Akil et al., 1984).

## 19.5   PHARMACOKINETICS OF OPIOIDS

The first natural opioid extracted from opium plant was morphine which is, used for general anesthetic purposes (Beaver, 1980) but, due to the extent of dependency and abuse of this compound people search for its alternatives which lead to the discovery of the first semi-synthetic opium derivatives known as heroin. But, it was eventually discovered to have effects as morphine for its abuses and addictions. So, this is how the first fully synthetic opioid meperidine was discovered and worked upon in 1942 (Duarte, 2005).

Pharmacokinetics of some of the majorly used opioids is described briefly as below:

### 19.5.1   MEPERIDINE

In 1986, Naquib and his research team conducted a study on five male patients going through surgery to examine the pharmacokinetics of meperidine. They found that in all five patients, anesthesia was produced and time of peak concentration of meperidine in plasma was 90 min after the epidural administration of 30 mg of meperidine with a half-life of 198 min. The mean time at which sensory and motor block occurred was 5.8 and 6.2 min, respectively. The minimum concentration of meperidine in the plasma for analgesic effect was found to be 400–500 ngmL$^{-1}$ (Naquib et al., 1986).

### 19.5.2   MORPHINE

In a study done on seven healthy volunteers, morphine, and metabolites were found to be present in urine after 72 h of single oral (5 mg) and intravenous (20 mg) dose administration (separately, at interval of 72 h) and it was found that the average systemic plasma clearance for morphine was 21.1 ± 3.4 mL/min/kg (Hasselstrom and Sawe, 1993). Also, 43% and 48% of the oral and intravenously administered morphine was recovered in the form of morphine in the two cases, respectively, while, 2.6% and 8.2% of the same was recovered as unchanged morphine from urine (Hasselstrom and Sawe, 1993).

### 19.5.3   HEROIN

In a study performed by Rook et al. (2006), it was found that after intravenous administration of heroin, its levels in blood declined very rapidly and became undetectable after 10–40 min with an average half-life of about 1.3–7.8 min. The peak concentration of heroin in plasma was found after absorption via intranasal way, after 2–15 min of inhalation. The mean heroin clearance estimation was around 128–1939 L/h which was very much than the hepatic (80 L/h) and renal blood flow (60 L/h). This very high rate of heroin clearance from plasma was mainly because heroin gets hydrolyzed in the basic body fluid environment and also esterase work on this opioid and eliminates it very rapidly (Rook et al., 2006).

## 19.6 MECHANISM OF ACTION OF OPIOIDS

Opioids exert their effects on neurons by activating the opioid receptors. These receptors are widely distributed in a large number throughout the CNS; cerebral cortex, the nuclei of tractus solitarius, locus coeruleus, thalamus, periaqueductal gray area, and substantia gelatinosa of the spinal cord (Adrian et al., 1983). They can also be found in other various other locations in the body but, at a lesser extent. Such locations are: the peripheral afferent nerve terminals and various organs including the knee joints, immune system, gastrointestinal tract, heart and so on (Stein et al., 2003).

Opioid receptors are found to be coupled with inhibitory G-proteins and always the case with G proteins, their activation depends upon a series of events. G-proteins comprises three subunits namely, $\alpha$, $\beta$, and $\gamma$. During the resting phase, that is, in the absence of opioid, guanosine diphosphate (GDP) binds to the $\alpha$-subunit. Once opioids bind to their specific receptors, GDP and $\alpha$-subunit get dissociated along with the substitution of GDP with GTP. This results in a conformational change which is responsible for the decoupling of opioid and the receptor, as

well. Further dissociation of GTP and $\alpha$ complex occurs from rest of the two subunits leading to association with the intracellular signalling pathway to produce the effect (Chahl, 1996).

If the focus toward the analgesic effect produced by opioids, then it is because of their action at nervous system, specifically the inhibition of neurotransmitter release from afferent nerve terminals of spinal cord. They inhibit the entry of calcium via voltage sensitive calcium channels and this is achieved by the increase in exit of potassium ions from neurons and inhibition of an enzyme which converts of ATP to cAMP that is, adenylate cycliase (Harrison et al., 1998).

Thus, all three receptors for opioid are linked to this enzyme and inhibit release of neurotransmitter. So, the analgesic effects are the result of a complex series of neuronal interactions as shown in (Fig. 19.1).

### 19.6.1 OPIOID MEDIATED ANALGESIA

Strong thermal and mechanical stimuli induce increase activity in primary sensory activity causing the stimulus of pain. It can also occur due to chemicals released by tissue damage or inflammation. $\mu$ receptors are believed to be present in the midbrain and are involved in the opioid mediated analgesia (Mogil et al., 2005). They are the site for binding $\mu$ agonists which stimulate deceasing inhibitory pathways. $\mu$ agonists act upon the nucleus reticularis para-gigantocellularis (NRPG) and periaqueductal gray (PAG) and result into an activation of descending inhibitory neurons reducing the pain. In result, greater neuronal traffic increases through the nucleus raphe magnus (NRM) which stimulates 5-hydroxytryptamine and enkephalin-containing neurons. This involves the substantia gelatinosa of the dorsal horn

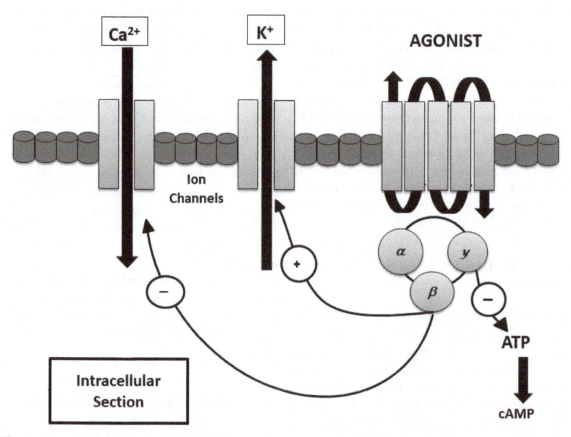

**FIGURE 19.1** Intracellular events after the binding of an agonist to a G-protein-coupled opioid receptor. Inhibition of adenylate cyclase with reduced production of cyclic adenosine monophosphate (cAMP) and other second messengers; stimulation of potassium conductance and target cells hyperpolarization which, makes the opioids less responsive to depolarizing pulse and close the voltage-sensitive calcium channel. These actions reduce the neuronal cell excitability and decrease the generation of the postsynaptic impulse adenosine triphosphate (ATP) and cAMP.

*Source:* Adapted from Zhang et al. (2013).

and leads to reduction of nociceptive transmission from the periphery to the thalamus. The same mechanism is followed by endogenous and exogenous opioids having inhibiting effects upon the peripheral nociceptive afferent neurones and substantia gelatinosa (in the dorsal horn). This results in reduction in nociceptive transmission from the periphery (Pathan and Williams, 2012).

## 19.7  OPIOIDS ANALGESICS

As described earlier, opioids analgesics are efficacious in treatment of acute pain as well as, with post-surgical pain and so on. They also help in reducing the pain associated with biliary and renal diseases or even cancer. Furthermore, these are used during the obstetric labor widely, but, as it can traverse the placental barrier, care and caution must be taken to reduce the chances of neonatal depression (Dews and Mekhail, 2004). Following are some examples of opioids analgesics:

### 19.7.1 MORPHINE

Morphine is a widely known opioid analgesic. After its administration, morphine gets modified by glucuronidation (described later in detail) resulting it into its metabolites: morphine-6 glucuronide (M6G) and morphine-3 glucuronide (M3G) in a ratio of 6:1, where M6G is responsible for some of the additional analgesic effects of morphine (Lotsch and Geisslinger, 2001) while M3G is responsible for hyper analgesia with increased agitation, pain, and myoclonus (Smith, 2000).

#### 19.7.1.1 METABOLISM

In the process of metabolization of morphine, it undergoes phase-I metabolism and phase-II metabolism. After the morphine has been administered to the patients via nebulizer or is taken orally, subcutaneously or intravenously, under phase-I metabolism, it is majorly metabolized via oxidation or hydrolysis in the liver and leads to the alteration in plasma levels. After this stage only, 40–50% of the morphine spreads to the CNS and within 72 h. 80% morphine dose is excreted via urine (Trescot et al., 2008). Then it undergoes, phase-II metabolism where it is metabolized by glucuronidation that is, the transfer of glucuronic acid component of uridine diphosphate glucuronic acid to a substrate in the presence of UDP-glucuronosyltransferase. The metabolizing process leads to synthesis of morphine-3-glucuronide and morphine-6-glucuronide in a ratio of 6:1 (Lotsch and Geisslinger, 2001), which are the two important metabolic morphine pathways in humans. The principal

enzyme for morphine metabolism is UGT2B7 [uridine diphosphoglucuronosyl transferases (UGT)] which is primarily released by liver, but can be released by brain and kidneys, as well (Chau et al., 2014) However, the meaningful clinical effect of hydromorphone metabolite of morphine is still unknown (Fig. 19.2).

CYPD26, cytochrome P450 2D6; CYP3A, cytochrome P450 3A4; UGT2B7, UDP-glucuronosyltransferase-2B7; C6G, codeine-6-glucuronide; M3G, morphine-3-glucuronide; M6G, morphine-6-glucuronide; ABCC2, ATP-binding cassette sub-family C member 2; ABCC3, ATP-binding cassette sub-family C member 3.

### 19.7.2 CODEINE

Codeine, a synthetic opioid analgesic, is a weak opioid. But, it has been reported to be very advantageous for suppressing cough. It is combined with non-opioid drugs for its uses. This drug has less sedation, reduced addiction, and respiratory depression if compared with morphine (Whittaker, 2013).

Figure 19.2 describes that how codeine gets converted to morphine and to another derivative as codeine-6-glucuronide and gets eliminated from the system (Chau et al., 2014).

### 19.7.3 HYDROCODONE

Hydrocodone and codeine are structurally similar but it, is a weak μ receptor agonist. However, after its demethylation by hydrogen bromide (HBr), it results into hydromorphone, having a comparatively stronger affinity for μ opioid receptor (Otton et al., 1993). It also possesses unique anti-sedative effect owing to its affinity toward multiple opioid receptors (Poyhia et al., 1992). It is metabolized by two cytochrome P450 (CYP) enzymes, that

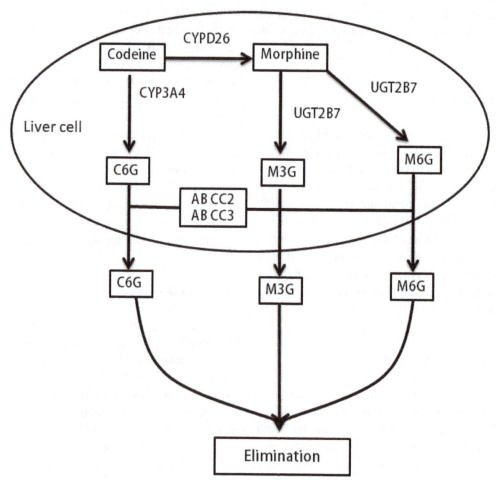

**FIGURE 19.2**   The metabolism of morphine and codeine.
*Source:* Adapted Gammal et al. (2016).

is, CYP3A4 and CYP2D6, which leads to synthesis of noroxycodone and oxymorphone metabolites, respectively, by demethylation (Prude Pharma, 2007). The drug hydrocodone and oxycodone are potent analgesics and are closer to morphine. They cause more respiratory depression and addiction liability than morphine, but ventricular standstill is resolved, preceded by nausea after first time oral administration of hydrocodone (Sudhakaran et al., 2014).

### 19.7.4   FENTANYL

Fentanyl is a potent agonist of μ-opioid receptors and has rapid onset and short duration of action. It has 80 times analgesic potency and respiratory depression as morphine. It is highly lipophilic in nature with strong binding toward plasma proteins (Feierman et al., 1996). Combination of fentanyl with droperidol, produces dissociative analgesia. Its metabolization by CYP3A4 results into the production of various inactive and nontoxic metabolites. High doses of fentanyl cause muscular rigidity (Kubat, 2013).

### 19.7.5   METHADONE

Methadone is similar to morphine but, is a synthetic analgesic and, more effective orally. Its main use is in the suppression of the withdrawal symptoms due to opiate addiction. Withdrawal of methadone is less severe than the withdrawal from other opiates. In a report, methadone was used for the treatment of patients suffering from the chronic back pain and scrotal pain along with the development of bilateral peripheral edema. The results indicated the reduction of edema following the cessation of methadone (Lynch, 2005). Traditionally, it has been used to treat heroin addicts and to cure neuropathic pain. However, its low cost and dosing flexibility have led to an increase in its usage. So far, its metabolism is unclear and, it has a long half-life. An exponential increase in the deaths has also been observed in the patients using this, for cure (Dawson et al., 2014).

### 19.8   OPIOID ANTAGONISTS

A drug that dampens the biological activity by blocking some or all the subtypes of opioid receptor(s) is referred as opioid antagonists. Majorly, they are utilized to treat acute opioid overdose management, addiction treatment, and for the management of that patient population which cannot obtain adequate analgesia from opioids (e.g., Crohn's disease, fibromyalgia, sickle cell anaemia, and neuropathy etc.). The most widely studied opioid antagonists are naloxone and naltrexone, which have higher affinity toward the receptors than agonists, but do not activate them. They effectively block the receptor, preventing the body from responding to exogenously administered opioids like morphine, heroin, meperidine, endorphins, and so on (Leavitt, 2009). These drugs have higher affinity for µ receptors in comparison to others and can pass through membranes (including blood–brain barrier) easily as; they have similar structure and equivalent lipophilicity. As shown in Figure 19.3, the evolution of opioid antagonists is also based on the morphine core structure.

The first pure opioid antagonist which was synthesized from morphine is Naloxone. It was made by the addition of C14 hydroxy group further exchanging of N-methyl group with alkyl group in 1993 (Fig. 19.3). If, the N-methyl group is formally exchanged with methylcyclopropyl then, it leads to formation of naltrexone (Rinner and Hudlicky, 2012), while the methylated analogue of naltrexone is referred as methylnaltrexone.

Naltrexone is a non-specific opioid receptor antagonist but, has half-life of 4 h as an advantage over naloxone as; it will have longer receptor dissociation constant. It is effective in its oral form and withstands with first-pass metabolism also (Miller, 2000). When managed systematically, methylnaltrexone does not cross the blood-brain barrier. As per many trials, it does not reverse centrally mediated effects shown by opioids but, is a vigorous antagonist of the side effects of peripheral opioids. Its oral bioavailability is lesser than 1% (Foss et al., 1997).

Nalmefene antagonizes opioids at other receptors but, do have a greater affinity for µ receptors like naloxone. It antagonizes µ receptors agonist effects four times vigorously as naloxone. If, given parenterally, it needs at least 4 h to act (dose dependent) but, it lasts up to 48 h if, given orally (Dixon et al., 1986).

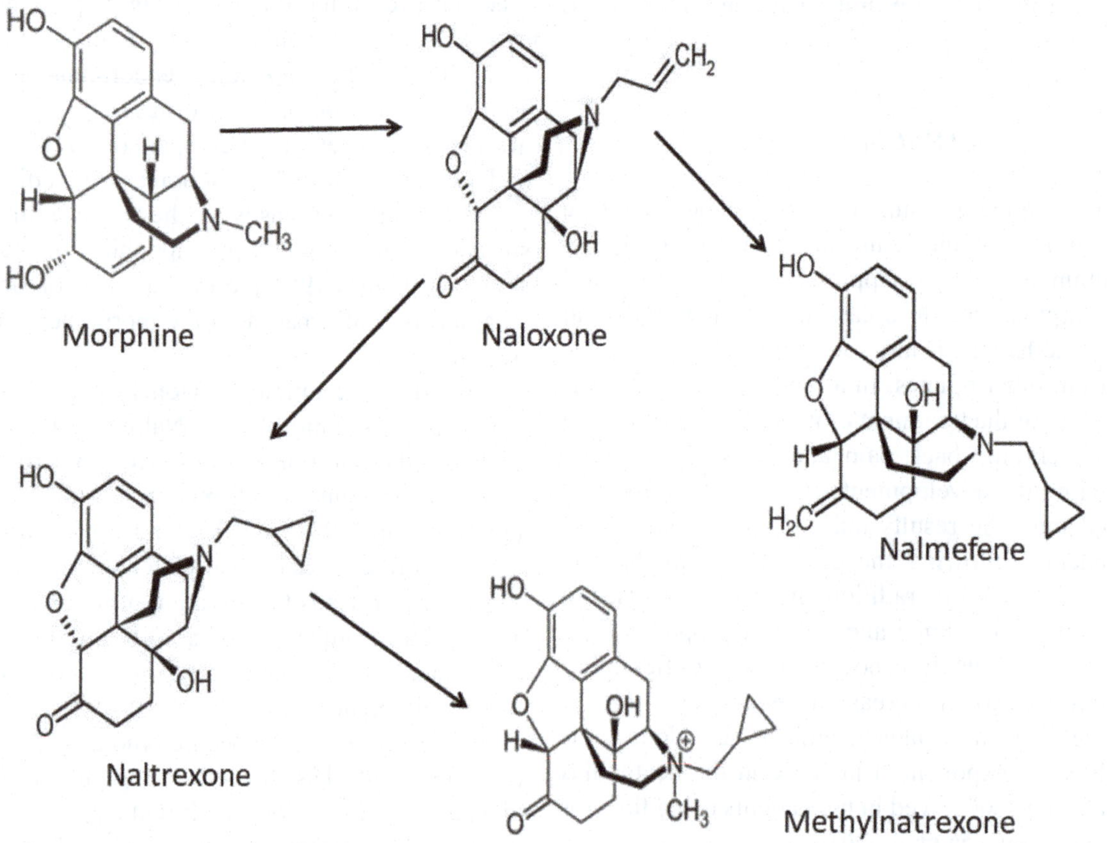

**FIGURE 19.3** Synthesis of opioid antagonists from morphine. Replacement of N-methyl group with methylcyclopropyl of naloxone, a synthetic morphinan derivative, results in the naltrexone and its methylation generates the methylnaltrexone.

*Source:* Reprinted from Strydom (2007).

## 19.8.1 EFFECTS IN THE PRESENCE OF AGONISTS

Combinations of opioid antagonists with agonist have been observed to help to restore the ability of opioid agonist to provide pain relief hindered by specific pathophysiology of these pain conditions (Taylor et al., 2013).

In a study, the combination of naloxone and morphine was administered in the sixteen children suffering from the sickle cell pain, which results in the improvement of pruritus score without adversely affecting the pain relief (Koch et al., 2008). One more double-blind study, with the combination of naloxone and morphine was also conducted for 90 hysterectomy surgery patients in which morphine was also administered with placebo rather than with the antagonist (ultra-low-dose). It was observed that the naloxone-treated patients had declined in the morphine usage within the first 24 h as compared with placebo. It was also observed that there was a decrease in vomiting and nausea incidences but, the reduction in pain score and prevalence of pruritus were considerably same (Movafegh et al., 2012).

Furthermore, in the presence of ultra-low-dose naltrexone, it has been shown that effect of buprenorphine for antinociception can be enhanced. Buprenorphine, an opioid agonist, has antinociceptive effect that is, blocks the pain detection stimuli by sensory neurons. To check its effect in the combination with ultra-low-dose naltrexone, 10 healthy volunteers were randomly administered with its combination as well as, naltrexone alone. The drugs were administered to either of the two arms and then subjected their hand to the ice water (cold pressor test). The results obtained, indicated a huge difference with the increase of 30.9% in cold pressor tolerance when both drugs are given at a dose ratio of 166:1 with the similar side effects (sedation, nausea, and vomiting; Hay et al., 2011).

In another study, when naloxone alone or with combination of several opioid agonists (morphine, nalbuphine, buprenorphine, oxycodone, and pentazocine) during various surgical interventions was administered, positive results were obtained regarding reduction in postsurgical pain, with the increase in analgesia with lengthier duration of analgesic effect (Leavitt, 2009).

### 19.8.2 EFFECTS IN THE ABSENCE OF AGONISTS

The profile of opioid antagonist effect also depends on the absence/presence of exogenous agonist and the degree to which physical dependence has been developed. Several researches have been conducted which reflects that in the absence of agonist, antagonists have advantages as well as, disadvantages for humans (Morgan and Christie, 2011).

A randomized, double-blind study was conducted over Crohn's disease patients. They were subjected to low-dose naltrexone alone, for 12 weeks and results were analyzed based upon Crohn's Disease Activity Index (CDAI), Crohn's disease endoscopy index severities score

(CDEIS), inflammatory bowel disease questionnaire (IBDQ) and 36-Item Short Form Survey (SF-36). They reported that CDAI scores were declined by 70% while CDEIS were greatly improved, whilst no noteworthy difference between the placebo and naltrexone patients regarding the SF-36 and IBDQ scores was observed. Apart this, few side effects were also noticed including insomnia, diarrhea, headache, nausea, and constipation (Smith et al., 2007).

There were some improvements shown in another case study, in the back, neck and spot pain after the low dose naltrexone treatment in the fibromyalgia patient. The patient had musculoskeletal pain along with the mood issues and insomnia before the treatment, which were further improved, leading toward a better quality of life (Ramanathan et al., 2012).

In addition to opioid antagonist, methylnaltrexone is also found to be effective when administered subcutaneously, intravenously or orally for the treatment of bowel dysfunction induced by opioids (Yuan et al., 1197, 2000, 2002). This indicates that an analgesic response of an opioid can be enhanced if given alone with a low dose antagonist.

## 19.9 SIDE EFFECTS OF OPIOIDS

As mentioned above, opioids have been found to be associated with various side effects. Some of these side effects are described as below:

### 19.9.1 CONSTIPATION

Constipation is the very common side effect for the patients treated with opioids (Swegle and Logemann, 2006). Opioids stimulate the receptors for opioids present in the gut, which subsequently leads to reduction in gastrointestinal motility, with an increase in fecal fluid absorption leading to hard and dry stool formation.

### 19.9.2 NAUSEA

A total of 25% of the patients treated with opioids are reported to suffer from nausea (Meuser et al., 2001). It may be due to the direct stimulation of the chemotactic trigger zone (CTZ), increased vestibular sensitivity, and reduced gastrointestinal motility, leading to gastric distention (Flake et al., 2004).

### 19.9.3 PRURITUS

Very less number of percentages of population develops pruritus if treated with opioids, which is often a result of a direct release of histamine, and not because of usual antigen/antibody interaction (McNicol et al., 2003). Therefore, it is considered as an adverse reaction rather than the allergic reaction. Its treatment generally includes the administration of antihistamines such as diphenhydramine.

### 19.9.4 SEDATION AND COGNITIVE DYSFUNCTION

Using opioids in increasing amount can lead to incidences of sedation in 20–60% of the population of patients (Cherny et al., 2001), while cognitive dysfunction can be compounded by dehydration, infections, metabolic abnormalities, or other advanced diseases (Cherny, 2000).

### 19.9.5 ENDOCRINE EFFECTS

Endorphins (naturally occurring opiates) are generally considered as the regulators of gonadotropins (Howlett and Rees, 1986). But, according to the various studies, 52% female patients who were on intrathecal opioids for chronic pain undergoes ardent inhibition of sex hormones and adrenal androgen production which is accompanied by amenorrhea (Daniell, 2008) while among the opioid-consuming males, subnormal testosterone levels were reported, which consequently leads to the poor life quality along with the persistent illness (Daniell, 2002).

### 19.9.6 IMMUNOLOGIC EFFECTS

Acute and chronic opioid administration can lead to inhibitory effects on cytokine expression, phagocytic activity, and natural killer functions. The proliferation capacity of macrophages and lymphocytes is decreased by the chronic administration of opioids (Roy and Loh, 1996).

### 19.9.7 OTHER EFFECTS

Use of opioids can lead to a condition called hyperalgesia. Hyperalgesia is a condition whereby a patient who is receiving opioids for pain treatment could actually become more sensitive to certain painful stimuli. It is a state of nociceptive sensitization caused by exposure to opioids (Marion et al., 2011).

Several case studies suggested that, long term administration of opioids leads to their accumulation and results in a neuroexcitatory side effect, that is, myoclonus (Hofmann et al., 2006). Myoclonus, a brief involuntary twitching of a muscle(s), is usually caused by the sudden muscle contractions (positive myoclonus) or brief lapses of contraction (negative myoclonus; Kojovic et al., 2011).

## 19.10 APPLICATION OF OPIOIDS

Opioids are administered for the treating of both acute and chronic pain. They are now being used to control pain for even cancer sufferers, as well (Vadivelu et al., 2016). As they affect the motility of the intestine they are also being used for managing diarrhea. Another use of opioid is as to reduce blood pressure and also known to modulate our immunity (Cruz et al., 2015). Endogenous opioids control body's response during stress times (Valentino, 2015). Autonomous nervous systems contain endorphins opioids which, regulate homeostasis in the body and perform many other functions, as well (Akil et al., 1984). Predictions are that along with their curative values, in future, stress-induced opioid peptides can also be useful as markers for diagnosis of tumors as, these peptides are found to be immunoreactive (Carr and Serou, 1995).

## 19.11 FUTURE OPPORTUNITIES AND CHALLENGES

Pain management is a very important aspect of medicines and people have been looking for various options to deal with it. Pain management is required for a tiny wound, headache, till managing the pain for a cancer patient. In spite of having many therapeutic applications various side effects are also associated the uses of opioids. Scientists have been working and developing new molecules or their combinations to avoid those side effects but, the challenges which still remained to be solved are:

1. Newly developed drug might not have equivalent efficacy as the original opioids.
2. Their solubility might change and so, it may affect their bioavailability and raise an additional problem to be handled.
3. Newly developed molecules might be efficacious but they might have similar side effects.

4. There are no strict rules or protocols which can be followed to design these combinations and they have been totally based on hit and trial basis.
5. As, distribution of the opioids receptor is different in different organs, we might have to develop different strategies to target these specific molecules.

Further elaborating the process of developing of a new drug, we may have to target various signaling molecules and pathways while acting on the same receptor subtype. This strategy has been a very useful and crucial way of developing pathway specific ligands. Studying the pathways and the interactions between different targets and the drug molecules, ligand binding can be modified to show minimum side effects along with specific therapeutic applications thus, efficiency of the drug can be improved along with increasing the effectiveness of the therapy (Jamshidi et al., 2015).

According to recent studies, opioid-induced analgesia is a result of MOR signaling through the G protein, while the side effects, are conferred via β-arrestin pathway (Manglik et al., 2016). Some of these various new structure-based alternatives to morphine opioid analgesics are BU08028, PZM21, and NFEPP and so on. These compounds act only on the analgesics pathway and therefore, have shown minimal side effects like respiratory regression, sedation, constipation, addictiveness, and opioid dependence (Ding et al., 2016; Manglik et al., 2016; Chan et al., 2017). These compounds are still in trials and might become great alternatives to the conventional opioid drugs in future.

Similarly, targeting various other pathways and strategies can also be used to reduce the toxic effects of opioids. For example, selective κOR agonists can be developed which acts on the peripheral sensory neurons and avoid most of the adverse effects brought about by the CNS (Chan et al., 2017). Safety and long-term efficacy of these compounds require further time, clinical studies and evaluation but it can be a start for the development of safe pain killers.

## 19.12   CONCLUSION

In the above description, the authors have discussed about the development of opioids, their uses, mechanism of action, and side effects. Although opioids are being very useful for various kinds of treatments but, the natural opioids like codeine and morphine, and semi-synthetic opioids like heroin, retain the liabilities of raw opium like addiction and many other side effects. In the process to avoid these side effects, various synthetic opioids are developed and after facing many failures and successes, now various new molecules or drug combinations have been developed to obtain the best curative effects. Thus, despite having many new drugs in hands, further investigations are required to establish their efficacy and safety.

## KEYWORDS

- illegal drug
- opioids
- opiates
- opium
- opioid receptors

## REFERENCES

Adrian, T. E.; Allen, J. M.; Bloom, S. R.; Ghatei, M. A.; Rossor, M. N.; Roberts, G. W.; Crow, T. J.; Tatemoto, K.; Polak, J. M. Neuropeptide Y Distribution in Human Brain. *Nature* **1983**, *306*, 584–586.

Akhondzadeh, S.; Kashani, L.; Mobaseri, M.; Hosseini, S. H.; Nikzad, S.; Khani, M. Passion Flower in the Treatment of Opiates Withdrawal: A Double–Blind Randomized Controlled Trial. *J. Clin. Pharm. Ther.* **2001**, *26*, 369–373.

Akil, H.; Watson, S. J.; Young, E.; Lewis, M. E.; Khachaturian, H.; Walker, J. M. Endogenous Opioids: Biology and Function. *Ann. Rev. Neurosci.* **1984,** *7*, 223–255.

Al-Hasani, R.; Bruchas, M. R. Molecular Mechanisms of Opioid Receptor-Dependent Signaling and Behavior. *Anesthesiology* **2011**, *115*, 1363–1381.

Attali, B.; Gouarderes, C.; Mazarguil, H.; Audiger, Y.; Cros, J. Evidence for Multiple "Kappa" Binding Sites by use of Opioid Peptides in the Guinea-Pig Lumbosacral Spinal Cord. *Neuropeptides* **1982**, *3*, 53–64.

Beaver, W. T. Analgesic Development: A Brief History and Perspective. *J. Clin. Pharmacol.* **1980**, *20*, 213–215.

Bilkei-Gorzo, A.; Berner, J.; Zimmermann, J.; Wickström, R.; Racz, I.; Zimmer, A. Increased Morphine Analgesia and Reduced Side Effects in Mice Lacking the Tac1 Gene. *Br. J. Pharmacol.* **2010**, *160*, 1443–1452.

Blakemore, P. R.; White, J. D. Morphine, The Proteus of Organic Molecules. *Chem. Commun.* **2002**, *11*, 1159–1168.

Brissett, D. I.; Whistler, J. L.; Rijn, R. M. Contribution of Mu and Delta Opioid Receptors to the Pharmacological Profile of Kappa Opioid Receptor Subtypes. *Eur. J. Pain.* **2012,** *16*, 327–337.

Brownstein, M. J. A Brief History of Opiates, Opioid Peptides, and Opioid Receptors. *Proc. Natl. Acad. Sci.* **1993**, *90*, 5391–5393.

Bushak, L. How Did Opioid Drugs Get to be so Deadly? A Brief History of its Transition From Trusted Painkiller to Epidemic. *Medical Daily* [Online], July 26, 2016. http://www.yalescientific.org/2017/01/ancient-analgesics-a-brief-history-of-opioids/.

Carr, D. J.; Serou, M. Exogenous and Endogenous Opioids as Biological Response Modifiers. *Immunopharmacology* **1995**, *31*, 59–71.

Charlton, J. E. Opioids: Core Curriculum for Professional Education in Pain, 3rd Ed.; IASP Press: Seattle, 2005.

Chau, N.; Elliot, D. J.; Lewis, B. C.; Burns, K.; Johnston, M. R.; Mackenzie, P. I.; Miners, J. O. Morphine

Glucuronidation and Glucosidation Represent Complementary Metabolic Pathways that are Both Catalyzed by Udp-Glucuronosyltransferase 2b7: Kinetic, Inhibition, and Molecular Modeling Studies. *J. Pharmacol. Exp. Ther.* **2014**, *349*, 126–137.

Chan, H. S.; Mccarthy, D.; Li, J.; Palczewski, K.; Yuan, S. Designing Safer Analgesics via M-Opioid Receptor Pathways. *Trends Pharmacol. Sci.* **2017**, *38*, 1016–1037.

Chahl, La. Opioids-Mechanisms of Action. *Aust. Prescr.* **1996**, *19*, 63–65.

Chen, Y.; Mestek, A.; Liu, J.; Hurley, J. A.; Yu, L. Molecular Cloning and Functional Expression of a L-Opioid Receptor From Rat Brain. *Mol. Pharmacol.* **1993**, *44*, 8–12.

Cherny, N. I. The Management of Cancer Pain. *CA: Cancer J. Clin.* **2000**, *50*, 70–116.

Cherny, N.; Ripamonti, C.; Pereira, J.; Davis, C.; Fallon, M.; Mcquay, H.; Mercadante, S.; Pasternak, G.; Ventafridda, V. Expert Working Group of the European Association of Palliative Care Network. Strategies to Manage the Adverse Effects of Oral Morphine: An Evidence-Based Report. *J. Clin. Oncol.* **2001**, *19*, 2542–2554.

Choi, Y. S.; Billings, J. A. Opioid Antagonists: A Review of their Role in Palliative Care. *J. Pain Symp. Manage* **2002**, *4*, 71–90.

Corbett, A.; Mcknight, S.; Henderson, G. Opioid Receptors. *In Vivo* **1999**, *1*, 1–10.

*Opioids and Opiates: Pharmacology, Abuse, and Addiction. Neuroscience in the 21st Century*, 2nd Ed.; Springer: New York, 2015; Vol. 1, pp. 1–33.

Daniell, H. W. Hypogonadism in Men Consuming Sustained-Action Oral Opioids. *J. Pain.* **2002**, *3*, 377–384.

Daniell, H. W. Opioid Endocrinopathy in Women Consuming Prescribed Sustained Action Opioids for Control of Nonmalignant Pain. *J. Pain.* **2008**, *9*, 28–36.

Dawson, C.; Paterson, F.; Mcfatter F.; Buchanan, D. Methadone and Oedema in the Palliative Care Setting: A Case Report and Review of the Literature. *Scott. Med. J.* **2014**, *59*, E11–E13.

Dews, T. E.; Mekhail, N. Safe Use of Opioids in Chronic Noncancer Pain. *Cleve. Clin. J. Med.* **2004**, *71*, 897–904.

Ding, H.; Czoty, P. W.; Kiguchi, N.; Cami-Kobeci, G.; Sukhtankar, D. D.; Nader, M. A.; Husbands, S. M.; Ko, M. C. A Novel Orvinol Analog, Bu08028, as a Safe Opioid Analgesic Without Abuse Liability in Primates. *PNAS* **2016**, *113*, E5511–E5518

Dixon, R.; Howes, J.; Gentile, J.; Hsu, H. B.; Hsiao, J.; Garg, D.; Weidler, D.; Meyer, M.; Tuttle, R. Nalmefene: Intravenous Safety and Kinetics of a New Opioid Antagonist. *Clin. Pharmacol. Ther.* **1986**, *39*, 49–53.

Duarte, D. F. Opium and Opioids: A Brief History. *Rev. Bras. Anestesiol.* **2005**, *55*, 135–146.

Evans, C. J.; Keith, D. E.; Morrison, H.; Magendzo, K.; Edwards, R. H. Cloning of a Delta Opioid Receptor by Functional Expression. *Science* **1992**, *258*, 1952–1952.

Feierman, D. E.; Lasker, J. M. Metabolism of Fentanyl, A Synthetic Opioid Analgesic, by Human Liver Microsomes. Role of Cyp3a4. *Drug Metab. Dispos.* **1996**, *24*, 932–939.

Flake, Z. A.; Scalley, R. G.; Bailey, A. G. Practical Selection of Antiemetics. *Am. Fam. Phys.* **2004**, *69*, 1169–1174.

Foss, J. F.; O'connor, M. F.; Yuan, C. S.; Murphy, M.; Moss, J.; Roizen, M. F. Safety and Tolerance of Methylnaltrexone in Healthy Humans: A Randomized, Placebo-Controlled, Intravenous, Ascending-Dose, Pharmacokinetic Study. *J. Clin. Pharmacol.* **1997**, *37*, 25–30.

Gammal, R.S.; Crews, K.R.; Haidar, C.E.; Hoffman, J.M.; Baker, D.K.; Barker, P.J.; Estepp, J.H.; Pei, D.; Broeckel, U.; Wang, W.; Weiss, M.J. Pharmacogenetics For Safe Codeine Use in Sickle Cell Disease. *Pediatrics* **2016**, *138*, e20153479.

Harrison, C.; Smart, D.; Lambert, D. G. Stimulatory Effects of Opioids. *Br. J. Anaesth.* **1998**, *81*, 20–28.

Hasselström, J.; Säwe, J. Morphine Pharmacokinetics and Metabolism in Humans. *Clin. Pharmacokinet.* **1993**, *24*, 344–354.

Hay, J. L.; La Vincente, S. F.; Somogyi, A. A.; Chapleo, C. B.; White, J. M. Potentiation of Buprenorphine Antinociception with Ultra-Low Dose Naltrexone in Healthy Subjects. *Eur. J. Pain.* **2011**, *15*, 293–298.

Helm, I. I.; Trescot, A. M.; Colson, J.; Sehgal, N.; Silverman, S. Opioid Antagonists, Partial Agonists, and Agonists/Antagonists: The Role of Office-Based Detoxification. *Pain Phys.* **2008**, *11*, 225–235.

Hofmann, A.; Tangri, N.; Lafontaine, A. L; Postuma, R. B. Myoclonus as an Acute Complication of Low-Dose Hydromorphone in Multiple System Atrophy. *J. Neurol. Neurosurg. Psychiatry* **2006**, *77*, 994–995.

Howlett, T. A.; Rees, L. H. Endogenous Opioid Peptides and Hypothalamo-Pituitary Function. *Ann Rev. Physiol.* **1986**, *48*, 527–536.

Inturrisi, C. E. Clinical Pharmacology of Opioids for Pain. *Clin. J. Pain.* **2002**, *18*, S3–S13.

Jamshidi, R. J.; Jacobs, B. A.; Sullivan, L. C.; Chavera, T. A.; Saylor, R. M.; Prisinzano, T. E.; Clarke, W. P.; Berg, K. A. Functional Selectivity of Kappa Opioid Receptor

Agonists in Peripheral Sensory Neurons. *J. Pharmacol. Exp. Ther.* **2015**, *355*, 174–182.

Kojovic, M.; Cordivari, C.; Bhatia, K. Myoclonic Disorders: A Practical Approach for Diagnosis and Treatment. *Ther. Adv. Neurol. Disord.* **2011**, *4*, 47–62.

Kubat, B. Drugs, Muscle Pallor, and Pyomyositis. *Forensic Sci. Med. Pathol.* **2013**, *9*, 564–567.

Leavitt, S. B. Opioid Antagonists in Pain Management. *Pract. Pain Manag.* **2009**, *9*, 1–7.

Lee, M.; Silverman, S.; Hansen, H.; Patel, V. A Comprehensive Review of Opioid-Induced Hyperalgesia. *Pain Phys.* **2011**, *14*, 145–161.

Lotsch, J.; Geisslinger, G. Morphine-6-Glucuronide: An Analgesic of the Future? *Clin. Pharmacokine.* **2001**, *40*, 485–499.

Lynch, M. E. A Review of the Use of Methadone for the Treatment of Chronic Noncancer Pain. *Pain Res. Manag.* **2005**, *10*, 133–144.

Manglik, A.; Lin, H.; Aryal, D. K.; Mccorvy, J. D.; Dengler, D.; Corder, G.; Levit, A.; Kling, R. C.; Bernat, V.; Hübner, H.; Huang, X. P. Structure-Based Discovery of Opioid Analgesics with Reduced Side Effects. *Nature* **2016**, *537*, 185.

Mccarberg, B.; Barkin, R. Long-Acting Opioids for Chronic Pain; Pharmacotherapeutic Opportunities to Enhance Compliance, Quality of Life, and Analgesia. *Am. J. Ther.* **2001**, *8*, 181–186.

Mcdonald, J.; Lambert, D. G. Opioid Receptors. *Cont. Educ. Anaesth. Crit. Care Pain* **2005**, *5*, 22–25.

Mcnicol, E.; Horowicz-Mehler, N.; Fisk, R. A.; Bennett, K.; Gialeli-Goudas, M.; Chew, P. W.; Lau, J.; Carr, D. Management of Opioid Side Effects in Cancer-Related and Chronic Noncancer Pain: A Systematic Review. *J. Pain.* **2003**, 231–256.

Meng, F.; Xie, G. X.; Thompson, R. C.; Mansour, A.; Goldstein, A.; Watson, S. J.; Akil, H. Cloning and Pharmacological Characterization of a Rat Kappa Opioid Receptor. *Proc. Natl. Acad. Sci.* **1993**, *90*, 9954–9958.

Meuser, T.; Pietruck, C.; Radbruch, L.; Stute, P.; Lehmann, K. A.; Grond, S. Symptoms During Cancer Pain Treatment Following Who-Guidelines: A Longitudinal Follow-Up Study of Symptom Prevalence, Severity and Etiology. *Pain Med.* **2001**, *93*, 247–257.

Miller, R. D. *Anesthesia*, 5th Ed.; Churchill Livingstone: New York, 2000; pp 273–376.

Mogil, J. S.; Ritchie, J.; Smith, S. B.; Strasburg, K.; Kaplan, L.; Wallace, M. R.; Romberg, R. R.; Bijl, H.; Sarton, E. Y.; Fillingim, R. B.; Dahan, A. Melanocortin-1 Receptor Gene Variants Affect Pain and M-Opioid Analgesia in Mice and Humans. *J. Med.Genet.* **2005**, *42*, 583–587.

Morgan, M. M.; Christie, M. J. Analysis of Opioid Efficacy, Tolerance, Addiction and Dependence from Cell Culture to Human. *Br. J. Pharmacol.* **2011**, *164*, 1322–1334.

Movafegh, A.; Shoeibi, G.; Ansari, M.; Sadeghi, M.; Azimaraghi, O.; Aghajani, Y. Naloxone Infusion and Post-Hysterectomy Morphine Consumption: A Double-Blind, Placebo-Controlled Study. *Acta Anaesthesiol. Scand.* **2012**, *56*, 1241–1249.

Naguib, M.; Famewo, C. E.; Absood, A. Pharmacokinetics of Meperidine in Spinal Anaesthesia. *Can. Anaesth. Soc. J.* **1986**, *33*, 162–166.

Niewijk, G. Ancient Analgesics: A Brief History of Opioids. *Yale Scientific* [Online], 2017, 19, p 27 http://www.yalescientific.org/2017/01/ancient-analgesics-a-brief-history-of-opioids/

Otton, S. V.; Schadel, M.; Cheung, S. W.; Kaplan, H. L.; Busto, U. E.; Sellers, E. M. Cyp2d6 Phenotype Determines the Metabolic Conversion of Hydrocodone to Hydromorphone. *Clin. Pharmacol. Ther.* **1993**, *54*, 463–472.

Oxycontin (Oxycodone Hcl Controlled-Release Tablets) [Package Insert] Stamford, Ct: Purdue Pharma Lp; 2007.

Pasternak, G. W. Opioids and their Receptors: Are we there yet? *Neuropharmacology* **2014**, *76*, 198–203.

Pathan, H.; Williams, J. Basic Opioid Pharmacology: An Update. *Br. J. Pain.* **2012**, *6*, 11–16.

Poyhia, R.; Seppala, T.; Olkkola, K. T.; Kalso, E. The Pharmacokinetics and Metabolism of Oxycodone after Intramuscular and Oral Administration to Healthy Subjects. *Br. J. Clin. Pharmacol.* **1992**, *33*, 617–621.

Pradhan, A. A.; Befort, K.; Nozaki, C.; Gavériaux-Ruff, C.; Kieffer, B. L. The Delta Opioid Receptor: An Evolving Target for the Treatment of Brain Disorders. *Trends Pharmacol. Sci.* **2011**, *32*, 581–590.

Ramanathan, S.; Panksepp, J.; Johnson, B. Is Fibromyalgia an Endocrine/Endorphin Deficit Disorder? Is Low Dose Naltrexone a New Treatment Option? *Psychosomatics* **2012**, *53*, 591–594.

Rinner, U.; Hudlicky, T. Synthesis of Morphine Alkaloids and Derivatives. *Top. Curr. Chem.* **2012**, *309*, 33–66.

Riters, L. V.; Cordes, M. A.; Stevenson, S. A. Prodynorphin and Kappa Opioid Receptor mRNA Expression in the Brain Relates to Social Status and Behavior in Male European Starlings. *Behav. Brain Res.* **2017**, *320*, 37–47.

Rook, E. J.; Huitema, A. D.; Ree, J. M. V.; Beijnen, J. H. Pharmacokinetics and Pharmacokinetic Variability of Heroin and its Metabolites: Review of the Literature. *Curr. Clin. Pharmacol.* **2006**, *1*, 109–118.

Roy, S.; Loh, H. H. Effects of Opioids on the Immune System. *Neurochem. Res.* **1996**, *21*, 1375–1386.

Smith, J. P.; Stock, H.; Bingaman, S.; Mauger, D.; Rogosnitzky, M.; Zagon, I. S. Low-Dose Naltrexone Therapy Improves Active Crohn's Disease. *Am. J. Gastroenterol.* **2007,** *102,* 820–828.

Smith, M. T. Neuroexcitatory Effects of Morphine and Hydromorphone: Evidence Implicating the 3-Glucuronide Metabolites. *Clin. Exp. Pharmacol. Physiol.* **2000,** *27,* 524–528.

Standiford Helm, I. I.; Trescot, A. M.; Colson, J.; Sehgal, N.; Silverman, S. Opioid Antagonists, Partial Agonists, and Agonists/Antagonists: The Role of Office-Based Detoxification. *Pain Physic.* **2008,** *11,* 225–235.

Stein, C.; Schafer, M.; Machelska, H. Attacking Pain at its Source: New Perspectives on Opioids. *Nat. Med.* **2003,** *9,* 1003–1008.

Strydom, J. Opioid Antagonists and their Therapeutic Role in Anaesthesia and Chronic Pain Management. *South Afr. J. Anaesth. Analg.* **2007,** *13,* 61–68.

Sudhakaran, S.; Surani, S. S.; Surani, S. R. Prolonged Ventricular Asystole: A Rare Adverse Effect of Hydrocodone Use. *Am. J. Case Rep.* **2014,** *15,* 450–453.

Swegle, J. M.; Logemann, C. Opioid-Induced Adverse Effects. *Am. Fam. Phys.* **2006,** *74,* 1347–1352.

Taylor Jr, R.; Pergolizzi Jr, J. V.; Porreca, F.; Raffa, R. B. Opioid Antagonists for Pain. *Exp. Opin. Invest. Drugs* **2013,** *22,* 517–525.

Tompkins, D. A.; Smith, M. T.; Bigelow, G. E.; Moaddel, R.; Venkata, S. V.; Strain, E. C. The Effect of Repeated Intramuscular Alfentanil Injections on Experimental Pain and Abuse Liability Indices in Healthy Males. *Clin. J. Pain* **2014,** *30,* 36–45.

Traynor, J. R.; Elliott, J. Δ-Opioid Receptor Subtypes and Cross-Talk with M-Receptors. *Trends Pharmacol. Sci.* **1993,** *14,* 84–86.

Trescot, A. M.; Datta, S.; Lee, M.; Hansen, H. Opioid Pharmacology. *Pain Phys.* **2008,** *11,* S133–S153.

Vadivelu, N.; Schermer, E.; Kodumudi, G.; Berger, J. M. The Clinical Applications of Extended-Release Abuse-Deterrent Opioids. *CNS Drugs* **2016,** *30,* 637–646.

Valentino, R. J.; Van Bockstaele, E. Endogenous Opioids: The Downside of Opposing Stress. *Neurobiol. Stress* **2015,** *1,* 23–32.

Rijn, Van R. M.; Whistler, J. L. The Δ 1 Opioid Receptor is a Heterodimer that Opposes the Actions of the Δ 2 Receptor on Alcohol Intake. *Biol. Psychiatry* **2009,** *66,* 777–784.

Waldhoer, M.; Bartlett, S. E.; Whistler, J. L. Opioid Receptors. *Ann. Rev. Biochem.* **2004,** *73,* 953–990.

Whittaker, M. R. Opioid use and the Risk of Respiratory Depression and Death in the Pediatric Population. *J. Pediatr. Pharmacol. Ther.* **2013,** *18,* 269–276.

Yuan, C. S.; Foss, J. F.; O'connor, M.; Osinski, J.; Karrison, T.; Moss, J.; Roizen, M. F. Methylnaltrexone for Reversal of Constipation due to Chronic Methadone Use. *JAMA* **2000,** *283,* 367–372.

Yuan, C. S.; Foss, J. F.; Osinski, J.; Toledano, A.; Roizen, M. F.; Moss, J. The Safety and Efficacy of Oral Methylnaltrexone in Preventing Morphine-Induced Delay in Oral-Cecal Transit Time. *Clin. Pharmacol. Ther.* **1997,** *61,* 467–475.

Yuan, C. S.; Wei, G.; Foss, J. F.; O'connor, M.; Karrison, T.; Osinski, J. Effects of Subcutaneous Methylnaltrexone on Morphine-Induced Peripherally Mediated Side-Effects: A Double-Blind Randomized Placebo-Controlled Trial. *J. Pharmacol. Exp. Ther.* **2002,** *300,* 118–123.

Zhang, X.; Eggert, U.S. Non-Traditional Roles of G Protein-Coupled Receptors In Basic Cell Biology. *Mol. BioSyst* **2013,** 9, 586–595.

Zhou, Y., Eds. Principles of Pain Management. In *Bradley's Neurology In Clinical Practice*; Saunders Elsevier: PA, Philadelphia, 2012.

# CHAPTER 20

# Drugs of Abuse and Addiction

SHALINI MANI,* CHAHAT KUBBA, and AARUSHI SINGH

*Jaypee Institute of Information Technology, Uttar Pradesh, India*

*Corresponding author. E-mail: mani.shalini@gmail.com*

## ABSTRACT

Majority of the drug molecules affects the circuit of the brain by releasing one or the other type of chemical signals. This effect on brain wiring may alter the ability of an individual to feel pleasure and might motivate the person to repeat the behavior needed to thrive. These physical changes may last for a long time and make a person lose self-control and drive the person toward damaging behaviors. Though different psychosocial factors are very crucial in case of drug addiction and their abuses, however, the phenomenon of drug addiction involves a biological process at its core. Due to some important biological consequences, the repeated exposure of an abusive drug molecule induces changes in a vulnerable brain. The current chapter introduces the abusive drugs, their different classes, and also summarizes the different mechanism of action of these drugs that leads to their addiction and abuse. Authors further details about the mechanism of action of different types of common drugs and put forth an explanation for their addictive nature.

## 20.1 INTRODUCTION

Drug addiction is best described as the uncontrollable or repeated use of drugs and an urge to seek it persistently despite being aware of the adverse consequences the drug could have on an individual (Uddin et al., 2017). The National Institute on Drug Abuse defines this disease as being chronic and relapsing and leading to an alteration in the structure and working of brain; the changes could endure and accelerate harmful behavior in individuals who abuse drugs. The concepts of drug addiction and drug abuse are nearly indistinguishable but a subtle distinction can be drawn between the two. Addiction commences when frequentative use of a drug or medication is done, which has a noticeable impact on the reward center of the brain, thereby affecting its functionality. On contrary, drug induced abuse can be defined as an inappropriate use of an illicit compound or a medication, which might eventually lead to the process of addiction. Once habituated, it can become a condition for entire life. Even in absence of the drug for a year long time, there can be an increased risk for relapse (Uddin et al., 2017).

For centuries, drug addiction has afflicted our society. The 2014 data released by National Survey on Drug Use and Health (NSDUH) revealed that 27.0 million people from 12 years or older demonstrated the use of an illegal drug in the past 30 days; the figure corresponds to approximately 1 in 10 Americans (Hedden et al., 2015). The data generated by NSDUH also highlights that 6.5 million Americans between the age of 12 years or older were currently illegal users of psychotherapeutic drugs such as pain relievers, stimulants, tranquilizers, and sedatives, which are being manufactured and distributed without prescription (Hedden et al., 2015).

Drug addiction as well as drug abuse is economically exacting the United States with an annual cost exceeding $600 billion, which includes health care expenses, crime and loss of work productivity associated with it. Moreover, it is particularly alarming to note that this alarming issue of drug addiction poses a global burden of 5.4%, as estimated by the World Health Organization (WHO) (American Addiction Centers, 2018). This statistic reflects that the toll drug addiction and abuse implicate medically, emotionally, and financially in immense. As a result, there is a substantiating need to understand the mechanism of abusive and addictive nature of a drug as well the neuronal effects of the same. As it may help in determining novel targets to treat and prevent addictive disorders (Uddin et al., 2017). In particular, the exigent components of drug addiction, including sensitization, craving, counter adaptation, relapse and abstinence, need to be explored from both neurobiological as well as non-neurobiological perspective. This chapter emphasizes the classes of such drugs, their mode of action, neuroadaptation, and the progress so far in terms of its underlying neuropsychological pathways.

## 20.2  CLASSIFICATION OF ADDICTIVE DRUGS

Under the Controlled Substances Act of implemented 1990, drugs were categorized into a legal classifications system, based on their use, safety profiles, and the potential for abuse. Table 20.1 gives the categorical illustration of addictive drugs.

**TABLE 20.1**  Classes of Addictive Drugs.

| Class | Level of drug abuse | Medical use | Safety level/ dependency risk | Examples |
|---|---|---|---|---|
| Class 1 | High drug abuse | No medical use is known | Lack of accepted safety levels | Cannabis, methaqualone, ecstasy, heroin, GHB, LSD, mescaline |
| Class 2 | High drug abuse | Little medical use | Potential risk of dependency | Cocaine, hydrocodone, fentanyl, hydromorphone, oxycodone |
| Class 3 | Lower level of drug abuse | Has a medical use | Risk of dependency is moderate | Ketamine, steroids, buprenorphine |
| Class 4 | Limited level of abuse | High medical use | Lower risk of dependency | Modafinil, benzodiazepines, tramadol |
| Class 5 | Lowest drug abuse | Accepted medical use | Limited dependence | Diphenoxylate, lacos amide, pregabalin |

In 2006, Luscher et al. also came up with a mechanistic classification of drugs that are considered to be addictive (Luscher et al., 2006). The bases of their classification were all those mechanisms of drug addiction, which had a defining commonality of enhancing the dopamine concentrations in specific parts of the human brain. The three distinguished groups were:

### 20.2.1 CLASS I (DRUGS ACTIVATING G PROTEIN-COUPLED RECEPTORS)

This class of drugs includes opioids, delta-9-tetrahydrocannabinol (THC), and γ-hydroxy butyrate (GHB). Though each bind to a different receptor (opioids to μ-opioid receptors, THC binds to cannabinoid receptors type I and GHB to $GABA_B$ receptor), in general, they lead to neuronal disinhibition (Luscher et al., 1997; Vaughan et al., 1997; Luscher et al., 2006).

### 20.2.2 CLASS II (DRUGS THAT ACT THROUGH IONOTROPIC RECEPTORS)

This class includes drugs such as benzodiazepines, nicotine, and ethanol. Nicotine binds to receptors for nicotinic acetylcholine (nAChRs) present in the brain and increases the permeation rate of cations, thereby leading to depolarization of cells (Pidoplichko et al., 1997; Fagen et al., 2003). Benzodiazepines increase dopamine present in mesocorticolimbic and positively regulate the $GABA_A$ receptors. Next is ethanol, which has a complex pharmacology; it binds to numerous receptors such as $GABA_A$ receptors, Kir3/GIRK, receptors for N-methyl-D-aspartate (NMDA), potassium channels, etc. but causes an overall disinhibition of neuronal cells (Mansvelder et al., 2002).

### 20.2.3 CLASS III (DRUGS BINDING BIOGENIC AMINES TRANSPORTERS)

The third category includes drugs such as cocaine, amphetamines (AMPH), and ecstasy. Cocaine is known to block the action of dopamine, noradrenaline, and serotonin in the central nervous system (CNS) by inhibiting their respective transporters thereby preventing their uptake. Blocking eventually causes a spike in dopamine levels in the nucleus accumbens (Kelz et al., 1999; Brown et al., 2011; Grueter et al., 2013). Amphetamine exerts its effect by causing an alteration in the action of transporters for biogenic amine located at the plasma membrane and inhibiting the normal vesicular release of dopamine while increasing its the nonvesicular release. Lastly, ecstasy binds to its respective receptor and causes the release of biogenic amines; it strongly increases dopamine concentration (Ali et al., 1999; Bartu et al., 2004; Silva and Yonamine, 2004).

## 20.3 ADDICTIVE DRUGS: MODE OF ACTION

The mechanism underlying the process of addiction is complex, as repeated exposure of drug to the brain transmutes the neuronal circuits. The interplay of events are governed by various genetic, nongenetic, and environmental factors which, once formed, can eventually lead to the manifestation of complex behaviors like dependence, craving, tolerance, and sensitization (Koob and Moal, 1997; Kendler et al., 2003). The mode of action of some abusive drugs involves some common modification, as follows.

## 20.3.1  EPIGENETIC MODIFICATIONS

The link between genotypic and environmental elements indicates a significant role of epigenetics in the significant response toward different drugs and the development of addiction. The term "epigenetics" can broadly be defined as a cumulative of all those processes that are obligatory for unwinding the genetic programmes, without altering the DNA sequences (Holliday, 2006). It can be considered as a mediator through which environment interacts with the genome to induce susceptibility to addiction as well as drug-mediated mal adaptations that are elementary to the addiction process. The epigenetic inheritance is primarily regulated by DNA methylation, modifications in histone, and regulatory RNAs (Nestler, 2014).

### 20.3.1.1  DNA METHYLATION

DNA methylation is one of the important and stable modifications out of all the epigenetic changes. It occurs due to the addition of a methyl group at the C5 position of one of the four DNA bases, namely cytosine, predominately at the CpG sites (Bird and Klose, 2006). Methylation negatively regulates the gene transcription by inhibiting the attachment of DNA binding transcription factors to the target sequence, thereby inducing the nearby chromatin into a silenced state (Moore et al., 2013). CpG-rich promoters are methylated in mammals to prevent the transcriptional initiation, silence genes, inactivate X chromosome, and aid genomic imprinting (Jones and Takai, 2001).

Evidence demonstrates a dynamic regulation of DNA methylation in the brain of an adult. The catalyzation and maintenance of the events are mediated by a class of enzymes called DNA methyltransferases (*Dnmts*), being highly expressed in post-mitotic nerve cells (Feng et al., 2010). Given the significant role of DNA methylation in sustaining the transcriptional mechanism, various studies have been conducted to reveal its link with addiction; Table 20.2 gives the summary of the same.

### 20.3.1.2  HISTONE MODIFICATION

Modifications in histone modification are another type of epigenetic change that modulates the expression of gene, post-translationally. In eukaryotic cells, the genomic material is packed into chromatin, in which the DNA wraps the nucleosome and the histone proteins (H2A, H2B, H3, and H4) becomes the basic unit for structure of the nucleosome along with H1, which spans the non-nucleosomal DNA (Luger et al., 1997). The noncovalent modification of histone proteins at the N-terminal modulates the gene expression via alteration of the chromatin structure to create either a transcriptionally active state (euchromatin) or transcriptionally repressive state (heterochromatin) (Cedar and Bergman, 2009). These dynamic modifications such as phosphorylation, acetylation, sumoylation, ubiquitylation, and methylation are actively mediated by two key enzymes: histone acetyltransferases (HATs) and histone deacetylases (HDACs) (Narlika et al., 2002).

The process of deciphering the transcriptional regulation mechanisms has widened our understanding of histone modification processes among which histone methylation, acetylation, and phosphorylation forms the framework for epigenetic maintenance. Table 20.3 summarizes some of the central aspects of the modification process.

**TABLE 20.2** Drug-Induced DNA Methylation Leading to Stable Epigenetic Changes.

| Drugs | Epigenetic factors involved | Effects | Experimental model | References |
|---|---|---|---|---|
| Fluoxetinc, cocaine | Methyl-CpG-binding protein (MBD1) and methyl-CpG-binding protein 2 (MeCP2) and | Gene silencing by recruiting Histone deacetylase (HDAC) | Adult rat brain | Cassel et al., 2006 |
| Amphetamine | – | Induces hyperlocomotion | Rats nucleus accumbens (NAc) | Kim et al., 2008 |
| Cocaine | MeCP2, HDAC2 and HDAC11 | Increased synthesis of MeCP2, HDAC2, HDAC11 and decreased nuclear localization of HDAC5 | Rat | Host et al., 2009 |
| Cocaine | – | Increase in number of thin dendritic spines on the Nac neurons | Mouse nucleus accumbens | LaPlant et al., 2010 |
| Amphetamine | MeCP2 | MeCP2 modulated AMPH-induced behaviors by enhancing amphetamine reward | Mice | Deng et al., 2010 |
| Ethanol | MeCP2 | Glutamatergic presynaptic mechanisms affected in the ventral segmental area and the striatal parts of rat brain | Rodent | Vrettou et al., 2015 |
| Metham-phetamine, modafinil | | Methamphetamine administration impaired object recognition memory whereas modafinil did not | Mice | Gonzalez et al., 2017 |

Evidence indicates that these modifications cause certain of the behavioral and functional defects associated with multiple drug abuse or addiction models. The cocaine exposure is known to enhance the degree of acetylated H3 and H4 histone in NAc, which is the key reward part of the brain (Schroeder et al., 2008). In another study, it was reported that H3 acetylation was particularly increased at the gene responsible for coding of DAD1-interacting protein and acetylation of H4 increased at the gene *DRD3* (Renthal et al., 2009). Furthermore, experiments demonstrate that the reward properties of cocaine and AMPH are also regulated by acetyltransferase modifications and HDAC regulations (Kalda et al., 2007; Wang et al., 2010; Malvaez et al., 2013).

Histone methylation is also directly controlled by psychostimulants. Chronic exposure to cocaine decreased the levels of H3K9me2 in mouse model through the repression of histone methyl transferase G9a (Maze et al., 2010). Reports highlight that H3K9me2 is known to regulate the transcription of FosB in the mouse model and affects behavioral aspects of cocaine (Heller et al., 2014). Similarly, phosphorylation of histone proteins is further a crucial epigenetic change due to psychostimulants. Drugs induce robust phosphorylation of H3 at the c-fos and c-Jun promoters proteins present within the NAc, which are positively regulated by extracellular signal-regulated kinase (ERK) or MSK1 signaling pathway (Gonzalez et al., 2008; Brami-Cherrier et al., 2009).

**TABLE 20.3**  Major Types of Histone Modifications Induced by Drugs Along with Their Key Features.

| Type of modification | Key features | Gene regulation mechanisms | Change in chromatin | Enzymes involved | References |
|---|---|---|---|---|---|
| Histone acetylation | Occurs on different histone proteins | Neutralizes the positively charged proteins on histones. Consequently, ionic interaction with DNA decreases and access to transcriptional activators increases | Activated | Histone acetyltransferases | Turner, 2000 |
| | In the brain, acetylation occurs on lysines 9, 14, 18, and 23 of H3 and at lysines 5, 8, 12, and 16 of H4. | Histones with acetylation act as an important site of binding for bromodomain containing proteins | | Histone deacetylases | |
| Histone methylation | Transfer of methyl groups to different amino acid residues in histone proteins | Di-methylation at lysine 9 of H3 (H3K9me2) causes DNA inactivation | Activation or repression | Histone methyltransferases | Fuks et al., 2003 |
| | DNA repression or activation or depends on the level of methylation and the nature of methylated amino acid | Tri-methylation of H3 at three different lysine residues (K4, K36 and K79 cause DNA activation | – | | |
| Histone phosphorylation | Histone tails are phosphorylated by protein kinases and dephosphorylated by phosphatases | Phosphorylation of serines of H3 at 10 and 28 and serine 32 of H2B causes gene transcription | Activated | Kinases | Fischle et al., 2005 |
| | Can occur on tyrosine threonine and serine residues | Dephosphorylation of H2AX modulates apoptosis | – | | |

### 20.3.1.3  REGULATORY RNAs

The revelation of mammalian genome sequence has depicted the distribution of noncoding RNAs (ncRNAs), which contribute majorly to the genome complexity and occupy important nodes in a variety of physiological networks, regulatory processes, and genome stability (Berezikov, 2011; Rinn and Chang, 2012). Considering these observations, it can be clearly stated that ncRNAs are also involved in the chief aspects of neuroplasticity and homeostasis of the nervous system and chronic drug use or any addictive stimuli can perturb the structural or functional coherence of neurons (Im et al., 2010; Saba et al., 2012). Some research outcomes that highlight the significance and

involvement of ncRNAs in drug addiction and abuse are highlighted in Table 20.4.

### 20.3.2  TRANSCRIPTIONAL FACTORS

Major types of changes at molecular and cellular level take place during drug-induced adaptations. Such changes are stable in most cases of addiction and even after years of abstinence these changes can lead to a relapse. Addiction process also goes through alterations in gene expression and linked to these expressions are the transcription factors which moderate the expression by binding to respective regulatory regions (Robison and Nestler, 2012). Transcription factors are large

**TABLE 20.4**   Drug-induced Changes in ncRNA and Their Involvement in Drug Addiction Process.

| Drug | ncRNAs involved | Expression level | Effects | References |
|---|---|---|---|---|
| Cocaine | MiR-181a, MiR-124 and Let-7d | MiR-181a - increases MiR-124 and Let-7d - decreases | The ncRNAs led to behavioral changes and altered transcription factors, receptors, etc. involved in cocaine-induced plasticity | Chandrasekar and Dreyer, 2011 |
| | MicroRNA-212 | MiR-212 - decreases | Increased motivational salience for cocaine-paired cues during periadolescence | Viola et al., 2016 |
| | MiR-124, MiR-212, MiR-132 and MiR-134 | MiR-212, MiR-132 - increases; MiR-134, MiR-124 - no change | Involved in controlling the synaptic plasticity of neurons and motivational learning | Chudy et al., 2017 |
| Nicotine | MiR-140 | MiR-140 - increases | Regulated *Dnm1* (dynamin 1 gene) expression, which in turn plays an important role in regulating synaptic endocytosis in the central nervous system | Huang and Li, 2009 |
| | MicroRNA | 78% microRNA (e.g., MiR-80, MiR-79, MiR-80 and MiR-79, etc.) - increases 22% (e.g., MiR-230, Mir-58, Mir-58, etc.) - decreases | MicroRNAs cumulatively mediated regulatory hormesis manifested in physiological phenotypes and biphasic behavior | Taki et al., 2014 |
| Alcohol | MiR-9 | MiR-9 - increases | Contributed to alcohol tolerance, adaptation and neuronal plasticity | Pietrzykowski et al., 2008 |
| | MiR-382 | MiR-382 - decreases | Modulated the expression of ΔFosB, involved in alcohol abuse | Li et al., 2013 |
| Opioids | Let-7 | Let-7 - increases | Repressed MOR thereby playing an integral role integral role in opioid tolerance | He et al., 2010 |
| | MiR-339-3p | MiR-339-3p - increases | Inhibited the production of μ-opioid receptor (MOR) proteins and was involved in opioid receptor regulation | Wu et al., 2013 |

protein complexes which have specificity toward the promoter regions of respective target genes and their binding to these regions is the mediating source to the gene expression, these factors include some prominent ones like cyclic adenosine monophosphate (cAMP) response element binding (CREB) and ΔFosB and several others such as nuclear factor kappa B (NF-κB) and MEF2 (Robison and Nestler, 2012; Nestler, 2013).

### 20.3.2.1   *CAMP RESPONSE ELEMENT BINDING PROTEIN*

In assorted regions of brain responsible for addiction, the CREB protein gets activated by the drugs and most prominently in the nucleus accumbens. This CREB activation represents a negative feedback mechanism as it reduces the satisfying effects of the drug by actually lessening the sensitivity of an individual and mediates dependence on drug at the time of drug withdrawal (Robison and Nestler, 2012; Nestler, 2013). This negative reinforcement leads to self-administering of drugs by the people and probably relapse. This action of CREB in the regions of amygdala and hippocampus is contemplated as a vital factor in behavioral memory. For example, in case of opioid, the initiation of dynorphin (peptides of opioid) expression in NAc neurons is regulated via CREB (Shaw-Lutchman et al., 2002; Shaw-Lutchman et al., 2003; Robison and Nestler, 2012; Nestler, 2013). CREB is associated with the increasing dynorphin activation of κ-opioid receptors present on the DA neurons which lie in the VTA region and henceforth suppressing dopamine transmission to NAc and thus impairing reward related behavior. Target genes for CREB are known only in some drugs, about its mantle in other drugs of abuse still remains undisclosed (Shaw-Lutchman et al., 2002; Shaw-Lutchman et al., 2003; Carlezon et al., 2005; Edwards et al., 2007; Altarejos and Montminy, 2011).

### 20.3.2.2   ΔFosB

Continous or acute exposure to any drug leads to induction of all Fos family transcription factors in NAc and other areas of brain. Most of these Fos protein levels come to normal after 8–12 h of exposure except ΔFosB which is a truncated product of a *FosB* gene (Robison and Nestler, 2012; Nestler, 2013). This product accumulates as it is unusually stable and with course of repeated drug exposure more Fos protein expressed lead to more accumulation. Their stability is to such an expanse that they persist for weeks after drug withdrawal (Nestler, 2013). This chronic induction of ΔFosB is studied almost in every drug and it is established to be more selective for D1-type NAc

neurons. It increases an individual's sensitivity toward the natural and satisfying rewards of drug and promotes self-administration. While CREB induces dynorphin, ΔFosB suppresses it and contribute to pro-reward effects. Functional effects of ΔFosB in other assorted areas of brain are still unrevealed except the orbito frontal cortex (Nestler, 2013). It was found that exposure to cocaine at chronic levels leads to cognitive-disrupting effects which guides to the incidence of ΔFosB-mediated tolerance and this adaptation leads to a rise in cocaine self-administration. Role of ΔFosB in behavioral memory is still under exploration (Nestler, 2013; Robison et al., 2012).

### 20.3.2.3   NF-κB

NF-κB transcription factor is linked to synaptic plasticity and evocation in drug addiction. Repeated cocaine administrations lead to induction of this nuclear factor in the nuclear accumbens and also these are analogous to nicotine dependence (Meffert et al., 2003; Russo, 2009; Christoffel, 2011; Robison and Nestler, 2012; Nestler, 2013). Several researches are presently going on to get acquainted with the cellular basis of NF-κB action and the plasticity changes occuring. Many researches have scrutinized that ΔFosB plays a prominent role in mediating these nuclear factors thus it is practicable that there exists a complex transcriptional cascade in this whole drug addiction process (Meffert et al., 2003; Russo, 2009; Christoffel, 2011).

### 20.3.2.4   MEF2

There are multiple myocyte enhancing factor-2 (MEF2) proteins which get expressed in nucleus accumbens of brain and these proteins form homo or heterodimers to regulate gene transcription (Pulipparacharuvil, 2008; Robison and Nestler, 2012). Cocaine administration suppresses MEF2 protein activity through some novel mechanisms involving cAMP. Increased dendritic spine density induced by cocaine is also a result of this decreased activity of MEF2 proteins. Still a lot is to be known about the MEF2 activity in other drugs of abuse (Pulipparacharuvil, 2008; Chen, 2010).

### 20.3.3   EFFECT ON G PROTEIN-COUPLED RECEPTORS

G protein-coupled receptors (GPCRs) are an extensive class of receptors, bound to G-proteins that are hetrotrimeric in nature (Beck-Sickinger and Böhme, 2009). It mediates a wide variety of cellular functions including the alteration in the levels of cAMP, which acts as secondary messengers, mobilization of intracellular calcium, modifying the release of neurotransmitter, and reorganization or alteration of cellular gene expression (Kiselyov et al., 2003; Betke et al., 2012). Various preclinical models have depicted a correlation between drug abuse and addiction with GPCRs. Repeated exposure to psychostimulants such as cocaine, nicotine, alcohol, etc. produces irreversible changes in the ability of GPCRs to modulate neurotransmitter release; such changes lead to an alteration in the behavioral responses by either promoting or constraining the stimulating effects of a range of addictive drugs (Navarro et al., 2013). Three well-known classes of GPCRs are known to mediate the effect of abused drugs on CNS most effectively. These include dopamine receptors ($D_1$–$D_5$), cannabinoid type 1 receptors (CB1), and group II metabotropic glutamate receptors (mGlus).

Dopamine is a neurotransmitter that binds to GPCRs and produces neurochemical effects (Sidhu, 1998). It is a direct target of a number of stimulating drugs such as methylphenidate, cocaine, and AMPH, which interacts with dopamine transporters thereby inhibiting the dopaminergic transmission and reuptake (Nestler, 2005; Calipari and Ferris, 2013). The next class of receptors is CB1 receptors, which are predominantly activated by endocannabinoids (eCBs) (Mackie, 2006). The rewarding effect of abused drugs is strongly mediated through the eCB system. CB1 receptors restore the drug seeking behavior, which is previously extinguished, when an individual is re-exposed to drug associated cues (Filip et al., 2006). It has been validated in experimental studies conducted on animal model systems that blocking CB1 receptors decreases the rewarding effect of opiates, nicotine, and alcohol and can therefore be considered as an effective therapy for drug abusers via preventing the relapse to drug seeking behavior (Foll and Goldberg, 2004). Another important class of receptors is mGlus, which are predominantly expressed on the glutamatergic terminals at presynaptic membranes; the receptor negatively regulates the glutamate release (Conn and Niswender, 2010). It has been demonstrated that mGlus antagonist reduces the active drug self-administration and drug-seeking behavior, as observed in the case of cocaine (Hao et al., 2010), nicotine (Liechti and Markou, 2007), and ethanol (Sidhpura et al., 2010) and thus, it might act as an effective tool to prevent the dependence on abused drugs.

### 20.3.4   NEUROTRANSMITTER TRANSPORTERS AS MOLECULAR TARGETS

Addictive drugs such as cocaine, AMPH (class III drugs) hinder the monoamine transporters of the neurons mainly which are present in the ventral tegmental area (VTA) and blocks the dopamine uptake by them leading to extracellular accumulation of dopamine as illustrated in Figure 20.1. These elevated dopamine concentrations may lead to anterior pituitary hypoplasia, inability to lactate owing to the reduced hypothalamic content of growth hormone-releasing hormone (Amara and Sonders, 1998).

In CNS, cocaine blocks the uptake of neurotransmitters like dopamine, serotonin, epinephrine by impeding their respective transporters. The restrained dopamine transporters lead to an elevation in dopamine concentrations in several regions specifically the nucleus accumbens (Brodie and Dunwiddie, 1990; Rocha et al., 1998; Rocha, 2003; Chen et al., 2006).

While the AMPH reverse the manoeuvre of the biogenic amine transporters which exist on the plasma membrane, AMPH are the substrates of these amine transporters. These enter the cell and accumulate in the synaptic vesicle and disturb the protein electrochemical gradient disrupting monoamine transport through VMAT. Thus, dopamine uptake by VMAT is prevented. In this way, amphetamine when administered to body leads to neurotoxic increase in intracellular dopamine and dopamine delivery from vesicles is decreased (Sulzer et al., 2005; Ingram et al., 2002).

Translocation of AMPH as substrates of DAT increases the probability of DAT binding sites to come in contact with the cytosol. Dopamine which is in higher concentration in the cytosol attaches to internalized binding sites and reverses transport of dopamine takes place. A dopamine molecule released by reverse transport is followed by uptake of an amphetamine molecule (Floor and Meng, 1996; Floor et al., 1995; Hughes and Brodie, 1959; Hughes et al., 1958).

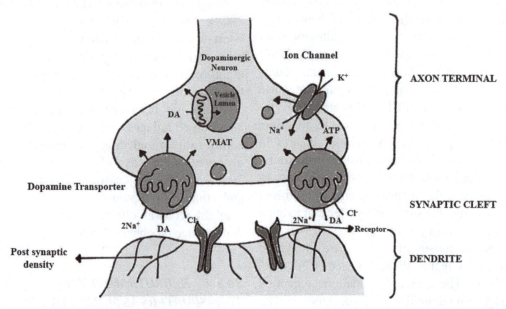

**FIGURE 20.1   (See color insert.)** Transporter proteins involved in uptake of dopamine (DA). The DA transporter uses the energy stored across the plasma membrane due to the occurrence of sodium gradient. Neurotransmitters diffuse through synaptic cleft and reach their respective receptors. Neurotransmitter once interior to the cell is taken to the vesicles through vesicular monoamine transporter (VMAT).

*Source*: Adapted from Amara and Sonders, 1998.

## 20.4 NEUROADAPTATION RELATED TO DRUG ADDICTION

Most of the research studies are centralized upon dopamine as per its participation in drug addiction, the increased dopamine in limbic regions of brain is considered decisive for the fortifying effects of the drugs (Rosenkranz and Grace, 2001). This increased concentration leads to some configurational changes in the frontal lobe of the brain; these changes are a cause of structural and functional modulations done in frontal cortex by dopamine. For example, cocaine administration acts on the prefrontal cortex and changes in dendrites present in this region takes place which further makes them morphologically different and also certain changes in the nucleus accumbens occur as shown in Figure 20.2 (Rosenkranz and Grace, 2001; Miller and Cohen, 2001).

Drug addiction is scrutinized as a syndrome of blunt response impediment and distinctive attribution which is actually known as iRISA—impaired response inhibition and salience attribution. The operation of the prefrontal circuits (modulated due to dopamine) decides the four behaviors linked with iRISA syndrome as depicted in Figure 20.3. The prefrontal cortex, orbitofrontal cortex, and anterior cingulate are all involved in these behaviors of drug addiction (Goldstein and Volkow, 2002). Because these circuits are operating simultaneously and are interacting with one another, thus any behavior (linked to I-RISA) which takes place involves the joint participation of all of these circuits as can be seen in Figure 20.4.

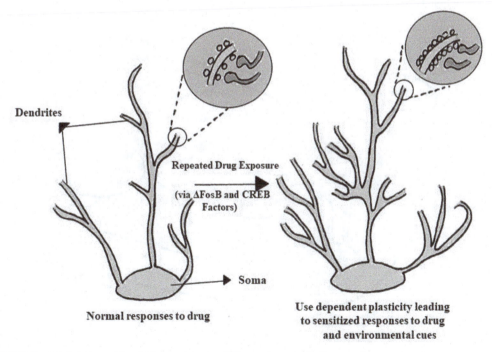

**FIGURE 20.2** Modulation of dendritic structure by different drugs including cocaine.
*Source*: Adapted from Nestler et al., 2001.

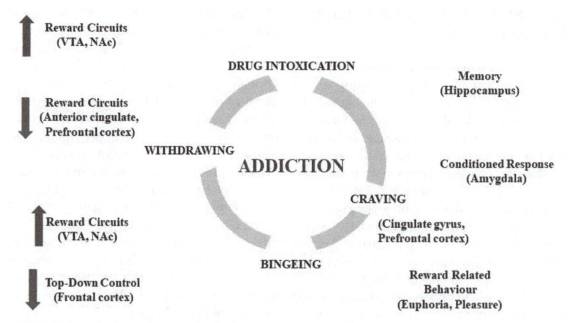

**FIGURE 20.3**    Behaviors and events related to iRISA syndrome of drug addiction. VTA, Ventral tegmental area; NAc, nucleus accumbens.

*Source*: Adapted from Goldstein and Volkow, 2002.

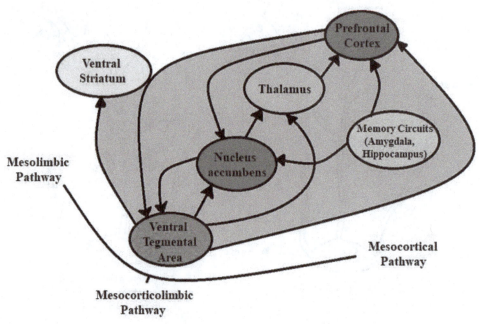

**FIGURE 20.4**    Synergy of the involved circuits in drug addiction.

*Source*: Adapted from Goldstein and Volkow, 2002.

## 20.4.1 DRUG INTOXICATION

Short-term drug administration affects the limbic brain regions where the extracellular concentration of dopamine becomes higher. Increased dopamine in the frontal regions of brain is also apparent. During drug intoxication, there is an exhilarating effect on the prefrontal cortex and anterior cingulate gyrus which are considered as the main drastic occurrences of drug administration (Goldstein and Volkow, 2002). Few studies have depicted that lower glucose metabolism in certain assorted regions of brain associated to cocaine intoxication while marijuana intoxication has a reverse effect of increased glucose metabolism levels in the prefrontal cortex (Ritz et al., 1987; Hurd and Ungerstedt, 1989; Goeders and Smith, 1986).

## 20.4.2 DRUG CRAVING

Reward-related behavior like euphoria, pleasure are some of the learned responses to which craving is associated with. Consolidation of this memory is likely to involve some neuroanatomical substrates like amygdala and hippocampus (Brown and Fibiger, 1993; Meil, 1997). Triggering of the prefrontal cortex and anterior cingulate gyrus are the definite events related to the episode of craving. To activate the fronto limbic circuits, acute drug administration is not required because of foregoing exposure to any drug, craving itself is sufficient and efficient for such activation (Brown and Fibiger, 1993; Meil, 1997; Franklin and Druhan, 2000; Volkow et al., 1999; Childress et al., 1999; Garavan et al., 2000; Grant et al., 1996; Maas et al., 1998).

## 20.4.3 COMPULSIVE DRUG ADMINISTRATION

Overpowering drug self-administration occurs even in cases where no rewards of pleasure are further provided. Craving being the chief reason for such a response. Dopaminergic, serotonergic circuits are associated with this loss of control and drug sessions and exhilaration of thalamo-orbitofrontal circuit and the anterior cingulate gyrus occurs (Fischman et al., 1985; Loh and Roberts 1990; Cornish et al., 1999).

## 20.4.4 DRUG WITHDRAWAL

Subsequent drug withdrawal after the recurrent drug administration leads to disruption of the behavioral circuits probably the disruption of those morphological changes which occurred in frontal cortical circuits including the disruption of the neurotransmitters. This culminates to dysphoria and irritability (Willner et al., 1992; Hodgins et al., 1995). Dysthymia is an elemental symptom of withdrawal which reflects the adaptation responses to replicated dopamine intensification by drugs. Withdrawal can be considered as a symptom caused due to the turning around of the neuroadaptive adaptations to a drug (Willner et al., 1992; Hodgins et al., 1995; Johnson and Fiscgman, 1989; Koob and Le Moal, 2001).

Along with other adaptations, alterations in dendrites also take place with chronic exposures of drug. Expansion of dendritic tree occurs (as observed in nucleus accumbens and prefrontal cortex). Conversions in dendritic structures which are homogenous to other examples of synaptic plasticity can mediate long-term sensitized responses to drugs of abuse (Nestler et al., 2001).

Specific adaptations occur with time-dependent changes in synaptic function and behavioral features of addiction. Such as in chronic morphine administration, VTA dopamine soma size is decreased as shown in Figure 20.5, while neuronal excitability increases and dopamine transmission to NAc is decreased (Mazei et al., 2011). Downregulation of IRS2 in VTA is the key factor mediating effects of morphine. α-amino butyric acid (GABA) currents are reduced and suppression of $K^+$ channel expression takes place, both these events control the effect of excitability (Mazei et al., 2011).

Some people get addicted while others in the same set of conditions do not, this still remains the most exacting issue in drug abuse research. Vulnerability to drug addiction is linked through several neurobiological mechanisms though these mechanisms are not yet clearly brought into light. The studies upto date possibly did not rule out the exact mechanism of the changed frontal activity in different individuals for different drug exposures (Hyman et al., 2006). The neurological adaptation is probably distinct with regard to every individual depending on

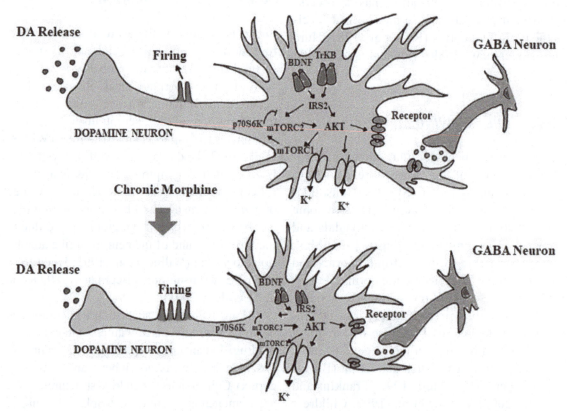

**FIGURE 20.5   (See color insert.)** Chronic morphine decreases VTA dopamine soma size. DA, dopamine; BDNF, brain-derived neutrophic factor; IRS, insulin receptor substance; mTORC, mammalian target of rapamycin complex; TrkB, tropomyosin receptor kinase B; AKT, protein kinase.

*Source*: Adapted from Nestler, 2013.

ones tolerance and involvement of different regions of brain (Wang et al., 2012).

Few studies suggest that genetic factors account for 50% of this individual variability in addiction vulnerability and this heritability is true for mostly all drugs (Kandel et al., 2006; Nestler, 2013). Hundreds of genetic variations have been a great setback in recognizing the exact genes involved in these addiction vulnerabilities. The other 50% of addiction is presumed to be from environmental factors which secondarily influence on the genetic composition of an individual (as per the epigenetic mechanisms discussed above) (Kalivas and O'Brien, 2008). Many of the "gateway" drugs like nicotine have been considered to impact upon the vulnerability (increase) of an individual to another drug. There is also increasing evidence that sufficiently high portions of a drug for longer time can transform an individual with lower genetic loading into an addict (Wang et al., 2012; Kandel et al., 2006; Kalivas and O'Brien, 2008).

## 20.5 MOLECULAR AND CELLULAR MECHANISMS OF ADDICTION IN SOME OF THE COMMON DRUGS

From increase in dopamine concentration to the neuroadaptive changes that occur all throughout the drug addiction process, many molecular and cellular mechanisms are involved due to the drugs acting upon their various molecular targets in brain (Nestler, 2002). Though the fledgling effects of most of the drugs are identical but as drug consumption moves toward addiction phase different mechanisms come into effect. Such mechanisms are discussed below with respect to some of the most common drugs of abuse (Nestler, 2002).

### 20.5.1 OPIOIDS

Opioids belong to the first class of mechanistic classification of drugs that is they act by activating GPCRs (Luscher et al., 2006). G-proteins, cAMP second messenger, and protein phosphorylation pathway mediate important aspects of opioids. Opioids increase dopamine concentration by acting on MORs which is manifested on GABA neurons of the VTA region. These MORs act with a dual action on these GABA neurons, first by hyperpolarising the GABA neurons this way the plasma membrane potential becomes more negative and action potential is never reached. Secondly, it decreases the release of GABA (Pickel et al., 2002; Luscher et al., 1997). Mediation of post-hyperpolarization is done by G-protein coupled inwardly rectifying K$^+$ channels (GIRK). MORs inhibit Ca$^{2+}$ channels and activate voltage-gated K$^+$ channels (Pickel et al., 2002; Luscher et al., 1997; Vaughan et al., 1997; Johnson and North, 1992). Opioids also act on the locus coeruleus (LC) which is the largest nonadrenergic nucleus in brain. Opioids decrease firing rate of LC neurons by activation of inward rectifying K$^+$ channels and inhibition of slowly depolarizing cation channel as you can see in Figure 20.6 and Figure 20.7. Both actions occur through G-proteins. cAMP levels of neurons are reduced and cAMP-dependent protein kinase is activated. Adenylate cyclase activity is acutely reduced in LC (Aghajanian, 1978; Andrade et al., 1983; Wang and Aghajanian, 1990; Alreja and Aghajanian, 1991).

Repeated uptake of opioids leads to diminished effects which are due to the tolerance. This tolerance is probably due to the availability of opioid receptors and neurotransmitter system (such as noradrenaline) changes (Maldonado, 1997). Tolerance can vary dependent on the environment as well (Siegel et al., 1982).

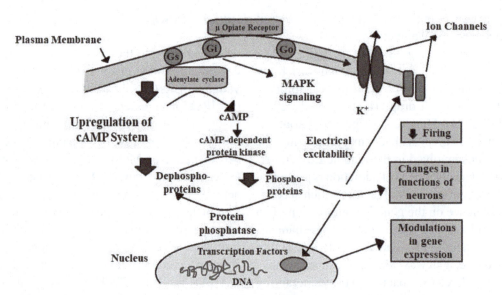

**FIGURE 20.6** Symbolic representation of opioid action on locus coeruleus. Gi/Go, Gs, G proteins; cAMP, cyclic adenosine monophosphate; MAPK, mitogen-activated protein kinase.
*Source*: Adapted from Nestler, 1992.

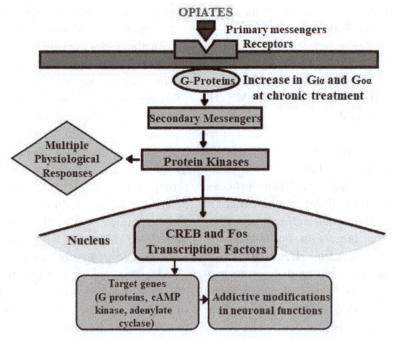

**FIGURE 20.7** Intracellular mechanism of the addictive changes brought by opioid action. Gia, Goa, G protein subunit; cAMP, cyclic adenosine monophosphate; CREB, cAMP response element binding; Fos, transcription factor.
*Source*: Adapted from Nestler, 1992.

Withdrawal from opioids leads to signs and symptoms including stomach cramps, pupil dilation, irritability, sweating, tachycardia, dysphoria, insomnia, rhinorrhoea, increased heart rate, increased blood pressure (Cushman and Dole, 1973). Dysphoria persists for much longer. Repeated dosing of opioids leads to more profound tolerance and dependence. Distinct neural systems are involved in addiction and dependence (Koob and Le Moal, 1997).

Adjunctive medications are used to reduce the symptoms of opioid withdrawal. These medications are formulated to target the noradrenaline system of brain as well as the regions of brain involved in withdrawal symptoms (Day et al., 2005).

### 20.5.2 COCAINE

Cocaine administration leads to blockage of catecholamine reuptake and hence an extracellular increase in catecholamine levels. In CNS, cocaine blocks the uptake of neurotransmitters like dopamine, epinephrine, and serotonin by inhibiting their respective transporters (Brodie and Dunwiddie, 1990; Rocha et al., 1998; Rocha, 2003; Chen et al., 2006). In striatum and NAc, the dopamine transients reach higher concentrations due to reuptake blockade at the electrode. Dopamine in the dorsal striatum is implicated with craving and is a rudimentary component for addiction. Craving being the main reason for relapse thus decreasing dopamine concentrations can be a beneficial step against cocaine addiction (Rocha, 2003; Whitman et al., 2007). Figure 20.8 shows how chronic exposure to cocaine leads to transient reorganization of α-amino-3-hydroxy-5-methyl-4-isoxazolepropionic acid (AMPA) and NMDA glutamate receptors at nucleus accumbens medium spiny neuron

(MSN) synapses as well as structural changes occur (Kelz et al., 1999; Brown et al., 2011; Grueter et al., 2013).

Psychological withdrawal symptoms of cocaine persist for long. Cocaine withdrawal is analogous to the changes in the mesocorticolimbic dopamine system with symptoms including fatigue, depression or anxiety, muscle aches, chills (Kampman et al., 2000). There has been no victorious progress of any pharmacotherapeutic intervention in cocaine addiction treatments (Kreek et al., 2012).

### 20.5.3 AMPHETAMINES

AMPH belong to class III of mechanistic classification of drugs like cocaine. AMPH is a common drug of abuse in Sweden and other northern European countries. AMPH acts by binding to dopamine transporters (Ingram et al., 2002; Sulzer et al., 2005). This reverses the action of the biogenic amine transporters. These enter the cell and accumulate in the synaptic vesicle and disturb the protein electrochemical gradient disrupting monoamine transport through VMAT. Thus, dopamine uptake by VMAT is prevented (Brodie and Dunwiddie, 1990; Rocha et al., 1998; Rocha, 2003; Chen et al., 2006). Redistribution of dopamine from synaptic vesicles to the cytosol is also induced by AMPH. This dopamine is released extracellularly by reverse transport and DA levels rise in the synaptic cleft region. Dopamine metabolism leads to formation of hydroxyl and super oxide radicals which cause the toxic effect of drug. These toxic effects are a result of free radical mediated destruction of monoaminergic terminals (Ali et al., 1999; Bartu et al., 2004; Silva and Yonamine, 2004). AMPH also induces reactive oxygen species (ROS) which regulate the transcription factors, their induction or suppression, and also the activation and repression of genes related to some neuronal functions is modulated by ROS as well. These might be the critical steps of AMPH-induced toxic events (Bartu et al., 2004; Ali et al., 1999; Silva and Yonamine, 2004).

Dependence on AMPH lead to increased extracellular noradrenaline concentrations including several effects such as increase in heart rate, respiration rate, blood pressure, body temperature, reduced appetite (Bönisch and Brüss, 2006; Cruickshank and Dyer, 2009).

Currently there is no pharmacological therapy which has proved its efficacy toward the addiction of amphetamines (Vocci and Appel, 2007).

### 20.5.4 NICOTINE

Nicotine belongs to class II of mechanistic classification thus it acts upon ionotropic receptors unlike opioids and cocaine which act on GPCRs and DAT, respectively. nAChRs are present on the GABA neurons (Pidoplichko et al., 1997;

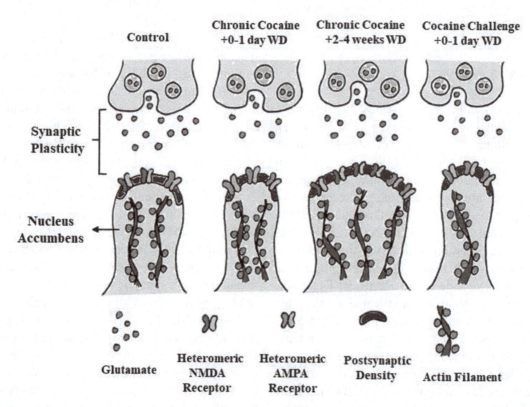

**FIGURE 20.8** Chronic cocaine induces surface expression of NMDA receptors, silent synapse formation, and long-term depression at early withdrawal (WD). With more days of withdrawal, AMPA receptors start increasing. These effects revert with a challenge dose of cocaine and depression of synaptic strength takes place. NMDA, N-methyl-D-aspartate; AMPA, α-amino-3-hydroxy-5-methyl-4-isoxazolepropionic acid.

*Source*: Adapted from Russo et al., 2010.

Fagen et al., 2003). Nicotine binds to these and makes them cation permeable and depolarize the cell. Prolonged exposure leads to rapid receptor desensitization. Desensitization of β2-nAChR on GABA neurons leads to decrease of GABA leading to a more extended disinhibition of dopamine neurons. Gratifying effects of nicotine are due to this β2-nAChR. Nicotine also acts upon homomeric α7 containing nAChR which are present at the synaptic terminals of DA neurons in the VTA region thus facilitating glutamate release (Picciotto et al., 1998; Mansvelder et al., 2002). Both of these actions are considered to be reasons of nicotine addiction. It is also studied that nicotine also effects dopamine release by NAc (Pidoplichko et al., 1997; Picciotto et al., 1998; Mansvelder et al., 2002; Fagen et al., 2003) as shown in Figure 20.9.

Dependence in nicotine is due to the activation of mesolimbic dopaminergic reward system. Exposure to nicotine leads to elevation in nAChRs which further lead to an increased nicotine tolerance (Cummings and Hyland, 2005).

In amygdala, an upregulation in adenylyl cyclase activity is considered to be linked with nicotine withdrawal. Symptoms including changes in cognition, restlessness, anxiety, depressed mood, insomnia, increased appetite. Participation of specific nAChRs is supposed to be linked with this withdrawal (McLaughlin et al., 2015).

To reduce the withdrawal symptoms, nicotine replacement medications are prominent. Nicotine replacement products presently available are nicotine nasal sprays, vapor inhaler, sublingual tablet, etc. (Henningfield et al., 2005).

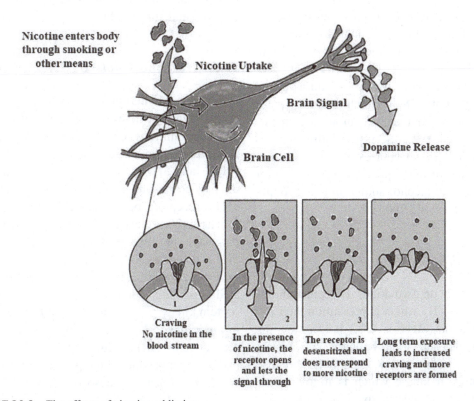

**FIGURE 20.9** The effects of nicotine addiction.

## 20.5.5  ALCOHOL

Alcohol is the most addictive drug used. Like nicotine, alcohol is also a part of second class of mechanistic classification of drugs (Morrow, 1995; Koob et al., 1998). No single receptor is involved in effect of alcohol. Alcohol alters quite a range of receptors like GABA receptors, nAChR, NMDA receptors, GIRK channels (Morrow, 1995; Koob et al., 1998; Luscher et al., 2006). Apart from these, adenosine re-uptake is also influenced by alcohol by obstructing the equilibrative nucleoside transporter (ENT1). Alcohol also increases dopamine release like all other drugs but the exact mechanism behind this is not known in the case of alcohol (Morrow, 1995; Koob et al., 1998). This increase can be due to direct excitation of DA neurons or may be due to any other altered receptor. Involvement of all the different receptors in the action pathway of alcohol make it unclear what is the actual reason of its addiction (Koob et al., 1998; Morrow, 1995).

Physiological aspects of withdrawal of alcohol last upto 48 h which include motor abnormalities, convulsions, autonomic disturbances, nausea, seizures, visual hallucinations (Majchrowicz, 1975). Naltrexone and acamprosate are some of the adjuvant interventions for the treatment of alcohol addiction (Swift et al., 1996; Johnson et al., 2003).

## 20.5.6  CANNABINOIDS

Cannabis acts on two kinds of cannabinoids receptors, the CB1 which is present on the CNS system and the other is CB2 present on the immune cells (Gupta and Kulhara, 2007). CB1 receptors stimulate the $K^+$ channels and also activate protein kinase which is mitogen-activated and both of these activations occur when CB1 inhibits adenylate cyclase and calcium channels. Development of forbearance and severe effects of cannabinoids are mediated by G-protein coupled cannabinoid receptors (Gupta and Kulhara, 2007).

Cannabinoids addiction and tolerance is developed due to the uncoupling and down-regulation of the brain CB1Rs (CB1 receptors) (Fratta and Fattore, 2013). Abstinence from the continuous uptake leads to various withdrawal symptoms such as decreased appetite, weight loss, as well as emotional changes such as irritability, anxiety, restlessness (Lichtman and Martin, 2005).

Several pharmacological interventions are going on, where CB1 receptor remains the primary target. Agonist medications (similar neurobiological mechanisms as cannabis) are beneficial to diminish the withdrawal symptoms (Cooper and Haney, 2009).

## 20.5.7  LYSERGIC ACID DIETHYLAMIDE

Lysergic acid diethylamide (LSD) influences diverse types of neurotransmitter systems, thus its functioning is complex and still remains unexplored. Activation of $5HT_{2A}$ receptors, $5HT_{2C}$, and $5HT_{1A}$ receptors lead to occurrence of psychosensory effects where $5HT_{1A}$ and $5HT_{2C}$ receptors be the significant modulators (Nichols, 2004). Several intracellular signaling pathways get involved on the activation of $5HT_{2A}$ receptors. Inositol triphosphate–diacylglycerol pathway gets activated by the Gq-mediated signaling, leading to activation of the protein kinase C (Garcia et al., 2007). Signaling through the GPCRs cause the expression of several genes such as egr-1 and egr-2 which are necessary in producing the psychotropic effects of LSD (Gonzalez-Maeso et al., 2007). LSD-induced $5HT_{2A}$ receptors activation leads to the breakdown of hippocampal prefrontal cortex. This reduces brain activity. LSD also effects the expression of BDNF. LSD-induced neuroplastic changes are the basis of persistent behavior changes. No FDA approved LSD assisted therapies exist today (Das et al., 2016).

Continuous ingestion of LSD is almost impossible as the "good trip" or high remains for a longer time. The dependence on LSD is not due to cravings or any physical effects but only due to personal need and psychological aspects (Dolder et al., 2017).

## 20.6 FUTURE OPPORTUNITIES AND CHALLENGES

A great number of studies have depicted that genes have transcriptional potential which play a role in regulation and are connected in the malfunctioning of brain due to abusive drugs. The mechanisms of transcriptional and epigenetic regulation are varied and complex. Thus, vast number of regulatory events has to be researched and studied upon in the upcoming years to be more clear of the regulatory mechanisms. The existing literature on epigenetic changes and transcriptional regulations in case of drug addiction is not enough in different key aspects. Till date different studies have used the conditioned place preference and locomotors sensitization paradigms (Xu et al., 2016). Though the behavioral assays seem to provide more meaningful insight in to an animal's sensitivity toward the activity of abusive/addictive drugs on the circuit; however, they do not provide direct measures of drug addiction. Work is also needed to conduct beyond the short time span of most recent experiments to examine transcriptional and epigenetic endpoints. Thus, experiment needs to be performed after longer periods of drug exposure and longer periods of

withdrawal from drug exposure. Future studies will incorporate more experimental paradigms which will be an explanatory approach and a better model for describing human addiction.

Induction of adaptive processes needs to be more clearly understood so that more of novel pharmacological treatment strategies can be discovered for more effective responses. Since the activation of dopamine system is found to be a vital factor for development of addiction, it makes it necessary for the researchers to find some strategies to inhibit the mesocorticolimbic DA system. These drugs produce different types of events within individual neurons, which may likely end up in influencing the behavior. Thus, a systems biology approach will be probably important to understand the biological phenomenon of addiction.

## 20.7 CONCLUSION

Based upon the findings of different studies revealed so far, it can be proposed that the precise mechanism of action of addiction and abuse has still not been clearly depicted. As most of the drugs are known to influence the dopamine system through a common and single process, thus this neurotransmitter serves as the basis of most of the recent theories and hypothesis of drug addictions and abuses. Many biochemical and molecular biology studies conducted to elucidate the basic mechanism of drug addiction are clinically very important, as it may help in designing strategies against drug abuse and addictions. Apart from molecular and biochemical studies, a better understanding of the neurobiological mechanisms underlying the addictive actions of drugs of abuse and of the genetic factors that contribute to drug addiction is also imperative for developing therapeutic agents that prevent/reverse the actions of the drugs affecting the neuronal wiring or targeting specific neurons. The designing of such novel therapeutic molecules will serve as a revolutionary step in our battle against drug addiction.

## KEYWORDS

- **addiction**
- **abusive drugs**
- **recreational drugs**
- **brain circuit**
- **opioids**

## REFERENCES

Aghajanian, G. K. Tolerance of Locus Coeruleus Neurons to Morphine and Suppression of Withdrawal Response by Clonidine. *Nature* **1978**, *267*, 186–188.

Ali, S. F.; Martin, J. L.; Black, M. D.; Itzhak, Y. Neuroprotective Role of Melatonin in Methamphetamineand1-methyl-4-phenyl-1,2,3,6-tetrahydropyridine-induceddopaminergic Neurotoxicity. *Ann. NY Acad. Sci.* **1999**, *890*, 119.

Alreja, M.; Aghajanian, G. K. Pacemaker Activity of Locus Coeruleus Neurons: Whole-Cell Recordings In Brain Slices Show Dependence on cAMP and Protein Kinase A. *Brain Res.* **1991**, *556*, 339–343.

Altarejos, J. Y.; Montminy, M. CREB and the CRTC co-activators: Sensors For Hormonal And Metabolic Signals. *Nat. Rev. Mol. Cell Biol.* **2011**, *12*, 141–151.

Amara, S. G.; Sonders, M. S. Neurotransmitter Transporters as Molecular Targets for Addictive Drugs. *Drug and Alcohol Dependence* **1998**, *51*, 87–96.

American Addiction Centers. Statistics on Drug Addiction. https://americanaddictioncenters.org/rehab-guide/addiction-statistics/ (Accessed June 23, 2018).

Andrade, R.; VanderMaelen, C. P.; Aghajanian, G. K. Morphine Tolerance and Dependence in the Locus Coeruleus: Single Cell Studies in Brain Slices. *Eur. J. Pharmacol.* **1983**, *91*, 161—169.

Bartu, A.; Freeman, N. C.; Gawthorne, G. S.; Codde, J. P.; Holman, C. D. Mortality in a Cohort Of Opiate And Amphetamine Users in Perth, Western Australia. *Addiction* **2004**, *99*, 53–60.

Berezikov, E. Evolution of Micro RNA Diversity and Regulation in Animals. *Nat. Rev. Genet.* **2011**, *12*, 846–860.

Bertran-Gonzalez, J.; Bosch, C.; Maroteaux, M.; Matamales, M.; Herve, D.; Valjent, E.; Girault, J. Opposing Patterns Of Signaling Activation In Dopamine D1 And D2 Receptor-Expressing Striatal Neurons In Response To Cocaine And Haloperidol. *J. Neurosci.* **2008**, *28*, 5671–5685.

Betke, K.; Wells, C.; Hamm, H. GPCR Mediated Regulation of Synaptic Transmission. *Prog. Neurobiol.* **2012**, *96*, 304–321.

Böhme, I.; Beck-Sickinger, A. Illuminating the Life of Gpcrs. *Cell Commun. Signal* **2009**, *7*, 16.

Bönisch, H.; Brüss, M. The Norepinephrine Transporter in Physiology and Disease. *Handb Exp. Pharmacol.* **2006**, *175*, 485–524.

Brami-Cherrier, K.; Roze, E.; Girault, J.; Betuing, S.; Caboche, J. Role of the ERK/MSK1 Signaling Pathway In Chromatin Remodelling And Brain Responses To Drugs Of Abuse. *J. Neurochem.* **2009**, *108*, 1323–1335.

Brodie, M. S.; Dunwiddie, T. V. Cocaine Effects in the Ventral Tegmental Area: Evidence for an Indirect Dopaminergic Mechanism of Action. *Naunyn Schmiedebergs Arch Pharmacol.* **1990**, *342*, 660–665.

Brown, E. E.; Fibiger, H. C. Differential Effects of Excitotoxic Lesions of the Amygdala on Cocaine-Induced Conditioned Locomotion And Conditioned Place Preference. *Psychopharmacology (Berl).* **1993**, *113*, 123–130.

Brown, T. E.; Lee B. R.; Mu, P. A Silent Synapse-Based Mechanism For Cocaine-Induced Locomotor Sensitization. *J. Neurosci.* **2011**, *31*, 8163–8174.

Calipari, E.; Ferris, M. Amphetamine Mechanisms and Actions at the Dopamine Terminal Revisited. *J. Neurosci.* **2013**, *33*, 8923–8925.

Carlezon, W. A. Jr.; Duman, R. S.; Nestler, E. J. The Many Faces of CREB. *Trends Neurosci.* **2005**, *28*, 436–445.

Cassel, S. Fluoxetine And Cocaine Induce The Epigenetic Factors Mecp2 And MBD1 In Adult Rat Brain. *Mol. Pharmacol.* **2006**, *70*, 487–492.

Cedar, H.; Bergman, Y. Linking DNA Methylation and Histone Modification: Patterns and Paradigms. *Nat. Rev. Genet.* **2009**, *10*, 295–304.

Chandrasekar, V.; Dreyer, J. Regulation of Mir-124, Let-7D and Mir-181A in the Accumbens Affects the Expression, Extinction and Reinstatement of

Cocaine-Induced Conditioned Place Preference. *Neuropsychopharmacology* **2011**, *36*, 1149–1164.

Chen, L. Chronic Ethanol Feeding Impairs AMPK and MEF2 Expression and Is Associated with GLUT4 Decrease in Rat Myocardium. *Exp. Mol. Med.* **2010**, *42*, 205–215.

Chen, R.; Tilley, M. R.; Wei, H.; Zhou, F.; Zhou, F. M. Abolished Cocaine Reward in Mice with a Cocaine-insensitive Dopamine Transporter. *Proc. Natl. Acad. Sci. U. S. A.* **2006**, *103*, 9333–9338.

Childress, A. R.; Mozley, P. D.; McElgin, W.; Fitzgerald, J.; Reivich, M.; O'Brien, C.P. Limbic Activation During Cue-induced Cocaine Craving. *Am. J. Psychiatry.* **1999**, *156*, 11–18.

Christoffel, D. J.; IkappaB Kinase Regulates Social Defeat Stress-induced Synaptic and Behaviouralplasticity. *J. Neurosci.* **2011**, *31*, 314–321.

Cooper, Z. D.; Haney, M. Actions of delta-9-tetrahydrocannabinol in cannabis: Relation to Use, Abuse, Dependence. *Int. Rev. Psychiatry* **2009**, *21* (2), 104–112.

Cornish, J. L.; Duffy, P.; Kalivas, P. W. A Role for Nucleus Accumbens Glutamate Transmission in the Relapse to Cocaine-seeking Behavior. *Neuroscience* **1999**, *93*, 1359–1367.

Cruickshank, C. C.; Dyer, K. R. A Review of the Clinical Pharmacology of Methamphetamine. *Addiction* **2009**, *104* (7), 1085–1099.

Cummings, K. M.; Hyland, A. Impact of Nicotine Replacement Therapy on Smoking Behavior.
*Annu. Rev. Public Health* **2005**, *26*, 583–599.

Cushman, P.; Dole, V. P. Detoxification of Rehabilitated Methadone-maintained Patients. *JAMA* **1973**, *226* (7), 747–752.

Das, S.; Barnwal, P.; Ramasamy, A.; Sen, S.; Mondal, S. Lysergic Acid Diethylamide: A Drug of 'Use'? *Ther. Adv. Psychopharmacol.* **2016**, *6* (3), 214–228.

Day, E.; Ison, J.; Keaney, F. A National Survey of Inpatient Drug Services in England. London: NTA, 2005.

Deng, J.; Rodriguiz, R.; Hutchinson, A.; Kim, I.; Wetsel, W.; West, A. Mecp2 in the Nucleus Accumbens Contributes to Neural and Behavioral Responses to Psychostimulants. *Nat. Neurosci.* **2010**, *13*, 1128–1136.

Dolder, P. C.; Schmid, Y.; Steuer, A. E.; Kraemer, T.; Rentsch, K. M.; Hammann, F.; Liechti, M. E. Pharmacokinetics and Pharmacodynamics of Lysergic Acid Diethylamide in Healthy Subjects. *Clin. Pharmacokinet.* **2017**, *56* (10), 1219–1230.

Drug Abuse Statistics. https://drugabuse.com/library/drug-abuse-statistics/ (accessed Apr 12, 2018).

Edwards, S.; Graham, D. L.; Bachtell, R. K.; Self, D. W. Region-specific Tolerance to Cocaine-regulated cAMP-dependent Protein Phosphorylation Following Chronic Self-administration. *Eur. J. Neurosci.* **2007**, *25*, 2201–2213.

Fagen, Z. M.; Mansvelder, H. D.; Keath, J. R.; McGehee, D. S. Short- and Long-term Modulation of Synaptic Inputs to Brain Reward Areas by Nicotine. *Ann. N Y Acad. Sci.* **2003**, *1003*, 185–195.

Feng, J.; Zhou, Y.; Campbell, S.; Le, T.; Li, E.; Sweatt, J.; Silva, A.; Fan, G. Dnmt1 And Dnmt3a Maintain DNA Methylation and Regulate Synaptic Function in Adult Forebrain Neurons. *Nat. Neurosci.* **2010**, *13*, 423–430.

Filip, M.; Golda, A.; Zaniewska, M.; McCreary, A.; Nowak, E.; Kolasiewicz, W.; Przegaliński, E. Involvement of Cannabinoid CB1 Receptors in Drug Addiction: Effects of Rimonabant on Behavioral Responses Induced by Cocaine. *Pharmacol. Rep.* **2006**, *58*, 806–819.

Fischle, W.; Tseng, B.; Dormann, H.; Ueberheide, B.; Garcia, B.; Shabanowitz, J.; Hunt, D.; Funabiki, H.; Allis, C. Regulation of HP1–chromatin Binding by Histone H3 Methylation and Phosphorylation. *Nature* **2005**, *438*, 1116–1122.

Fischman, M. W.; Schuster, C. R.; Javaid, J.; Hatano, Y.; Davis, J. Acute Tolerance Development to the Cardiovascular and Subjective Effects of Cocaine. *J. Pharmacol. Exp. Ther.* **1985**, *235*, 677–682.

Floor, E.; Leventhal, P. S.; Wang, Y.; Meng, L.; Chen, W. Dynamic Storage of Dopamine in Rat Brain Synaptic Vesicles In Vitro. *J. Neurochem.* **1995**, *64*, 689–699.

Floor, E.; Meng, L. Amphetamine Releases Dopamine from Synaptic Vesicles by Dual Mechanisms. *Neurosci. Lett.* **1996**, *215*, 53–56.

Franklin, T. R.; Druhan, J. P. Expression of Fos-related Antigens in the Nucleus Accumbens and Associated Regions Following Exposure to a Cocaine-paired Environment. *Eur. J. Neurosci.* **2000**, *12*, 2097–2106.

Fratta, W.; Fattore, L. Molecular Mechanisms of Cannabinoid Addiction. *Curr. Opin. Neurobiol.* 2013, *23*, 1–6.

Fuks, F.; Hurd, P.; Wolf, D.; Nan, X.; Bird, A.; Kouzarides, T. The Methyl-CpG-binding Protein MeCP2 Links DNA Methylation to Histone Methylation. *J. Biol. Chem.* **2002**, *278*, 4035–4040.

Garavan, H.; Pankiewicz, J.; Bloom, A.; Cho, J-K.; Sperry, L.; Ross, T. J.; Salmeron, B. J.; Risinger, R.; Kelley, D.; Stein, E. A. Cue-induced Cocaine Craving: Neuroanatomical Specificity for Drug Users and Drug Stimuli. *Am. J. Psychiatry* **2000**, *157*, 1789–1798.

Garcia, E.; Smith, R.; Sanders-Bush, E. Role of G(q) Protein in Behavioral Effects of the Hallucinogenic Drug 1-(2,5-dimethoxy-4-iodophenyl)-2-aminopropane. *Neuropharmacology* **2007**, *52*, 1671–1677.

Goeders, N. E.; Smith, J. E. Reinforcing Properties of Cocaine in the Medial Prefrontal Cortex: Primary Action on Presynaptic Dopaminergic Terminals. *Pharmacol. Biochem. Behav.* **1986**, *25*, 191–199.

Goldstein, R. Z.; Volkow, N. D. Drug Addiction and Its Underlying Neurobiological Basis: Neuroimaging Evidence for the Involvement of the Frontal Cortex. *Am. J. Psychiatry* **2002**, *159*, 1642–1652.

González, B.; Jayanthi, S.; Gomez, N.; Torres, O.; Sosa, M.; Bernardi, A.; Urbano, F.; García-Rill, E.; Cadet, J.; Bisagno, V. Repeated Methamphetamine and Modafinil Induce Differential Cognitive Effects and Specific Histone Acetylation and DNA Methylation Profiles in the Mouse Medial Prefrontal Cortex. *Prog. Neuro-Psychopharmacol. Biol. Psychiatry* **2018**, *82*, 1–11.

Gonzalez-Maeso, J.; Weisstaub, N.; Zhou, M.; Chan, P.; Ivic, L.; Ang, R. Hallucinogens Recruit Specific Cortical 5-HT(2A) Receptor-mediated Signaling Pathways to Affect Behavior. *Neuron* **2007**, *53*, 439–452.

Grant, S.; London, E. D.; Newlin, D. B.; Villemagne, V. L.; Liu, X.; Contoreggi, C.; Phillips, R. L.; Kimes, A. S.; Margolin, A. Activation of Memory Circuits During Cue-elicited Cocaine Craving. *Proc. Natl. Acad. Sci. U. S. A.* **1996**, *93*, 12040–12045.

Grueter, B. A.; Robison, A. J.; Neve, R. L.; Nestler, E. J.; Malenka, R. C. ΔFosB Differentially Modulates Nucleus Accumbens Direct and Indirect Pathway Function. *Proc. Natl. Acad. Sci. U. S. A.* **2013**, *110*, 1923–1928.

Gupta, S.; Kulhara, P. Cellular and Molecular Mechanisms of Drug Dependence. *Ind. J. Psychiatry.* **2007**, *49* (2), 85–90.

Hao, Y.; Martin-Fardon, R.; Weiss, F. Behavioral and Functional Evidence of Metabotropic Glutamate Receptor 2/3 and Metabotropic Glutamate Receptor 5 Dysregulation in Cocaine-escalated Rats: Factor in the Transition to Dependence. *Biol. Psychiatry.* **2010**, *68*, 240–248.

He, Y.; Yang, C.; Kirkmire, C.; Wang, Z. Regulation of Opioid Tolerance by Let-7 Family Microrna Targeting the Opioid Receptor. *J. Neurosci.* **2010**, *30*, 10251–10258.

Hedden, S.; Kennet, J.; Lipari, R.; Medley, G.; Tice, P. Behavioral Health Trends In The United States: Results From The 2014 National Survey On Drug Use And Health, 2015.

Heller, E.; Cates, H.; Peña, C.; Sun, H.; Shao, N.; Feng, J.; Golden, S.; Herman, J.; Walsh, J.; Mazei-Robison, M. et al. Locus-specific Epigenetic Remodeling Controls Addiction- and Depression-related Behaviors. *Nat. Neurosci.* **2014**, *17*, 1720–1727.

Henningfield, J. E.; Fant, R. V.; Buchhalter, A. R.; Stitzer, M. L. Pharmacotherapy for Nicotine Dependence. *CA Cancer J. Clin.* **2005**, *55* (5), 281–299.

Hodgins, D. C.; el-Guebaly, N.; Armstrong, S. Prospective and Retrospective Reports of Mood States Before Relapse to Substanceuse. *J. Consult. Clin. Psychol.* **1995**, *63*, 400–407.

Holliday, R. Epigenetics: A Historical Overview. *Epigenetics* **2006**, *1*, 76–80.

Host, L.; Dietrich, J.; Carouge, D.; Aunis, D.; Zwiller, J. Cocaine Self-administration Alters the Expression of Chromatin-remodelling Proteins; Modulation by Histone Deacetylase Inhibition. *J. Psychopharmacol.* **2009**, *25*, 222–229.

Huang, W.; Li, M. Nicotine Modulates Expression of Mir-140*, Which Targets the 3'-Untranslated Region of Dynamin 1 Gene (Dnm1). *Int. J. Neuropsychopharmacol.* **2008**, *12*, 537.

Hughes, F. B.; Shore, P. A.; Brodie, B. B. Serotonin Storage Mechanism and Its Interaction with Reserpine. *Experientia* **1958**, *14*, 178–180.

Hughes, F. B.; Brodie, B. B. The Mechanism of Serotonin and Catecholamine Uptake by Platelets. *J. Pharmacol. Exp. Ther.* **1959**, *127*, 96–102.

Hurd, Y. L.; Ungerstedt, U. Cocaine: An In Vivo Micro Dialysis Evaluation of Its Acute Action on Dopamine Transmission in Rat Striatum. *Synapse* **1989**, *3*, 48–54.

Hyman, S. E.; Malenka, R. C.; Nestler, E. J. Neural Mechanisms of Addiction: The Role of Reward-related Learning and Memory. *Annu. Rev. Neurosci.* **2006**, *29*, 565–598.

Im, H.; Hollander, J.; Bali, P.; Kenny, P. Mecp2 Controls BDNF Expression and Cocaine Intake Through Homeostatic Interactions with Microrna-212. *Nat. Neurosci.* **2010**, *13*, 1120–1127.

Ingram, S. L.; Prasad, B. M.; Amara, S. G. Dopamine Transporter-mediated Conductance Increase Excitability of Midbrain Dopamineneurons. *NatNeurosci.* **2002**, *5*, 971–978.

Johanson, C. E.; Fischman, M. W. The Pharmacology of Cocaine Related to Its Abuse. *Pharmacol. Rev.* **1989**, *41*, 3–52.

Johnson, S. W.; North, R. A. Opioids Excite Dopamine Neurons by Hyperpolarization of Local Interneurons. *J. Neurosci.* **1992**, *12*, 483–488.

Johnson, B. A.; Ait-Daoud, N.; Bowden, C. L.; DiClemente, C. C.; Roache, J. D.; Lawson, K.; Javors, M. A. Oral

Topiramate for Treatment of Alcohol Dependence: A Randomised Controlled Trial. *The Lancet* **2003**, *361*, 1677–1685.

Jones, P. The Role of DNA Methylation in Mammalian Epigenetics. *Science* **2001**, *293*, 1068–1070.

Kalda, A.; Heidmets, L.; Shen, H.; Zharkovsky, A.; Chen, J. Histone Deacetylase Inhibitors Modulates the Induction and Expression of Amphetamine-induced Behavioral Sensitization Partially Through an Associated Learning of the Environment in Mice. *Behav. Brain Res.* **2007**, *181*, 76–84.

Kalivas, P. W.; O'Brien, C. Drug Addiction as a Pathology of Staged Neuroplasticity. *Neuropsychopharmacology* **2008**, *33*, 166–180.

Kampman, K. M.; Volpicelli, J. R.; Alterman, A. I.; Cornish, J.; O'Brien, C. P. Amantadine in the Treatment of Cocaine-dependent Patients with Severe Withdrawal Symptoms. *Am. J. Psychiatry.* **2000**, *157*, 2052–2054.

Kandel, D. B.; Yamaguchi, K.; Klein, L. C. Testing the Gateway Hypothesis. *Addiction* **2006**, *101*, 470–472.

Kelz, M. B.; Chen, J.; Carlezon, W. A. Jr. Expression of the Transcription Factor ΔFosB in the Brain Controls Sensitivity to Cocaine. *Nature* **1999**, *401*, 272–276.

Kendler, K.; Jacobson, K.; Prescott, C.; Neale, M. Specificity of Genetic and Environmental Risk Factors for Use and Abuse/Dependence of Cannabis, Cocaine, Hallucinogens, Sedatives, Stimulants, and Opiates in Male Twins *Am. J. Psychiatry.* **2003**, *160*, 687–695.

Kim, W.; Kim, S.; Kim, J. Chronic Microinjection of Valproic Acid into the Nucleus Accumbens Attenuates Amphetamine-induced Locomotor Activity. *Neurosci. Lett.* **2008**, *432*, 54–57.

Kiselyov, K.; Shin, D.; Muallem, S. Signaling Specificity in GPCR-dependent Ca2+ Signaling. *Cell. Signal.* **2003**, *15*, 243–253.

Klose, R.; Bird, A. Genomic DNA Methylation: The Mark and Its Mediators. *Trends Biochem. Sci.* **2006**, *31*, 89–97.

Koob, G.; Moal, M. Drug Abuse: Hedonic Homeostatic Dysregulation. *Science* **1997**, *278*, 52–58.

Koob, G. F.; LeMoal, M. Drug Addiction, Dysregulation of Reward, and Allostasis. *Neuropsychopharmacology* **2001**, *24*, 97–129.

Koob, G. F.; Roberts, A. J.; Schulteis, G.; Parsons, L. H.; Heyser, C. J. Neuro Circuitry Targets in Ethanol Reward and Dependence. *Alcohol Clin Exp Res.* **1998**, *22*, 3–9.

Kreek, M. J.; Zhou, Y.; Butelman, E. R. Opiate Addiction and Cocaine Addiction: Underlying Molecular Neurobiology and Genetics. *J. Clin. Invest.* **2012**, *122* (10), 3387–3393.

LaPlant, Q.; Vialou, V.; Covington, H.; Dumitriu, D.; Feng, J.; Warren, B.; Maze, I.; Dietz, D.; Watts, E.; Iñiguez, S. et al. Dnmt3a Regulates Emotional Behavior and Spine Plasticity in the Nucleus Accumbens. *Nat. Neurosci.* **2010**, *13*, 1137–1143.

Le Foll, B. Cannabinoid CB1 Receptor Antagonists as Promising New Medications for Drug Dependence. *J. Pharmacol. Exp. Ther.* **2004**, *312*, 875–883.

Li, J.; Li, J.; Liu, X.; Qin, S.; Guan, Y.; Liu, Y.; Cheng, Y.; Chen, X.; Li, W.; Wang, S. et al. Microrna Expression Profile and Functional Analysis Reveal That Mir-382 is a Critical Novel Gene of Alcohol Addiction. *EMBO Mol. Med.* **2013**, *5*, 1402–1414.

Lichtman, A. H.; Martin, B. R. Cannabinoid Tolerance and Dependence. *HEP* **2005**, *168*, 691–717.

Liechti, M.; Markou, A. Interactive Effects of the Mglu5 Receptor Antagonist MPEP and the Mglu2/3 Receptor Antagonist LY341495 on Nicotine Self-administration and Reward Deficits Associated with Nicotine Withdrawal in Rats. *Eur. J. Pharmacol.* **2007**, *554*, 164–174.

Loh, E. A.; Roberts, D. C. Break-points on a Progressive Ratio Schedule Reinforced by Intravenous Cocaine Increase Following Depletion of Forebrain Serotonin. *Psychopharmacology (Berl).* **1990**, *101*, 262–266.

Luger, K.; Mäder, A.; Richmond, R.; Sargent, D.; Richmond, T. Crystal Structure of the Nucleosome Core Particle at 2.8 Å Resolution. *Nature* **1997**, *389*, 251–260.

Luscher, C.; Jan, L. Y.; Stoffel, M.; Malenka, R. C.; Nicoll, R. A. G Lüscher, C.; Ungless, M. The Mechanistic Classification of Addictive Drugs. *PLOS Med.* **2006**, *3*, e437.

Luscher, C.; Jan, L. Y.; Stoffel, M.; Malenka, R. C.; Nicoll, R. A. G Protein-coupled Inwardly Rectifying K+ Channels (Girks) Mediate Postsynaptic But Not Presynaptic Transmitter Actions in Hippocampal Neurons. *Neuron* **1997**, *19*, 687–695.

Maas, L. C.; Lukas, S. E.; Kaufman, M. J.; Weiss, R. D.; Daniels, S. L.; Rogers V. W.; Kukes, T. J.; Renshaw, P. F. Functional Magnetic Resonance Imaging of Human Brain Activation During Cue-induced Cocaine Craving. *Am. J. Psychiatry.* **1998**, *155*, 124–126.

Mackie, K. Mechanisms of CB1 Receptor Signaling: Endocannabinoid Modulation of Synaptic Strength. *Int. J. Obes.* **2006**, *30*, S19–S23.

Maldonado R. Participation of Noradrenergic Pathways in the Expression of Opiate Withdrawal: Biochemical and Pharmacological Evidence. *Neurosci. Biobehav. Rev.* **1997**, *21*, 91–104.

Malvaez, M.; McQuown, S.; Rogge, G.; Astarabadi, M.; Jacques, V.; Carreiro, S.; Rusche, J.; Wood, M. HDAC3-Selective Inhibitor Enhances Extinction of Cocaine-seeking Behavior in a Persistent Manner. *Proc. Natl. Acad. Sci.* **2013,** *110,* 2647–2652.

Majchrowicz, E. Induction of Physical Dependence Upon Ethanol and the Associated Behavioral Changes in Rats. *Psychopharmacologia* **1975,** *43* (3), 245–254.

Mansvelder, H. D.; Keath, J. R.; McGehee, D. S. Synaptic Mechanisms Underlie Nicotine Induced Excitability of Brain Reward Areas. *Neuron* **2002,** *33,* 905–919.

Maze, I.; Covington, H.; Dietz, D.; LaPlant, Q.; Renthal, W.; Russo, S.; Mechanic, M.; Mouzon, E.; Neve, R.; Haggarty, S. et al. Essential Role of the Histone Methyltransferase G9a in Cocaine-induced Plasticity. *Science* **2010,** *327,* 213–216.

Mazei-Robison, M. S.; Koo, J. W.; Friedman, A. K. Role for mTOR signaling and Neuronal Activity in Morphine-induced Adaptations in Ventral Tegmental Area Dopamine Neurons. *Neuron* **2011,** *72,* 977–990.

McLaughlin, I.; Dani, J. A.; De Biasi, M. Nicotine Withdrawal. *Curr. Top Behav. Neurosci.* **2015,** *24,* 99–123.

Meffert, M. K.; Chang, J. M.; Wiltgen, B. J.; Fanselow, M. S.; Baltimore, D. NF-kappa B Functions in Synaptic Signaling and Behavior. *Nat. Neurosci.* **2003,** *6,* 1072–1078.

Meil, W. M. Lesions of the Basolateral Amygdala Abolish the Ability of Drug Associated Cues to Reinstate Responding During Withdrawal from Self Administered Cocaine. *Behav. Brain. Res.* **1997,** *87,* 139–148.

Miller, E. K.; Cohen, J. D. An Integrative Theory of Prefrontal Cortex Function. *Annu. Rev. Neurosci.* **2001,** *24,* 167–202.

Moore, L.; Le, T.; Fan, G. DNA Methylation and Its Basic Function. *Neuropsychopharmacology* **2012,** *38,* 23–38.

Morrow, A. L. Regulation of GABA Receptor Function and Gene Expression in the Central Nervous System. *Int. Rev. Neurobiol.* **1995,** *38,* 1–41.

Narlikar, G.; Fan, H.; Kingston, R. Cooperation Between Complexes That Regulate Chromatin Structure and Transcription. *Cell* **2002,** *108,* 475–487.

Navarro, G.; Moreno, E.; Bonaventura, J.; Brugarolas, M.; Farré, D.; Aguinaga, D.; Mallol, J.; Cortés, A.; Casadó, V.; Lluís, C. et al. Cocaine Inhibits Dopamine D2 Receptor Signaling Via Sigma-1-D2 Receptor Heteromers. *PLoS One* **2013,** *8,* e61245.

Nestler, E. Epigenetic Mechanisms of Drug Addiction. *Neuropharmacology* **2014,** *76,* 259–268.

Nestler, E. J.; Barrot, M.; Self, D. W. ΔFosB: A Sustained Molecular Switch for Addiction. *PNAS* **2001,** *98* (20), 11042–11046.

Nestler, E. J. Common Molecular and Cellular Substrates of Addiction and Memory. *Neurobiol. Learn Memory* **2002,** *78,* 637–647.

Nestler, E. J. Molecular Mechanisms of Drug Addiction. *J. Neurosci.* **1992,** *12* (7), 2439–2450.

Nestler, E. The Neurobiology of Cocaine Addiction. *Sci. Pract. Perspect.* **2005,** *3,* 4–10.

Nestler, E. J. Transcriptional Mechanisms of Addiction: Role of Delta FosB. *Philos. Trans. R Soc. London B Biol. Sci.* **2008,** *363,* 3245–3255.

Nestler, E. J. Cellular Basis of Memory for Addiction. *Dialogues Clin. Neurosci.* **2013,** *15,* 431–443.

Nichols, D. Hallucinogens. *Pharmacol. Ther.* **2004,** *101,* 131–181.

Niswender, C.; Conn, P. Metabotropic Glutamate Receptors: Physiology, Pharmacology, and Disease. *Annu. Rev. Pharmacol. Toxicol.* **2010,** *50,* 295–322.

Pickel, V. M.; Garzon, M.; Mengual, E. Electron Microscopic Immunolabeling of Transporters and Receptors Identifies Transmitter-specific Functional Sites Envisioned in Cajal's Neuron. *Prog. Brain Res.* **2002,** *136,* 145–155.

Picciotto, M. R.; Zoli, M.; Rimondini, R.; Lena, C.; Marubio, L. M. Acetylcholine Receptors Containing the Beta2 Subunit are Involved in the Reinforcing Properties of Nicotine. *Nature* **1998,** *391,* 173–177.

Pidoplichko, V. I.; DeBiasi, M.; Williams, J. T.; Dani, J. A. Nicotine Activates and Desensitizes Midbrain Dopamine Neurons. *Nature* **1997,** *390,* 401–404.

Pietrzykowski, A.; Friesen, R.; Martin, G.; Puig, S.; Nowak, C.; Wynne, P.; Siegelmann, H.; Treistman, S. Posttranscriptional Regulation of BK Channel Splice Variant Stability by Mir-9 Underlies Neuroadaptation to Alcohol. *Neuron* **2008,** *59,* 274–287.

Pulipparacharuvil, S. Cocaine Regulates MEF2 to Control Synaptic and Behavioral Plasticity. *Neuron* **2008,** *59,* 621–633.

Renthal, W.; Kumar, A.; Xiao, G.; Wilkinson, M.; Covington, H.; Maze, I.; Sikder, D.; Robison, A.; LaPlant, Q.; Dietz, D. et al. Genome-wide Analysis of Chromatin Regulation by Cocaine Reveals a Role for Sirtuins. *Neuron* **2009,** *62,* 335–348.

Rinn, J.; Chang, H. Genome Regulation by Long Noncoding Rnas. *Annu. Rev. Biochem.* **2012,** *81,* 145–166.

Ritz, M. C.; Lamb, R. J.; Goldberg, S. R.; Kuhar, M. J. Cocaine Receptors on Dopamine Transporters are

Related to Self-administration of Cocaine. *Science* **1987**, *237*, 1219–1223.

Robison, A. J.; Vialou, V.; Mazei-Robison, M. Behavioral and Structural Responses to Chronic Cocaine Require a Feed-forward Loop Involving ΔFosb And CaMKII in the Nucleus Accumbens Shell. *J. Neurosci.* **2013**, *33*, 4295–4307.

Robison, A. J.; Nestler, E. J. Transcriptional and Epigenetic Mechanisms of Addiction. *Nat. Rev. Neurosci.* **2012**, *12* (11), 623–637.

Rocha, B. A. Stimulant and Reinforcing Effects of Cocaine in Monoamine Transporter Knockout Mice. *Eur. J. Pharmacol.* **2003**, *479*, 107–115.

Rocha, B. A.; Fumagalli, F.; Gainetdinov, R. R.; Jones, S. R.; Ator, R. Cocaine Self Administration in Dopamine-transporter Knock Out Mice. *Nat. Neurosci.* **1998**, *1*, 132–137.

Rosenkranz, J. A.; Grace, A. A. Dopamine Attenuates Prefrontal Cortical Suppression of Sensory Inputs to the Basolateral Amygdala of Rats. *J Neurosci.* **2001**, *21*, 4090–4103.

Russo, S. J. Nuclear Factor Kappa B Signaling Regulates Neuronal Morphology and Cocaine Reward. *J Neurosci.* **2009**, *29*, 3529–3537.

Russo, S. J.; Dietz, D. M.; Dumitriu, D; Morrison, J. H.; Malenka, R. C.; Nestler, E. J. The Addicted Synapse: Mechanisms of Synaptic and Structural Plasticity in Nucleus Accumbens. *Trends Neurosci.* **2010**, *33*, 267–276.

Saba, R.; Storchel, P.; Aksoy-Aksel, A.; Kepura, F.; Lippi, G.; Plant, T.; Schratt, G. Dopamine-regulated Microrna Mir-181A Controls Glua2 Surface Expression in Hippocampal Neurons. *Mol. Cell. Biol.* **2011**, *32*, 619–632.

Sadakierska-Chudy, A.; Frankowska, M.; Miszkiel, J.; Wydra, K.; Jastrzębska, J.; Filip, M. Prolonged Induction of Mir-212/132 and REST Expression in Rat Striatum Following Cocaine Self-administration. *Mol. Neurobiol.* **2016**, *54*, 2241–2254.

Schroeder, F.; Penta, K.; Matevossian, A.; Jones, S.; Konradi, C.; Tapper, A.; Akbarian, S. Drug-induced Activation of Dopamine D1 Receptor Signaling and Inhibition of Class I/II Histone Deacetylase Induce Chromatin Remodeling in Reward Circuitry and Modulate Cocaine-related Behaviors. *Neuropsychopharmacology* **2008**, *33*, 2981–2992.

Shaw-Lutchman, S. Z.; Impey, S.; Storm, D.; Nestler, E. J. Regulation of CRE Mediated Transcription in Mouse Brain by Amphetamine. *Synapse* **2003**, *48*, 10–17.

Shaw-Lutchman, T. Z.; Barrot, M.; Wallace. Regional and Cellular Mapping of CRE-mediated Transcription During Naltrexone-precipitated Morphine Withdrawal. *J. Neurosci.* **2002**, *22*, 3663–3672.

Sidhpura, N.; Weiss, F.; Martin-Fardon, R. Effects of the Mglu2/3 Agonist LY379268 and the Mglu5 Antagonist MTEP on Ethanol Seeking and Reinforcement are Differentially Altered in Rats with a History of Ethanol Dependence. *Biol. Psychiatry.* **2010**, *67*, 804–811.

Sidhu, A. Coupling of D1 and D5 Dopamine Receptors to Multiple G Proteins. *Mol. Neurobiol.* **1998**, *16*, 125–134.

Siegel, S.; Hinson, R. E.; Krank, M. D. Heroin 'Overdose' Death: Contribution of Drug-associated Environmental Cues. *Science* **1982**, *216*, 436–437.

Silva, O. A.; Yonamine, M. Drug Abuse Among Workers in Brazilian Regions. *Rev. Saude Publica.* **2004**, *38*, 552–556.

Sullivan, P. F. Candidate Genes for Nicotine Dependence via Linkage, Epistasis, and Bioinformatics. *Am. J. Med. Genet. B Neuropsychiatr. Genet.* **2004**, *126B*, 23–36.

Sulzer, D.; Sonders, M. S.; Poulsen, N. W.; Galli, A. Mechanisms of Neurotransmitter Release by Amphetamines. *Prog.Neurobiol.* **2005**, *75*, 406–433.

Swift, R. M.; Davidson, D.; Whelihan, W.; Kuznetsov, O. Ondansetron Alters Human Alcohol Intoxication. *Biol Psychiatry.* **1996**, *40*, 514–521.

Taki, F.; Pan, X.; Zhang, B. Chronic Nicotine Exposure Systemically Alters Micro RNA Expression Profiles During Post-embryonic Stages in Caenorhabditis Elegans. *J. Cell. Physiol.* **2014**, *229*, 78–89.

Turner, B. Histone Acetylation and an Epigenetic Code. *Bioessays* **2000**, *22*, 836–845.

Uddin, M. S.; Sufian, M. A.; Kabir, M. T.; Hossain, M. F.; Nasrullah, M.; et al. Amphetamines: Potent Recreational Drug of Abuse. *J. Addict. Res. Ther.* **2017**, *8*, 1–12.

Vaughan, C. W.; Ingram, S. L.; Connor, M. A.; Christie, M. J. How Opioids Inhibit GABA-mediated Neurotransmission. *Nature* **1997**, *390*, 6+11–614.

Viola, T.; Wearick-Silva, L.; De Azeredo, L.; Centeno-Silva, A.; Murphy, C.; Marshall, P.; Li, X.; Singewald, N.; Garcia, F.; Bredy, T. et al. Increased Cocaine-induced Conditioned Place Preference During Peri adolescence in Maternally Separated Male BALB/C Mice: The Role of Cortical BDNF, Microrna-212, and Mecp2. *Psychopharmacology* **2016**, *233*, 3279–3288.

Vocci, F. J.; Appel, N. M. Approaches to the Development of Medications for the Treatment of Methamphetamine Dependence. *Addiction* **2007**, *102*, 96–106.

Volkow, N. D.; Wang, G. J.; Fowler, J. S.; Hitzemann, R.; Angrist, B.; Gatley, S. J.; Logan, J.; Ding, Y. S.; Pappas,

N. Association of Methylphenidate-induced Craving with Changes in Right Striato-Orbito Frontal Metabolism in Cocaine Abusers: Implications in Addiction. *Am. J. Psychiatr.* **1999,** *156,* 19–26.

Vrettou, M.; Granholm, L.; Todkar, A.; Nilsson, K.; Wallén-Mackenzie, Å.; Nylander, I.; Comasco, E. Ethanol Affects Limbic and Striatal Presynaptic Glutamatergic and DNA Methylation Gene Expression in Outbred Rats Exposed to Early-life Stress. *Addict. Biol.* **2015,** *22,* 369–380.

Wang, J. C.; Kapoor, M.; Goate, A. M. The Genetics of Substance Dependence. *Annu. Rev. Genomics Hum. Genet.* **2012,** *13,* 241–261.

Wang, L.; Lv, Z.; Hu, Z.; Sheng, J.; Hui, B.; Sun, J.; Ma, L. Chronic Cocaine-induced H3 Acetylation and Transcriptional Activation of Camkiiα in the Nucleus Accumbens is Critical for Motivation for Drug Reinforcement. *Neuropsychopharmacology* **2009,** *35,* 913–928.

Wang, Y. Y.; Aghajanian, G. K. Excitation of Locus Coeruleus Neurons by Vasoactive Intestinal Peptide: Role of cAMP and Protein Kinase A. *J Neurosci.* **1990,** *10,* 3335–3343.

Wightman, R. M.; Heien, M. L.; Wassum, K. M.; Sombers, L. A.; Aragona, B. J.; Khan, A. S.; Ariansen, J. L; Cheer, J. F.; Phillips, P. E.; Carelli, R. M. Dopamine Release is Heterogeneous Within Microenvironments of the Rat Nucleus Accumbens. *Eur. J. Neurosci.* **2007,** *26,* 2046–2054.

Willner, P.; Muscat, R.; Papp, M. Chronic mild stress-induced anhedonia: a realistic animal model of depression. *Neurosci Bio behav Rev.* **1992**. 16, 525–534.

Wu, Q.; Hwang, C.; Zheng, H.; Wagley, Y.; Lin, H.; Kim, D.; Law, P.; Loh, H.; Wei, L. Microrna 339 Down-regulates M-opioid Receptor at the Post-transcriptional Level in Response to Opioid Treatment. *FASEB J.* **2013,** *27,* 522–535.

Xu, P.; Qiu, Y.; Zhang, Y.; Bai, Y.; Xu, P.; Liu, Y.; et al. The Effects of 4-methylethcathinone on Conditioned Place Preference, Locomotor Sensitization, and Anxiety-like Behavior: A Comparison with Methamphetamine. *Int. J. Neuropsychopharmacol.* **2016,** *19* (4), 120.

# CHAPTER 21

# Therapeutic Gases for Neurological Disorders

RACHANA*, TANYA GUPTA, SAUMYA YADAV, and MANISHA SINGH

*Jaypee Institute of Information Technology, Uttar Pradesh, India*

*Corresponding author. E-mail: rachana.dr@gmail.com*

## ABSTRACT

The neuroprotective properties for certain thera-peutic gases have been observed for decades, leading to extensive research that has been widely reported and continues to attract the scientific interests. Medically important gases include oxygen ($O_2$), carbon dioxide ($CO_2$), nitric oxide (NO), helium (He), etc., and have been supported by the evidence alluding to their use as potential neuroprotective agents. They are found to be useful in several neural disorders including traumatic brain injury, ischemic injury, autism, stroke, Alzheimer's diseases, and subarachnoid hemorrhage, etc. as they have various neuropro-tective effects. Until now, none of these has been widely used clinically except oxygen, followed by nitric oxide. The present chapter summarizes the therapeutic roles and mechanism of action of these four gases, and further discusses the possible hindrances and future perspectives.

## 21.1 INTRODUCTION

Gases have been used as pharmaceuticals, playing vital role in treating various disease conditions. Some of these gases are impor-tant physiological regulatory factors having antioxidant, anti-inflammatory, and anti-apop-totic protective properties, and have been shown to have pharmacological effects on various neurological disorders like multiple scle-rosis, traumatic brain injury, and aneurysmal subarachnoid hemorrhage, etc. (Nakao et al., 2009; Deng et al., 2014). The four therapeutic gases, oxygen, carbon dioxide, nitric oxide (NO), and helium (He) are important in various ways to treat several therapeutic conditions.

Oxygen plays a vital role in the develop-ment and growth of multicellular organisms. If we focus on its role for curing specifically the neurological disorders, then hyperbaric oxygen therapy (HBOT) is the emerging method for the treatment of various pathological condi-tions including traumatic brain injury, multiple sclerosis, stroke, and glioblastoma multiforme, etc. In case of carbon dioxide, therapeutic hypercapnia holds a promising approach to treat ischemic brain injury patients, highlighted later in the chapter (Brambrink and Orfanakis, 2010).

NO is naturally involved in various signaling pathways as a biological messenger or neurotransmitter (Garry et al., 2014). Its role as a neuroprotective agent depends on the production of NO from three different isoforms of nitric oxide synthase (NOS), that is, neuronal nitric oxide synthase, inducible nitric oxide synthase, and endothelial nitric oxide synthase (nNOS, iNOS, and eNOS) (Garry et al., 2014). The chapter discusses the potential of NO to manipulate the signaling pathway as a treatment for various brain injuries. At the end, neuroprotective effects of He on brain tissue have been discussed. Very few studies have been conducted for He and till yet its mechanism of action is not very well known (Berganza and Zhang, 2013). In this chapter, authors are focusing only on the available evidence of neurological aspects of He.

## 21.2   OXYGEN

Oxygen ($O_2$) is a colorless, odorless, tasteless, and transparent gas, which combines with all other elements except inert gases to form oxides. It was discovered by the Joseph Priestley of England and Carl Wilhelm Scheele of Germany. It was named as oxygen by Antoine Lavoisier in 1774 (Partington et al., 1989). It is the third most abundant element on the earth and is utilized by the body for oxidation and combustion processes. As a therapeutic gas, it was first experimented on the mouse by the Priestley who reported that the mice were more active and lived longer breathing it (Emsley et al., 2001).

The credit for the development of oxygen administration devices goes to the "oxygen box" for infants and young children, created by Dr. Julius Hess (Hess, 1934). Oxygen box is a special type of incubator which provides a stable thermal environment along with the oxygen administration. The only drawback of this box is the uncontrolled flow of oxygen in the incubator. Furthermore, advancement in science has led to the development of various oxygen administrating devices which not only reduces the death rates and proved to be essential for infants and children but also provides various

health benefits for several other conditions (Robertson, 2003; Saugstad, 1990). The rate of oxygen prescription, its accurateness, and consequent monitoring of the therapy has also been discussed by various researchers (Cook et al., 1996; Fitzgerald et al., 1988; Holbourn et al., 2014; Rudge et al., 2014; Small et al., 1992).

Inappropriate use of oxygen, especially in absence or insufficient humidification, can lead to sepsis, resulting due to the development of cracks in mucosa and drying of the same (Kopelman et al., 2003). If oxygen administration is not controlled while delivering, its higher concentrations can result in the aggravation of hypercapnia too (Bosson et al., 2014).

### 21.2.1   OXYGEN THERAPY

Oxygen therapy is a treatment in which supplemental oxygen is given to the patient in chronic or acute disorders. Normally, our lungs take the oxygen from the air we breathe but the patients suffering from severe diseases like asthma, pneumonia, chronic obstructive pulmonary disorder (COPD) require this gas as a therapy (Suzzane et al., 2005). It is also referred in case of hypoxemia, where the concentration of blood oxygen in the arteries goes below 60 mmHg, and tissue hypoxia occurs (Schreiner et al., 1982). Before delivering oxygen to the patients, one has to determine the dose, time duration, and pressure of oxygen by assessing the patients age, his/her need, technological capability, and matching technology (Walsh et al., 2017).

When mechanical support is also required to meet oxygen demand, then ventilators or resuscitators are used but during low dose requirements, nasal cannula and simple mask technique is sufficient for delivery (Lamb et al., 2016; Myers et al., 2002).

Oxygen therapy can be categorized on the basis of the use of oxygen. If oxygen is used at high percentage of normal atmospheric pressure, delivered via a mask connected to an oxygen cylinder, it is referred as normobaric oxygen therapy (NBOT), while when 100% oxygen is administered at pressures above one atmosphere in a specialized vessel it is known as HBOT (Bennet et al., 2015).

### 21.2.1.1 HYPERBARIC OXYGEN THERAPY

HBOT is a medication in which a patient breathes 100% oxygen from time to time, inside a chamber, in which pressure is higher than at sea level (Gessel et al., 2008). In this therapy, the diffusion of oxygen into the plasma is 10 times more than the normal concentration. Researchers started evaluating the clinical applications of HBOT since 1895 and by late 1950s there were sufficient evidences available regarding HBOT to work in various medical and surgical conditions. Experiments included use of oxygen in reversing the effects of effect on the carbon monoxide (CO) toxicosis in humans (Weaver, 2014), central nervous system (CNS) toxicity in cancer patients (Gray et al., 1953), and decompression illness (Gray et al., 1953; Moon, 2014), etc. Furthermore, various surgical operations were also been performed under hyperbaric oxygen environment (Kindwall, 2008).

In a report by Weaver (2014), HBOT was suggested as a good treatment for acute CO poisoning. While breathing hyperbaric oxygen, carboxyhemoglobin elimination was found to be hastened which in turn was helpful in regulating the inflammatory processes occurred during CO poisoning. Moreover, hyperbaric oxygen is also reported to improve mitochondrial function, inhibits lipid peroxidation, impairs leukocyte adhesion, and reduces brain inflammation (Weaver, 2014).

Another study on decompression illness reviewed that the treatment with HBOT immediately reduces the bubble volume which leads to an increase in the diffusion gradient for inert gas, and along with this, oxygenation of ischemic tissue increases and CNS edema decreases. HBOT is considered as the chief support for the treatment of this disease (Moon, 2014).

HBOT is given to the patients according to diseases and with a specific treatment protocol. In general, hyperbaric oxygen is given at 2.0–3.0 atmospheric absolute (ATA) for 90 min. To the patients suffering from clostridial infection of muscles or gas gangrene, oxygen is administrated with hyperbaric oxygen for three times a day at 3.0 ATA for the first 24 h, followed by twice per day for the next 2–5 days. If toxicity still persists, the treatment profile can be extended. For compartment syndrome, it is given twice at 2.0 to 2.4 ATA for 10 days (Gessel, 2008).

This therapy is also beneficial to cure malignancy as it increases the intra-tumor reactive oxygen species (ROS) level which not only induces the tumor destruction but also removes the hypoxic stimulus which is responsible for angiogenesis. The only reason of this therapy for not being 100% successful in this condition is that the treatment depends on the type and stage of tumor and has been found to be ineffective therapy if used alone (Daruwalla et al., 2006).

### 21.2.1.2 NORMOBARIC OXYGEN THERAPY

NBOT is referred as the delivery of high flow oxygen via facemask at 1 atm pressure. It has several advantages over HBOT including ease to administration, noninvasiveness, economic, widely available, and can be started promptly after stroke onset (Shi et al., 2016).

The condition after pediatric drowning can be treated with normobaric 100% oxygen to maintain adequate systemic oxygenation. Normobaric hyperoxia decreases infarct volume and magnetic resonance imaging (MRI) abnormalities (Singhal et al., 2002). Intermittent use of hyperoxia provides protection against ischemia reperfusion (IR) injury by promoting ischemic tolerance in rat model (Bigdeliet al., 2007) and provides stability for the short term which results in the thriving treatment of ailments such as thrombolysis (Kim et al., 2005). Interestingly, in another study in 2008, it was demonstrated that NBO also improves aerobic metabolism by playing a neuroprotective role after decreasing the lactate levels and preserving N-acetylcysteine levels (Bigdeliet al., 2008).

In 2012, a study was conducted over eight patients; their eyes were suffering from scleral ischemia and were poorly responsive to conventional medical and surgical therapy. After several days administration of NBO (10 L/min) for 1 h, twice a day, they noticed a therapeutic response suggesting NBO to provide therapeutic benefit as an adjunctive measure to standard therapy for scleral necrosis (Sharifipour et al., 2012).

### 21.2.2 OXYGEN FOR NEUROLOGICAL DISORDERS

#### 21.2.2.1 MULTIPLE SCLEROSIS

Multiple sclerosis is a chronic neurological disorder in which patchy inflammation, demyelination, and gliosis are seen in the CNS (Compston, 1998). Inflammation appears at distinct areas within the white matter which further extends to gray matter and proliferates deliberately within the CNS, showing a marked perivenular distribution (Prineas et al.,

1979). Perivascular smacking with lymphocytes follows breaking of the blood–brain barrier, and inflammatory cells exoduses intravascular compartment, further headed for cascading inflammatory activation (Prineas et al., 1979).

Multiple sclerosis currently is an inoperable and fatal disorder. Attempts to cure it are limited to treatments such as plasmapheresis and corticosteroids administration, etc., which results in similar effects like modification of disease succession and declining in degeneration rate by HBOT (Bennett et al., 2004). Majority of such measures include immunosuppressive or immunomodulatory agents. Drugs including interferon-$\beta$ (IFN-$\beta$), glatiramer acetate (GA), intravenous immunoglobulin, mitoxantrone, methotrexate, and corticosteroids are used for the treatment of multiple sclerosis (Goodin et al., 2002).

Additionally, HBOT possesses some adverse effects as well, including damage to the ears, sinuses, lungs, temporary worsening of short-sightedness, claustrophobia, and oxygen poisoning. Despite rare probability of such serious adverse events, HBOT cannot be considered as an entirely benign intervention and also requires further clinical trials for further validation (Bennett et al., 2004; James 1982).

#### 21.2.2.2 AUTISM SPECTRUM DISORDER

Autism spectrum disorder (ASD) is a neurodevelopmental disorder and is characterized by persistent communication and socialization impairments along with the restricted and repetitive behavioral patterns (American Psychiatric Association, 2013). It is also associated with some physiological and metabolic abnormalities including cerebral hypoperfusion, brain inflammation, immune dysregulation, and oxidative stress (Rossignol and Frye, 2012). Meanwhile, behavioral therapies are the only efficacious treatment for ASD, else there is no known cure till date. Thus, there is an urgent requirement to investigate an alternative treatment for ASD with the known benefits.

With several past studies, it was observed that HBOT is efficacious for the treatment of ASD, as it improves the behavioral and physiological flaws in ASD suffering patients. Before the randomized and multicentered studies, several nonrandomized trials have been conducted. One of them is the study with 18 children (3–16 years) who were given 40 hyperbaric sessions of 45 min either with 1.3 atm and 24% oxygen or with 1.5 atm and 100% oxygen. The results indicated a significant improvement in speech, motivation, and physiological/cognitive awareness with no adverse side effects in both the cases (Rossignol et al., 2007).

In 2009, first controlled study was conducted by Centre for Autism Research and Education along with the International Child Development Resource Center in USA, which demonstrated that children from the age group of 3–7 years when administered with HBOT at 1.3 atm for 40 h, 24% oxygen showed a significant improvement in social interaction, overall functioning, cognitive awareness, and receptive language. They reported that it was may be due to the decrease in brain inflammation and amelioration of cerebral hypoperfusion and modulating immune dysregulation but the exact mechanism of this improvement is still unknown (Rossingol et al., 2009).

Another study was also conducted in 2009 which reviews the single-photon emission computed tomography (SPECT) scan of basal and control brain perfusion in ASD suffering children. In this study, patients were treated with HBOT at 1.5 atm for 60 min daily. It was continued for 50 sessions and the results indicated a significant improvement in temporal,

frontal, and other areas of brain which leads to the correction in speech, sociability, cognitive awareness, and physical behavior (Kinaci et al., 2009). While in 2010, another randomized controlled trial was performed in which 16 autistic children received 24–28% supplemental oxygen and HBOT at 1.3 atm and 18 were subjected to control treatment (free airflow at ambient pressure per hour, 80 sessions), results obtained for both the groups showed no significant difference on direct observations as well as, via Autism Diagnostic Observation Schedule, generic tool indicating HBOT not to be an effective tool for treating autism (Granpeesheh et al., 2010).

Thus, very few evidences are present to relate HBOT with autism and its use as a treatment is not recommended due to controversial evidences and the risks and benefits of HBOT for autistic children. In 2017, a warning has also been published by FDA for parents to be cautious for the false and misleading declarations of HBOT authentication for the treatment of autism (FDA, 2017).

### 21.2.2.3 CEREBRAL PALSY

The injury in immature brain leads to cerebral palsy (CP) (Bax, 1964). It comprises the extensive range of motor function disorders accompanied by abnormal perception, speech impairment, mental retardation, sensation and cognition disturbance, or seizures (Pineau and Moqadem, 2007). Currently, there is no known treatment to cure CP except the supportive care (occupational training, speech therapy, etc.). But, there is a growing interest for the use of HBOT in CP affected children.

Some researches were done in late 90s which suggested the role of hyperbaric oxygenation in treating CP. A pilot study was conducted in 1999 on the children suffering from spastic diplegic CP. The children were given 20 HBOT therapies at 1.7 atm, 95% oxygen for 60 min, and were evaluated revealing that there was an improvement in fine and gross motor function as well as reduction in spasticity (Montgomery et al., 1999). In another study as well, similar results were obtained along with the reduction in tissue inflammation and normalization of glucose metabolism in brain injured cells (Cronje, 1999). Thus, these successful treatments open up a new way in managing the CP among children.

The most controversial study for the use of HBOT in CP was a double-blind multicenter trial by Collet et al. (2001). In this case study, cerebral palsied children were divided into two groups. One group received the HBOT treatment (100% oxygen, 1.75 atm, 1 h) while another was subjected to the pressurized air, that is, 1.3 atm. After the treatment, it was concluded that both the treatments were equally effective, that means, HBOT has no significant advantage over pressurized air in the cure of CP. Therefore, rampant demeanor rises among the researchers which breaks the ground and led to further investigations (Collet et al., 2001).

In 2002, a study was conducted by Chavdarov over large group of CP suffering patients. They were encountered with hyperbaric oxygenation for 40–50 min at 1.5–1.7 atm and after the treatment, the therapy had been proved successful with the positive results (recovery in mental, speech, and motor dysfunction) in 86% of children, with no side effects. But the limitation for this study is the lack of control group, thus further investigations are required to examine the benefits of HBOT in CP patients (Chavdarov, 2002).

Another study conducted in 2012 was found to be a contradictory study which suggests that neither hyperbaric oxygen (100% oxygen at 1.5 atm) nor hyperbaric air (14% oxygen at 1.5 atm)

improves the gross motor function in spastic cerebral palsied patients because no change had been observed in gross motor function measure (GMFM) global score from pre- to posttreatment (Lacey et al., 2012). Several more compelling evidences are required to prove HBOT as an effective therapy for the treatment of CP as compared with hyperbaric air.

### 21.2.2.4 GLIOBLASTOMA MULTIFORME

Glioblastoma multiforme (GBM) is the most common and malignant brain tumor which is originated from the glial cells in brain. The prognosis for GBM patients is very poor and the tumor is incurable even after the surgical sessions followed by chemotherapy. Its aggressive biology (i.e., apoptosis resistance) and neoplasm's infiltrative growth are its major characteristics (Stępien et al., 2016). It is believed that tumor hypoxia is the major reason for the treatment failure of GBM as it generally eliminates the cell function and becomes radiotherapy and chemotherapy resistant. Thus, HBOT can help to increase the efficacy of the treatment because it increases the oxygen content in tumor area, several studies are also conducted to support this statement.

In 2007, a study was conducted on mice with transplanted glioma cells. It was observed that when the levels of $pO_2$ is increased via normobaric therapy or moderate HBOT, it inhibits tumor progression due to the induction of proliferation inhibitors and proapoptotic genes and reduction of anti-apoptotic and proangiogenic genes. Although intracranial glioma cells are different from the subcutaneous transplanted glioma cells and even no cytological experiments were done, it was suggested that before the recommendation of this therapy further investigations are required (Stuhr et al., 2007).

Several more studies were conducted which demonstrated HBOT as chemo/radiotherapy adjunctive treatment for GBM. The median survival time was found to be increased to 6–7 weeks while survival rate was increased to 28% from 10%, when GBM patients were given HBOT during irradiation, as compared with the radiotherapy alone (Chang, 1997). Despite of these benefits, late side effects were also observed including seizures and radiation injury. The treatment of these side effects from HBOT along with different neuro-oncology treatment settings has been discussed in the review by Kohshi et al. in 2013. They analyzed the data till 2012 and concluded that HBOT not only repairs the radiation necrosis but also prevents the tissue from the injury caused by radiotherapy for brain tumors (Kohshi et al., 2013). Recently, a study was conducted over human glioma stem cells bearing mice. They were divided into the four groups on the basis of the therapy they were administered (control, HBOT, nimustine [ACNU], HBOT+ACNU). After the specified time interval, the results they obtained suggested that HBOT enhances the oxygen level in tumor tissues which subsequently results in the reduction of glioma cells proliferation and neoplasm infiltration. Also, it lowers the expression of HIF-1$\alpha$, TNF-$\alpha$, VEGF, NF-$\kappa$B, IL-1$\beta$, and MMP9, which indicates HBOT as an effective therapy for treating GBM. They also deduce that HBOT with ACNU exhibits a significant reduction in glioma growth, thus, it can also be a promising approach to treat GBM (Lu et al., 2016).

### 21.2.3 SIDE EFFECTS AND TOXICITY

There are various risks associated with oxygen therapy. In late 19th century, Paul Bert was the first who reported the oxygen toxicity in patients who are administered with hyperbaric oxygen for 3 h (Donald, 1947). He stated that after 3 h of exposure, patients start experiencing the chest pain and prolonged exposure led to bronchopneumonia. Exposure to high concentrations of oxygen first damage the capillary endothelium, followed by interstitial edema (0–12 h), worsening acquiescence and vital capacity (12–30 h), and lastly the thickening of the alveolar-capillary membrane (30–72 h) (Kacmarek et al., 2013).

As stated earlier, if it is used in nonhumidified way, it may cause drying, crusting, and may lead to various injuries to the nose and face by all types of administration methods including masks (Tamir et al., 2007). Oxygen therapy can also cause carbon dioxide narcosis chronic obstructive pulmonary disease (COPD) patients due to excessive administration (New, 2006). It can also cause structural damage to the lungs and during necropsy in COPD patients, proliferative and fibrotic changes of oxygen toxicity are also observed (Davis et al., 1983; Jenkinson, 1998). There is also a risk of retrolental fibrosis and bronchopulmonary dysplasia, which may occur in newborn after oxygen administration (Higgins et al., 2007; Saugstad, 2005).

Due to the rapid pressure changes, few patients may experience mild to severe pain from rupture of the middle ear, the cranial sinuses, and, in rare cases, barotraumas (Clark et al., 1971). When amount of oxygen inhaled is high (under the pressure), it may precipitate generalized seizures though it does not cause permanent damage (Clark et al., 1971). With repeated exposure to hyperbaric oxygen, some patients may set reversible trachea-bronchial symptoms including chest tightness, a sub-sternal burning sensation, and cough (Kindwall et al., 1993).

Oxygen is toxic to the tissues in high doses and this can be marked acutely as a wide variety of preseizure symptoms. These events are observed to be self-terminating with the falling in brain oxygen tension (Bennett et al., 2004).

Major problems using oxygen therapy are high costs and feasibility obtaining oxygen supply. There is also a lack of medical expertise and sources of supply. Patient's resources to meet the expense of the treatment are also scarce. Moreover, no apparent guidelines are available regarding reimbursement of costs on oxygen and the apparatus (Walsh et al., 2017).

## 21.3    CARBON DIOXIDE

Carbon dioxide ($CO_2$), a colorless, nonflammable, odorless gas which was first described as a discrete substance by van Helmont. It was first discovered by Joseph Black in 1750s who named it as "fixed air" (West, 2014). During the ancient period of Roman times, hot springs with carbon dioxide were said to have medical properties. The antibacterial activity of carbon dioxide first came in light in 17–18 century described by Robert Boyle and Antoine Lavoisier (Pennazio, 2005). They reported the treatment of chronic ulcers using this gas in 1777 (Zupanets et al., 2015).

Carboxy therapy used carbon dioxide-rich water bath, and was originated in France in 1930 which was found to help in wound healing and lipolysis of accumulated fat (Shalan et al., 2015). Carbon dioxide was first used by injectional application for Balneotherapy in 1932 in Medical Spa Roy, France (Koutna, 2006). During the late 1990s, Italian Ministry of Health approved devices utilizing carbon dioxide therapy (Brandi et al., 2010). In the 1980s, laser resurfacing to treat photodamaged and aging skin originated utilizing carbon dioxide (Ortiz et al., 2014). During the mid-2000, fractional $CO_2$ lasers were discovered which allowed the surgeons to leave small areas of untreated skin

intact between the irradiated skin (Moreira et al., 2014). 2012 was announced as the European Year Carboxy therapy.

Another effective method to use carbon dioxide as a therapeutic gas is therapeutic hypercapnia, as it not only improves the tissue perfusion but also enhances oxygenation (Zhou et al., 2010). It is the increase in arterial carbon dioxide tension ($PaCO_2$) by addition of carbon dioxide to inspired gas and has been reported as a new treatment for various diseases, including, lung, intestinal, myocardial, and CNS injuries (Chonghaile et al., 2008; Laffey et al., 2003 and Normura et al., 1994). The levels of increased arterial carbon dioxide tension vary from 50 mmHg to more than 100 mmHg. It has been reported that levels of $PaCO_2$ between 50 and 70 mmHg (mild hypercapnia) protects the brain from hypoxic-ischemic injury while $PaCO_2$ levels > 100 mmHg (severe hypercapnia) are found to be detrimental (Vannucii et al., 1997). This may be due to the impairment of cellular calcium homeostasis or by extra- and intracellular acidification (Akca, 2006). It has been hypothesized that it was due to the activation of the hypothalamic–pituitary–adrenal axis, enhancement of neurotransmitter function, and exertion of anti-inflammatory and antioxidant effects via mild hypercapnia (Zhou et al., 2010).

## 21.3.1    *CARBON DIOXIDE FOR NEUROLOGICAL DISORDERS*

### 21.3.1.1    *CEREBRAL ISCHEMIA*

Brain ischemia or cerebral ischemia is the insufficient supply of blood to the brain to meet its metabolic demand (Jonathon, 2008). It results in cerebral hypoxia which ultimately leads to the ischemic stroke, death of brain tissues, and cerebral infarction (Cure Hunter, 2008). It can be confined to a specific portion of brain majorly due to the blockage of cerebral vessels and referred as focal cerebral ischemia or it affects the wide regions of brain tissues by sudden reduction of blood flow to brain and termed as global cerebral ischemia (Jonathon, 2008).

In 2010, a study was conducted over the global cerebral ischemic mice, who were exposed to different $CO_2$ arterial tensions, that is, mild (60–80 mm Hg), moderate (80–100 mm Hg), and severe (100–120 mm Hg) for 2 h. Results were evaluated after 24 and 72 h of recovery for neurologic deficit score, protein expression, histopathological changes, and tissue edema formation. The results indicated that moderate hypercapnia possesses best neuroprotective effects while severe hypercapnia results in increased brain edema, thus, worsening the condition. They reported that mild and moderate hypercapnia modulates the apoptosis regulating protein, thus fewer ultrastructural histopathological changes were observed (Zhou et al., 2010).

Another study was conducted in 2013, which shows the improvement in focal cerebral ischemia after the treatment with mild hypercapnia. The hypercapnia treated rats exhibit decrease in infarct volume, increase in survival rate, and the upregulation of anti-apoptotic proteins including Bcl-2 which resulted in the inhibition of Bax translocation. Thus, indicating hypercapnia as an adjunctive therapy in the treatment of focal cerebral ischemia (Tao et al., 2013).

A study conducted in 2016 demonstrated the inhibition of hypoxia-induced disruption of cortical cerebral blood flow and blood–brain barrier permeability in ischemic brain injury by $CO_2$ inhalation. It was the first study which exhibited a correlation between hypercapnia and hypoxia exposure extent. This study evaluated the role of moderate hypercapnia in ischemic brain injured rats after the exposure to mild to severe hypoxia. Results indicated that hypercapnia efficacy depends on the systemic hypoxia as well as on cerebral pH. Thus, further studies are required to evaluate the actual mechanism of hypercapnia in cerebral ischemia (Yang et al., 2016).

### 21.3.1.2 CORPUS CALLOSOTOMY

Corpus callosotomy is a neurosurgical procedure used to control nonfocal epilepsy, drop attacks, atonic seizures, tonic–clonic seizures, Lennox–Gastaut syndrome, and recurrent status epilepticus (Maehara and Shimizu, 2001; Oguni et al., 1991).

There are many types of procedures to achieve corpus callosotomy like surgical, endoscopic, and radiosurgical, however, laser systems have a significant part in multiple neurosurgical procedures from the late 1960s in order to reduce damage to normal brain tissue, working in a narrow surgical passage at the same time (Choudhri et al., 2015).

Corpus callosum is one of the most important commissural pathways linking the cerebral hemispheres. The procedure of callosotomy works by separating the interhemispheric seizures spread. The main goal of this process is to reduce the frequency of seizure because the outcome of seizures is hardly ever a cure. This process makes only the quality of life better and therefore patients are made to stay on an antiepileptic treatment. Patients often continue to have drop attacks and generalized seizures, which are thought to be due to other commissural pathways, such as anterior, posterior, and hippocampal commissures (Rougier et al., 1997). Thus, the degree of sectioning in callosotomy is to attain the equilibrium between good seizure control and minimizing consequences to reduce the disconnection syndrome.

In most cases, an anterior corpus callosotomy is enough but if the patient still has seizures very repeatedly, complete callosotomy is important (Choudhri et al., 2015).

Corpus callosotomy achieved by means of an interhemispheric approach is linked with perils and neurological consequences which includes paresis of the nondominant limb, gait difficulty, and urinary incontinence and are temporary (Choudhri et al., 2015).

Of all the types of lasers used in corpus callosotomy, $CO_2$ laser has features like absorption by water and minimal thermal penetration, which permits its use for diverse functions like tumor resection, cord detethering, dural closure, and vascular anastomosis (Kilori et al., 2010).

Recently, growth of the OmniGuide fiber was achieved by Fink and associates at the Massachusetts Institute of Technology (MIT) and they have reintroduced the $CO_2$ laser to neurosurgery. In 2015, the first successful use of a fiberoptic $CO_2$ laser for corpus callosotomy in six children with intractable epilepsy has been reported, two of whom had the initial anterior two-thirds callosotomy completed with bipolar electrocautery (Choudhri et al., 2015). Among all the six patients, there has been an improvement in the drop attacks frequency after the complete callosotomy, along with this, no side effects/complications had been encountered. Cerebral imaging demonstrated a clean sectioning of callosal fibers with preservation of normal ventricular anatomy indicating that the low-profile laser fiber tip was effective for working in the depths of the interhemispheric fissure with minimal brain retraction.

### 21.3.2 SIDE EFFECTS AND TOXICITY

If $CO_2$ exposure goes high (60–67% $CO_2$), the displacement of oxygen occurs and it notably leads to toxicity. In this case, patient's show symptoms of asphyxia when the atmospheric oxygen falls below 16% and when $O_2$ levels fall between 10% and 13% instant unconsciousness and can cause death (Watanabe and Morita, 1998). Some of the patients exhibit seizures if the levels of carbon dioxide increase (>600 ppm) and can lead to no cerebral functioning. Hypopigmentation, because of carbon dioxide laser, is the most frequently reported long-term side effect (Manuskiatti et al., 1999).

### 21.4 NITRIC OXIDE

Ever since the discovery of NO in 1772 by Joseph Priestly, who also gave its composition consisting of single oxygen atom and a nitrogen atom, it was thought to be an atmospheric pollutant until the 1980s when nitric oxide (NO) synthesis was discovered in mammalian cells (Anacak et al., 2006). This discovery was important as NO has many physiological roles which were not identified till then. Biologically, L-arginine is used by a family of enzymes (iNOS and endothelial NOS) to produce NO (Mali et al., 2017).

During 1980s, numerous investigations were going on to find out the process of vasodilation, which is important for controlling blood pressure. During and before this time, patients were given nitroglycerin for heart disorders to induce vasodilation and normalization of blood pressure (Anacak et al., 2006) without knowing the exact mechanism. In 1977, independent investigations on the mechanism of nitroglycerin were conducted by Ferid Murad, who found that it could cause vasodilation because of the release of gas called NO (Steinhorn et al., 2015). In 1980, another independent investigation was carried out by Robert Furchgott on the mechanism of a drug acetylcholine on the process of vasodilation and he discovered that relaxation of blood vessels only occurs if a special class of cells (today known as endothelial cells) are present (Loscalzo, 2013). He and his group later discovered that endothelial cells produce a factor that is important for the relaxation of smooth muscle cells. They later termed the factor as endothelial-derived relaxing factor (EDRF). Since its discovery, various studies were carried out to find and identify EDRF and it was finally identified by Louis Ignaro in 1987 to be identical to the gas NO (Steinhorn et al., 2015).

So, the very first role of NO identified as a therapeutic gas was as a vasodilation regulator. And since this first discovery, it has been found to be useful in various other therapeutic conditions including the treatment of pulmonary disorders and cancer, maintenance of vascular tone and blood pressure, modulating immunity and inflammation, neurotransmission, and kidney and reproductive functions (Giles, 2006). NO is neuroprotective (Fig. 21.1) and plays an important role in a number of vital processes in the developing brain, including myelination and reduction in the size of excitotoxic and ischemic lesions. It also has a central position in deciding how an immature brain response to hypoxic–ischemic insults (Charriaut-Marlangue et al., 2012).

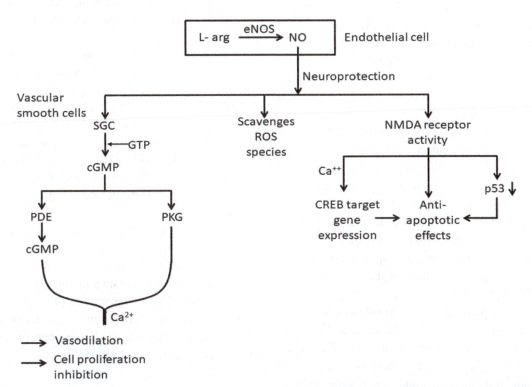

**FIGURE 21.1** The neuroprotective effect of nitric oxide. L-arg, L-arginine; eNOS, endothelial NOS; ROS, reactive oxygen species; NMDA, N-methyl-D-aspartate; CREB, cAMP response element binding protein; p53, tumor supressor protein 53; sGC, Soluble guanylate cyclase; cGMP, cyclic guanosine monophosphate; GTP, guanosine triphosphate; PDE, phosphodiesterase; PKG: cGMP dependent protein kinase.

*Source:* Adapted from Calabrese et al., 2009.

### 21.4.1  MECHANISM OF ACTION

Role of NO in the CNS was firstly character-ized as an intercellular messenger that acti-vates the glutamate receptors by mediating the increase in levels of cyclic GMP. NO is produced by three different types of NO synthases. Neuronal nitric oxide synthase (nNOS) has the highest activity in brain and is mainly found in neurons. The highest density of these nNOS-containing neurons is present in the accessory olfactory bulb and granule cells of the cerebellum (Esplugues, 2002). Despite the fact that nNOS neurons constitute only about 1% of cell bodies in the cerebral cortex, nearly every neuron in the cortex is open to nNOS nerve terminals. The location of nNOS can be any presynaptic or postsynaptic. It is chiefly concerned with signaling of neurons, neurotoxicity, synaptic plasticity, and behav-ioral pathways modulation like expression of pain (Esplugues, 2002).

Endothelial NOS (eNOS) is also present in some of the neurons and glia but is largely located in the endothelial cells of cerebral vessels and primarily involved in vascular func-tion regulation (Li et al., 2006).

Finally, stimulation of inducible nitric oxide synthase (iNOS) in glial cells is involved in the unspecific brain immune response and is typi-cally linked with pathological conditions (Saha and Pahan, 2006).

NO plays most important role in neurotrans-mitter inhibition of peripheral nonadrenergic, noncholinergic (NANC) nerves. Peripheral nitrergic nerves are widely distributed throughout the body and are mainly vital in producing smooth muscle relaxation of the gastrointestinal, respiratory, vascular, and urogenital systems (Esplugues, 2002).

### 21.4.1.1   HEME OXYGENASE OVEREXPRESSION

Production of ROS occurs in all living being including human body. ROS are formed as an outcome of regular cellular metabolism and environmental causes like air pollutants, cigarette smoke, etc. (Birben et al., 2012). ROS being extremely reactive molecules have the power to damage cellular structures and can also modify their functions. Uncontrolled or extreme production of these ROS can result in oxidative damage to the body. A reduction in production of endogenous antioxidants can also lead to the oxidative stress. This shifting in the balance of oxidants and antioxidants toward favoring oxidants is oxidative stress (Sun et al., 2006; Birben et al., 2012).

An early cellular response to oxidative stress is heme oxygenase-1 induction which has been shown to have neuroprotective func-tion (Calabrese et al., 2007). It has been proven that heme oxygenase-1 is induced by NO in rat astrocytes, microglia, and hippocampus (Calabrese et al., 2007). Excess NO and reactive nitrogen species (RNS) are produced as a result of upregulation of iNOS during pro-oxidant conditions.

When the upregulation of heme oxygenase 1 protein occurs, an increase in biliverdin also occurs. This biliverdin molecule is then reduced into the antioxidant and anti-nitrosative molecule, bilirubin by the activity of an enzyme biliverdin reductase. This is another mechanism which is also seen to have shown neuroprotec-tive effects (Calabrese et al., 2007).

### 21.4.1.2   S-NITROSYLATION

S-nitrosylation is an essential posttranslational protein modification (Sun et al., 2006). It was demonstrated by experimental evidences that the process of S-nitrosylation of the proteins forms regulates the effects of NO on the functions of the cells and various S-nitrosylated proteins (SNO proteins) were identified recently and studies have shown that hypo-S-nitrosylation or hyper-S-nitrosylation of specific proteins resulting in protein function alterations are directly connected to the causes and the symptoms of a number of diseases, mainly including nervous system disorders (Foster et al., 2009). Under oxidative stress, the process of *S*-nitrosylation of protein thiols occurs and prevents the cells from more oxidative damage by increased level of ROS and RNS. These reactive species induce stress-signaling pathways engaged in dysfunction of mitochondria, overload of intracellular $Ca^{2+}$, heart failure, necrosis, and apoptosis (Sun et al., 2006).

NMDA receptors, the $Ca^{2+}$-favoring glutamate-gated ion channels are majorly responsible for the neuronal injury. Their activation (especially the NR1/NR2B-subtype) mediates damage to cells and their overstimulation can lead to cell death by excitotoxicity (Dong et al., 2009). In this case also protection against such excitotoxicity is done by NO molecule by the process of S-nitrosylation. The NR1 and NR2 subunits of the NMDA receptor undergo S-nitrosylation which reduces the intracellular $Ca^{2+}$ influx responsible for neuronal death or it can also diminish the formation of peroxynitrite (Calabrese et al., 2007).

NO also inhibits the activity of caspases which are the enzyme induced during apoptosis, via S-nitrosylation of cysteines present in their catalytic site and thus provide cytoprotection. This process of S-nitrosylation reduces the caspases activity in neurons. Various studies have confirmed that when NO donors like S-nitrosothiols are used to treat cortical neurons that had been treated with several NO donors, it is noticed that there is remarkable reduction in staurosporine-induced caspase 3 and caspase 9 activation (Mannick et al., 2001; Melino et al., 1997).

## 21.4.2 NITRIC OXIDE FOR NEUROLOGICAL DISORDERS

### 21.4.2.1 ANEURYSMAL SUBARACHNOID HEMORRHAGE

Aneurysmal subarachnoid hemorrhage (SAH) is a fatal neurological disorder which accounts for 27% stroke related deaths before the age of 65 years (Johnston et al., 1998). It may occur as a result of the ruptured cerebral aneurysm due to head injury which results in aneurysm re-bleeding, large aneurysm size, cerebral ischemia, and cerebral infarction from vasospasm (Komotar et al., 2009). SAH results in the alteration of various pathological mechanisms. It alters the cerebral physiology by raising the intracranial pressure and displacing the cerebrospinal fluid with the blood releases during the aneurysm rupture. It also results in the reduction of cerebral blood flow and cerebral perfusion pressure which leads to the development of ischemic brain injury as well (Sehba et al., 2010).

Cerebral hemodynamics is regulated by the NO/NOS pathways which also get altered during the SAH and results in pathological consequence. NO regulates the cerebral blood flow and blood pressure by dilating the blood vessels, thus decrease in the NO levels after SAH give rise to the constriction of cerebral vessels, platelets aggregation, and adherence of leukocytes to vascular endothelium (Flores et al., 2009; Sehba et al., 2005; Sehba et al., 2000). Thus, alterations in NO pathways results in poor clinical outcomes after SAH.

Studies revealed that early use of NO donors like S-nitrosoglutathione and nitroglycerin, NONOate, glyceroltrinitrate, and diazenium-diolate after SAH helps to recover the cerebral blood flow, dilates the cerebral vessels, prevents the excitotoxic glutamate release, and decrease the chances of developing cerebral infarction and vasospasm (Sehba et al., 1999; Sehba et al., 2007). Another approach is by altering the NO signaling pathway. During its vasodilation mechanism, inhibition of cyclic guanosine 3',5'-monophosphate (cGMP) degradation by phosphodiesterase improves the NO-mediated vasodilation (Sobey and Quan, 1999).

In 1980s, the first evidence over cats proved that cortical spreading depression (CSD) occurs during SAH (Hubschmann and Kornhauser, 1980 and 1982). CSD consists of an electro-encephalogram (EEG) wave moving across the cortical surface and have an important role in several brain injuries as well as in strokes. Depolarization spread, in a healthy brain tissue, is linked with vasodilatation. However, vasoconstriction and tissue ischemia occur in damaged brain tissue because of inverse blood brain flow response (Garry et al., 2014).

When NO level reduces, the brain becomes more vulnerable to cortical spreading depression. However, optimum levels of NO take actions for the restoration of impaired cere-brovascular reactivity that occurs after CSD. Therefore, while NO signaling has a key responsibility for the blood flow changes occurring after SAH, its reduction has a significant function in early brain injury, microthrombus formation, and cortical spreading depression. To prevent secondary neuronal injury, targeting this pathway may be beneficial and thus may partially restore physiological blood flow (Garry et al., 2014).

### 21.4.2.2 ALZHEIMER'S DISEASE

Alzheimer's disease (AD) is a complex, multi-factorial, and the most common neurodegen-erative disorder. It involves the loss of specific synapses and neurons in the brain which leads to progressive memory loss, changes in cogni-tive ability, and ultimately to death (Reitz et al., 2011). Till date, no complete cure is known for AD. The two key hallmarks of AD are the accu-mulation and aggregation of $\beta$-amyloid (A$\beta$) protein, in the form of extracellular plaques, and hyperphosphorylated tau protein, as intra-cellular neurofibrillary tangles (Querfurth and LaFerla, 2010). This results in the disrup-tion of normal signaling pathways, mitochon-drial dysfunction, and neuroinflammation by inducing oxidative stress in the cell (Balez and Ooi, 2016). A$\beta$ protein also leads to the exces-sive formation of ROS via NMDAR activation. Thus, the disruption of NMDAR activity and oxidative stress are found to be harmful in AD (De Felice et al., 2007).

Use of NO in AD may serve neuroprotective roles through cGMP pathway leading to vasodi-lation which is known to increase cerebral blood supply to neurons. This ultimately might result in reducing oxidative stress. It also can mini-mize influx of excess Ca$^{2+}$ through inhibition of NMDA receptors at glutamatergic synapses (Aranico et al., 1996 and Kohgami et al., 2010).

### 21.4.2.3 COGNITIVE DISORDERS

Evaluation of the role of NO in cognitive disorders and experimental AD was done by Manukhina et al. (2008) using mice in hypo-baric pressure chamber. AD was induced by injecting fragment of β-amyloid into nucleus basalis magnocellularis. NO-synthase inhibitor-N ω-nitro-L-arginine (L-NNA), or

NO-donor–dinitrosyl iron complex was administered to rats (Manukhina et al., 2008). The results indicated that that L-NNA aggravated the harmful effects of β-amyloid, stopped the shielding effect of hypoxia adaptation, and also caused memory disorders in rats. On the other hand when the NO-donor could increase the NO level in the body and could prevent the progression of disease due to β-amyloid (Manukhina et al., 2008).

### 21.4.3   SIDE EFFECTS AND TOXICITY

NO shows cytotoxic effects if produced in excess amount or if the cell is in a pro-oxidant state. Since NO can react with superoxide anions (possess strong oxidant properties) produced by iNOS in inflammatory conditions or nNOS during excitotoxicity to form peroxynitrite, therefore cellular components damaging can take place because of protein nitration cause by the interaction of peroxynitrite and cellular components (Calabrese et al., 2007).

The NO-mediated S-nitrosylation (S-NO) of certain substrates like matrix metalloproteinase 9 (MMP9), GAPDH, and protein-disulfide isomerase (PDI) has been anticipated to be a mechanism by which NO can become neurotoxic (Calabrese et al., 2007).

Also, the activation of hemoprotein cyclooxygenase (COX) by NO and the generation of free radicals (FRs) and prostaglandins (PGs) during its catalytic cycle results in strong pro-inflammatory effects (Calabrese et al., 2007).

### 21.5   HELIUM

Helium (He) was named after the Greek god of the Sun Helios. It is a colorless, odorless, tasteless, monatomic, inert, and nontoxic gas

and is included in the group of noble gas in periodic table (Weber et al., 2015). In 1895, a Scottish chemist, William Ramsey was the first on Earth to discover He by the treatment of mineral cleveite with mineral acids (Harris et al., 2010). Its unique physical and chemical properties generated an increase interest in experimental and clinical investigations for its ever-expanding applications in medicine (Oei et al., 2010). Its inertness can be explained by its outer shell which is completely filled with electrons (Preckel et al., 2005). It has the lowest melting and boiling points among all elements and is present in the ample amount in nature. It is the least dense element in the universe after the hydrogen (Schmidt et al., 1937). Regardless of this, most of the He in use is formed due to the radioactive decay of α-particles and extracted during natural gas production by fractional distillation (Banks et al., 2010). Due to the various unusual physical and chemical properties of He, it always remains as a field of interest for researchers.

In 1926, Sayers and Yant found that helium has lower solubility than nitrogen, due to which there is no discomfort caused to humans and animals while breathing helium–oxygen mixtures. This helium–oxygen mixture not only decreases the formation of nitrogen bubbles but also reduces the decompression sickness or caisson disease in deep-sea divers (Sayers et al., 1926). Helium does not have anesthetic properties but possess various other biological effects (Koblin et al., 1998). Below is the brief description of knowledge about the uses of He in medicine.

### 21.5.1   HELIUM FOR NEUROLOGICAL DISORDERS

From last so many years, scientists are proving the neuroprotective effects of xenon in in vitro as well as in in vivo models (Maze et al., 2016). But, the exact mechanism of action for xenon is still unknown. Supply for xenon is limited and it was found that He has similar chemical characteristics scientists concern toward this noble gas from xenon (Dickinson et al., 2010).

### 21.5.1.1 BRAIN INJURY

In an in vitro mouse model of traumatic brain injury, increased He pressure (i.e., upto 2 atm) could reduce the cell damage and also could protect the brain as well. It was further proved by replacing the He pressure with nitrogen pressure, which resulted in the worse injury. Thus, it was suggested that protection was caused not only because of the beneficial effect of He alone but also by eradicating the detrimental effects of nitrogen. Despite this, the only drawback of this model was they were not able to distinguish between the pharmacological and pressure per se effects of He (Coburn et al., 2008).

Some contradictory results were also observed when the neuronal cultures are exposed to normobaric He and normobaric nitrogen apoxia, respectively. When the results were compared, nitrogen injured cells show more functional damage as compared with xenon and argon while He injured cells worsen the damage, and these results leads to the further findings to clear (Jawad et al., 2009) the contradiction.

### 21.5.1.2 NEUROPROTECTION

There has been an interesting study on rats in which He (75% volume) could provide neuroprotection against acute ischemic stroke when they are administered with middle cerebral artery occlusion (MCAO) after reperfusion. The study was conducted with mice and authors concluded that He could show neuroprotection by producing the hypothermia and it takes the benefit of high thermal conductivity of He when compared with air (David et al., 2009).

In a recent in vitro study (2016) on mice scientists reported that He could be an efficient neuroprotectant if given after plasminogen activator-induced reperfusion. In thromboembolic brain ischemic mice, both intra and postischemic He (75% volume) for 20 min reduces the brain damage and risk of brain hemorrhage as well (Haelewyn et al., 2016).

With the evolution of He, another experimental phase in neurosurgery has evolved the plasma technology (Beckley et al. 2004). The plasma coagulator is a device, which uses ionized He gas to enhance the visualization of bleeding sites, in which jet flow of gas is used to clear the bleeding and other liquids away from the surgical field and the chances of electric current dissipation is eliminated within the blood.

### 21.5.2 SIDE EFFECTS AND TOXICITY

He sometimes even leads to more fatal complications than $CO_2$ Due to the high solubility of He in blood (approximately 50 times more) than $CO_2$, it can lead to venous gas embolism (Schob et al., 1996). Another major concern with He is, during laparoscopic anti-reflux procedures, it has the potential to cause He pneumothorax (Crabtree et al., 1999; Ludemann et al., 2003). He also increases the risk of hypoxia due to its high thermal conductivity which in severe cases cause high pressure nervous syndrome (HPNS), a condition with a congregation of complications related to autonomic and neurological (Bennet, 1982).

## 21.6 FUTURE OPPORTUNITIES AND CHALLENGES

It has been discussed above that so far these gases have been used in an unorganized manner and there are no sets of fixed protocols which are accepted worldwide. So an efficient planning is required for the future trials/uses of these gases before establishing their utility for such treatments. We need to design the experiments in a scientific manner and take appropriate sample sizes of the patients along with the manpower needed. Also, the inclusion criteria and outcome measures are to be analyzed carefully. We would also need high-end instruments like MRI and others to validate quality of life and diseases during and after the therapy.

Safety and cost-effectiveness of the therapies also need to be taken in account by all upcoming future uses/trials. Also, we need to make sure that these treatments and diagnostic tests are accessible to the entire community. In total, there is a lot of things which still have to be explored in the case of gases to be utilized for treatments and clinicians and scientists can work together to establish their utility in the treatment of neural and other diseases and disorders.

## 21.7 CONCLUSION

The use of the discussed gases in the present chapter for the treatment of various neurological disorders is very wide and have very less side effects in comparison to various other drugs used for the similar treatments. There are numerous types of disorders related to brain, and treatment for each disorder is handled in a different manner. A single gas can work on different disorders utilizing different mechanism of action. For example, NO shows neuroprotection by the process of S-nitrosylation and overexpression of heme oxygenase-1 or it can work as a second messenger and modulate signaling mechanism. Oxygen can work though modulating the hyperoxic status or can modulate the apoptotic mechanism. Carbon dioxide gas can treat microbial infections utilizing its antimicrobial nature and on other hand it has been utilized in lasers too. Similarly, in case of He it can be mixed with oxygen and aid oxytherapy and, on the other hand it is useful for various diagnostic procedures. In conclusion, it can be said that these and many other gases can further be explored to establish their different roles for different treatments.

## KEYWORDS

- **oxygen**
- **carbon dioxide**
- **nitric oxide**
- **helium**
- **neurological disorders**

## REFERENCES

Akca, O. Optimizing the Intraoperative Management of Carbon Dioxide Concentration. *Curr. Opin. Anaesthesiol.* **2006,** *19,* 19–25.

Arancio, O.; Lev-Ram, V.; Tsien, R. Y.; Kandel, E. R.; Hawkins, R. D. Nitric Oxide Acts as a Retrograde Messenger During Long-term Potentiation in Cultured Hippocampal Neurons. *J. Physiol. Paris.* **1996,** *90,* 321–322.

Balez, R.; Ooi, L. Getting to NO Alzheimer's Disease: Neuroprotection Versus Neurotoxicity Mediated by Nitric Oxide. *Oxidative Med. Cell Longev.* **2016,** *2016,* 1–8.

Bandali, K. S.; Belanger, M. P.; Wittnich, C. Does Hyperoxia Affect Glucose Regulation and Transport in the Newborn? *J. Thorac. Cardiovasc. Surg.* **2003,** *126,* 1730–1735.

Banks, M. Helium Sell-off Risks Future Supply, PhysicsWorld.com. Institute of Physics, 2010. (accessed Jan 1, 2012.

Bax, M. C. Terminology and Classification of Cerebral Palsy. *Dev. Med. Child Neurol.* **1964,** *6,* 295–297.

Beckley, M. L.; Ghafourpour, K. L.; Indresano, A. T. The Use of Argon Beam Coagulation to Control Hemorrhage: A Case Report and Review of the Technology. *J. Oral Maxillofac. Surg.* **2004,** *62,* 5–8.

Bennett, M. H.; Heard, R. Hyperbaric Oxygen Therapy for Multiple Sclerosis. *Cochrane Database Syst. Rev.* **2004,** *1.*

Bennett, M. H.; French, C.; Schnabel, A.; Wasiak, J.; Kranke, P.; Weibel, S. Normobaric and Hyperbaric Oxygen Therapy for the Treatment and Prevention of Migraine and Cluster Headache. *Cochrane Database Syst. Rev.* **2015,** *12.* doi: 10.1002/14651858.CD005219.pub3.

Bennett, P. B. The High-pressure Nervous Syndrome in Man. In: *The Physiology and Medicine of Diving and Compressed Air Work*; Bennett, P. B., Elliot, D. H., Eds; London: Balliere-Tindall, 1982, pp. 262–296.

Berganza, C. J.; Zhang, J. H. The Role of Helium Gas in Medicine. *Med. Gas Res.* **2013,** *3,* 18.

Bigdeli, M. R.; Hajizadeh, S.; Froozandeh, M.; Rasulian, B.; Heidarianpour, A.; Khoshbaten, A. Prolonged and Intermittent Normobaric Hyperoxia Induce Different Degrees of Ischemic Tolerance in Rat Brain Tissue. *Brain Res.* **2007,** *1152,* 228–233.

Birben, E.; Sahiner, U. M.; Sackesen, C.; Erzurum, S.; Kalayci, O. Oxidative Stress and Antioxidant Defense. *World Allergy Organ. J.* **2012,** *5,* 9.

Bosson, N.; Gausche-Hill, M.; Koenig, W. Implementation of a Titrated Oxygen Protocol in the Out-of-hospital Setting. *Prehosp. Disaster Med.* **2014,** *29,* 403–408.

Brain Ischemia (Cerebral Ischemia). Cure Hunter Incorporated, 2003. (accessed Nov 11, 2008).

Brambrink, A.; Orfanakis, A. Therapeutic Hypercapnia" after Ischemic Brain Injury. Is There a Potential for Neuroprotection?. *Anesthesiology* **2010,** *112,* 274–276.

Brandi, C.; Grimaldi, L.; Nisi, G.; Brafa, A.; Campa, A.; Calabro, M.; Campana, M.; D'Aniello, C. The Role of Carbon Dioxide Therapy in the Treatment of Chronic Wounds. *In Vivo* **2010,** *24,* 223–226.

Calabrese, V.; Cornelius, C.; Rizzarelli, E.; Owen, J.B.; Dinkova-Kostova, A.T.; Butterfield, D.A. Nitric Oxide in Cell Survival: A Janus Molecule. *Antioxid. Redox Signal* **2009,** *11,* 2717–2739.

Calabrese, V.; Mancuso, C.; Calvani, M.; Rizzarelli, E.; Butterfield, D. A.; Stella, A. M. G. Nitric Oxide in the Central Nervous System: Neuroprotection Versus Neurotoxicity. *Nat. Rev. Neurosci.* **2007,** *8,* 766–775.

Chang, C. H.; Hyperbaric Oxygen and Radiation Therapy in the Management of Glioblastoma. *Natl. Cancer Inst. Monogr.* **1977,** *46,* 163–169.

Charriaut-Marlangue, C.; Bonnin, P.; Gharib, A.; Leger, P. L.;Villapol, S.; Pocard, M.; Gressens, P.; Renolleau, S.; Baud, O. Inhaled Nitric Oxide Reduces Brain Damage by Collateral Recruitment in a Neonatal Stroke Model. *Stroke* **2012,** *43,* 3078–3084.

Chavdarov, I. In *The Effects of Hyperbaric Oxygenation on Psycho-motor Functions by Children with Cerebral Palsy*, Proceedings of the 2nd International Symposium on Hyperbaric oxygenation in Cerebral Palsy and the Brain-injured Child, 2002.

Choudhri, O.; Lober, R. M.; Camara-Quintana, J.; Yeom, K. W.; Guzman, R.; Edwards, M. S. Carbon Dioxide Laser for Corpus Callosotomy in the Pediatric Population. *J. Neurosurg. Pediatr.* **2015,** *15,* 321–327.

Clark, J. M.; Lambertsen, C. J. Pulmonary Oxygen Toxicity: A Review. *Pharmacol. Rev.* **1971,** *23,* 37–133.

Coburn, M.; Maze, M.; Franks, N. P. The Neuroprotective Effects of Xenon and Helium in an *In Vitro* Model of Traumatic Brain Injury. *Crit. Care Med.* **2008,** *36,* 588–595.

Collet, J. P.; Vanasse, M.; Marois, P.; Amar, M.; Goldberg, J.; Lambert, J.; Lassonde, M.; Hardy, P.; Fortin, J.; Tremblay, S. D.; Montgomery, D. Hyperbaric Oxygen for Children with Cerebral Palsy: A Randomised multicentre Trial. *The Lancet* **2001,** *357,* 582–586.

Compston, D. The Genetic Epidemiology of Multiple Sclerosis. In: *McAlpine's Multiple Sclerosis*; Compston, D., Ebers, G. C., Lassmann, H., McDonald, W. I., Matthews, W. B., Wekerle, H., Eds; 3rd ed, London: Churchill Livingstone, 1998 45–142.

Cook, D. J.; Reeve, B. K.; Griffith, L. E.; Mookadam, F.; Gibson, J. C. Multidisciplinary Education for Oxygen Prescription: A Continuous Quality Improvement Study. *JAMA Intern. Med.* **1996,** *156,* 1797–1801.

Crabtree, J. H.; Fishman, A. Videosocpic Surgery Under Local and Regional Anesthesia with Helium Abdominal Insufflation. *Surg. Endosc.* **1999,** *13,* 1035–1039.

Cronje, F. Hyperbaric Oxygen Therapy for Children with Cerebral Palsy. *S. Afr. Med. J.* **1999,** *89,* 359–361.

Daruwalla, J.; Christophi, C. Hyperbaric Oxygen Therapy for Malignancy: A Review. *World J. Surg.* **2006,** *30,* 2112–2131.

David, H. N.; Haelewyn, B.; Chazalviel, L.; Lecocq, M.; Degoulet, M.; Risso, J. J.; Abraini, J. H. Post-ischemic Helium Provides Neuroprotection in Rats Subjected to Middle Cerebral Artery Occlusion-induced Ischemia by Producing Hypothermia. *J. Cereb. Blood Flow Metab.* **2009,** *29,* 1159–1165.

Davis, W. B.; Rennard, S. I.; Bitterman, P. B.; Crystal, R. G. Pulmonary Oxygen Toxicity: Early Reversible Changes in Human Alveolar Structures Induced by Hyperoxia. *N. Engl. J. Med.* **1983,** *309,* 878–883.

De Felice, F. G.; Velasco, P. T.; Lambert, M. P.; Viola, K.; Fernandez, S. J.; Ferreira, S. T.; Klein W. L. Aß Oligomers Induce Neuronal Oxidative Stress Through an *N*-methyl-*D*-aspartate Receptor-dependent Mechanism That is Blocked by the Alzheimer Drug Memantine. *J. Biol. Chem.* **2007,** *282,* 11590–11601.

Deng, J.; Lei, C.; Chen, Y.; Fang, Z.; Yang, Q.; Zhang, H.; Cai, M.; Shi, L.; Dong, H.; Xiong, L. Neuroprotective Gases–Fantasy or Reality for Clinical Use? *Prog. Neurobiol.* **2014,** *115,* 210–245.

Diagnostic and Statistical Manual of Mental Disorders. American Psychiatric Association, 5th ed.; VA: American Psychiatric Association, 2013.

Dickinson, R.; Franks, N. P. Bench-to-bedside Review: Molecular Pharmacology and Clinical Use of Inert Gases in Anesthesia and Neuroprotection. *Crit. Care.* **2010,** *14,* 229.

Donald, K. W. Oxygen Poisoning in Man (I,II). *Br. Med. J.* **1947,** 667–672, 712–717.

Dong, X. X.; Wang, Y.; Qin, Z. H. Molecular Mechanisms of Excitotoxicity and Their Relevance to Pathogenesis of Neurodegenerative Diseases. *Acta Pharmacol. Sin.* **2009,** *30,* 379–387.

Emsley, J. *Nature's Building Blocks: An A-Z Guide to the Elements.* Oxford, England: Oxford University Press, 2001, pp. 297–304.

Esplugues, J. V. NO as a Signaling Molecule in the Nervous System. *Br. J. Pharmacol.* **2002,** *135,* 1079–1095.

Fitzgerald, J. M.; Baynham, R.; Powles, A. C. Use of Oxygen Therapy for Adult Patients Outside of the Critical Care Areas of a University Hospital. *Lancet* **1988,** *1,* 981–983.

Flores, R.; Muller, A.; Friedrich, V.; Sehba, F. A. Neutrophils in Early Vascular Injury After Subarachnoid Hemorrhage (SAH). *Soc. Neurosci. Abs.* **2009** Poster 51.5/I20.

Food and Drug Administration [website]. Beware of False or Misleading Claims for Treating Autism. Silver Spring, MD: Food and Drug Administration, 2017.

Foster, M. W.; Hess, D. T.; Stamler, J. S. Protein S-nitrosylation in Health and Disease: A Current Perspective. *Trends Mol. Med.* **2009,** *15,* 391–404.

Garry, P. S.; Ezra, M.; Rowland, M. J.; Westbrook, J.; Pattinson, K.T.S. The role of the nitric oxide pathway in brain injury and its treatment—from bench to bedside. *Exp. Neurol.* **2015,** *263,* 235–243

Gesell, L. B. *Hyperbaric Oxygen Therapy Indications,* 12th ed; Durham, NC: Undersea and Hyperbaric Medical Society, 2008.

Giles, T. D. Aspects of Nitric Oxide in Health and Disease: A Focus on Hypertension and Cardiovascular Disease. *J. Clin. Hypertens.* **2006,** *8,* 2–16.

Goodin, D. S.; Frohman, E. M.; Garmany, G. P.; Halper, J.; Likosky, W. H.; Lublin, F. D. Disease Modifying Therapies in Multiple Sclerosis. *Neurology* **2002,** *58,* 169–178.

Granpeesheh, D.; Tarbox, J.; Dixon, D. R.; Wilke, A. E.; Allen, M. S.; Bradstreet, J. J. Randomized Trial of Hyperbaric Oxygen Therapy for Children with Autism. *Res. Autism Spectr. Disord.* **2010,** *4,* 268–275.

Gray, L. H.; Conger, A.; Ebert, M.; Hornsey, S.; Scott, O. C. The Concentration of Oxygen Dissolved in Tissues at the Time of Irradiation as a Factor in Radiotherapy. *Br. J. Radiol.* **1953,** *26,* 638–648.

Guix, F. X.; Uribesalgo, I.; Coma, M.; Munoz, F. J. The Physiology and Pathophysiology of Nitric Oxide in the Brain. *Prog. Neurobiol.* **2005,** *76,* 126–152.

Haelewyn, B.; David, H. N.; Blatteau, J. E.; Vallée, N.; Meckler, C.; Risso, J. J.; Abraini, J. H. Modulation by the Noble Gas Helium of Tissue Plasminogen Activator: Effects in a Rat Model of Thromboembolic Stroke. *Crit. Care Med.* **2016,** *44,* e383–e389.

Harris P. D., Barnes, R. The Uses of Helium and Xenon in Current Clinical Practice. *Anaesthesia* **2008,** *63,* 284–293.

Hess, J. Oxygen Unit for Premature and Very Young Infants. *Am. J. Disease. Child.* **1934,** *47,* 916–917.

Higgins, R. D.; Bancalari, E.; Willinger, M.; Raju, T. N. Executive Summary of the Workshop on Oxygen in Neonatal Therapies: Controversies and Opportunities for Research. *Pediatrics* **2007,** *119,* 790–796.

Holbourn, A.; Wong, J. Oxygen Prescribing Practice at Waikato Hospital Does Not Meet Guideline Recommendations. *Int. Med. J.* **2014,** *44,* 1231–1234.

Hubschmann, O. R.; Kornhauser, D. Cortical Cellular Response in Acute Subarachnoid Hemorrhage. *J. Neurosurg.* **1980,** *52,* 456–462.

Hubschmann, O. R.; Kornhauser, D. Effect of Subarachnoid Hemorrhage on the Extracellular Microenvironment. *J. Neurosurg.* **1982,** *56,* 216–221.

Iscoe, S.; Fisher, J. A. Hyperoxia-induced Hypocapnia: An Underappreciated Risk. *Chest* **2005,** *128,* 430–433.

Iwase, K.; Miyanaka, K.; Shimizu, A.; Nagasaki, A.; Gotoh, T.; Mori, M.; Takiguchi, M. Induction of Endothelial Nitric-oxide Synthase in Rat Brain Astrocytes by Systemic Lipopolysaccharide Treatment. *J. Biol. Chem.* 275 (16), 11929–11933.

James, P. B. Evidence for Subacute Fat Embolism as the Cause of Multiple Sclerosis. *Lancet* **1982,** 380–386.

Jawad, N.; Rizvi, M.; Gu, J.; Adeyi, O.; Tao, G.; Maze, M.; Ma, D. Neuroprotection (and Lack of Neuroprotection) Afforded by a Series of Noble Gases in an *In Vitro* Model of Neuronal Injury. *Neurosci. Lett.* **2009,** *460,* 232–236.

Jenkinson, S. G. Oxygen Toxicity. *J. Intensive Care Med.* **1988,** *3,* 137–152.

Johnston, S. C.; Selvin, S.; Gress, D. R. The Burden, Trends, and Demographics of Mortality from Subarachnoid Hemorrhage. *Neurology* **1998,** *50,* 1413–1418.

Kacmarek, R. M.; Stoller, J. K.; Heuer, A. J.; Egan, D. F. *Egan's Fundamentals of Respiratory Care.* St. Louis, Missouri: Elsevier/Mosby, 2013, pp. 868–870.

Kety, S.; Schmidt, C. The Effects of Altered Arterial Tensions of Carbon Dioxide and Oxygen on Cerebral Blood Flow and Cerebral Oxygen Consumption of Normal Young Men. *J. Clin. Invest.* **1948,** *27,* 484–492.

Killory, B. D.; Chang, S. W.; Wait, S. D.; Spetzler, R. F. Use of Flexible Hollow-core $CO_2$ Laser in Microsurgical Resection of CNS Lesions: Early Surgical Experience. *Neurosurgery* **2010,** *66,* 1187–1192.

Kim, H. Y.; Singhal, A. B.; Lo, E. H. Normobaric Hyperoxia Extends the Reperfusion Window in Focal Cerebral Ischemia. *Ann. Neurol.* **2005,** *57,* 571–575.

Kinaci, N.; Kinaci, S.; Alan, M.; Elbuken, E. The Effects of Hyperbaric Oxygen Therapy in Children with Autism Spectrum Disorders. *Undersea Hyperbaric Med.* **2009,** *36,* 4.

Kindwall, E. P. Hyperbaric Oxygen's Effect on Radiation Necrosis. *Clin. Plast. Surg.* **1993,** *20,* 473–483.

Kindwall, E. P. A History of Hyperbaric Medicine. In: *Hyperbaric Medicine Practice;* Kindwall, E. P., Wheelan, H. T., Eds; 3rd ed, Flagstaff, AZ: Best Publishing, 2008, pp. 3-22.

Koblin, D. D.; Fang, Z.; Eger, E. I. I. I.; Laster, M. J.; Gong, D.; Ionescu, P.; Halsey, M. J.; Trudell, J. R. Minimum Alveolar Concentrations of Noble Gases, Nitrogen, and Sulfur Hexafluoride in Rats: Helium and Neon as Nonimmobilizers (Nonanesthetics). *Anesth. Analg.* **1998,** *87,* 419–424.

Kohgami, S.; Ogata, T.; Morino, T.; Yamamoto, H.; Schubert, P. Pharmacological Shift of the Ambiguous Nitric Oxide Action from Neurotoxicity to Cyclic GMP-mediated Protection. *Neurol. Res.* **2010,** *32,* 938–944.

Kohshi. K.; Beppu, T.; Tanaka, K.; Ogawa, K.; Inoue, O.; Kukita, I.; Clarke, R. E. Potential Roles of Hyperbaric Oxygenation in the Treatments of Brain Tumors. *Undersea Hyperb. Med.* **2013,** *40,* 351–362.

Komotar, R. J.; Schmidt, J. M.; Starke, R. M.; Claassen, J.; Wartenberg, K. E.; Lee, K.; Badjatia, N.; Connolly Jr, E. S.; Mayer, S. A. Resuscitation and Critical Care Of Poor-Grade Subarachnoid Hemorrhage. *Neurosurgery* **2009,** *64,* 397–410.

Kopelman, A. E.; Holbert, D. Use of Oxygen Cannulas In Extremely Low Birthweight Infants Is Associated With Mucosal Trauma and Bleeding, and Possibly with Coagulase-negative Staphylococcal Sepsis. *J. Perinatol.* **2003,** *23,* 94–97.

Koutna, N. Carboxytherapy—A New Non-invasive Method in Aesthetic Medicine. *Cas. Lek. Cesk.* **2006,** *145,* 841–843.

Lacey, D. J.; Stolfi, A.; Pilati, L. E. Effects of Hyperbaric Oxygen on Motor Function in Children with Cerebral Palsy. *Annals Neurol.* **2012,** *72,* 695–703.

Lamb, K.; Piper, D. Southmedic Oxy Mask Compared with the Hudson RCI Non-rebreather Mask: Safety and Performance Comparison. *Can. J. Respir. Ther.* **2016,** *52,* 13–15.

Lambertsen, C. J.; Dough, R. H.; Cooper, D. Y.; Emmel, G. L.; Loeschcke, H. H.; Schmidt, C. F. Oxygen Toxicity; Effects in Man of Oxygen Inhalation at 1 And 3.5 Atmospheres Upon Blood Gas Transport, Cerebral Circulation and Cerebral Metabolism. *J. Appl. Physiol.* **1953,** *5,* 471–486.

Li, H.; Witte, K.; August, M.; Brausch, I.; Gödtel-Armbrust, U.; Habermeier, A.; Gloss, E. I.; Oelze, M.; Munzel, T.; Forstermann, U. Reversal of eNOS Uncoupling and Upregulation of eNOS Expression Lowers Blood Pressure in Hypertensive Rats. *J. Am. Coll. Cardiol.* **2006,** *47,* 2536–2544.

Liu, L.; Stamler, J. S. NO: An Inhibitor of Cell Death. *Cell Death Differ.* **1999,** *6,* 937–942.

Lu, Z.; Ma, J.; Liu, B.; Dai, C.; Xie, T.; Ma, X.; Li, M.; Dong, J.; Lan, Q.; Huang, Q. Hyperbaric Oxygen Therapy Sensitizes Nimustine Treatment for Glioma in Mice. *Cancer Med.* **2016,** *5,* 3147–3155.

Ludemann, R.; Krysztopik, R.; Jamieson, G. G.; Watson, D. I. Pneumothorax During Laparoscopy. *Surg. Endosc.* **2003**, *17*, 1985–1989.

Macey, P. M.; Woo, M. A.; Harper, R. M. Hyperoxic Brain Effects are Normalized by Addition of $CO_2$. *PLoS Med.* **2007**, *4*, 828–836.

Maehara, T.; Shimizu, H. Surgical Outcome of Corpus Callosotomy in Patients with Drop Attacks. *Epilepsia* **2001**, *42*, 67–71.

Mannick, J. B.; Schonhoff, C.; Papeta, N.; Ghafourifar, P.; Szibor, M.; Fang, K.; Gaston, B. S-Nitrosylation of Mitochondrial Caspases. *J. Cell Biol.* **2001**, *154*, 1111–1116.

Manukhina, E. B.; Pshennikova, M. G.; Goryacheva, A. V.; Khomenko, I. P.; Mashina, S. Y.; Pokidyshev, D. A.; Malyshev, I. Y. Role of Nitric Oxide in Prevention of Cognitive Disorders in Neurodegenerative Brain Injuries in Rats. *Bull. Exp. Biol. Med.* **2008**, *146*, 391–395.

Manuskiatti, W.; Fitzpatrick, R. E.; Goldman, M. P. Long-term Effectiveness and Side Effects of Carbon Dioxide Laser Resurfacing for Photoaged Facial Skin. *J. Am. Acad. Dermatol.* **1999**, *40*, 401–411.

Maze, M. Preclinical Neuroprotective Actions Of Xenon And Possible Implications For Human Therapeutics: A Narrative Review. *Can. J. Anaesth.* **2016**, *63*, 212–226.

Melino, G.; Melino, G.; Bernassola, F.; Knight, R. A.; Corasaniti, M. T.; Nistic, G.; Finazzi-Agr, A. S-nitrosylation Regulates Apoptosis. *Nature* **1997**, *388*, 432–433.

Montgomery. D.; Goldberg. J.; Amar, M.; Lacroix, V. Effects of Hyperbaric Oxygen Therapy on Children with Spastic Diplegic Cerebral Palsy: A Pilot Project. *Undersea Hyperbar Med.* **1999**, *26*, 235.

Moon, R. E. Hyperbaric Oxygen Treatment for Decompression Sickness. *Undersea Hyperb. Med.* **2014**, *41*, 151–157.

Moreira, A. C.; Moreira, M.; Motta, R. L.; Moreira, Y. C.; Bettoni, A. P.; Tokunaga, H. H.; Fagundes, D. J. The Combination of Rhytidoplasty and Fractional CO2 Laser Therapy in the Treatment of Facial Aging. *Aesthetic Plast. Surg.* **2014**, *38*, 839–848.

Murphy, S. Production of Nitric Oxide by Glial Cells: Regulation and Potential Roles in the CNS. *Glia* **2000**, *29*, 1–14.

Myers, T. R. American Association for Respiratory Care. AARC Clinical Practice Guideline: Selection of an Oxygen Delivery Device For Neonatal And Pediatric Patients: 2002 Revision And Update. *Respir. Care.* **2002**, *47*, 707–716.

Nakao, A.; Sugimoto, R.; Billiar, T. R.; McCurry, K. R. Therapeutic Antioxidant Medical Gas. *J. Clin. Biochem. Nutr.* **2009**, *44*, 1–13.

New, A. Oxygen: Kill or Use? Prehospital Hyperoxia in COPD Patients. *Emerg. M. J.* **2006**, *23*, 144–146.

Oei, G. T.; Weber, N. C.; Hollmann, M. W.; Preckel, B. Cellular Effects of Helium in Different Organs. *Anesthesiology* **2010**, *112*, 1503–1510.

Oguni, H.; Olivier, A.; Andermann, F.; Comair, J. Anterior Callosotomy in the Treatment of Medically Intractable Epilepsies: A Study of 43 Patients with a Mean Follow-up of 39 months. *Ann. Neurol.* **1991**, *30*, 357–364.

Ortiz, A. E.; Goldman, M. P.; Fitzpatrick, R. E. Ablative $CO_2$ Lasers for Skin Tightening: Traditional Versus Fractional. *Dermatol. Surg.* **2014**, *40*, S147–S151.

Partington, J. R. *A Short History Of Chemistry*, 3rd ed; New York: Dover Publications, 1989, p. 90.

Pennazio, S. Air and the Origin of the Experimental Plant Physiology. *Biol. Forum/Rivista di Biologia* **2005**, 98.

Pineau, G.; Moqadem, K.; Obadia, A.; Perron, S. Place of Hyperbaric Oxygen Therapy in the Management of Cerebral Palsy; AETMIS 07-01; Montréal: AETMIS, 2007.

Preckel, B.; Schlack, W. Inert Gases as the Future Inhalational Anaesthetics? *Best Pract. Res. Clin. Anaesthesiol.* **2005**, *19*, 365–379.

Prineas, J.; Connell, F. Remyelination in Multiple Sclerosis. *Ann. Neurol.* **1979**, *5*, 22–31.

Querfurth, H. W.; LaFerla, F. M. Alzheimer's Disease. *N. Engl. J. Med.* **2010**, *362*, 329–344.

Rathnasiri Bandara, S.M. Paranasal Sinus Nitric Oxide and Migraine: A New Hypothesis on the Sinorhinogenic Theory. *Med. Hypotheses* **2013**, *80*, 329–340.

Reitz, C.; Brayne, C.; Mayeux, R. Epidemiology of Alzheimer disease. *Nat. Rev. Neurol.* **2011**, *7*, 137–152.

Rice, S. A. Human Health Risk Assessment of CO2: Survivors of Acute High-Level Exposure And Populations Sensitive To Prolonged Low-Level Exposure. *Environments* **2014**, *3*, 7–15.

Robertson, A. F. Reflections on Errors in Neonatology: I. The "Handsoff" Years, 1920 to 1950. *J. Perinatol.* **2003**, *23*, 48–55.

Rossignol, D. A.; Frye, R. E. A Review of Research Trends in Physiological Abnormalities in Autism Spectrum Disorders: Immune Dysregulation, Inflammation, Oxidative Stress, Mitochondrial Dysfunction and Environmental Toxicant Exposures. *Mol. Psychiatry* **2012**, *17*, 389–401.

Rossignol, D. A.; Rossignol, L. W.; James, S. J.; Melnyk, S.; Mumper, E. The Effects of Hyperbaric Oxygen Therapy on Oxidative Stress, Inflammation, and Symptoms in Children with Autism: An Open-label Pilot Study. *BMC Pediatr.* **2007,** *7,* 36.

Rossignol, D. A.; Rossignol, L. W.; Smith, S.; Schneider, C.; Logerquist, S.; Usman, A.; Neubrander, J.; Madren, E. M.; Hintz, G.; Grushkin, B.; Mumper, E. A. Hyperbaric Treatment For Children with Autism: A Multicenter, Randomized, Double-blind, Controlled Trial. *BMC Pediatr.* **2009,** *9,* 21.

Rougier, A.; Claverie, B.; Pedespan, M. J.; Marchal, C.; Loiseau, P. Callostomy for Intractable Epilepsy: Overall Outcome. *J. Neurosurg. Sci.* **1997,** *41,* 51–57.

Rudge, J.; Odedra, S.; Harrison, D. A New Oxygen Prescription Produces Real Improvements in Therapeutic Oxygen Use. *BMJ Quality Improvement Reports* **2014,** *3,* 1–5.

Saha, R. N.; Pahan, K. Regulation of Inducible Nitric Oxide Synthase Gene in Glial Cells. *Antioxid. Redox Signal.* **2006,** *8,* 929–947.

Saugstad, O. D. Oxygen Toxicity in the Neonatal Period. *Acta. Paediatr. Scand.* **1990,** *79,* 881–892.

Saugstad, O. D.; Ramji, S.; Vento, M. Resuscitation of Depressed Newborn Infants with Ambient Air or Pure Oxygen: A Meta-analysis. *Biol. Neonate* **2005,** *87,* 27–34.

Sayers, R. R.; Yant, W. P. The Value of Helium-oxygen Atmosphere in Diving and Caisson Operations. *Anesth. Analg.* **1926,** *5,* 127–138.

Schmidt, G.; Keesom, W. H. New Measurements of Liquid Helium Temperatures: I. The Boiling Point of Helium. *Physica* **1937,** *4,* 963–970.

Schob, O. M.; Allen, D. C.; Benzel, E.; Curet, M. J.; Adams, M. S.; Baldwin, N. G.; Largiader, F.; Zucker, K. A.; Fulton, R. L. A Comparison of the Pathophysiologic Effects of Carbon Dioxide, Nitrous Oxide and Helium Pneumoperitoneum on Intracranial Pressure. *Am. J. Surg.* **1996,** *172,* 248–253.

Schreiner, R. L.; Kisling, J. A. *Practical Neonatal Respiratory Care.* New York: Raven Press, 1982.

Sehba, F. A.; Ding, W. H.; Chereshnev, I.; Bederson, J. B. Effects of S-nitrosoglutathione on Acute Vasoconstriction and Glutamate Release after Subarachnoid Hemorrhage. *Stroke* **1999,** *30,* 1955–1961.

Sehba, F. A.; Friedrich, V. Jr.; Makonnen, G.; Bederson, J. B. Acute Cerebral Vascular Injury after Subarachnoid Hemorrhage and Its Prevention by Administration of a Nitric Oxide Donor. *J. Neurosurg.* **2007,** *106,* 321–329.

Sehba, F. A.; Mustafa, G.; Friedrich, V.; Bederson, J. B. Acute Microvascular Platelet Aggregation after

Subarachnoid Hemorrhage. *J. Neurosurg.* **2005,** *102,* 1094–1100.

Sehba, F. A.; Pluta, R. M.; Zhang, J. H. Metamorphosis of Subarachnoid Hemorrhage Research: from Delayed Vasospasm to Early Brain Injury. *Mol. Neurobiol.* **2011,** *43,* 27–40.

Sehba, F. A.; Schwartz, A. Y.; Chereshnev, I.; Bederson, J. B. Acute Decrease in Cerebral Nitric Oxide Levels after Subarachnoid Hemorrhage. *J. Cereb. Blood Flow Metab.* **2000,** *20,* 604–611.

Sharifipour, F.; Panahi-Bazaz, M.; Idani, E.; Malekahmadi, M.; Feizi, S. Normobaric Oxygen Therapy for Scleral Ischemia or Melt. *J. Ophthalmic Vis. Res.* **2012,** *7,* 275–280.

Shi, S. H.; Qi, Z. F.; Luo, Y. M.; Ji, X. M.; Liu, K. J. Normobaric Oxygen Treatment in Acute Ischemic Stroke: A Clinical Perspective. *Med. Gas Res.* **2016,** *6,* 147–153.

Singhal, A. B.; Dijkhuizen, R. M.; Rosen, B. R.; Lo, E. H. Normobaric Hyperoxia Reduces MRI Diffusion Abnormalities and Infarct Size in Experimental Stroke. *Neurology* **2002,** *58,* 945–952.

Small, D.; Duha, A.; Wieskopf, B.; Dajczman, E.; Laporta, D.; Kreisman, H.; Wolkove, N.; Frank, H. Uses and Misuses of Oxygen in Hospitalized Patients. *Am. J. Med.* **1992,** *92,* 591–595.

Sobey. C. G.; Quan, L. Impaired Cerebral Vasodilator Responses to NO and PDE V Inhibition after Subarachnoid Hemorrhage. *Am J. Physiol.* **1999,** *277,* 1718–1724.

Steinhorn, B. S.; Loscalzo, J.; Michel, T. Nitroglycerin and Nitric Oxide—A Rondo of Themes in Cardiovascular Therapeutics. *N. Engl. J. Med.* **2015,** *373,* 277–280.

Stępień, K.; Ostrowski, R. P.; Matyja, E. Hyperbaric Oxygen as an Adjunctive Therapy in Treatment of Malignancies, Including Brain Tumours. *Med. Oncol.* **2016,** *33,* 101.

Stuhr, L. E.; Raa, A.; Oyan, A. M.; Kalland, K. H.; Sakariassen, P. O.; Petersen, K.; Bjerkvig, R.; Reed, R. K. Hyperoxia Retards Growth and Induces Apoptosis, Changes in Vascular Density and Gene Expression in Transplanted Gliomas in Nude Rats. *J. Neurooncol.* **2007,** *85,* 191–202.

Sullivan, J. "Two Flavors of Ischemia." Brain Ischemia 101. Emergency Medicine Cerebral Resuscitation Lab. (accessed Oct 13, 2008).

Sullivan, J. "What is Brain Ischemia?" WSU Emergency Medicine Cerebral Resuscitation Laboratory. (accessed Nov 11, 2008).

Sun, J.; Steenbergen, C.; Murphy, E. S-nitrosylation: NO-related Redox Signaling to Protect Against Oxidative Stress. *Antioxid. Redox. Signal.* **2006,** *8,* 1693–1705.

Suzanne, C.; Lareau, R. N.; Bonnie Fahy, R. N. Oxygen Therapy. *Am. J. Respir. Crit. Care Med.* **2005,** *171,* P1–P2.

Tamir, G.; Issa, M.; Yaron, H. S. Mobile Phone-triggered Thermal Burns in the Presence of Supplemental Oxygen. *J. Burn Care Res.* **2007,** *28,* 348–350.

Tao, T.; Liu, Y.; Zhang, J.; Xu, Y.; Li, W.; Zhao, M. Therapeutic Hypercapnia Improves Functional Recovery and Attenuates Injury via Antiapoptotic Mechanisms in a Rat Focal Cerebral Ischemia/Reperfusion Model. *Brain Res.* **2013,** *1533,* 52–62.

Tomita, R.; Kurosu, Y.; Munakata, K. Relationship Between Nitric Oxide and Non-adrenergic Non-cholinergic Inhibitory Nerves in Human Lower Esophageal Sphincter. *J. Gastroenterol.* **1997,** *32,* 1–5.

Walsh, B. K.; Smallwood, C. D. Pediatric Oxygen Therapy: A Review and Update. *Respir. Care.* **2017,** *62,* 645–661.

Weaver, L. K. Hyperbaric Oxygen Therapy for Carbon Monoxide Poisoning. *Undersea Hyperb. Med.* **2014,** *41,* 339–354.

Weber, N. C.; Smit, K. F.; Hollmann, M. W.; Preckel, B. Targets Involved in Cardioprotection by the Non-anesthetic Noble Gas Helium. *Curr. Drug Targets* **2015,** *16,* 786–792.

West, J. B. Carbon Dioxide, Latent Heat, and the Beginnings of the Discovery of the Respiratory Gases. *Am. J. Physiol. Lung Cell. Mol. Physiol.* **2014,** *306,* L1057–L1063.

Wong, T. T.; Kwan, S. Y.; Chang, K. P.; Hsiu-Mei, W. Yang, T. F.; Chen, Y. S.; Yi-Yen, L. Corpus Callosotomy in Children. *Childs Nerv. Syst.* **2006,** *22,* 999–1011.

Yetik-Anacak, G.; Catravas, J. D. Nitric Oxide and the Endothelium: History and Impact on Cardiovascular Disease. *Vascul. Pharmacol.* **2006,** *45,* 268–276.

Zhou, Q.; Cao, B.; Niu, L.; Cui, X.; Yu, H.; Liu, J.; Li, H.; Li, W. Effects of Permissive Hypercapnia on Transient Global Cerebral Ischemia–Reperfusion Injury in Rats. *Anesthesiology* **2010,** *112,* 288–297.

Zupanets, M. V.; Lagutina, A. S.; Pavliy, E. K. The History of Carboxytherapy. The National University of Pharmacy, Ukraine, 2015.

# PART III
## Miscellaneous Drugs

# CHAPTER 22

# Pharmacotherapy of Neurochemical Imbalances

RUPALI PATIL[1*], AMAN UPAGANLAWAR[2], and SUVARNA INGALE[3]

[1]GES's SDMSG College of Pharmaceutical Education and Research, Maharashtra, India

[2]SNJB's SSDJ College of Pharmacy, Maharashtra, India

[3]SCES's Indira College of Pharmacy, Maharashtra, India

*Corresponding author. E-mail: ruupalipatil@rediffmail.com

## ABSTRACT

Human brain is made up of billions of neurons which communicate with each other through chemical messengers which are referred to as neuroactive substances. These neuroactive substances include neurotransmitters, neuromodulators, neurohormones, and neuromediators. Some neurotransmitters also act as neuromodulators and neurohormones. It would be cynical if there will ever be general covenant about the meanings of these neuroactive substances including neurotransmitters, since the term neurotransmitter has traditionally been used very loosely indeed which includes neurotransmitters, neurohormones, neuromodulators, and neuromediators. Any alterations in functioning of these neuroactive substances can cause diseases. At the same time, these neuroactive substances also act as best targets for drug research in the treatment of various brain diseases. Hence, the main objective of this chapter is to differentiate and understand neurotransmitters, neuromodulators, neurohormones, and neuromediators, to focus on different neurochemicals and their role in neuropsychiatric disorders, and treatment of such central nervous system (CNS) ailments mainly by managing neurochemical imbalances.

## 22.1 INTRODUCTION

The human brain is the most multifaceted component in the body with specific structural and functional properties. It is made up of a huge network of billions of interconnected neurons that exceed any social network. With the help of this huge network of billions of interconnected neurons, brain collects and stores huge information exceeding a supercomputer (Uddin et al., 2017; Guyton and Hall, 2011). This super ability of brain empowers humans to accomplish amazing milestones such as reaching the moon, identifying human genome, creating artistic sculptures, composing musical bits, and script literature (Brookhart and Mountcastle, 1977). The brain controls each and every biological process in our body like heart rate, sexual activity, emotion, thoughts, hopes, dreams, imaginations, learning, and memory. Brain plays an important role in evoking immunological reactions to disease. Brain is the house of billions of neurons that convey signal to other neurons by secreting chemical messengers known as neurotransmitters, at synapse—the contact point where one neuron communicates with another. Neurotransmitter substance after releasing at neuronal terminals moves across the synapse. It then combines with the receptors on the target cell and thus regulates every facet of our body (Guyton and Hall, 2011).

Chemical substance, which is produced in a neuron and alters the functions of other neurons or muscle cells is known as natural neuroactive substance. These include neurotransmitters, neuromodulators, and neurohormones. Some natural neuroactive constituents play dual role as transmitters and hormones (Dismukes, 1979). Several authors favor to entitle all neuroactive substances under the single title neurotransmitters. For them, neuromodulators are a type of transmitter, and neurohormones are a type of neuromodulator. In fact, it would be doubtful if there will ever be general agreement about the definitions of these neuroactive substances including neurotransmitters as the term, neurotransmitter has historically been used very lightly (Hoyle, 1985).

Although it is unlikely that a set of definitions will seem satisfactory, the following definitions could prove useful.

- A substance synthesized and discharged in a neuron at structurally and functionally specific junction (synapse), which diffuses through a synaptic cleft to activate or stimulate postsynaptic neurons or muscle or effector cell, is known as neurotransmitter. Example includes: Acetylcholine (ACh), Noradrenaline/ Norepinephrine (NA/NE), Adrenaline/Epinephrine (A/EPN), Dopamine (DA), histamine, gamma-aminobutyric acid (GABA), glycine, serotonin or 5-hydroxytryptamine (5-HT), and glutamate (Oja and Saransaari, 2009).

- A neuromodulator interacts with the suitable receptors affecting neurons or effector cells and is released from neuronal terminal in the CNS or peripheral nervous system (PNS). It is not always secreted in synapse, but frequently works through second messengers and thereby generates long-term actions. Sometimes, neuromodulator is released in close proximity affecting only adjacent neurons or effectors or sometimes release may be more widespread (Dismukes, 1979). Opioid peptides like enkephalins, dynorphins and endorphins, circulating steroid hormones, neurosteroids, locally released adenosine, and other purines, eicosanoids, Carbon monoxide (CO) and nitric oxide (NO) are all now regarded as neuromodulators (Barchas et al., 1978).

- A neurohormone is a substance released from neuronal terminals into the blood and lymph and thereby acts on distant peripheral targets. They vary from a neuromodulator only with respect to extent of its action. The most common examples of neurohormones are oxytocin and vasopressin (Barker, 1977; Barker, 1978).

- A neuromediator is a substance that is involved in evoking response when a transmitter binds with receptor site. Examples of such neuromediators include inositol phosphates, adenosine 3',5'-cyclic monophosphate (cAMP), guanosine 3',5'-cyclic monophosphate (cGMP) which are known as second messengers (Kulinsky and Kolesnichenko, 2005; Burrows, 1996; Bloom, 2006).

From the above-mentioned definitions, it was understood that the functions of natural neuroactive substances overlap freely, especially of neuromodulators and neurohormones. It has been revealed that neuromodulators and neurohormones do not necessarily bind with specific receptors present on cell membranes and so, do not always associate with adjustable ion channels. They produce their effects after passing into the specific cells. Moreover, it is increasingly pragmatic that more than one chemical substance with only one of these be a transmitter and the other a modulator, may be secreted at any given synapse synchronously or asynchronously (Hoyle, 1985). Utmost all neuropsychiatric disorders occur due to neurochemical imbalances. Therefore, the main objective of this chapter is to emphasize different neurochemicals, their utility in the management of neuropsychiatric disorders mainly by restoring the neurochemical imbalances.

## 22.2 NEUROTRANSMITTERS

Neurotransmitter functions as a facilitator for the neurotransmission from neuron to neuron through a synapse (Oja and Saransaari, 2009). TR Elliott; a graduate student at Cambridge established the first belief in 1904 that chemical entities may be accountable for neuronal communication (Sembulingam and Sembulingam, 2012). According to Elliott, a chemical substance is produced from nerve terminals upon stimulation in small amount to elicit actions on the effect or cell or organ (Barchas et al., 1978). However, the first convincing evidence for involvement of chemical substance in neurotransmission was demonstrated by Loewi's experiment in 1921, wherein the observed release of vagusstoff subsequent to vagal stimulation. Similarly, Cannon and Uridil

in 1921 demonstrated that through sympathetic activation of nerves liver cells release "sympathin." These two compounds namely vagusstoff and sympathin later were recognized as ACh and EPN, respectively. ACh and EPN are two well-studied neurotransmitters that received widespread investigation (Barchas et al., 1978).

### 22.2.1 CRITERIA

A substance/chemical should fulfill following standards to qualify as a neurotransmitter molecule:

- The chemical should be located presynaptically, probably distributed throughout the brain.
- The precursors and enzymes needed for biosynthesis should be present in the cytoplasm of nerve cell.
- Depolarization of nerve terminal should discharge the substance in substantial amount.
- The effect produced by direct administration or by application of the chemical to the synapse should be similar to that evoked by stimulation of nerve terminal.
- The substance should bind with specific receptors that are located on presynaptic neuron.
- The substance after binding with receptor should alter postsynaptic membrane potential to elicit the effect.
- The effects produced by substance should be washed off in reasonable timeframe through an inactivating mechanism or through enzymatic degradation.
- The effects of direct administration or application of the substance to the synapse must be equally responsive to inactivating mechanisms and likewise

must be altered by interventions at postsynaptic sites.

- The effects of direct administration or application of the substance and normal neurotransmission at synaptic cleft must be blocked or reversed by the same antagonist (Ayano, 2016; Smith, 2002; Barchas et al., 1978; Werman, 1966).

## 22.2.2  CLASSIFICATION

### 22.2.2.1  BASED ON FUNCTION

Neurotransmitters functions as either excitatory or inhibitory on postsynaptic neuron. Accordingly, neurotransmitters may be classified as excitatory and inhibitory.

#### 22.2.2.1.1  *Excitatory Neurotransmitters*

It is a chemical substance, which is accountable for the conduction of nerve impulse from presynaptic to postsynaptic neuron. Alterations in the resting membrane potential of postsynaptic neuron through altered membrane permeability to sodium ions by a neurotransmitter secreted from the nerve terminal is called slow excitatory postsynaptic potential (EPSP). However, it does not directly alter the postsynaptic potential. Most important excitatory neurotransmitters are ACh and NA (Sadock, 2009).

#### 22.2.2.1.2  *Inhibitory Neurotransmitters*

It is a chemical substance, accountable for inhibiting the conduction of impulse from presynaptic to postsynaptic neuron. Alteration in the membrane potential of postsynaptic neuron through altered membrane permeability to potassium ions by a neurotransmitter

secreted from the nerve terminal is called inhibitory postsynaptic potential (IPSP). The most important inhibitory neurotransmitters are GABA and DA (Stahl, 2008; Sadock, 2009).

### 22.2.2.2  BASED ON CHEMICAL NATURE

Many substances of different chemical nature are identified as neurotransmitters. Depending upon their chemical nature, neurotransmitters are classified into five groups:

- **Amino Acids:** These neurotransmitters mediate fast synaptic transmission and functions as inhibitory and excitatory in nature. Examples include GABA, glycine, glutamate, and aspartate (Ayano, 2016).
- **Amines:** These are chemically transformed amino acids. They mediate slow synaptic transmission. These also functions as inhibitory and excitatory in action. Examples include NA, adrenaline, DA, serotonin, and histamine (Ayano, 2016).
- **Trace Amines:** For example, phenethylamine, N-methylphenethylamine, tyramine, 3-iodothyronamine, octopamine, tryptamine, etc.
- **Peptides:** For example opioid peptides, substance P, etc.
- **Purines:** For example adenosine triphosphate (ATP), adenosine, etc.

There are some substances which do not fall in any of above category. One such substance is ACh. It is formed from the choline and acetyl coenzyme A (CoA) in the presence of the enzyme called choline acetyltransferase (CAT or ChAT). Another substance included in this category is the soluble gas NO (Ayano, 2016; Bloom, 2006).

### 22.2.2.3   BASED ON SIZE

Based on size neurotransmitters are classified into two groups:

#### 22.2.2.3.1   Small-molecule Neurotransmitters

These include ACh, biogenic amines-NA, histamine, DA, serotonin, amino acids, ATP, and NO (Tortora and Derrickson, 2009).

#### 22.2.2.3.2   Neuropeptides

These include enkephalins, endorphins, dynorphins, substance P, angiotensin II, cholecystokinin (CCK), and hypothalamic releasing and inhibiting hormones. Neuroactive substances, composed of 3–40 amino acids connected by peptide linkage are called neuropeptides, are found plentifully in both the CNS and the peripheral nervous system. They are synthesized in the cell body of neuron, stored into synaptic vesicles, and conveyed to axon terminals (Tortora and Derrickson, 2009).

### 22.2.3   LIFE CYCLE

To realize the concept of neurotransmitter, it is utmost necessary to understand the dynamics of neurotransmitter and neurotransmission, as each process involved in it may reveal a probable target of action for the numerous classes of drugs acting as potentiating or inhibiting agents at a given neurotransmitter pathway (Lodish et al., 2000).

Life cycle of neurotransmitter involves its biosynthesis, storage, release, receptor stimulation, and inactivation and reuptake from the synaptic cleft as depicted in Figure 22.1.

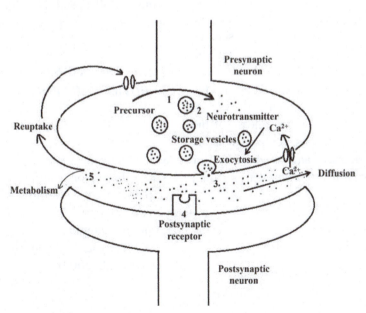

**FIGURE 22.1**   Life cycle of neurotransmitters.

### 22.2.3.1   BIOSYNTHESIS

The cell body of the neuron is the major site for production of many neurotransmitters in presence of enzymes which are found in the cell body and are conveyed through axon to the nerve terminals. They are produced from modest and abundant amino acids precursors. These precursors freely obtained from the diet undergo synthetic conversion reactions in presence of necessary enzymes to form neurotransmitter substances (Ayano, 2016).

### 22.2.3.2   STORAGE AND RELEASE

Neurotransmitters are stored in small membrane lined organelles called synaptic vesicles located at the axon terminal. Both small (around 50 nm) and large (around 100 nm) vesicles containing granules are present in monoaminergic neurons. NE, DA, and 5-HT are stocked in cell bodies, axons, and nerve terminals (Tortora and Derrickson, 2009). However, enzymes such as DA-β-hydroxylase and quite small amounts of amines are stored in large vesicles in the axons. The arrival of a nerve action potential at an axonal terminal, discharge enough amount of the transmitter into the synaptic cleft. Influx of $Ca^{2+}$ ions into axonal cytoplasm stimulates the release of the transmitter into the synaptic cleft by exocytosis which is the utmost noteworthy phase responsible for release of the transmitter. After exocytosis, empty synaptic vesicles are reprocessed and used again for recharging them with neurotransmitter (Webster, 2001).

### 22.2.3.3   RECEPTOR ACTIVATION

After release from the storage vesicles, the first initiating event in neurotransmission is binding of released quanta of neurotransmitter with specific postsynaptic receptors which then produce either an excitatory or inhibitory postsynaptic potential and thereby regulates every physiological process in the body. It is all possible and observed that one neurotransmitter characteristically stimulates several different subtypes of receptors like ligand-gated ion channels or G-protein coupled receptors (GPCR) (Nestler et al., 2009).

### 22.2.3.4   INACTIVATION AND REUPTAKE

After the execution of the action, neurotransmitter is inactivated by different mechanisms such as diffusion out of synaptic cleft, inactivation or disintegration by specific enzymes, engulfment by astrocytes (macrophages) and reuptake into the axon terminal. Many neurotransmitters are actively taken back (reuptake) into the nerve terminals with the help of membrane proteins known as neurotransmitter transporters and repacked into new synaptic vesicles for further actions (Muller and Nistico, 1989; Edwards, 2007).

### 22.2.4   IMPORTANT NEUROTRANSMITTERS

### 22.2.4.1   ACETYLCHOLINE

ACh is chemically acetyl ester of choline. ACh is the major neurotransmitter at all parasympathetic nerve terminals except at some sympathetic nerve like neuromuscular junction, autonomic ganglia, basal forebrain complex, and pontomesencephalic cholinergic complex (Nestler et al., 2009).

### 22.2.4.1.1  Biosynthesis, Storage, and Release

ACh is produced in cholinergic neurons from choline and acetate as depicted in Figure 22.2. Acetate is derived from acetyl coenzyme A which is formed from glucose during glycolysis or from citrate. Choline is derived by reuptake from the synaptic cleft and partly from the blood. The synthetic step converting choline and acetate into ACh is catalyzed by ChAT (Oja and Saransaari, 2009).

### 22.2.4.1.2  Receptor Activation

ACh (cholinergic) receptors can be delineated into two type such as muscarinic acetylcholine receptors (mAChRs) and nicotinic receptors (nAChRs) depending on their pharmacologic properties. The mAChRs are G-protein-coupled receptors. Muscarinic receptors (Table 22.1) can be further divided according to G-protein coupling as $M_1$, $M_3$, and $M_5$ receptors which are attached with Gq protein, and $M_2$ and $M_4$ receptors which are attached with Gi. The nAChRs (Table 22.2) are ligand gated ion channel. They can further be divided into three main classes, the muscle, ganglionic, and CNS types according to their location (Nestler et al., 2009).

### 22.2.4.1.3  Inactivation

ACh is quickly deactivated by hydrolysis to choline and acetate by an enzyme Acetylcholinesterase (AChE) as depicted in Figure 22.2. This process is rapid enough to elucidate the pragmatic alterations in sodium ion conductance and thereby electrical activity during synaptic transmission (Smith, 2002; Barrettet et al., 2009; Oja and Saransaari, 2009; Nestler et al., 2009).

**TABLE 22.1**  Types of mAChRs.

| Receptors | $M_1$ (Neural) | $M_2$ (Cardiac) | $M_3$ (Glandular) |
|---|---|---|---|
| Locations | Ganglia (autonomic and enteric), gastric paracrine cells, CNS (cortex and hippocampus) | SA node, AV node, atrium, ventricle; neural; presynaptic terminals | Exocrine glands, smooth muscles, vascular endothelium |
| G-protein | $G_{q/11}$ | $G_{i/o}$ | $G_{q/11}$ |
| Second messengers | Inositol 1,4,5-triphosphate ($IP_3$), diacylglycerol (DAG) | cAMP | $IP_3$, DAG |
| Functions | Gastric acid secretion, gastrointestinal tract (GIT) motility, CNS excitation | Decrease rate of impulse generation, conduction velocity and contractility, vagal bradycardia | Increase exocrine secretions, smooth muscle contraction |
| Agonists | $M_c$N-A-343, oxotremorine | Methacholine | Bethanechol |
| Antagonists | Pirenzepine, telenzepine | – | 1,1-dimethyl-4-diphenyl-acetoxypiperidinium iodide (4-DAMP), Hexahydrocele difenidol (HHSD), Darifenacin |

*Source*: Adapted from Sharma and Sharma (2017) and Nestler et al. (2009).

**TABLE 22.2** Types of Nicotinic Receptors.

| Receptors | Muscle type ($N_M$) | Neuronal type ($N_N$) | CNS type |
|---|---|---|---|
| Locations | At skeletal NMJ (postsynaptic) | At all autonomic ganglia and at adrenal medulla (postsynaptic) | At sensory nerve terminals and in other parts of brain but mostly located presynaptically. |
| Types | Ligand gated ion channel family receptors—opening for cation channel (end plate depolarization) | | |
| Functions | Muscle contraction | Conduction of electrical impulse through autonomic ganglia and stimulating postganglionic neuron, secretion of NE, and EPN from adrenal medulla | Presynaptic facilitation of the release of DA and glutamate |
| Agonists | ACh, succinylcholine, PTMA, nicotine | Nicotine, DMPP, epibatidine | – |
| Antagonists | D-tubocurarine, α-bungarotoxin | – | – |

*Source*: Adapted from Sharma and Sharma (2017) and Nestler et al. (2009).

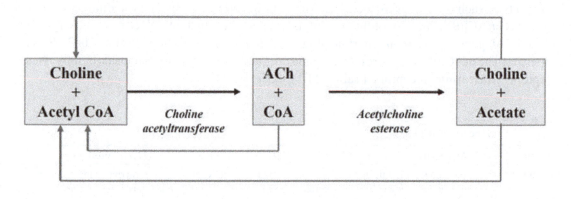

**FIGURE 22.2** Biosynthesis and degradation of ACh. Where, CoA, Coenzyme A.

### 22.2.4.2 MONOAMINES—DA, EPINEPHRINE, NE, SEROTONIN, AND HISTAMINE

Neurotransmitters with one –NH$_2$ group attached to an aromatic ring by a –CH$_2$-CH$_2$ chain are called monoamine neurotransmitters, for example, DA, NE, and epinephrine (EPN; synthesized in one biosynthetic pathway), as well as 5-HT or serotonin and histamine. As monoamine neurotransmitters sometimes modify the responses of neurons to excitatory and inhibitory amino acids via G-protein linked receptor system, they are also referred to as neuromodulators (Webster, 2001).

### 22.2.4.2.1 Biosynthesis, Storage, and Release

Biosynthesis of catecholamine (CA) neurotransmitters, DA, NE, and EPN initiates with amino acid tyrosine through sequential biosynthetic enzyme catalyzed reaction as depicted in Figure 22.3. CA biosynthesis starts with amino acid tyrosine obtained from diet, which moves by active transport into the brain (or peripheral sympathetic neurons). Within neurons, it undergoes hydroxylation in presence of an enzyme tyrosine hydroxylase (TH) to form di-hydroxyphenylalanine (DOPA). Ferrous ion acts as a cofactor for TH. This reaction also needs presence of molecular oxygen and a hydrogen donor—tetrahydrobiopterin (Oja and Saransaari, 2009).

It is important to note that synthesis of DA in dopaminergic neurons needs another enzyme L-aromatic amino acid decarboxylase to convert di-hydroxyphenylalanine to DA. L-aromatic amino acid decarboxylase is an enzyme present in neuronal cytoplasm and requires pyridoxal phosphate as a cofactor derived from vitamin $B_6$. While an enzyme dopamine-β-hydroxylase converts DA to NE in noradrenergic neurons (Rang et al., 2011). Dopamine-β-hydroxylase (DBH) requires $Cu^{2+}$ and ascorbic acid (vitamin C) as cofactors. DBH is bound to synaptic vesicles containing NE. In addition, another enzyme, phenylethanolamine-*N*-methyltransferase, converts NE to EPN in the adrenal medulla and in brainstem using *S*-adenosyl-l-methionine (SAM), a methyl donor. In these cells, NE leaves the storage vesicles by a process of diffusion, is methylated in adrenal medulla to epinephrine, and then gets stored in newly recruited storage vesicles (Nestler et al., 2009; Kuhar et al., 1999; Barrett et al., 2009). In synaptic storage vesicles, NE and EPN are stored linked with ATP and chromogranin A. CAs are stored into the granulated synaptic vesicles by transportation with two vesicular transporters. CAs are discharged from adrenergic neurons and adrenal medulla cells by the process of

exocytosis along with ATP, chromogranin A, and DBH as they are also stored in vesicles (Sharma and Sharma, 2017).

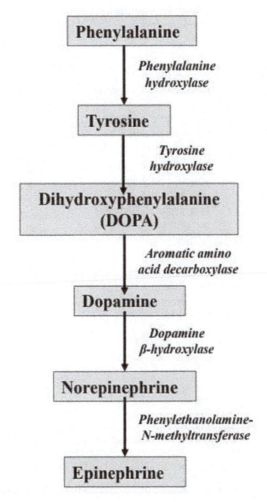

**FIGURE 22.3**  Biosynthesis of DA, NE, and epinephrine.

### 22.2.4.2.2  Receptor Activation

Once NE is secreted from adrenergic neuron, it combines with postsynaptic adrenergic receptors and stimulates the receptors and thereby evokes various physiological responses. The adrenoceptors (Table 22.3) are

**TABLE 22.3**   Types and Functions of Adrenergic Receptors.

| Receptors | Locations | G-protein | Second messengers | Functions | Agonists | Antagonists |
|---|---|---|---|---|---|---|
| $\alpha_1$ | Postsynaptic, most smooth muscles, salivary glands, liver cells | $G_q$ | ↑ IP3; ↑ DAG | ↑ $Ca^{2+}$ conc., contractions of smooth muscle ↑ Secretions of glands | Phenylephrine, methoxamine | Prazosin, indoramin |
| $\alpha_2$ | Presynaptic on adrenergic or cholinergic nerve terminals; postsynaptic in brain; ß-Pancreatic cells vascular smooth muscle | $G_i$ | ↓ cAMP | ↓ NE release ↓ Central sympathetic outflow ↓ Insulin release | Clonidine | Yohimbine, prazosin |
| $\beta_1$ | Postsynaptic at cardiac muscle, Juxtaglomerular apparatus' also presynaptic at adrenergic and cholinergic nerve terminal (but mainly postsynaptic) | $G_s$ | ↑ cAMP | ↑ Heart rate ↑ Force of contraction ↑ Renin release | Isoproterenol, terbutaline | Alprenolol, betaxolol, propranolol |
| $\beta_2$ | Post as well as presynaptic in bronchi, coronary arteries, uterus and smooth muscles; also in myocardium | $G_s$ | ↑ cAMP | ↑ NE release, relaxation of smooth muscle ↑ Glycogenolysis ↑ Heart rate ↑ Force of contractions | Procaterol, zinterol | Propranolol |
| $\beta_3$ | Postsynaptic at adipocytes | $G_s$ | ↑ cAMP | ↑ Lipolysis Thermogenesis | Propranolol | Pindolol, bupranolol |

Where, ↑, Increase; ↓, Decrease.

*Source*: Adapted from Sharma and Sharma (2017) and Nestler et al. (2009).

**TABLE 22.4**   Types and Functions of DA Receptors.

| Receptors | Locations | G-protein | Second messengers | Functions | Agonists | Antagonists |
|---|---|---|---|---|---|---|
| $D_1$ | Postsynaptic, brain, renal, mesenteric, coronary vascular bed and myocardium | Gs | ↑cAMP and/or $PIP_2$ hydrolysis | ↑ Myocardial contractility Vasodilatation of renal vasculature | SKF38393, fenoldopam | Bromocriptine, lisuride |
| $D_2$ | Mainly in CNS, also presynaptically at nerve terminals, also in renal vascular bed | Gi | ↓cAMP ↑$K^+$ conductance, Blocks $Ca^{2+}$ channels | ↓ NE release Modulates DA release | SCH23390, SKF83566, haloperidol | Sulpiride |

↑, Increase; ↓, Decrease.

*Source*: Adapted from Sharma and Sharma (2017) and Nestler et al. (2009).

classified into two classes: α and β (Rang et al., 2011). α-adrenoceptors are further subdivided into two types: $α_1$ and $α_2$. β-adrenoceptors are further subdivided into three subtypes: $β_1$, $β_2$, and $β_3$ (Muller et al., 1978). EPN and NE produce actions on α and β receptors. However, NE has predominant actions on α-adrenergic receptors, and epinephrine has predominant actions on β-receptors. As studied extensively earlier, α and β receptors are characteristic GPCRs existing in numerous forms (Nestler et al., 2009). All β-adrenoceptors are Gs-protein-coupled, and most of $α_1$ adrenoceptors are Gq-protein-coupled; $α_2$ receptors, which are generally Gi-coupled, function as inhibitory autoreceptors and as postsynaptic receptors. DA receptors (Table 22.4) are categorized into two classes: $D_1$ and $D_2$ family. $D_1$ family includes $D_1$ and $D_5$ receptors, coupled to Gs or the related $G_{olf}$. $D_2$ family includes $D_2$, $D_3$, and $D_4$ receptors, coupled to Gi/Go. $D_2$ and $D_3$ receptors function as inhibitory presynaptic autoreceptors and as postsynaptic receptors (Stanford, 2001; Muller and Nistico, 1989).

### 22.2.4.2.3 Inactivation

Monoamine neurotransmitters released from the synaptic cleft are destroyed by reuptake into the nerve terminal via specialized transporter proteins situated on the presynaptic neuronal membrane, or catabolism. NE is mainly inactivated by reuptake. Monoamine oxidase (MAO) and catechol-O-methyltransferase (COMT) are the two major enzymes involved in inactivation of these monoamines (Rang et al., 2011).

MAO is expressed in two forms namely—intracellular and extracellular forms. The intracellular form of MAO is most predominant in the outer membrane of mitochondria found abundant in presynaptic terminals. The intracellular form inactivates monoamines once they are taken up into presynaptic terminals (Rang et al., 2011). While, the extracellular form inactivates monoamine neurotransmitter when present in the synapse. MAO enzymes exist in two types: MAO-A and MAO-B. CAs are also catabolized by COMT. COMT also exists in two forms: peripherally major soluble isoform and central membrane-bound isoform. CAs undergo methylation reaction in presence of COMT using S-adenosylmethionine as a methyl donor. Homovanillic acid and vanillyl mandelic acid are the principal metabolites of DA and NE formed from the enzymatic degradation of CAs by MAO and COMT are as shown in Figures 22.4 and 22.5 (Webster, 2001).

Generally, measurement of cerebrospinal fluid (CSF), blood, and urine level of these metabolites was used as an indirect measure of brain catecholaminergic dysfunction in depression and schizophrenia. However, activity of the sympathetic nervous system and adrenal medulla and by many other factors complicates interpretation of these metabolite levels significantly thereby limiting its practicality as markers of CNS CA function (Nestler et al., 2009).

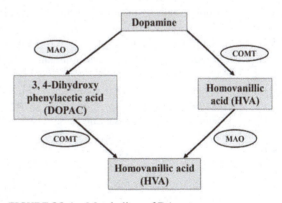

**FIGURE 22.4** Metabolism of DA.

### 22.2.4.3  SEROTONIN

Serotonin exists abundantly in blood platelets and GIT. In GIT, it is found in the enterochromaffin cells and the myenteric plexus. It is also found within the brain stem. 5-HT is produced from an essential amino acid, tryptophan. Tryptophan is hydroxylated in presence of an enzyme tryptophan hydroxylase to produce 5-hydroxyl tryptophan, which further undergoes decarboxylation to form 5-HT by aromatic L-amino acid decarboxylase (Oja and Saransaari, 2009).

When 5-HT is released from serotonergic neurons, some amount of serotonin released is recollected back in serotonergic neurons by an active reuptake mechanism and some amount is inactivated by MAO and aldehyde dehydrogenase into 5-hydroxyindole acetic acid as shown in Figure 22.6 (Barrett et al., 2009; Nestler et al., 2009).

At present, there are seven types of 5-HT receptors (5-HT$_1$ to 5-HT$_7$) as stated in Table 22.5. Of which 5-HT$_1$ and 5-HT$_2$ are further divided. 5-HT$_1$ receptors have 5-HT$_{1A}$, 5-HT$_{1B}$, 5-HT$_{1D}$, 5-HT$_{1E}$, and 5-HT$_{1F}$ subtypes. 5-HT$_1$ receptors are inhibitory presynaptic receptors, located mainly in CNS. 5-HT$_2$ receptors are postsynaptic receptors present mainly in periphery and to some extent in CNS. There are three subtypes of 5-HT$_2$ receptors: 5-HT$_{2A}$, 5-HT$_{2B}$, and 5-HT$_{2C}$. There are two subtypes of 5-HT$_5$: 5-HT$_{5A}$ and 5-HT$_{5B}$. The serotonergic receptors are GPCR coupled to adenylyl cyclase or phospholipase C (PLC), except 5-HT$_3$

**TABLE 22.5**  Types of Serotonin Receptors.

| Receptors | Locations | G-protein | Agonists | Antagonists |
|---|---|---|---|---|
| 5HT$_{1A}$ | Hippocampus, septum, amygdala, dorsal raphe, cortex | Gi/o | 8-OH-DPAT, buspirone, gepirone | WAY-100135 |
| 5HT$_{1B}$ | Substantia nigra, basal ganglia | Gi/o | Sumatriptan and related triptans | – |
| 5HT$_{1D}$ | Substantia nigra, striatum nucleus, Accumbens, hippocampus | Gi/o | Sumatriptan and related triptans | GR-127935 |
| 5HT$_{1E}$ | – | Gi/o | – | – |
| 5HT$_{1F}$ | Dorsal raphe, hippocampus, cortex | Gi/o | – | – |
| 5HT$_{2A}$ | Cortex, olfactory tubercle, claustrum | Gq/11 | DMT and related psychedelics | Ketanserin, MDL900239 |
| 5HT$_{2B}$ | Not located in brain | Gq/11 | DMT | – |
| 5HT$_{2C}$ | Basal ganglia, choroid plexus, substantia nigra | Gq/11 | DMT, MCPP | Mesulergine, fluoxetine |
| 5HT$_3$ | Spinal cord, cortex, hippocampus, brainstem nuclei | Ligand-gated channel | Ondansetron, granisetron | – |
| 5HT$_4$ | Hippocampus, nucleus accumbens, striatum substantia nigra | Gs | Metoclopramide | GR 113808 |
| 5HT$_{5A}$ | Cortex, hippocampus, cerebellum | Gs | – | Methiothepin |
| 5HT$_{5B}$ | Habenula, hippocampal CA1 | Unknown | – | Methiothepin |
| 5HT$_6$ | Striatum, olfactory tubercle, cortex, hippocampus | Gs | – | Methiothepin, clozapine, amitriptyline |
| 5HT$_7$ | Hypothalamus, thalamus, cortex, suprachiasmatic nucleus | Gs | – | Methiothepin, clozapine, amitriptyline |

*Source*: Adapted from Sharma and Sharma (2017) and Nestler et al. (2009).

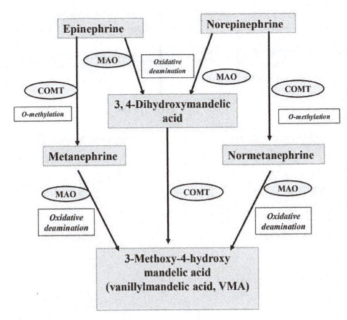

**FIGURE 22.5** Metabolism of epinephrine and NE.

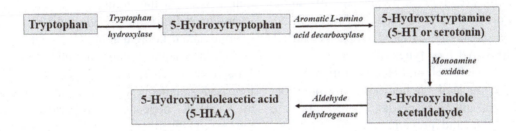

**FIGURE 22.6** Synthesis and inactivation of serotonin.

receptors which are ligand-gated ion channel receptors (Sharma and Sharma, 2017).

### 22.2.4.4   HISTAMINE

Histamine—a naturally occurring imidazole derivative is found abundantly in the gastric mucosa, mast cells located in the anterior and posterior pituitary gland as well as at body surfaces, brain, and bone marrow (Rang et al., 2011).

An amino acid histidine, undergoes decarboxylation in presence of histidine decarboxylase enzyme to form Histamine. It is converted to methyl histamine in presence of an enzyme imidazole N-methyl transferase or, alternatively, to imidazole acetic acid in presence of an enzyme diamine oxidase (histaminase). MAO facilitates the oxidation of methyl histamine to methyl imidazole acetic acid (Fig. 22.7; Oja and Saransaari, 2009).

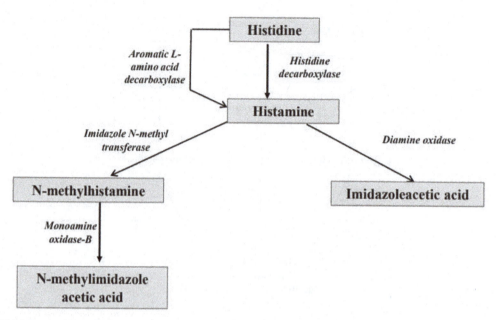

**FIGURE 22.7** Synthesis and metabolism of histamine.

There are four subtypes of histamine receptors—$H_1$, $H_2$, $H_3$, and $H_4$ (Table 22.6). They are widely scattered centrally and in periphery. All four types of histamine receptors are GPCRs. $H_1$ and $H_2$ receptors are post-synaptic, located on neuroeffector junction. Unlike $H_1$ and $H_2$ receptors, $H_3$ receptors are presynaptic, and prevent secretion of hista-mine and other transmitters via a G-protein. $H_4$ receptors are located on blood cells like eosinophils, neutrophils, and CD4 T cells. Of the four types, $H_1$ and $H_2$ receptors, agonists, and antagonists have been studied extensively for targets of drug research (Barrett et al., 2009; Nestler et al., 2009).

### 22.2.4.5 AMINO ACID NEUROTRANSMITTERS

#### 22.2.4.5.1 Glutamate

Glutamate, an excitatory neurotransmitter in the CNS is responsible for 75% of the neuronal transmission in the brain. In brain, glutamate is produced from glucose via the Krebs cycle intermediate α-ketoglutarate. Glutamate is then housed in synaptic storage sacs and secreted by exocytosis which is a $Ca^{2+}$-dependent process. Released glutamate is recycled back into cells by $Na^+/H^+/K^+$ dependent transporters and then further conveyed into synaptic vesicles, by a $H^+$ dependent transporters across the vesicular membrane (Barrett et al., 2009; Nestler et al., 2009).

Glutamate receptors (Table 22.7) is composed of two families of receptors—iono-tropic and metabotropic receptors. Ionotropic glutamate receptors are ion channels gated receptors which open/close upon agonist binding. Ionotropic glutamate receptors are of three types that are: N-methyl-D-aspartate (NMDA), α-amino-3-hydroxy-5-methyl-4-isoxazole propionic acid (AMPA), and kainate receptors. The metabotropic glutamate receptors are: G-protein-coupled receptors. There are 11 subtypes identified (Sharma and Sharma, 2017).

**TABLE 22.6**  Types of Histamine Receptors.

| Receptors | Locations | G-protein | Agonists | Antagonists |
|---|---|---|---|---|
| $H_1$ | Cortex, hippocampus, nucleus accumbens, thalamus | Gq/11 | – | Mepyramine, triprolidine, diphenhydramine, dimenhydrinate |
| $H_2$ | Basal ganglia, hippocampus; amygdala, cortex | Gs | Dimaprit | Ranitidine, cimetidine |
| $H_3$ | Basal ganglia, hippocampus, cortex | Gi/o | R-α-methylhistamine, Imetit | Thioperamide |
| $H_4$ | Eosinophils, neutrophils, and CD4 T cells | Gi | Imetit, clozapine | Thioperamide |

*Source*: Adapted from Sharma and Sharma (2017) and Nestler et al. (2009).

**TABLE 22.7**  Types of Glutamate Receptors.

| Receptors | NMDA | AMPA | Kainate |
|---|---|---|---|
| Locations | Postsynaptic | Postsynaptic | Pre- and postsynaptic |
| **Effects/ Mechanisms** | Ligand-gated ion channel | Ligand-gated ion channel | Ligand-gated cation channel |
| **Functions** | Slow EPSP, Synaptic plasticity (long-term potentiation, long-term depression), Excitotoxicity | Fast EPSP | Fast EPSP |
| **Agonists** | NMDA | AMPA | Kainate |
| **Antagonists** | AP5/APV ((2R)-amino-5- phosphonovaleric acid), 2-amino-5-phosphonopentanoic acid; CPP, 3-(2-carboxypirazin-4-yl)-propyl-1-phosphonic acid | NBQX | NBQX, ACET |

ACET, kainate receptor antagonist; AMPA, α-amino-3-hydroxy-5-methyl-4-isoxazole propionic acid; EPSP, slow excitatory postsynaptic potential; NBQX, AMPA receptor antagonist; NBQX, AMPA receptor antagonist; NMDA, N-Methyl-D-aspartate.

*Source*: Adapted from Rang et al. (2011) and Nestler et al. (2009).

Glutamate is primarily inactivated by carrier mediated reuptake mechanism which under certain conditions such as brain ischemia functions in reverse manner and thereby leads to glutamate release. Glutamate that enters astrocytes is transformed back to glutamine. It is then recycled back to the neurons via transporters, which convert the glutamine back to glutamate (Rang et al., 2011).

### 22.2.4.5.2  *Gamma-aminobutyric Acid*

GABA is the major inhibitory neurotransmitter in the brain, though the role of glycine is also imperative. Perhaps, 25–45% of all nerve terminals contain this transmitter. GABA is present abundantly in brain tissue (about 10 µmol/g tissue) in the nigrostriatal system and in lower concentrations (2–5 µmol/g) in the gray matter (Rang et al., 2011).

The main neurotransmitter fraction of GABA is produced by a metabolic pathway called GABA shunt. Glucose is the starting material and pyruvate also can act as a precursor for GABA synthesis. First, there is conversion of α-ketoglutarate into glutamate by an enzyme α-oxoglutarate transaminase. Glutamic acid then undergoes decarboxylation in presence of an enzyme glutamic acid decarboxylase (GAD) to form GABA which is stored into small synaptic vesicles and then released into the synaptic cleft. GABA is inactivated in the synaptic cleft by the actions of several types of plasma membrane GABA transporters (Bowery and Smart, 2006).

GABA receptors are two types—$GABA_A$ and $GABA_B$ receptors. $GABA_A$ receptors (Hosie et al., 2003) are ligand-gated ion channel whereas the other, $GABA_B$ receptors, is G-protein-coupled (Barnard et al., 1998).

GABA is inactivated by transamination to succinic semialdehyde in presence of an enzyme, GABA-T and then to succinate in the citric acid cycle. Pyridoxal phosphate acts as a cofactor for GAD and GABA-T. In addition, GABA is inactivated by active reuptake of GABA. A vesicular GABA transporter (VGAT) transports GABA and glycine into secretory vesicles (Rang et al., 2011; Barrett et al., 2009; Nestler et al., 2009; Smith, 2002).

### 22.2.4.5.3 *Glycine*

Unlike GABA, glycine is smallest, well-established inhibitory amino acid neurotransmitter, mainly found in the spinal cord and brain stem with its abundance in gray matter of spinal cord (5 μmol/g) (Bowery and Smart, 2006). It is produced in the brain by an enzyme serine hydroxyl methytransferase, which is a pyridoxal phosphate-dependent enzyme. By taking into consideration the distribution pattern of mitochondrial serine hydroxymethyltransferase and distribution of glycine, this conversion occurs in the mitochondrial compartment (Rang et al., 2011). Glycine is perhaps packed in small elliptical presynaptic vesicles. Upon release, it shows rapid hyperpolarization of the subsynaptic membrane and thereafter is taken up from the synaptic cleft into nerve terminals by high-affinity $Na^+$ dependent reuptake system. Glycine is mainly involved in protein synthesis. Glycine is packaged into small synaptic vesicles in small fractions for release as neurotransmitter (Webster, 2001).

Like with other neurotransmitters, glycine is released into the synaptic cleft on the receipt of an action potential in the presynaptic glycinergic neuron which promotes a cascade of reaction leading to vesicular fusion and diffuses and binds with its receptors located on the postsynaptic cell membrane (Rang et al., 2011).

Glycine receptors facilitate the speediest inhibitory neurotransmission in the mammalian brain and spinal cord. The potency of various amino acids at glycine receptors is in the following order: glycine > β-alanine > taurine > L-alanine > L-serine > proline (Vannier and Triller, 1997).

Like $GABA_A$ receptors, the glycine receptor is a receptor ionophore that contains a $Cl^-$ channel. It is also similar in size to the $GABA_A$ receptor and is believed to possess a quasi-symmetrical pentameric structure that surrounds a water-filled ion conduction pore. Glycine receptors belong to the same super family as nicotinic cholinergic, $GABA_A$, and $5HT_3$ receptors (Kuhse and Betz, 1995).

Glycine is inactivated from the synaptic cleft by reuptake transporters located on the plasma membranes of glial cells and of presynaptic nerve terminals. The degradation of glycine occurs mainly by means of the glycine cleavage system (GCS) in the inner mitochondrial membrane (Rang et al., 2011; Nestler et al., 2009; Smith, 2002; Barrett et al., 2009).

## 22.3 NEUROMODULATORS

Neuromodulator is the chemical messenger, which modifies and regulates activities that take place during the synaptic transmission. They do not propagate nerve impulses like neurotransmitters.

### 22.3.1 NEUROMODULATORS VERSUS NEUROTRANSMITTERS

Neuromodulators are distinct from neurotransmitters (Table 22.8). However, both the terms are wrongly interchanged as few peptides like substance P act as neurotransmitters as well as neuromodulators. Generally, the neuromodulators are peptides. So neuromodulators are often referred as neuropeptides. Almost all the peptides found in nervous tissues are neuromodulators (Butcher and Talbot, 1978). During the historical period between 1960s and 1970s, it became apparent that the peptidergic substances that regulate the pituitary gland and some enteric peptides are synthesized in the CNS. This observation established first breakthrough in the discovery of neuropeptides. Like neurotransmitter, some neuropeptides are also produced in the neurons, released from their presynaptic nerve terminals and act at postsynaptic sites. Still, they are not considered as neurotransmitter as they do not meet all standards to qualify as a neurotransmitter, they are often categorized as "putative neurotransmitters" (e.g., endorphins; Halbach and Dermietzel, 2002). Some neuropeptides are released from neurons like

neurotransmitters, but devoid of any effects on neuronal activity, rather they act on peripheral tissues and hence be named neurohormones [e.g., follicle-stimulating hormone (FSH)]. Thus, all neuropeptides do not work as neuromodulators (Bloom, 2006). Based on these facts, de Wied in 1987 formulated a hypothesis which states that "neuropeptides are synthesized and secreted from neuronal origin and exert effects on neuronal activities itself." At present, it was found that almost all neuropeptides affect neurons as well as non-neuronal tissues. Hence, the term neuropeptide is not referred in this limited way but now it has become more elaborate. Additionally, most neuropeptides connect neuronal activities with other body functions and thereby regulate various physiological reactions, such as growth and development, evolution, body homeostasis, behavioral performance, and immune reactions (de Wied, 1987).

**TABLE 22.8** Difference Between Neuromodulators and Neurotransmitters.

| Neuromodulators | Neurotransmitters |
|---|---|
| Modify and regulate the activities of synaptic transmission | Propagate nerve impulses through synapse |
| Packed in large synaptic vesicles, which are present in all parts of neuron like soma, dendrite, axon, and nerve endings | Packed in small vesicles in axon terminals only |
| Neurons have one or more neuromodulators | Neurons have one conventional neurotransmitter |
| Chemically, neuromodulators are only peptides | Chemically, neurotransmitters are amino acids, amine or others |
| Neuromodulators have diverse actions | Acts by producing depolarization or hyperpolarization through the receptors |

*Source*: Adapted from Phillis and Kostopoulos (1977).

Neuropeptides are synthesized from large precursor molecules leading to the formation of one or more peptides. During biosynthesis, a cascade of cellular processes decode the genetic information and produce biological active substance, neuropeptide which is stored in vesicles and released once stimulus is reached at the neuronal terminal. It has been found that a single neuron houses different neuropeptides. Even sometimes neuropetides co-localize with neurotransmitters, whereby neuropeptides can alter the effects of the neurotransmitter co-released with it (Burnstock, 1976).

Till date, above 50 different neuropeptides are discovered and studied. This list of neuro-modulators and neuropeptides expected to grow continuously in coming years. Some derivatives of fatty acids are functionally similar to that of neuropeptides (Herlenius and Lagercrantz, 2001). Like neuropeptides, they are synthesized from precursors through different enzymatic steps, bind to membrane bound receptors, which lead to signal transduction producing biological responses. Some diffusible gases, like NO and CO, also work as neuromodulators. However, they work in the close proximity of neurons as they have short half-life and diffusion-dependent range of activity, (Werman, 1966; Hokfelt, 2003).

### 22.3.2   CRITERIA

Like neurotransmitters, a substance should fulfill following standards to qualify as a neuromodulator:

- The substance must not act as a neurotransmitter.
- The substance must be distributed in biological fluids and must reach the modulatory site in substantial amount.

- Alterations in concentrations of the substance in body must modify the neuronal activity almost certainly and constantly.
- Direct administration of the substance must produce the same effect as with increase in its endogenous concentrations.
- The substance must act through precise targets of action with which it can modu-late neuronal activity.
- There must be inactivating mechanisms which will determine its duration of neuronal effects produced with changes in concentrations of the substance endog-enously or exogenously.
- Interference of the neuronal effects of increased endogenous concentrations or exogenous administration of the substance must be equal (Barchas et al., 1978).

### 22.3.3   ACTIONS

Neuromodulators have diverse actions such as:

- Regulation of synthesis, breakdown, or reuptake of neurotransmitter
- Excitation or inhibition of membrane receptors by acting independently or together with neurotransmitter
- Control of gene expression
- Regulation of local blood flow
- Promotion of synaptic formation
- Control of glial cell morphology
- Regulation of behavior

### 22.3.4   CLASSIFICATION

Neuromodulators are classified into two types:

### 22.3.4.1 NON-OPIOID PEPTIDES

Non-opioid neuropeptides act by binding with GPCRs. These neuropeptides are also called non-opioid neuromodulators. The most common examples are prostaglandins, ATP, adenosine, carbon dioxide, ammonia, steroid hormones, and thyroid releasing hormone (TRH; Barrett et al., 2009).

#### 22.3.4.1.1 *Adenosine and ATP*

Both adenosine and ATP act as transmitters and/or modulators in the CNS as they do in the periphery. Mapping the pathways is difficult, because purinergic neurons are not easily identifiable histochemically, but it is likely that adenosine serves as a very widespread neuromodulator, while ATP has more specific synaptic functions as a fast transmitter and as a local modulator (Webster, 2001).

Adenosine is produced intracellularly from ATP. It is not packaged into vesicles but is released mainly by carrier mediated transport. Because the intracellular concentration of ATP greatly exceeds that of adenosine, conversion of a small proportion of ATP results in a large increase in adenosine. ATP is stored into vesicles and released by exocytosis as a conventional transmitter, but can also leak out of cells in large amounts under conditions of tissue damage. Like glutamate, ATP can behave as an excitotoxin in high concentrations leading to neuronal damage. However, it is rapidly converted back to adenosine exerting protective effect. This neuroprotective characteristic of adenosine proposes that it can be used as a security tool keeping the neurons safe from damage in conditions like ischaemia or seizure activity in which their viability is at risk (Edward and Gibb, 1992).

Adenosine also acts as a neuromodulator and exerts CNS depressant action and also exhibits its extensive actions all over the body. There are four types of adenosine receptors such as $A_1$, $A_{2A}$, $A_{2B}$, and $A_3$. All are GPCRs. $A_{2A}$ and $A_{2B}$ act by increasing AMP concentrations while $A_1$ and $A_3$ decrease it. CNS stimulants, caffeine, and theophylline block adenosine receptors in CNS and thus produce stimulatory effects of coffee and tea. Adenosine and various adenosine receptor agonists are inhibitory in nature producing depressant effects such as drowsiness and sedation, motor incoordination, analgesia, and anticonvulsant activity. Xanthines, such as caffeine, which are antagonists at $A_2$ receptors, produce arousal and alertness. At present, use of $A_1$ antagonists to reduce the risk and effects of strokes is noteworthy (Burnstock et al., 2011).

ATP has now also established its role as a transmitter through its widespread receptor-mediated actions in the body. ATP binds with two types of receptors, P2X and P2Y receptors. P2X receptors are ligand-gated ion channel receptors subdivided into seven subtypes (P2X1 to P2X7). P2X receptors are widely distributed all over the body. P2X1 and P2X2 receptors are found in the dorsal horn, and hence play an important role in sensory transmission. P2Y receptors are GPCRs and there are eight subtypes of P2Y receptors such as P2Y1, P2Y2, P2Y4, P2Y6, P2Y11, P2Y12, P2Y13, and P2Y14 (Rang et al., 2011; Edward and Gibb, 1993; Barrett et al., 2009; Webster, 2001).

#### 22.3.4.1.2 *Endogenous Cannabinoids (Endocannabinoids)*

Historically, cannabinoids have been known for its consumption for mainly frivolous and medical purpose. Cannabinoids are mainly obtained from *Cannabis sativa.* Delta-9-tetrahydrocannabinol

(THC) is a main active phytoconstituent of *Cannabis sativa.* Cannabinoids being psycho-active, produce euphoria, augmentation of sensory perception, increased heart rate, loss of pain perception, lack of concentration, and impaired memory. The important pharmacological actions of cannabinoids and tolerance are facilitated through G-protein-coupled cannabinoid receptors (CB) which are of types CB1 and CB2 (Mechoulamand Fride, 1995).

CB1 receptors are coupled via Gi/o protein which inhibits adenylyl cyclase and voltage sensitive calcium channels, and opens G-protein-sensitive inward-rectifying potassium (GIRK) channels, causing hyperpolarization. CB2 are coupled via Gi/o to adenylyl cyclase, G-protein-sensitive inward-rectifying potassium channels and mitogen activated protein kinase similarly to CB1, but not to voltage operated calcium channels. Other as-yet-unidentified G-protein-coupled receptors are also implicated because cannabinoids exhibit analgesic actions and activate G-proteins in the brain of CB1 knockout mice despite the absence of CB1 receptors (Iversen, 2003).

The interesting fact observed was that though cannabinoids exist only naturally in plant with no biological connection in humans, many parts of brain, namely cerebral cortex, basal ganglia, cerebellum, and hippocampus express huge numbers of receptors for cannabinoids. This made scientific workers to think of endogenous substances which may be selectively interacting with CB and whose action is facilitated by Delta-9-tetrahydrocannabinol. Thus in 1992, the first endogenous ligand of CB1 receptors later labeled as Anandamide was discovered in porcine brain. The name, Anandamide was derived from the Sanskrit word 'Ananda' meaning 'Bliss.' With the discovery of anandamide, many other metabolites collectively termed as endocannabinoids, were characterized

and discovered to act as useful agonists of CB in the brain, however they were not superior in efficacy than anandamide (Devane et al., 1992). The endocannabinoids are found in the brain or other tissues only in small amounts. Similar to other lipid mediators, they are formed and released locally on call. Anandamide and endocannabinoids are rapidly inactivated by reuptake through transporter and by metabolism through the enzyme fatty acid amide hydrolase. The anandamide is formed from the precursor N-arachidonic phosphatidyl ethanolamine by hydrolysis in presence of an enzyme phosphodiesterase enzyme phospholipase D (Iversen, 2003).

Anandamide bind with G-protein-coupled CB1 receptors on presynaptic nerve terminal, thereby altering membrane potential through changes in neuronal membrane permeability of calcium and potassium ions and adenylate cyclase (AC) leading to neurotransmitter release or action, or both. Neuromodulatory actions of endocannabinoids could influence a wide range of physiological activities including nociception, cardiovascular, respiratory, and gastrointestinal function. Hypothalamic hormone interactions could influence food intake and reproductive function. These physiological effects mentioned above intensively recommended cannabinoids as modulators of neurotransmitter release and action causing possible ultimate effect in various brain regions (Marzo et al., 1998; Rang et al., 2011; Iversen, 2003).

### 22.3.4.1.3   Nitric Oxide

NO is a new neurotransmitter as well as neuromodulator. It plays an important role in signaling functions in almost all animals. However, NO being gaseous and very small in size and is a predominantly fascinating case of a neuromodulator. NO is produced from arginine by the enzymatic reactions wherein arginine is converted to citrulline and NO in presence of an enzyme, NO synthase (NOS). Like neurotransmitter substances, it is not readily stored but is produced on demand. It is a water-soluble gas that diffuses spontaneously through membranes and binds with target molecules through covalent bonds. For example, it binds to guanylyl cyclase and haemoglobin, through iron and also to other molecules through metals (Jacklet, 2005).

Conventionally, NO signaling mainly operates through the activation of guanylyl cyclase which forms cGMP. However, sometimes NO signaling operated through redox chemistry reactions wherein NO binds to the sulfur in membrane proteins (O'Shea and Husbands, 2013).

NO functions as a diverse neuromodulator which acts as an orthograde transmitter, as a retrograde transmitter, as a cotransmitter, and as a paracrine messenger. It modulates neurotransmitter release, membrane ionic movements as well as receptor function (Baranano et al., 2001). Based on the situation, NO may behave as either neurotoxic or neuroprotective. It plays an important role in olfaction, sensation, chemosensation, and neuronal integrative activities. The various physiological functions of NO are vasodilatation, vasorelaxation, neuroprotection, neurotoxicity, cytotoxicity, control of secretions, intestinal smooth muscle relaxation, penile erection, regulation of developmental processes, neurotransmission, and neuromodulation. The improper release of NO is responsible for several diseases. There is significant indication that confirms the role of NO as a neurotransmitter and neuromodulator in the CNS (O'Shea and Husbands, 2013; Nestler et al., 2009; Baranano et al., 2001).

### 22.3.4.1.4 Carbon Monoxide

It is a highly toxic gas when breathed in large amount. However, it also functions as a neurotransmitter. Its function in the brain is still being clarified. In neurons, enzyme heme oxygenase (HO) degrades heme to form biliverdin, iron, and finally CO. CO stimulates soluble guanylyl cyclase enzyme like NO. In enteric nervous system and olfactory receptor neurons, CO acts as strong neurotransmitter. The neurons of the myenteric plexus contain HO2 and neuronal nitric oxide synthase (nNOS) in abundant amount. Any genetic deletion or pharmacologic inhibition of HO2 considerably decreases noradrenergic neurotransmission in the gut. Similarly, there is reduction in the generation of cGMP in olfactory neurons due to inhibition of heme oxygenase, which normally produce adequate CO to stimulate guanylyl cyclase. Production of CO also plays an important role in the conservation of circadian rhythms by influencing the DNA binding activity of key circadian transcription factors (Nestler et al., 2009; Baranano et al., 2001).

Other non-opioid neuromodulators include bradykinin, secretin, cholecystokinin, gastrin, motilin, neurotensin, CRH, GH-RH, GH-IH, GnRH, TRH, vasoactive intestinal polypeptide, neuropeptide Y, ghrelin etc. (Sembulingam and Sembulingam, 2012).

### 22.3.4.2 OPIOID PEPTIDES

Peptides, which bind to opioid receptors, are called opioid peptides. Opioid peptides are also called opioid neuropeptides or opioid neuromodulators. Opioid receptors are the membrane proteins located in nerve endings in brain and GIT. Opioid receptors are of three types: μ, κ, and δ. These proteins are called opioid receptors because of their affinity toward the opiate or morphine, which are derived from opium. Opiate and morphine act by binding with the receptor proteins (opioid receptors) for the natural neuropeptides. Natural neuropeptides are called endogenous opioid peptides. Endogenous opioid peptides have opiate like activity and inhibit the neurons in the brain involved in pain sensation (Froehlich, 1997). Opioid peptides are of three types:

### 22.3.4.2.1 Enkephalins

Enkephalins are the natural opiate peptides recognized first in pig's brain. Derived from the precursor proenkephalin, these peptides are present in the nerve endings in many parts of forebrain, substantia gelatinosa of brainstem, spinal cord, and GIT. Two types of enkephalins are known, leucine-enkephalin (YGGFL) and methionine enkephalin (YGGFM) (Sharma and Sharma, 2017).

### 22.3.4.2.2 Dynorphins

Dynorphins are derived from prodynorphin. Dynorphins are found in hypothalamus,

posterior pituitary, and duodenum. Dynorphins are of two types: α- and β-dynorphins (Barrett et al., 2009).

### 22.3.4.2.3 Endorphins

Endorphins are the large peptides derived from theprecursor pro-opiomelanocortin. Endorphins are predominant in diencephalic region particularly hypothalamus and anterior and intermediate lobes of pituitary gland. Three types of endorphins are recognized: α-, β- and γ-endorphins (Barrett et al., 2009).

## 22.4 NEUROHORMONES

Nervous tissue possesses a superior capability to expend some chemical substances for communication, fairly numerous from that of neurotransmitters, that is proscribed to specific neurons. These chemical substances are known as neurohormones, substances produced from neurosecretory cells of the nervous systems of vertebrates and invertebrates. Neurohormones have ability to travel to distant non-neuronal locations like endocrine messengers through blood and lymph. Unlike neurotransmitters, they are inactivated slowly and they are used by cells only once for the same action. Thus, neurohormones are the major

**TABLE 22.9** Types and Functions of Vasopressin Receptors.

| Receptors | Locations | G-protein | Second messengers | Functions |
|---|---|---|---|---|
| $V_1$ | Vascular and other smooth muscles, platelets, hepatocytes | $G_q$ | $IP_3$; DAG | VSM contraction, visceral smooth muscle contraction, platelet aggregation, glycogenolysis |
| $V_2$ | Collecting tubules, vascular endothelium | $G_s$ | cAMP | Water retention, vasodilation |
| $V_3$ | Anterior pituitary | $G_q$ | $IP_3$; DAG | Adrenocorticotropic hormone (ACTH) release |

*Source*: Adapted from Sharma and Sharma (2017).

mediators between nervous and non-nervous system (Scharrer, 1969). Yet, using this term is ambiguous because many times hypothalamic neurons also form synapses with central neurons. Cytochemical evidence indicates that the same substances that are secreted as hormones from the posterior pituitary, mediate transmission at these sites (Bloom, 2006).

Mostly cells involved in the synthesis and release of neurohormones are glandular in nature and are chiefly present in nervous systems of invertebrates and vertebrates. Hence, they are referred to as neurosecretory cells. These cells predominantly contain membrane-bound granular vesicles of varying densities called synaptoids. Neurosecretory granules move through axonal cytoplasm to terminal and gather there in a spherical sac. The exact mechanism of release is still not entirely known. However, it seems to take place similar to that of hormonal release from endocrine cells (Scharrer, 1967; Scharrer, 1968). Vasopressin and oxytocin were the first peptide neurohormones known historically. They are nonapeptides chemically (Jan et al., 1979).

## 22.4.1 IMPORTANT NEUROHORMONES

### 22.4.1.1 VASOPRESSIN

Vasopressins are nonapeptide hormones. They are stored in the neurosecretory cells of neurohypophysis. When plasma osmolarity raises or blood pressure falls, it is released from posterior pituitary gland cells into the blood. Vasopressin facilitates water reabsorption from the collecting tubules of kidney and thereby reduces plasma osmolarity. It also causes vasoconstriction (hence named vasopressin) and thus increases blood pressure (Sharma and Sharma, 2017). Pain, stress,

anxiety, ACh, nicotine, and carbamazepine also facilitate release of vasopressin. In addition, they are well-known compounds to function as neurotransmitters as well. However, the significance of their neurotransmitter function is not yet cleared (Craig, 2004). Table 22.9 represents the classes and roles of vasopressin receptors.

### 22.4.1.2 OXYTOCIN

Oxytocin is a peptide hormone released from posterior pituitaty gland. It plays an important role in milk ejection in lactating mothers and also in initiation of labor and parturition mainly through oxytocin receptors. Oxytocin receptors are $G_q/G_{11}$ protein-coupled receptors which produce their actions through $IP_3$ and $Ca^{2+}$ mobilization from intracellular stores (Sharma and Sharma, 2017).

Currently the role of oxytocin is extended to a hormone released from corpus luteum and as a neuroregulator in regulation of behavior and affiliation (Higuchi, 1995).

### 22.4.1.3 OTHER NEUROHORMONES

In addition to vasopressin and oxytocin, there are some other substances which are referred to as neurohormones. However, they are well-known as endocrine hormones. They include GnRH, TRH, (GHRH), somatostatin, DA, CRH (Sembulingam and Sembulingam, 2012).

## 22.5 NEUROMEDIATORS

A neuromediator is one of the main messengers that mediate production of the postsynaptic response via neurotransmitter. The important examples of such neuromediators are second messengers such as cAMP, cGMP, and $IP_3$ (Greengard, 1976). However, it would be exaggeration to state that only cAMP and cGMP level variation in brain can generation of the synaptic potential. Rather Second messenger such as cAMP, cGMP, and IP causes phosphorylation of intracellular protein and initiate a complex cascade of molecular reactions that alter membrane and cytoplasmic proteins which are an important event to induce neuronal excitability (Bloom, 2006).

Second messengers refer to small intracellular molecules that are produced after the first messenger (i.e., neurotransmitter, neurohormone, or neuromodulator)-dependent receptor activation. Second messengers are required to stimulate intracellular signaling pathways inducing a cellular response. The second messengers are chemically diverse in nature. They are cyclic nucleotides, lipid derivatives and small active compounds, and some ions. The most studied second messengers are cyclic AMP or cyclic GMP, calcium, DAG, $IP_3$, and reactive oxygen and nitrogen species (ROS, NOS; Kulinsky and Kolesnichenko, 2005).

### 22.5.1 CYCLIC ADENOSINE MONOPHOSPHATE

cAMP is a low-molecular-weight, hydrophilic second messenger formed by ATP hydrolysis by a membrane bound enzyme AC which is coupled with G-proteins. As AC is activated, there is rise in intracellular cAMP leading to stimulation of a number of proteins, for example, cAMP can modulate the opening of some cationic channels that possess a binding domain for cyclic nucleotide binding domain (CNBD). cAMP also activates the protein kinase A (PKA) (Greengard and Kebabian, 1974).

### 22.5.2 INOSITOL 1, 4, 5-TRIPHOSPHATE

Lipid-derived messengers are essential components in signal transduction and its augmentation besides to cyclic nucleotides (Kulinsky and Kolesnichenko, 2005). $IP_3$ and DAG are two important intracellular messengers formed by hydrolysis of phosphatidylinositol-4,5 bisphosphate in presence of an enzyme PLC. $IP_3$ released from the cell membrane under the action of PLC, being water-soluble disseminates into the cytoplasm and binds with endoplasmic reticulum (ER) specified tetrameric receptors. This interaction prompt to discharge of $Ca^{2+}$ ion from intracellular storage sites which is also an important second messenger. (Kulinsky and Kolesnichenko, 2005). DAG formed by the action of PLC, which is an important activator of proteins of Protein kinase C family, which possess requisite sites for DAG and $Ca^{2+}$ ions and enhance the phosphorylation of other proteins and amplifying the signal (Kulinsky and Kolesnichenko, 2005).

### 22.5.3 CALCIUM

Calcium ($Ca^{2+}$) is a second messenger possessing special particularities. $Ca^{2+}$ presents in the cytoplasm may be released from either intracellular stores or the extracellular space; through activation of $IP_3$ or ryanodine receptors located at intracellular membranes (such as endoplasmic reticulum). Calcium is released from intracellular and enters to cells through specific channels, extracellularly divided into three main categories:

- Voltage-operated channels (VOCs),
- Receptor-operated channels (ROCs), and
- Store-operated channels (SOCs).

Cells are very sensitive to identify even minor changes in the cytoplasmic $Ca^{2+}$ concentrations and couple those changes to particular activation of regulatory proteins and transcription factors, regulating cell motility, apoptosis, protein synthesis, and exocytosis, among other functions (Meech, 1978).

Distinct proteins in cell cytosol are able to bind $Ca^{2+}$ modifying its activity, for example, calmodulin (CaM), calcineurin, and synaptotagmin. Upon $Ca^{2+}$ binding, CaM undergoes a structural change and binds to numerous proteins to activate them. For example, $Ca^{2+}$-CaM is able to bind to the calcium-calmodulin kinase IV (CaMKIV), which is forming an inactive complex in cell cytoplasm.

Other molecule which is able to bind $Ca^{2+}$ is the phosphatase calcineurin that is a serine/threonine phosphatase. Occupation of the calcium-binding site makes a structural change that exposes a $Ca^{2+}$-CaM binding site. Once $Ca^{2+}$-CaM binds to a subunit, it becomes active, dephosphorylation a number of intracellular targets. Exocytosis process of neurotransmitter release from synaptic vesicles is known to be mediated via Synaptotagmin present in all neurons (Claudia and Fabiola, 2014).

## 22.6 DISORDERS ASSOCIATED WITH DEFECTS IN NEUROCHEMICALS AND POTENTIAL DRUGS FOR NEUROCHEMICAL IMBALANCES

The brain is the ultimate center that regulates all neurological and behavioral aspects of body through neuronal communications mediated via various neurochemicals as discussed above. Thus, neuropsychiatric disorders are mostly result of disturbed neurochemical balance. Besides the multifaceted involvement of billions of neuronal cells, CNS is a multifaceted organization with its diverse number of neurotransmitter systems as compared to ANS,

in which the parasympathetic system works on "rest and digest" phenomenon and sympathetic system on "fight-or-flight" phenomenon. There are more than 20 neurotransmitter systems and multiple receptors for each neurotransmitter as discussed in above sections. Any alterations in neurochemical balance are experienced as neurological or psychiatric disorders like epilepsy, Parkinson's disease (PD), Alzheimer's disease (AD), psychosis, depression, and so on (Choudhury et al., 2018). ACh, NA, DA, and 5-HT are of utmost importance amongst neurotransmitters for their profound role in pathogenesis of various neurologic and psychiatric disorders in humans. Yet, the involvement of various proteins and peptides such as neurotrophic factors, growth factors, and endogenous chemical compounds cannot be ignored (Garris, 2010).

DA known as "the molecule of happiness," regulates movement, memory, cognition, attention, pleasure, reward, motivation, sleep, creativity, and personality. Long-lasting decline in DA levels is one of the major events involved in the neurodegeneration in PD (Wise, 2004). In addition to neurodegeneration, shortage of DA causes depression and mood swings (Gaspar et al., 2003). 5-HT is one more important neurotransmitter which regulates social behavior, mood, sleep, digestion, appetite, memory, and sexual desire. Abundance of serotonin in the gastrointestinal tract controls the appetite and bowel movement (Garris, 2010). In addition, it also mediates breast milk production, liver regeneration, blood clotting, vasoconstriction, and bone metabolism. Reduction in blood serotonin levels is reported to cause mainly depression (Garris, 2010). NE acts as a neurohormone as well as neurotransmitter. NE regulates stress response through fight-or-flight mechanism. It is one of the major mediators in attention deficit hyperactivity disorder (ADHD), depression, and blood pressure. It regulates blood pressure, heart rate, blood flow to the muscles and gastrointestinal motility as and when required (Choudhury et al., 2018). ACh is an important regulator of memory and cognitive aptitude related tasks. Lack of ACh is main pathological event in AD, PD, and myasthenia gravis (Brown, 2006) which is expressed in the form of low energy levels, fatigue, memory loss, cognitive decline, learning disabilities, muscle aches, nerve damage, and frequent mood swings (Rand, 2007; Ohmura et al., 2012).

Some of the important neurological and psychiatric disorders accompanied by neurochemical imbalance are as follows:

### 22.6.1  ALZHEIMER'S DISEASE

AD is chronic, progressive, and irreversible neurodegenerative disorder of aged brain in which brain cells deterioration cause cognitive impairment. It primarily affect hippocampal-dependent learning, memory, judgment and reasoning, movement coordination, and recognition (Parihar and Hemnani, 2004). Cognitive deficit taking place in Alzheimer's patient is accompanied by advanced loss of cholinergic neurons and subsequently decrease in levels of ACh and ChAT activity in brain. It was reported to be a most common characteristic feature of senile dementia of the AD (Keverne and Ray, 2005). In addition, senile dementia of AD is also related to loss of glutamatergic neurons and glutamate activity by affecting the NMDA glutamate receptors in brain. Pathologically, AD is characterized by deposition of senile plaque and neurofibrillary tangles in brain (Francis et al., 1999). Other neurotransmitter is also involved in pathogenesis of AD. However, little information is known about their relationship to the underlying neuropathology of AD.

Neurochemical changes associated with AD are summarized in Table 22.10.

Based on the understanding of neurochemical changes evident during AD, following classes of drugs are used in the management of AD (Table 22.11):

### 22.6.2  PARKINSON'S DISEASE

PD is a gradually worsening neurodegenerative disease. In PD, there is destruction of dopaminergic neurons projecting from substantia nigra pars compacta ($SN_{PC}$) to neostriatum causing the imbalance between DA and ACh (Sharma and Sharma, 2017). The three cardial signs of PD are tremors, rigidity, and bradykinesia. In large part, Parkinson's signs become evident when more than 75–85% of dopaminergic neurons in the midbrain are destroyed. As a result of this, the brain areas that functionally depend on dopaminergic inputs, that is specifically the post-commissural putamen and other basal ganglia regions suffer from DA deficiency (Galvan and Wichmann, 2008).

**TABLE 22.10**  An Outline of Reported Neurochemical Changes in Alzheimer's Disease.

| Neurotransmitters | Parameters | Changes |
|---|---|---|
| Ach | ChAT | ↓ |
| | AChE | ↓ |
| | BuChE | ↑ |
| | ACh release | ↓ |
| | Muscarinic receptors | ↓ |
| | Nicotinic receptors | ↓ |
| | High affinity choline uptake | ↓ |
| | Nerve growth factor (NGF) receptors | ↓ |
| | Number of neurons | ↓ |

**TABLE 22.10** *(Continued)*

| Neurotransmitters | Parameters | Changes |
|---|---|---|
| Glutamate | Glutamate | ↓ |
| | Cerebrospinal fluid | ↓ |
| 5-HT | Serotonin | ↓ |
| | 5-HT uptake sites | ↓ / − |
| | Number of neurons | − |
| NE | NE | ↓ |
| | DBH | ↓ |
| | Number of neurons | ↓ |
| DA | DA | − |
| | Number of neurons | − |
| GABA | Glutamic acid decarboxylase (GAD) | ↓ |
| Somatostatin | Receptors for somatostatin, neuropeptide, adenosine, corticotrophin-releasing factor | − |
| Neuropeptide Y | | ↓ |
| Adenosine | | ↓ |
| Corticotrophin-releasing factor | | ↓ |

BuChE, butyrylcholinesterase; ↑, increase; ↓, decrease; −, no change.

*Source*: Adapted from Keverne (2005).

Most of the anti-parkinsonian drugs act by countering shortage of DA in basal ganglia or by blocking muscarinic receptors (Table 22.12).

**TABLE 22.11** Drugs Used in Treatment of AD.

| Class | Examples | Mechanism |
|---|---|---|
| Cholinergic agents | Donepezil, galantamine, rivastigmine | AChE inhibitor |
| Glutaminergic agents | Ampakine, memantine | Glutamate signaling enhancement |
| Nicotinic agonist | Nicotine | Increased release of ACh |
| Monoamine and agents acting on them | Methyl-penidate, modafinil | Effects on CA, 5-HT, glutamate, GABA, and histamine system |

*Source*: Adapted from Ingole et al. (2008).

**TABLE 22.12** Drugs Used for Management of PD.

| Class | Examples | Mechanism |
|---|---|---|
| DA precursor | Carbidopa/ levodopa | Increase DA |
| COMT inhibitors | Entacapone | Inhibit DA inactivation |
| DA agonists | Bromocriptine, pergolide, pramipexole | Increase DA |
| MAO-B inhibitors | Selegiline | Inhibit DA inactivation |
| Anticholinergics | Benztropine, trihexyphenidyl | Decrease cholinergic overactivation |

*Source*: Adapted from Goetz et al. (2004) and Samii and Ransom (2005).

### 22.6.3 *PSYCHOSIS AND SCHIZOPHRENIA*

Psychosis is a psychological illness which occurs as a result of some biological disease states or due to changes in the neurotransmitter concentration (mainly DA and 5-HT) of the brain. The most common symptoms of psychosis include delusions (i.e., a wrong faith which cannot be justified with intellectual reasons) or illusions (i.e., mental beliefs derived from misunderstanding of an actual existing visual sensory stimulus) or phantasms (i.e., a fabricated sensitivity occurring even in absence of a sensory stimulus) which may be in the form of visual or auditory or olfactory hallucinations. Schizophrenia is a very specific term used for special type of functional psychosis which occurs mainly due to increased DA level in the limbic area of brain. Patients suffering from Schizophrenia are typically reserved dual characters; they display withdrawal from realism with no consistency between their feelings, thoughts, dialogue, and action (Sharma and Sharma, 2017). However, many medical co-workers have proved serotonin hypothesis and glutamate hypothesis of psychosis, and Schizophrenia which states that other neurotransmitters are also involved in genesis of psychological illness. Based on the DA and serotonin hypothesis of schizophrenia and psychosis, drugs used in treatment of psychosis known as antipsychotic or neuroleptic drugs are antagonists or partial agonists at D2 DA receptors, and atypical antipsychotics are 5-HT2 receptor antagonists (Crow and Harrington, 1994; Table 22.13).

**TABLE 22.13**   Drugs Used in the Treatment of Psychosis.

| Class | Examples | Mechanism |
|---|---|---|
| Typical antipsychotics | Chlorpromazine, haloperidol, fluphenazine, flupentixol, clopentixol | D2 DA receptors antagonists |
| Atypical antipsychotics | Clozapine, risperidone, sertindole, quetiapine, amisulpride, aripiprazole, zotepine, ziprasidone | 5-HT2 receptor antagonists |

*Source*: Adapted from Rang et al. (2011).

### 22.6.4   EPILEPSY AND SEIZURE

Epilepsy is brain's erratic functional anomaly characterized by persistent unpredictable spontaneous seizures. Convulsions are unstructured, violent, and asymmetrical or long standing muscle contractions. A seizure means abnormal, coordinated, and recurring electrical firing of neuronal populations in the CNS leading to behavioral changes (Shin et al., 1994; Sharma and Sharma, 2017). The major pathogenic events accountable for seizures are augmented glutamate excitatory neurotransmission, compromised GABA inhibitory neurotransmission or anomalous electrical activity of affected epileptic foci in brain (Rang et al., 2011).

Most commonly used antiepileptic drugs (Table 22.14) in current practice act by enhancing GABA-mediated inhibition or decreasing excitatory glutaminergic neurotransmission (Dwivedi, 2001).

### 22.6.5   DEPRESSION

Depression is an affective disorder characterized by grief, loss of attention or pleasure, guilty feelings, sleep disturbances, and cognitive impairments worsening the quality of life. DA, NE, or 5-HT are the most important players in pathophysiology of depressive disorders (Rang et al., 2011). Table 22.15 summarizes neurochemical/hormonal abnormalities in depression.

**TABLE 22.14**   Drugs Used in Treatment of Epilepsy.

| Drugs | Mechanism |
|---|---|
| Phenobarbital | ↑ in GABA-mediated inhibition |
| Valproic acid; Divalproex sodium | ↑ in GABA-mediated inhibition; ↓ in rapid repetitive firing |
| Clonazepam, diazepam, lorazepam, clorazepate | ↑ in GABA-mediated inhibition |
| Felbamate | ↓ in glutaminergic neurotransmission |
| Gabapentin, Lamotrigine | ↑ in GABA-mediated inhibition |
| Tiagabine | ↑ in GABA-mediated inhibition |
| Topiramate | ↓ in rapid repetitive firing; ↑ in GABA-mediated inhibition |
| Vigabatrin | ↑ in GABA-mediated inhibition |
| Felbamate | ↑ in glutaminergic neurotransmission |

↑, increase; ↓, decrease.

*Source*: Adapted from Dwivedi (2001).

**TABLE 22.15** Neurochemical/hormonal Abnormalities in Depression.

| Neurochemical | Change |
|---|---|
| 5-HT | ↓ |
| NA | ↓ |
| Brain derived neurotrophic factors | ↓ |
| Cortisol, corticotropin-releasing hormone (CRH) | ↑ |
| Proinflammatory cytokines | ↑ |

↑, increase; ↓, decrease.

*Source*: Adapted from Rang et al. (2011).

Although the monoamine hypothesis is insufficient to justify the sole component of depression, pharmacological manipulation of monoamine transmission is the most commonly used therapeutic approach in the management of depression. Main types of antidepressants include:

### 22.6.5.1 MONOAMINE UPTAKE INHIBITORS

These drugs work through preventing reuptake of NA and/or 5-HT by monoaminergic nerve terminals (tricyclic antidepressants, selective serotonin reuptake inhibitors, newer inhibitors of NA and 5-HT reuptake).

### 22.6.5.2 MONOAMINE RECEPTOR ANTAGONISTS

$\alpha_2$-adrenoceptor antagonists can indirectly elevate 5-HT release.

### 22.6.5.3 MONOAMINE OXIDASE INHIBITORS

These drugs work through preventing one or both forms of brain MAO, thus increasing the cytosolic stores of NA and 5-HT in nerve terminals (Rang et al., 2011).

Table 22.16 provides a simplified outline of key neurotransmitters involved in clinical manifestations of various disorders believed to

**TABLE 22.16** Disorders Associated with Defects in Neurotransmitters, Neuromodulators, and Neurohormones with their Treatment Options.

| Disorders | Neurotransmitters involved | Pathophysiology | Drug class and examples |
|---|---|---|---|
| Myasthenia gravis | Ach | Production of IgG antibody that binds to ACh receptors ($N_M$) at neuromuscular junction (NMJ) | AChE inhibitors: Neostigmine, pyridostigmine, ambenonium |
| AD | Ach | Deposition of extracellular β-amyloid plaques and intracellular neurofibrillary tangles in cholinergic neurons that produce and utilize Ach | Cholinesterase inhibitors: Donepezil, rivastigmine, galantamine |
| PD | ACh, DA | Degeneration of dopaminergic neurons in the substantianigra, disturbing DA/ACh equilibrium which causes ACh predominance in striatal cells | Anticholinergic drugs: Benztropine |
| Anxiety | NE, GABA, 5-HT | Reduced activity of GABA imbalances in NE and 5-HT responses in brain | GABA-mimetic drugs: Diazepam, alprazolam, clonazepam, phenobarbital; SSRIs: Fluoxetine, venlafaxine |

**TABLE 22.16**    *(Continued)*

| Depression | NE, DA, 5-HT, ACh, GABA, substance P | Irregularities in NE, DA, 5-HT, ACh, GABA transmission; Contribution of other hormones and neuropeptides like substance P, DA, ACh, GABA | NE and 5-HT reuptake inhibitors: Imipramine, doxepine; NE reuptake inhibitors: Desipramine, nortriptyline; Selective serotonin reuptake inhibitors: Fluoxetine, citalopram; MAO inhibitors: Moclobemide |
|---|---|---|---|
| Mania | NE, DA, 5-HT, glutamate | Augment NE and DA activity, decrease 5-HT levels, and abnormal glutamate neurotransmission | Lithium, topiramate, valproate and lamotrigine |
| Schizophrenia | DA, 5-HT | Increased secretion and production of DA and increased number of DA receptors | $D_2$ receptors antagonists: Haloperidol, chlorpromazine; $5\text{-HT}_2$ receptors antagonists: Clozapine, Olanzapine, and risperidone |
| Hypertension | NE, renin | Increased sympathetic activation (increased NE release) | Sympatholytic drugs: Clonidine, α-methyldopa, prazosin, atenolol, metoprolol |
| Epilepsy | GABA, glutamate | Enhanced actions of glutamate or decrease defects of GABA in epileptic foci in brain areas | Phenytoin, lamotrigine, phenobarbital and benzodiazepines, tiagabine, valproate, topiramate |
| Pain | Prostaglandin, 5-HT, histamine; Neuropeptides: substance P, neurokinin A | Increased stimulation of sensory nerve endings by excessive prostaglandin, 5-HT and histamine release causing pain. | Opioid analgesics: Morphine, codeine, naloxone; NSAIDs: Diclofenac, aceclofenac, nimesulide, aspirin |
| Hyperacidity, duodenal and gastric ulcers, Zollinger-Ellison syndrome | Histamine | Increased histamine release increases gastric acid secretion | $H_2$ receptor antagonists: Cimetidine, ranitidine, nizatidine |
| Allergy | Histamine | Increased stimulation of sensory nerve endings by excessive histamine release causing itching, urticarial | $H_1$ receptor antagonists: Cetirizine, chlorpheniramine |
| Hypersensitivity reactions | Histamine, prostaglandins and NE/E | Increased release of histamine, bradykinin, prostaglandins in response to antigen | Adrenaline |
| Bronchial asthma | Histamine, 5-HT, prostaglandins, NE and ACh; Neuropeptides: substance P, neurokinin A | Increased release of histamine, bradykinin, prostaglandins from activated mast cells; Increased ACh release causing bronchoconstriction | $β_2$-receptor agonists: Salbutamol, terbutaline; Sympathomimetics: Isoprenaline, ephedrine; Anticholinergics: Ipratropium; Leukotriene modulator: Montelukast |

ACh, acetylcholine; DA, dopamine; NE, norepinephrine; GABA, gamma-aminobutyric acid; 5-HT, 5-hydroxytryptamine; AChE, acetylcholinesterase; MAO, monoamine oxidase; IgG, immunoglobulin G.

*Source*: Adapted from Sharma and Sharma (2017).

occur as a result of chemical imbalances with their treatment options.

## 22.7 CONCLUSION

In this chapter, the authors described biosynthesis, storage, release, receptor activation, and inactivation of important neurotransmitters. Various neuromodulators are described as opioid and non-opioid peptides. It has been observed that vasopressin and oxytocin play an important role as neurohormones while second messengers like cAMP, cGMP, DAG, IP$_3$, and calcium facilitate the responses of neurotransmitters at their postsynaptic receptors. Throughout the chapter, it was evident that some neurotransmitters like adrenaline also work as neurohormone. Adenosine and ATP act as transmitters and/or modulators in the CNS. NO also acts as a neurotransmitter and neuromodulator in the CNS. Thus, it can be concluded that it is very complex to clearly delineate the role of neurotransmitters, neuromodulators, and neurohormones. However, imbalance of these neurochemicals is one of the major pathological events in genesis of various CNS disorders like AD, PD, psychosis, and depression. It is also evident that drugs used in management of such neuropsychiatric disorders mainly target on correcting the neurochemical imbalance.

## KEYWORDS

- **neuroactive substances**
- **neurohormones**
- **neuromediators**
- **neuromodulators**
- **neurotransmitters**

## REFERENCES

Ayano, G. Common Neurotransmitters: Criteria for Neurotransmitters, Key Locations, Classifications and Functions. *Adv. Psychol. Neuro.* **2016**, *1* (1), 1–5.

Baranano, D. E.; Ferris, C. C.; Snyder, S. H. Atypical Neural Messengers. *Trends Neurosci.* **2001**, *24* (2), 99–106.

Barchas, J. D.; Akil, H.; Elliott, G. R.; Holman, R. B.; Watson, S. J. Biochemical Neurochemistry: Neuroregulators and Behavioral States. *Science* **1978**, *200*, 964–973.

Barrett, K.; Brooks, H.; Boitano, S.; Barman, S. Neurotransmitters & Neuromodulators. In Ganong's Review of Medical Physiology, 23rd ed; McGraw-Hill: New York, 2009; pp 129–146.

Barker, J. L. Neurosciences Research Program Bulletin In Peptides and Behavior: A Critical Analysis of Research Strategies. *Evid. Div. Cell. Roles Peptides Neuronal Funct.* **1978**, *16* (4), 535–553.

Barker, J. L. Physiological Roles of Peptides in the Nervous System. In *Peptides in Neurobiology* Gainer, H., Ed.; Academic Press: New York: Plenum, 1977; pp 295–343.

Barnard, E.A., Skolnick, P. and Olsen, R.W..mohler H, Sieghart W, Biggio G, Braestrup C, Bateson AN, Langer Z: International union of pharmacology. XV. Subtypes of (aminobutyric acid A receptors: classification on the basis of subunit structure and receptor function. *Pharmacol Rev.* **1998**, *50*, 291–313.

Barrett, K.; Brooks, H.; Boitano, S.; Barman, S. Neurotransmitters & Neuromodulators. In *Ganong's Review of Medical Physiology*, 23rd ed; McGraw-Hill: New York, 2009; pp 129–146.

Bowery, N. G.; Smart, T. G. GABA and Glycine as Neurotransmitters: A Brief History. *Br. J. Pharmacol.* **2006**, *147*, S109–S119.

Brookhart, J. M.; Mountcastle, V. B. *The Nervous System. Handbook of Physiology, Vol 1, Cellular Biology of Neurons, Part 2*; American Physiological Society: Bethesda, 1977.

Brown, D. A. Acetylcholine. *Br. J. Pharmacol.* **2006**, *147*, S1.

Burnstock, G., Purinergic Receptors in the Nervous System. *Curr. Top. Membr.* **2003**, *54*, 307–368.

Burnstock, G. Do Some Nerve Cells Release More Than One Transmitter? *Neuroscience* **1976**, *1*, 239–248.

Butcher, L. L., Talbot, K. Chemical Communication Processes Involving Neurons: Vocabulary and Syntax. In *Cholinergic- Monoaminergic Interactions in the Brain*; Butcher, L. L., Ed.; Academic Press: New York, 1978; pp 3–22.

Bloom, F. E.; Neurotransmission and the Central Nervous System. In *Goodman & Gilman's: The Pharmacological Basis of Therapeutics*; Brunton, L. L.; Lazo, J. S.; Parker K. L., Eds.; McGraw-Hill: New York, 2006; pp 317.

Burrows, M. Neurotransmitters, neuromodulators and, neurohormones. The neurobiology of an insect brain. Oxford University Press, Oxford, 1996, pp.168–228.

Choudhury, A.; Sahu, T.; Ramanujam, P. L.; Banerjee Kumar, A.; Chakraborty, I.; Arora, N. Neurochemicals, Behaviours and Psychiatric Perspectives of Neurological Diseases. *Neuropsychiatry (London)*. 2018, 8 (1), 395–424.

Claudia, G. E.; Fabiola, G. M. Basic Elements of Signal Transduction Pathways Involved in Chemical Neurotransmission. In *Identification of Neural Markers Accompanying Memory*; Meneses, A., Ed.; Elsevier, 2014; Vol. 8, pp 127.

Craig, C. R. Introduction to Central Nervous System Pharmacology. In *Modern Pharmacology with Clinical Applications*; Craig, C. R.; Stitzel, R. E., Eds.; Lippincott Williams and Wilkins: Philadelphia, 2004, pp 287.

Crow, T. J.; Harrington C. A. Etiopathogenesis and Treatment of Psychosis. *Annu. Rev. Med.* 1994, 45, 219–234.

De wied, D. The Neuropeptide Concept. *Prog. Brain. Res.* 1987, 72, 93–108.

Devane, W. A.; Hanus, L.; Breuer, A. Isolation and Structure of a Brain Constituent that Binds to the Cannabinoid Receptor. *Science* 1992, 258, 1946–1949.

Dismukes, P. L. K. New Concepts of Molecular Communication Among Neurons. *Behav. Brain Sci.* 1979, 2, 409–448.

Dwivedi, C. Antiepileptic Drugs. *Am. J. Pharm. Educ.* 2001, 65, 197–202.

Edwards, R. H. The Neurotransmitter Cycle and Quantal Size. *Neuron* 2007, 55 (6), 835–858.

Edward, F. A; Gibb, A. J. ATP - A Fast Neurotransmitter. *FEBS Lett.* 1993, 325 (1–2), 86–89.

Edwards, F. A.; Gibb, A. J.; Colquhoun, D. ATP Receptor Mediated Synaptic Currents in the Central Nervous System. *Nature* 1992, 359, 144–147.

Francis, P. T.; Palmer, A. M.; Snape, M.; Wilcock, G. K. The Cholinergic Hypothesis of Alzheimer's Disease: A Review of Progress. *J. Neurol. Neurosurg. Psychiatry* 1999, 66, 137–147.

Froehlich, J. C. Opioid peptides. Neurotransmitter Review. *Alcohol Res. Health* 1997, 21 (2), 132–136.

Galvan A.; Wichmann, T. Pathophysiology of Parkinsonism. *Clin. Neurophysio.* 2008, 119, 1459–1474.

Garris, P. A. Advancing Neurochemical Monitoring. *Nat. Methods* 2010, 7, 106–108.

Gaspar, P.; Cases, O.; Maroteaux L. The Developmental Role of Serotonin, News from Mouse Molecular Genetics. *Nat. Rev. Neurosci.* 2003, 4, 1002–1012.

Geoffrey, B.; Bertil, B. Fredholm and Alexei Verkhratsk. Adenosine and ATP Receptors in the Brain. *Curr. Top. Med. Chem.* 2011, 11, 973–1011.

Goetz, C. G.; Poewe, W.; Rascol, O.; Sampaio, C. Evidence-based Medical Review Update: Pharmacological and Surgical Treatments of Parkinson's Disease: 2001 to 2004. *Mov. Disord.* 2005, 20, 523–539.

Greengard, P. Possible Role for Cyclic Nucleotides and Phosphorylated Membrane Proteins in Postsynaptic Actions of Neurotransmitters. *Nature* 1976, 260, 101–108.

Greengard, P.; Kebabian, J. W. Role of Cyclic AMP in Synaptic Transmission in the Mammalian Peripheral Nervous System. *Fed. Proc.* 1974, 33, 1059–1067.

Halbach, O. B.; Dermietzel, R. *Neurotransmitters and Neuromodulators*; Wiley-VCH Verlag GmbH & Co., Weinheim, Germany, 2002.

Hall, J. E.; Guyton, A. C. *Guyton and Hall Textbook of Medical Physiology*, 12th ed; Saunders Elsevier: Philadelphia, 2011.

Henley, J. M. Amino Acid Neurotransmitters. In *Encyclopedia of Life Sciences*; Nature Publishing Group: New York, 2001; pp 1–7.

Herlenius E.; Lagercrantz H. Neurotransmitters and Neuromodulators During Early Human Development. *Early Hum. Dev.* 2001, 65, 21–37.

Higuchi, T. Oxytocin: A Neurohormone, Neuroregulator, Paracrine Substance. *Jpn. J. Physiol.* 1995, 45, 1–21.

Hokfelt, T. Neuropeptides: Opportunities for Drug Discovery. *Lancet. Neurol.* 2003, 2, 463–472.

Hosie, A. M.; Dunne, E. L.; Harvey, R. J.; Smart, T. G. Zinc-mediated Inhibition of GABAA Receptors: Discrete Binding Sites Underlie Subtype Specificity. *Nat. Neurosci.* 2003, 6, 362–369.

Hoyle, G. Neurotransmitters, Neuromodulators, and Neurohormones. In *Neurobiology. Proceedings in Life Sciences*; Gilles, R.; Balthazart, J. Eds.; Springer: Heidelberg, Berlin, 1985.

Ingole, S. R.; Rajput, S. K.; Sharma, S. S. Cognition Enhancers: Current Strategies and Future Perspectives. *CRIPS* 2008, 9 (3), 42–48.

Iversen, L. Cannabis and the Brain. *Brain.* 2003, 126, 1252–1270.

Jacklet, J. W. Nitric Oxide as a Neuronal Messenger. In *Encyclopedia of Life Sciences*; Wiley J. & Sons, United States, 2005; pp 1–8.

Jan, Y. N.; Jan, L. Y.; Kuffler, S. W. *A Peptide as a Possible Transmitter in Sympathetic Ganglia of the Frog*. Proc. Natl. Acad. Sci. USA 1979, 76, 1501–1505.

Juif, P.; Poisbeau, P. Neurohormonal Effects of Oxytocin and Vasopressin Receptor Agonists on Spinal Pain Processing in Male Rats. *Pain* **2013,** *154,* 1449–1456.

Keverne J.; Ray M. Neurochemistry of Alzheimer's Disease. *Psychiatry* **2005,** *4* (1), 40–42.

Kulinsky, V. I.; Kolesnichenko, L. S. Molecular Mechanisms of Hormonal Activity. I. Receptors. Neuromediators. Systems with Second Messengers. *Biochemistry (Moscow)* **2005,** *70* (1), 24–39.

Kuhar, M. J., Couceyro, P. R., Lambert, P. D. Biosynthesis of Catecholamines. In *Basic Neurochemistry: Molecular, Cellular and Medical Aspects;* Siegel, G. J.; Agranoff, B. W.; Albers, R. W. et al., Eds.; Lippincott - Raven: Philadelphia, 1999.

Kuhse, J; Betz, H. The Inhibitory Glycine Receptor: Architecture, Synaptic Localization and Molecular Pathology of a Postsynaptic Ion—Channel Complex. *Curr. Opin. Neurobiol.* **1995,** *5,* 318–323.

Langer, S. Z. Subtypes of γ-aminobutyric acid a Receptors: Classification on the Basis of Subunit Structure and Receptor Function. *Pharmacol. Rev.* **1998,** *50,* 291–313.

Lodish, H.; Berk, A.; Zipursky, S. L. Molecular Cell Biology: Section 21.4 Neurotransmitters, Synapses, and Impulse Transmission, 4th ed; W. H. Freeman: New York, 2002.

Marzo, V. D.; Melck, D.; Bisogno, T.; Petrocellis, L. D. Endocannabinoids: Endogenous Cannabinoid Receptor Ligands with Neuromodulatory Action. *Trends Neurosci.* **1998,** *21* (12), 521–528.

Mechoulam, R.; Fride, E. Cannabinoid Receptors. Pertwee, R. G., Ed.; Academic Press, United States, 1995; pp 233–258.

Meech, R. W. Calcium-dependent Potassium Activation in Nervous Tissue. *Ann. Rev. Biophys. Bioeng.* **1978,** *7,* 1–18.

Muller, E. E.; Nistico, G. *Brain Messengers and the Pituitary;* Academic Press: San Diego, 1989.

Nestler, E. J.; Hyman, S. E.; Malenka, R. C. *Molecular Neuropharmacology: A Foundation for Clinical Neuroscience,* 2nd ed.; McGraw-Hill Companies: USA, 2009.

Ohmura, Y.; Tsutsui-Kimura, I.; Yoshioka, M. Impulsive Behaviour and Nicotinic Acetylcholine Receptors. *J. Pharmacol. Sci.* **2012,** *118* (4), 413–422.

Oja, S. S; Saransaari, P. Neurotransmitters and Modulators. In *Physiology and Maintenance, Encyclopedia of Life Support Systems:* UK, 2009; Vol. 5.

O'Shea, M.; Husbands, P.; Philippides, A. Nitric Oxide Neuromodulation. In *Encyclopedia of Computational Neuroscience;* Jaeger, D.; Jung R., Eds.; Springer: New York: NY, 2013.

Parihar, M. S.; Hemnani, T. Alzheimer's Disease Pathogenesis and Therapeutic Interventions. *J. Clin. Neurosci.* **2004,** *11* (5), 456–467.

Phillis, J. W.; Kostopoulos, G. K. Activation of a Noradrenergic Pathway from the Brain Stem to Rat Cerebral Cortex. *Gen. Pharmacol.* **1977,** *8,* 207–211.

Rand, J. B. Acetylcholine. *WormBook.* **2007,** *131* (1), 1–21.

Rang, H. P.; Dale, M. M.; Ritter, J. M.; Flower, R. J.; Henderson, G. Rang & Dale's Pharmacology, 7th ed; Churchill Livingstone: Edinbergh, **2011**.

Sadock, B. J.; Sadock, V. A.; Ruiz, P. *Kaplan and Sadock's Comprehensive Textbook of Psychiatry,* 9th Ed; Lippincott Williams & Wilkins: Philadelphia, 2009.

Samii, A.; Ransom, B. R. Movement Disorders: Overview and Treatment Options. *P&T* **2005,** *30* (4), 228–238.

Scharrer, B. Neurohumors and Neurohormones: Definitions and Terminology. *J. Neural. Rel.* **1969,** Suppl. IX, 1–20.

Scharrer, B.: The Neurosecretory Neuron in Neuroendocrine Regulatory Mechanisms. *Amer. Zool.* **1967,** *7,* 161–169.

Scharrer, B.; Neurosecretion, X. I. V. Ultrastructural Study of Sites of Release of Neurosecretory Material in Blattarian Insects. *Zschr. Zellforsch.* **1968,** *89,* 1–16.

Sembulingam, K.; Sembulingam P. *Essentials of Medical Physiology,* 6th ed; Jaypee Brothers Medical Publishers: New Delhi, 2012.

Sharma, H. L.; Sharma, K. K. *Principles of Pharmacology,* 3rd ed; Paras Medical Publisher: Hyderabad, 2017.

Shin, C.; McNamara, J. O. Mechanism of Epilepsy. *Annu. Rev. Med.* **1994,** *45,* 379–389.

Smith, C. M. Elements of Molecular Neurobiology, 3rd ed; John Wiley & Sons Ltd: England, 2002.

Stanford, S. C. Noradrenaline. In *Neurotransmitters, Drugs and Brain function;* Roy, W., Ed.; John Wiley and Sons: UK and Chichester, 2001.

Stahl, S. M. Stahl's Essential Psychopharmacology: Neuroscientific Basis and Practical Applications, 3rd ed; Cambridge University Press: New York, 2008.

Tortora, G. J.; Derrickson, B. *Principles of Anatomy and Physiology,* 12th ed; John Wiley & Sons: USA, 2009.

Uddin, M. S.; Mamun, A. A., Kabir M. T. et al. Neurochemistry of Neurochemicals: Messengers of Brain Functions. *J. Intellec. Disab. Diag. Treat.* **2017,** *5,* 137–151.

Vannier, C; Triller, A. Biology of the Postsynaptic Glycine Receptor. *Int. Rev. Cytol.* **1997,** *176,* 201–244.

Webster, R. A. Other Transmitters and Mediators. In *Neurotransmitters, Drugs and Brain Functions,* Roy,

W., Ed.; John Wiley and Sons: UK and Chichester, 2001, pp 265–285.

Werman, R. Criteria for Identification of a Central Nervous Transmitter. *Comp. Biochem. Physiol.* **1996**, *18*, 745–766.

Wise, R. A. Dopamine, Learning and Motivation. *Nat. Rev. Neurosci.* **2004,** *5*, 483–494.

# CHAPTER 23

# Evaluation Models for Drug Transport Across the Blood–Brain Barrier

CHINNU SABU and K. PRAMOD*

*Government Medical College, Kerala, India*

*Corresponding author. E-mail: pramodkphd@yahoo.com*

## ABSTRACT

The blood-brain barrier (BBB) acts as a merging channel between peripheral and central nervous system (CNS). This barrier enacts a major part in the regulation of substances out and into the brain. The specialized structure of the barrier act as a hindrance to the advancement of novel drugs targeted to CNS. In vitro BBB models are easy as well as valuable aiding tools that perform a great role in regulating and controlling brain homeostasis. Basically, models are cultures of brain endothelial cells. Various models like monolayer, co-culture, microfluidic, dynamic, and to the most recently developed three-dimensional models are reviewed. In vivo methods for evaluation of drug movement across BBB are discussed. This chapter reviews various evaluation models for drug transport across the BBB which covers both in vivo and in vitro models. The chapter also focuses on the selection criteria, validation, and optimization of parameters in the development of an ideal model.

## 23.1 INTRODUCTION

The brain is a labile organ which safeguards itself independently from toxic substances and injuries, at the same time this complex structure hinders the drug development. The major obstacle to the discovery of pharmaceuticals is the capability to go across the blood–brain barrier (BBB) which is often a challenging problem. BBB is vital for the regular functioning of the central nervous system (CNS) as well as to treat them. The mechanism by which the barrier is altered during pathological studies is still undergoing studies and the solution to the problem is not yet discovered. Understanding of basic of the structure leads to the evolvement of various in vitro models (Palmiotti et al., 2014).

Besides the barrier function, it acts as a carrier in the transfer of nutrients to brain and removal of metabolites. Also, it plays a crucial role in clinical practice. The permeability of the barrier is increased in various neurological conditions like cerebral ischemia, brain trauma, tumors, and neurodegenerative disorders (Weiss et al., 2009). However, the impermeability of the barrier to different drugs limits the presence at therapeutically relevant concentration is one of the major obstacles in the treatment of CNS disorders (Jeffrey and Summerfield, 2010). The barrier permeability is characterized with the help of in vitro models.

It has been seen that physiological responses are difficult to obtain in vivo (Xu et al., 2016). To overcome this issue, a good number of in vitro models are developed which helps to figure out the functioning associated with various pathological conditions (He et al., 2014). The barrier acts as an interface performing various specialized functions like protection from injuries as well as for the passage of molecules. The in vitro BBB simulation is a major hurdle in mimicking the physiological characteristics, which have to correlate with the in vivo condition. A high throughput screening can be achieved with the help of combinatorial chemistry. The structure of the barrier and its changes associated with various pathological conditions can be studied. In this chapter authors presented, the different types of in vitro and in vivo models. In addition,

the various criteria in the selection of an ideal model are discussed.

## 23.2 BLOOD–BRAIN BARRIER

### 23.2.1 *CELLULAR STRUCTURE*

BBB is an anatomical structure which prevents the influx and efflux of substances by forming tight junctions (TJs) between the cells making it protective in action. The barrier acts as a blocking membrane for the transport of substances into CNS. The core components that make up BBB include endothelial cells, astrocytes, and pericytes (Fig. 23.1). The tight junction of endothelial cells at BBB is made distinct from endothelial cells present in other parts of the human body by characteristics like

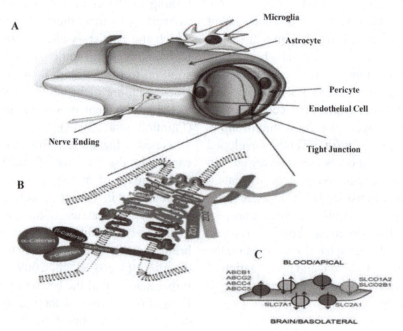

**FIGURE 23.1**   Schematic representation of the cellular structure of brain capillary. (A) Represents the brain capillary lined by endothelial cells, astrocytes, and pericytes. (B) Illustrates the tight and adherence junction proteins. (C) Transporters of blood–brain barrier.

*Source*: Reprinted with permission from Wilhelm and Krizbai (2014). © American Chemical Society.

lack of openings and sparse vesicular transport (Praveen et al., 2014).

TJs and adherence junctions (AJs) mainly constitute the BBB junction complex. TJs are an array of ceaseless intracellular membranous strands or fibrils on the peripheral phase and grooves on endothelial phase. The three integral membrane proteins that form the backbone are claudin, occludin, junction adhesion molecules (JAMs), and numerous cytoplasmic accessory proteins. Claudin 1 and claudin 2 are mainly found in BBB junction, where these proteins interplay with one another to frame a tight seal (Furuse et al., 1998, 1999). Occludin is a regulatory protein with four transmembrane domains. The paracellular barrier of TJ is contributed by the extracellular loop of claudin and occludin originating from the neighboring cells essential for the development of BBB (Hirase et al., 1997; Sonoda et al., 1999). JAMs are immunoglobulin having a single membrane trans-domain. Mainly JAM 1 and JAM 2 are revealed in BBB. They are responsible for cell to cell adhesion, paracellular permeability, and monocyte transmigration (Bazzoni et al., 2000). Major cytoplasmic accessory proteins responsible for the formation of TJs include zonula occludens (ZO) 1, ZO 2, ZO 3, cingulin, and many others. Interaction of transmembrane protein with cytoplasmic protein imposes skeletal rigidity to the endothelial cells (Haskin et al., 1998). Cadherin-Catenin complex and further associated proteins constitute AJ.

Astrocyte end-feet help in inducing tightening of the endothelium (Janzer el al., 1987). The formation of optimal BBB results from the interaction of endothelial cells with astrocyte. Apart from it helps in down-regulation of tetraethylenepentamine (TEPA) and thrombomodulin (TM) expression (Rubin et al., 1991). Pericytes (PCs) are cells of microvessel that have a significant role in imparting the structural stability to endothelial cells, also in cerebral auto-regulation, angiogenesis, differentiation of BBB, reducing capillary diameter, and blood flow (Ramsauer et al., 2002; Winkler et al., 2011).

## 23.2.2 CHARACTERISTICS

Normally, the serum protein leakage into CNS is restricted by BBB. The modification of the barrier thus creates a greater reduction in the demand of pinocytosis and the absence of intracellular fenestration. The substances are transported across the membrane through different mechanisms like diffusion, adsorptive endocytosis, saturable transport, and extracellular pathways (Banks, 2009).

## 23.2.3 FUNCTIONS

The functions of BBB are not fixed but gets adapted according to the needs. BBB has a significant role in the retainment of brain homeostasis, control of influx and efflux transport, and protection of brain. It regulates the entry and secretion of substance into CNS. The barrier plays a significant role in the exchange and control of informational molecules between CNS and blood (Quan and Banks, 2007). Specific efflux transporter system at the BBB mainly contribute to the protection of brain from harmful amphipathic and hydrophobic substance.

## 23.2.4 BLOOD–BRAIN BARRIER IN DISEASED CONDITION

Interruption of BBB may be a result of trauma as well as various inflammatory responses which cause entry of various components into the CNS. The permeability of the barrier may be helpful as well as, to that same extent, can be dangerous. In one way, increase in permeability allows the substance to enter the cell by targeting the molecule. Disruption of the barrier leads to gradual deterioration, injury, neuronal dysfunction and finally it leads to the death (Daneman, 2012). The BBB disruption in various pathological conditions like Parkinson disease, Alzheimer's and epilepsy have been studied using various in vitro models on endothelial cell cultures using imaging tools and histopathological studies (Mihaly and Bozoky, 1984; Zlokovic, 2005).

## 23.3 IN VITRO BLOOD–BRAIN BARRIER MODELS

Relevant model helps in studying the brain pathologies without which it is often difficult. Some of the developed tissue culture models have characteristics similar to in vivo, but the junctions may not be as strong when compared to in vivo (Butt et al., 1990). There is no perfect in vitro model that mimics characteristics similar to in vivo. The limitation encountered

in the build-up of barrier models is the reduced availability of primary endothelium which is compensated by stems derived brain endothelial cells (Lippmann et al., 2012). The functional and structural requirement for an ideal in vitro model includes restricted paracellular diffusion between endothelial cells, selective and asymmetric permeability to crucial ions, expression of efflux systems, transport mechanisms and drug-metabolizing enzymes, exposure to laminar shear stress and finally responsiveness to permeation modulators. A schematic representation of in vitro and in vivo transport of reported polyethylene glycolated (PEGylated) silica nanoparticles across BBB is shown in Figure 23.2.

### 23.3.1 CLASSIFICATION OF IN VITRO BLOOD–BRAIN BARRIER MODELS

#### 23.3.1.1 STATIC MODELS

Static models do not imitate the shear push produced by the stream of blood under in vivo

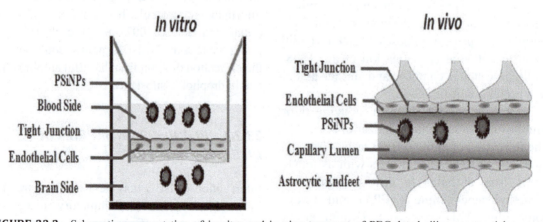

**FIGURE 23.2** Schematic representation of in vitro and in vivo transport of PEGylated silica nanoparticles across blood–brain barrier.

*Source*: Reprinted with permission from Liu et al. (2014). © American Chemical Society.

conditions. Based on the number of cell sorts included, they are sub-classified as follows.

### 23.3.1.1.1 Endothelial Monoculture Models

A simple BBB model is a monolayer of endothelial cell grown on Trans well insert. The insert shows the same properties as that of blood side, whereas the well on which the insert is positioned depicts the abluminal side mimicking a side-by-side diffusion system across a microporous semipermeable membrane. The microporous membrane like polycarbonate filter permits the movement of molecules. Low passage of brain microvascular endothelial cells (BMEC) is used that shows the same characteristics and properties similar to in vivo conditions (Cecchelli et al., 2007). They are widely used because of convenience but face problems like poor agitation and temperature control which shows significant differences compared to in vivo conditions (Berezowski et al., 2004; Patlak et al., 1981). The major challenges faced in isolating BMEC are low yield and contamination. High yield of BMEC can be obtained by the involvement of larger species like bovine, porcine and so on (Lippmann et al., 2013). However, they are nontransgenic and only a few antibodies are available in case of larger species. This limitation results in the use of rodents in various studies. Recently, various studies have reported that incorporation of pluromycin in the culture medium improves the purity of BMECs (Perriere et al., 2005, 2007; Yao and Tsirka, 2011).

Human endothelial cells are taken when the human-specific receptor is used. However, due to ethical reasons, they are not currently used. In order to balance the impediment, immortalized cells are utilized which is derived either from a human cerebral vascular endothelial cell line (Weksler, 2013) and immortalized human cerebral endothelial cells. Immortalized cell lines from rodents are for the most part utilized in view of their expanded capacity to multiply in culture. Reduced articulation of transporters and compound (Urich, 2012), a decrease in γ glutamyl transpeptidase and antacid phosphate movement (Hafny, 1996; Hosoya, 2000) prompts loss of specific qualities which make it different from in vivo (Roux et.al. 1994; Roux and Couraud, 2005; Weksler, 2005). These limitations lead to the formation of a loose monolayer. Another way to deal with the advancement of BMECs is by the utilization of human pluripotent immature microorganisms for the most part communicated at different tight intersections, transporters, and receptors (Lippmann et al., 2012). The principal limitation associated with monolayer model is that they contain only one cell type and lacks communication between various cells. Because of its effortlessness, they are for the most part utilized as a part of the flagging pathway, transport energy, restricting proclivity estimation, and high-throughput screening (Berezowski et al., 2004).

### 23.3.1.1.2 Co-Culture Models

Introduction of various cells like glial cells and astrocytes help in the maintenance and proper functioning of BBB essential for the design of a co-culture model (Deli et al., 2005).

#### 23.3.1.1.2.1 Co-Culture Model of Endothelial Cells with Glial Cells

Glial cells are primary cells used in the study of gliomas. Glial cells have the property to stimulate the BBB characteristics. Thus, incorporation of endothelial cells with glial cells creates a model that shows the same may be obtained from different sources. A major draw-back encountered is the availability of tissue of surgical origin (Hutamekalin et al., 2008; Stamatovic et al., 2005; Veszelka et al., 2007).

### 23.3.1.1.2.2   Co-Culture of Endothelial Cells with Astrocytes

Recent studies proved that a direct contact of endothelial cells and astrocytes is necessary to obtain expected transendothelial electrical resistance (TEER) value compared to other systems which show negligible TEER values (Malina et al., 2009). Astrocytes make an important part in the maintenance of tight junction as well as in the up-regulation of efflux transporters (Gaillard et al., 2000; Hori, et al., 2004, Wolburg, et al., 1994). The effect of astrocytes on endothelium can be altered by various experimental arrangements. One method includes culturing of endothelial cells on the top of a filter with astrocytes grown at the bottom of the filter which enables a direct contact. The second method involves the growth of astrocytes at the bottom of the filter which permits the soluble factors to exert their effect. The last arrangement involves the use of astrocyte conditioned medium, but the effect of factors released by astrocytes on the endothelial cell cannot be determined.

### 23.3.1.1.2.3   Co-Culture of Endothelial Cells using Pericytes

Pericytes act as a root of adult pluripotent stem cells and act as an important factor in the development of basement membrane (Sa-Pereira et al., 2012). These cells serve various contractile, immune, phagocytic, and antigenic functions. Endothelial co-culture model with pericyte can mediate functions like occludin gene expression (Hori et al., 2004), regulation of functions of TGF-β (Dohgu et al., 2005), reducing

permeability (Nakagawa et al., 2007), induction of MRP6 expression (Berezowski et al., 2004), and functional transporter activity (Ahmad et al., 2011).

### 23.3.1.1.2.4   Co-Culture of Endothelial Cells with Neuronal Precursors or Other Cells

Since this culture model shows low TEER value and high permeability, are mainly used to study the effect of drugs on embryonic life (Weidenfeller et al., 2007). A recent study showed that human model showing high TEER value has been developed using neural progenitor cell-derived astrocytes and neurons (Lippman et al., 2014). Neurons can have the capability to induce blood–brain-related enzymes in an endothelial cell culture (Tontsch and Bauer, 1991). Various studies concluded that direct contact with endothelial cell culture is not necessary for increased TEER.

### 23.3.1.1.2.5   Triple Cell Co-Culture Models

Various experimental set up can be designed with the use of different types of cells. A fundamental setup incorporates refined of endothelial cells on the highest point of channel embed and pericytes and astrocytes on the lower side of the channel and at the base of the well. The difference in BBB functions at various stages of life can have been differentiated since the endothelial cells retained properties in vivo (Takata et al., 2013). The main benefit of the model is the use of primate endothelial cells which result in low species difference.

New means to the advancement of triple cell culture models employing endothelial cells, astrocytes, and neuronal cells have been recently introduced. Studies were conducted to establish the relation on contact between endothelial cells and others. It is shown that both in contact and non-contact mode, they show no difference in transporter expression, but the permeability was much lower in non-contact mode (Vandenhaute et al., 2011).

### 23.3.1.2 DYNAMIC MODELS

The fact that a shear stress is able to affect the endothelial and barrier properties has led to the development of dynamic in vitro model (Tarbel, 2010). The model mimics the in vivo blood circulation by culturing of endothelial cells in hollow fibers. Similarly, the induction of variable stress is attributed by circulating the culture media.

#### 23.3.1.2.1 Cone and Plate Apparatus

In cone and plate apparatus model, endothelial cells are cultured on the bottom of the plate which is exposed to shear stress generated by the cone. Thereby, the force is transferred to the cells by the rotation of the cone. The amount of stress generated is directly proportional to the cone angle and angular velocity. Back-flow and pulsatile shear stress can be achieved by reversing and pulsing the angular velocity of the cone. Laminar shear stress has a great impact on the endothelial physiology. In addition, it also leads to the production of vasoactive substances and in the modulation of the bioenergetic behavior (Bussolari et al., 1982; Dewey et al.,1981). Moreover, turbulent shear stress has a different effect when compared with laminar shear stress. The model is widely used in biochemical studies and physiological investigations. The model lacks the ability to characterize the flow experienced by endothelium in vivo. As a result, decreased permeability is observed due to lack

of contact of endothelial cells with other cells. This limitation makes a way to the development of a new model which gives importance to the flow.

#### 23.3.1.2.2 Dynamic In Vitro Blood–Brain Barrier Model

Under the influence of laminar shear stress, brain endothelial cells are co-cultured with astrocytes in dynamic in vitro BBB model which shows exact characteristics as in situ (Cucullo et al., 2002; Santaguida et al., 2006). In this model, the cerebral endothelial cells are cultured on the lumen and astrocyte co-cultured on the Poly-D-lysine coat on the outer surface. A large number of hollow fibers are suspended inside a sealed chamber. The artificial capillaries are in contact with the medium through silicone tubing. The exposure to both luminal and abluminal compartment can be achieved by various ports. A variable pulsate pump can be used to generate similarly in vivo conditions (Dewey et al., 1981; Koutsiaris et al., 2007). The system allows continuous circulation of blood making the studies related to various diseased conditions easier. A recently developed model allows cell migration studies which help to evaluate the neuroinflammation of brain tissue (Takeshita and Ransohoff, 2012). Here, the morphological changes of the endothelium cannot be determined since the model does not allow the visualization of intraluminal components. Besides this, the system requires a lot of time and skill. Moreover, the limited ability of cell characterization makes the system less applicable.

### 23.3.1.3 MICROFLUIDIC MODEL

The microfluidic model generates shear stress and dynamic flow by the incorporation of two perpendicular crossing channels. The model consists of a semi-porous membrane placed at the interface of two channels with the co-culture of endothelial and glial cells on both sides of the membrane (Fig. 23.3). Silver/silver chloride (Ag/AgCl) electrodes connected at both side of surfaces are then used for TEER measurement. A further modification of the model can be done by using platinum electrodes instead of Ag/AgCl electrode resulting in the use of less number of cells and reliable estimation of TEER (Griep et al., 2013). The model finds promising application in permeability studies.

Generation of a model with two side-by-side compartments having separation by a continuous improved TJ protein resulted in inaccurate TEER measurement (Prabhakarpandian et al., 2013). Another microfluidic model makes use of a micro-hole structure for trapping cells as a result of pressure gradient (Yeon et al.,

2012). The model consists of two horizontally oriented chambers with micro-hole structure incubated with astrocyte conditioned medium. The permeability of compounds is validated by use of several drugs like antipyrine, atenolol and so on. A limitation accounted for these models is that it lacks cell to cell contact. In addition, it does not allow replication of microvasculature in vivo.

Further, the modification of above model by a synthetic microvasculature model was developed. The model consists of two microchannels separated by microfabricated pillar infused with endothelial cells and astrocyte co-cultured medium through various ports (Prabhakarpandian et al., 2013).

The microfluidic BBB model offers several advantages in comparison with the dynamic in vitro BBB model. Non-destructive microscopy, transmigration or trafficking studies, use of less

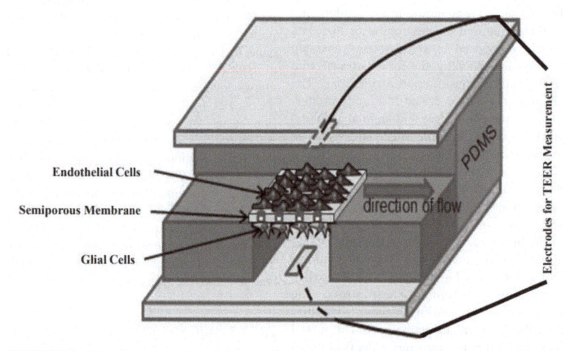

**FIGURE 23.3**  Schematic representation of a microfluidic model.

*Source*: Reprinted with permission from Wilhelm and Krizbai (2014). © American Chemical Society.

amount of cells, fewer technical skills and less time to reach steady state TEER are the main advantages associated with the model. Apart from the advantages, the model has certain limitations. Short time in culture results in low TEER value than expected. Moreover, the model contains only two types of cell. Microfluidic models find extensive application in new drug research and development and neurovascular research (Booth and Kim, 2012).

### 23.3.1.4 THREE-DIMENSIONAL MODEL

The three-dimensional (3D) model of human BBB in vitro indicates cell structure like that of the CNS and it can be utilized to break down the conveyance of particles to the brain. The model consists of human astrocytes incorporated in a collagen gel, overlaid by a layer of human brain endothelium (Peddagangannagari et al., 2015). The endothelial cells arranged in a cylindrical monolayer permits leukocyte transmission across BBB. The model is developed by a coating an adhesive molecule like PDL and collagen gel on microstructures and glass substrates followed by a coating of brain endothelial cell on top and bottom surfaces to form a tight monolayer (Hansang et al., 2015). Characterization of the model can be done by transmission electron microscopy (TEM), immunofluorescence microscopy, and flow cytometry. The results can be relevant to human conditions since the cell used is human cells. Monitoring of the whole process provides crucial information at various time points.

### 23.3.1.5 CELL-BASED IN VITRO MODEL OF NON-CEREBRAL ORIGIN

The use of noncerebral peripheral epithelial cell line has come into focus due to lack of ideal barrier properties associated with the immortalized brain endothelial cell line. The various models include Madin–Darby canine kidney (MDCK) cell line, Caco-2 cell line, and ECV304/C6 cell line (Hellinger et al., 2012).

#### 23.3.1.5.1 MDCK Cell Line

MDCK which is easy to grow with low permeability to sucrose have appreciable TEER and can be transfected with multidrug resistance gene (MDCR1) resulting in polarized expression of P-glycoprotein (Pgp) (Gumbleton and Audus, 2001; Pastan et al., 1988). MDCR1 MDK cell line can be used as a model to aid drug discovery and for studying the in vitro permeability of compounds (Wang et al., 2005). Paracellular transport has been overestimated with this type of cell line due to a highly intercellular transverse area since kidney cells are cubical in shape having small surface area results in large cell density (Gumbleton and Audus, 2001). The effect of other efflux transporters should also need to be considered when studies are conducted using MDCK cell line.

#### 23.3.1.5.2 Caco-2 Cell Line

Caco-2 cells are epithelial cells derived from human colon adenoma cell line having distinct morphological characteristics are mainly used in BBB permeability studies (Lundquist and Renftel, 2002). Caco-2 cells are a poor model for permeability studies (Lundquist, 2002; Lohmann et al., 2002).

#### 23.3.1.5.3 ECV304/C6 Cell Line

ECV304/C6 is combinations of ECV304 co-cultured with C6 cells have both epithelial and endothelial properties. The major draw-backs like low TEER value and lack of Pgp expression make the application of the cell line limited (Hurst and Fritz, 1996).

### 23.3.1.6 NON-CELL BASED IN VITRO MODEL

#### 23.3.1.6.1 Immobilized Artificial Membrane Chromatography

The membrane acting as interface consists of phosphatidyl choline residue bound with an inert silica support that can be used as an alternative for screening drug permeability through the cell membrane (Pidgeon et al., 1995; Stewart and Chan,1998). The permeability of drug across the membrane can be determined drug partitioning into JAMs (Nasal et al., 1995; Ong et al., 1996). The major draw-back of the model is that it does not mimic the mechanism of fluid membrane especially in case of passive diffusion and the prediction power is low when the brain uptake is affected by other factors like plasma protein binding, active transport, and so on.

#### 23.3.1.6.2 Parallel Artificial Membrane Permeability Assay

Parallel artificial membrane permeability assay (PAMPA) model was proposed for studying CNS permeability. Porcine brain lipid extract dissolved in n-dodecane serves as PAMPA membrane barrier. The permeability characteristics of chemical compounds can be determined by a combined solubility permeability assay method (Wexler et al., 2005). In this method, it involves the determination of solubility at different pH values. The filtered

saturated solution acts as an input material for BBB permeability. This method offers certain merits like reduced of sample usage and preparation time, removal of interference, maximization of input concentration and optimization of the sample to track. A combined approach of both active and passive transporter studies helps in differentiation between paracellular and transcellular components of transport (Kerns et al., 2004).

### 23.3.2 SELECTION OF MODELS

Selection of appropriate model is based upon the purpose of the study which eliminates the effort of time and money spent on it. Co-culture models and dynamic in vitro models are used when permeability studies are needed to be carried out. The microfluidic model finds its application when cell migration study is involved. Monolayer models can be an appropriate choice when the objective is to study transport kinetics, for exploring signaling pathway and in the measurement of binding affinities. The results are then validated when they are correlated in vivo (Yarong et al., 2014).

In vitro models find a suitable role in new drug research and development process which includes various stages like target identification, hit identification, lead identification, and finally optimization of the product. The first stage involves screening numerous compounds by high throughput screening when a target is identified. Simple models like monolayer and co-culture models are mainly used in the first step. The validation of identified compound and SAR are usually carried out in optimization stage which utilizes in vitro models like static co-culture and dynamic models that are sensitive to in vivo conditions. The correlation with human cells is required to be carried out to avoid interspecies variability at various stages of development (Paradis et al., 2016).

### 23.3.3 VALIDATION MARKERS FOR IN VITRO BLOOD–BRAIN BARRIER MODELS

Validation markers selected should be such that it makes the model important. The tightness of the barrier is usually measured by TEER. This feature is validated by permeability studies with hydrophilic tracer molecules like sucrose, mannitol, and sodium fluorescein. The experimental design and size of the molecule play an important role in the correlation of TEER with permeability studies. Claudin 5 acts as major tightening claudin with high BBB expression (Furuse, 2010). The permeability of small molecules can be assessed by the efflux transporter of ABC family and SLC. Validation can be mainly performed by protein or mRNA expression studies. The delivery of compounds conjugated to ligands can be assessed by receptors like leptin, insulin, glutathione and transferrin receptor. Among this transferring transferrin receptor is significant in CNS delivery of drugs. High expression of the receptor in brain endothelial cells in vivo, makes it an excellent marker for the study of a new model. A brief outline of the validation markers for in vitro BBB model is enlisted in Table 23.1.

### 23.3.4 OPTIMIZATION OF IN VITRO BLOOD–BRAIN BARRIER MODELS

Isolation of primary endothelial cells is a difficult procedure which often requires the removal of contaminating cells which can be accomplished by use of puromycin-containing medium having the property of killing cells with no Pgp activity (Perriere et al., 2005). The optimization of culture shows great significance in the selection of appropriate in vitro model. The model having a polycarbonate filter is mainly used for the growth of endothelial cells. Before the incubation, endothelial cells are coated with materials having properties same as that of the basement membrane. Inserts having the pore size of 0.4 μm are used for drug testing (Wuest et al., 2013). The cell trafficking studies are conducted using inserts having much larger pore size.

**TABLE 23.1** Validation Markers for In Vitro Blood–Brain Barrier Model.

| Category | Validation |
|---|---|
| Validation of cell lineage | Visualization; F-actin staining |
| Tight junctions | mRNA and protein expression-localization; TEER and permeability measurements |
| Efflux transporters | mRNA and protein expression; Bidirectional transport studies |
| SLC expression | mRNA and protein expression; Trans endothelial transport studies |
| Receptor systems | mRNA and protein expression; Trans endothelial transport of iron |

The formation of TJ is contributed by components like astrocytes, pericytes present in the barrier shows an increase in the intra-endothelial cyclic adenosine monophosphate (cAMP) level (Rubin et al., 1991). Wingless-type MMTV (mouse mammary tumor virus) integration site family (Wnt) signaling pathway activation plays a major role in the formation of the barrier (Hoheisel et al., 1998).The tightness of the barrier can be greatly increased by various factors like increased the buffer capacity of the medium, and the presence of insulin, sodium selenite, and so on.

## 23.4 IN VIVO METHODS

### 23.4.1 INVASIVE IN VIVO METHODS

#### 23.4.1.1 BRAIN/PLASMA RATIO $K_p$ (LOG BB)

It is the ratio of steady-state concentration of the compound in the brain to that in plasma. Passive diffusion properties, transporter effects, metabolism, and drug binding affinity are some of the main factors influencing log BB (Kalvass JC, 2002; Liu et al., 2008). The non-availability of a standard protocol for the determination of log BB results in variations. Compounds having a ratio of 0.3–0.5 have great access to CNS, values greater than 1 can easily cross BBB whereas a value of less than 0.1 finds difficult to enter CNS. This method is very costly since it requires more animals per unit time points. The partitioning of a compound into the lipid matrix can be determined by log BB (Van et al., 2001).

#### 23.4.1.2 BRAIN UPTAKE INDEX

It is one of the earliest methods, used to determine the uptake of a compound into the brain (Oldendorf, 1970). The carotid artery single pass technique involves the injection of both test and reference compound in the physiological buffer into carotid artery. Thereby internal standard determines the amount that reaches the brain (Bonate, 1995). The method is so fast and various compounds can be evaluated in less amount of time. Major limitation associated with this method is that the capillary transit time is short and only 20% of the injected dose enters the internal carotid artery (Bickel, 2005).

#### 23.4.1.3 IN SITU PERFUSION TECHNIQUE

The method is an extension of brain uptake index which longer requires experimental time followed by sampling at the defined time (Takasato et al., 1984). Before the initiation of the experiment, common carotid artery is ligated to prevent the mixing of perfusion fluid with the systemic blood. It possesses a great advantage since the systemic exposure is avoided. The effect of pH, ionic content, and flow rate can be monitored by this technique (Bonate, 1995). The method cannot be used for high throughput screening as the experimental set up is much complicated. Prolonged diffusion time possess a major limitation to this method (Bickel et al., 1993).

#### 23.4.1.4 INTRAVENOUS INJECTION TECHNIQUE

This technique also referred to as gold standard method involves cannulation of the femoral vein. At multiple time points, the brain levels are determined (Ohno et al., 1978). Brain concentration after correcting of intravascular content can be determined by co-administration of a marker or by a washing technique via ascending aorta. Both plasma and brain pharmacokinetics can be estimated using this sensitive method. This technique is much easier to be carried out compared to other ones (Bickel et al., 1993).

#### 23.4.1.5 INTRACEREBRAL MICRO DIALYSIS

The method involves incorporation of a micro-dialysis probe which is semi-permeable hollow fiber membrane mixed with perfusate into the extracellular matrix of the brain. The concentration of the molecule is determined by the collection of dialysate and estimated by an appropriate analytical method. In this method, concentrations are determined at different time intervals resulting in use of less number of animals for investigations. It also results in the reduction in issues associated with inter-animal variability.

The most interesting feature is that the probe can be placed anywhere in the brain. However, the insertion of the probe into the brain can affect BBB functionality. The sensitivity of the technique is mainly based on the assay method employed (Lange et al., 1999). It is seen that in vitro extraction of fiber wall measurement is greater than in vivo (Terasaki et al., 1992).

### 23.4.2 NON-INVASIVE EXTERNAL METHODS

#### 23.4.2.1 QUALITATIVE AUTORADIOGRAPHY

The technique involves the injection of the radioactive compound into the brain. Radioactivity is measured by collection of samples at defined time intervals. Soon after the procedure, the animal is sacrificed and isolated brain is kept in frozen condition. Sections are made from this frozen brain, placed inside X-ray case and autoradiographed. In conjugation with an image analysis method, the radioactivity can be measured (Sakurada et al., 1978; Sokoloff et al., 1977). The technique is cheap, easy, and offers a high spatial resolution. However, the distribution of parent compound and metabolites cannot be carried out by this method.

#### 23.4.2.2 MAGNETIC RESONANCE IMAGING

It is a sensitive technique used for the detection and assessment of BBB in various pathological conditions. The technique offers both qualitative and quantitative determination. It involves quantification of a tracer after intravenous administration. BBB permeability surface area product (PS) can be determined with the help of graphical analysis method (Ewing et al., 2003; Patlak et al., 1983). Above all, the method is

expensive and limited in application for routine screening.

#### 23.4.2.3 POSITRON EMISSION TOPOGRAPHY

The method is a non-invasive quantitative technique to determine BBB PS product in humans under normal and diseased condition (Brooks et al., 1984). The method involves incorporation of positron emitting radionuclide into the subject followed by placing of the subject in a counter which detects the positrons emitted. The two dimensional (2 D) image of the brain can be obtained by a computerized imaging technique. Currently, PET is used as a tool to measure the in vivo P-gp function at BBB both in rodents and animals (Bart et al., 2005; Sasongko et al., 2005). Moreover, the technique is expensive and both the preparation and stability of the tracers are a matter of concern. Also, it provides no distinction between parent compound and metabolites. Single photon emission topography finds its application in characterizing the efflux transporters and in determining BBB permeability in diseased states.

### 23.5 FUTURE OPPORTUNITIES AND CHALLENGES

Increasing cost of experimental studies, pre-screening, and validation studies are the major factors concerning the development of an in vitro model. Factors like stability, predictability, and permeability are major problems in the design of an ideal model; researches are still continuing to find the solution to tackle this problem (Ruck et al., 2015). A newly developed system with real-time monitoring and computer controlled system need to be considered which helps in reducing the time required for monitoring and maintenance. The physiological properties of BBB can be achieved with the use of advanced cell-based models but lack the ability in providing information on drug molecules when it passes the BBB. However, all in vitro models lacks certain characteristics. Thus, they are not an ideal substitute for in vivo models (Helms et al., 2016). The delivery of drug to CNS can be better studied with the potent human cell line (Wilhelm and Krizbai, 2014). The development of an ideal in vitro model will be a promising approach, but the lack of quality and quantity of in vivo BBB permeability data sets possess a major limitation. Irrespective of the experimental setup if the model offers good characteristics, it will be highly challenging. The challenge encountered is bringing together all the technologies under a group to cover the various aspect related to CNS drug discovery.

## 23.6  CONCLUSION

In vitro model plays a significant role in studying the transport kinetics across the BBB and for testing the interaction of various drugs on barrier properties. The need for the pre-screening of the drugs in CNS disorder forces researchers for the development of a suitable model. High predictability and scalability are some of the important factors to be considered in the development of a suitable model. Automation of the various factors help in reduces the daily maintenance and for the optimization of the operating characteristics. Great advancement has been made in the recent years. It should be noted that there is not a perfect in vitro model that mimics in vivo properties. More accurate results can be obtained with in situ model that exhibits physiological characteristics similar to BBB. The reliability of a model depends upon its usage.

## KEYWORDS

- **adherence junctions**
- **blood–brain barrier**
- **drug transport**
- **models**
- **tight junctions**

## REFERENCES

Ahmad, A. A.; Taboada C. B.; Gassmann, M.; Ogunshola, O. O. Astrocytes And Pericytes Differentially Modulate Blood-Brain Barrier Characteristics During Development and Hypoxic Insult. *J. Cereb. Blood Flow Metab.* **2011**, *31*, 693–705.

Ballabh, P.; Braun, A.; Nedergaard. M. The Blood–Brain Barrier: An Overview Structure, Regulation, and Clinical Implications. *Neurobiol. Dis.* **2014**, *16*, 1–14.

Banks, W. A. Characteristics of Compounds that Cross the Blood-Brain Barrier. *BMC Neurol.* **2009**, *9* (Suppl 1), S3.

Bart, J.; Dijkers, E. C.; Wegman, T. D.; De Vries, E. G.; Van Der Graaf, W. T.; Groen, H. J.; Vaalburg, W.; Willemsen, A. T.; Hendrikse, N.H. New Positron Emission Tomography Tracer [11C] Carvedilol Reveals P-Glycoprotein Modulation Kinetics. *Br. J. Pharmacol.* **2005**, *145*, 1045–1051.

Bazzoni, G.; Martinez-Estrada, O. M.; Mueller, F.; Nelboeck, P.; Schmid,G.; Bartfai, T.; Dejana, E.; Brockhaus, M. Homophilic Interaction of Junctional Adhesion Molecule. *J. Biol. Chem.* **2000**, *275*, 30970–30976.

Berezowski, V.; Landry, C.; Dehouck, M. P.; Cecchelli, R.; Fenart, L. Contribution of Glial Cells and Pericytes to the mRNA Profiles of P-Glycoprotein and Multidrug Resistance-Associated Proteins in an in vitro Model of the Blood-Brain Barrier. *Brain Res.* **2004**, *1018*, 1–9.

Berezowski, V.; Landry, C.; Lundquist, S.; Dehouck, L.; Cecchelli, R.; Dehouck, M. P., Fenart, L. Transport Screening of Drug Cocktails Through an in vitro Blood-Brain Barrier: Is it a Good Strategy for Increasing the Throughput of the Discovery Pipeline? *Pharm. Res.* **2004**, 756–760.

Bickel, U. How to Measure Drug Transport Across the Blood-Brain Barrier. *Neurorx* **2005**, *2* (1), 15–26.

Bickel, U.; Yoshikawa, T.; Pardridge, W. M. Delivery of Peptides and Proteins Through the Blood-Brain Barrier. *Adv. Drug. Deliv. Rev.* **1993**, *10*, 205–245.

Bonate, P. L. Animal Models for Studying Transport Across the Blood-Brain Barrier. *J. Neurosci. Methods.* **1995**, *56*, 1–15.

Booth, R.; Kim, H. Characterization of a Microfluidic in vitro Model of the Blood-Brain Barrier. *Lab Chip.* **2012**, 1784–1792.

Brooks, D. J.; Beaney, R. P.; Lammertsma, A. A.; Leenders, K. L.; Horlock, P. L.; Kensett, M. J.; Marshall, J.; Thomas, D. G.; Jones, T. Quantitative Measurement of Blood-Brain Barrier Permeability Using Rubidium-82 and Positron Emission Tomography. *J. Cereb. Blood Flow Metab.* **1984**, *4*, 534–545.

Bussolari, S. R.; Dewey, C. F.; Gimbrone, M. A. Apparatus for Subjecting Living Cells to Fluid Shear Stress. *Rev. Sci Instrum.* **1982**, *53*, 1851–1854.

Butt, A. M.; Jones, H. C; Abbott, N. J. Electrical Resistance Across the Blood-Brain Barrier in Anaesthetized Rats: A Developmental Study. *J. Physiol.* **1990**, 47–62.

Cecchelli, R.; Berezowski, V.; Lundquist, S.; Culot, M.; Renftel, M.; Dehouck, M. P. Modelling of the Blood-Brain Barrier in Drug Discovery and Development. *Nat. Rev. Drug Discov.* **2007**, 250–261.

Cucullo, L.; Mcallister, M. S.; Kight, K.; Bengez, L. K.; Marroni, M.; Mayberg, M. R.; Stanness, K. A.; Janigro. D. A New Dynamic in vitro Model for the Multidimensional Study of Astrocyte–Endothelial Cell Interactions at the Blood–Brain Barrier. *Brain Res.* **2002**, *951*, 243–254.

Daneman, R. The Blood-Brain Barrier in Health and Disease. *Ann. Neurol.* **2012**, *72*, 648–672.

De Lange, C. M.; De Boer, A. G.; Breimer, D. D. Microdialysis for Pharmacokinetic Analysis of Drug Transport to the Brain. *Adv. Drug Deliv. Rev.* **1999**, *36*, 211–227.

Deli, M. A.; Abraham, C. S.; Kataoka, Y.; Niwa, M. Permeability Studies on in vitro Blood-Brain Barrier Models: Physiology, Pathology, and Pharmacology. *Cell Mol. Neurobiol.* **2005**, *25*, 59–127.

Dewey, C. F.; Bussolari, S. R.; Gimbrone, M. A.; Davies, P. F. The Dynamic Response of Vascular Endothelial Cells to Fluid Shear Stress. *J. Biomech Eng.* **1981**, *103*, 177–185.

Dohgu, S.; Takata, F.; Yamauchi, A.; Nakagawa, S.; Egawa, T.; Naito, M.; Tsuruo, T.; Sawada, Y.; Niwa, M.; Kataoka, Y. Brain Pericytes Contribute to the Induction and Up-Regulation of Blood-Brain Barrier Functions Through Transforming Growth Factor-Beta Production. *Brain Res.* **2005**, *1038*, 208–215.

Ewing, J. R.; Knight, R.; Nagaraja, T. N.; Yee, J. S.; Nagesh, V.; Whitton, P. A.; Li, L.; Fenstermacher, J. D. Patlak Plots of Gd-DTPA MRI Data Yield Bloodbrain Transfer Constants Concordant with those of 14C-Sucrose in Areas of Blood-Brain Opening. *Magn. Reson. Med.* **2003**, *50*, 283–292.

Furuse, M. Molecular Basis of the Core Structure of Tight Junctions. *Cold Spring Harb. Perspect. Biol.* **2010**, *2*, A002907.

Furuse, M.; Fujita, K.; Hiiragi, T.; Fujimoto, K.; Tsukita, S. Claudin-1 and -2: Novel Integral Membrane Proteins Localizing at Tight Junctions with no Sequence Similarity to Occludin. *J. Cell Biol.* **1998**, *141*, 1539–1550.

Furuse, M.; Sasaki, H.; Tsukita, S. Manner of Interaction of Heterogeneous Claudin Species Within and Between Tight Junction Strands. *J. Cell Biol.* **1999**, *147*, 899–903.

Gaillard, P. J.; Van Der Sandt, I. C.; Voorwinden, L. H.; Vu. D. Astrocytes Increase the Functional Expression of P-Glycoprotein in an in vitro Model of the Blood-Brain Barrier. *Pharm. Res.* **2000**, *17*, 1198–1205.

Griep, L. M.; Wolbers, F.; Wagenaar, B. D.; Ter Braak, P. M.; Weksler, B. B.; Romero, I. A. BBB on Chip: Microfluidic Platform to Mechanically and Biochemically Modulate Blood-Brain Barrier Function. *Biomed. Microdevices* **2013**, *15* (1), 145–150.

Gumbleton, M.; Audus, K. L. Progress and Limitations in the use of in-vitro Cell Cultures to Serve as a Permeability Screen for the Blood-Brain Barrier. *J. Pharm Sci.* **2001**, *90*, 1681–1698.

Hafny, B. E.; Bourre, J. M.; Roux, F. Synergistic Stimulation of Gamma-Glutamyl Transpeptidase and Alkaline Phosphatase Activities by Retinoic Acid and Astroglial Factors in Immortalized Rat Brain Microvessel Endothelial Cells. *J. Cell Physiol.* **1996**, 451–460.

Hansang, C.; Seo, J. H.; Keith, Wong, H. K.; Terasaki, Y.; Park, J.; Bong, K.; Arai, K.; Lo, E. H.; Irimia, D. Three-Dimensional Blood-Brain Barrier Model for in vitro Studies of Neurovascular Pathology. *Sci. Rep.* **2015**, *5*, 15222.

Haskins, J.; Gu, L.; Wittchen, E. S.; Hibbard, J.; Stevenson, B. R. ZO-3, A Novel Member of the MAGUK Protein Family Found at the Tight Junction, Interacts with ZO-1 and Occludin. *J. Cell Biol.* **1998**, *141*, 199–208.

He, Y.; Yao, Y.; Tsirka, S. E.; Cao, Y. Cell-Culture Models of the Blood–Brain Barrier. *Stroke* **2014**, *45* (8), 2514–2526.

Hellinger, E.; Veszelka, S.; Toth, A. E.; Walter, F.; Kittel, A.; Bakk, M. L.; Tihanyi, K.; Hada, V.; Nakagawa, S.; Duy, T. D.; Niwa, M.; Deli, M. A; Vastag, M. Comparison of Brain Capillary Endothelial Cell-Based and Epithelial (MDCK-MDR1, Caco-2, and VB-Caco-2) Cell-Based Surrogate Blood–Brain Barrier Penetration Models. *Eur. J. Pharm. Biopharm.* **2012**, *82* (2), 340–351.

Helms, H. C.; Abbott, N. J.; Burek, M.; Cecchelli, R.; Couraud, P. O.; Deli, M. A.; Brodin, B. In Vitro Models of the Blood–Brain Barrier: An Overview of Commonly Used Brain Endothelial Cell Culture Models and Guidelines for their Use. *J. Cereb. Blood Flow Metab.* **2016**, *36* (5), 862–890.

Hirase, T.; Staddon, J. M.; Saitou, M.; Akatsuka, Y. A.; Itoh, M.; Furuse. Occludin as a Possible Determinant of Tight Junction Permeability in Endothelial Cells. *J. Cell Sci.* **1997**, *110*, 1603–1613.

Hoheisel, D. Hydrocortisone Reinforces the Blood-Brain Properties in a Serum Free Cell Culture System. *Biochem. Biophys. Res. Commun.* **1998**, *247*, 312–315.

Hori, S.; Ohtsuki, S.; Tachikawa, Masanori; Kimura, N.; Kondo, T.; Watanabe, M.; Nakashima, E.; Terasaki, T. Functional Expression of Rat ABCG2 on the Luminal Side of Brain Capillaries and Its Enhancement by Astrocyte-Derived Soluble Factor(S). *J. Neurochem.* **2004**, *90*, 526–536.

Hori, S., Ohtsuki, S.; Hosoya, K.; Nakashima, E.; Terasaki, T. A. Pericyte-Derived Angiopoietin-1 Multimeric Complex Induces Occludin Gene Expression in Brain Capillary Endothelial Cells Through Tie-2 Activation in vitro. *J. Neurochem.* **2004**, *89*, 503–513.

Hosoya, K. I.; Takashima, T.; Tetsuka, K.; Nagura. T.; Ohtsuki. S.; Takanaga, H. mRNA Expression and Transport Characterization of Conditionally Immortalized Rat Brain Capillary Endothelial Cell Lines; A New In Vitro BBB Model for Drug Targeting. *J. Drug Target.* 2000, 357–370.

Hurst, R. D.; Fritz, I. B. Properties of an Immortalized Vascular Endothelial/Glioma Cell Co-Culture Model of the Blood-Brain Barrier. *J. Cell Physiol.* **1996**, *167*, 81–88.

Hutamekalin. P.; Farkas, A. E.; Orbiok, A.; Wilhelm, I.; Nagyoszi, P.; Veszelka, S.; Deli, M. A.; Buzas, K.; Gulyas. E. H.; Medzihradszky, K. F.; Meksuriyen, D.; Krizbai, I. A. Effect of Nicotine and Polyaromtic Hydrocarbons on Cerebral Endothelial Cells. *Cell Biol. Int.* **2008**, 198–209.

Janzer, R. C.; Raff, M. C. Astrocytes Induce Blood–Brain Barrier Properties in Endothelial Cells. *Nature* **1987**, *325*, 253–257.

Jeffrey, P; Summerfield, S. Assessment of the Blood-Brain Barrier in CNS Drug Discovery. *Neurobiol. Dis.* **2010**, *37*, 33–37.

Kalvass, J. C.; Maurer, T. S. Influence of Nonspecific Brain and Plasma Binding on CNS Exposure: Implications for Rational Drug Discovery. *Biopharm. Drug Disp.* **2002**, *23*, 327–338.

Kerns, E. H.; Di, L.; Petusky, S.; Farris, M.; Ley, R.; Jupp, P. Combined Application of Parallel Artificial Membrane Permeability Assay and Caco-2 Permeability Assays in Drug Discovery. *J. Pharm. Sci.* **2004**, *93*, 1440–1453.

Koutsiaris, A. G.; Tachmitzi, S. V.; Batis, N.; Kotoula, M. G.; Karabatsas, C. H.; Tsironi, E.; Chatzoulis, D. Z. Volume Flow and Wall Shear Stress Quantification in the Human Conjunctival Capillaries and Post-Capillary Venules in vivo. *Biorheology* **2007**, *44*, 375–386.

Lippmann, E. S.; Ahmad, A. A.; Azarin, S. M.; Palecek, S. P.; Shusta, E. V. A Retinoic Acid-Enhanced, Multicellular Human Blood Brain Barrier Model Derived from Stem Cell Sources. *Sci. Rep.* **2014**, *4*, 4160.

Lippmann, E. S.; Ahmad, A. A.; Palecek, S. P.; Shusta, E. V. Modeling the Blood Brain Barrier Using Stem Cell Sources. *Fluids Barr. CNS* **2013**, *10* (1), 2.

Lippmann, E. S.; Azarin, S. M.; Kay, J. E.; Nessler, R. A.; Wilson, H. K.; Ahmad, A. Derivation of Blood-Brain Barrier Endothelial Cells from Human Pluripotent Stem Cells. *Nat Biotechnol.* **2012**, 783–791.

Liu, D.; Lin, B.; Shao, W.; Zhu, Z.; Ji.; Yang, C. In vitro and in vivo Studies on the Transport of Pegylated Silica Nanoparticles Across the Blood–Brain Barrier. *ACS Appl. Mater. Interfaces* **2014**, *6* (3), 2131–2136.

Liu, X.; Chen, C.; Smith, B. J. Progress in Brain Penetration in Drug Discovery and Development. *J. Pharmacol. Exp. Ther.* **2008**, *325*, 349–356.

Lohmann, C.; Huwel, S.; Galla, H. J. Predicting Blood-Brain Barrier Permeability of Drugs: Evaluation of Different in-vitro Assays. *J. Drug Target* **2002**, *10*, 263–276.

Lundquist, S.; Renftel, M. The Use of in Vitro Cell Culture Models for Mechanistic Studies and as Permeability Screens for the Blood-Brain Barrier in the Pharmaceutical Industry–Background and Current Status in the Drug Discovery Process. *Vascul. Pharmacol.* **2002**, *38*, 355–364.

Malina, C. K.; Cooper, I.; Teichberg, V. I. Closing the Gap Between the in-vivo and in-vitro Blood-Brain Barrier Tightness. *Brain Res.* **2009**, *1284*, 12–21.

Mihaly, A.; Bozoky, B. Immunohistochemical Localization of Extravasated Serum Albumin in the Hippocampus of Human Subjects with Partial and Generalized Epilepsies and Epileptiform Convulsions. *Acta Neuropathol.* **1984,** *65,* 25–34.

Nakagawa, S.; Nakao, S.; Honda, M.; Hayashi, K.; Nakaoke, R.; Kataoka, Y.; Niwa, M. Pericytes from Brain Microvessels Strengthen the Barrier Integrity in Primary Cultures of Rat Brain Endothelial Cells. *Cell Mol. Neurobiol.* **2007,** *27,* 687–694.

Nasal, A.; Sznitowska, M.; Bucinski, A.; Kaliszan, R. Hydrophobicity Parameter from Highperformance Liquid Chromatography on an Immobilized Artificial Membrane Column and its Relationship to Bioactivity. *J. Chromatogr. A.* **1995,** *692,* 83–89.

Ohno, K.; Pettigrew, K. D.; Rapoport, S. I. Lowerlimits of Cerebrovascular Permeability to Nonelectrolytes in the Conscious Rat. *Am. J. Physiol.* **1978,** *235,* 299–307.

Oldendorf, W. H. Measurement of Brain Uptake of Radiolabeled Substances Using Tritiated Water Internal Standard. *Brain Res.* **1970,** *24,* 372–376.

Ong, S.; Hanlan, L.; Pidgeon, C. Immobilized Artificial-Membrane Chromatography: Measurements of Membrane Partition Coefficient and Predicting Drug Membrane Permeability. *J. Chromatogr. A* **1996,** *728,* 113–128.

Palmiotti, C. A.; Prasad, S.; Naik, P.; Abul, K. M.; Sajja, R. K.; Achyuta, A. H.; Cucullo, L. In Vitro Cerebrovascular Modeling in the 21st Century: Current and Prospective Technologies. *Pharm. Res.* **2014,** *31* (12), 3229–3250.

Paradis, A.; Leblanc, D.; Dumais, N. Optimization of an in vitro Human Blood–Brain Barrier Model: Application to Blood Monocyte Transmigration Assays. *Methodsx* **2016,** *3,* 25–34.

Pastan, I.; Gottesman, M. M.; Ueda, K.; Lovelace, E.; Rutherford, A. V.; Willingham, M. C. A Retrovirus Carrying an MDR1 cDNA Confers Multidrug Resistance and Polarized Expression of P-Glycoprotein in MDCK Cells. *Proc. Natl. Acad. Sci. USA* **1988,** *85,* 4486–4490.

Patlak, C. S.; Blasberg, R. G.; Fenstermacher, J. D. Graphical Evaluation of Blood-To-Brain Transfer Constants from Multiple-Time Uptake Data. *J. Cereb. Blood Flow Metab.* **1983,** *3,* 1–7.

Patlak, C. S.; Paulson, O. B. The Role of Unstirred Layers for Water Exchange Across the Blood–Brain Barrier. *Microvasc. Res.* **1981,** *21,* 117–127.

Peddagannagari, S.; Gromnicova, R.; Davies, H.; Phillips, H.; Romero, I. A. A Three-Dimensional model of the Human Blood-Brain Barrier to Analyse the Transport of Nanoparticles and Astrocyte/Endothelial Interactions. *F1000 Res.* **2015,** 1279.

Perriere, N.; Demeuse, P.; Garcia, E.; Regina, A.; Debray, M.; Andreux, J. P.; Couvreur, P.; Scherrmann, J. M.; Temsamani, J.; Couraud, P. O.; Deli, M. A.; Roux, F.

Puromycin-Based Purification of Rat Brain Capillary Endothelial Cell Cultures. Effect on the Expression of Blood-Brain Barrier-Specific Properties. *J. Neurochem.* **2005,** *93,* 279–289.

Perriere, N.; Yousif, S.; Cazaubon, S.; Chaverot, N.; Bourasset, F.; Cisternino, S.; Decleves, X.; Hori, S.; Terasaki, T.; Deli, M.; Scherrmann, J. M.; Temsamani, J.; Roux, F.; Couraud, P. O. A Functional in vitro Model of Rat Blood-Brain Barrier for Molecular Analysis of Efflux Transporters. *Brain Res.* **2007,** 1–13.

Pereira, S. I.; Brites, D.; Brito, M. A. Neurovascular Unit: A Focus on Pericytes. *Mol. Neurobiol.* **2012,** *45,* 327–347.

Pidgeon, C.; Ong, S.; Liu, H.; Qiu, X.; Pidgeon, M.; Dantzig, A. H.; Munroe, J.; Hornback, W. J.; Kasher, J. S.; Glunz, L.; Czereba, T. J. IAM Chromatography: An in vitro Screen for Predicting Drug Membrane Permeability. *J. Med. Chem.* **1995,** *38,* 590–594.

Prabhakarpandian, B.; Nichols, J. B.; Mills, I. R.; Wegrzynowicz, M. S.; Aschner, M.; Pant, K. Sym-BBB: A Microfluidic Blood Brain Barrier Model. *Lab Chip.* **2013,** 1093–1101.

Quan, N.; Banks, W. A. Brain-Immune Communication Pathways. *Brain Behav. Immun.* **2007,** *21,* 727–735.

Ramsauer, M.; Krause, D.; Dermietzel, R. Angiogenesis of the Blood–Brain Barrier in vitro and the Function of Cerebral Pericytes. *FASEB J.* **2002,** *95,* 11981–11986.

Roux, F.; Couraud, P. O. Rat Brain Endothelial Cell Lines for the Study of Blood-Brain Barrier Permeability and Transport Functions. *Cell Mol. Neurobiol.* **2005,** *25* (1), 41–58.

Roux, F.; Trautmann, D. O.; Chaverot, N.; Claire, M.; Mailly, P.; Bourre, J. M; Strosberg, A. D.; Couraud, P. O. Regulation of Gamma-Glutamyl Transpeptidase and Alkaline Phosphatase Activities in Immortalized Rat Brain Microvessel Endothelial Cells. *J. Cell Physiol.* **1994,** *159* (1), 101–113.

Rubin, L. L.; Hall, D. E.; Porter, S.; Barbu, K.; Cannon, C.; Horner, H. C.; Janatpour, M.; Liaw, C. W.; Manning, K.; Morales, J. A Cell Culture Model of the Blood-Brain Barrier. *J. Cell Biol.* **1991,** *115,* 1725–1735.

Rubin, L. L.; Barbu, K.; Bard, F.; Cannon, C.; Hall, D. E.; Horner, H.; Janatpour, M.; Liaw, C.; Manning, K.; Morales, J. et al. Differentiation of Brain Endothelial Cells in Cell Culture. *Ann. NY Acad. Sci.* **1991,** *633,* 420–425.

Ruck, T.; Bittner, S.; Meuth, S. G. Blood-Brain Barrier Modeling: Challenges and Perspectives. *Neural Regen Res.* **2015,** *10* (6), 889–891.

Sakurada, O.; Kennedy, C.; Jehle, C.; Brown, J. D.; Carbin, G. L.; Sokoloff, L. Measurement of Local Cerebral

Blood Flow with Iodo[14C]-Antipyrine. *Am. J. Physiol.* **1978**, H59.

Santaguida, S.; Janigro, D.; Hossain, M.; Oby, E.; Rapp, E.; Cucullo, L. Side by Side Comparison Between Dynamic Versus Static Models of Blood–Brain Barrier in vitro: A Permeability Study. *Brain Res.* **2006**, *1109*, 1–13.

Sasongko, L.; Link, J. M.; Muzi, M.; Mankoff, D. A.; Yang, X.; Collier, A. C.; Shoner, S. C.; Unadkat, J. D. Imaging Pglycoprotein Transport Activity at the Human Blood-Brain Barrier with Positron Emission Tomography. *Clin. Pharmacol. Ther.* **2005**, *77*, 503–514.

Sokoloff, L.; Reivich, M.; Kennedy, C.; Desroseiers, M. H.; Patlak, C. S.; Pettigrew, K. D.; Sakurada, O.; Shinohara, M. The [14C]Deoxyglucose Method for the Measurement of Local Cerebral Glucose Utilization: Theory, Procedure, and Normal Values in the Conscious and Anesthetized Albino Rat. *J. Neurochem.* **1977**, *28*, 897.

Sonoda, N.; Furuse, M.; Sasaki, H.; Yonemura, S.; Katahira, J.; Horiguchi, Y.; Tsukita, S. Clostridium Perfringens Enterotoxin Fragment Removes Specific Claudins from Tight Junction Strands. *J. Cell Biol.* **1999**, *147*, 195–204.

Stamatovic, S. M.; Shakui, P.; Keep, R. F.; Moore, B. B.; Kunkel, S. L.; Rooijen, V. N.; Andjelkovic, A. V. Monocyte Chemoattractant Protein-1 Regulation of Blood-Brain Barrier Permeability. *J. Cereb. Blood Flow Metab.* **2005**, *25*, 593–606.

Stewart, B. H.; Chan, O. H. Use of Immobilized Artificial Membrane Chromatography for Drug Transport Applications. *J. Pharm. Sci.* **1998**, *87*, 1471–1478.

Takasato, Y.; Rapoport, S. I.; Smith, Q. R. An in situ Brain Perfusion Technique to Study Cerebrovascular Transport in the Rat. *Am. J. Physiol.* **247**, 484–493.

Takata, F.; Dohgu, S.; Yamauchi, A.; Matsumoto, J.; Machida, T.; Fujishita, K.; Shibata, K.; Shinozaki, Y.; Sato, K.; Kataoka, Y.; Koizumi, S. In vitro Blood-Brain Barrier Models Using Brain Capillary Endothelial Cells Isolated from Neonatal and Adult Rats Retain Age Related Barrier Properties. *PLoS One* **2013**, *8*, E55166.

Takeshita, Y.; Ransohoff, R. M. Inflammatory Cell Trafficking Across the Blood-Brain Barrier (BBB): Chemokine Regulation and *in vitro* Models. *Immunol. Rev.* **2012**, *248* (1), 228–239.

Tarbel, J. M. Shear Stress and the Endothelial Transport Barrier. *Cardiovasc. Res.* **2010**, *87*, 320–330.

Terasaki, T.; Deguchi, Y.; Kasama, Y.; Pardridge, W. M.; Tsuji, A. Determination of in vivo Steady-State Unbound Drug Concentration N the Brain Interstitial Fluid by Microdialysis. *Int. J. Pharm.* **1992**, *81*, 143–152.

Tontsch, U.; Bauer, H. C. Glial Cells and Neurons Induce Blood-Brain Barrier Related Enzymes in Cultured Cerebral Endothelial Cells. *Brain Res.* **1991**, *539*, 247–253.

Urich, E.; Lazic, S. E.; Molnos, J.; Wells, I.; Freskgård, P. O. Transcriptional Profiling of Human Brain Endothelial Cells Reveals Key Properties Crucial for Predictive in vitro Blood-Brain Barrier Models. *PLoS One* **2012**, E38149.

Van De Waterbeemd, H.; Smith, D. A.; Jones, B. C. Lipophilicity in PK Design: Methyl, Ethyl, Futile. *J. Comput. Aided Mol. Des.* **2001**, *56*, 273–286.

Vandenhaute, E.; Dehouck, L.; Boucau, M. C.; Sevin, E.; Uzbekov, R.; Tardivel, M.; Gosselet, F.; Fenart, L.; Cecchelli, R.; Dehouck, M. P. Modelling the Neurovascular Unit and the Blood-Brain Barrier with the Unique Function of Pericytes. *Curr. Neurovasc. Res.* **2011**, *8*, 258–269.

Veszelka, S.; Pasztoi, M.; Farkas, A. E.; Krizbai, I.; Ngo, T. K.; Niwa, M.; Abrahám, C. S.; Deli, M. A. Pentosan Polysulfate Protects Brain Endothelial Cells Against Bacterial Lipopolysaccharide-Induced Damages. *Neurochem. Int.* **2007**, *50*, 219–228.

Wang, Q.; Rager, J. D.; Weinstein, K.; Kardos, P. S.; Weinstein, K.; Kardos, P. S.; Dobson, G. L.; Li J.; Hidalgo, I. J. Evaluation of the MDR-MDCK Cell Line as a Permeability Screen for the Blood-Brain Barrier. *Int. J. Pharm.* **2005**, *288*, 349–359.

Weidenfeller, C.; Svendsen, C. N.; Shusta, E. V. Differentiating Embryonic Neural Progenitor Cells Induce Blood-Brain Barrier Properties. *J. Neurochem.* **2007**, *101*, 555–565.

Weiss, N; Miller, F; Cazaubon, S; Couraud, P. O. The Blood-Brain Barrier in Brain Homeostasis and Neurological Diseases. *Biochim. Biophys. Acta* **2009**, *1788*, 842–857.

Weksler, B.; Romero, I. A.; Couraud, P. O. The Hcmec/D3 Cell Line as a Model of the Human Blood Brain Barrier. *Fluids Barr. CNS* **2013**, *10* (1), 16.

Weksler, B. B.; Subileau, E. A.; Perriere, N.; Charneau, P.; Holloway, K.; Leveque, M.; Leignel, H. T.; Nicotra, A.; Bourdoulous, S.; Turowski, P.; Male, D. K.; Roux, F.; Greenwood, J.; Romero, I. A.; Couraud, P. O. Blood-Brain Barrier-Specific Properties of a Human Adult Brain Endothelial cell Line. *FASEB. J.* **2005**, *19*, 1872–1874.

Wexler, D. S.; Gao, L.; Anderson, F.; Ow, A.; Nadasdi, L.; Mcalorum, A.; Urfer, R.; Huang, S. G. Linking Solubility and Permeability Assays for Maximum Throughput and Reproducibility. *J. Biomol. Screen* **2005**, *10*, 383–390.

Wilhelm, I.; Krizbai, I. A.; In vitro Models of the Blood–Brain Barrier for the Study of Drug Delivery to the Brain. *Mol. Pharm.* **2014,** *11* (7), 1949–63.

Winkler, E. A.; Bell, R. D.; Zlokovic, B. V. Central Nervous System Pericytes in Health and Disease. *Nat. Neurosci.* **2011,** *14*, 1398–1405.

Wolburg, H.; Neuhaus, J.; Kniesel, U.; Krauss, B.; Schmid, E. M.; Ocalan, M.; Farrell, C.; Risau, W. Modulation of Tight Junction Structure in Blood-Brain Barrier Endothelial Cells Effects of Tissue Culture, Second Messengers and Cocultured Astrocytes. *J. Cell. Sci.* **1994,** *107*, 1347–1357.

Wuest, D. M.; Wing, A. M.; Lee, K. H. Membrane Configuration Optimization for a Murine in vitro Blood-Brain Barrier Model. *J. Neurosci. Methods.* **2013,** *212*, 211–221.

Xu, H.; Li, Z.; Yu, Y.; Sizdahkhani, S.; Ho, W. S.; Yin, F.; Qin, J. A Dynamic *in vivo*-Like Organotypic Blood-Brain Barrier Model to Probe Metastatic Brain Tumors. *Sci. Rep.* **2016,** *6*, 36670.

Yao, Y.; Tsirka, S. E. Truncation of Monocyte Chemoattractant Protein 1 by Plasmin Promotes Blood-Brain Barrier Disruption. *J. Cell Sci.* **2011,** *124*, 1486–1495.

Yarong, H.; Yao, Y.; Tsirka, S. E. Cell-Culture Models of the Blood–Brain Barrier. *Stroke* **2014,** 2514–2526.

Yeon, J. H.; Na, D.; Choi, K.; Ryu, S. W.; Choi, C.; Park, J. K. Reliable Permeability Assay System in a Microfluidic Device Mimicking Cerebral Vasculatures. *Biomed. Microdevices* **2012,** *14* (6), 1141–1148.

Zlokovic, B. V. Neurovascular Mechanisms of Alzheimer's Neurodegeneration. *Trends Neurosci.* **2005,** *25*, 202–208.

# CHAPTER 24

# Therapeutic Potentials of Nanomedicine for Brain Disorders

HIMANI CHAURASIA*, VISHAL K. SINGH, VIVEK K. CHATURVEDI, RICHA MISHRA, and VISHAL SRIVASTAVA

*University of Allahabad, Allahabad, Uttar Pradesh, India*

*Corresponding author. E-mail: himaniangel13@gmail.com*

## ABSTRACT

The incidence of brain disorders (BDs) of unknown etiology is increasing, including well-studied diseases, such as Alzheimer's disease (AD), Parkinson's disease (PD), and multiple sclerosis. The blood–brain barrier (BBB) provides protection to the brain, but hinders the treatment and diagnosis of these neurological diseases, because the drugs must cross the BBB to reach the lesions. Nanotechnology utilizes designed materials or gadgets that connect with natural frameworks at an atomic level and could alter the treatment of BDs by allowing, reacting to, and connecting with the target locales to prompt physiological reactions while limiting symptoms. Traditional medication conveyance frameworks don't give satisfactory cytoengineering innovation and association designs that are basic for useful recuperation in BDs, because of constraints posed by the prohibitive blood–cerebrum boundary. Still, the treatment of numerous BDs is enigmatic. In this chapter, authors discussed in details about the present uses of nanoempowered drug conveyance frameworks for the treatment of BDs and also investigate the future uses of nanotechnology in clinical neuroscience to create inventive remedial modalities for the treatment of BDs.

## 24.1   INTRODUCTION

For quick and efficient working, brain should be healthy but different kinds of disorders affect the brain at every stage of life. Consequently the results can be devastating. Genetic variations, illness, or a traumatic injury can affect the brain and cause brain disorder. Nowadays Alzheimer's disease (AD), Parkinson's disease (PD), stroke, amyotrophic lateral sclerosis (ALS), and many others have become more common brain disorders. The incidence of these neurodegenerative disorders is increasing continuously as the population ages, with enormous economic and human costs. Genetic variations are the main factors responsible for some brain disorders like Huntington's disease (HD) or familial ALS. Otherwise, genetic and environmental interactions can also affect the brain in many ways (Chaturvedi et al., 2017). PD is marked by deterioration of mesencephalon, which results in depletion in the amount of brain dopamine

(DA) leads to typical manifestation concerning bradykinesia, rigidity, and resting tremor (Shukla et al., 2010). Worldwide, approximately 24 million people are affected with dementia, out of that 60% of them suffer from AD (Giacobini et al., 2007). According to current scenario 1% of individuals aged from 50 to 70 years are suffering from AD and this ratio increases drastically to 50% in 70-years old people. AD typified defined by learning problem and memory loss whereas pathologically through stout cerebral atrophy symptomatic of neuronal reduction, amidst plenty of extracellular neuritic amyloid patches and intracellular neurofibrillary labyrinth arise prominently with in frontal and temporal lobes, together with hippocampus (Prabhakaran et al., 2011).

Nanotechnology plays a crucial part to abolish BDs. Delivery of traditional drug to the central nervous system (CNS), prevails the cardinal provocation for therapeutic section of every BDs. Due to the presence of many types of defensive barriers enclosing the CNS interrupt the conveyance of the drugs. In this contrast, nanotechnology utilizes designed materials or gadgets with its little practical associating nanometre scale (1–100 nm) that can cooperate with organic frameworks at the subatomic level. Nanoparticles can enter the BBB (Kim et al., 2007). Nanotechnology, in this manner can be employed to frame newer evocative techniques and apparatuses, as well as additionally nano-empowered conveyance frameworks, which can sidestep the BBB all together to expedite standard and modish neuroanaleptic arbitrations, such as medicational treatment, genetic engineering, and tissue reclamation (Gabathuler, 2010; Lee, et al., 2010).

The progress of early diagnosis and treatment of BDs have been deeply investigated but still there are limitations. Due to the BBB, majority concerning drugs prevent the penetration of

barrier and causing vast peripheral side effects. Nanomaterials (NMs) are proposed to deal with the obstacles to facilitate the penetration of BBB cells at subatomic region, by using current physiological delivery systems, deprived in meddling with the ordinary work of BBB. Nanotechnology also promotes transcytosis mechanisms, which are mediated by some biological receptors to facilitate the conveyance of NM from the blood to the mesencephalon and then to the CNS (Bhaskar et al., 2010). Till date, many types of NMs have been designed (in the form of tubes, small particles, fibers, gels, etc., in nanorange), which are accessible for clinical as well as biological purposes while targeting the loaded drug with different physical substances, highlights, and applications. In this chapter, authors present the therapeutic potentials of nanomedicine for BDs.

## 24.2 NANOTECHNOLOGY FOR NEUROPROTECTION AND NEURONAL TISSUE REGENERATION

### 24.2.1 NEUROPROTECTIVE NATURE

Several compounds are proposed as neuroprotective agents are classified into different categories, such as fullerenes and nanoceria. Fullerenes, the radical scavenger also known as radical "sponges," have proficiency toward assimilation of numerous radicals/molecule, as fullerenes have double bond system in which $\pi$-electrons can take part in delocalization (Cho et al., 2011). It has a profound efficiency for removal of superoxide radicals across a dismutation catalytic mechanism. Fullerene being a radical scavenger, works well during increased in level of reactive oxygen-species perceive when CNS was injured. Tris-malonic acid derivative (Ali et al., 2004), derived from fullerene C60 molecule (C3) has also a potency to upsurge the viability extent in mice paucity mitochondrial superoxide dismutase by 300%. Cerium oxide nanoparticles (NPs) are another category of nanoparticles that are used in the role of neuroprotective substances possessing antioxidant properties (Das et al., 2007), provide cushions to the cells of isolated rat spinal cord neurons, from ruination by cause of oxidative stress, while performing experiments in artificial insemination.

## 24.2.2 NEURONAL TISSUE REGENERATION

In artificial insemination (in vitro), experiments neuronal tissue (NT) and nanofibers (NF) are utilized to repair damaged neuronal tissues and also successfully put to use for regenerative medicine on grounds of analogy of their structure to those of cells. The rehabilitation of damaged neuronal tissues have been attained by employing the extracellular arena for promotion of neuronal linkage and for the buttress of axonal growth. An artificial-insemination experiment had been recently executed by Jin et al. (2011), who fabricated multiwalled carbon neuronal tissue smeared with electrospun poly(1-lactic acid-*co*-caprolactone) nanofiber, that was facile toward the enhancement of the neurite outgrowth. Deploy of NT/NF, as frame and conveyance across their activity with neurotrophins, stimulating propagation, and discrimination of neurons is a spare approach of nanotechnology toward neuronal tissue regeneration (Liu et al., 2011). The proficiency of NF in consolidating nerve growth factor to efficaciously boost up sciatic nerve regeneration could be shown by artificial insemination (in vitro) experiments. In promoting regenerative techniques alongside future ambition in perspective

of brain curvet function, the construction of electrically conducted NT/NF frame is another tactic. The electrically transmitting nanoframe had broadly been employed for conduction of signals from neuronal regenerative cells by virtue of their reduced impedance and elevated charge transmission (Prabhakaran et al., 2011). Poly (lactic acid-*co*-glycolic acid) NF layer smeared alongside electrically relegating polypyrrole was employed by Lee et al. (2011), along with combined strategy and accompanied by neuronal extensive factor. In vitro experiment of nanofibers promoted to neurite regeneration and proliferation.

## 24.3 NANOMEDICINE-MEDIATED BRAIN DRUG DELIVERY

Nanotechnology mediated targeting of nondrug in BBB mainly by two types, like adsorptive mediated transcytosis and receptor mediated transcytosis.

## 24.3.1 ADSORPTIVE-MEDIATED TRANSCYTOSIS

Transcytosis mechanism, which is mediated by adsorption phenomenon is the most vitalizing mechanism out of different approaches to assist the conveyance of nanomedicine to the brain across cellular channel all the way through blood devoid of intrusive among the normal work of barrier (Bhaskar et al., 2010). Other techniques are also come to conquer the BBB (a vibrant substantial and natal obstacle between blood flow and CNS) extending from intrusive modus operandi, toward chemical refinement of medicinal compounds and computational modus operandi for depiction of molecules in addition to improved penetrability (Abbott et al., 1999).

"Trojan horse" method has been deployed to penetrate the substances those are not crossing the BBB by integrating them to one that does. Passing the substances through intranasal via olfactory lobe is also another method to go across the BBB. To go across the BBB, nanomedicines should exhibit perpetuate half-life in blood averting the reticulo-endothelial organization as well as should have at least, a surface operation toward intending the BBB (Patel et al., 2009). To avoid any side effects nanomedicines ought to be non-poisonous, nonerythrogenic, noninsusceptible, biodegradable, and biocompatible. The electrostatic synergistic effects of ligand in the midst of charges revealed by the side of luminal facet belonging to endothelial cells are the backbone of adsorptive-mediated transcytosis (Bhaskar et al., 2010). The surface of nanomedicines have been upgraded with cell invading peptides those are

derived from TAT along with cationic proteins for instance albumin that accelerate penetration of medicinal substance to go across the BBB which is mentioned in Figure 24.1.

### 24.3.2 RECEPTOR-MEDIATED TRANSCYTOSIS

An alternative physiological mechanism is also proposed to overcome the BBB, named as receptor-mediated transcytosis, a profoundly probed mechanism. Some receptors are present on the BBB cells which are highly expressed by BBB cells and facilitate the nanomedicine to permeate across the BBB endothelium and the presence of these receptors on BBB cells is the backbone of this mechanism. Different type of proteins, such as insulin, transferring,

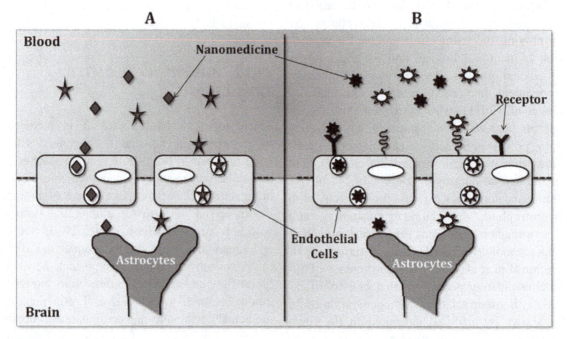

**FIGURE 24.1** Transcytosis of macromolecules at the blood–brain barrier. (A) Adsorptive-mediated transcytosis and (B) receptor-mediated transcytosis.

apolipoprotein E, α2-macroglobulin are also gamut the brain through this mechanism (Gabathuler et al., 2010) and activate the nanomedicine surface to accelerate their BBB crossing (Kim et al., 2007). Receptors present on the BBB that take part in receptor-mediated transcytosis, are opposed by specific monoclonal antibodies. Monoclonal antibodies that (Ulbrich et al., 2009) worked counter to insulin receptors present on BBB may take into account in creating a brain-target drug transporting device (Boado et al., 2008).

## 24.4 NANOMEDICINE FOR BRAIN DISORDERS

### 24.4.1 NANOPARTICLE-BASED DRUGS EFFECTIVE FOR CONTROLLING ALZHEIMER'S DISEASE

A senescent derangement commonly known by AD that gradually ruins memory and other important mental abilities causes several negative effects, such as confusion and mood swings. Neuronal and synaptic loss throughout the brain (Cherubini et al., 2010) is the characteristic features of AD. Its symptoms customarily advanced by time and get deteriorated over time, seemly inexorable ample to intervene with daily chore. This disease accounts for 60–80% of dementia cases. Rudimental cerebral cortex cholinergic system, limbic brain, hippocampus along with many cortical section are those parts in the brain that eventually lose their functions when affected by AD (Oliveira et. al., 2005). Hippocampal CA1 and CA2, entorhinal cortex and the locus coeruleus neuronal loss prevails during AD (Ryu et al., 2009). This derangement has some noticeable peculiarity, such as extracellular amyloid β (Aβ) plaques, intracellular neurofibrillary labyrinth and neuron adrift.

The loss of neuronal cells caused by AD can be superseded by the neuronal regenerative tissues. Dementia therapeutic target can be achieved by stem cell therapy, which is the most accepted strategy where various cells like adult neural stem cell (Blurton-Jones et al., 2009), ESC-derived neural cells (Moghadam et al., 2009), mesenchymal stem cells (Lee et al., 2010a,b) and astrocytes (Pihlaja et al., 2008) are most commonly employed. Finally, it is presumed that transplanted cells ransom the inadequacy caused in animals due to this disease. Stem cells have high transient capability; therefore they can be transmogrify genetically and can be employed in place of fibroblasts to facilitate the functionalization for transmission of nerve extensive factor which obstruct declension of nerve cells. Stem cells are also employed as conveyance of factors which control the impacts of disease and negate the cholinergic neuronal mortality, in turn trigger the cells for their activity and simultaneously enhance the brain function (Giacobini et al., 2007).

Today, methods using NP for AD treatment have been principally used to intervene with Aβ accumulation or to keep apart the peptide, centrally as well as in the blood, with the thought of reduction in its brain level. Gobbi et al. (2010) developed phosphatidic acid or cardiolipin functionalized nanoliposomes (NL), utilized for sink effect. These functionalized NL have a high attraction for Aβ and decreases its toxicity in vitro (Gobbi et al., 2010). Canovi et al. (2011) also developed NL adorned with an anti-Aβ monoclonal antibody, with high attraction for Aβ either in vitro or ex vivo along necropsy AD brain sampling. Curcumin derivative adorned NL developed by Mourtas et al. (2011 binds Aβ with high attraction and suppress its collection in vitro. Due to toxic effect especially hepatotoxicity and neurotoxicity, metal chelators are less commonly used to protect against oxidative scathe.

## 24.4.2 NANOPARTICLE-BASED DRUGS EFFECTIVE FOR CONTROLLING PARKINSON'S DISEASE

Another neurodegenerative disorder that results in cessation of dopaminergic nerve cells have its existence in substantia nigra pars compacta (SNpc) that simultaneously causes alleviation of striatal dopamine is commonly known as Parkinson's disease (Glass et al., 2010). It is characterized to kaputs transmission of informative signals toward basal ganglia. Symptoms that seems in PD are as follows tremor in one hand, slow movement, stiffness in muscles, loss of balance, muscular rigidity, problems in coordination, restless sleep, dizziness, amnesia, dementia, blunt voice and damaged voice box leading to to shiver, consternation, distorted olfactory sense, drizzling sensation in urination, constraint jaws, blank stare, constipation, melancholy, slobbering, scared of falling, impoverishment of contradistinction susceptivity, and unintentional writhing. It is the most common movement disorder (Cova et al., 2010). Several dysfunctional cellular processes, stress and inflammations contribute to cell damage (Kim et al., 2002). Diversified therapies are accessible which can obstruct the commencement of motor syndrome which can improve the life span. Major therapies include the oral administration of levodopa and dopamine receptor agonist that have the ability to convert itself into dopamine in the affected site of brain and deep stimulation to the brain in the subthalamic nucleus. Carbidopa is also prescribed with levodopa as cotreatment to protect the levodopa from disintegration before reaching at the action center (Prokai-Tatrai et al., 2009). Due to the development of

some opportunistic infections and side effects, pharmacological treatments have some limitations. To overcome these drawbacks other strategies are also developed by employing stem cells, which can revitalize the damaged dopaminergic neuron cells with replacing the dopaminergic synthesizing cells. The treatment of brain disorders will become more accessible through stem-cell based therapy in future. Adult bone marrow-derived mesenchymal stem cells (BM-MSC) along with olfactory ensheathing cells (OEC) are most commonly employed for therapeutic purposes in PD by using techniques of regenerative cells (Kim et al., 2002; Kriks et al., 2011). A drawback of using these cells are also occurs as these cells are not completely able to distinguish into dopamine neurons. To resolve this somatic cells are remodeled into nerve regenerative cells along with dopamine nerve fibers through prenatal brain tissue, embryonic stem cells, incited pluripotent stem cells including precisely procured dopamine neuronal cells had been commonly utilized (Blandini et al., 2010; Shukla et al., 2010). Replacement of anthropoid fetal DA neurons in patients can engender long-lasting development and improve quality of life span (Cova et al., 2010; Meyer et al., 2010; Tsunemoto et al., 2015).

For the treatment of PD nano-based methods are reported for the delivery and liberation of DA in the brain. Chitosan NP with DA adsorbed onto the outer surface is inclined by Trapani et al. (2011). After intraperitoneal (IP) administration, in vivo experiments with rats demonstrated that NP-loaded DA is less cytotoxic, compasses the brain and step up in the striatum more than DA alone. As substitute to viruses, for gene therapy of PD, NP has been utilized as transfection vehicles in order to bind the peril of mutagenesis, as well as extravagant immune response (Trapani et al., 2011). Huang et al. (2010) demonstrated importantly enhanced locomotor activity, decreased dopaminergic neuronal loss, and increased DA levels in PD rat brain utilizing lactoferrin-modified NP encapsulating human neurotrophic factor gene.

### 24.4.3 NANOPARTICLE-BASED DRUGS EFFECTIVE FOR CONTROLLING AMYOTROPHIC LATERAL SCLEROSIS

ALS is a broad category of rare brain disorder also known by motor neuron disorders, generally generated by restrained retrogression and demise of motor nerve cells. Motor-nerve cells lengthen from the brain to spinal cord and moves all over the body through muscles. Mandatory transmissions connections between brain and voluntary muscles are commenced through motor neurons. ALS is a lethal neurological disorder and it has no cure till date and nor functional coverage or therapy to stop or reverse the advancement of disease. This disease results in paralytic impact on voluntary muscular system. Some cases related to ALS may occurs due to mutation appears at genetic level in superoxide dismutase-1 (*SOD1*) gene inscribe an enzyme named as superoxide dismutase that produces free radicals that are highly toxic in nature and can abrupt the normal transmissions through motor neurons from brain to spinal cord, as well as in different muscles (Kiernan et al., 2011). This mutant *SOD1* gene accumulates at intracellular region resulting in inhibition of chaperone or proteasome work ability that lead toward the abruption in several protein functions. For the treatment of ALS solid-lipid nanoparticles consist of riluzole was prepared specifically. These nanoparticles have high permeability, high drug absorbing ability, a greater capability than free riluzole, high efficiency as a conveyor to transmit drug into brain and first applying on rats that give an idea to use of nanoparticles for ALS treatment (Bondi et al., 2010). ALS can

be detected by counting the $CD^{4+}$ lymphocytes (Batavelji et al., 2011) and $CD^{8+}$ cells in the different part of brain. Going by this idea Machtoub et al. (2011) injected USPIO NP besides anti-CD4 antibodies, which have the capability of diagnosing the morbid zone inside the ALS obsessive rat's brain.

ALS is a calamitous BDs affecting 1–2/100,000 person/year. Mutations in *SOD1* gene, encoding superoxide dismutase enzyme causes approximately 20% of familial ALS cases. Mutated *SOD1* produces toxic free radicals. Extra, mutant *SOD1* produces intracellular repository that subdue chaperone and/or proteasome activity, with subsequent misfolding and insufficient clearance of numerous proteins. Only one report has been published regarding the synthesis of solid-lipid NP having riluzole for the particular treatment of ALS at the best of our knowledge. These NP displayed high-drug loading, a higher efficiency than free riluzole, a greater ability to deliver the drug into the brain and a lower indiscriminate biodistribution in rats, initiate the way to the use of NP for ALS therapy. Machtoub et al. (2011), injected IV USPIO NP conjugated with anti-CD4 antibodies, and were able to detect the pathological regions in ALS rat brain.

### 24.4.4 NANOPARTICLE-BASED DRUGS EFFECTIVE FOR CONTROLLING PRION DISEASE

Prion disease (PrD) is a club of escalating contagious brain disorders which was engendered by hoarding of misfolded isoform of the prion protein (PrP) in which PrP misleads and turn to mutate form gives the first prion disease that was recognized in human prions was Creutzfeldt–Jakob disease raised in scattered instances and episodically of about one case per million individuals/year. PrD is incorporated with diarrhea and autonomic neuropathy is usually familial. Other forms of PrD in humans are named as fatal familial insomnia (FFI), Gerstmann–Straussler–Scheinker syndrome (GSS), Kuru and variably protease-sensitive prionopathy (VPSPr). The forms belong to prion diseases that results in quite variable trait of the same protein which has two similar or dissimilar large alpha helical forms including pathogenic, protease-resistant isoform (PrPSc), primarily β-sheets, awfully configure noxious amyloid clump. If a pathogenic trait has been originated, prion protein having robust cellular isoform (PrPC) is transmutes to prion protease-resistant isoform (PrPSc) through protein–protein interaction (Norrby, 2011). Prion spreads silently throughout brain of patient and its development continues for a long time unaccompanied to any harm. Ultimately prions begin with assassinate neurons, further when symptoms appear, individual gets tremendous intellectual dwindle. Therefore, this disease becomes fatal within few months. Prion disease can also be communicated by ways of infected operative instruments, human growth hormone supplements, or transplanting tissue that circumjacent the brain. Such arising pathological conditions had been named as "iatrogenic infections." Therefore, prion disease had to be classified into acquired, sporadic, and genetic, based on from where they arise. It is further classified into other three categories: Creutzfeldt–Jakob disease (CLD), fatal familial insomnia (FFI), and GSS syndrome. Throughout its treatment the focal key object regarding antiprion approach is the morbid trait of the cellular prion protein (PrPC) named PrPSc, that is always allied with the disorder. A lot of diversification has been seen for the availability of different compounds which can influence the formation of PrPSc or amplify its demolition in vitro replica and extended life durability in experimental animals. Some of antiprion agents, such as quinacrine and pentosan polysulfate, as well

as dioxycycline are subject to clinical trial but no satisfactory result was obtained. A resistant and nonresistant immunization escorted by antibodies antagonistic toward PrP along with mucosal inoculation have been manifesting toward fortify amid tangential septicemia during experiments. Moreover, the probability to impede with PrPC to PrPSc transmutation by an active command of PrPC is other riveting strategy arising from experiments. Transmutation of PrPC to PrPres with outcome progression and agglomeration results in neuronal death and amyloidogenesis. The potential of various polyamine dendrimers to abolish morbific prion traits from damaged cells through hauling those traits to lysosomal mortifications (Lim et al., 2010). On the account of consequences attributed to polyamines that pretend to be hinging over positive charges, Lim et al. (2010) developed perpetually charged polyamines, thanks to be reduced toxicity. A polyelectrolyte is a polymer with multiple repeating unit and veiled by gold (Au) NP along with two functional groups over surface that are sulfonates and primary amines where these functional groups are clinically pertinent and adequate to impede the aggregation of PrPSc at highly minimal concentration in neuroblastoma cells, had been established by Ai Tran et al. (2010) Au NP surface-operationalized with the same groups are also capable to crucifix the murine BBB and can invade the delineate neuronal structures that virtual to regions in which PrP accumulation occurs, was demonstrated by Sousa et al. (2010) and this information perhaps further employed to diagnostic scope. The diagnosis program that hinge on two aptamers was established by Xiao et al. (2010) that perceive two peculiar epitopes of PrPSC to differentiate it from PrPC in serum as well as in brain homogenate. In presence of PrPSC, aptamers affiliated to the surface of magnetic microparticles and quantum dot (QD), respectively, frame a sandwich structure which is intensely fluorescent in aqueous mediums and conceivably discriminated through external magnetic field. For the site specific tagging of PrP which is asserted over the cell surfaces in vitro (Xie et al., 2010), the PEG-interspersed nitrilotriacetic acid-functionalized QD could be employed.

The ability of different polyamine dendrimers to take away pathogenic prion isoform (PrPSc) from infected cells by dragging them to lyso-somal degradation indicated by in vitro studies. Lim et al. synthesized permanently charged polyamines, which resulted less toxic, since the effect of polyamines appears to be dependent on positive charges. Tran et al. (2010) synthesized polyelectrolyte multilayer-coated Au NP with two therapeutically relevant functional groups on the surface, sulfonates and primary amines, able to shackle the deposition of PrPSc in neuroblastoma cells at very low concentration. Sousa et al. (2010) showed that Au NP surface-functionalized with the same groups, are able to cross the murine BBB and enter defined neuronal structures close to areas in which PrP aggregation takes place. This method could be also used for diagnostic purposes (Sousa et al., 2010).

## 24.5 NANOFORMULATIONS INVESTIGATED TO COMBAT BRAIN DISORDERS

Nanoparticles (NPs) are believed as one of the most promising and flexible drug-delivery systems into unreachable regions like the brain (Palmer, 2010). NPs can vary from 1 to 1000 nm in size and they are synthetic origin. Micelles and liposome are the example of NPs. They studied properly for drug delivery to the brain. Synthetic NPs can be made from synthetic NPs may be prepared from polymeric

materials, such as poly(ethylenimine) (PEI), poly(alkyl cyanoacrylates), poly(amidoamine), dendrimers (PAMAM), poly(ε-caprolactone) (PCL), poly(lactic-co-glycolic acid) (PLGA), polyesters (poly(lactic acid) (PLA), or from inorganic materials, such as AU, silicon dioxide (silica) (Barbu et al., 2009; Cupaioli et al., 2014; Lockman et al., 2009). Such carriers can perform by covalently to them. Inorganic NPs are advantageous because of their function, simple preparation method, shape, and control size. Inorganic NPs use on account of tracking is easy by analytic technique (e.g., ICP-MS) or by microscopy techniques (e.g., MRI, TEM). There is lots of parameter of which affect the productivity of BBB passage, NPs system-atic circulation and cellular conveyance. The shape and size of NPs affects body circulation and cellular level. The different shaped NPs exist like cubic, spherical, rod-like, and so on, though spherical NPs are easy to prepare. There is many specifications, which affects the NPs passage, among them zeta potential is important one (Lockman et al., 2004). Lots of ligands have been adjoining to NPs for the facilitation of BBB entrance. Many parameters affect the transportation of NPs by mean of BBB at different ranges. Those NPs adjoining with ligand, capable to collaborate with BBB receptor at comparatively low validity, have the matchless performance. Table 24.1 repre-sents nanoparticles based formulation for drug delivery to the brain.

## 24.6 PATENTS RELATED TO BRAIN TARGETING TO CONTROL BRAIN DISORDERS

BDs are increasing at a great rate as popu-lace senility which affects the populace with medical, as well as financial burden. BDs

**TABLE 24.1**  List of Nanoformulations Investigated for Better Brain Delivery.

| Formulation | API/modelmolecule | Materials used | Advantages |
|---|---|---|---|
| Dendrimers | Docetaxel | Polypropyleneimine | Higher targeting efficiency and biodistribution to the brain |
| | Paclitaxel | Polyamidoamine | 12-fold greater permeability across porcine brain endothelial cells |
| Nanoemulsion | Saquinavir | Flax-seed, safflower oil | Improved brain uptake |
| | Risperidone | Glycerylmonocaprylate | Higher drug transport efficiency and increased direct nose to brain drug transport |
| Polymeric nanoparticles | Doxorubicin | PBCA-tween 80 | Augmented accumulation of NP in the tumor site and in the contralateral hemisphere |
| | Venlafaxine | Chitosan | Better brain uptake, higher direct transport percentage |
| | Etoposide | PLGA and PCL | Selective distribution with higher brain permeability |
| | Amphotericin B | PLA–PEG-tween 80 | Drug concentration in mice brain greatly enhanced, reduced the toxicity of amphotericin B to liver, kidney etc. |
| | Imatinib mesylate | PLGA | Increased the extent of drug permeation to brain |
| Micelles | Olanzapine | Block copolymers of ethylene oxide/propylene oxide | Demonstrated higher drug targeting index (5.20), drug targeting efficiency (520.26%) and direct transport percentage (80.76%) |
| Liposomes | Citicoline | DSPC, cholesterol, DSPE | Considerable increase (10-fold) in the bioavailability of the drug in the brain parenchyma |
| | Oregon green | DPPC, DC, cholesterol, DOPE | Liposomes were strongly internalized in cultured cell lines within 6 h |
| Solid lipid nanoparticles/nanostructured lipid carriers | Baicalein | Tripalmitin, gelucire, vitamin E | Brain-targeting efficiency of baicalein was greatly improved by NLCs |
| | Camptothecin | Cetyl palmitate, dynasan, witepsol | Higher affinity to the porcine brain capillary endothelial cells as compared to macrophages |
| | Idarubicin | Stearic acid | Drug and its metabolite were detected in the brain only after IDA–SLN administration |
| | Clozapine | Trimyristin, tripalmitin, tristearin | The AUC and MRT of clozapine SLNs were significantly higher in brain |

AUC, area area under the curve; DC, dendritic dendritic cell; DOPE, dioleoyl dioleoyl phosphatidylethanolamin; DPPC, dipalmitoyl dipalmitoyl phosphatidylcholine; DSPC, distearoyl distearoyl phosphatidylcholine; DSPE, 1,2-distearoyldis-tearoyl-*sn*-glycero-3-phosphoethanolamine; IDA-SLN, idarubicinidarubicin-loaded solid lipid nanoparticles; MRT, mean mean residence time; NLCs, nanostructured lipid carriers; PBCA, poly(butyl cyanoacrylate); PCL, poly ε-caprolactone; PLGA, poly(dl-lactic-*co*-glycolic acid).

**TABLE 24.2** List of Patents Claiming for Neuroprotective Applications.

| Claimed product | Patent name | Patent no. | Activity claimed | Inventors |
|---|---|---|---|---|
| Gold nanoparticles | Gold nanoparticles coated with polyelectrolytes and albumin | US20110262546 | Drug delivery | Legname et al., 2011 |
| Cerium oxide | Nanoparticles of cerium oxide targeted to an amyloid beta antigen of AD and associated methods | US9463253 | Drug delivery | Sudipta et al., 2016 |
| Cerium oxide | Nanoparticles of cerium oxide targeted to an amyloid-beta antigen of AD and associated methods | US20120070500 | Drug targeting | Annamaria et al., 2012 |
| Carbon nanotubes coated with cyclodextrin | Method of diagnosing, prognosing, and monitoring AD | US20120245854 | Diagnosing, monitoring | Hossam et al., 2012 |
| Carbon nanotubes coated with cyclodextrin | Method of diagnosing, prognosing, and monitoring PD | US20120245434 | Diagnosin, monitoring | Hossam et al., 2012 |
| Cerium oxide nanoparticles | Neuronal protection by cerium oxide nanoparticles | US20160038537 | Disease treatment | William et al., 2016 |
| Carbon nanotubes or metal nanoparticles | Diagnosing, prognosing, and monitoring multiple sclerosis | US8945935 | Diagnostic | Hossam et al., 2015 |
| Gold-creatinine nanoparticle | Polyelectrolyte-encapsulated gold nanoparticles capable of crossing blood–brain barrier (BBB) | US20110111040 | Treatment of stroke drug delivery | Silke et al., 2010 |
| Mitochondria containing nanoparticles | Therapeutic nanoparticles for accumulation in the brain | US20170216219 | An antioxidant and an antiinflammatory brain injury | Shanta et al., 2017 |
| Lipid nanoparticle | Lipid-derived nanoparticles for brain-targeted drug delivery | US20100076092 | Drug delivery | Jayanth et al., 2010 |
| Chitosan-polyethylene oxide nanoparticle | Nanoparticles for brain tumor imaging | US2010260686 | Imaging | Miqin et al., 2010 |
| Magnetic nanoparticle | Magnetic nanodelivery of therapeutic agents across the BBB | US2011213193 | Drug delivery and imaging | Madhavan et al., 2011 |

*Source*: FreePatentsOnline.com, 2018

have complicated nature that mostly results from combination of environmental, pathological and genetic constituent. This study based on compiling different patented documents for exploration of BDs (i.e., EP2282779A1, US2011022955A1), to deliver information in term of research and development in the form of technological alteration. The strategies for the treatment of various BDs are based on different patent approaches. This based on age-grouped BDs such as a novel remedy of technique for the treatment of AD and incorporated disorder by modulated cell stress reaction EP22827791, via joint therapies that revamp angiogenesis US2012005899A1 (Kumar and Sharma, 2017). A list of some patent published on nanomedicine mediated diagnosis and targeting of brain disorders are depicted in Table 24.2.

## 24.7 NEUROTOXICITY OF NANOMATERIALS

As the nanotechnology, and NMs have wide applications in many industries and shows a significant impact on medicinal field and also have some potential threat for human health, particularly having feasible virulent impact over CNS (Kim et al., 2007; Sousa et al., 2010; Ulbrich et al., 2009). Number of data are available to demonstrate the virulent of NMs which are based upon several aspects, such as chemical conformation, shape, surface area, size, surface charge, absorbance, and so many others. A system that can act as conveyance to transport the drug at target site is the requisite. Therefore, superparamagnetic iron oxide nanoparticles (SPIONS) as novel drug conveyance had been prepared and guide it with the assistance of an extramural magnetic field to its goal. Finally, impact of surface chemistry on the neurotoxin behavior of SPIONS has been explored in inseminational condition in human neuroblastoma cell culture at cellular level, as well as molecular levels. SPIONS having stark $-NH_2$ or $-COOH$ functional groups have higher toxicity as compare to those encrust SPIONS because of propensity to assimilate amino acids, proteins, vitamins, and ions that causes variations in pH, as well as compositions in the cellular medium. Toxicity of bare SPIONS can be assessed by MTT assay. It is also evaluated that the SPIONS–COOH persuade the upregulation of oxidative strain empirical gene by modulating it, was reported by DNA microarray experiments (Mahmoudi et al., 2011). Au nanoparticles have also been investigated as neurotoxic agents in vitro or in vivo examinations and the shape (spheres, rods, and urchins), size, and surface coating (PEG or acetyltriethyl ammonium bromide) of Au nanoparticles affect the neurotoxicity. Usually, it had been observed in vitro studies that electropositive charged gold nanoparticles shows virulent effect on neurons and microglia and it does not depends upon the shape of Au nanoparticles (Hutter et al., 2010). Ephemeral microglia activation and impulsion of TLR-2 promoter bustle in transgenic

mice was caused by intranasal insertion of Au nanoparticles, in a shape and surface-contingent manner was examined in vivo studies. Results obtained when PC12 cells are used to assess the toxic effectiveness of nanotechnologies were not satisfactory as nanotechnologies persuade a decrease in PC12 cells growth, as well as activity and mitochondrial membrane potential in dose-dependent and time dependent manner whereas an increment of reactive oxygen species (ROS) productivity arrests the cell cycle at G2/M phase and the apoptotic rate was depend on dose (Wang et al., 2011). Titanium dioxide ($TiO_2$) nanoparticles exposure also shows toxicity as it can cause configurationally variation of cortex neurons and also cause interference in level of the monoamine neurotransmitter which is found in subbrain division when administered intranasally in mice and it also produced potential neurological lesion in brain. Due to these toxic impacts of nanoparticles, their use in future for clinical purpose will be expectedly limited to in vitro diagnostics. So, for manufacturing newer nanoparticles planned for brain-specific drug transportation and diagnostic purposes, benefit-risk calculation must be in balance (Win-Shwe et al., 2011; Zhang et al., 2011).

## 24.8   FUTURE OPPORTUNITIES AND CHALLENGES

Synchronically, brain disorders have shown its significant impact on number of individuals including all age groups. The race of disorders that distresses older people also increases its number of cases throughout the world with increment in life-span. Currently, with available data, approximately 18 million people suffer from AD and a regular increase in numbers, which will lead to 34 million, till 2025. A number of contradictions had been represented by BDs. Some of them have direct genetic evolutions, whereas others are somewhat related to genetic mutations. Onset of the disease before time or increment in severity of disease is the result of such mutations that occurs in aforementioned diseases. Some common features are usually present in multiple brain related diseases like some deformities in protein mortification pathways, malfunction in mitochondria, copout axonal transmission process, and finally introduction of morbidity of cells. The elective unproductivity of distinct cell populations in various brain disorders is presently sorely understood but these information's may give some necessary clues toward disease-customize therapeutic approaches. It devotes to various practical concerns that represent symbolic challenges in twain, the diagnosis and therapeutic advancement phases of highly potent therapeutic agents. Till date the reductionist approaches of drug productivity has not been rewarding toward delivering workable disease-transmogrifying therapies regarding BDs and mostly drug experiment campaigns depends on animal models that sufficiently recapitulate the human disease. Various bottlenecks have come into picture at clinical trials stage related to BDs like as recognition of patients at initial phase of disease, categorization of populations for clinical trials in a proper manner and the inadequacy of sufficient biomarkers toward appraising treatment potency. In this series various novel strategies have been proposed and are associated to nanotechnology for effective treatments (Cuny et al., 2012). Ultimately, the challenges posed by BDs along with opportunities to gain the target regarding efficient disease-customize therapeutic approaches related to these disastrous situations (Voss et al., 2012).

## 24.9 CONCLUSION

This chapter comprises an overview of studies about BDs, their effective treatments through nanotechnology by employing nanomedicines, supportive and advanced recovery of harmed neurons, neuronal protections, and invigorates the qualities and conveyance of drugs toward damaged cells over the BBB, subscribing fundamentally to the improvement of drug delivery through nanoframeworks for the treatment of BDs. Though, a few studies have proposed that NMs can show dangerous results for different human system organs and frameworks, counting the CNS, here all mechanical and logical fields are working vigorously to furnish their nonmaterial's and looks into with the excellent. Therefore, it is the backbone of the current therapeutic strategy to recover from destructive and lethal BDs. Nanomedicines have high potential to cross the BBB which was the major challenge earlier during the treatment of BDs. They not only facilitate to crucifix the BBB but also facilitate the conveyance of drug toward target, as well as gene delivery. Nanotechnology provides newer vision in the development of novel contrivance but additional efforts are required in the field of nanotechnology to transform the theoretical education to clinical approaches.

## KEYWORDS

- **brain disorders**
- **blood–brain barrier**
- **nanotechnology**
- **nanomedicine**

## REFERENCES

Abbott, N. J.; Chugani, D. C.; Zaharchuk, G.; Rosen, B. R.; Lo, E. H. Delivery of Imaging Agents into Brain. *Adv. Drug Deliv. Rev.* **1999**, *37* (1–3), 253–277.

Ali, S. S.; Hardt, J. I.; Quick, K. L.; Kim-Han, J. S.; Erlanger, B. F.; Huang, T.-T.; Epstein, C. J.; Dugan, L. L. A Biologically Effective Fullerene (C60) Derivative with Superoxide Dismutase Mimetic Properties. *Free Rad. Biol. Med.* **2004**, *37* (8), 1191–1202.

Barbu, E.; Molnàr, É.; Tsibouklis, J.; Górecki, D. C. The Potential for Nanoparticle-Based Drug Delivery to the Brain: Overcoming the Blood–Brain Barrier. *Ex. Opin. Drug Deliv.* **2009**, *6* (6), 553–565.

Bataveljić, D.; Stamenković, S.; Bačić, G.; Andjus, P. Imaging Cellular Markers of Neuroinflammation in the Brain of the Rat Model of Amyotrophic Lateral Sclerosis. *Acta Physio. Hung.* **2011**, *98* (1), 27–31.

Bhaskar, S.; Tian, F.; Stoeger, T.; Kreyling, W.; de la Fuente, J. M.; Grazú, V.; Borm, P.; Estrada, G.; Ntziachristos, V.; Razansky, D. Multifunctional Nanocarriers for Diagnostics, Drug Delivery and Targeted Treatment Across Blood–Brain Barrier: Perspectives on Tracking and Neuroimaging. *Particle Fib. Toxic.* **2010**, *7* (1), 3.

Blandini, F.; Cova, L.; Armentero, M.-T.; Zennaro, E.; Levandis, G.; Bossolasco, P.; Calzarossa, C.; Mellone, M.; Giuseppe, B.; Deliliers, G. L. Transplantation of Undifferentiated Human Mesenchymal Stem Cells Protects Against 6-Hydroxydopamine Neurotoxicity in the Rat. *Cell Transplant.* **2010**, *19* (2), 203–218.

Blurton-Jones, M.; Kitazawa, M.; Martinez-Coria, H.; Castello, N. A.; Müller, F.-J.; Loring, J. F.; Yamasaki, T. R.; Poon, W. W.; Green, K. N.; LaFerla, F. M. Neural Stem Cells Improve Cognition via BDNF in a Transgenic Model of Alzheimer Disease. *Proc. Nat. Acad. Sci* **2009**, *106* (32), 13594–13599.

Boado, R. J. A New Generation of Neurobiological Drugs Engineered to Overcome the Challenges of Brain Drug Delivery. *Drug News Per.* **2008**, *21* (9), 489–503.

Bondì, M. L.; Craparo, E. F.; Giammona, G.; Drago, F. Brain-Targeted Solid Lipid Nanoparticles Containing Riluzole: Preparation, Characterization and Biodistribution. *Nanomedicine* **2010**, *5* (1), 25–32.

Canovi, M.; Markoutsa, E.; Lazar, A. N., Pampalakis, G.; Clemente, C.; Re, F.; Sesana, S.; Masserini, M.; Salmona, M.; Duyckaerts, C.; Flores, O. The Binding Affinity of Anti-Aβ1-42 MAb-Decorated Nanoliposomes to Aβ1-42 Peptides In Vitro and to Amyloid Deposits in Post-Mortem Tissue. *Biomaterials* **2011**, *32* (23), 5489–5497.

Chaturvedi, V. K. Hangloo, A.; Pathak, R. K.; Gupta, K. K.; Singh, V.; Singh, A. K.; Verma, V.; Singh, M. P. Stem Cell Therapy for Neurodegenerative Disorders. In *Stem Cells from Culture Dish to Clinic;* Nova Science Publish, Inc.: New York, 2017.

Cherubini, A.; Spoletini, I.; Péran, P.; Luccichenti, G.; Di Paola, M.; Sancesario, G.; Gianni, W.; Giubilei, F.; Bossù, P.; Sabatini, U. A Multimodal MRI Investigation of the Subventricular Zone in Mild Cognitive Impairment and Alzheimer's Disease Patients. *Neurosc. Lett.* **2010**, *469* (2), 214–218.

Cho, Y.; Borgens, R. B. Polymer and Nano-Technology Applications for Repair and Reconstruction of the Central Nervous System. *Experi. Neurol.* **2012**, *233* (1), 126–144.

Cova, L.; Armentero, M.-T.; Zennaro, E.; Calzarossa, C.; Bossolasco, P.; Deliliers, G. L.; Polli, E.; Nappi, G.; Silani, V.; Blandini, F. Multiple Neurogenic and Neurorescue Effects of Human Mesenchymal Stem Cell After Transplantation in an Experimental Model of Parkinson's Disease. *Brain Res.* **2010**, *1311*, 12–27.

Cuny, G. D. Neurodegenerative Diseases: Challenges and Opportunities. *Future Sci.* **2012**, *4* (13), 1647–1649.

Cupaioli, F. A.; Zucca, F. A.; Boraschi, D.; Zecca, L. Engineered Nanoparticles. How Brain Friendly is this New Guest? *Prog. Neurobiol.* **2014**, *119*, 20–38.

Das, M.; Patil, S.; Bhargava, N.; Kang, J.-F.; Riedel, L. M.; Seal, S.; Hickman, J. J. Auto-Catalytic Ceria Nanoparticles Offer Neuroprotection to Adult Rat Spinal Cord Neurons. *Biomaterial* **2007**, *28* (10), 1918–1925.

Ferri, C. P.; Prince, M.; Brayne, C.; Brodaty, H.; Fratiglioni, L.; Ganguli, M.; Hall, K.; Hasegawa, K.; Hendrie, H.; Huang, Y. Global Prevalence of Dementia: A Delphi Consensus Study. *Lancet* **2005**, *366* (9503), 2112–2117.

Gabathuler, R. Approaches to Transport Therapeutic Drugs Across the Blood–Brain Barrier to Treat Brain Diseases. *Neurobiol. Dis.* **2010**, *37* (1), 48–57.

Gaur, A.; Midha, A.; Bhatia, A. Significance of Nanotechnology in Medical Sciences. *Asian J. Pharma.* **2008**, *2* (2), 80.

Giacobini, E.; Becker, R. E. One Hundred Years After the Discovery of Alzheimer's Disease. A Turning Point for Therapy? *J. Alzheimer Dis.* **2007**, *12* (1), 37–52.

Gobbi, M.; Re, F.; Canovi, M.; Beeg, M.; Gregori, M.; Sesana, S.; Sonnino, S.; Brogioli, D.; Musicanti, C.; Gasco, P.; Salmona, M. Lipid-Based Nanoparticles with High Binding Affinity for Amyloid-β1-42 Peptide. *Biomaterial* **2010**, *31*, 6519–6529. FreePatentsOnline.com. http://www.freepatentsonline.com (accessed June 27, 2018).

Huang, R.; Ke, W.; Liu, Y.; Wu, D.; Feng, L.; Jiang, C.; Pei, Y. Gene therapy Using Lactoferrin-Modified Nanoparticles in a Rotenone-Induced Chronic Parkinson Model. *J. Neurol. Sci.* **2010,** *290* (1), 123–130.

Hutter, E.; Boridy, S.; Labrecque, S.; Lalancette-Hébert, M.; Kriz, J.; Winnik, F. M.; Maysinger, D. Microglial Response to Gold Nanoparticles. *ACS Nano.* **2010,** *4* (5), 2595–2606.

Jin, G.-Z.; Kim, M.; Shin, U. S.; Kim, H.-W. Neurite Outgrowth of Dorsal Root Ganglia Neurons is Enhanced on Aligned Nanofibrous Biopolymer Scaffold with Carbon Nanotube Coating. *Neurosci. Lett.* **2011,** *501* (1), 10–14.

Kiernan, M. C.; Vucic, S.; Cheah, B. C.; Turner, M. R.; Eisen, A.; Hardiman, O.; Burrell, J. R.; Zoing, M. C. Amyotrophic Lateral Sclerosis. *Lancet* **2011,** *377* (9769), 942–955.

Kim, H. R.; Andrieux, K.; Gil, S.; Taverna, M.; Chacun, H.; Desmaële, D.; Taran, F.; Georgin, D.; Couvreur, P. Translocation of Poly (Ethylene Glycol-*co*-Hexadecyl) Cyanoacrylate Nanoparticles into Rat Brain Endothelial Cells: Role of Apolipoproteins in Receptor-Mediated Endocytosis. *Biomacromolecules* **2007,** *8* (3), 793–799.

Kim, J.-H.; Auerbach, J. M.; Rodríguez-Gómez, J. A.; Velasco, I.; Gavin, D.; Lumelsky, N.; Lee, S.-H.; Nguyen, J.; Sánchez-Pernaute, R.; Bankiewicz, K. Dopamine Neurons Derived from Embryonic Stem Cells Function in an Animal Model of Parkinson's Disease. *Nature* **2002,** *418* (6893), 50.

Kostarelos, K.; Miller, A. D. Synthetic, Self-Assembly ABCD Nanoparticles; a Structural Paradigm for Viable Synthetic Non-Viral Vectors. *Chem. Soc. Rev.* **2005,** *34* (11), 970–994.

Kriks, S.; Shim, J.-W.; Piao, J.; Ganat, Y. M.; Wakeman, D. R.; Xie, Z.; Carrillo-Reid, L.; Auyeung, G.; Antonacci, C.; Buch, A. Dopamine Neurons Derived from Human ES Cells Efficiently Engraft in Animal Models of Parkinson's Disease. *Nature* **2011,** *480* (7378), 547.

Kumar, B.; Sharma, D. Recent Patent Advances for Neurodegenerative Disorders and its Treatment. *Recent Pat. Drug Deliv. Formul.* **2017,** *11*(3), 158–172.

Leary, S. P.; Liu, C. Y.; Apuzzo, M. L. Toward the Emergence of Nanoneurosurgery: Part II—Nanomedicine: Diagnostics and Imaging at the Nanoscale Level. *Neurosurgery* **2006,** *58* (5), 805–823.

Leary, S. P.; Liu, C. Y.; Apuzzo, M. Toward the Emergence of Nanoneurosurgery: Part III—Nanomedicine: Targeted Nanotherapy, Nanosurgery, and Progress Toward the Realization of Nanoneurosurgery. *Neuroscience* **2006,** *58* (6), 1009–1026.

Lee, H. J.; Lee, J. K.; Lee, H.; Shin, J.-W.; Carter, J. E.; Sakamoto, T.; Jin, H. K.; Bae, J.-S. The Therapeutic Potential of Human Umbilical Cord Blood-Derived Mesenchymal Stem Cells in Alzheimer's Disease. *Neuro. Lett.* **2010a,** *481* (1), 30–35.

Lee, J. K.; Jin, H. K.; Endo, S.; Schuchman, E. H.; Carter, J. E.; Bae, J. S. Intracerebral Transplantation of Bone Marrow-Derived Mesenchymal Stem Cells Reduces Amyloid-Beta Deposition and Rescues Memory Deficits in Alzheimer's Disease Mice by Modulation of Immune Responses. *Stem Cells* **2010b,** *28* (2), 329–343.

Lee, J. Y.; Bashur, C. A.; Milroy, C. A.; Forciniti, L.; Goldstein, A. S.; Schmidt, C. E. Nerve Growth Factor-Immobilized Electrically Conducting Fibrous Scaffolds for Potential use in Neural Engineering Applications. *IEEE Trans. Nanobiosci.* **2012,** *11* (1), 15–21.

Liu, J. J.; Wang, C. Y.; Wang, J. G.; Ruan, H. J.; Fan, C. Y. Peripheral Nerve Regeneration Using Composite Poly (Lactic Acid-Caprolactone)/Nerve Growth Factor Conduits Prepared by Coaxial Electrospinning. *J. Biomed. Mater. Res.* **2011,** *96* (1), 13–20.

Lockman, P. R.; Koziara, J. M.; Mumper, R. J.; Allen, D. D. Nanoparticle Surface Charges Alter Blood–Brain Barrier Integrity and Permeability. *J. Drug Target.* **2004,** *12* (9–10), 635–641.

Lockman, P. R.; Mumper, R. J.; Khan, M. A.; Allen, D. D. Nanoparticle Technology for Drug Delivery Across the Blood-Brain Barrier. *Drug Dev. Ind. Pharma.* **2002,** *28* (1), 1–13.

Machtoub, L.; Bataveljic, D.; Andjus, P. Molecular Imaging of Brain Lipid Environment of Lymphocytes in Amyotrophic Lateral Sclerosis Using Magnetic Resonance Imaging and SECARS Microscopy. *Physio. Res.* **2011,** *60*, S121.

McLeod, M.; Hong, M.; Mukhida, K.; Sadi, D.; Ulalia, R.; Mendez, I. Erythropoietin and GDNF Enhance Ventral Mesencephalic Fiber Outgrowth and Capillary Proliferation Following Neural Transplantation in a Rodent Model of Parkinson's Disease. *Eur. J. Neurosci.* **2006,** *24* (2), 361–370.

Meyer, A. K.; Maisel, M.; Hermann, A.; Stirl, K.; Storch, A. Restorative Approaches in Parkinson's Disease: Which Cell Type Wins the Race? *J. Neurol. Sci.* **2010,** *289* (1), 93–103.

Moghadam, F. H.; Alaie, H.; Karbalaie, K.; Tanhaei, S.; Esfahani, M. H. N.; Baharvand, H. Transplantation of Primed or Unprimed Mouse Embryonic Stem Cell-Derived Neural Precursor Cells Improves Cognitive Function in Alzheimerian Rats. *Differentiation* **2009,** *78* (2–3), 59–68.

Mourtas, S.; Canovi, M.; Zona, C.; Aurilia, D.; Niarakis, A.; La Ferla, B.; Salmona, M.; Nicotra, F.; Gobbi, M.; Antimisiaris, S. G. Curcumin-Decorated Nanoliposomes with very High Affinity for Amyloid-β1-42 Peptide. *Biomaterial* **2011**, *32* (6), 635–1645.

Oliveira, J.; Alcyr, A.; Hodges, H. M. Alzheimer's Disease and Neural Transplantation as Prospective Cell Therapy. *Curr. Alz. Res.* **2005**, *2* (1), 79–95.

Palmer, A. M. The Role of the Blood–CNS Barrier in CNS Disorders and Their Treatment. *Neurobiol. Dis.* **2010**, *37* (1), 3–12.

Patel, M. M.; Goyal, B. R.; Bhadada, S. V.; Bhatt, J. S.; Amin, A. F. Getting into the Brain. *CNS Drugs* **2009**, *23*(1), 35–58.

Pihlaja, R.; Koistinaho, J.; Malm, T.; Sikkilä, H.; Vainio, S.; Koistinaho, M. Transplanted Astrocytes Internalize Deposited β-Amyloid Peptides in a Transgenic Mouse Model of Alzheimer's Disease. *Glia* **2008**, *56* (2), 154–163.

Popovic, N.; Brundin, P. Therapeutic Potential of Controlled Drug Delivery Systems in Neurodegenerative Diseases. *Int. J. Pharm.* **2006**, *314* (2), 120–126.

Prabhakaran, M. P.; Ghasemi-Mobarakeh, L.; Ramakrishna, S. Electrospun Composite Nanofibers for Tissue Regeneration. *J. Nanosci. Nanotechnol.* **2011**, *11* (4), 3039–3057.

Prokai-Tatrai, K.; Prokai, L. Prodrugs of Thyrotropin-Releasing Hormone and Related Peptides as Central Nervous System Agents. *Molecules* **2009**, *14* (2), 633–654.

Ryu, J. K.; Cho, T.; Wang, Y. T.; McLarnon, J. G. Neural Progenitor Cells Attenuate Inflammatory Reactivity and Neuronal Loss in an Animal Model of Inflamed AD Brain. *J Neuroinflam.* **2009**, *23* (6), 39.

Shukla, S.; Trivedi, A.; Singh, K.; Sharma, V. Pituitary Tuberculoma. *J. Neuro. Rural Pract.* **2010**, *1* (1), 30.

Singh, N. V.; Pillay, Y. E. Advances in the Treatment of Parkinson's Disease. *Prog. Neurobiol.* **2007**, *81*, 29–44.

Sousa, F.; Mandal, S.; Garrovo, C.; Astolfo, A.; Bonifacio, A.; Latawiec, D.; Menk, R. H.; Arfelli, F.; Huewel, S.; Legname, G.; Galla, H. J. Functionalized Gold Nanoparticles: A Detailed In Vivo Multimodal Microscopic Brain Distribution Study.*Nanoscale* **2010**, *2* (12), 2826–2834.

Tran, H. N. A.; Sousa, F.; Moda, F.; Mandal, S.; Chanana, M.; Vimercati, C.; Morbin, M.; Krol, S.; Tagliavini, F.; Legname, G. A Novel Class of Potential Prion Drugs: Preliminary In Vitro and In Vivo Data for Multilayer Coated Gold Nanoparticles. *Nanoscale* **2010**, *2* (12), 2724–2732.

Trapani, A.; De Giglio, E.; Cafagna, D.; Denora, N.; Agrimi, G.; Cassano, T.; Gaetani, S.; Cuomo, V.; Trapani, G. Characterization and Evaluation of Chitosan Nanoparticles for Dopamine Brain Delivery. *Int. J. Pharm.* **2011**, *419* (1–2), 296–307.

Tsunemoto, R. K.; Eade, K. T.; Blanchard, J. W. Baldwin, K. K. Forward Engineering Neuronal Diversity Using Direct Reprogramming. *EMBO J.* **2015**, *34* (11), 1445–1455.

Ulbrich, K.; Hekmatara, T.; Herbert, E.; Kreuter, J. Transferrin- and Transferrin Receptor-Antibody-Modified Nanoparticles Enable Drug Delivery Across The Blood–Brain Barrier (BBB). *Eur. J. Pharm. Biopharm.* **2009**, *71*, 251–256.

Voss, J.; Ebert, A.; Wolfe, M.; Lindsley, C.; Cookson, C.; Deane, R. Ask the Experts. *Fut. Med. Chem.* **2012**, *4* (13), 1661–1669.

Wang, J.; Sun, P.; Bao, Y.; Liu, J. An L. Cytotoxicity of Single-Walled Carbon Nanotubes on PC12 Cells. *Toxicol. In Vitro* **2011**, *25*, 242–250.

Win-Shwe, T.T.; Fujimaki, H. Nanoparticles and Neurotoxicity. *Int. J. Mol. Sci.* **2011**, *12*, 6267–6280.

Xiao, S.J.; Hu, P. P.; Wu, X. D. et al. Sensitive Discrimination and Detection of Prion Disease-Associated Isoform with a Dual-Aptamer Strategy by Developing a Sandwich Structure of Magnetic Microparticles and Quantum Dots. *Anal Chem.* **2010**, *82,* 9736–9742.

Xie, M.; Luo, K.; Huang, B. H. et al. PEG-Interspersed Nitrilotriacetic Acid Functionalized Quantum Dots for Site-Specific Labeling of Prion Proteins Expressed on Cell Surfaces. *Biomaterial* **2010**, *31,* 8362–8370.

Zhang, L.; Bai, R.; Li, B.; et al. Rutile $TiO_2$ Particles Exert Size and Surface Coating Dependent Retention and Lesions on the Murine Brain. *Toxicol. Lett.* **2011**, *207,* 73–81.

# Index

616

Index

blood–brain barrier (BBB), 32
classification, 33
depolarizing neuromuscular blockers
    abnormal plasma cholinesterase, 46
    cardiovascular side effects, 45
    fasciculation, 45
    hyperkalemia, 46
    intragastric pressure, 45–46
    intraocular and intracranial pressure, 46
    mechanisms of action, 44–45
    muscle pains, 45
    succinylcholine, 45
exocrine glands
    acute nicotine poisoning, 54
    pharmacokinetics, 53–54
ganglion blockers, 54
    adverse drug reactions, 56
    mechanism of action, 55
    pharmacokinetics, 55–56
    pharmacological actions, 55
    therapeutic uses, 56
ganglionic blockers
    classification of, 50
    individual nondepolarizing neuromuscular blockers, properties, 49
    mechanism of action, 48, 50
    neurotransmission, 50, 52
    stimulants, 51
neuromuscular blocking, 43
    classification, 44
    drug interaction, 50
nicotine, 51
    cardiovascular system, 53
    central nervous system (CNS), 53
    gastrointestinal tract, 53
    mechanism of action, 52
    peripheral nervous system, 52–53
nondepolarizing neuromuscular blockers
    drug interactions, 48
    mechanisms of action, 46–47
    pharmacokinetics, 47
nonselective muscarinic receptor antagonist
    absorption, 34
    airways, 37
    atropine, 33–34
    cardiovascular system, 36
    distribution, 34–35
    elimination, 35
    eye, 37
    gastrointestinal tract, 36

mechanism of action, 35
    metabolism, 35
    pharmacodynamics, 35–38
    prostate glandular epithelium, 37–38
    sinoatrial (SA), 36
    urinary bladder, 36–37
selective muscarinic receptor antagonists
    darifenacin, 38–39
    fesoterodine, 39
    pirenzepine, 38
    solifenacin, 39
    telenzepine, 38
    tolterodine, 38
therapeutic applications
    cardiovascular disorders, 40–41
    gastrointestinal disorders, 41
    motion sickness, 39–40
    ophthalmoscopic examination, 40
    Parkinson's disease (PD), 39
    respiratory disorders, 40
    sweat gland, 41
    urinary urgency, 41
urinary incontinence (UI), 32
Cognitive enhancement, 447
    amphetamines and methylphenidate (MPH), 448
    downside, 458
    ghrelin
        mechanism of action, 456
    methods
        D-serine, 455
        glycine, 455
    signaling pathways, 449
        Akt/PKB signaling pathway, 454
        approaches, 450
        C1 domain proteins, 451
        inhibitors of apoptosis, 454
        in learning, 450
        mitogen-activated protein kinase (MAPK) signaling, 453
        in neuronal repair, 452
        neurotrophins (NTs), 452–453
        NMDA signaling, 453
        NT receptors, activators of, 453
        PKC activators as, 454–455
        PKC substrates, 451–452
        protein kinase C (PKC), 450–451
    strategies, 456
        genes, 457
        protein degradation, 457–458
        protein translation, 457
    unconventional ways, 448